The Biology of Wetas, King Crickets and Their Allies

The Biology of Wetas, King Crickets and Their Allies

Edited by

L.H. Field

Department of Zoology
University of Canterbury
New Zealand

CABI *Publishing*

CABI *Publishing* is a division of CAB *International*

CABI Publishing
CAB International
Wallingford
Oxon OX10 8DE
UK
Tel: +44 (0)1491 832111
Fax: +44 (0)1491 833508
Email: cabi@cabi.org
Web site: http://www.cabi.org

CABI Publishing
10 E 40th Street
Suite 3203
New York, NY 10016
USA
Tel: +1 212 481 7018
Fax: +1 212 686 7993
Email: cabi-nao@cabi.org

A catalogue record for this book is available from the British Library, London, UK.

A catalogue record for this book is available from the Library of Congress, Washington DC, USA

ISBN 0 85199 408 3

Typeset by Columns Design Limited, Reading
Printed and bound in the UK by Biddles Ltd, Guildford and King's Lynn

Contents

Contributors

Andrés O. Angulo, Departamento de Zoologia, Facultad Ciencias Naturales y Oceanográficas, Universidad de Concepción, Casilla 4040, Correo 3, Concepción, Chile. email: aangulo@udec.cl

Robert S. Bigelow (deceased), *Formerly of:* Department of Zoology, University of Canterbury, Private Bag 4800, Christchurch, New Zealand

Paul R.L. Cary, Ecology Group, Institute of Natural Resources, Massey University, PO Box 11222, Palmerston North, New Zealand

Neil A. Deans, Fish and Game Council, Nelson, New Zealand. email: zool146@it.canterbury.ac.nz

Laurence H. Field, Department of Zoology, University of Canterbury, Private Bag 4800, Christchurch, New Zealand. email: zool146@it.canterbury.ac.nz

George W. Gibbs, School of Biological Sciences, Victoria University, Box 600, Wellington, New Zealand. email: george.gibbs@vuw.ac.nz

Stephen Glasgow, Department of Zoology, University of Canterbury, Private Bag 4800, Christchurch, New Zealand. email: zool146@it.canterbury.ac.nz

Andrej V. Gorochov, Zoological Institute, Russian Academy of Sciences, Universitetskaya nab. 1, St Petersburg 199034, Russia. email: gor@zisp.spb.su

Roderick J. Hale, Department of Zoology, University of Western Australia, Nedlands, Western Australia 6907, Australia. email: rjhale@barclays.net

Thomas H. Jarman, Department of Zoology, University of Canterbury, Private Bag 4800, Christchurch, New Zealand. email: zool146@it.canterbury.ac.nz

Tania King, Zoology Department, University of Otago, PO Box 56, Dunedin, New Zealand. Present address: School of Biology, University of Leeds, Leeds LS2 9JT, UK.

John Leader, Department of Physiology, University of Otago, Dunedin 913, New Zealand. email: jleader@gandalf.otago.ac.nz

Robert D. Lewis, School of Biological Sciences, University of Auckland, Private Bag 92019, Auckland, New Zealand. email: b.lewis@auckland.ac.nz

Mary McIntyre, School of Biological Sciences and Island Biology Research Programme, Victoria University, Box 600, Wellington, New Zealand. email: mary.mcintyre@vuw.ac.nz

Geoffrey B. Monteith, Queensland Museum, Brisbane, Queensland 4101, Australia. email: geoffm@qm.qld.gov.au

Mary Morgan-Richards, Zoology Department, University of Otago, PO Box 56, Dunedin, New Zealand. email: mary.morgan-richards@stonebow.otago.ac.nz

Doug Neufeld, Department of Physiology, University of Otago, Dunedin 913, New Zealand. email: jleader@gandalf.otago.ac.nz

Barry O'Brien, Department of Biological Sciences, University of Waikato, Private Bag 3105, Hamilton, New Zealand. email: b.obrien@waikato.ac.nz

Robert Poulin, Department of Zoology, University of Otago, PO Box 56, Dunedin, New Zealand. email: zoodw@zoo.latrobe.edu.au

David C.F. Rentz, Division of Entomology, CSIRO, Canberra, ACT 2601, Australia. email: dave.rentz@ento.csiro.au

Graham R. Sandlant, Department of Zoology, University of Canterbury, Private Bag 4800, Christchurch, New Zealand. email: zool146@it.canterbury.ac.nz

Greg Sherley, Science and Research Division, Science Technology and Information Service, Department of Conservation, PO Box 10420, Wellington, New Zealand. email: gsherley@doc.govt.nz

Ian A.N. Stringer, Ecology Group, Institute of Natural Resources, Massey University, PO Box 11222, Palmerston North, New Zealand. email: i.stringer@massey.ac.nz

Rob B. Toms, Department of General Entomology, Transvaal Museum, Northern Flagship Institution, PO Box 413, Pretoria 0001, South Africa. email: toms@nfi.co.za

Steve Trewick, Department of Zoology, University of Otago, PO Box 56, Dunedin, New Zealand. email: mary.morgan-richards@stonebow.otago.ac.uk

Claudine L. Tyrrell, Department of Zoology, University of Otago, PO Box 56, Dunedin, New Zealand. email: zoodw@zoo.latrobe.edu.au

Thomas J. Walker, Department of Entomology and Nematology, University of Florida, Gainesville, FL 32611, USA. email: tjw@gnv.ifas.ufl.edu

David B. Weissman, Department of Entomology, California Academy of Sciences, Golden Gate Park, San Francisco, California 94118, USA. email: gryllus1@juno.com

David A. Wharton, Department of Zoology, University of Otago, PO Box 56, Dunedin, New Zealand. email: zoodw@zoo.latrobe.edu.au

Anna York, School of Biological Sciences, University of Auckland, Private Bag 92019, Auckland, New Zealand. email: b.lewis@auckland.ac.nz

Foreword

The wetas, king crickets and their allies are some of the most fascinating members of the insect world. Some are the heaviest of insects, dwarfing even some small mammals in weight. Their large size and often ferocious nature always engenders public interest. This book has been long awaited. Orthopterists have always had 'trouble' dealing with these insects because they appear to bear no characters distinctive for species separation or secondary sexual characters have been misinterpreted. The insects included in this book are for the most part members of the orthopterous superfamilies Stenopelmatoidea and Gryllacridoidea, collectively known as king crickets. However, the common name 'bugaboo' has helped to confuse the situation with 'king cricket', being the common name for quite different insects, depending on geography. This is the reason for including the review of the Gryllacrididae here. All these insects are of Gondwanan origin and are important 'flagship species' for biogeographical studies. But, as a result of human-initiated activities, including introduced animals, such as rats and cane-toads, and development, a number of species are threatened with extinction.

A peculiarity of king crickets is the bizarre production of secondary sexual characters, such as enlarged heads, mandibles and hind legs. These can be asymmetrical as well. Extensive behavioural studies are reported here. Odd weaponry, such as tusks and mandibular extensions and protrusions, suggest bizarre courtship rituals unlike those found in other orthopteroids.

Larry Field has assembled a most interesting array of topics dealing with the morphology, reproduction, physiology, behaviour, biogeography, systematics and ecology of these insects. The chapter on conservation should prove valuable for the continuity of measures leading to the continued existence of threatened species. The review of the Gryllacrididae, noted above, should prove useful in relating their biology, behaviour, etc. to those of the true king crickets. This book will go far in helping to answer questions that have long puzzled orthopterists. If it stimulates further investigations of these insects, it will have more than fulfilled its purpose.

D.C.F. Rentz
Canberra, January 2000

Preface

This book was conceived at the Hilo conference of the Orthopterists Society, where, owing to the enthusiastic interest shown, I became strongly aware that no previous synthesis of knowledge existed for this primitive group of orthopterans. Encouraged by Win Bailey and Dan Otte, I set about gathering together all the published research, as well as work buried in government coffers, and unpublished research on the wetas of New Zealand. It soon became obvious that the scope should be extended worldwide, since additional unpublished and ongoing research existed in other southern hemisphere countries and the USA. The resulting book therefore represents a timely foundation for the ever increasing research on these stenopelmatoid insects.

I owe much to the contributing authors and to past students and colleagues for their efforts in making this book possible. I am particularly grateful to Dan Otte for offering to illustrate the male secondary sexual characters of many of the strange anostostomatids in the collections of the Philadelphia Academy of Science, and to John Edwards for use of unpublished figures of tree weta cercal sensory anatomy and histology. I thank the following for reviewing chapters and for proofreading: David Rentz, Ian Stringer, Richard Holdaway, George Gibbs, John Edwards, Piotr Naskrecki, Paul Edwards, Hiroshi Nishino, Neil Deans, Frank Howorth and Barry O'Brien. My thanks also to Tim Hardwick and Emma Critchley of CAB *International* for their enduring patience while the book was being written. The figures were digitized by Catherine Collins and Craig Russell, two young and talented digital artists, who never thought that they would be confronted with the exacting labour of making every last scientific detail correct. Finally, I owe especial gratitude to my wife Mitsuyo for her patience and endurance of the stresses involved in this project.

This book was funded by a grant from the University of Canterbury and supported by the facilities of its Department of Zoology.

Introduction

Laurence H. Field
Department of Zoology, University of Canterbury, Private Bag 4800, Christchurch, New Zealand

Experiences and Background

In every biologist there may be kindled a sense of wonderment and discovery, something akin to that which fired the excitement of the 18th- and 19th-century naturalists, when first encountering a large king cricket, weta or Jerusalem cricket. Add to that a night-time setting in a far-flung South Pacific island reserve, offshore from New Zealand, where your torchlight is roving across the low-hanging fronds of Nikau palms and suddenly picks up the enormous tell-tale antennae of a giant weta the size of a small rat, perched on the green blades. If you're careful and don't move suddenly, you can coax the nearly extinct giant insect on to your palm, which it fills entirely with its leg span. In another land, you press your memory into action as you walk carefully at night, trying to recall the vague trail through the wet and overhanging vegetation of the North Queensland montane rainforest where you laid the rotting pineapple bait stations the previous afternoon. Suddenly your light beam encounters a bait covered with a swarm of waving antennae on the muddy ground up ahead. They belong to a group of long-legged 'white-knees', the flightless king cricket *Penalva*. But just on the edge is lurking a huge-headed male of the giant Australian king cricket *Anostostoma*, looking prehistoric and bizarre, with great curved mandibles, which make the whole head seem nearly as long as the body.

These are the real-life experiences that make the study of this superfamily of orthopteran insects a fascinating pursuit for both biologist and lay person alike. Their attraction lies not in any cuddly or virtuous appearance but, instead, in their eccentric and often bizarre features and behaviour. You immediately sense that here is something unusual indeed, photogenic because the insect's enormous head and threatening mandibles are all out of proportion with its body, or because it has stout, curved tusks projecting from its head, like some mythical beast in miniature, or because it emerges from a shovelful of desert sand into the daylight like a hidden monster living a subterranean life, unknown and unseen by the world at large.

The Stenopelmatoidea is a superfamily of ensiferan Orthoptera, commonly known as wetas, king crickets, Jerusalem crickets and raspy crickets. The group encompasses a huge variety of cricket-like insects, many of which are flightless and which seem rather closer to their ancient origins than do the more modern grasshoppers, crickets and katydids (long-horned crickets). Most are nocturnal and infrequently encountered, unless attracted by bait or sought by excavation or night lighting. In New Zealand, Australia and North America, certain species are abundant enough to be studied in their habitats or maintained in laboratory colonies. Much of the present volume stems from 25 years of research by the editor, and his students and colleagues, on many aspects of the biology of New Zealand wetas.

An explanation of the term 'weta' is necessary at the outset. This Maori word refers generically to New Zealand species of *Hemideina* (tree wetas), *Hemiandrus* (ground wetas) and *Deinacrida* (giant wetas), all of which were recognized as similar-looking flightless orthopterans distinct from crickets, bush crickets and tettigoniids. Because the Maori language lacks the plural noun form, as well as the 's' sound, some English-speaking authors in New Zealand use 'weta' as a singular and plural noun, thus retaining the Maori grammar. More commonly, 'weta' has become an English cognate and is used in the plural form, 'wetas'. Both usages are found in this book, wherein the reader, alerted by the above, will perceive clear meaning in the context.

Audience

The book's audience is viewed in a broad sense; thus chapters are included on basic anatomy, behaviour and ecology, as well as information on conservation-orientated practices, such as captive breeding of the rare giants and experimental reintroduction of endangered species to island habitats. Much of the material is presented or reviewed for the first time in a publicly available volume, and the book is expected to be a fundamental information source for research scientists who deal with this insect group. Because many of the wetas and king crickets, in particular, are large and spectacular in appearance, some chapters will be a resource for schools (science projects and teaching modules on these 'exciting' insects, in New Zealand, Australia, California, South Africa, South America), zoos (display, breeding and maintenance of large insects as a popular public attraction), museums (background information for explanations in displays of these unusual insects) and government conservation agencies concerned with the protection and fostering of this unique insect heritage, particularly in New Zealand, Australia, South Africa, Chile, Madagascar and North America.

Overview

No single treatise has attempted to gather together the known biology of the Stenopelmatoidea. Probably because of the disparate distribution of the families into regions remote from the populated centres of research, the literature remains sparse, localized and little published. Species have radiated especially in the southern hemisphere: Africa, Australia, New Zealand. Yet, within each region, a core of scientists has quietly pursued the fascinating biological stories that may be gleaned from this unusual insect fauna. To set the scene of that research, regional overviews of stenopelmatoid fauna are offered in five chapters of the book's first section, 'Systematics and Biogeography' (omission of some families in this treatment simply indicates the great wealth of biological research awaiting future workers). Most of the section deals with the family Anostostomatidae (wetas and king crickets), while lesser coverage, certainly not indicative of a lesser fauna, is devoted to the Gryllacrididae (leaf-rolling crickets) and to the Stenopelmatidae (Jerusalem crickets). The family Rhaphidophoridae (cave and camel-backed crickets) has received little attention other than taxonomic treatments, and is dealt with here on a comparative anatomical basis (O'Brien and Field, Chapter 8). Stenopelmatoid phylogeny has presented, and in some respects still remains, a tangled history of evolutionary changes inevitably bound to the crustal breakup and wanderings of the great Gondwana terranes. Every taxonomic attempt to resolve the confusing history of these insects has led to contentious results, although, as Gorochov explains in Chapter 1, recent efforts have shed greater light than ever before on understanding the interrelationships. The most detailed knowledge, based upon molecular genetic techniques, is being developed in New Zealand for generic and specific relationships, as explained by Morgan-Richards *et al.* (Chapter 7).

The 'Morphology and Anatomy' section is treated in a broad sense. Historically, initial scientific efforts concentrated upon museum collections, which brought attention to the plethora of primitive characters expressed in the anatomy of especially the Anostostomatidae and Rhaphidophoridae. Chapter 8 not only places these into perspective, but also serves as an extensive guide to the anatomy and morphology of the group, using the New Zealand tree and giant wetas as examples. This work is complemented by Angulo's treatment of the cratomeline 'red cricket' of Chile, in Chapter 11.

The early work on the Stenopelmatoidea also highlighted the diversity of extraordinary secondary sexual characters found particularly in males, and presumably underlying intense agonistic competition for females. Elephantine tusks mounted on mandibles can reach 44% of the body length in the New Zealand *Motuweta*; lance-like spines project forward from the greatly enlarged head (a condition termed megacephaly) in the African *Henicus*, and long, arched, sharp-toothed mandibles extend far out from the head to meet only at the tips in African and Australian *Nasidus* and *Anostostoma*. Perhaps the most bizarre is the grossly asymmetric head capsule found in the megacephalic Caribbean species of *Licodia*, a delight and challenge for studies of fluctuating asymmetry. In this genus, the left side of the head is greatly distended laterally, pulling the left mandibular articulation far from the median line and resulting in a left mandible up to 138% longer than the right one. These are a few of the marvellous workings of sexual selection, illustrated by Toms (Chapter 4) and Field and Deans (Chapter 10). Most of the male characters are apparently weapons of combat, but we are just beginning to understand the ways in which these structures are used behaviourally (Field, Chapter 18). The escalating battles between male tree wetas with enormous sharp mandibles turn out to be highly ritualized and part of a defence-resource polygyny mating system. The above groups are well suited to provide rich rewards for studies in the behavioural and evolutionary mechanisms of sexual selection.

The 'Ecology' section reviews how radio-tracking techniques have revealed the habits of some highly protected giant wetas of New Zealand (McIntyre, Chapter 12), as well as the tightly knit relationship between *Hemideina* wetas and the tree galleries which they opportunistically seek, but do not make. Field and Sandlant (Chapter 13) describe how galleries create a limiting mechanism to prevent runaway Fisherian sexual selection of megacephaly in males, by limiting accessibility only to males with head capsules that fit through the entrance. Larger males cannot obtain refuge from diurnal predators and will be selected against. This principle appears to operate most harshly on *Hemideina femorata*, which depends upon galleries created with little variation in diameter by a monospecific host-provider beetle.

The stenopelmatoid families are susceptible to a large variety of parasites, but the most spectacular are the enormously long gordian worms (Nematomorpha), as discussed by Wharton *et al.* in Chapter 14. The authors present experimental research that reveals an impressive survival ability of another weta parasite, the nematode *Wetahula*, co-adapted with its freeze-tolerant alpine weta host to recover from freezing at −61°C!

Some of the greatest forward strides in biological knowledge of the stenopelmatoids are reviewed in the 'Behaviour' section. In contrast to more than a century of earlier research dealing with preserved insects, recent research has generated a wealth of remarkable discoveries about living wetas, Jerusalem crickets and African king crickets. This is especially true for sound production and communication, as described by Field in Chapter 15. While the stridulatory mechanisms of most species have long been known to involve simple, somewhat primitive, femoro-abdominal pegs, the variety and embellishment of structures uncovered in New Zealand and Australian species exceeds that found in the more celebrated modern ensiferan songsters, the Tettigoniidae and Gryllidae. A rich diversity of kinds of stridulatory teeth and pegs has evolved in this island-bound group: spikes, tubercles, molar-like teeth, double-lipped crescents, club-like articulated pegs on abdominal tergites, files, slats and microscopic conical spines. The associated behaviour in wetas includes several different means of moving abdomen against femur, abdominal tergo-tergal stridulation and mandibular stridulation in the tusked wetas, while rudimentary tegminal stridulation is found in some winged stenopelmatoid species. An additional non-stridulatory behaviour, tremulation, is produced during agonistic fights between male tree wetas.

The sound patterns, however, remain primitive and less sophisticated than those of the modern ensiferans, and do not conform to the gryllid and tettigoniid 'rules' played out under conditions of intense interspecific competition for mates and territory. In tree wetas, there are no courtship sound patterns, nor are sounds of calling males evidently used to attract mates. While such calling sounds may serve as territorial markers, they exhibit far more variation than those seen in gryllids and tettigoniids. As a whole, the signalling system suggests a more primitive, intermediate stage in the evolution of social acoustic signalling in orthopterans. And, finally, by some unknown physical mechanism, several species of giant wetas tick very regularly, like a clock, when alarmed.

Another mode of social signalling, transmission of substrate vibratory signals, appears to be widespread in New Zealand tree wetas and ground wetas (both Anostostomatidae), the abundant Australian raspy crickets (Gryllacrididae) and the North American Jerusalem crickets (Stenopelmatidae). Thus, it has come to light that highly species-specific drumming signals are transmitted through the desert sands of California and Mexico to bring together male and female subterranean Jerusalem crickets for prolonged mating sessions (Weissman, Chapters 3 and 19). Drumming, whether by the abdomen against the substrate (Jerusalem crickets) or by rapid tapping of the hind tibiae on bush stems or soil (raspy crickets, ground wetas), is used in genera with and without ears, and the signals are picked up by sensitive vibration detectors in the tarsi, as explained by Field in Chapters 15 and 22.

Stridulatory behaviour is also used as a defence mechanism. In Chapter 16, Field and Glasgow show for the first time that the spectacular display of raised hind legs with colossal spines of tree and giant wetas, combined with sound production, effectively deters avian and reptilian predators. While the latter groups are likely to have evolved together with wetas and exerted selection pressure on the development of such defence mechanisms, introduced mammalian predators clearly played no such role, and the defence behaviours of wetas against them are quite useless. Thus the terrestrial giant wetas have become nearly extinct due to introduced rats, cats and mustelids. Other species have evolved an entirely different defence ploy of flipping over on their back with legs outspread. An experimental behavioural analysis has shown that this is especially suitable for protection against reptilian predators. A further, highly unusual form of insect defence behaviour is reported for two unrelated riverine anostostomatids from Australia and New Zealand. These insects, which live adjacent to streams, jump into the stream and crawl underwater along the rocks as a means of escape. The New Zealand example is an unnamed species of tusked weta which has been observed to stay submerged for up to 5 minutes.

Many exciting discoveries are presented in the chapters on mating behaviour by Weissman (Chapter 3), Field and Jarman (Chapter 17) and Stringer (Chapter 20). For example, New Zealand tree weta males fight for possession of harems of females, with associated sexual selection leading to exaggerated weaponry (such as the enormous mandibles described in Field and Deans, Chapter 10). However the mating systems evidently evolved in allopatry because there appear to be no cross-species mating barriers. The normal chemical and/or acoustic species-specific clues are lacking and males of one species of *Hemideina* readily mate (successfully) with females of other species (Morgan-Richards *et al.*, Chapter 7). Although hybrid zones have been demonstrated in two cases, apparently most matings produce sterile offspring, indicating that postmating barriers exist at genetic and developmental levels. Another curious behaviour complicates the tree weta mating system and remains to be fully investigated. Males show an apparent conflict in parental investment by displaying aggression toward females after mating or when mating attempts are rebuffed by choosy females.

In contrast to tree wetas, the closely related giant wetas have retained a primitive mating behaviour, and multiple males aggregate around a female for mating, rather than showing aggressive competition. In addition, Stringer (Chapter 20) describes how males of the giant tusked weta and at least several tree weta species show early maturation and can become imagines up to three instars earlier than normal. Such undersized, but sexually mature males engage in sneak mating tactics and thus circumvent the aggressive competition system with an alternative mating approach.

However, the ultimate form of male parental investment was discovered in Jerusalem crickets, where the male is eaten by the female in a bizarre act of postmating cannibalism in about 10% of matings. Weissman in Chapter 19 presents the first described cases of male complicity in postcoital cannibalism, in which the male lets the female attack him after completing spermatophore transfer. In his words:

> These six laboratory males had the opportunity to escape but made no such effort. In one laboratory mating, the female left the male for a full 15 min, during which time he did not move, before returning to consume him. Once attacked by their female, no male made any effort to escape or repel her onslaught.

Although the male offers the ultimate parental investment in his offspring (his body), questions remain to be answered about the pay-off for such an investment. Weissman suggests that cannibalism may be necessary to trigger some reproductive process, such as egg formation or oviposition, which somehow ensures the eaten male's paternity.

The 'Reproduction and Development' section focuses on studies of New Zealand tree wetas and giant wetas. A wealth of information is presented on sex ratios, seasonality, copulation and spermatophores, egg structure, development and embryology by Stringer in Chapter 20. Knowledge of postembryonic development is rich in many aspects, but critically lacking in others, such as the elucidation of life-tables for species. Most of these large insects undergo a relatively long development without a diapause, and they are unusually long-lived as adults. The early-maturing phenomenon in tree weta males (*Hemideina*) provides an unusual and inviting avenue for further research in many aspects of sexual development, behaviour and sexual competition.

The unique island geography and evolution of New Zealand has been the justification for studies of many biological aspects of wetas. The expectation has been that isolation has either allowed retention of primitively derived characters or promoted development of interesting features specific to the isolated fauna. This conjecture is supported by the finding of numerous primitive characters in wetas, no less so in the physiological systems than in the anatomy and morphology described by O'Brien and Field in Chapter 8. While most knowledge of insect physiology is based upon on several 'model' species commonly available in the northern hemisphere, the section on 'Physiology' presents a surprisingly full account of many unusual mechanisms found in the relatively primitive gryllacridoid Orthoptera. Not only is this useful material for comparative physiologists, but it also provides interesting insights into how newly described mechanisms evolved in the face of physiological challenges in the isolation of the New Zealand islands. We see in Field's Chapter 22 on sensory physiology, for example, that tree wetas possess remarkably swollen forelegs, which contain tympanal structures (eardrums) and large inflated air sacs to specifically amplify intraspecific sound frequencies; this foreleg amplifier is unique in the insect world. Moreover, the giant wetas possess the largest foreleg ears of any orthopteran, but they are unsophisticated in terms of directional hearing when compared with those of crickets and tettigoniids. Most internal stretch receptor organs in the exoskeletal joints resemble those of other ensiferan Orthoptera, but unusual differences show up. Thus the tibiotarsal chordotonal organ (which normally monitors tarsal movement and angle) is attached to entirely different moving points than described in other insects, and consequently raises a new set of questions about its function. In the massive male mandibles of alpine tree wetas, the main stretch receptor organ (ventral muscle receptor organ (VMRO)) is strikingly different. Up to 170 sensory axons have been counted in the sensory nerve to the VMRO, making this the most richly innervated muscle receptor organ (MRO) ever reported. A more typical number of sensory neurons, eight to ten, is seen in the mandibular MROs of beetles.

The search for unusual and/or primitive features in stenopelmatoids has also been amply rewarded in the realm of neuromuscular physiology and motor control, as described by Field in Chapter 23. Such features are inextricably bound to the anatomy: for example, the coxal leg musculature is more highly divided and dispersed than that of modern orthopterans, and even somewhat recalls features seen in apterygote leg musculature. The dispersed array of muscles may be specifically related to supporting the characteristic raised leg defence display of wetas, for they have been found to have unique neuromuscular synaptic properties. Thus, although the leg muscles are driven by the typical orthopteran array of motor neurons, the discovery that these muscles can be locked into a catch contraction (i.e. a frozen state), whereby no further nerve impulses are required to maintain prolonged muscle tension while the legs are raised, is unprecedented in any other insect group. Wetas appear to utilize catch contraction not just in maintenance of the defence posture, but also in thanatosis, where all limbs are frozen in position. The latter is easily elicited experimentally, and is likely to be a natural mode of resting vertically in tree galleries during the daytime without prolonged electrical neuromuscular activation.

Both behaviours above appear to be mediated by the neurosecretory product, octopamine, produced by a special class of motor neurons called dorsal unpaired motor (DUM) neurons. Because there are many differences in the physiological responses to octopamine between locusts and wetas, it is important to investigate the extent to which octopamine plays a role in the behaviour of ensiferan Orthoptera compared with that of the caeliferan Orthoptera. Is the lack of catch contraction in the locust simply a reflection of this species' behavioural repertoire, or is it a characteristic of the entire suborder of grasshoppers and locusts?

Another fascinating story includes the role of octopamine in mediating the ability of tree wetas to undergo conditioned learning of leg position. Wetas can be trained to set the leg into a narrow joint angle window. However, contrary to the results found in more modern insects, such as locusts, which can retain a memory of the learned behaviour within the central nervous system (CNS) for at least an hour, tree wetas do not retain a memory of window position after the tibia enters it. Therefore the weta must be setting the position by a peripheral locking mechanism, the catch contraction (which holds the tibia in the window for over 2 h), but is not able to reset the tibia into the window once it drifts outside; it must relearn the task. Apparently the more modern caeliferan Orthoptera (locusts and grasshoppers) have developed a sophisticated memory ability in the CNS, while the more primitive wetas have not.

The chapter also describes how the seemingly simple mandibular motor control system consists of an elaborate evolutionary response to solve the engineering challenge of controlling the position and force of two asymmetrical and independently articulated structures. The singularly large and powerful mandibles of male tree wetas have made possible a detailed neural control study, which has heretofore been too difficult in other insects. A startling fact to emerge is that the huge closer muscles can develop several thousand pounds per square inch of force on the cutting edges of the mandibles, but this potentially self-destructive set of forces is balanced dynamically by a complex feedback servo-controlled system involving the multineuronal VMRO described above.

The mechanisms controlling the highly predictable evening emergence of tree wetas from within the dark confines of their galleries have been explored and modelled by Lewis and York in Chapter 24 on circadian rhythms and the underlying control oscillators. Surprisingly, the light-sensitive resetting mechanism of the biological clock is not found in the compound eyes or in the ocelli (the insects' only presumed photoreceptors). The precise timing and rich variety of locomotory rhythms have enabled the authors to develop a sophisticated dual pacemaker, time-delay model for the clock, and to set the scene for a molecular approach to analysing biological clocks.

The final chapter on physiology explores control of ionic concentrations in the weta haemolymph. Neufeld and Leader (Chapter 25) explain its importance in alpine wetas, which are subject to winter freezing conditions. When subjected to freezing temperatures, *Hemideina maori* does not become supercooled, but can tolerate freezing to at least $-10°C$ for 30 minutes or more, provided that cooling is slow, of the order of 1°C per minute. *Hemideina maori* is thus the largest insect known to survive freezing. Survival appears to involve the ability of the cells to tolerate the concentrated medium formed by the precipitation of water.

The section on 'Conservation of Endangered Species' focuses on New Zealand's pioneering efforts in the conservation of the rare giant wetas, as well as a number of other anostostomatid species. In this country, as well as Australia and South Africa, the larger and more spectacular anostostomatids have become flagship species in respect of the preservation of each country's unique natural heritage. The public in these countries is becoming more aware of this interesting insect heritage, through magazine and newspaper articles, live displays in zoos and school science projects. Moreover, the weta fauna of New Zealand has featured in New Zealand, British and American television documentaries. Australia's largest king cricket is the subject of a popular display in the Queensland Museum. Southern Africa's tusked king cricket (known locally as the 'Parktown prawn' from its predilection for wandering into swimming pools in a wealthy suburb of Johannesburg) has been given its own web page by Pretoria's Transvaal Museum. In addition, a recent centenary celebration of the scientifically historical description of this bizarre insect led to an enhanced public awareness of biodiversity and natural heritage in South Africa.

Sherley summarizes in Chapter 26 how the combined action of New Zealand government legislation and of his own Department of Conservation offers landmark approaches to captive breeding and release of endangered land invertebrates, and wetas in particular, as well as practical reports of successful and unsuccessful experimental trials. It is hoped that this material serves as an instructive nucleus to motivate international efforts along the same track.

Part I
Systematics and Biogeography

1
The Higher Classification, Phylogeny and Evolution of the Superfamily Stenopelmatoidea

Andrej V. Gorochov
Zoological Institute, Russian Academy of Sciences, Universitetskaya nab. 1, St Petersburg 199034, Russia

Introduction and Historical Background

The classification by Burmeister (1838) is the first reasonably detailed classification of the suborder Ensifera. This author considered the Ensifera as two families ('Locustina' and 'Gryllodea') of the group Saltatoria. His 'Locustina' was divided into seven tribes, six of which comprised the different Tettigonioidea, while the seventh ('Stenopelmatidae') contained the Stenopelmatoidea. The following authors (Blanchard, 1845; Saussure, 1859; Thomas, 1872; Stål, 1876; Brunner-Wattenwyl, 1888; Jacobson, 1905) proposed the name Stenopelmatoidea for Burmeister's seventh group, some dividing it into several tribes and groups of genera. The classification of Stenopelmatoidea from the above-mentioned monograph by Brunner-Wattenwyl has remained the most important contribution from this period. The subsequent development of the classification of this superfamily (1910–1937) has been connected with the numerous works by Karny. His final classification (Karny, 1937) included all Stenopelmatoidea known at that time into one family, the 'Gryllacrididae', consisting of seven subfamilies and eight tribes. He also included in this family the representatives of Prophalangopsinae (true recent Hagloidea) and Cyphoderrinae (recent Hagloidea or perhaps primitive Tettigonioidea). In this period, the first attempts at a phylogenetic study of this superfamily appeared (Karny, 1921, 1930; Handlirsch, 1925, 1929), but they were either eclectic or oversimplified.

The more interesting phylogenetic schemes for the Stenopelmatoidea were those proposed by Ander (1939) and Zeuner (1939). They supposed that the Stenopelmatoidea or 'stenopelmatoid forms' are the primitive group ancestral for the Hagloidea and Tettigonioidea. Ander considered the Grylloidea to be unrelated to the Stenopelmatoidea and their 'descendants', while Zeuner considered the Grylloidea to be descendants of 'stenopelmatoid forms'. Ragge's (1955) opinion about the phylogeny of the Stenopelmatoidea is similar to Zeuner's, but he considered that one of the stenopelmatoid groups, the 'Schizodactylidae', was closer to the Grylloidea than to the other Ensifera. In the famous book by Sharov (1968), the Stenopelmatoidea and Grylloidea were considered to be descendants of the Hagloidea, and the Tettigonioidea were placed as an unrelated ensiferan group.

The subsequent period saw numerous reviews of classification, including two important ones by orthopteran specialists (Beier, 1972; Kevan, 1982), descriptions of new supergeneric taxa (Rentz and Weissman, 1973; Rentz, 1980; Kevan and Wighton, 1981, 1983), discussions about the ranks of higher taxa (Kevan, 1976, 1977), and special (Rhaphidophoridae only) or very general phylogenetic studies (Hubbel and Norton, 1978; Hennig, 1981). Hennig presented reasonable critical remarks about the phylogenetic schemes by Zeuner and Sharov.

In 1984–1997, several new publications about the classification, phylogeny and evolution of Stenopelmatoidea appeared (Gorochov, 1984, 1988a, 1995; Gwynne, 1995; Johns, 1997). The paper by Johns contains mainly nomenclatural corrections. Gwynne's phylogenetic scheme is similar partly to Ander's scheme and partly to Ragge's scheme; it is based on the formalistic 'cladistic analysis', which contains the logically incorrect usage of the parsimony principle (Pesenko, 1989, 1991). Gorochov's views, with some changes, are expounded below.

Systematic Position and Origin of Stenopelmatoidea

The order Orthoptera is usually divided into two superorders: Ensifera or long-horned Orthoptera and Caelifera or short-horned Orthoptera; the first suborder is considered more ancient (from the Carboniferous) and ancestral to the second one (from the Triassic). A review and critique of differing opinions regarding this scheme are given by Gorochov (1995). The suborder Ensifera is divided by the different authors into infraorders or only superfamilies. Gorochov suggests the division into four infraorders: Oedischiidea and Elcanidea (the very primitive Palaeozoic–Mesozoic groups still without tibial tympana and tegminal stridulatory apparatus or, in some Mesozoic Oedischiidea, with tegminal stridulatory apparatus appearing to be non-homologous to that of the Recent Ensifera), Tettigoniidea and Gryllidea (the Mesozoic–Recent groups with the following general synapomorphies: presence of tibial tympana and characteristic tegminal stridulatory apparatus).

The infraorder Tettigoniidea *sensu* Gorochov consists of three superfamilies: Hagloidea (the most ancient Triassic–Recent taxon of Tettigoniidea with primitive tegminal stridulatory apparatus), Tettigonioidea (the Cretaceous–Recent superfamily with originally specialized tegminal stridulatory apparatus) and Stenopelmatoidea (the Palaeocene–Recent superfamily with loss of tegminal stridulatory apparatus and originally more or less parallel longitudinal venation of tegmina). The infraorder Oedischiidea is possibly ancestral to the Hagloidea. The early (Triassic) Hagloidea were possible ancestors of the infraorder Gryllidea (its autapomorphies are obviously connected with the adaptations to the more or less digging mode of life of its ancestor). The late (Jurassic–Cretaceous) Hagloidea appear to be ancestors of the Tettigonioidea and Stenopelmatoidea.

Sharov (1968) first proposed the mainly Mesozoic superfamily Hagloidea as ancestral for Stenopelmatoidea. He described the fossil genus *Zeuneroptera* (Palaeocene of Scotland), which is intermediate between Hagloidea (Fig. 1.1-1, 2) and Stenopelmatoidea. The tegmina of Zeuneroptera (the only imprint of this genus possibly belongs to a male (Fig. 1.1-3)) preserve the traces of the tegminal stridulatory apparatus characteristic for males of Hagloidea, but the tegminal venation of *Zeuneroptera* is also similar to that of the Stenopelmatoidea because of their almost parallel longitudinal veins (synapomorphy of Stenopelmatoidea).

In addition, Kevan and Wighton (1981, 1983) described the very similar genus *Albertoilus* (Palaeocene of Canada) as a representative of the Hagloidea, and included Sharov's only genus in the subfamily Zeuneropterinae. Gorochov (1988a) provisionally included *Albertoilus* in the Zeuneropterinae and placed this subfamily closer to the family Anostostomatidae, especially to Cratomelinae, because the tegmina of *Cratomelus* Blanchard have a characteristic venation, which is similar to the rudiments of a stridulatory apparatus in the Zeuneropterinae (Fig. 1.1-6). Now, due to insufficient data, I prefer not to include this subfamily in the general classification of Stenopelmatoidea.

Nevertheless, the Zeuneropterinae is the most ancient and most primitive group of all known Stenopelmatoidea. The venation of the hind wings of *Albertoilus* (Fig. 1.1-4) is almost identical to that of representatives of the Aboilinae (the Jurassic–Cretaceous subfamily of Hagloidea) (Fig. 1.1-2) and is characterized by the well-developed MP+CuA_1; this vein in all other Stenopelmatoidea is partly or almost completely lost (Fig. 1.1-5). This fact suggests that the Aboilinae are possible ancestors of the Stenopelmatoidea (and the Tettigonioidea, as shown by Gorochov (1988b, 1995)). It is possible that the time of separation of Stenopelmatoidea from Aboilinae is not later than Early Cretaceous (the supposed time of origin of Tettigonioidea). In that case, only future finds of fossil Stenopelmatoidea from that age could confirm or refute whether the presence of a rudimentary tegminal stridulatory apparatus in the

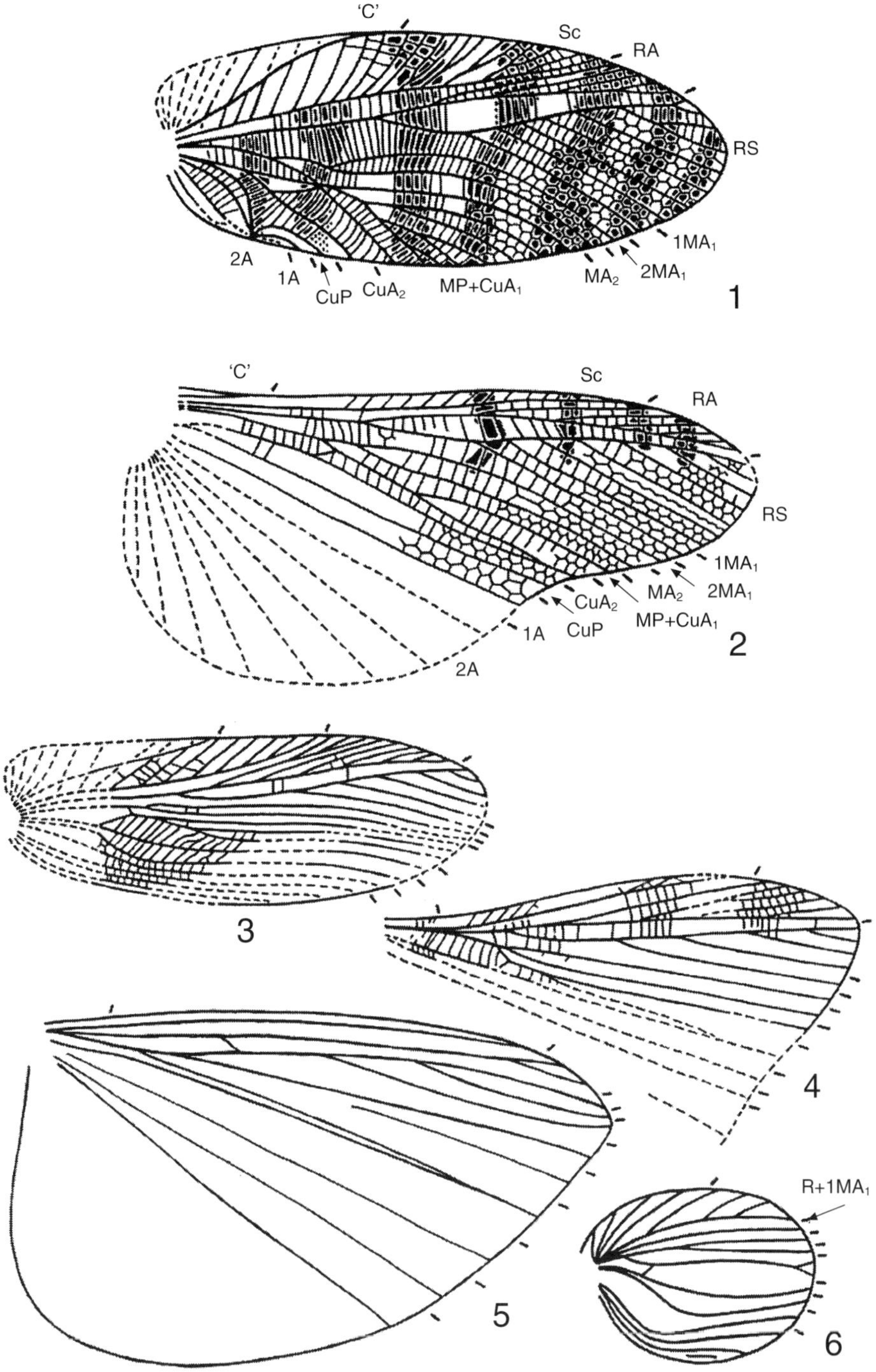

Fig. 1.1. Hagloidea and Stenopelmatoidea (1–4, fossil; 5, 6, Recent). 1, *Aboilus columnatus* Mart., tegmen of male; 2, *A. maculatus* (Mart.), hind wing; 3, *Zeuneroptera scotica* (Zeun.), tegmen of male (?); 4, *Albertoilus cervirufi* Kev. et Wight., hind wing; 5, *Transaevum landatum* Johns, hind wing of male (schematic); 6, *Cratomelus armatus* Blanch., tegmen of male (schematic).

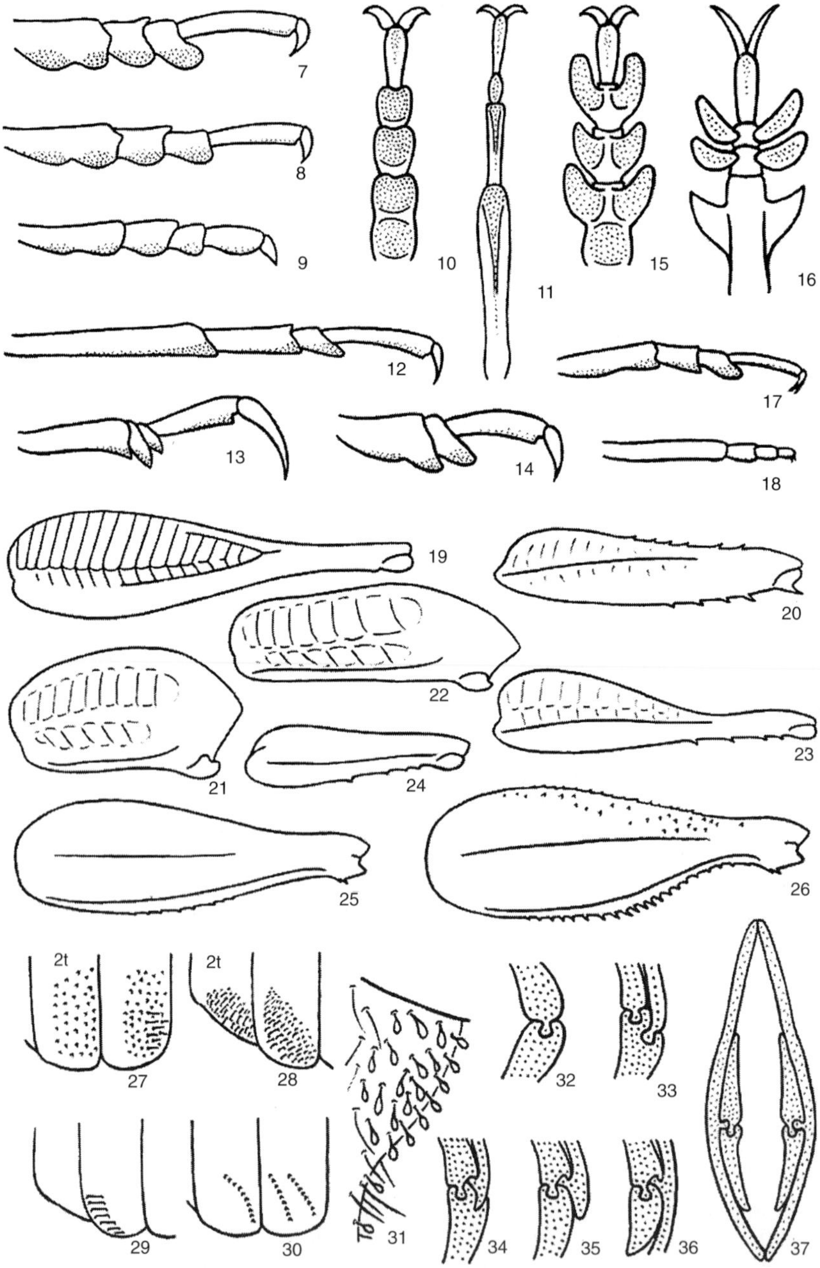

Fig. 1.2. Recent Stenopelmatoidea (7–31, 34–37), Grylloidea (32) and Tettigonioidea (33) (schematic). 7–18, Hind tarsus: 7, *Carcinopsis*, 8, 9, *Cooloola* (8, male, 9, female), 10, *Pteranabropsis*, 11, 12, *Dolichopoda*, 13, *Macrobaenetes*, 14, *Daihinia*, 15, *Gryllacris*, 16, *Schizodactylus*, 17, *Oryctopterus*, 18, *Oryctopus*; 19–26, hind femur: 19, *Apteranabropsis*, 20, *Hemideina*, 21, 22, *Cooloola* (21, female, 22, male), 23, *Lezina*, 24, *Sia*, 25, 26, *Macrobaenetes* (25, female, 26, male); 27–30, abdominal stridulatory area (2t, second tergite): 27, *Lezina*, 28, *Deinacrida*, 29, *Hemideina*, 30, *Diaphanogryllacris*; 31, *Maxentius*, inner part of cercal base; 32–37, transverse section of ovipositor (32–36, only place of articulation of right valvae): 32, Acheta, 33,*Tettigonia*, 34, *Cratomelus*, 35, *Diestrammena*, 36, *Gryllacris*, 37, *Diaphanogryllacris*. View from side (7–9, 12–14, 17–30), from below (10, 11, 15, 16), from above (31) and from behind (32-37).

Zeuneropterinae and in *Cratomelus* is a very old relict character for both these taxa.

Classification of Recent Stenopelmatoidea

Most finds of fossil Stenopelmatoidea do not allow for more or less correct placement into the distinct higher taxa of this superfamily. Therefore these fossils, excepting Protroglophilinae (see below), are not discussed here, and the proposed classification deals almost entirely with Recent taxa. The present classification is based on the works by Gorochov (1988a, 1995), but with some nomenclatural changes by Johns (1997).

Superfam. Stenopelmatoidea Burmeister, 1838

Diagnosis

Tegmina with more or less parallel longitudinal venation (CuP and 1A usually (originally) ended in distal half of tegmen); tegminal stridulatory apparatus absent (sometimes only its traces preserved); costal area with Sc branches usually (originally) crossed by long false costal vein (sometimes this 'C' likened to Sc branch); MA_2 developed (originally) or fused with $2MA_1$. Tarsi with four segments (sometimes with only three segments, but place of fusion of two proximal segments distinct); fore tibiae with two open tympana (originally) or without them. Femoro-abdominal stridulatory apparatus consisting of rough stridulatory areas at lower part of fore abdominal tergites (proximal region of inner surface of hind femora rubs up against these areas) either developed (maybe originally) or absent (maybe lost); cerci soft, usually not adapted for clasping, but very rarely hooked in male (see Fig. 1.10-119). Upper and middle valvae of ovipositor articulated with lower ones and with each other (Fig. 1.2-34–37) (in Tettigonioidea, upper and middle valvae articulated only with lower valvae (Fig. 1.2-33); in Grylloidea, upper and lower valvae articulated only with each other (middle valvae reduced) (Fig. 1.2-32)).

Composition

Three recent families and one extinct subfamily (Zeuneropterinae Kevan et Wighton, 1983) possibly related to Anostostomatidae (see above).

Key to families

1a. Tarsi always with four segments and not widened pulvilli at sole: first segment with two pulvilli (Fig. 1.2-7–10); tympana developed or lost; outer side of hind femora with distinct chevron ridges (Fig. 1.2-19) or only traces of them (Fig. 1.2-20–23). Abdominal stridulatory areas of femoro-abdominal stridulatory apparatus (if developed) only irregularly denticulated (Fig. 1.2-27) or with small ridges (Fig. 1.2-28, 29); male abdominal apex: tenth tergite divided into lateral halves or undivided, with pair of hooked processes, paraprocts usually with specialized process or sometimes without it (see Fig. 1.7-77–85); ovipositor (if well-developed) with lower part of upper valvae inserted in special furrow of lower valvae (Fig. 1.2-34)..............................(I) **Anostostomatidae**

1b. Tarsi with four segments, but very rarely with three segments, and with varied pulvilli at sole (Fig. 1.2-11–18); tympana always absent; hind femora always without chevron ridges (Fig. 1.2-24–26). Abdominal stridulatory areas of femoro-abdominal stridulatory apparatus (if developed) with rows of rather large denticles (Fig. 1.2-30); male abdominal apex: tenth tergite usually undivided, with pair of varied processes or without them, paraprocts without specialized process or sometimes with it (see Fig. 1.15-176–180); ovipositor (if well-developed) with lower valvae partly or completely covered with lower part of upper valvae on outside (Fig. 1.2-35–37) **2a**

2a. Wings always absent. Tarsi with four segments, but very rarely with three segments, and with non-widened pulvilli: first segment with only one pulvillus (Fig. 1.2-11–13) unless two proximal segments fused (Fig. 1.2-14). Femoro-abdominal stridulatory apparatus lost; cerci without any bulb-like sensillae; ovipositor always developed, with lower part of upper valvae partly (sometimes almost completely) covering lower valvae on outside (Fig. 1.2-35) ....(II) **Rhaphidophoridae**

2b. Wings developed or lost. Tarsi always with four segments and with varied pulvilli: first segment with two pulvilli, distal of which sometimes paired (Fig. 1.2-15, 17), or without any distinct pulvilli (Fig. 1.2-16, 18). Femoro-abdominal stridulatory apparatus developed or lost; cerci with bulb-like sensilla at inner proximal part (Fig. 1.2-31) or without them; ovipositor developed or reduced; if well-developed, lower part of its upper valvae completely covering lower valvae on outside (Fig. 1.2-36, 37) ....(III) **Stenopelmatidae**

(I) Fam. Anostostomatidae Saussure, 1859

Composition

Eight recent subfamilies and one recent American tribe with unclear subfamilial position (Glaphyrosomatini Rentz and Weissman, 1973). The systematic position of the genera *Gryllacropsis* Br.-W. and *Hypocophoides* Karny (Hindustan), *Coccinellomima* Karny (Kalimantan), *Anisoura* Ander (Nicobar Islands), *Hypocophus* Br.-W. (Madagascar), *Dolichochaeta* Phill. (South America), *Aistus* Br.-W. (New Caledonia), *Hemiandrus* Ander (New Zealand), *Gryllotaurus* Karny, *Penthoplophora* Tepper and *Transaevum* Johns (Australia) belonging, or perhaps belonging, to this family is also unclear.

Key to subfamilies

1a. Eyes and antennal cavities in female slightly smaller than in male (Fig. 1.3-42, 43); mandibles without sexual specialization (similar in both sexes), somewhat pricking (Fig. 1.3-53); laciniae without any teeth (Fig. 1.3-54); antennae shorter than pronotum. Wings strongly shortened in male and lost in female; male tegmina with almost straight CuP and area between MP+CuA_1 and CuA_2 not widened (Fig. 1.4-56); male hind wings strongly reduced. All legs strongly adapted for digging, thick in male and very thick in female; tibiae almost without subapical spines (Fig. 1.5-69, 70); fore coxae without spine; hind femora not adapted for jumping, with only traces of chevron ridges (Fig. 1.2-21, 22). Tympana and femoro-abdominal stridulatory apparatus absent; cerci very short; male abdominal apex: genital plate with styles, tenth tergite distinctly divided into lateral halves, paraprocts with specialized process (Fig. 1.7-77); ovipositor strongly reduced ....(1) **Cooloolinae** Rentz, 1980

Composition

The genus *Cooloola* Rentz (Australia).

1b. Eyes and antennal cavities similar in both sexes (Fig. 1.3-44–47); mandibles varied, but not pricking (Fig. 1.3-50, 51, 55); laciniae with three teeth (Fig. 1.3-52); antennae much longer than pronotum. Wings varied, similar in both sexes, often lost in both sexes. Legs without strong adaptation for digging, more or less thin and similar in both sexes (Figs 1.5-71, 72; 1.6-73–76); tibiae varied; fore coxae with spine; hind femora varied. Tympana and femoro-abdominal stridulatory apparatus developed or absent; cerci distinctly longer; male abdominal apex varied (Fig. 1.7-78–85); ovipositor developed or reduced **2a**

2a. Upper side of all tibiae with subapical spines; outer side of hind femora with distinct chevron ridges (Fig. 1.2-19). Femoro-abdominal stridulatory apparatus (if developed) with irregularly denticulated abdominal stridulatory areas (see Fig. 1.15-170); male genital plate with styles; male paraprocts with specialized process (Fig. 1.7-78–80, 83, 84); ovipositor always developed..........................**3a**

2b. Upper side of fore tibiae without subapical spines; middle and hind tibiae varied; outer side of hind femora with only traces of chevron ridges (Fig. 1.2-20, 23). Femoro-abdominal stridulatory apparatus (if developed) with more or less varied abdominal stridulatory areas (Fig. 1.2-27–29); male genital plate and male paraprocts varied (Fig. 1.7-81, 82, 85); ovipositor developed or reduced **6a**

3a. Mandibles of male without sexual specialization (Fig. 1.3-38) (similar in both sexes). Pronotum and body not arched in profile (Fig. 1.5-72). Wings (if developed) with almost straight CuP and area between MP+CuA_1 and CuA_2 not widened in tegmina, with remains of MP+CuA_1 not fused with CuA_2 in hind wings (Fig. 1.4-58, 59). Tympana and femoro-abdominal stridulatory apparatus usually developed. Abdominal apex of male: ninth tergite with more or less distinct paired small hind lobes over the hooks of tenth tergite, tenth tergite usually undivided and with wide distance between abovementioned hooks (Fig. 1.7-78, 79) (apterous females practically indistinguishable from those of Anostostomatinae). ....(2) **Anabropsinae** Rentz et Weissman, 1973

Composition

The genera *Anabropsis* Rehn, *Paterdecolyus* Griff., *Pteranabropsis* Gor., *Apteranabropsis* Gor. (tropical

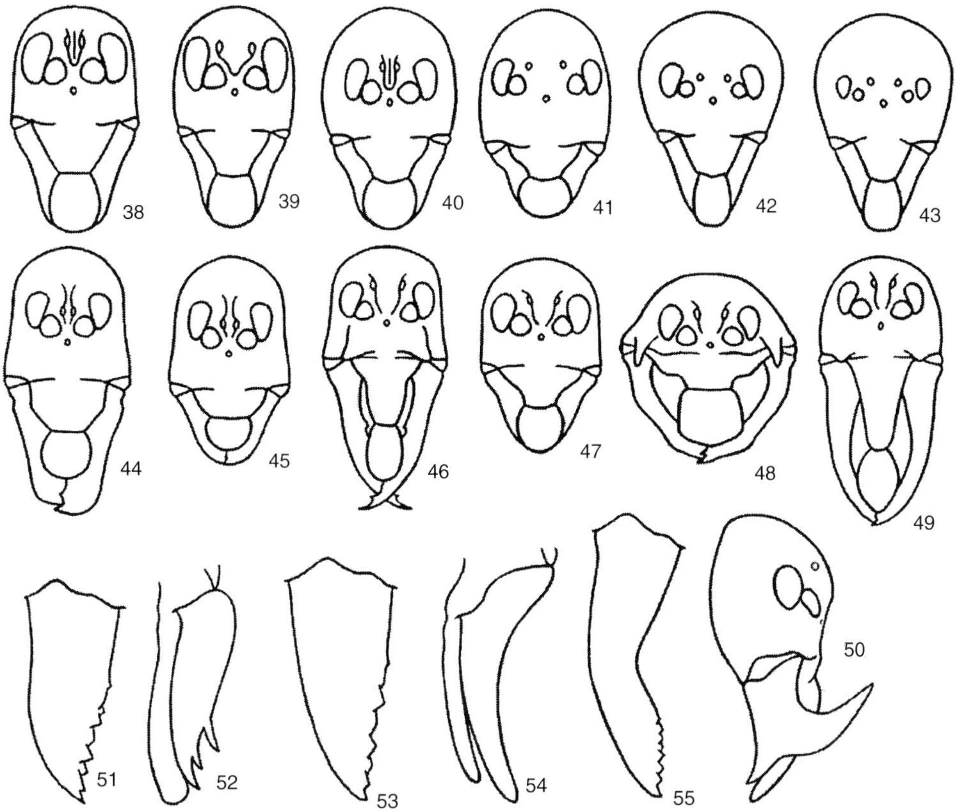

Fig. 1.3. Recent Anostostomatidae (schematic). 38–50, Head without antennae: 38, *Pteranabropsis*, 39, *Lutosa*, 40, *Cratomelus*, 41, *Lezina*, 42, 43, *Cooloola* (42, male, 43, female), 44, 45, *Hemideina* (44, male, 45, female), 46, 47, *Spizapterus* (46, male, 47, female), 48, *Henicus* (male), 49, *Anostostoma* (male), 50, *Libanasidus* (male); 51, 52, *Anabropsis* (51, mandible, 52, distal part of maxilla); 53, 54, *Cooloola* (53, mandible, 54, distal part of maxilla); 55, *Mimnermus*, mandible of male. View from front (38–49, 51, 53, 55), from side (50) and from behind (52, 54).

America and Africa, South-East Asia; the three latter genera were recently synonymized by Johns, but his opinion is illogical, because, in that case, *Paterdecolyus* (*sensu* Johns) lacks any distinction from *Anabropsis*) and possibly *Brachyporus* Br.-W., *Exogryllacris* Will., *Penalva* Walk., *Papuaistus* Griff., and *Onosandrus* Stål (Madagascar, Australia, New Guinea, South Africa).

3b. Mandibles of male, shape of pronotum and body, venation of wings (if developed) varied. Tympana and femoro-abdominal stridulatory apparatus developed or lost. Abdominal apex of male: ninth tergite without any paired lobes, tenth tergite divided into two lateral halves and usually with short distance between hooks (Fig. 1.7-80, 83, 84)..**4a**

4a. Mandibles of male without sexual specialization (Fig. 1.3-40) (similar in both sexes). Pronotum and body not arched in profile (Fig. 1.5-71). Wings shortened, with distinctly S-shaped CuP and widened area between MP+CuA_1 and CuA_2 in tegmina, with unclear homology of veins in hind wings (these veins modified in connection with appearance of tegmino-alary stridulatory apparatus) (Figs 1.1-6; 1.4-57). Tympana and femoro-abdominal stridulatory apparatus absent............(3) **Cratomelinae** Brunner-Wattenwyl, 1888

Composition

The genus *Cratomelus* Blanchard (South America).

4b. Mandibles of male, shape of pronotum and body varied. Wings (if developed) with almost straight CuP and area between MP+CuA_1 and CuA_2 not widened in tegmina, with remains of MP+CuA_1 almost completely fused with CuA_2 in hind wings (Fig. 1.4-60, 61). Tympana and femoro-abdominal stridulatory apparatus developed or lost **5a**

5a. Mandibles of male without sexual specialization (Fig. 1.3-39) (similar in both sexes). Pronotum and body arched in profile (in connection with enlargement of mesonotum and metanotum almost as in Rhaphidophoridae) (Fig. 1.6-74). Wings lost. Tympana and femoro-abdominal stridulatory apparatus usually developed......................(4) **Lutosinae** Gorochov, 1988

Composition

The genera *Lutosa* Walk., *Apotetamenus* Br.-W. and possibly *Licodia* Walk. (all from tropical America).

5b. Mandibles of male usually with sexual specialization (increased and arched or with horn-like process) (Fig. 1.3-46, 48–50, 55); mandibles of female without sexual specialization (Fig. 1.3-47). Pronotum and body not arched in profile (Fig. 1.6-73). Wings, tympana and femoro-abdominal stridulatory apparatus developed or lost (apterous females practically indistinguishable from those of Anabropsinae) ....(5) **Anostostomatinae** Saussure, 1859 (= Mimnermidae Stål, 1876; = Henicinae Karny, 1928)

Composition

The genera *Anostostoma* Gray, *Henicus* Gray, *Mimnermus* Stål (the two latter names were synonymized by Johns, but they contain very different representatives), *Nasidius* Stål, *Libanasidus* Pering., *Spizaphilus* Kirby, *Spizapterus* Karny (the two latter names were also synonymized by Johns, but it seems to me this synonymy is questionable), *Carcinopsis* Br.-W. (tropical and South Africa, Madagascar, Australia, New Caledonia) and possibly *Libanasa* Walk., *Bochus* Pering., *Onosandridus* Pering., *Borborothis* Br.-W., *Motuweta* Johns (tropical and South Africa, New Zealand).

6a. Mandibles of male without sexual specialization (Fig. 1.3-41) (similar in both sexes). Wings lost. Upper side of middle and hind tibiae with subapical spines (Fig. 1.6-75). Tympana and femoro-abdominal stridulatory apparatus developed; cerci rather long; male abdominal apex: genital plate without styles, paraprocts without specialized process (Fig. 1.7-82); ovipositor strongly reduced ....(6) **Lezininae** Karny, 1932

Composition

The genus *Lezina* Walk. (North and East Africa, South-West and Central Asia).

6b. Mandibles of male varied. Wings developed or lost. Upper side of middle tibiae without subapical spines (Figs 1.4-64; 1.6-76); hind tibiae varied. Tympana and femoro-abdominal stridulatory apparatus developed or absent; cerci slightly shorter; male abdominal apex: genital plate with styles, paraprocts with specialized process (Fig. 1.7-81, 85); ovipositor developed (or female unknown) **7a**

7a. Mandibles of male without sexual specialization (female unknown). Wings developed (tegmina with almost straight CuP and no widened area between MP+CuA_1 and CuA_2; hind wings with remains of MP+CuA_1 almost completely fused with CuA_2) (Fig. 1.4-62, 63). Upper side of hind tibiae with only almost indistinct subapical denticles (Fig. 1.4-65, 66). Tympana and femoro-abdominal stridulatory apparatus absent ....(7) **Leiomelinae** subfam. n.

Composition

The genus *Leiomelus* Ander (South America).

7b. Mandibles of male with sexual specialization (increased length) (Fig. 1.3-44) or without it as well as those of female (Fig. 1.3-45). Wings lost. Upper side of hind tibiae with large subapical spines (Figs 1.4-67, 68; 1.6-76). Tympana and femoro-abdominal stridulatory apparatus developed; abdominal stridulatory areas with ridges (Fig. 1.2-28, 29) ....(8) **Deinacridinae** Karny, 1932

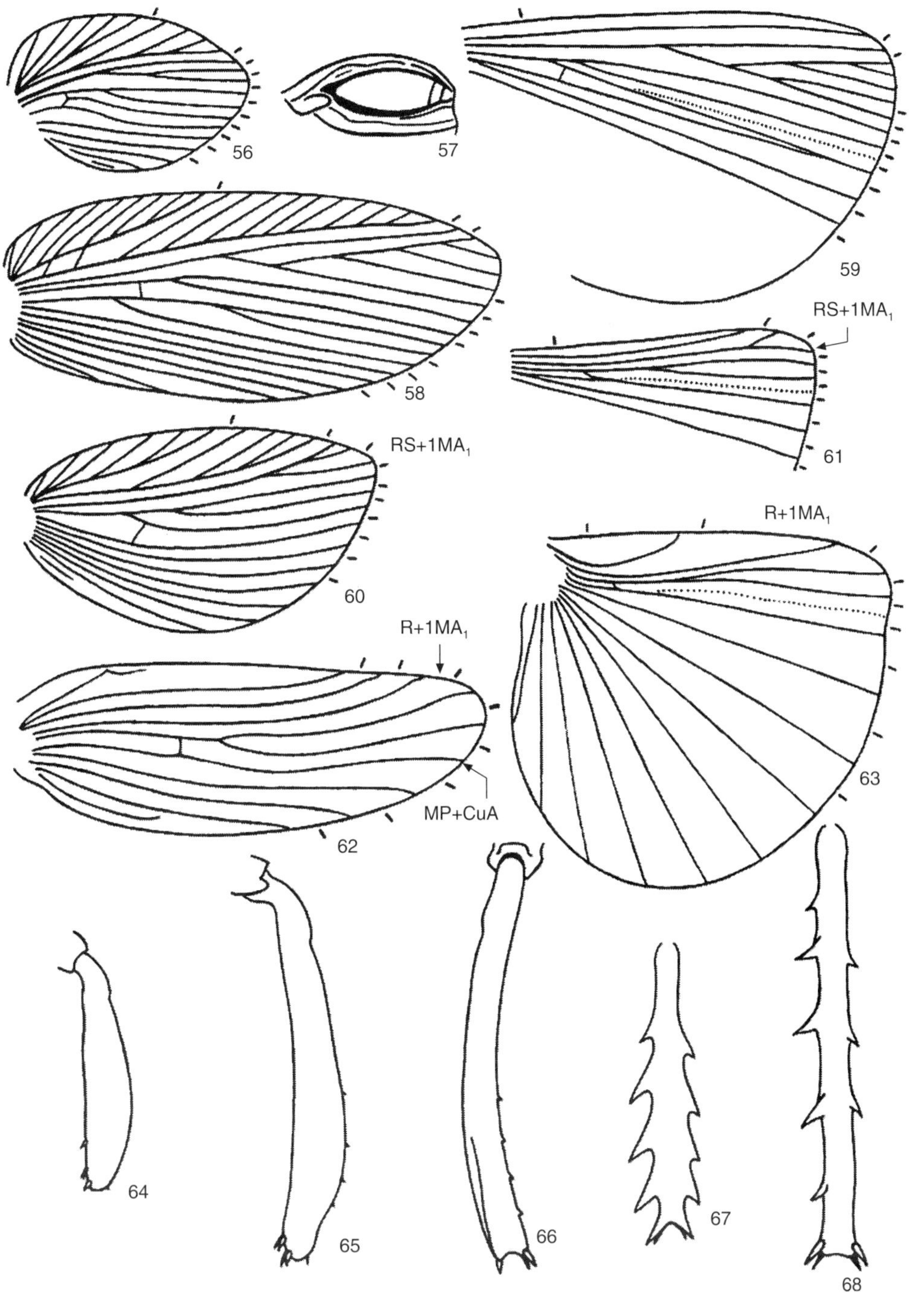

Fig. 1.4. Recent Anostostomatidae (schematic). 56, *Cooloola*, tegmen of male; 57, *Cratomelus*, costal part of hind wing; 58, 59, *Pteranabropsis*: 58, tegmen, 59, hind wing; 60, 61, *Spizaphilus*: 60, tegmen, 61, costal part of hind wing; 62–66, *Leiomelus*: 62, tegmen, 63, hind wing, 64, outer side of middle tibia, 65, outer side of hind tibia, 66, upper side of left hind tibia; 67, 68, upper side of right hind tibia: 67, *Hemideina*, 68, *Deinacrida*.

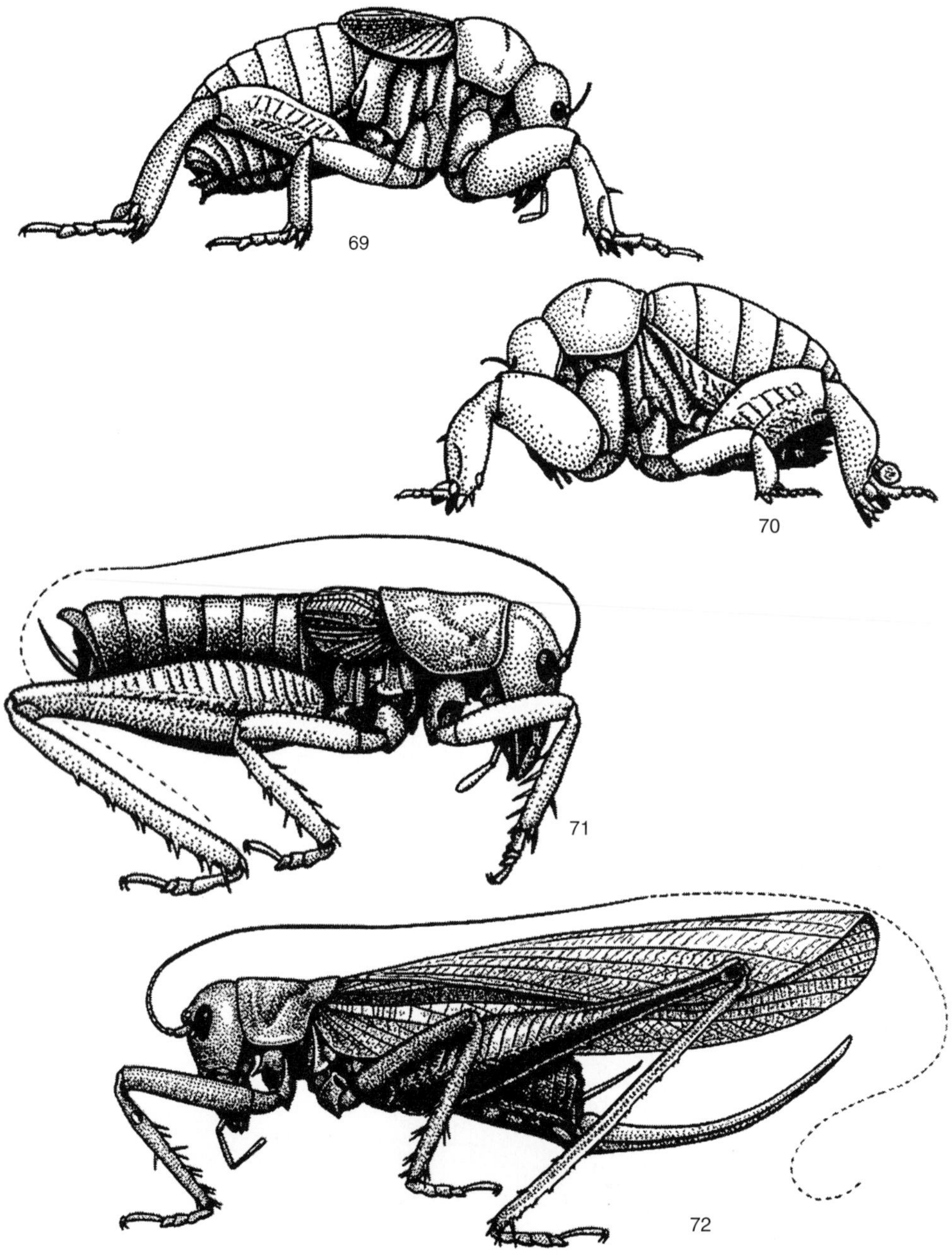

Fig. 1.5. Recent Anostostomatidae, general view. 69, 70, *Cooloola ziljan* Rentz: 69, male, 70, female; 71, *Cratomelus armatus* Blanch., male; 72, *Pteranabropsis carli* (Griff.), female.

Composition

The genera *Deinacrida* White and *Hemideina* Walk. (New Zealand).

(II) Fam. Rhaphidophoridae Thomas, 1872

Composition

Eight recent subfamilies and one extinct subfamily. The extinct subfamily (Protroglophilinae) is

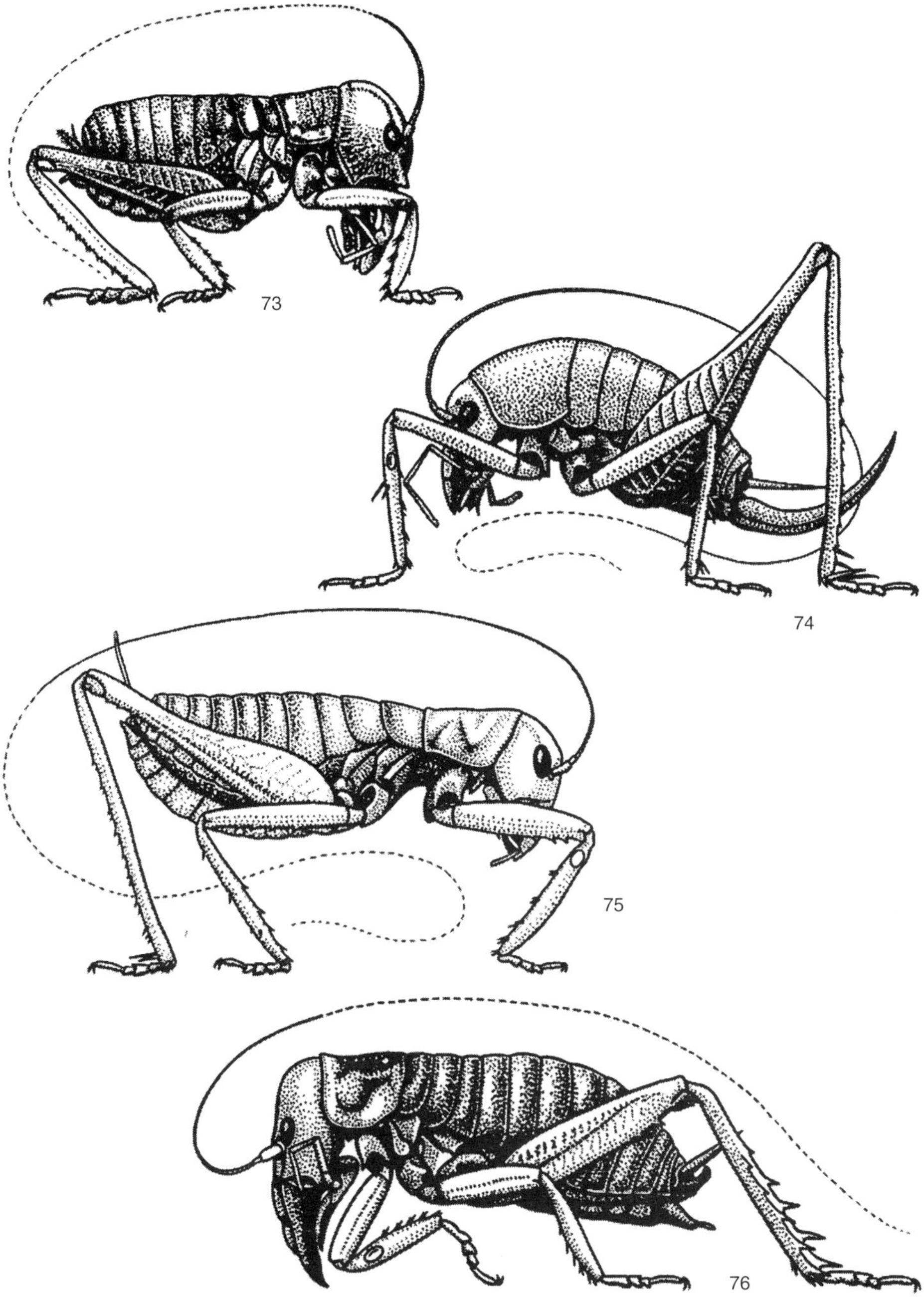

Fig. 1.6. Recent Anostostomatidae, general view. 73, *Mimnermus cephalotes* Bol., male; 74, *Lutosa marginalis* Walk., female; 75, *Lezina mutica* (Br.-W.), male; 76, *Hemideina maori* (Pict. et Sauss.), male.

known from amber inclusions; therefore it may be included in the key. The systematic position of the genera *Anoplophilus* Karny and *Alpinanoplophilus* Ishik. (Japan), united by Ichikawa (1997) in the invalid group Anoplophilinae (without description and diagnosis), of this family is rather unclear.

Key to subfamilies

1a. Shape of body and armaments of legs rather varied (Figs 1.8-101, 104; 1.9-114, 115), but first and second segments of hind tarsi with paired apical spines (spurs) (Fig. 1.8-86, 87). Distal part of male genital plate usually with styles, more or less divided into upper and lower parts (Fig. 1.10-118) ....(1)
Macropathinae Karny, 1929

Composition

Two tribes: **Macropathini** Karny and **Talitropsini** Gorochov, stat. n. These tribes contain nearly 30 genera (Australia, southern part of South America and South Africa, New Zealand and some other islands of southern hemisphere).

1b. Shape of body and armaments of legs very varied, but first and/or second segments of hind tarsi with unpaired upper apical spine (spur) or small tubercle (Fig. 1.8-88–98). Distal part of male genital plate varied, but undivided into upper and lower parts (Fig. 1.10-119–121) **2a**

2a. Legs short (Fig. 1.9-116); fore coxae without spine or tubercle; hind tibiae with only non-articulated upper subapical spines (Fig. 1.8-111). Upper valvae of ovipositor with upper denticles (Fig. 1.10-128). Male genital plate without styles (Fig. 1.10-119).............(2) **Gammarotettiginae** Karny, 1934

Composition

The genus *Gammarotettix* Br.-W. (North America).

2b. Legs distinctly longer, but sometimes rather short (Fig. 1.9-117); fore coxae with spine or tubercle; hind tibiae varied. Upper valvae of ovipositor without denticles (Fig. 1.10-122–127, 129, 130). Male genital plate varied **3a**

3a. Hind tibiae with articulated upper subapical spines and non-articulated upper subapical denticles together (Figs 1.8-102, 103, 105, 106; 1.9-117). Male genital plate usually without styles ....(3)
Ceuthophilinae Brunner-Wattenwyl, 1888

Composition

Five tribes: **Ceuthophilini** Brunner-Wattenwyl, **Daihiniini** Karny, **Hadenoecini** Ander, **Pristoceutophilini** Hubbel and **Argyrtini** Hubbel. These tribes contain around 20 genera (North America).

3b. Hind tibiae with only articulated upper subapical spines or with non-articulated upper subapical spines and/or denticles (Fig. 1.8-107–110, 112, 113). Male genital plate varied **4a**

4a. Middle femora with one inner and one outer apical spines (spurs); hind tibiae with non-articulated upper subapical spines and/or denticles (Fig. 1.8-109, 110) **5a**

4b. Middle femora with only inner apical spine (spur) or without any apical spines (spurs); hind tibiae with varied armaments (Fig. 1.8-107, 108, 112, 113) **6a**

5a. Fore femora with one inner apical spine (spur); longest apical spine (spur) of hind tibiae longer than first and second segments of hind tarsi together; hind tarsi short, with more or less high first segment bearing large apical spine (spur), with rather short second segment (Fig. 1.8-94, 95). Male genital plate with styles (Fig. 1.10-120)...............(4)
Rhaphidophorinae Thomas, 1872

Composition

The genera *Rhaphidophora* Serv., *Stonychophora* Karny, *Eurhaphidophora* Gor., *Pararhaphidophora* Gor. and *Neorhaphidophora* Gor. (from Hindustan and East Asia to Australia and Samoa).

5b. Fore femora with only outer apical spine (spur); longest apical spine (spur) of hind tibiae shorter than first and second segments of hind tarsi together; hind tarsi long, with low first segment bearing small apical spine (spur) or without this spine (spur), with more or less long second segment (Fig. 1.8-90, 91). Male genital plate without styles (Fig. 1.10-121) ....(5) **Aemodogryllinae** Jacobson, 1905

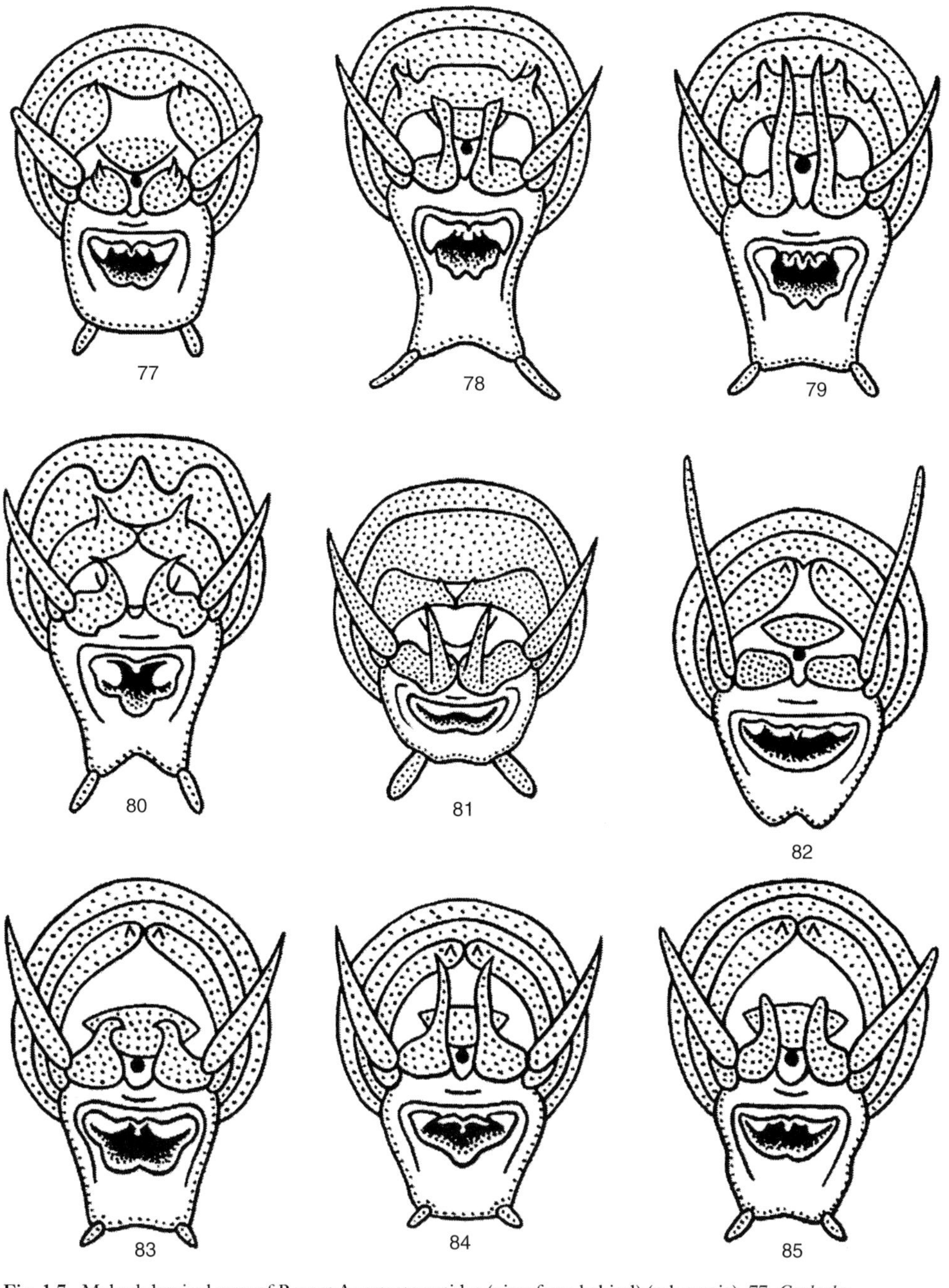

Fig. 1.7. Male abdominal apex of Recent Anostostomatidae (view from behind) (schematic). 77, *Cooloola*; 78, *Pteranabropsis*; 79, *Apteranabropsis*; 80, *Cratomelus*; 81, *Leiomelus*; 82, *Lezina*; 83, *Lutosa*; 84, *Mimnermus*; 85, *Hemideina*.

Composition

Two tribes: **Aemodogryllini** Jacobson and **Diestramimini** Gorochov. These tribes contain around ten genera (from Hindustan and East Asia to New Guinea, one or two synanthropic species almost cosmopolitan).

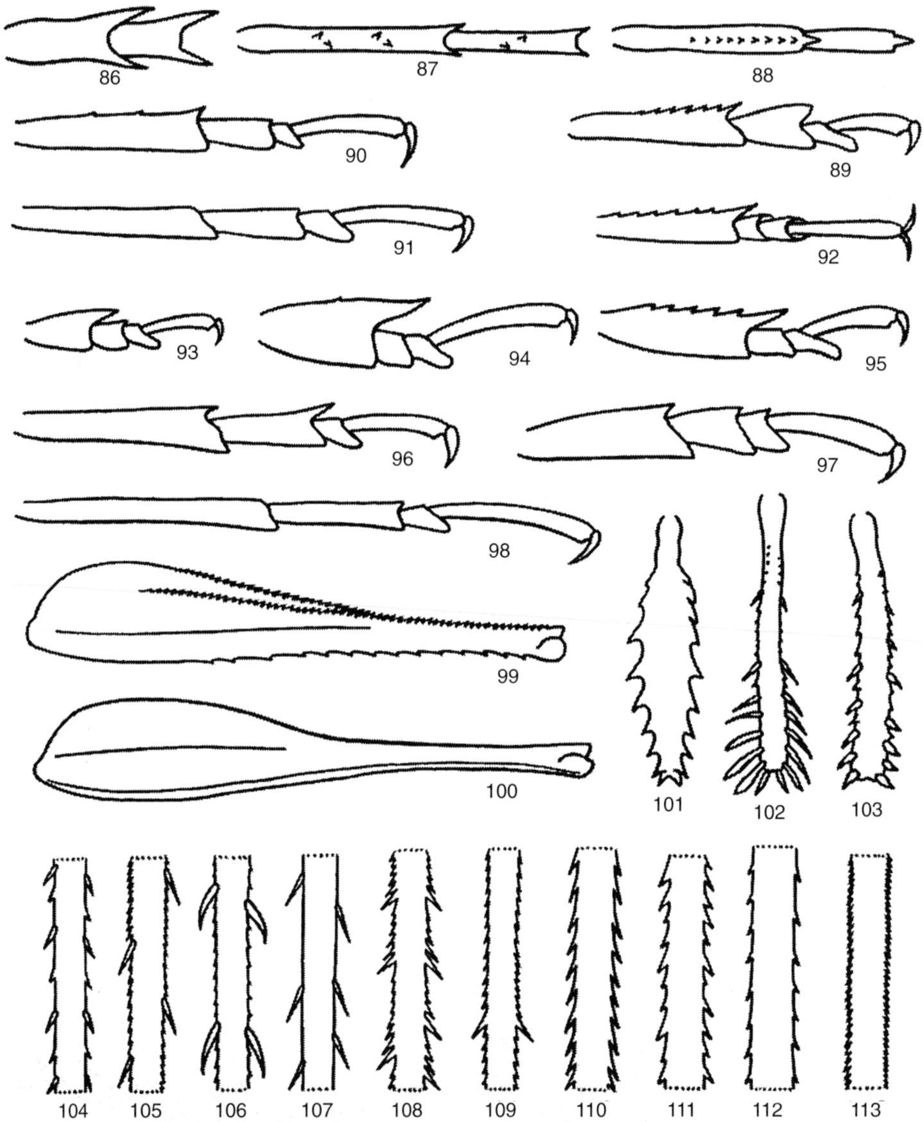

Fig. 1.8. Rhaphidophoridae (86–91, 93–111, 113, Recent; 92, 112, fossil) (schematic). 86–88, First and second segments of hind tarsus from above: 86, *Talitropsis*, 87, *Gymnoplectron*, 88, *Troglophilus*; 89–98, hind tarsus from side: 89, *Troglophilus*, 90, *Diestramima*, 91, *Diestrammena*, 92, *Prorhaphidophora*, 93, *Gammarotettix*, 94, *Stonychophora*, 95, *Rhaphidophora*, 96, *Tropidischia*, 97, *Argyrtes*, 98, *Hadenoecus*; 99, 100, hind femur from side: 99, *Tropidischia*, 100, *Dolichopoda*; 101–103, hind tibia from above: 101, *Talitropsis*, 102, *Macrobaenetes*, 103, *Daihinia*; 104–113, middle part of hind tibia from above: 104, *Gymnoplectron*, 105, *Hadenoecus*, 106, *Ceutophilus*, 107, *Dolichopoda*, 108, *Troglophilus*, 109, *Diestrammena*, 110, *Rhaphidophora*, 111, *Gammarotettix*, 112, *Protroglophilus*, 113, *Tropidischia*.

6a. Femora and tibiae with two rows of upper and two rows of lower non-articulated subapical denticles (excepting outer lower edge of fore femora, which are without any spines or denticles) (Fig. 1.8-99). Male genital plate with styles...............(6)

Tropidischiinae Scudder, 1897

Composition

The genus *Tropidischia* Scudd. (North America).

6b. Femora without any upper subapical spines or denticles (Fig. 1.8-100). Male genital plate varied..**7a**

7a. Hind tibiae with articulated upper subapical spines (Fig. 1.8-107); first segment of hind tarsi without any spines or denticles; second segment of hind tarsi long and low (see Fig. 1.2-12). Male genital plate without styles............(7) **Dolichopodinae** Brunner-Wattenwyl, 1888

Composition

The genus *Dolichopoda* Bol. (South Europe, Asia Minor, Caucasus, Transcaucasia).

7b. Hind tibiae with non-articulated upper subapical denticles or spines and denticles (Fig. 1.8-108, 112); first segment of hind tarsi with rather large apical spine (spur) and unpaired row of subapical denticles; second segment of hind tarsi varied (Fig. 1.8-89, 92). Male genital plate with styles (or this part of body unknown)...........................**8a**

8a. Second segment of hind tarsi long and high, with large apical spine (spur) (Fig. 1.8-88, 89). Male genital plate with styles ....(8) **Troglophilinae** Karny, 1929

Composition

The genus *Troglophilus* Krauss (from South Europe to Syria).

8b. Second segment of hind tarsi short and low, practically without apical spine (spur) (Fig. 1.8-92) (male genital plate unknown) ....(9) **Protroglophilinae** Gorochov, 1988

Composition

The fossil genera *Protroglophilus* Gor. and *Prorhaphidophora* Chop. (Eocene of East Europe).

(III) Fam. Stenopelmatidae Burmeister, 1838

Composition

Five recent subfamilies. They form two groups, which are considered by some authors as two families ('Stenopelmatidae' and 'Gryllacrididae'). Sometimes the subfamily Schizodactylinae is considered as an additional family (or even superfamily!). But all these taxa are well united by the loss of tympana and the characteristic stridulatory areas of the femoro-abdominal stridulatory apparatus (with more or less vertical rows of large stridulatory denticles) (Fig. 1.15-171–173). The good development of this stridulatory apparatus and the simultaneous loss of tympana are possibly connected with the disappearance of intraspecific acoustic communication and the employment of sound for the scaring of predators. This supposition is corroborated by the great variability of number of stridulatory denticles in the same species and the presence of the femoro-abdominal stridulatory apparatus in the nymphs of Schizodactylinae.

Key to subfamilies

1a. Fore thoracic stigmata with more or less horizontal slit-like aperture and without any hind projections (Fig. 1.11-148–150). Tarsi with not widened pulvilli at sole (Fig. 1.11-153). Wings (if developed) with non-fused $2MA_1$ and MA_2 (Fig. 1.12-156–159). Cerci with bulb-like sensillae at inner proximal part (Fig. 1.15-174) **2a**

1b. Fore thoracic stigmata with more or less vertical slit-like aperture and characteristic hind projection (Fig. 1.11-151, 152). Tarsi with two or three pairs of widened (from slightly to strongly widened) pulvilli at sole (see Figs 1.2-15, 16; 1.11-154, 155). Wings (if developed) with fused $2MA_1$ and MA_2 (Fig. 1.12-160–163). Cerci without any bulb-like sensillae (Fig. 1.15-175).......................**4a**

2a. Sexual dimorphism present (Figs 1.11-134, 135; 1.13-166, 167). Mandibles pricking (Fig. 1.11-134–136, 144, 146); laciniae with two or three teeth, longer than galeae (Fig. 1.11-145, 147). Fore thoracic stigmata with only two lobules forming slit-like aperture (Fig. 1.11-148). All or only middle and hind legs adapted for digging (Fig. 1.13-165–167). Femoro-abdominal stridulatory apparatus absent; first abdominal sternite not widened and distinctly not fused with metathoracic sternite; male abdominal apex without any hooks (Fig. 1.15-178); ovipositor strongly reduced.....................(1) **Oryctopinae** Kevan, 1986

Composition

Two tribes: **Oryctopini** Kevan and **Oryctopterini** Gorochov. These tribes contain two genera (Hindustan, Sri Lanka).

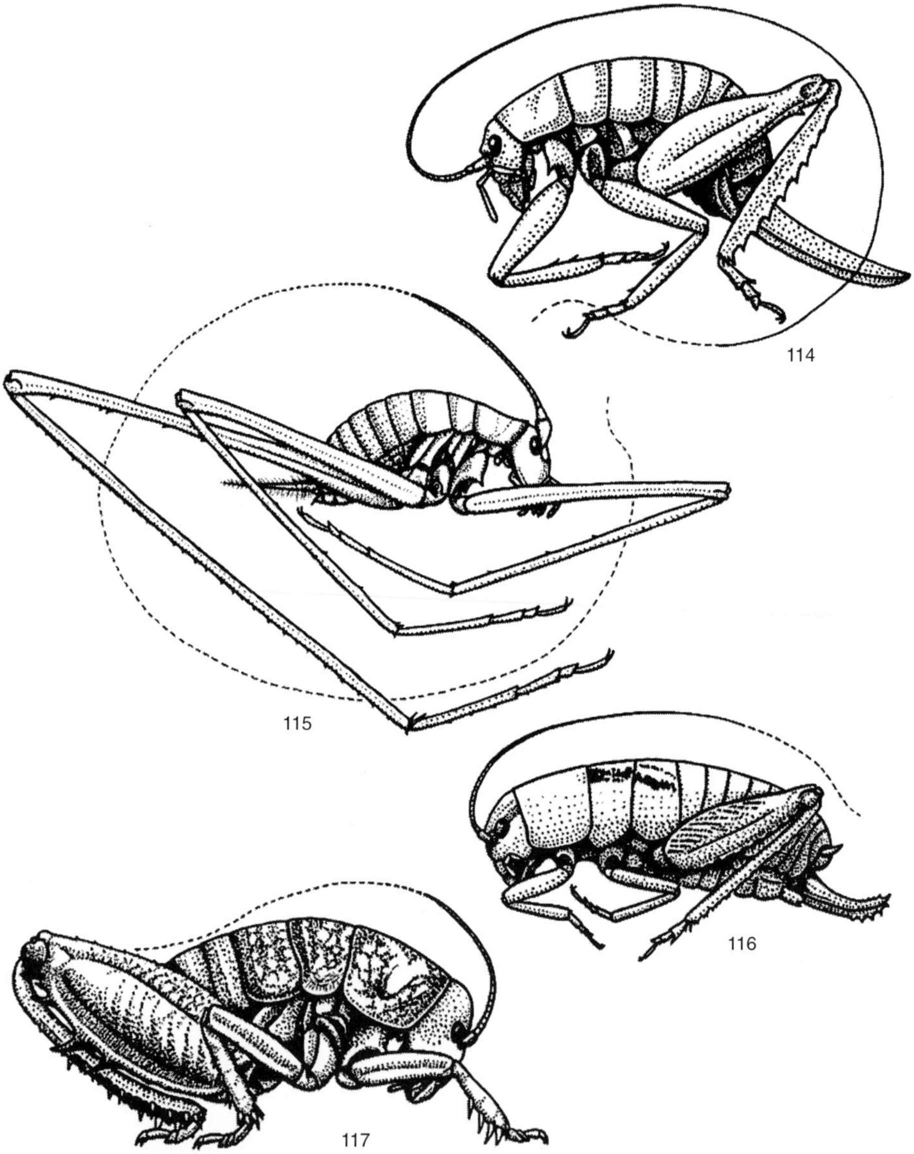

Fig. 1.9. Recent Rhaphidophoridae, general view. 114, *Talitropsis crassicruris* Hutt., female; 115, *Macropathus filifer* Walk., male; 116, *Gammarotettix bovis* Rehn, female; 117, *Daihinia brevipes* Hald., male.

2b. Sexual dimorphism not developed. Mandibles not pricking (Fig. 1.11-131–133, 140, 142); laciniae with three teeth, not longer than galeae (Fig. 1.11-141, 143). Fore thoracic stigmata with somewhat varied lobules forming slit-like aperture (Fig. 1.11-149, 150). All or only fore legs more or less adapted for digging (Fig. 1.13-164). Femoro-abdominal stridulatory apparatus developed or absent; first abdominal sternite varied; male abdominal apex with hooks (Fig. 1.15-176, 177); ovipositor only shortened (Fig. 1.15-181–186) **3a**

3a. Head with slight, but distinct, rostrum between antennae (Fig. 1.11-131, 132). Fore thoracic stigmata with two normal and third rudimentary lobules forming slit-like aperture (Fig.

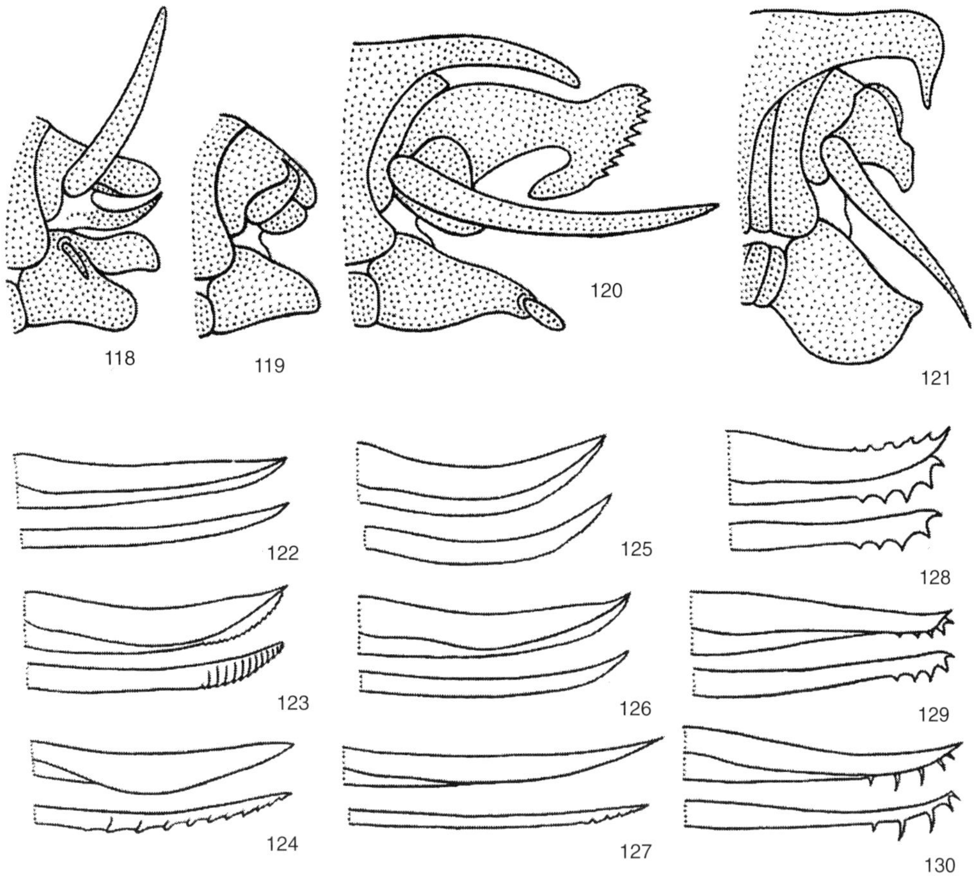

Fig. 1.10. Recent Rhaphidophoridae (schematic). 118–121, Male abdominal apex from side: 118, *Heteromallus*, 119, *Gammarotettix*, 120, *Pararhaphidophora*, 121, *Gigantettix*; 122–130, ovipositor and its lower valva from side: 122, *Tropidischia*, 123, *Dolichopoda*, 124, *Troglophilus*, 125, *Rhaphidophora*, 126, *Diestrammena*, 127, *Hadenoecus*, 128, *Gammarotettix*, 129, *Ceutophilus*, 130, *Daihinia*.

1.11-149). Femoro-abdominal stridulatory apparatus developed (Fig. 1.15-171); first abdominal sternite not widened and distinctly not fused with metathoracic sternite; male abdominal apex with hooks at upper part of anal plate (tenth tergite + epiproct) (Fig. 1.15-176); lower and inner valvae of ovipositor fused with each other only at base (Fig. 1.15-182, 184)..(2) **Siinae** Gorochov, 1988

Composition

The genera *Sia* Gieb. and *Maxentius* Stål (Malay archipelago, South Africa).

3b. Head without distinct traces of rostrum between antennae (Fig. 1.11-133). Fore thoracic stigmata with only two lobules forming slit-like aperture (Fig. 1.11-150). Femoro-abdominal stridulatory apparatus absent; first abdominal sternite widened and almost fused with metathoracic sternite; male abdominal apex with hooks at lateral parts of anal plate (tenth tergite + epiproct) (Fig. 1.15-177); lower and inner valvae of ovipositor fused with each other at base and middle part (Fig. 1.15-186) ....(3) **Stenopelmatinae** Burmeister, 1838

Composition

The genera *Stenopelmatus* Burm., *Ammopelmatus* Tinkh., *Viscainopelmatus* Tinkh. and *Stenopelmatopterus* Gor. (North and Central America).

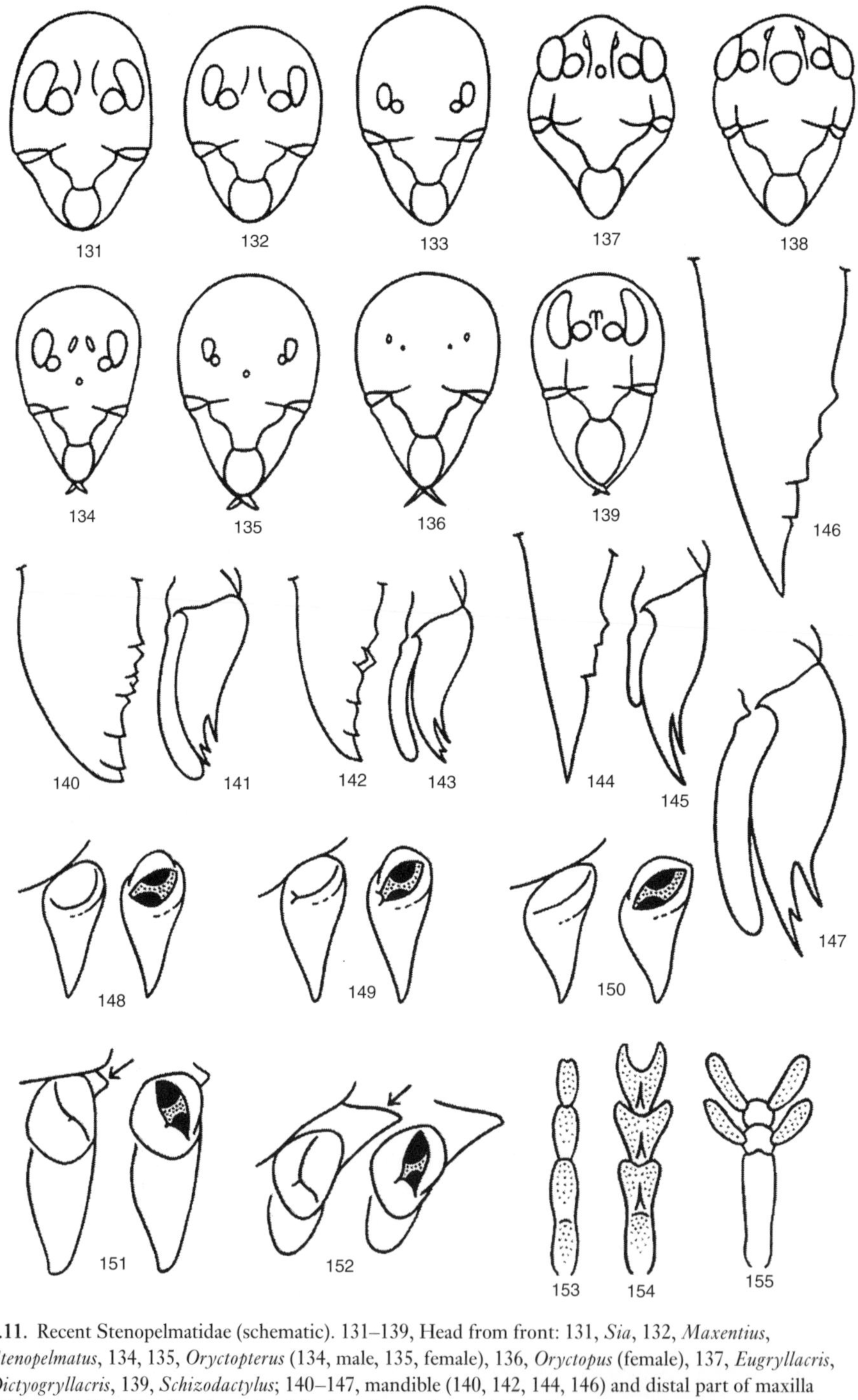

Fig. 1.11. Recent Stenopelmatidae (schematic). 131–139, Head from front: 131, *Sia*, 132, *Maxentius*, 133, *Stenopelmatus*, 134, 135, *Oryctopterus* (134, male, 135, female), 136, *Oryctopus* (female), 137, *Eugryllacris*, 138, *Dictyogryllacris*, 139, *Schizodactylus*; 140–147, mandible (140, 142, 144, 146) and distal part of maxilla (141, 143, 145, 147): 140, 141, *Maxentius*, 142, 143, *Stenopelmatus*, 144, 145, *Oryctopus* (female), 146, 147, *Oryctopterus* (female); 148–152, fore thoracic stigma with closed aperture (and part of pronotal edge; arrow points at hind projection) and with open aperture: 148, *Oryctopterus*, 149, *Maxentius*, 150, *Stenopelmatus*, 151, *Schizodactylus*, 152, *Diaphanogryllacris*; 153–155, three proximal segments of hind (153, 154) and fore (155) tarsi from below: 153, *Sia*, 154, *Borneogryllcris*, 155, *Schizodactylus*.

4a. Rostrum of head between antennae usually rather wide and not tubercle-like (Fig. 1.11-137, 138). Fore thoracic stigmata with distinct rudiment of third lobule and rather large hind projection (Fig. 1.11-152). Tegmina (if wings well-developed) with median part not folded like a fan (distal part of wing cannot be rolled into a ring) (Figs 1.12-162; 1.14-168). Fore coxae with spine; tarsi with three pairs of lateral lobes of more or less widened pulvilli (see Figs 1.2-15; 1.11-154). Male abdominal apex with enlarged ninth tergite (Fig. 1.15-180); ovipositor usually well-developed (Fig. 1.14-168).......(4) **Gryllacridinae** Blanchard, 1845

Composition

This largest subfamily of Stenopelmatoidea contains around 80 genera, but has not been divided into any tribes up to now (almost all tropical and many subtropical countries).

4b. Rostrum of head between antennae narrow and consists of one pair of small rounded tubercles (Fig. 1.11-139). Fore thoracic stigmata with indistinct rudiment of third lobule and small hind projection (Fig. 1.11-151). Tegmina (if wings well-developed) with very large fan-like folded median part (distal part of wings can be rolled into a ring shape) (Figs 1.12-160; 1.14-169). Fore coxae without spine; tarsi with two pairs of lateral lobes of strongly widened pulvilli (hind tarsi with one additional pair of lateral processes at first segment) (see Figs 1.2-16; 1.11-155). Male abdominal apex with non-enlarged ninth tergite (Fig. 1.15-179); ovipositor strongly reduced................................(5) **Schizodactylinae** Blanchard, 1845

Composition

Two tribes: **Schizodactylini** Blanchard and **Comicini** Handlirsch. These tribes contain two genera (from Asia Minor to Hindustan and Burma, South Africa).

Phylogeny of Stenopelmatoidea

(Fig. 1.16)

The phylogenetic relations of the three families of Stenopelmatoidea are more or less clear. The superfamily has the following autapomorphies: the parallel or almost parallel longitudinal venation of tegmina, the loss of tegminal stridulatory apparatus and possibly the rise of the femoro-abdominal stridulatory apparatus. The autapomorphies of Rhaphidophoridae are also rather dependable: the presence of only one pulvillus at the sole of first tarsal segments, the absence of wings and possibly the loss of the distinct femoro-abdominal stridulatory apparatus. Stenopelmatidae have only one possible original autapomorphy – the abdominal stridulatory areas with the more or less vertical rows of large denticles (see Figs 1.2-30; 1.15-171–173). It is necessary to emphasize that the strong development of the femoro-abdominal stridulatory apparatus in the latter family is not accompanied by the restoration of tibial tympana. Two of the above-mentioned families are united by only one dependable synapomorphy in the structure of the ovipositor – the lower valvae are covered with the lower parts of upper valvae on the outside (see Fig. 1.2-35–37). This peculiarity is present in all Rhaphidophoridae and some Stenopelmatidae (Gryllacridinae, Stenopelmatinae, Siinae). The ovipositor of other Stenopelmatidae is strongly reduced. Anosto-stomatinae are characterized by a dissimilar ovipositor – the lower parts of upper valvae are inserted in special furrows of the lower valvae (see Fig. 1.2-34). The latter type of ovipositor is possibly plesiomorphic and also characteristic of Hagloidea (the female of only one Recent genus of Hagloidea is unknown; two other possible Recent genera of the same superfamily have a strongly reduced ovipositor) as the ovipositor of Recent Tettigonioidea (and even Grylloidea) may have derived from it (see Fig. 1.2-32, 33). The other general character for Rhaphidophoridae and Stenopelmatidae – the loss of tympana (and apparently the loss of intraspecific acoustical communication) – is also a possible synapomorphy.

The phylogenetic position of Anostostomatidae is more complex. I cannot exclude that some extinct primitive group of Anostostomatidae was ancestral for the two other families of Stenopelmatoidea, as Anostostomatidae are characterized by mainly primitive characteristics (it is not quite clear whether the presence of chevron ridges at the outer sides of the hind femora is the original synapomorphy of only Recent subfamilies of Anostostomatidae, or the peculiarity of a general ancestor of all Stenopelmatoidea, as these ridges are easily reduced and even disappear in some subfamilies of Anostostomatidae).

Two opinions about the most primitive group of the Recent Anostostomatidae are possible.

1. The tenth tergite of the male abdomen in most Anostostomatidae is divided into two halves by the membranous (or almost membranous) area (see Fig. 1.7-77, 80–85), but this tergite in Anabropsinae is usually undivided. The latter condition is rather characteristic of Rhaphidophoridae, Stenopelmatidae and other Recent Ensifera. It would be logical to consider that this condition is a plesiomorphy and to put Anabropsinae near the base of Recent Anostostomatidae stock, as their wing venation (see Fig. 1.4-58, 59) is also rather primitive.
2. But the venation of tegmina in Cratomelinae and of hind wings in the enigmatic genus *Transaevum* (see Fig. 1.1-5, 6) looks more primitive. The tenth tergite of the male abdomen in the Cratomelinae and (according to the description by Johns (1997)) *Transaevum* is divided. If this division is the primitive condition, the latter taxa must be put near the base of Recent Anostostomatidae stock (I also cannot exclude that *Transaevum* is the possible Recent representative of Zeuneropterinae).

The Anabropsinae have the following autapomorphies: the presence of paired small hind lobes of the ninth abdominal tergite over hooks of the tenth abdominal tergite in the male, possibly the originally (for only this subfamily) undivided tenth abdominal tergite of the male or (if the latter character is plesiomorphic) the straightened (independently from other Anostostomatidae) tegminal CuP.

The phylogenetic scheme of other Anostostomatidae (Fig. 1.16) is supported by the less contradictory (but single and scarcely dependable) synapomorphies. The membranous epiproct of the male (see Fig. 1.7-80, 81) is a possible synapomorphy of only Leiomelinae (their autapomorphy: the loss of chevron ridges of hind femora) and Cratomelinae (their autapomorphy: the development of the tegmino-alary stridulatory apparatus). The loss of the curved shape of CuP in the tegmina of Leiomelinae (see Fig. 1.4-62) was also probably passed independently from other Anostostomatidae.

The subfamilies Cooloolinae, Anostostomatinae, Lezininae, Lutosinae, and Deinacridinae are united by the combination of two derivative characters (synapomorphy): the straight tegminal CuP (see Fig. 1.4-56, 60) and the presence of a more or less membranous area between the tenth abdominal tergite and the epiproct in the male (see Fig. 1.7-77, 82–85). The comparatively short distance between the hooks of the tenth abdominal tergite of the male is a possible synapomorphy of Lezininae, Anostostomatinae, Lutosinae and Deinacridinae (see Fig. 1.7-82–85). The autapomorphies of Cooloolinae consist of the characters connected with the strong development of a digging mode of life.

The rather wide, more or less membranous area above the male epiproct is characteristic of Anostostomatinae, Lutosinae and Deinacridinae (the possible synapomorphy) (see Fig. 1.7-83–85). Lezininae have lost the wings, the male paraproctal processes, the styles of the male genital plate and the ovipositor (this loss is their possible autapomorphy). Anostostomatinae and Deinacridinae are united by the trend toward the development of sexual dimorphism in the shape of the mandibles (see Fig. 1.3-44–50, 55). The autapomorphies of Lutosinae and Deinacridinae are connected with jumping: the former are strongly adapted for jumping (the short arched body, the strong hind femora) and the latter have lost the ability to jump strongly (the hind femora are only slightly adapted to jumping, with chevron ridges strongly reduced). Anostostomatinae do not have any distinct autapomorphies. It is possible that this group was ancestral for the Deinacridinae.

The family Rhaphidophoridae is divided into two groups. One of them includes a single subfamily, Macropathinae, which is characterized by a rather dependable plesiomorphic peculiarity – the presence of paired apical spines (spurs) at the two basal segments of the hind tarsi (see Fig. 1.8-86, 87). The autapomorphies of Macropathinae include the male genital plate divided into upper and lower parts (Fig. 1.10-118) and the male genitalia with the rather complicated system of sclerites (partly analogous to that of Gryllidae: with epiphallus, endoparameres, slight guiding rod, and sometimes rami) (Fig. 1.17-188–191). All the other subfamilies of Rhaphidophoridae have the dependable original synapomorphy – the presence of the only unpaired upper apical spine (spur) at two basal segments of hind tarsi (see Fig. 1.8-88).

The subfamilies Ceuthophilinae and Gammarotettiginae are united by the possible original synapomorphy in the structure of the ovipositor: the lower edges of the lower valvae with

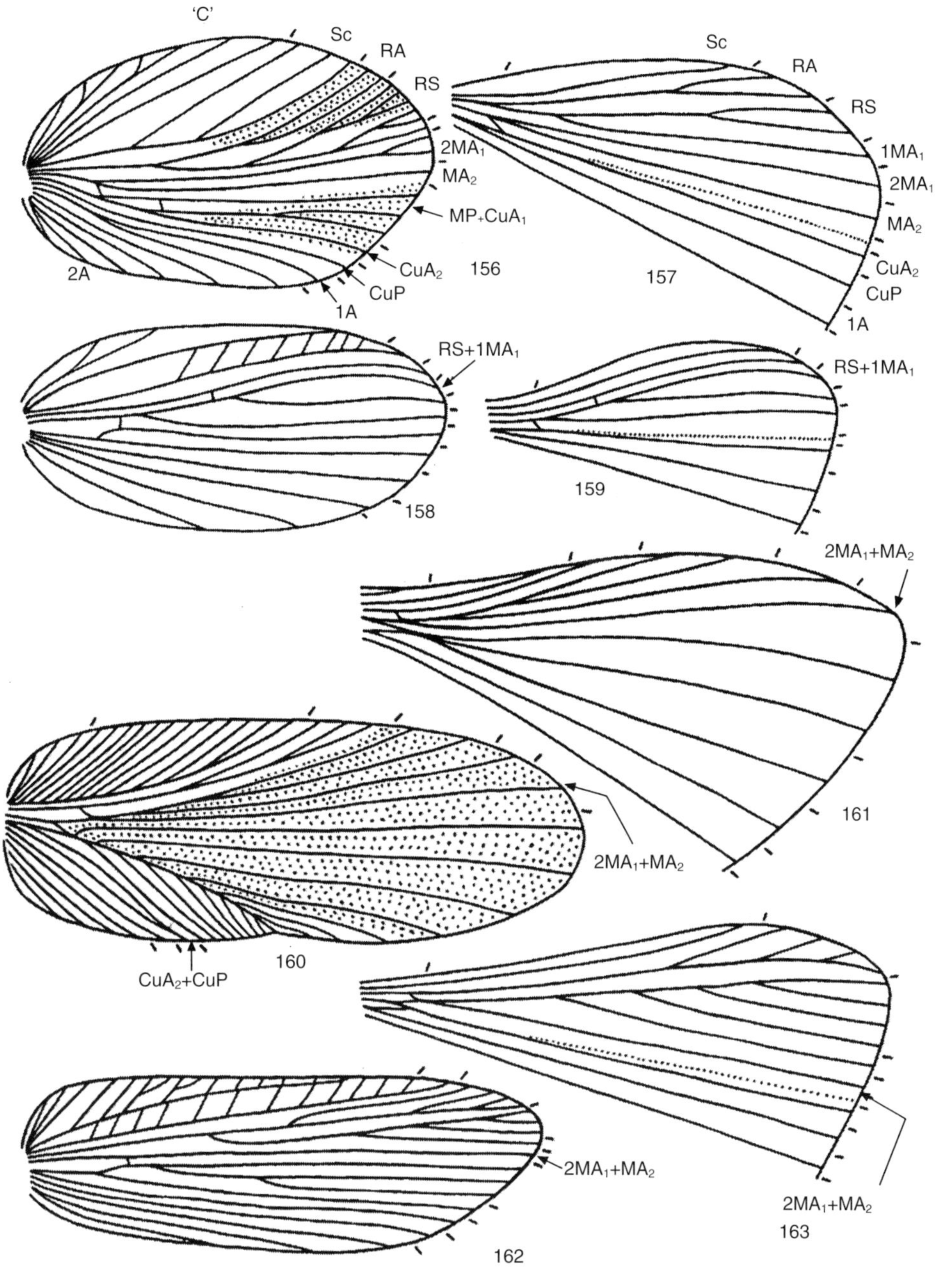

Fig. 1.12. Recent Stenopelmatidae (schematic) (fan-like parts of tegmina dotted). 156, 157, *Sia*: 156, tegmen, 157, costal part of hind wing; 158, 159, *Oryctopterus*, male: 158, tegmen, 159, costal part of hind wing; 160, 161, *Schizodactylus*: 160, tegmen, 161, costal part of hind wing; 162, *Hadrogryllacris*, tegmen; 163, *Paragryllacris*, costal part of hind wing.

the long, narrow teeth at the distal part (Fig. 1.10-128–130); it is necessary to emphasize that some representatives of Ceuthophilinae have the secondary modifications of the ovipositor, which may be independently rather similar to that of some other Rhaphidophoridae (see Fig. 1.10-123, 124,

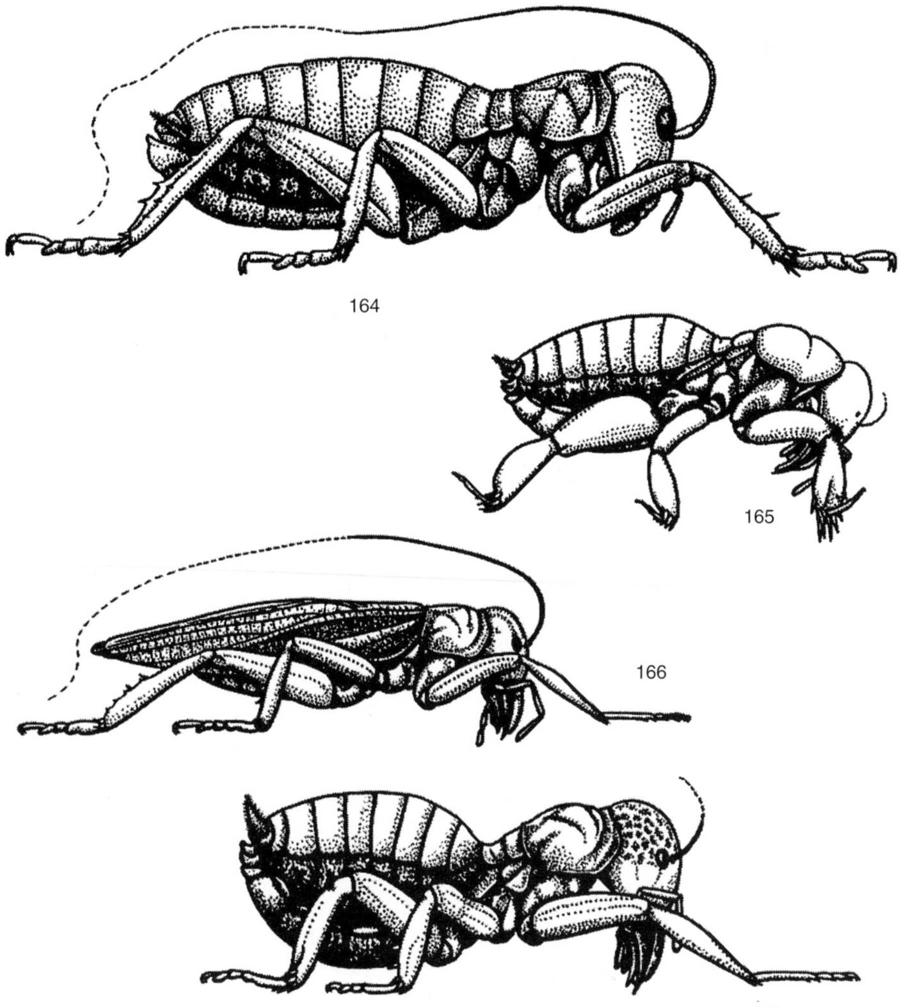

Fig. 1.13. Recent Stenopelmatidae, general view. 164, *Maxentius pinguis* (Walk.), male; 165, *Oryctopus prodigiosus* Bol., female; 166, 167, *Oryctopterus lagenipes* (Karny): 166, male, 167, female.

127). There are the following autapomorphies of Gammarotettiginae: the absence of a spine at the fore coxae and the denticulated upper edge of the distal part of upper valvae of the ovipositor. The more or less dependable autapomorphies of Ceuthophilinae are unknown. It is possible that Ceuthophilinae were the ancestral group for Gammarotettiginae.

I cannot find any dependable synapomorphies for the Protroglophilinae, Dolichopodinae, Troglophilinae, Rhaphidophorinae and Aemodogryllinae, but their general origin (after the separation of the above-mentioned American stock) is very possible, as three of these subfamilies (Dolichopodinae, Troglophilinae, Aemodogryllinae) have a trend to develop an epiphallic sclerite at the upper surface of the base of the dorsal lobe in the male genitalia (Ceuthophilinae and Gammarotettiginae have a trend to form an epiphallic sclerite at the distal part of the dorsal lobe in the male genitalia) (Fig. 1.17-192–202).

The Dolichopodinae and Troglophilinae are united by the rather similar transverse epiphallic sclerite (synapomorphy) (Fig. 1.17-201, 202).

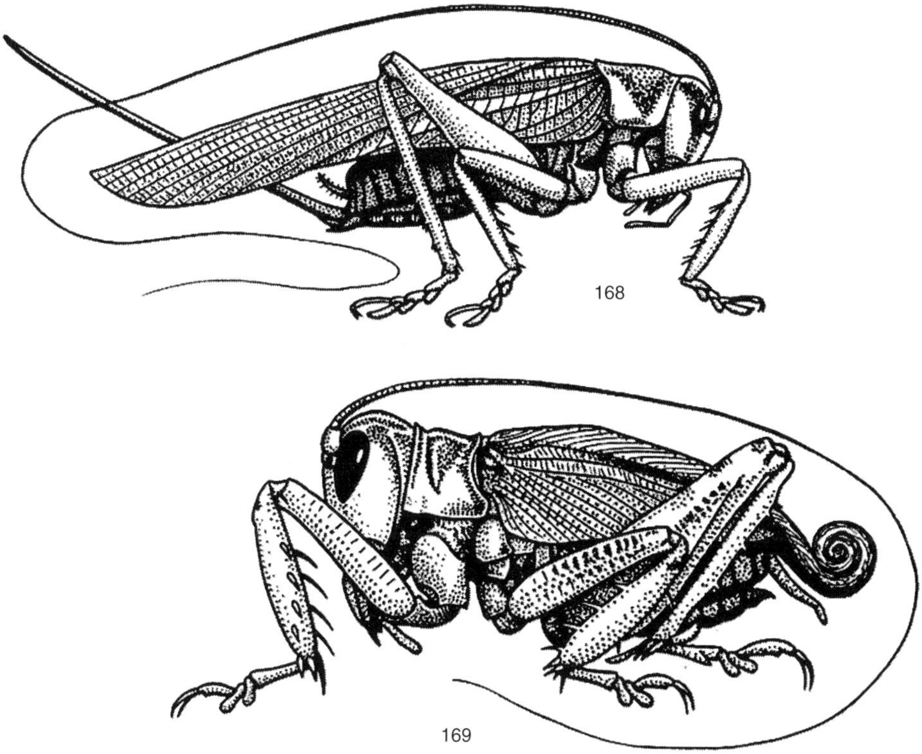

Fig. 1.14. Recent Stenopelmatidae, general view. 168, *Diaphanogryllacris laeta* (Walk.), female; 169, *Schizodactylus hesperus* B.-Bien., male.

Their autapomorphies include the structure of the tarsi: Dolichopodinae – first segment without any denticles, Troglophilinae – second segment rather high. Rhaphidophorinae and Aemodogryllinae are characterized by the presence of one pair of rather long apical spines (spurs) on the middle femora (the possible synapomorphy). The autapomorphies of Rhaphidophorinae are the presence of a distinct inner apical spine (spur) at the fore femora and the development of very long apical spines (spurs) on the hind tibiae. Aemodogryllinae have the following autapomorphies: the development of a long second segment of the tarsi, the presence of a rather long outer apical spine (spur) on the fore femora and the absence of styles on the male genital plate.

Protroglophilinae (extinct) are possibly the most primitive group of these five subfamilies. Their genital complex is unknown, but the all other characters look plesiomorphic. It is possible that this subfamily was ancestral for Rhaphidophorinae + Aemodogryllinae and Troglophilinae + Dolichopodinae (see Fig. 1.16). Moreover, I cannot exclude the possibility that the enigmatic Japanese genera *Anoplophilus* and *Alpinanoplophilus* may be Recent representatives of the Protroglophilinae, as the peculiarities of their structure are not opposed to the diagnosis of fossil Protroglophilinae.

The phylogenetic position of Tropidischiinae is not clear at present. It is necessary only to indicate that this subfamily belongs to the northern stock of the Rhaphidophoridae, which is characterized originally by the unpaired apical spine (spur) on two basal segments of the hind tarsi.

The family Stenopelmatidae consists of two Recent branches. One of them includes Stenopelmatinae, Oryctopinae and Siinae. These subfamilies are united by two more or less dependable synapomorphies: the fore thoracic stigmata with almost horizontal slit-like aperture (see Fig. 1.11-148–150), the base of the inner part

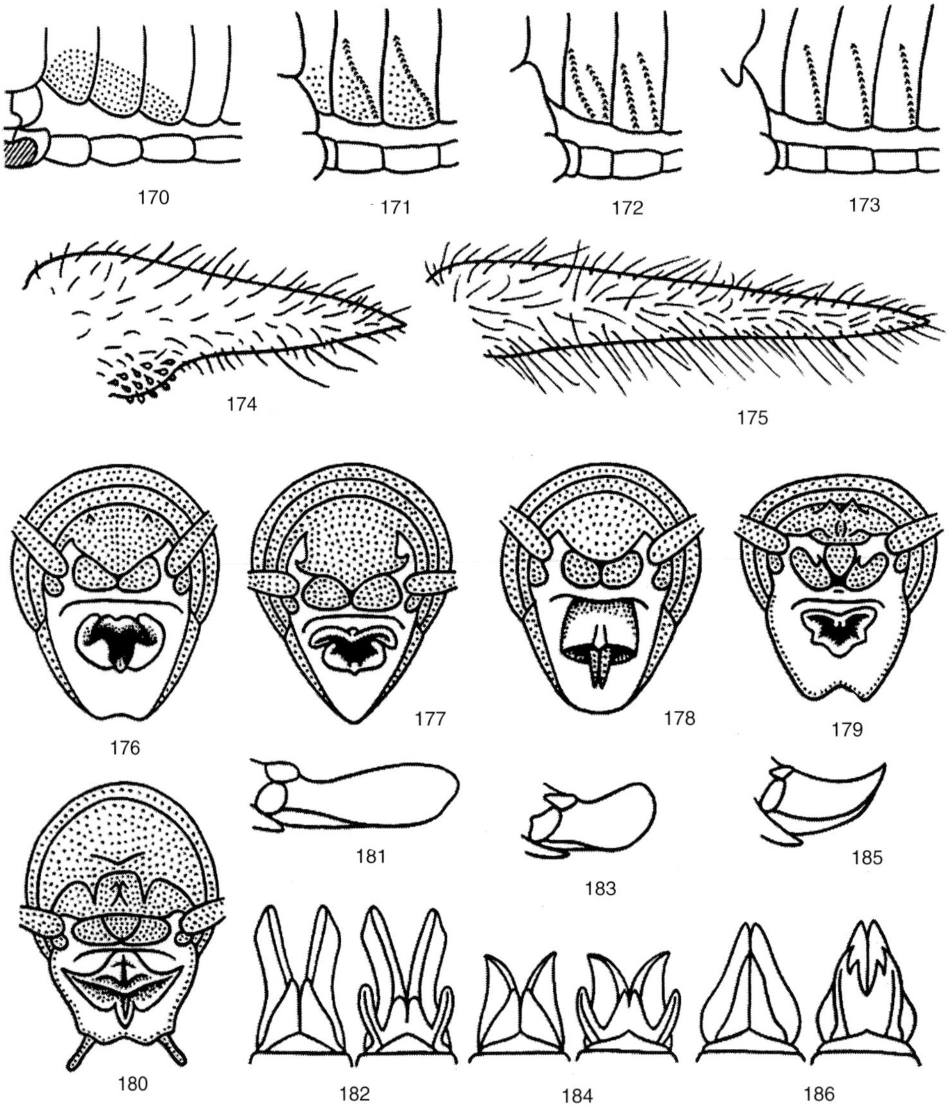

Fig. 1.15. Recent Anostostomatidae (170) and Stenopelmatidae (171–186) (schematic). 170–173, Abdominal stridulatory area of femoro-abdominal stridulatory apparatus: 170, *Lutosa*, 171, *Maxentius*, 172, *Borneogryllacris*, 173, *Schizodactylus* (nymph); 174, 175, cercus from above: 174, *Oryctopterus*, 175, *Gryllacris*; 176–180, male abdominal apex from behind: 176, *Maxentius*, 177, *Stenopelmatus*, 178, *Oryctopterus*, 179, *Schizodactylus*, 180, *Diaphanogryllacris*; 181–186, ovipositor from side (181, 183, 185) and from below (with lower valvae closed and moved aside) (182, 184, 186): 181, 182, *Sia*, 183, 184, *Maxentius*, 185, 186, *Stenopelmatus*.

of the cerci with characteristic bulb-like sensillae (see Figs 1.2-31; 1.15-174) (it is interesting that very similar sensillae are independently developed in Grylloidea). There are also the less dependable synapomorphies of these subfamilies: the digging or almost digging mode of life and the shortening of the ovipositor.

The Stenopelmatinae and Oryctopinae have rather slight synapomorphies: the more or less narrow (pricking or almost pricking) distal part of mandibles (Fig. 1.11-142, 144, 146) and the strong trend to the digging mode of life. The autapomorphy of the Stenopelmatinae is the widened first abdominal sternite almost fused with the metathoracic sternite. The autapomorphies of the Oryctopinae are the clearly pricking mandibles and the more or less shortened galeae (see Fig. 1.11-144–147). The Siinae differ from the above-mentioned subfamilies in the possible plesiomorphic structure of abdominal stridulatory areas, which are irregularly denticulated and provided with vertical rows of large denticles (see Fig. 1.15-171). Their autapomorphies are not clear. It is possible that this group was ancestral for Oryctopinae + Stenopelmatinae.

The second branch of Stenopelmatidae consists of the Gryllacridinae and Schizodactylinae. These have the following dependable synapomorphies: the fore thoracic stigmata with the more or less vertical slit-like aperture and a characteristic hind projection (see Fig. 1.11-151, 152) and the tarsi with widened (lobe-like or almost lobe-like) pulvilli (see Figs 1.2-15, 16; 1.11-154, 155). The possible autapomorphies of Gryllacridinae are the development of the enlarged ninth abdominal tergite in males (see Fig. 1.15-180) and a rather large hind projection of the fore thoracic stigmata (see Fig. 1.11-152). The autapomorphies of the Schizodactylinae are more dependable: the characteristic articulated pulvilli and non-articulated projections of the tarsi, the very peculiar structure of the tegmina (if the wings are not lost) which can be folded like a fan (the distal parts of the wings can be rolled into a ring) (see Figs 1.12-160; 1.14-169), the digging mode of life and the reduction of the ovipositor.

Evolutionary Tendencies in Mode of Life

The mode of life of the Aboilinae (the extinct subfamily of the Prophalangopsidae (Hagloidea) – the possible ancestral group for the Stenopelmatoidea and Tettigonioidea) was possibly somewhat similar to that of the Recent representatives of the Anabropsinae and/or some Tettigoniidae (*Tettigonia*, *Drymadusa*, *Hexacentrus*). The Aboilinae, judging by the fossil imprints, were predators (Gorochov, 1995) or almost phytophagous forms feeding on pollen grains (Krassilov *et al.*, 1997). It is possible that they had zoophagous and phytophagous habits simultaneously (as in the Recent species of *Drymadusa*). The Aboilinae probably spent most of their life on plants, but they also had adaptations for jumping on the ground (the well-developed apical spines (spurs) of the hind tibiae) and their oviposition was made in soil. They did not have any digging habits.

The origin of the Stenopelmatoidea was possibly connected with a transfer of one of the descendants of the predatory Aboilinae to the digging mode of life. Most Stenopelmatoidea have zoophagous or even predatory habits, and only some of them transferred to a mixed diet (for example, most Rhaphidophoridae) or even to phytophagy (for example, some Gryllacridinae and Deinacridinae). During this transfer the ancestor of Stenopelmatoidea may have acquired slight adaptations to digging and the partly parallel longitudinal venation of the tegmina (as in Gryllotalpidae). It is possible that the loss of a tegminal stridulatory apparatus and the rise of a femoro-abdominal stridulatory apparatus were connected with this process.

Further evolution of the modes of life in the Stenopelmatoidea includes the rather numerous trends toward strengthening the ability to dig or the weakening of this ability (up to the complete loss of any digging habits) (Fig. 1.18-203). It is proposed that an analogous (oscillatory) type of evolution is usual for the taxa during their assimilation of new adaptive zones (Gorochov, 1995).

If the genus *Transaevum* is representative of the Zeuneropterinae, the latter subfamily (according to Johns' description) had the tendency toward a weakening of the digging ability. The Cratomelinae and Cooloolinae preserve (or independently acquired again) the digging habit, but the Cooloolinae attained a stronger specialization for digging. The adaptations for digging in the Lezininae are rather slight, and developed independently, as the loss of the upper spines of their fore tibiae is probably the inheritance of a non-digging ancestor. Leiomelinae possibly have a trend toward the opposite mode of life, but with traces of adaptations for digging (the fore and middle tibiae are thickened). It is impossible to exclude that the origin of the practically unstudied mode of life in this group may be more complicated.

The complete loss of the digging mode of life is characteristic of the Anabropsinae and Lutosinae. This process possibly developed independently in both subfamilies. The loss of digging ability in Deinacridinae and Anostostomatinae is possibly only partial (and maybe also indepen-

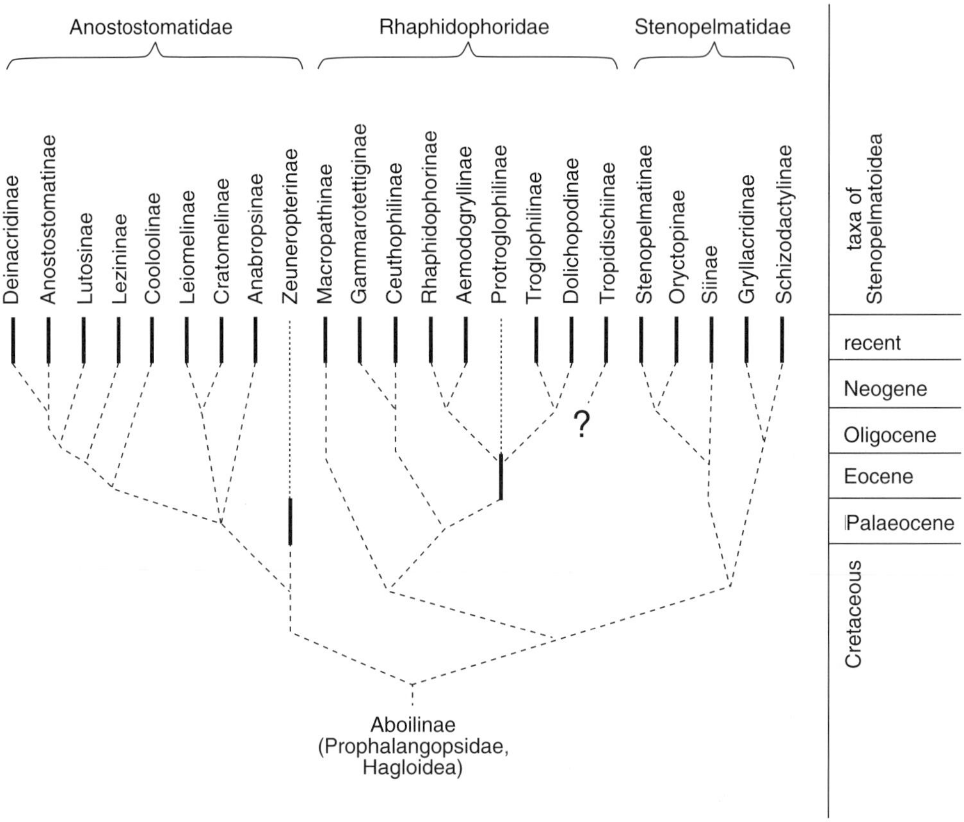

Fig. 1.16. Phylogenetic scheme of Stenopelmatoidea. Vertical line, taxon represented: bold line; dependably; line from dots; problematically interrupted line, hypothetically; oblique interrupted line, hypothetical phylogenetic connection (if three of these lines go from 1 dot, the order of their branching unknown).

dent). Some of the Anostostomatinae (*Mimnermus*) probably preserve or acquired again the slight adaptation for digging. It is possible that the general ancestor of the Rhaphidophoridae and Stenopelmatidae originally had a trend toward the loss of a digging ability, as its ovipositor acquired certain improvements in the articulation of the valvae (adaptation to the digging mode of life leads to a reduction of the ovipositor, not to its improvement). The Rhaphidophoridae originally preserved this trend and attained the complete loss of any digging ability. They were adapted for the more or less geophilous life, with a strong jumping ability, and had the ability to climb rocks and bark. This mode of life is preserved in most extant Rhaphidophoridae. Many of them live in hollows or caves. Sometimes obligatory troglobionts occur (their coloration is very light, without or almost without any spots, and their eyes are partly reduced or completely lost; the coloration of non-obligatory troglobionts is darkish or spotted, and their eyes are well-developed). Some Ceuthophilinae (Daihiniini) have returned to digging habits (Fig. 1.18-203).

The primitive Stenopelmatidae possibly returned to a more or less digging mode of life (their descendants are characterized by a large head with a strong mandibles and rather short, robust legs). Further specializations for digging were possibly the main evolutionary tendency of the general ancestor of Siinae, Stenopelmatinae and Oryctopinae (the shortening of the ovipositor and the development of bulb-like sensillae of the cerci (the latter character is also present in the Grylloidea; it possibly arose as one of the adaptations to life in burrows)). In the Siinae, the weaking of a digging ability may be observed (the legs of *Sia* have comparatively slight digging

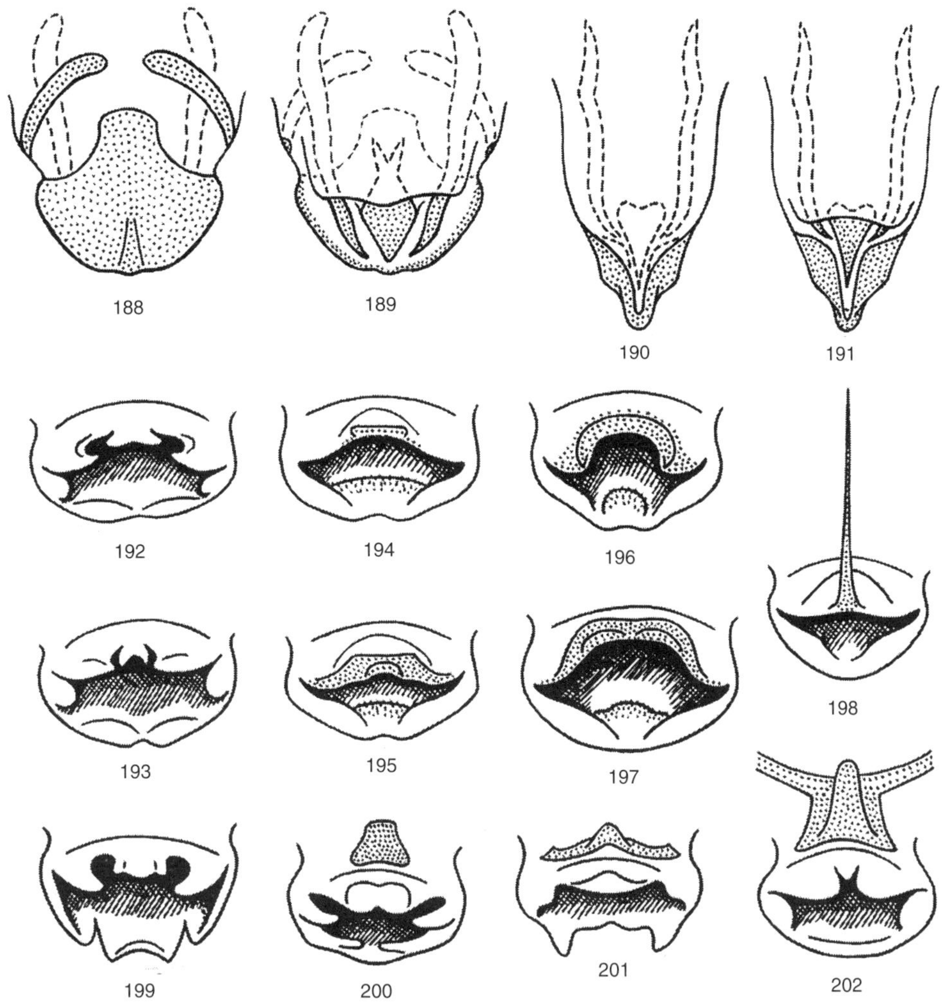

Fig. 1.17. Male genitalia of Rhaphidophoridae (schematic). 188–191, Macropathinae (188, 190, from above; 189, 191, from below): 188, 189, *Talitropsis*, 190, 191, *Heteromallus*; 192, Gammarotettiginae (*Gammarotettix*, from behind); 193–198, Ceuthophilinae (from behind): 193, *Hadenoecus*, 194, *Macrobaenetes*, 195, *Daihinia*, 196, *Ceuthophilus*, 197, *Geotettix*, 198, *Argyrtes*; 199, Rhaphidophorinae (*Rhaphidophora*, from behind); 200, *Aemodogryllinae* (*Diestrammena*, from behind); 201, Troglophilinae (*Troglophilus*, from behind); 202, Dolichopodinae (*Dolichopoda*, from behind).

spines, but in *Maxentius* these spines are partly reduced).

The Stenopelmatinae + Oryctopinae preserve the trend of strengthening digging habits (the development of pricking mandibles in both groups; this structure of the mandibles is a possible adaptation to feeding on slow-moving invertebrates in the soil) and independently intensify it (as shown by the strong development of pricking mandibles in all the Oryctopinae and the rise of almost fused metathoracic and first abdominal sternites in the Stenopelmatinae). Some Oryctopinae (Oryctopini) developed a strong ability to dig (their forelegs are well adapted to digging), but other Oryctopinae (Oryctopterini) are beginning to go in the opposite direction (their forelegs have lost digging specializations).

The general ancestor of the Gryllacridinae and Shizodactylinae may have developed the loss of a digging ability in connection with the transfer to a phytophilous mode of life (indicated by the development of rather large lobe-like pulvilli of the tarsi for attachment to leaves). One of their descendant lines probably returned to the digging, geophilous mode of life and gave rise to the ancestral Shizodactylinae (their lobe-like pulvilli on the tarsi are preserved, but they have lost their primary adaptive function). The other descendants evidently preserved the phytophilous mode of life and intensified it (Gryllacridinae), but a small number of them returned to a partly geophilous and even almost burrowing mode of life (the latter process possibly took place more than once) (Fig. 1.18-203).

Notes on Distribution

The fossil Zeuneropterinae were found in the Palaeocene of Scotland and Canada; the possible Recent representative of this taxon maybe lives in Australia (the relict area?). The rather wide distribution of Anabropsinae (from America and Africa to East Asia and probably Australia) is possibly a result of the comparatively recent range expansion, as the genera of this subfamily from the different regions are very similar (young). This conclusion is surprising, as this subfamily is rather primitive morphologically.

The endemic Recent subfamilies Cratomelinae, Leiomelinae and Lutosinae (South America), Cooloolidae (Australia) and Deinacridinae (New Zealand) were possibly formed within the limits of their recent areas (Cratomelinae and Leiomelinae – from the very primitive general ancestor maybe during the isolation of South America; the other subfamilies – from the less primitive ancestors, which arrived in these isolated lands possibly with the help of rafts of fallen trunks). Lezininae (which are characteristic of arid territories) were probably also formed within the limits of their recent area (South-West and Central Asia, North and East Africa). The area of Anostostomatinae includes Africa, Madagascar, Australia, New Caledonia, and maybe New Zealand, and is possibly narrowing now.

The distribution of the Recent Rhaphidophoridae suggests a more or less plausible hypothesis about their geographical history. The primitive Rhaphidophoridae with the paired spines of the hind tarsi originally disappeared from the tropical zone as a possible result of the rise of Recent types of tropical forests. Their descendants (which gave the beginnings of Recent forms) were originally preserved only in a southern and a northern refuge. The inhabitants of the southern one preserved the plesiomorphic tarsi. Formerly I proposed more than one southern refuge (Gorochov, 1995), but now, after study of the male genital complex of the Macropathinae from New Zealand, Australia and South America, my opinion is different, as the genera from these regions are very similar, and their ancestors must have arrived in most of these lands reasonably recently with the help of rafts (the only genus from South Africa usually included in Macropathinae is inadequately studied).

The northern refuge contained the ancestors of all other Recent Rhaphidophoridae united by the dependable synapomorphy, unpaired spines of the hind tarsi. The descendants of the latter arrived in North America (the general ancestor of Ceuthophilinae + Gammarotettiginae) or Eurasia (Protroglophilinae), probably across Beringia. The division into these two stocks took place not later than the Eocene, as the Protroglophilinae are known from Eocene Baltic amber. Then the North American stock was divided into the two above-mentioned Recent subfamilies living in only this region up to now. The Protroglophilinae possibly extended to East Asia (the Recent genera *Anoplophilus* and *Alpinanoplophilus*), generating two branches (western and eastern) as a result of the development of enormous arid territories without forests in the central part of Eurasia.

The western branch gave rise to the mainly European forest Troglophilinae and Dolichopodinae, and the eastern branch gave rise to the Rhaphidophorinae and Aemodogryllinae. The two latter subfamilies occupied the forests of the southern half of East Asia (including Hindustan). They also extended to the adjacent islands, including the Kuril Islands (Aemodogryllinae) and even Samoa (Rhaphidophorinae), presumably with the help of rafts (voyage by wood rafts is evidently the usual mode of distribution of the Rhaphidophoridae, as they frequently live in hollow trunks and may well oviposit in the mouldering wood). The history of the distribution of the Recent North American subfamily Tropidischiinae is unclear.

Two subfamilies of the Stenopelmatidae have relict distributions: the Siinae (South Africa and the Malay archipelago) and the Schizodactylinae

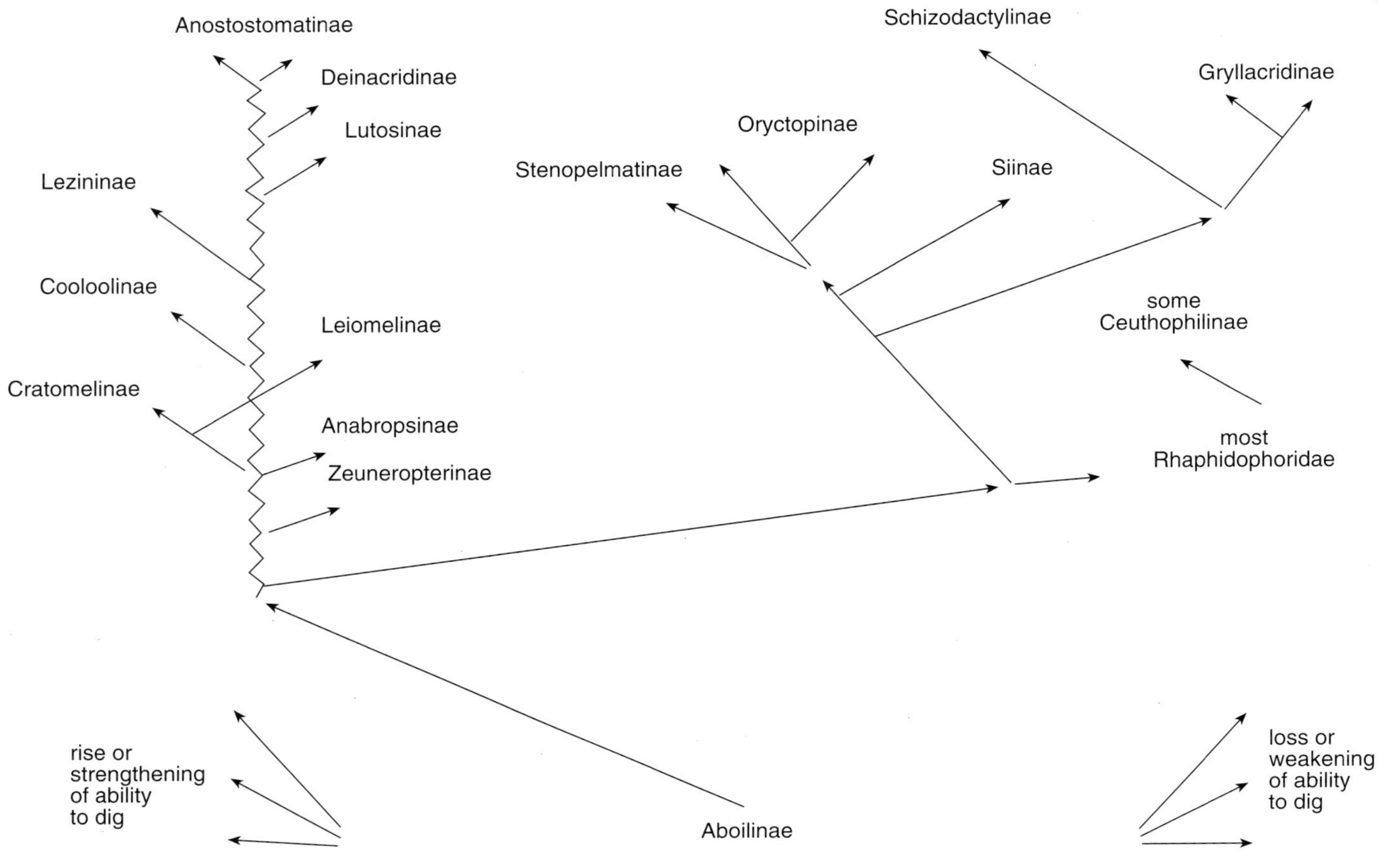

Fig. 1.18. 203, Hypothetical scheme of oscillations of evolutionary tendencies in mode of life of Stenopelmatoidea (the length of arrows has no connection with the degree of specialization or despecialization for digging).

(South Africa and Asia, from Turkey to Burma). Probably their earlier distributions were larger. Either the origins of the distributions of the Stenopelmatinae (North and Central America) and Oryctopinae (Hindustan and Sri Lanka) were probably more or less endemic, or else these distributions are also relict (it appears that the relict hypothesis is more likely for the Oryctopinae than for the Stenopelmatinae). The Gryllacridinae now occupy all tropical and subtropical regions (except some small islands). This distribution may have resulted from a relatively recent expansion, but I cannot exclude that the distribution was formerly somewhat wider, as this subfamily is known (but not dependably) from Miocene fossils of the European temperate zone.

Abbreviations Used in Figures

1A, first anal; $1MA_1$, first branch of media anterior 1; 2A, second anal; $2MA_1$, second branch of media anterior 2; 'C', false costa; CuA_1, branch 1 of cubitus anterior; CuA_2, branch 2 of cubitus anterior; CuP, cubitus posterior; MA_2, media anterior 2; MP, media posterior; R, radius; RA, radius anterior; RS, radius sector; Sc, subcosta.

References

Ander, K. (1939) Vergleicheud-Anatomische und Phylogenetische Studien uber die Ensifera (Saltatoria). *Opuscula Entomologica,* Supplement 2. Berlingska Boktryckeriet, Lund, 306 pp.

Beier, M. (1972) *Ordnung Saltatoria (Grillen und Heuschrecken).* Handbuch der Zoologie 4, Walter de Gruyter, Berlin, 217 pp.

Blanchard, E. (1845) *Histoire des Insectes,* Vol. 2. Libraire de Firmin Didot Frères, Paris, 524 pp.

Brunner-Wattenwyl, C. (1888) *Monographie der Stenopelmatiden und Gryllacriden.* Adolf Holzhausen, Vienna, 150 pp.

Burmeister, H. (1838) *Gymnognatha (Vulgo Orthoptera).* Handbuch der Entomologie 2, T.C.F. Guslin, Berlin, 1050 pp.

Gorochov, A.V. (1984) *On the Higher Classification of the Recent Ensifera (Orthoptera).* Verhandlungen des Zehnten Internationalen Symposiums uber Entomofaunistik Mitteleuropas (SIEEC 10), Muzsak Kozmuvelodesi Kiado, Budapest, 420 pp.

Gorochov, A.V. (1988a) System and phylogeny of the recent Orthoptera of the superfamilies Hagloidea and Stenopelmatoidea with a description of new taxa. Communications 1 and 2. *Zoologicheskij Zhurnal [Zoological Journal]* 67 (3, 4), 353–366, 518–529. [In Russian.]

Gorochov, A.V. (1988b) *Classification and Phylogeny of Tettigonioidea (Gryllida = Orthoptera). Melovoy biotsenoticheskiy krizis i evolyutsiya nasekomykh [Cretaceous Biocoenotic Crisis and Evolution of Insects].* Nauka, Moscow, 236 pp. [In Russian.]

Gorochov, A.V. (1995) System and evolution of the suborder Ensifera (Orthoptera). Parts 1 and 2. *Trudy Zoologicheskogo Instituta Rossijskoj Akademii Nauk [Proceedings of the Zoological Institute, Russian Academy of Sciences]* 260, 1–224, 1–213. [In Russian.]

Gwynne, D.T. (1995) Phylogeny of the Ensifera (Orthoptera): a hypothesis supporting multiple origins of acoustical signalling, complex spermatophores and maternal care in crickets, katydids, and weta. *Journal of Orthoptera Research* 4, 203–218.

Handlirsch, A. (1925) *Geschichte, Literatur, Technik, Palaontologie, Phylogenie, Systematik.* Handbuch der Entomologie 3, G. Fischer, Jena, 1202 pp.

Handlirsch, A. (1929) *5. Ordnung der Pterygogenea: Saltatoria oder Heuschrecken.* Handbuch der Zoologie 4, Walter de Gruyter, Berlin, 2456 pp.

Hennig, W. (1981) *Insect Phylogeny.* Wiley-Interscience, New York, 514 pp.

Hubbel, Th.H. and Norton, R.M. (1978) *The Systematics and Biology of the Cave-crickets of the North American Tribe Hadenoecini (Orthoptera Saltatoria: Ensifera: Rhaphidophoridae: Dolichopodinae).* Miscellaneous Publications of the Museum of Zoology, University of Michigan, Ann Arbor, no. 156, 124 pp.

Ichikawa, A. (1997) Camel crickets (Orthoptera: Rhaphidophoridae) of Japan, including the Ryukyu Islands. *Bulletin of the Hoshizaki Green Foundation* 1, 67–76. [In Japanese.]

Jacobson, G.G. (1905) *Orthoptera. Pryamokrylye i lozhnosetchatokrylye Rossijskoj Imperii i sopredelnykh stran.* A.F. Devrien, St Petersburg, 925 pp. [In Russian.]

Johns, P.M. (1997) The Gondwanaland weta: family Anostostomatidae (formerly in Stenopelmatidae, Henicidae or Mimnermidae): nomenclatural problems, world checklist, new genera and species. *Journal of Orthoptera Research* 6, 125–138.

Karny, H.H. (1921) Zur Systematik der Orthopteroiden Insekten. *Treubia* 1, 163–269.

Karny, H.H. (1930) Uber das Flugelgeader der Gryllacriden. *Archivio Zoologico Italiano* 15, 193–244.

Karny, H.H. (1937) *Orthoptera Fam. Gryllacrididae Subfamiliae Omnes.* Genera Insectorum, A.F. Devrien, Brussels, 317 pp.

Kevan, D.K.McE. (1976) Suprafamilial classification of

'Orthopteroid' and related insects, applying the principles of symbolic logic. *Notes from the Lyman Entomological Museum and Research Laboratory* 2, 1–24.

Kevan, D.K.McE. (1977) The higher classification of the orthopteroid insects: a general view. *Lyman Entomological Museum and Research Laboratory Memoirs* 4, 1–31.

Kevan, D.K.McE. (1982) *Orthoptera. Phasmatoptera.* Synopsis and Classification of Living Organisms 2, McGraw-Hill, New York, 1232 pp.

Kevan, D.K.McE. and Wighton, D.C. (1981) Paleocene orthopteroids from south-central Alberta, Canada. *Canadian Journal of Earth Sciences* 18, 1824–1837.

Kevan, D.K.McE. and Wighton, D.C. (1983) Further observations on North American tertiary orthopteroids (Insecta: Grylloptera). *Canadian Journal of Earth Sciences* 20, 217–224.

Krassilov, V.A., Zherikhin, V.V. and Rasnitsyn, A.P. (1997) Classopollis in the guts of Jurassic insects. *Palaeontology* 40, 1095–1101.

Pesenko, Yu.A. (1989) Methodological analysis of systematics I. Clarification of the problem, the main taxonomic schools. *Trudy Zoologicheskogo Instituta Akademii Nauk SSSR [Proceedings of the Zoological Institute, USSR Academy of Sciences]* 206, 8–119. [In Russian with English summary.]

Pesenko, Yu.A. (1991) Methodological analysis of systematics II. Phylogenetic reconstructions as scientific hypotheses. *Trudy Zoologicheskogo Instituta Akademii Nauk SSSR [Proceedings of the Zoological Institute, USSR Academy of Sciences]* 234, 61–155. [In Russian with English summary.]

Ragge, D.R. (1955) *The Wing-venation of the Orthoptera Saltatoria with Notes on Dictyopteran Wing-venation.* British Museum, London, 159 pp.

Rentz, D.C.F. (1980) A new family of ensiferous Orthoptera from the coastal sands of southeast Queensland. *Memoirs of the Queensland Museum* 2, 49–63.

Rentz, D.C.F. and Weissman, D.B. (1973) The origins and affinities of the Orthoptera of the Channel islands and adjacent mainland California. Part 1. The genus Cnemotettix. *Proceedings of the Academy of Natural Sciences of Philadelphia* 125, 89–120.

Saussure, H. (1859) Orthoptera nova Americana (diagnoses préliminaires). *Revue et Magasin de Zoologie* 11, 201–212.

Sharov, A.G. (1968) The phylogeny of the orthopteroid insects. *Trudy Paleontologicheskogo Instituta Akademii Nauk SSSR [Proceedings of the Paleontological Institute, USSR Academy of Sciences]* 118, 217 pp. [In Russian.]

Stål, C. (1876) Observations orthoptérologiques 2. Sur *Anostostoma* et quelques genres voisins. *Bihang till Konglige Svenska Vetenskaps-Akademiens Handlingar* 4, 47–53.

Thomas, C. (1872) Notes on the saltatorial Orthoptera of the Rocky Mountain regions. *Annual Report Progress United States Geological Survey Montana* 5, 423–466.

Zeuner, F.E. (1939) *Fossil Orthoptera Ensifera*, British Museum, London, 321 pp.

2
Habitats and Biogeography of New Zealand's Deinacridine and Tusked Weta Species

George W. Gibbs
School of Biological Sciences, Victoria University, Box 600, Wellington, New Zealand

This chapter reviews what is known about the habitats and distribution of 18 weta species in the sister genera *Hemideina* (seven species) and *Deinacrida* (11 species). Tusked weta (three species), which are not closely related to these two genera (M. Morgan-Richards, Dunedin, 1998, personal communication) are also included here. These large-bodied New Zealand weta occupy a wide range of habitats, from islands at sea level to high mountain scree slopes at 2000 m elevation. A recent phylogenetic hypothesis generated from morphological and allozyme data (Morgan-Richards and Gibbs, 2001) is used here for grouping the taxa in order to discuss their generalized habitat features. The same presumptive groupings are discussed in terms of their geographical patterns.

The accompanying maps show spot localities for all *Deinacrida* and tusked weta species. Generalized distribution areas only are depicted for *Hemideina* weta because many more localities are known for the common species and this review did not attempt to gather all of them. Detailed maps are presented by Trewick and Morgan-Richards (1995, figs 1–3) for the three North Island species of *Hemideina* and by Townsend *et al.* (1997, fig. 1) for the two species on Banks Peninsula. The maps are for biogeographical purposes and do not show locations of populations that have been translocated for the purpose of conservation.

The Stem Species: *Hemideina broughi* (Fig. 2.1)

This large (total length 85 mm), secretive, but rather aggressive species has been interpreted by both Field (1993) and Morgan-Richards and Gibbs (2001) as being closest to the ground-plan deinacridine weta. Evidence for this view comes from its lack of a stridulatory file on the second abdominal tergite (Field, 1993) and a number of presumptive ancestral morphological characters it shares with both *Hemideina* and *Deinacrida*. It is a unicolour reddish-brown species with the long legs and strong tarsi of an arboreal climber (Fig. 2.2). Since males of *H. broughi* do not exhibit the variation in head size that characterizes the other *Hemideina* species and their jaws are not sexually dimorphic, it is assumed that this species lacks the highly developed social behaviours of the latter. Nevertheless, they often aggregate in groups within suitably large galleries (P. Barrett, Raumati Beach, 1998, personal communication).

Known as the west coast bush weta, this species has been found mainly in old, partly hollow, red beech trees (*Nothofagus fusca*) and also mountain beech (*Nothofagus solandri* var. *cliffortiodes*). It is associated with high-rainfall, heavily forested areas of the northern west coast of the South Island between the Tasman Mountains and Greymouth (Fig. 2.1) at elevations up to 1100 m.

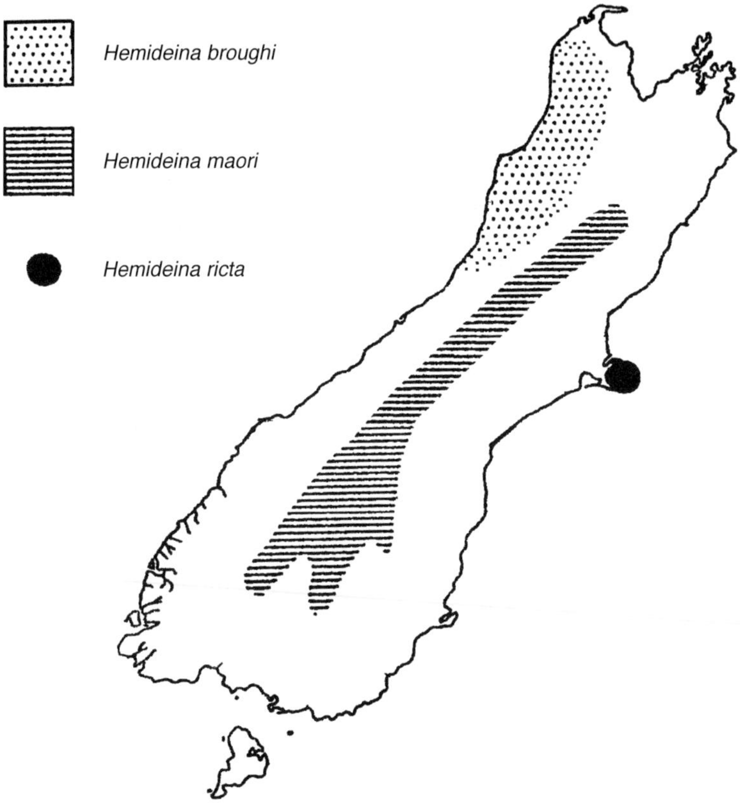

Fig. 2.1. Distribution of three South Island species of tree weta in the genus *Hemideina*.

Arboreal Tree Weta: *Hemideina thoracica, Hemideina trewicki, Hemideina crassidens, Hemideina femorata* (Fig. 2.3)

These four species (*c.* 40–60 mm) are the weta familiar to most New Zealanders, having successfully colonized urban hedges and gardens. Despite the fact that their key habitat feature is a secure tree-hole gallery – hence their common name – these weta are not wood-borers. Their galleries are normally tunnels that have initially been excavated by a large wood-boring larva, e.g. *Ochrocydus huttoni* (Coleoptera: Cerambycidae) or *Aenetus virescens* (Lepidoptera: Hepialidae), but many natural crevices and cavities in trees can serve the purpose, provided they have a narrow, tight-squeeze entrance. The relationship between hole size in manuka (*Leptospermum scoparium*) and weta head size in *H. femorata* and *H. crassidens* is positively correlated, with a significant difference between males and females (Spencer, 1995; see also Field and Sandlant, Chapter 13, this volume). The holes in manuka were longer (mean 14 mm, $n = 80$) than wide (mean 10 mm). Holes that are occupied by tree weta can be distinguished from unoccupied holes by their clean, neatly nibbled apertures (see Fig. 13.1-2, Field and Sandlant, Chapter 13, this volume). Tree weta enter the galleries head first and exit rear first. This orientation in the tunnel makes good use of the heavily spined hind tibiae, which effectively block the tunnel to intruders. They prefer living timber over dead wood (Asher, 1977) and generally avoid holes in rotten logs on the ground. Holes can be damp or dry but must not be susceptible to flooding.

Tree weta are in no way host tree-specific or confined to any particular forest/shrubland type, whether indigenous or exotic. Any suitable hole

Fig. 2.2. *Hemideina broughi:* male from Flora Saddle, Kahurangi National Park, in *Nothofagus fusca* forest at 950 m. Photo: G.W. Gibbs.

will serve as a gallery, with the provision that certain tree or shrub species tend to host the most appropriate wood-borers or to decay in such a way as to promote the development of secure cavities. The most favoured plants are manuka, kanuka (*Kunzea ericoides*), ngaio (*Myoporum laetum*) and mahoe (*Melicytus ramiflorus*), but many others can support large weta populations (Little, 1980; Moller, 1985; Rufaut, 1995; reviewed by Field and Sandlant, Chapter 13, this volume).

Tree weta are primarily herbivorous, feeding on the leaves, flowers and fruit of a wide range of trees and shrubs, but they will also take living or recently dead invertebrates when they get the opportunity. Their galleries and foraging activities normally keep them arboreal, with only gravid females having the need to descend on to the ground for oviposition in moist soil (Moller, 1985).

All species of *Hemideina*, except *H. broughi*, share a sexually dimorphic character, in which maturity can be reached over three different instars in the case of males but not females. This phenomenon results in a unique size polymorphism in mature males, which, in turn, is related to their social interactions (detailed discussion in Field and Jarman, Chapter 17, this volume). Although the size differences affect the whole body, they are most apparent in the allometric growth of head and jaw sizes. Males of *H. crassidens*, for example, can mature in their eighth, ninth or tenth instar (Spencer, 1995). Females always mature in their tenth instar, by which time their head capsule is about the same size as that of an eighth-instar male (for *H. crassidens*: h.l. = 12.1 mm), as seen in Fig. 2.4. The size of the male and his fighting apparatus establishes his position in the social dominance hierachy. Thus the largest megacephalic males can command the most desirable resources, which, in this case, are galleries and mature females to share them with (detailed discussion in Field, Chapter 18, this volume). A cost of size dominance is the time these males must spend in defence of their gallery. Spencer's (1995) study showed that in two populations sampled (one on a rat-free island, the other on the mainland), the mature males were made up of 31–41% in their tenth instar and 56–59% in their ninth instar. Only two eighth-instar males occurred in the overall sample of 60 sexually mature males.

The small-aperture galleries and social hierarchy are believed to be largely responsible for the

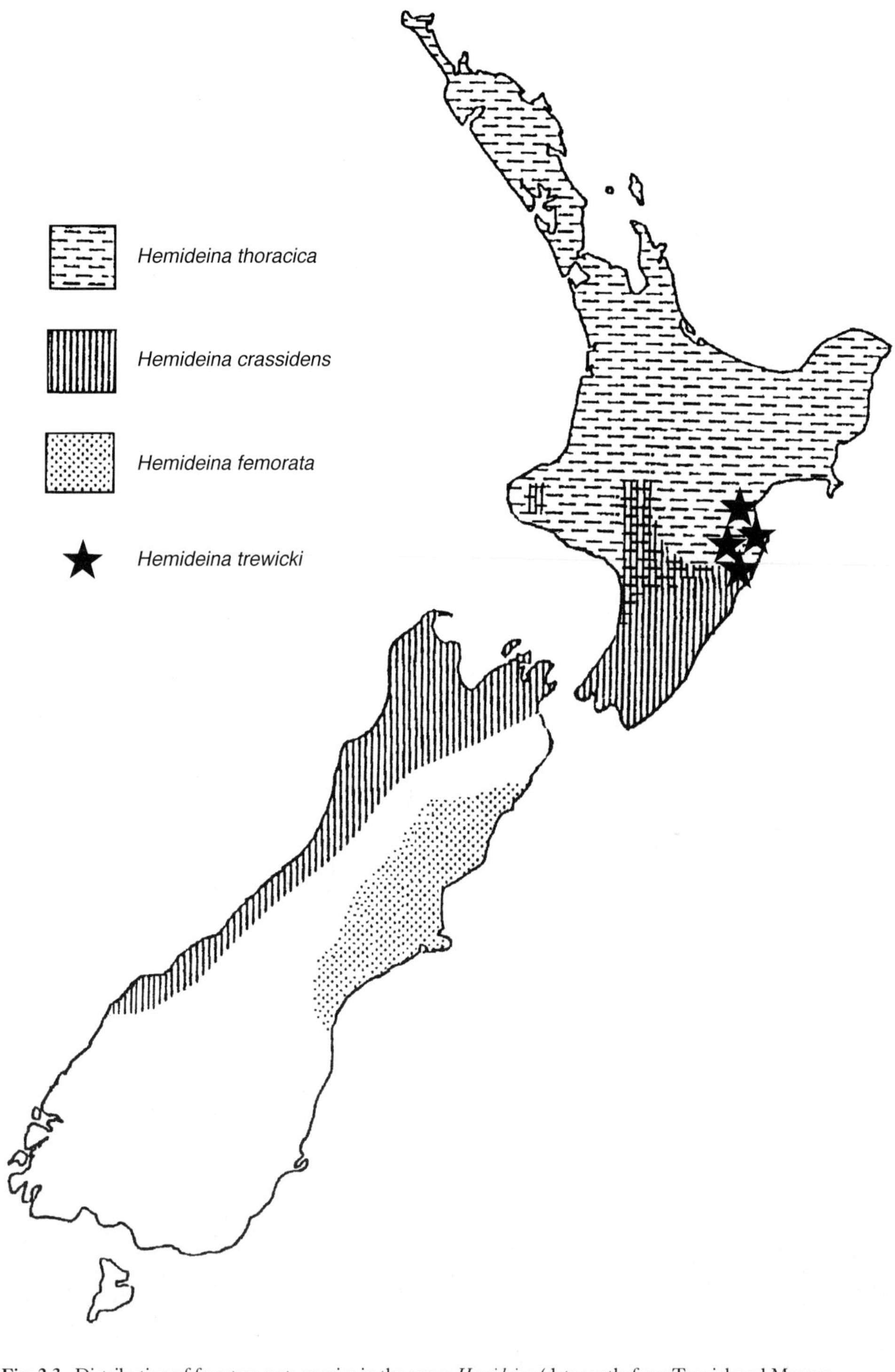

Fig. 2.3. Distribution of four tree weta species in the genus *Hemideina* (data partly from Trewick and Morgan-Richards, 1995, and P.M. Johns, Christchurch, 1998, personal communication).

success of this group of species, as compared with the solitary giant weta species, in the presence of alien predators (Gibbs, 1998). It has been suggested that selective pressures from rodent predation might affect the entrance hole size of tree weta galleries (Moller, 1985). A study by Rufaut (1995) of gallery statistics at locations with and without rats showed that the mean hole diameter is indeed significantly larger on rat-free islands (207 mm on Stephens Island; 237 mm on Maud Island) than on an island with kiore (*Rattus exulans*) (114 mm on Long Island) or on the mainland (181 mm at Cable Bay, near Nelson). This study showed that there was a preference for galleries that are 0.6 to about 4 m above ground level.

Tree weta are smooth and shiny, especially on the abdomen, which is ringed with bands of contrasting colours (see Figs 9.7 and 9.8 in Field and Bigelow, Chapter 9, this volume). The three North Island species comprise a stable monophyletic cluster in the phylograms, supported by both morphological and allozyme data sets (Morgan-Richards and Gibbs, 2001) and are distinguished morphologically from the South Island species by the lack of a longitudinal jaw ridge or carina and their unique eggshell microstructure (Morgan-Richards and Gibbs, 2001). Interestingly, the diploid number of chromosomes varies widely between 11 and 24 (XO males) in North Island species of *Hemideina* but is constant at 25 in the four South Island species (Morgan-Richards, 1995).

The distribution of this group of species covers low to medium altitudes in all parts of New Zealand where trees and shrubs are present, except in the south and south-eastern part of South Island (Fig. 2.3). They are absent from Three Kings, Poor Knights, Stewart and the Chatham Islands, but common on inshore islands adjacent to their respective distribution areas. Although broadly vicariant, notable overlaps occur. The North Island overlap zones have been the subject of a detailed study by Trewick and Morgan-Richards (1995) and are summarized by Morgan-Richards *et al.* (Chapter 7, this volume). They have shown that, where *H. crassidens* and *H. thoracica* co-occur on Mts Ruapehu and Taranaki, there is a zone of overlap at 800–900 m a.s.l., with *H. thoracica* extending below this and *H. crassidens* above it to 1100 m. The Canterbury tree weta, *H. femorata*, overlaps with *Hemideina maori* at Hanmer Springs (Little, 1980) and with *Hemideina ricta* on Banks Peninsula (Townsend *et al.*, 1997).

Fig. 2.4. *Hemideina crassidens:* megacephalic male from Wellington. Photo: Brett Robertson.

Tree Weta on the Ground: *Hemideina ricta, Hemideina maori* (Fig. 2.1)

Based on the evidence used for the phylogenetic hypotheses, these two species of *Hemideina* share a very recent common ancestor (Morgan-Richards and Gibbs, 2001). Both possess the mandibular carina, which they share with *H. femorata*, but together they are distinguished by having only a single apical spine on the hind femur and by the number of stridulatory ridges. Also, in contrast to all other *Hemideina*, in which the defence reaction involves a hind-leg-raising display, these two species share a unique defence reaction. It involves rolling over on to the back and lying motionless with jaws open and legs apart (described and illustrated by Field and Glasgow, Chapter 16, this volume). This seemingly vulnerable attitude enables the weta to instantly grasp an inquisitive prodding object (lizard snout?) with its claws and inflict a strong bite. These weta live on the ground in open country, where lizard predation would be significant, especially in prehuman times. A female is shown in Fig. 2.5.

The most notable feature of this pair of tree weta species is their ability to occupy treeless habitats, usually in tussock grassland. They utilize holes, crevices and galleries in rocky terrain and in fallen logs, in much the same way as other tree weta use tree holes, selecting cavities with tight-fitting entrances, which mammalian predators cannot access. *Hemideina maori* appears to be restricted to ground-level galleries even when shrubby vegetation is present, but *H. ricta* will occupy galleries in trees, where available (Townsend *et al.*, 1997). They share the sexual maturity dimorphism, social behaviour and diet of the tree-living species. Both species occur in relatively low-rainfall regions of the eastern South Island (Fig. 2.1). Their distributions are vicariant, with *H. ricta* restricted to the outer area of Banks Peninsula, beyond a line between Pigeon Bay and Akaroa Harbour, and *H. maori* ranging between the upper Clarence Valley in Marlborough and the Lammermore and Umbrella Ranges in southern Otago. Where *H. ricta* and *H. femorata* overlap on Banks Peninsula, *H. ricta* occupies sites at higher elevations (to 806 m), with *H. femorata* found only

Fig. 2.5. *Hemideina maori:* female from Rock and Pillar Range, Otago, at 1400 m. Head is much smaller in proportion to body than in male, and abdomen is more swollen. Both are characteristics of all *Hemideina* spp. Photo: G.W. Gibbs.

below 450 m. (Townsend *et al.*, 1997). Also, where *H. maori* and *H. femorata* overlap (e.g. Hanmer Springs), *H. maori* occurs at the higher elevations (Little, 1980). *Hemideina maori* extends to about 1500 m a.s.l. on the summits of many of the eastern ranges, but can equally well be found in the intermontane valleys. It is absent from the plains and coastal ranges.

Northern Arboreal Giant Weta: *Deinacrida fallai, Deinacrida heteracantha, Deinacrida mahoenui*

(Fig. 2.6)

These three species of giant weta, New Zealand's largest weta (up to 75 mm total length), comprise a

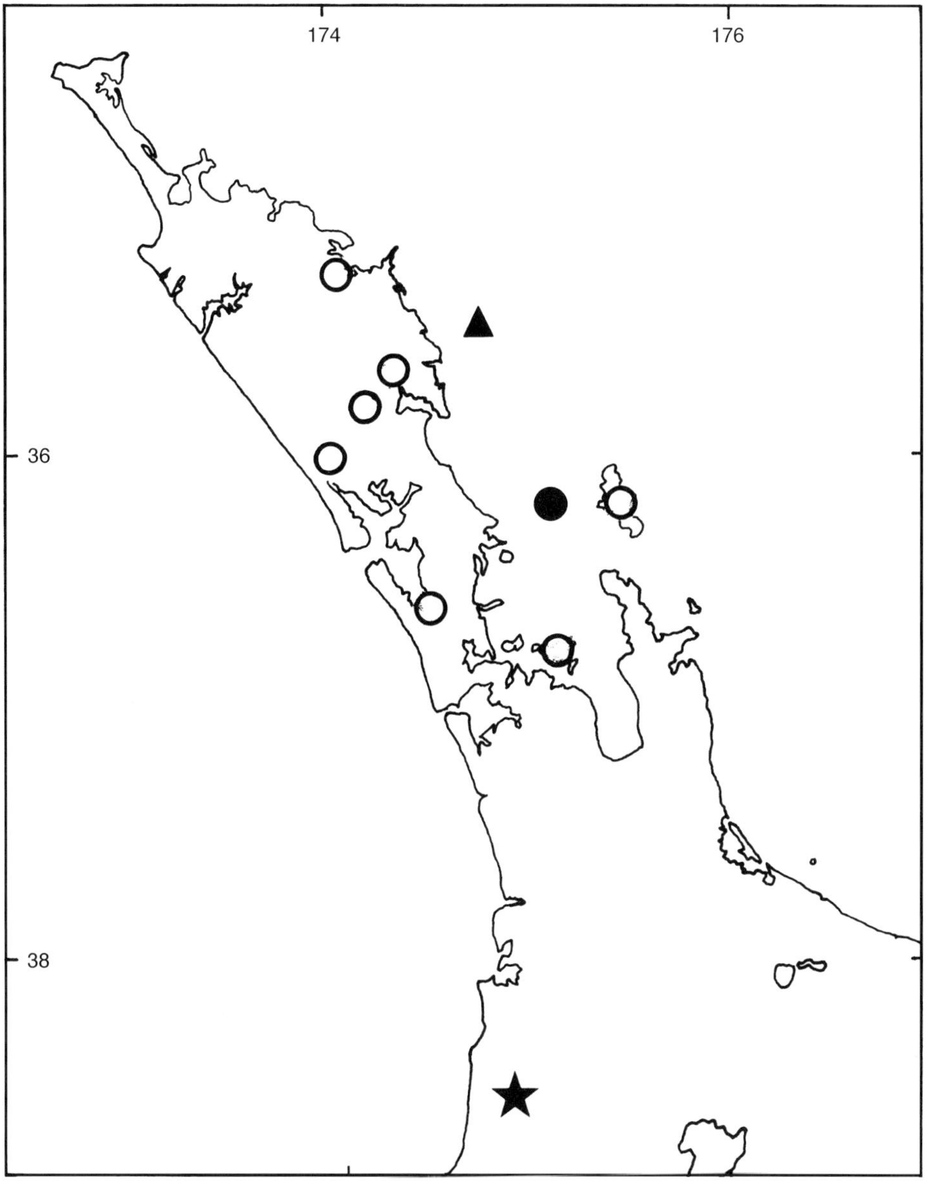

Fig. 2.6. Distribution of three northern species of *Deinacrida* (after Watt, 1963). ▲ *D. fallai*; ● *D. heteracantha*; ○ *D. heteracantha* extinct; ★ *D. mahoenui*.

tight monophyletic group in both morphological and genetic analysis (Morgan-Richards and Gibbs, 2001). These insect giants are all arboreal forest dwellers, at least in their original habitats, with a relatively nomadic lifestyle as adults (McIntyre, Wellington, 1997, personal communication). On Little Barrier Island, Richards (1973) records adults of *D. heteracantha* being found not only above ground level, under loose bark of kanuka and inside cavities in mahoe and pohutukawa (*Metrosideros excelsa*), but also at ground level under mats of pohuehue (*Muehlenbeckia complexa*). Gibbs and McIntyre (1997) note that, although no longer found in this ground-level situation, they make good use of dense 'epiphyte nests' to provide daytime refuges for adults and late instars. The sheer bulk (20–50 g) of these weta makes it difficult for them occupy cavity refuges that are secure from predators. The immatures are more site-specific than adults and can occupy holes and crevices under bark, although later instars are still extremely vulnerable, perhaps even more so because they keep returning to their insecure refuges. Richards (1973) kept *D. heteracantha* and *D. fallai* in captivity and noted that instars 1–5 tended to burrow into soft soil head first, where they remained by day, with only their hind tibiae and tarsi visible. On the other hand, she also found two second-instar nymphs on Little Barrier Island amongst foliage of pohutukawa, so it seems that daytime refuges of nymphs vary greatly. With *D. mahoenui*, Richards (1994) has shown that it spends much of its life in a small area of only 1–3 m^3, returning to rest in the same places on successive days (discussed in Sherley, Chapter 26, this volume). Adults can move across open ground to new bushes, up to 31 m in one night. By day, the weta rest 'tucked into junctions' of gorse branches or simply amongst foliage (Richards, 1994).

Present-day New Zealand is fortunate to still retain wild populations of these extremely vulnerable giant insects. The most northern species, *D. fallai*, illustrates the advantages of a habitat free from alien predators. It thrives in the pohutukawa forests of the Poor Knights group of islands (Fig. 2.6), where its main predators are the harrier hawk (*Circus approximans*) by day (Watt, 1982) and tuatara (*Sphenodon punctatus*) by night. Watt (1982) notes that their nocturnal activity is mostly arboreal and they are not often seen during the day. The other two species in this group have suffered severely since European rats arrived in New

Fig. 2.7. *Deinacrida heteracantha:* female in full raised-leg defence display. Individuals may reach 50 g body weight and 75 mm in length, easily covering one's outstretched hand. Little Barrier Island. Photo: C.R. Veitch.

Zealand. *Deinacrida heteracantha* (Fig. 2.7), which today survives solely as a declining population on Little Barrier Island (Gibbs and McIntyre, 1997), formerly inhabited most of North Auckland mainland, Waiheke and Great Barrier Islands (Watt, 1963). Buller (1871) noted that 'formerly, it was abundant in the forests north of Auckland; but of late years it has become extremely rare. The natives attribute its extermination to the introduced Norway rat' (*Rattus norvegicus*). The most southern species of this group is *D. mahoenui*, discovered in 1962 in a patch of gorse (*Ulex europaeus*) during gorse-cutting operations on Mahoenui Station in the southern King Country (Watt, 1963). Watt found subsequent specimens in remnants of native forest on farmland in the same area (from rotten tawhero (*Weinmannia silvicola*) and hollow ponga (*Cyathea dealbata*) and it is still known only from two small populations in the same district, where it persists in introduced gorse bushes. One of these is a 240 ha patch of gorse, surrounded by farmland, reserved by the Department of Conservation especially for the weta (Sherley and Hayes, 1993). This population has been the subject of ecological studies by Richards (1994) and Sherley and Hayes (1993) and is reviewed by Sherley (Chapter 26, this volume).

These largest weta are almost exclusively herbivorous and can be raised in captivity on plant matter alone. They are nocturnal and normally seen as solitary individuals (Richards, 1973). Copulation has been observed by day in captive populations but in the wild is known to occur on vegetation at night. Although capable of producing stridulatory sounds in defence (described by Field, Chapter 15, this volume), they have not been heard to communicate in this way in the wild.

Mountain-bluff Weta: *Deinacrida elegans* (Fig. 2.8)

The position of this highly distinctive and very athletic giant weta is not entirely clear from the phylogenetic analyses. Morphological analysis places it, along with two other smooth-bodied alpine giant weta, *Deinacrida pluvialis* and *Deinacrida talpa*, in a basal position within the *Deinacrida* clade, whereas the allozyme data place it as a sister taxon to the three northern arboreal giant weta (Morgan-Richards and Gibbs, 2001). Apart from its unique greyish-blue coloration, with red, black and white leg joints, this weta is also distinct in its occupation of a unique habitat, which combines the challenge of ground-dwelling with athletic climbing ability. Its habitat is essentially a cliff-face where the rock is very stable but jointed in such a way as to provide deep narrow crevices, into which the weta squeezes itself. Dry crevices, or at least those which are well drained, seem to be preferred, but one of the essential features is clean rock and freedom from plant matter. The weta are often so firmly wedged into the fissures that their abdomen is noticeably distorted (Meads and Notman, 1992). *Deinacrida elegans* has occurred on exposed rock-faces at elevations between 800 and 1700 m a.s.l. Although more than one weta may be found in a crevice, they display no interactive social behaviour. Meads and Notman (1991, 1992) observed on two occasions that this weta can emerge from crevices by day 'to sun itself'. The cliff habitat poses some climbing challenges, even to a weta, as evidenced by dry, sun-bleached carcasses of this species which are sometimes found on the river-beds below such cliffs. In several instances, these bleached remains are all we know of the presence of the species. The bluff weta's escape reaction is adapted to its cliff-face habitat and consists of a leap off the rock followed by an uncontrolled roll with its legs tucked up (Gibbs, 1999).

The bluff weta has a disjunct distribution (Fig. 2.8), with a northern population on the greywacke sandstone ranges of Marlborough, occupying the Seaward and Inland Kaikoura Ranges and the mountains around the head of the Wairau River. The southern population is on the Cretaceous rhyolite bluffs on the western faces of Mt Somers, Mid-Canterbury. Although bluff weta refuge requirements might serve to isolate them spatially from co-occurring species *Deinacrida connectens* and *Deinacrida parva* in the Seaward Kaikoura Ranges, various combinations of all three giant weta have been observed feeding in close proximity at night (Meads and Notman, 1992; G.W. Gibbs, unpublished data)

Cook Strait Giant Weta: *Deinacrida rugosa, Deinacrida parva* (Fig. 2.9)

There is some disagreement over the status of *D. parva*. With the type specimen evidently lost, Ramsay (1971) resurrected the species name on

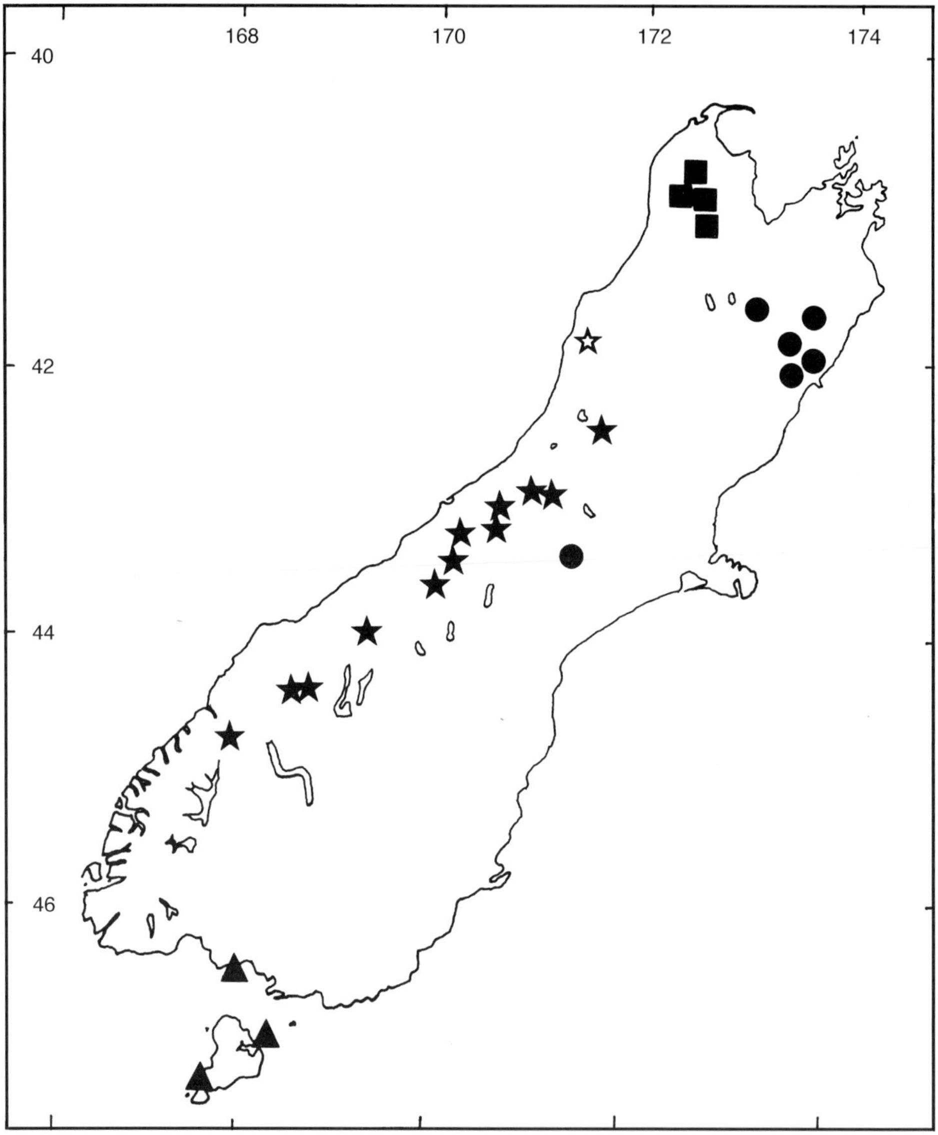

Fig. 2.8. Map showing known localities for five South Island species of *Deinacrida*. ■ *D. tibiospina;* ● *D. elegans;* ☆ *D. talpa;* ★ *D. pluvialis;* ▲ *D. carinata.*

the basis of two subadult males, found in the Upper Wairau Valley in 1966 and considered to resemble Buller's original description (Buller, 1895). Since then, giant weta closely resembling *D. rugosa*, but smaller, have been found in some numbers living in three valleys of the Seaward Kaikoura Range. Cameron (1996) compared morphometrics, allozymes and karyotypes of *D. rugosa* from Cook Strait Islands and the Kaikoura population. He found consistent differences but was unable to satisfactorily resolve whether *D. parva* and *D. rugosa* should be regarded as distinct species. Johns (1997) synonymizes *D. parva* with *D. rugosa* without presenting any justification. Pending further resolution of this fascinating taxonomic challenge, I shall refer to them as a pair of recently diverged sister species. The combined phylogenetic tree (Morgan-Richards and Gibbs,

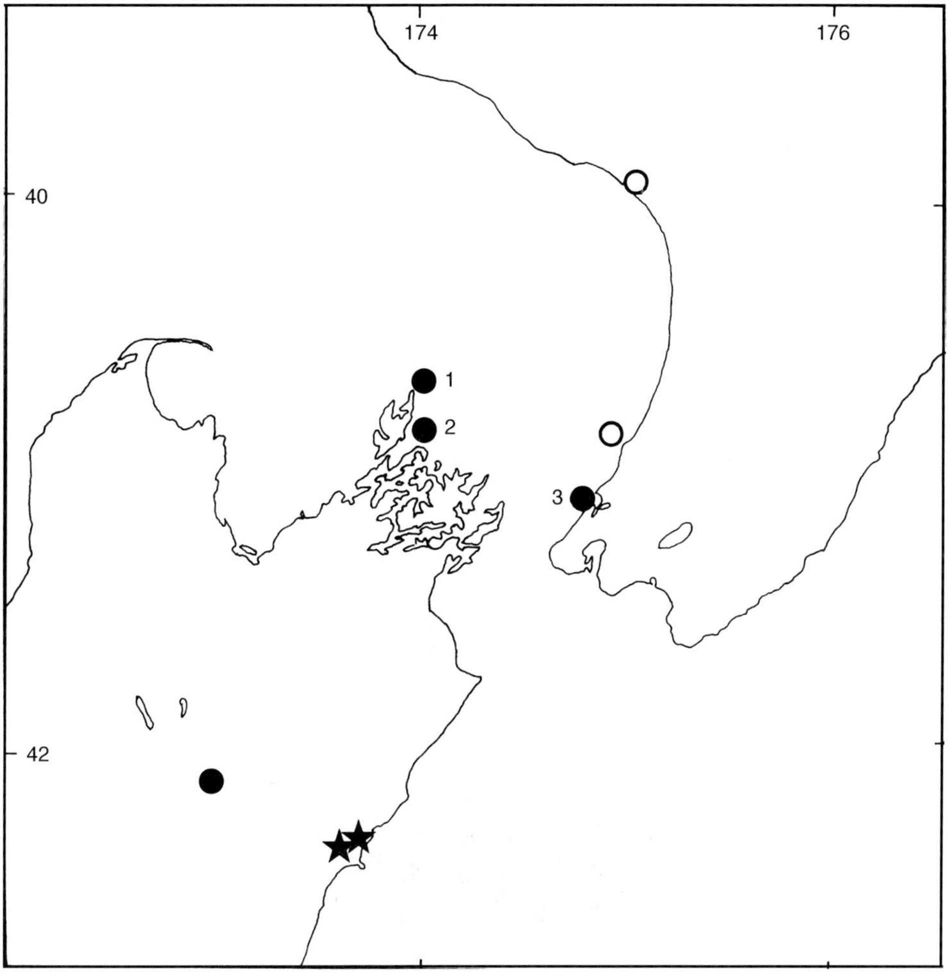

Fig. 2.9. Distribution of two central New Zealand species of *Deinacrida*. ● *D. rugosa*; ○ *D. rugosa* extinct or presumed extinct; ★ *D. parva*. Island localities as follows: 1, Stephens Is.; 2, Trios Is.; 3, Mana Is. The inland location for *D. rugosa* denotes two immature specimens found at Lake Sedgemere.

2001) supports this and places them in an unresolved trichotomy with all the *Deinacrida* included in the previous two sections. They are easily recognized from their cuticular character, which is brownish-ochreous or straw-coloured and exceptionally roughened. The abdominal tergites are emarginate posteriorly and granulose along their edges. These weta are primarily ground-dwelling, although they climb on to low bushes to browse on leaves and flowers. Their preferred habitat is grassland with scattered mat-forming plants and low shrubs. By day, they rest in temporary refuges, scooped out at soil level amongst the grass stems, or under stones and logs and, by night, they forage on available vegetation (McIntyre, 1993). Males locate females during the early evening, the pairs remaining in close proximity during the night but not copulating until the following day, when established in their daytime roost (McIntyre, 1993).

The larger of the species is *D. rugosa* (Fig. 2.10), in which females weigh 32 g. It was first described in 1871 from a specimen taken 'in the Wanganui district' (Fig. 2.9), on North Island mainland (Buller, 1871), but has subsequently been found only on islands around Cook Strait, including Stephens, Trios (two islands), Mana and

Kapiti (not seen since 1913 on the latter). It is not compatible with rats but has survived well in the presence of mice on Mana Island. In contrast, the entity known as *D. parva* is consistently smaller, with measurements that do not overlap those of the island populations of *D. rugosa* and a maximum adult female weight of about 14.5 g. It occurs in the Kowhai River, Hapuku River and Jordan Stream catchments of the eastern Seaward Kaikoura Ranges, where it is subalpine to alpine, seeming to prefer altitudes of about 900 m a.s.l. This habitat is, in many ways, very similar to that of the Cook Strait Islands. However, it has also occurred well above this level, at 1300 m on barren ridge tops, and well below it, on the farmed river-flats of the Kaikoura delta. Its preference for river terrace habitats close to swift mountain torrents results in many drownings. In fact, the best way to survey for this weta is to examine river margins for dead or dying specimens. Whether its fatal attraction to water is innate or whether it is mediated by a high frequency of gordian worm parasites, which plague this species, needs to be determined. The two immature weta found at Lake Sedgemere (1047 m a.s.l.), inland Marlborough, in 1966 (Ramsay, 1971) more closely resemble *D. rugosa* than *D. parva* but have some intermediate features (Cameron, 1996).

South Island Western Alpine Giant Weta: *Deinacrida pluvialis*, *Deinacrida talpa* (Fig. 2.8)

This pair of sister species occupies areas of subalpine tussock and herb field that receive very high rainfall to the west of the main divide. Both are smooth in texture, with an all-over brown pigmentation, giving them the appearance of a tree weta. Although closely related, there are two notable differences between these species, which relate to their respective habitats. *Deinacrida pluvialis* is the larger weta (54 mm total length). It uses loose stones as refuge sites, sometimes showing some evidence of continued occupation in the form of neatly scooped-out hollows in the underlying silt, whereas *D. talpa* is a slightly smaller (49 mm), burrow-forming weta. The latter excavates long tunnels, about 30 mm diameter, in the soft organic soil layer under carpet grass (*Chionochloa australis*), entering head first in the manner of a tree

Fig. 2.10. *Deinacrida rugosa:* female feeding on *Ozothamnus leptophylla* flowers on Mana Island, Cook Strait. Note coarsely tuberculated pronotum, characteristic of New Zealand *Deinacrida* spp. Photo: G.W. Gibbs.

weta, with its exceptionally spiny hind tibiae almost blocking the tunnel. A pair of adults may occur in the same 40 cm tunnel, with the male positioned nearer to the entrance than the female (Gibbs, 1999).

Deinacrida pluvialis has been recorded (Fig. 2.8) from Mt Alexander in the Taramakau Valley, south along the main divide of the Southern Alps to the Cleddau Cirque in Fiordland, with an altitude range of 700–1700 m. It seems to be most abundant in the western Otago mountains, where it overlaps with the range of *D. connectens* but occupies a less barren habitat. *Deinacrida talpa* is known only from the north-western flanks of three peaks in the Mt Faraday region of the Paparoa Range (Fig. 2.8) between 1200 and 1300 m.

These two *Deinacrida* species possess multiple stridulatory ridges on the second abdominal tergite, very similar to those of tree weta, which place them in a different clade from the remaining three species of the genus considered below (described by Field, Chapter 15, this volume).

South Island Giant Weta that Lack Stridulatory Ridges: *Deinacrida connectens, Deinacrida tibiospina, Deinacrida carinata* (Figs 2.8 and 2.11)

The phylogenetic analysis, based on the combined data set (Morgan-Richards and Gibbs, 2001), placed *D. connectens, D. tibiospina* and *D. carinata* in a monophyletic clade, defined by the presence of spines on the dorsum of the hind femur and the lack of a stridulatory file on the second abdominal tergite. The abdominal cuticle is finely rugose in *D. connectens*, *D. tibiospina* and *D. carinata* but smooth and shiny in *D. pluvialis* and *D. talpa*.

The scree weta, *D. connectens*, is an alpine specialist, occurring on extremely exposed rocky, barren slopes that are almost devoid of vegetation (Fig. 2.12). Its preferred altitude is about 1500 m, but it has been found as low as 770 m and as high as 3600 m. This moderate-sized (45 mm) greyish weta with white legs hides by day under stones on scree slopes, emerging at night to browse on scattered scree herbs, shrubs and lichens. It has the most widespread distribution of all giant weta (Fig. 2.11), extending from the Peel Range in the north-west or Blackbirch Range in the north-east, south along the eastern side of the main dividing ranges to St Mary's Range in Otago. It is also known from two isolated populations: in the vicinity of Headlong Peak in western Otago (Watt, 1980) and on Spence Peak, Takitimu Range, in Southland. Field (1980) described activity cycles and behaviour of *D. connectens.* Geographical, genetic and colour variation are described by Morgan-Richards and Gibbs (1996). In his checklist of Anostostomidae, Johns (1997) mistakenly lists both *D. connectans* (*sic*) (Ander, 1939) and *D. sonitospina* (Salmon, 1950) in his species list for *Deinacrida*, thus ignoring a previous synonymy by Ramsay (1961).

Another alpine giant weta, *D. tibiospina*, which lives in the subalpine tussock and herb-field zone between about 1000 and 1500 m, scarcely deserves the title 'giant' weta. This small (33 mm and 7 g weight), reddish-brown weta is associated with the vegetation, not the rocks of its mountain habitat. During the day, it occurs under the leaf bases of tufted tussocks (*Chionochloa*), *Astelia* and flax (*Phormium*) (Meads, 1990). It is only known from a number of scattered localities in the Tasman Mountains, north-western Nelson, from just south of the Heaphy Track to the Twins on the Arthur Range (Fig. 2.8).

The smallest and most southern of the giant weta, *D. carinata*, reaches only 6 g weight in females, with a length of 40 mm (Meads and Notman, 1993). Its habitat is uncertain, even to the extent of whether it is ground-dwelling or partly arboreal, since the only report (Meads and Notman, 1993) was made on Pig Island, where there is no tall vegetation. Here, it lives amongst herbaceous vegetation on the ground, but nothing is known about the nature of its refuges or whether it is site-specific. It has been found on three islands around Foveaux Strait (Fig. 2.8) – Herekopare, Kundy and Pig Island – which suggests a relict pattern, but there are no old records from the South Island mainland.

The Small Hokianga Tusked Weta: *Anisoura nicobarica* (Fig. 2.13)

Smallest of all weta in this review (20–23 mm), this one was described as an aberrant ground weta in the genus *Hemiandrus* (*H. monstrosus* Salmon, 1950) before it was realized that tusked weta are a valid taxonomic entity (Fig. 2.14). Johns (1997)

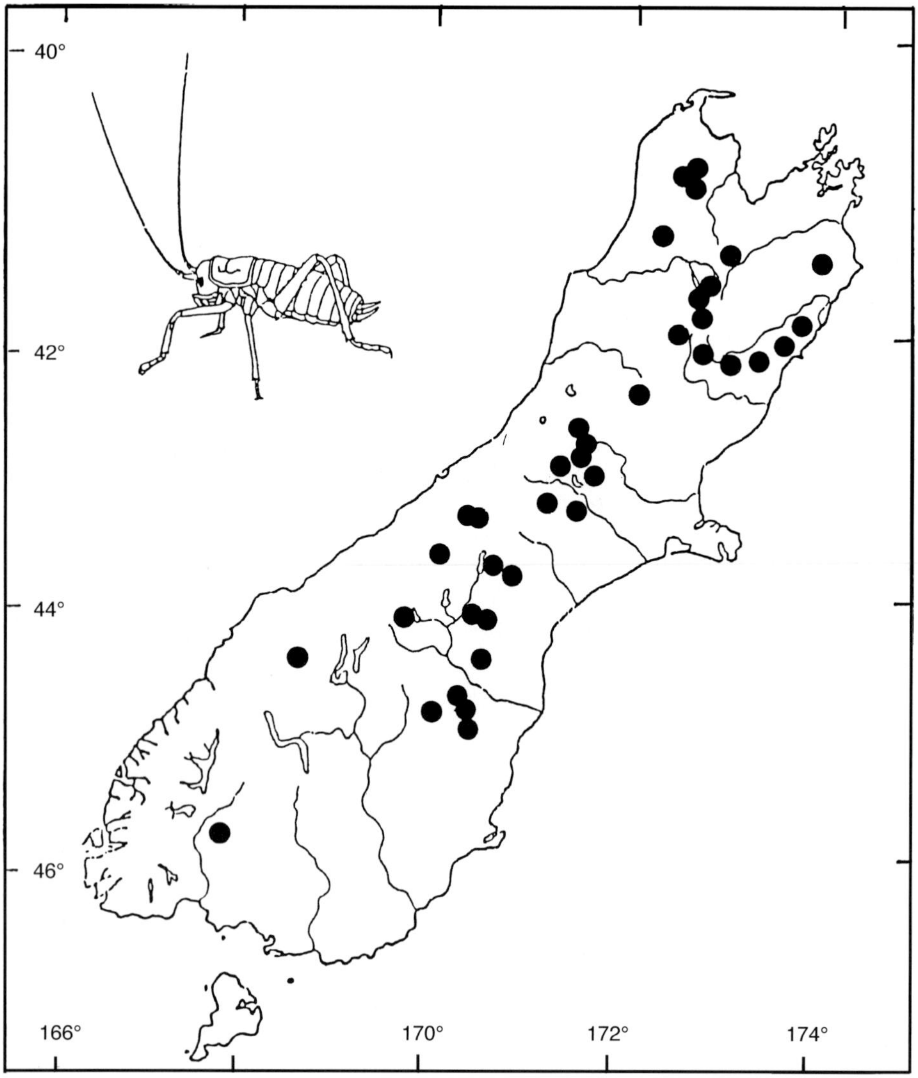

Fig. 2.11. Map of known localities for *Deinacrida connectens*. (Inset: from Field, 1980, with permission.)

regards *H. monstrosus* as a synonym of *A. nicobarica* Ander 1938 from the Nicobar Islands, but this is queried by Gorochov (Chapter 1, this volume). This small weta shares its habitat with the tree weta, *H. thoracica*. Both occupy galleries in manuka and other trees and shrubs, but with two important differences. The tusked weta orientates itself in its galleries so that it faces an intruder with its jaws agape in defence. It also frequently seals the entrance to the gallery, especially during a moult, with a mixture of saliva and wood scrapings, so as to render the aperture almost invisible. Both these characteristics are common to all three tusked weta species.

It is a forest weta but has been found in a number of rural localities well away from trees. One captive female *A. nicobarica* scooped out a 'nest' in soil, which was roofed over with debris, suggesting they can be quite flexible over their selection of refuges (J. Davidson, Kohukohu, 1991, personal communication). However, most have been found in tree holes. A wide range of plant and animal foods are consumed.

The Hokianga tusked weta is distributed

Fig. 2.12. *Deinacrida connectens:* female from rocky scree slope above Rees Saddle, Otago, at 1400 m. This southern population is characterized by very dark abdominal banding, which is lacking on the more northern mountain-range populations. Photo: G.W. Gibbs.

throughout the northernmost part of New Zealand as far south as 35° 35′ (Fig. 2.13).

Large Tusked Weta: *Motuweta isolata* and Raukumara Tusked Weta (Fig. 2.13)

These two large weta are almost certainly each other's nearest relatives (M. Morgan-Richards, Dunedin, 1998, personal communication), but at the time of writing this still has to be confirmed. They resemble each other in a number of morphological and behavioural features but *M. isolata* is much the larger (adults to 70 mm and 20 g versus 40 mm and 4.5 g) (Fig. 2.15). Adult males of both possess elongate, curved tusks projecting forward from, and fixed to, the mandibles which have been observed being used for jostling matches between males in *M. isolata*. They are essentially ground-dwelling weta in terms of refuge sites, but are capable climbers and ascend trees and bushes at night. Their habitat is undisturbed indigenous forest.

A single small population of *M. isolata* is known from the 13 ha Middle Island in the Mercury Group (Fig. 2.13), where it occurs in milk tree forest (*Streblus banksii*) amongst a diverse assemblage of burrowing seabirds (five species) and 11 species of reptiles, including tuatara (*Sphenodon punctatus*), but no alien predators. Their refuges are constructed in the soil, often amongst roots at the base of a tree. They excavate a chamber in the soil, which is roofed over and the entrance blocked when the weta is inside. Demographic and behavioural studies by McIntyre over a period of 4 years (McIntyre, 1994) have shown that the nocturnal emergence of the weta from its refuges is dependent upon rain and/or high humidity and also the amount of moonlight. *Motuweta isolata* consistently times its nocturnal forays to coincide with the totally dark period of the night, avoiding the moonlit period. This behaviour is assumed to be related to the visual hunting prowess of tuatara.

The undescribed Raukumara tusked weta is New Zealand's most recent large-bodied weta discovery. However, in just over 2 years since its discovery it has been found at ten different sites, covering a distance of 145 km, along the eastern Bay of Plenty ranges (Fig. 2.13), so it can be regarded as secretive and inaccessible, rather than

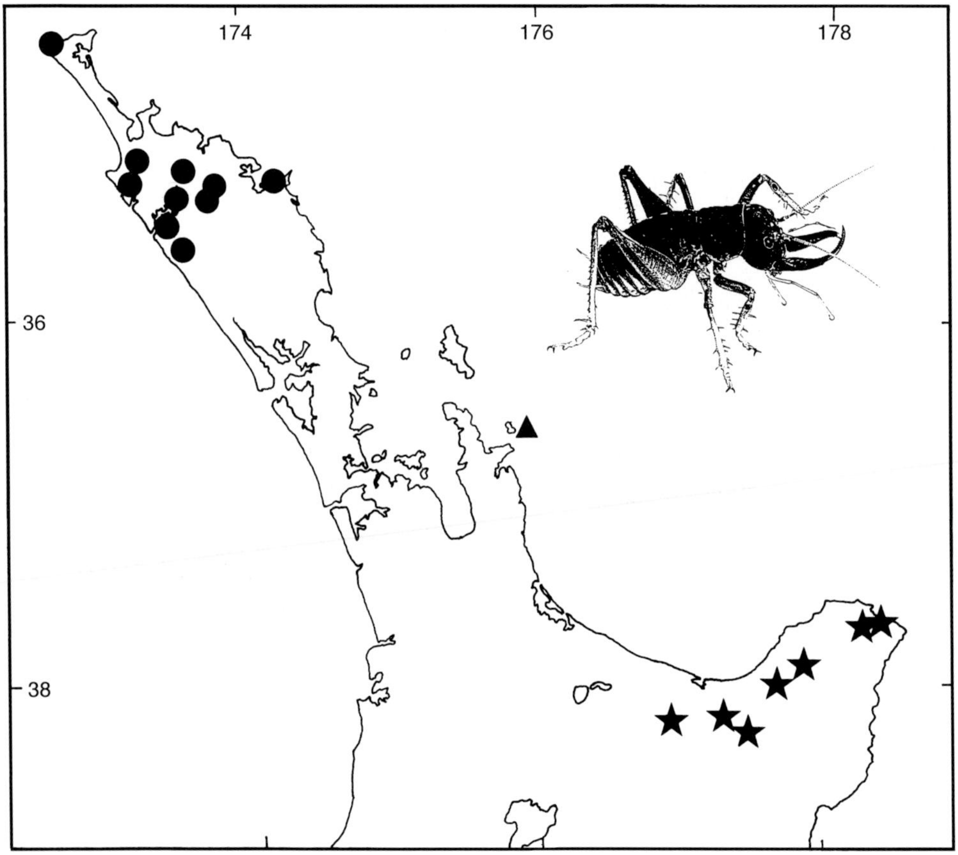

Fig. 2.13. Map of northern New Zealand showing known localities for three species of tusked weta. ● *Anisoura nicobarica;* ▲ *Motuweta isolata*; ★ Raukumara tusked weta. (Inset: Mercury Island tusked weta male, *Motuweta isolata*.)

rare. Apart from being smaller than *M. isolata*, males have distinctive tusks, which tend to droop at the tips instead of curving upwards like those of an elephant. Also, the tusks are devoid of stridulatory ridges, which are a feature of both *M. isolata* and *A. nicobarica*.

The Raukumara tusked weta habitat can be described with some precision, since all specimens have been found in close proximity to small first- or second-order streams in steep hill country clothed in undisturbed indigenous forest. Their refuges are under stones on gravelly stream terraces, not more than 3 m from the flowing water and clearly within the zone of spasmodic floods. Each weta excavates a cavity several times its own body size, with a narrow tunnel to the exterior. By day, the tunnel entrance is sealed with damp silt, which is kept in place for up to 4 weeks during a moult. This weta's intimate association with streams extends to its escape reaction, which involves leaping into the water and remaining on the stream bed for about 3 minutes before cautiously venturing across the bed to a position where part of the head and thorax can project above the surface. If the water flow is swift and the surface ruffled, the submerged weta is totally concealed. This escape reaction is reminiscent of the behaviour of a north Queensland mountain rainforest king cricket, *Transaevum laudatum*, described by Monteith and Field (Chapter 5, this volume). *Transaevum* nymphs, which feed on algae at night on the wet boulders of swift streams, will jump into the water when disturbed and drift downstream before climbing out.

Fig. 2.14. The small, reclusive tusked weta tentatively assigned to *Anisouris nicobarica* (= *Hemiandrus monstrosus* of New Zealand literature). Both species of tusked wetas have stridulatory ridges on the tusks and produce sound as a defence response. Male from Hokianga. Photo: G.W. Gibbs.

Biogeography

Satisfactory resolution of the higher-level systematics of the weta groupings has not yet been achieved, with the result that any rigorous examination of world biogeographical patterns is still premature.

The family Anostostomatidae (formerly Stenopelmatidae) is known from almost all parts of the world (Johns, 1997; reviewed by Gorochov, Chapter 1, this volume) and is especially diverse in the southern hemisphere land masses. It is thus no surprise that they are well represented in New Zealand. Present understanding of the New Zealand fauna seems to indicate that there are three distinct kinds of anostostomatids represented there: the large-bodied 'true weta', which comprises just *Hemideina* and *Deinacrida* (the Deinacridinae); the smaller (usually), more active, cricket-like burrowing weta, which we call 'ground weta', in the genus *Hemiandrus*; and the three species of 'tusked weta'.

New Zealand's 'true weta' (Deinacrininae), as defined above, have no close relatives elsewhere and appear to be a derived endemic group. Their key features are the heavy hind tibiae with a reduced number of very large spines (associated with their head-first occupation of narrow tunnels) and the development of socially interactive behaviour in *Hemideina*.

On the other hand, New Zealand's anostostomatine 'ground weta', *Hemiandrus*, are represented by species in both eastern Australia and New Zealand (Johns, 1997). All are apterous, suggesting that their separation is likely to date from at least the opening of the Tasman Sea in the late Cretaceous. Relationships of Australian and New Zealand taxa need to be investigated, especially since a few of the Anostostomatinae are winged.

Anostostomatids with fixed tusks on the mandibles in adult males are known from South Africa (*Henicus, Libanasidus*), Australia (*Australostoma*) and New Zealand (Johns, 1997; see also Gorochov, Chapter 1, this volume). Whether this feature constitutes evidence for the monophyly of 'tusked weta' has yet to be determined.

Within-New Zealand Distribution Patterns of Weta

This review, which considers 21 species that are distributed throughout virtually the whole of New Zealand, calls for an examination of patterns that could be biogeographically significant.

Apart from the broad distributional patterns of the presumptive phylogenetic groupings which have already been alluded to, we can identify three patterns of interest.

First, there is the position of the hypothetical stem species, *H. broughi*, in the north-western South Island (Fig. 2.1). This region is accepted as a centre of rich endemism for terrestrial biota (McGlone, 1985; Heads, 1989). Situated west of the median tectonic line, it is a component of the Tuhua terrane, having been part of the Pacific segment of Gondwana (Cooper, 1989). Moreover, the area remained above sea level during the Oligocene marine transgression (Cooper and Cooper, 1995). These factors are compatible with the long-term *in situ* survival of a stem taxon. Fiordland shares a similar history to that of north-western Nelson and also many biogeographical links (Heads, 1989). The surprising feature of weta distributions is that no deinacridine weta have been found there. Perhaps, if I could make a biogeographical prediction, it would be that a large and very significant anostostomatid weta might be discovered there one day.

Secondly, the occurrence of a pair of sister species, *D. pluvialis* and *D. talpa*, one on each side of the South Island Alpine Fault, provides us with yet another example of biological disjunction along this fault (for other examples, see Heads, 1989). Localities of *D. pluvialis* are all to the east of the fault on the greywacke of the main divide, whereas *D. talpa* is west of it on the older granite block mountains of the Paparoa Range (Fig. 2.8). On the basis of this distribution pattern, it could be predicted that perhaps further localities for *D. talpa* might be found on the Victoria Range. Indeed, there has been a hiker's report of a large weta there but, to date, no positive identification has been made. The Alpine Fault displacement is about 480

Fig. 2.15. Adult male of the spectacular tusked weta, *Motuweta isolata*, from Middle Mercury Island, the last protective refuge of this nearly extinct species. Large adults may reach 70 mm in body length, but like *Hemideina* males are polymorphic for size with a wide range of tusk lengths. Photo: Brett Robertson.

km between Nelson and Otago, with movement commencing in the Miocene (5 million years ago) and continuing today.

Thirdly, one of the most widely distributed species, *H. thoracica*, is apparently undergoing a mobilist phase in its development (*sensu* Croizat, 1964), with evidence of recent extension of range into the volcanic region of the North Island, as discussed by Morgan-Richards (1995); see also Morgan-Richards *et al.* (Chapter 7, this volume). According to geological evidence, the area around Lake Taupo would not have been habitable 2000 years ago and since then at least two other volcanic eruptions might have destroyed much of the weta habitat. A narrow (16 km) hybrid zone between two different chromosome races of this species now occupies this area (Morgan-Richards, 1995).

Morgan-Richards and Gibbs (1996) have drawn attention to the value of considering widespread alpine taxa such as *D. connectens* in the context of the age of the Alpine biota. This species occurs on many mountain-tops and displays geographical variations in colour patterns, karotypes and allozymes, which are not particularly concordant (Field, 1980). It is suggested that the minimal extent of allozyme differentiation between populations is in agreement with geological evidence that the Southern Alps of New Zealand are less than 5 million years old. The scree weta display either very recent evolution of their alpine habitat, or a fair degree of communication between populations after they became part of the alpine biota. Both are possible within the scenario of fluctuating Pleistocene climates. Diploid chromosome numbers from 16 sample localities ranged from 17 (XO) to 22 (XX). The distribution of three 'chromosome races' (17, 19 and 21 in XO males) shows that each 'race' occupies a coherent geographical area, which subdivides the species into a northern 21-karyotype population extending south and east to the narrowest portion of the South Island (limit at Fox Peak); a 19-karyotype population to the south and west between Mt Cook and the Takitimu Range; and a 17-karyotype population in the eastern Otago ranges. Congruence of allozyme and chromosome data indicate that the 19-karyotype form is likely to be ancestral (Morgan-Richards and Gibbs, 1996) and its distribution, known only from three widely spaced localities, could be interpreted as a relict one.

The phylogenetic hypothesis also enables us to examine the ecological shift from lowland to alpine habitat in the deinacridine weta taxa. Is adaptation to the alpine habitat a single event in their evolutionary history? The phylogenetic tree suggests otherwise, with no fewer than three independent clades of sister-species pairs having one member in a lowland habitat and the other alpine: *D. rugosa–D. parva*; *D. carinata–D. tibiospina*; *H. ricta–H. maori* (Morgan-Richards and Gibbs, 2001).

Acknowledgements

The new knowledge of New Zealand weta reviewed above would not have been possible without stimulation and input from Mary Morgan-Richards, Mary McIntyre and a number of dedicated Department of Conservation officers. Support, in the form of contracts and collecting permits, came from the New Zealand Department of Conservation, and financial support for the research came from the Internal Research Committee of Victoria University.

References

Ander, K. (1938) Diagnosen neuer Laubeuschrcken. *Opuscula Entomologica* 3, 50–56.

Ander, K. (1939) Vergleichend anatomische und phylogenetische Sudien. Ueber die Ensifera (Saltatoria). *Opuscula Entomologica Supplement* 2, 1–306.

Asher, G.W. (1977) Ecological aspects of the common tree weta (*Hemideina thoracica*) in native vegetation. Unpublished report, file no. 8/1/5, DSIR Ecology Division Lower Hutt.

Buller, W. (1871) Notes on the genus *Deinacrida* in New Zealand. *Transactions and Proceedings of the New Zealand Institute* 3, 34–37.

Buller, W. (1895) On the wetas: a group of orthopterous insects inhabiting New Zealand; with descriptions of two new species. *Transactions and Proceedings of the New Zealand Institute* 27, 143–147.

Cameron, W. (1996) A study of the taxonomic status of *Deinacrida parva* and *Deinacrida rugosa* (Orthoptera: Stenopelmatidae), two giant weta from central New Zealand. MSc thesis, Victoria University, Wellington, New Zealand.

Cooper, A. and Cooper, R.A. (1995) The Oligocene bottleneck and New Zealand biota: genetic record of past environmental crisis. *Proceedings of the Royal Society, London B* 261, 293–302.

Cooper, R.A. (1989) New Zealand tectonostratigraphic terranes and panbiogeography. *New Zealand Journal of Zoology* 16, 699–712.

Croizat, L. (1964) *Space, Time, Form: the Biological Synthesis*. Published by the author, Caracas.

Field, L.H. (1980) Observations on the biology of *Deinacrida connectens* (Orthoptera: Stenopelmatidae), an alpine weta. *New Zealand Journal of Zoology* 7, 211–220.

Field, L.H. (1993) Structure and evolution of stridulatory mechanisms in New Zealand wetas (Orthoptera: Stenopelmatidae). *International Journal of Insect Morphology and Embryology* 22, 163–183.

Gibbs, G.W. (1998) Why are some weta (Orthoptera: Stenopelmatidae) vulnerable yet others are common? *Journal of Insect Conservation* 2, 161–166.

Gibbs, G.W. (1999) Four new species of giant weta, *Deinacrida* (Orthoptera: Anostostomatidae: Deinacridinae) from New Zealand. *Journal of the Royal Society of New Zealand* 29, 307–324.

Gibbs, G.W. and McIntyre, M.E. (1997) Abundance and future options for wetapunga on Little Barrier Island. *Science for Conservation* (Wellington, New Zealand) 48, 5–24.

Heads, M. (1989) Integrating earth and life sciences in New Zealand natural history: the parallel arcs model. *New Zealand Journal of Zoology* 16, 549–585.

Johns, P.M. (1997) The Gondwanaland weta: family Anostostomatidae (formerly in Stenopelmatidae, Henicidae or Mimnermidae): nomenclatural problems, world checklist, new genera and species. *Journal of Orthoptera Research* 6, 125–138.

Little, G.A. (1980) Food consumption and utilisation in two species of weta (*Hemideina femorata, H. maori*: Stenopelmatidae). BSc Hons. project, University of Canterbury, Christchurch, New Zealand.

McGlone, M.S. (1985) Plant biogeography and the late Cenozoic history of New Zealand. *New Zealand Journal of Botany* 23, 723–749.

McIntyre, M.E. (1993) Dispersal and preliminary estimates of the giant weta, *Deinacrida rugosa*, following the eradication of mice from Mana Island. Unpublished report to Department of Conservation, Wellington, 16 pp.

McIntyre, M.E. (1994) The status and habitat of the Middle Island (Mercury Group) tusked weta with implications for management. Unpublished report to Department of Conservation, Wellington, 10 pp.

Meads, M. (1990) *Forgotten Fauna, the Rare, Endangered, and Protected Invertebrates of New Zealand*. DSIR Publishing, Wellington, New Zealand.

Meads, M. and Notman, P. (1991) A survey of the giant wetas (*Deinacrida parva* and *Deinacrida* sp.) in the north branch of the Hapuku River, Kaikoura, 29 April–3 May 1991. *DSIR Land Resources Technical Record (Lower Hutt, New Zealand)* 43, 1–34.

Meads, M. and Notman, P. (1992) Survey of the status of three species of giant wetas (*Deinacrida*) on the Seaward and Inland Kaikoura Ranges. *DSIR Land Resources Technical Record (Lower Hutt, New Zealand)* 89, 1–35.

Meads, M. and Notman, P. (1993) Giant weta (*Deinacrida carinata*) on Pig Island, Foveaux Strait. In: *Landcare Research Contract Report*, LC9293/116, for Department of Conservation, Wellington, pp. 1–14.

Moller, H. (1985) Tree wetas (*Hemideina crassicruris*) (Orthoptera: Stenopelmatidae) of Stephens Island, Cook Strait. *New Zealand Journal of Zoology* 12, 55–69.

Morgan-Richards, M. (1995) Weta karyotypes: the systematic significance of their variation. PhD thesis, Victoria University, Wellington, New Zealand.

Morgan-Richards, M. and Gibbs, G.W. (1996) Colour, allozyme and karyotype variation show little concordance in the New Zealand giant scree weta *Deinacrida connectens* (Orthoptera: Stenopelmatidae). *Hereditas* 125, 265–276.

Morgan-Richards, M. and Gibbs, G.W. (2001) A phylogenetic analysis of New Zealand giant and tree weta (Orthoptera: Anostostomatidae: *Deinacrida* and *Hemideina*) using morphological and genetic characters. *Invertebrate Taxonomy* 15 (in press).

Ramsay, G.W. (1961) The synonymy and systematics of a genus and two species of New Zealand weta (Orthoptera: Stenopelmatidae: Henicinae). *Proceedings of the Royal Entomological Society of London (B)* 30, 85–89.

Ramsay, G.W. (1971) Rediscovery of Sir Walter Buller's weta *Deinacrida parva* (Orthoptera: Gryllacridoidea; Henicidae). *New Zealand Entomologist* 5, 52–53.

Richards, A.M. (1973) A comparative study of the giant wetas *Deinacrida heteracantha* and *D. fallai* (Orthoptera: Henicidae) from New Zealand. *Journal of Zoology, London* 169, 195–236.

Richards, G.E. (1994) Ecology and behaviour of the Mahoenui giant weta, *Deinacrida* nov.sp. Unpublished MSc thesis, Massey University, Palmerston North, New Zealand.

Rufaut, C.G. (1995) A comparative study of the Wellington tree weta, *Hemideina crassidens* (Blanchard, 1851) in the presence and absence of rodents. Unpublished MSc thesis, Victoria University, Wellington, New Zealand.

Salmon, J.T. (1950) A revision of the New Zealand wetas (Anostostominae (Orthoptera: Stenopelmatidae). *Dominion Museum Records in Entomology* 1, 121–177.

Sherley, G.H. and Hayes, L.M. (1993) The conservation of a giant weta (*Deinacrida* n.sp. Orthoptera: Stenopelmatidae) at Mahoenui, King Country: habitat use, and other aspects of its ecology. *New Zealand Entomologist* 16, 55–68.

Spencer, A.M. (1995) Sexual maturity in the male tree weta *Hemideina crassidens* (Orthoptera: Stenopelmatidae). Unpublished MSc thesis, Victoria University, Wellington, New Zealand.

Townsend, J.A., Brown, B., Stringer, I.A.N. and Potter, M.A. (1997) Distribution, habitat and conservation status of *Hemideina ricta* and *H. femorata* on Banks Peninsula, New Zealand. *New Zealand Journal of Ecology* 21, 43–49.

Trewick, S.A. and Morgan-Richards, M. (1995) On the distribution of tree weta in the North Island, New Zealand. *Journal of the Royal Society of New Zealand* 25, 485–493.

Watt, J.C. (1963) The rediscovery of a giant weta, *Deinacrida heteracantha*, on the North Island mainland. *New Zealand Entomologist* 3, 9–13.

Watt, J.C. (1980) Notes on pitfall trapping on Headlong Peak, Mount Aspiring National Park. *New Zealand Entomologist* 7, 184–191.

Watt, J.C. (1982) Terrestrial arthropods from the Poor Knights Islands, New Zealand. *Journal of the Royal Society of New Zealand* 12, 283–320.

3

North and Central America Jerusalem Crickets (Orthoptera: Stenopelmatidae): Taxonomy, Distribution, Life Cycle, Ecology and Related Biology of the American Species

David B. Weissman

Department of Entomology, California Academy of Sciences, Golden Gate Park, San Francisco, California 94118, USA

Introduction

In 1980, while coordinating a project on the orthopteran fauna of Baja California, Mexico, I elected to do the Jerusalem crickets (*Stenopelmatus* Burmeister 1838, *Ammopelmatus* Tinkham 1965, and *Viscainopelmatus* Tinkham 1970) in the absence of other workers. The genera and species were supposedly well known, as were their distributions. Of the 35 species of *Stenopelmatus* in the literature, 19 were described from Mexico and Central America between 1838 and 1902. I have examined all type specimens (some determined to be suitable lectotypes and neotypes by T.H. Hubbell (Michigan, 1982, personal communication)) and confirm that many prior synonymies are erroneous. The US species of *Stenopelmatus* were last revised in 1916 (Hebard), with an additional five species subsequently described (Davis and Smith, 1926; Tinkham, 1968, 1979; Tinkham and Rentz, 1969; Rentz, 1978), giving 14 species in total. Two taxa have no specific locality. *Ammopelmatus* includes the type species described in 1965 (Tinkham, 1965) and a second in 1981 (Rentz and Weissman, 1981). *Viscainopelmatus* is monotypic (Tinkham, 1970). In this chapter, I refer to all species in all three genera as Jerusalem crickets (JCs).

Interestingly, the initial task proved more complicated than anticipated: the Baja California Peninsula has at least eight species, only one described; California has 35–50 species, only seven described; and the USA has 60–80 species in total, 14 described. Efforts in mainland Mexico and Central America have just begun and indicate a diverse fauna. In total, there may be 100 species of *Stenopelmatus*, making this genus the most significant component of the world's Stenopelmatidae, which Kevan (1982) noted is comprised of 'only about half a dozen genera and a little over 30 species' (but see Johns (1997) for a more extensive species list and taxa transferred to the Anostostomatidae).

After 20 years, I estimate my fieldwork and data collection for this revisionary study of the JCs are 70% complete. Thus my two chapters in this book will serve as both a progress report and a review of previous work. Unfortunately, most of the literature, unless dealing with a newly described, geographically limited or morphologically very unique species of JC, will be unassignable to taxa.

Described Species of Jerusalem Crickets and Type Locality

See Karny (1937) for bibliographic citations prior to 1937 and 'References' for subsequent ones.

Stenopelmatus Burmeister 1838 (type species *talpa*)
- *ater* Saussure & Pictet 1897 (Costa Rica)
- *cahuilaensis* Tinkham 1968 (California)
- *calcaratus* Griffini 1893 (Mexico)
- *californicus* Brunner von Wattenwyl 1888 (British Columbia, Canada)
- *cephalotes* F. Walker 1869 (west coast of North America)
- *comanchus* Saussure & Pictet 1897 (Mexico)
- *erythromelas* F. Walker 1869 (no locality)
- *fasciatus* Thomas 1872 (Idaho)
- *fuscus* Haldeman 1852 (New Mexico)
- *guatemalae* Brunner von Wattenwyl 1888 (Guatemala)
- *histrio* Saussure 1859 (Mexico)
- *hydrocephalus* Brunner von Wattenwyl 1888 (California)
- *intermedius* Davis & Smith 1926 (California)
- *irregularis* Brunner von Wattenwyl 1888 (Arizona)
- *lessonae* Griffini 1893 (Mexico)
- *longispina* Brunner von Wattenwyl 1888 (British Columbia, Canada)
- *lycosoides* F. Walker 1869 (Mexico)
- *mescaleroensis* Tinkham 1979 (New Mexico)
- *mexicanus* Saussure 1859 (Mexico)
- *minor* Saussure 1859 (Mexico)
- *navajo* Rentz 1978 (Arizona)
- *nieti* Saussure 1859 (Mexico)
- *nigrocapitatus* Tinkham & Rentz 1969 (California)
- *oculatus* Scudder 1876 (Utah)
- *piceiventris* F. Walker 1869 (no locality)
- *pictus* Scudder 1899 (California)
- *politus* F. Walker 1869 (Mexico)
- *sallei* Saussure 1859 (Mexico)
- *sartorianus* Saussure 1859 (Mexico)
- *sumichrasti* Saussure 1859 (Mexico)
- *talpa* Burmeister 1838 (Mexico)
- *terrenus* Rehn 1902 (Texas)
- *toltecus* (Saussure) 1861 (Mexico)
- *typhlops* Rehn 1902 (Mexico)
- *vicinus* Brunner von Wattenwyl 1888 (Guatemala)

Ammopelmatus Tinkham 1965
- *kelsoensis* Tinkham 1965 (California)
- *muwu* Rentz & Weissman 1981 (California)

Viscainopelmatus Tinkham 1970
- *davewerneri* Tinkham 1970 (Baja California, Mexico)

Collecting Techniques

Probably every entomological collection in the western USA includes JCs. Unfortunately, most specimens are juveniles, which do not show the few diagnostic (adult) characters. The rare adults present were dried on the pin, as opposed to preserved in alcohol, and are thus distorted for both colour and shape. Dead JCs cannot produce their calling drum (see Weissman, Chapter 19, this volume) nor can their testes be analysed for chromosome number. One can sometimes do mitochondrial DNA studies on alcohol-preserved specimens, but freezing of fresh material at −70°C is preferable. For all these reasons, I generally shun museum collections except to search for the few morphologically unusual specimens and to learn what time of year JCs were collected in certain areas. They are found in almost every imaginable habitat, from sea level to over 3000 m in the White Mountains, Inyo County, California, to over 4250 m on Mt Orizaba, Puebla, Mexico (considerably above the timberline (L. Swan, California, 1996, personal communication)).

Time of year and type of habitat determine how I collect JCs. Sand-dune habitats, either coastal or inland, that do not experience freezing temperatures are best collected during the wet winter and early spring months by sifting (Fig. 3.1) or turning over material (rubbish, boards, rocks, cow-pats) on the surface of the dunes. During the hot summer months (even in areas with summer monsoon rains), JCs are almost inaccessible during the day, apparently digging down to find the desired temperature and humidity. Late instars and adults of certain species do come to the surface at night and wander. They can then be collected by oatmeal trails, by pitfall traps baited with oatmeal or by driving along paved roads that traverse such dunes. Sand-dune habitats that experience winter freezes are best collected during the summer, using oatmeal trails, pitfall traps and road driving.

Fig. 3.1. Sifting for Jerusalem crickets in sand-dunes in winter, in Kern County, California, V.F. Lee, sifter. Inside measurements of sifter are 47.5 cm × 47.5 cm; mesh size is 6.3 mm × 6.3 mm. Entire sifter is suspended under a camera tripod. Highest yields usually occur when the sand is very moist and hard to sift.

Coastal, foothill and inland California habitats that experience, at most, mild freezing temperatures (no colder than −3°C at night), are easily collected during the winter months by finding damp, grassy, oak woodland gentle slopes covered with rocks and logs. Jerusalem crickets frequently seek daytime shelter under such objects. The remaining areas of California and most of the western USA, since they have deep winter freezes, are best collected during warming periods of spring by turning over objects and during the summer and autumn by nocturnal oatmeal trails, pitfall traps and road driving. Interestingly, during summer daylight hours, JCs in habitats above 1750 m, even if dry, are frequently found under objects, especially if such objects are shaded.

Jerusalem crickets living in cloud forests (usually above 1200 m) of Mexico and Central America, which include most of the taxa there, occupy a habitat completely different from that of any US species: they live in rotting logs and stumps, at least during the summer rainy season (June–September). These individuals are best collected by tearing apart such wood during the daytime, although rock and log turning can still be useful. Late autumn/early winter dry-season collecting was unproductive during my one trip to Honduras. Many tropical cloud forests are so moist that oatmeal on a trail quickly becomes saturated and may not attract many JCs.

Rearing Techniques

All collected JCs are kept separate, since they eat each other. They are returned to the laboratory and raised in 225 cm^3 (8 oz) plastic margarine tubs filled with damp, sterilized, fine-grained sand. They are fed shop-bought romaine lettuce and 'old-fashioned' oatmeal and exposed to ambient photoperiods. Summer temperatures are kept at 20°C and the JCs are fed every 7–10 days after cleaning their tubs. At higher temperatures, mite infestations become a problem. The laboratory passively cools during winter, reaching minimums of 6°C. At these cooler temperatures, JCs are fed every 10–12 days. The diet of Mexican and

Central American cloud forest JCs is supplemented with weekly feedings of greater wax moth larvae (*Galleria melonella* (Linnaeus)) obtained from a local bait shop. Without such supplements, these JCs never moult and they die within a few months.

Preservation Techniques

After I have recorded their calling drum (see Weissman, Chapter 19, this volume) and photographed many, adults are killed by injection with Williams' fixative (Williams, 1968) and their colours noted. They are then placed in fixative for 6 h after slitting the abdomen ventrally to let fixative into the cavity to 'sterilize' it and prevent bacterial discoloration. Fixed adults are transferred to 70% ethanol for permanent storage. While this process does harden the specimens, colour preservation is greatly improved. Jerusalem crickets have soft, unsclerotized internal genitalia of no taxonomic distinction (D.B. Weissman, unpublished).

Karyotyping

Testes were removed from penultimate or last-instar males (see 'Determination of Adults' below), incubated in insect saline and colchicine for 1–2 h, fixed in 3 : 1 100% ethanol : glacial acetic acid and stained with acetocarmine. All observations were made on squash preparations by Norihiro Ueshima, Matsusaka University, Japan, and will be reported in more detail elsewhere. Karyotyping is best done in newly moulted last instars, but it is difficult, in some species, to distinguish last from penultimate instars. The latter may not have any testes development. Most adult JCs, even if newly moulted, have no meiosis or mitosis and their testes, filled with mature sperm, are unsuitable for karyotyping. This absence of adult meiosis is a common phenomenon in Orthoptera taxa that overwinter as nymphs (e.g. *Gryllus* Linnaeus field crickets and *Cnemotettix* Caudell silk-spinning crickets), but probably does not limit fertility (see Henry, 1985, p. 28). In contrast, grasshoppers that overwinter as eggs still have meiosis 3 months after moulting to adults (D.B. Weissman, unpublished).

Overview of Results

When I started these studies, not only were adequate keys to species unavailable, but no documentation existed for determining when individuals were adult (Tinkham and Rentz, 1969; Rentz, 1978), although Tinkham and Rentz (1969) used 'size of adults' as the first couplet in their key. Two species were separated in that same key by unique head markings; yet these markings appear only in adults (D.B. Weissman, unpublished).

My subsequent efforts were successful because I started rearing thousands of field-captured nymphs, and also I discovered that many species of JCs have unique calling-song drumming patterns (see Weissman, Chapter 19, this volume). It is surprising to realize that two of the largest insect species (Fig. 3.2) in California, both weighing over 8 g as adults, are undescribed JCs; one species is actually common in populated areas of southern California. New JC species will be described elsewhere. Mitochondrial DNA studies have been done in the laboratory of Felix A.H. Sperling, under the direction of Michael S. Caterino, both then at the University of California, Berkeley, but are at a preliminary stage and are not discussed here.

Statements below refer to most, if not all, species of JCs. Where a specific (usually unnamed) species is involved, I refer to it by number (Table 3.1).

Distribution

Jerusalem crickets occur (Fig. 3.3) from southwestern Canada (including Vancouver Island (Buckell, 1930)), east through Montana and western North Dakota and south through the Great Plains as far east as Perkins, Oklahoma (Hebard, 1936, 1938; Vickery and Kevan, 1983; D.B. Weissman, unpublished). The peninsula of Baja California, Mexico, has JCs as far south as La Paz (D.B. Weissman, unpublished), while Mexico proper has JCs essentially confined to the highlands, although a species reaches the coast at Veracruz (unpublished records from Academy of Natural Sciences, Philadelphia (ANSP); University of Michigan, Museum of Zoology (UMMZ); V.F. Lee, California Academy of Sciences, San Francisco; J.K. Liebherr, Cornell

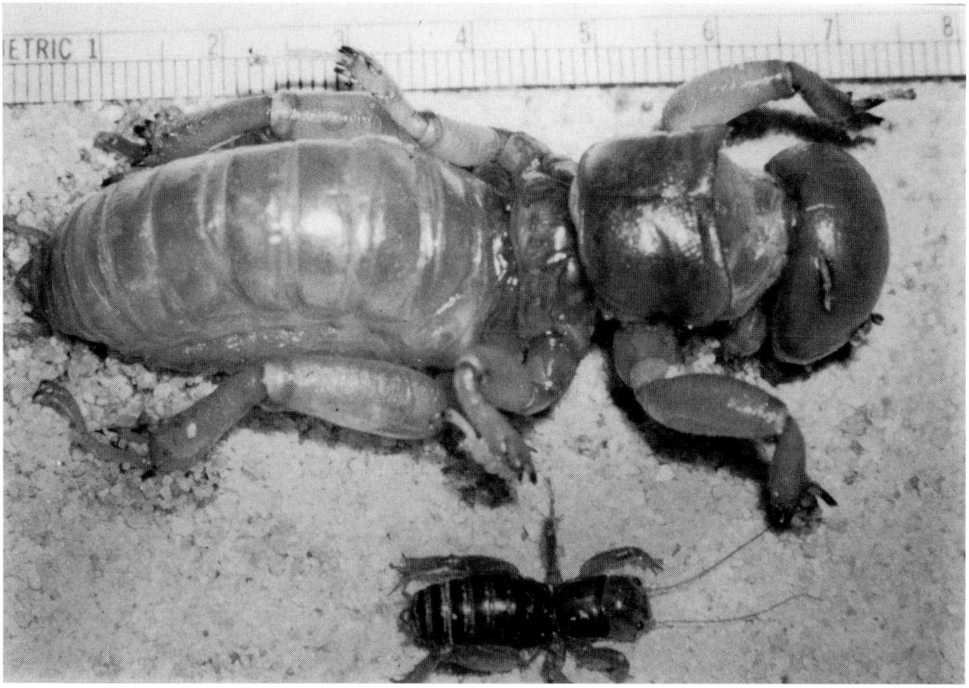

Fig. 3.2. Size range of adult Jerusalem crickets. Top, adult female of undescribed species from Kern County, California, live weight unknown. Other adults of this species, at half this female's volume, weigh over 8 g. Bottom, adult male of undescribed species from Monterey County, California, weighs around 750 mg.

University, Ithaca, New York). Jerusalem crickets have been captured in the Central American countries of Guatemala, Honduras, El Salvador and Costa Rica (unpublished records ANSP, UMMZ), and they undoubtedly occur in the highlands of Nicaragua. The southernmost record in Costa Rica is from Las Mellizas, Puntarenas Province, a few kilometres from Panama (unpublished records, Instituto Nacional de Biodiversidad, San Jose, Costa Rica).

Microsympatry of two species of JCs is rare and has been discovered at only 30 localities (out of hundreds), involving 20 different combinations of species. Microsympatry of three species of JCs is known from near Cherry Valley, Riverside County, California.

Table 3.1. Undescribed and described species of Jerusalem crickets referred to in text and their collection localities.

Species no.	Locality
1	Mexico, Baja California Norte, 34 km south of Ensenada
2	USA, California, San Bernardino County, Kelso Dunes, *A. kelsoensis*
3	USA, California, Monterey County, Arroyo Seco
4	USA, California, southern part of state, various locations
5	USA, California, San Bernardino County, Lake Arrowhead
6	USA, California, Los Angeles County, Santa Monica Mountains
7	USA, Idaho, Jefferson County, western part
8	Canada, British Columbia, Oliver
9	USA, Nevada, Humboldt County, Desert Valley Dunes
10	USA, Idaho, Owynee County, Bruneau Dunes

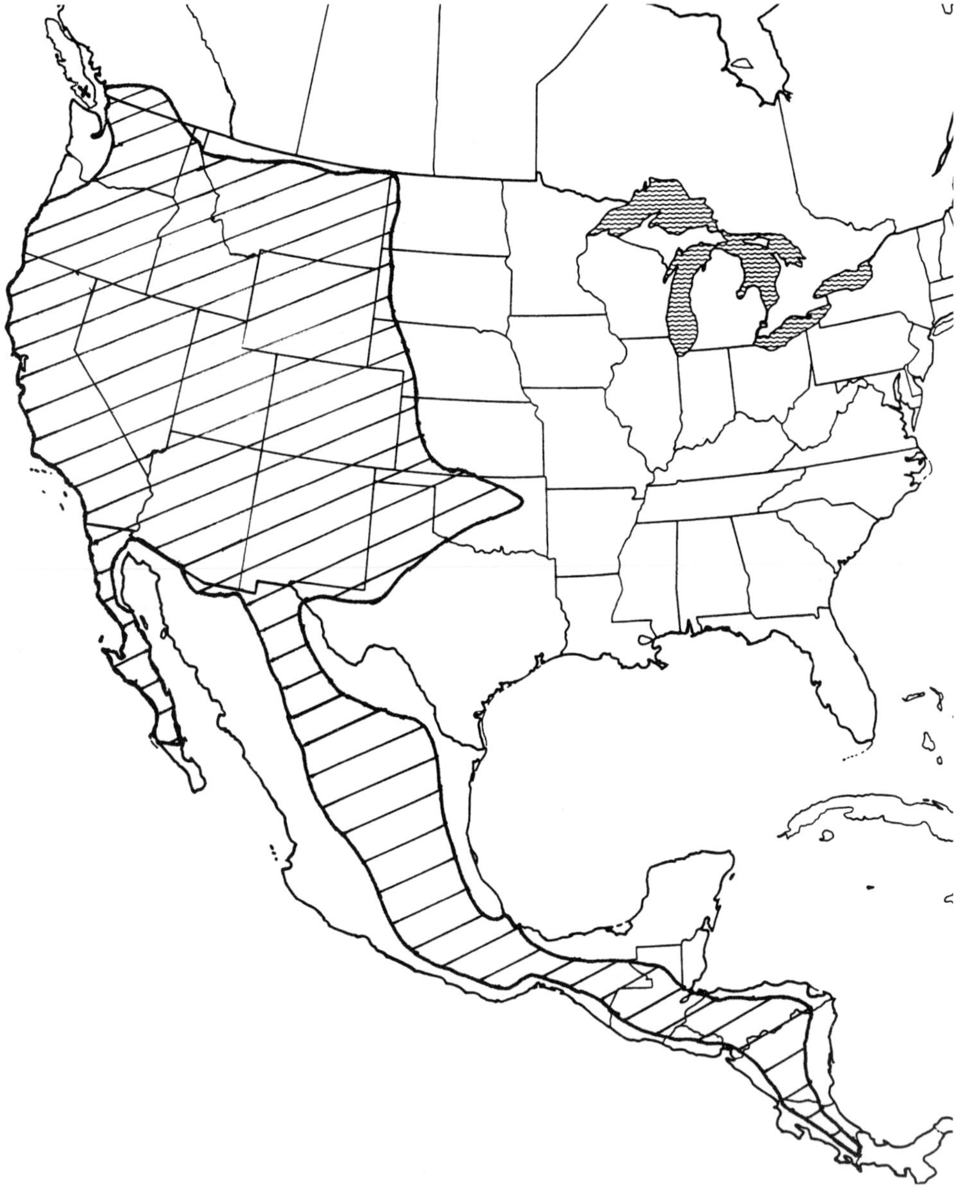

Fig. 3.3. Known distribution of *Stenopelmatus* Jerusalem crickets within stippled area.

Determination of Adults

Adult males are recognized by a pair of black, sclerotized, incurved hooks located on each side of the supra-anal plate between the cerci (Fig. 3.4). Contrary to Hebard's (1916) assertion, these hooks are found only in adult males, where they facilitate mating (see Weissman, Chapter 19, this volume). They are not part of a 'gin trap' (Gwynne, 1997, p. 297; see also Field and Jarman, Chapter 17, this volume). Earlier instars have unsclerotized ridges of varying development. Once males become adults, they moult no more and they mate and drum.

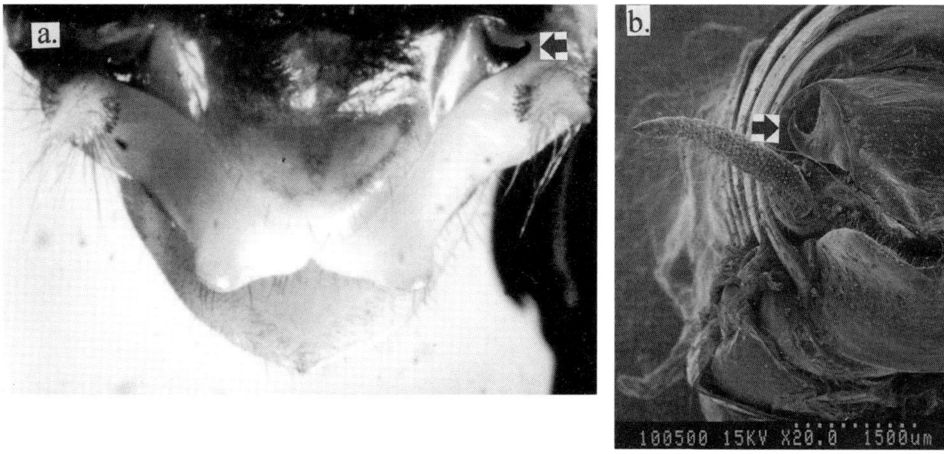

Fig. 3.4. (a) Adult male Jerusalem crickets are characterized by a pair of black, sclerotized hooks (arrow) on each side of the supra-anal plate, here seen in dorsal view. (b) Scanning electron micrograph (SEM) of left hook (arrow) seen from behind. Horizontal structure on left, is the male's left cercus.

Adult females have at least the tip and ventral surfaces of their ovipositor blades sclerotized (Fig. 3.5), although this can be difficult to see in entirely black Mexican and Central American species. Adult females cease moulting, mate, drum and may contain mature eggs.

Large-headed adult male and female JCs are associated with sandy or sand-dune-type habitats or loose soils that are easy to dig in, since they use mainly their heads for digging (Davis, 1927). This megacephalic condition is apparent in last instars and appears unrelated to aggression. Two megacephalic adult males enclosed with a virgin adult female show no antagonistic behaviour toward each other. This is similar to the behaviour of male giant wetas, but unlike the intense agonism between males of *Cratomelus* in Chile (see Angulo, Chapter 11, this volume) and male tree wetas in New Zealand (see Field, Chapter 18, this volume).

Life Cycle

I have been unable to study the complete life history of any JC species because of my inability to get females to oviposit. But collecting thousands of specimens at all times of the year still yields a comprehensive overall picture.

Individuals of most JC species become adult in midsummer in both the laboratory and nature, with individuals of some coastal California taxa maturing a month or two later. There may be some field protandry, where males mature before females, but no more than a few weeks are involved. There is no evidence of any summer adult aestivation period; individuals are merely underground and not obvious. Adults start drumming within 4 days to 2 weeks after their last moult and will mate then. Adult males disappear in the field before adult females, due I believe, to a combination of shorter lives, protandry and sexual cannibalism (see Weissman, Chapter 19, this volume), although the latter seems, by itself, too limited to account for the distorted adult sex ratios usually seen during late autumn–early winter.

Eggs are apparently laid soon after mating, judging by the appearance of tiny nymphs in autumn. These individuals could be first instars, as gauged by their size relative to the size of their eggs. I also find tiny nymphs in May. It may be that eggs of the first clutch, if laid by late summer or early autumn, can hatch before the cold of winter. Subsequent clutches may not hatch until spring or may be laid in spring by overwintering adult females. I believe eggs hatch as soon after laying as developmentally possible (i.e. there is no obligatory diapause) and that JCs, if deep underground, can safely pass the winter in either the egg or the nymphal stage. Eggs of all species are also apparently laid well below the surface (Davis, 1927), where they will not freeze, because I have never found a clutch while turning thousands of surface objects and sifting tons of sand.

From meagre records, JCs appear to have

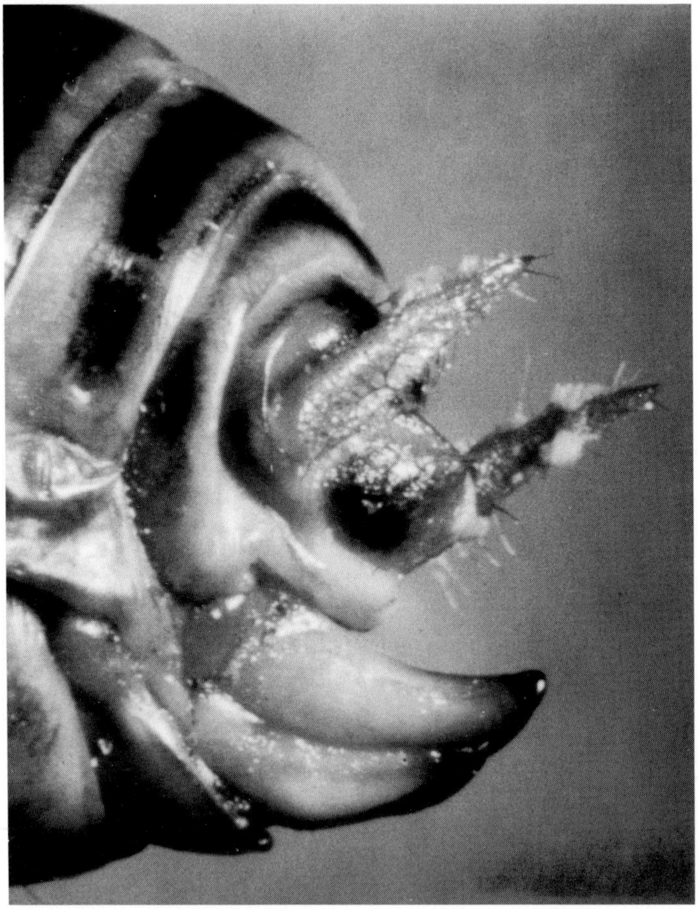

Fig. 3.5. Adult female Jerusalem crickets are characterized by a pointed, sclerotized, dark ovipositor. The two vertical structures dorsal to the ovipositor are cerci.

between nine and 11 moults, with differences between species likely. The following records are illustrative:

- Species no. 1. Male captured 8/1/1979, 4 mm in length, first or second instar. Molted 22/2/1979, 5/4/1979, 15/5/1979, 4/7/1979, 30/8/1979, 9/10/1979, 14/2/1980, 27/4/1980, 14/9/1980 adult. Died 25/4/1981.
- Species no. 2. Female captured 20/4/1963, length unrecorded but believed to be third instar (from unpublished notes of E.R. Tinkham, Indio, California, 1985). Moulted 6/1963, 5/8/1963, 8/10/1963, 13/12/1963, 13/5/1964, ?/9/1964, 16/7/1965, 18/9/1965, 8/3/1966 adult. Died 20/4/1967.

All JCs moult while on their backs (Fig. 3.6) and subsequently right themselves and eat the exuvium. In the laboratory, JCs usually moult while buried (housing them without moist sand results in high mortality rates) and, unless one examines them daily, moults can be missed.

Most JCs require $1\frac{3}{4}$ years from egg hatch to maturity (moult to adult), although populations from high elevations and/or northern latitudes, with their shorter growing seasons, probably require an extra year or two. In any group with such a long maturation period, one expects to find variation in growth. Nevertheless, collecting JCs in early winter in coastal California typically yields three age classes: adults who matured in the previous few months; mid-sized to large nymphs that will mature the next summer; and small, recently hatched individuals that will overwinter twice to

Fig. 3.6. All Jerusalem crickets moult on their back, with the exuvium ultimately positioned behind them. This adult male, from Santa Clara County, California, will right itself and eat the cast skin.

mature during their second summer. Collecting JCs in late spring on Mt Shasta, Siskiyou County, California, at almost 1500 m, also yields three age classes: large nymphs that will mature in 2–3 months; medium-sized nymphs that become adults in 14–15 months; and small nymphs that become adults in 26–27 months, probably having hatched the previous autumn. I have had isolated individuals of several species require unusually long periods to develop, possibly because of nutrient deficiencies or parasitic infections (Smith, 1929): a small female nymph of species no. 3, captured on 27 December 1992, moulted to adult on 22 July 1997. The individual of species no. 2, discussed above, required 4 years to become adult (see also Tinkham and Rentz, 1969). I have raised some 20 other individuals of species no. 3 and ten other individuals of species no. 2, and all matured in the 'normal' period of $1\frac{3}{4}$ years, thus indicating a possible abnormality in these two slow-maturing nymphs, although both drummed normally once adult. In the laboratory, most adults live from 2 to 6 months. A female of species no. 4, who had been an adult for 11 months, was readily mated by a newly moulted adult.

Gwynne (1995, p. 212), based on Essig's photograph (1926) of an egg clutch in '*S. longispina*' (without information as to the presence or absence of a parent), postulated, in an evolutionary phylogeny, that JCs have parental care. There is no evidence other than the photograph in support of, and much to argue against, this: (i) Davis (1927) discussed finding several such egg chambers, but none were noted to contain females; (ii) I have found tiny (presumably first-instar) nymphs together in the field on several occasions without any adults; (iii) adult females of several species can be common under rocks in California in late autumn and winter. Were there parental care, I would expect these females to be sequestered with their eggs and rarely seen near the surface; (iv) JCs are cannibalistic, especially large individuals on smaller ones; (v) adult female JCs are seen wandering with the advent of late autumn and winter rains, at exactly the time they should be engaged in parental care; (vi) I can raise the tiniest nymphs to maturity, indicating that an adult female is not necessary; (vii) JCs are such good diggers that they have no need for a long ovipositor to place eggs deep and protected; and (viii) many eggs apparently overwinter and females appear to be able to clutch more than once.

Parental care would thus severely limit fecundity and is essentially 'incompatible with an iteroparous life style' (Tallamy and Brown, 1999). I think the 'brood chamber' pictured by Essig (1926) probably represents the dirt cleared from the area by the large-bodied female, similar to the way JCs make 'burrows' under objects by packing the dirt with their heads as they dig.

Ecology

While widespread, JC species living in two particular habitats have a number of unique morphological and behavioural adaptations.

1. Obligate sand-dune-inhabiting JC species, when compared with species away from dunes, are lighter-coloured (for cryptic nocturnal movement on light-toned sand), have shorter antennae, megacephalic heads for digging, shorter and stouter legs and spatulate spines on the rear tibia for sand propulsion. In fact, these edaphic adaptations prompted Tinkham (1965, 1970) to erect two monotypic genera to encompass these changes. Additional work (D.B. Weissman, unpublished) has shown these supposed 'generic' adaptations to be widespread in sand-dune-inhabiting JC taxa. Along the Baja California–California coast, there is a succession of four localized, sand-obligate JC species with non-overlapping ranges. At several locations within these restricted ranges, a second, microsympatric species of non-sand-obligate JC is found. In all cases, the two species are readily distinguishable by the above physical criteria. This colour pattern is broken at one inland sanddune area where two sand-dune-obligate species co-occur: south-eastern New Mexico around the Mescalero sand-dunes to Monahans sand-dunes in western Texas. Dunes throughout this area can be densely covered with shin oak (*Quercus harvardi* Rydberg) and their fallen leaves. The black *S. mescaleroensis* inhabits these vegetated areas and is cryptic when moving about on the surface at night; while an undescribed, light coloured, megacephalic species inhabits areas with minimum ground cover (C. Anderson, New Mexico, 1993, personal communication).

2. Many JC species inhabiting mountains and cloud forests in Mexico and Central America

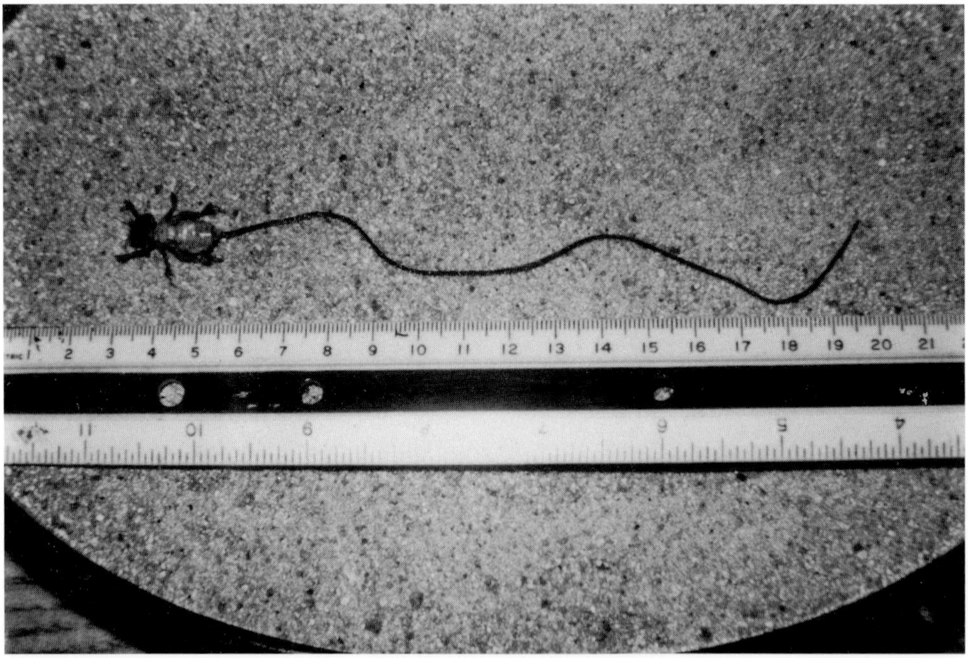

Fig. 3.7. This female nymph from Bernalillo County, New Mexico, was collected on 15 June 1988 and died the following 30 May, when the pictured 19.8 cm hair-worm emerged from her. She never moulted during this 11-month period, but was always active.

exhibit a behaviour never seen in US taxa: they jump 5–15 cm when disturbed. These taxa also live in rotting logs (including a fully winged species). The only two US occurrences of JCs living in logs were seen at Pt Conception, Santa Barbara County, and along coastal Santa Cruz County, both in California. I propose several possible reasons why inhabiting logs occurs in the tropics: (i) living in logs gets the JCs off the wet, frequently saturated ground and into an environment with a more predictable humidity (especially during dry conditions); (ii) the hollowed-out passageways in the logs provide easy access to the tunnel-making passalid, cerambycid and scarab larvae, which the JCs readily eat; (iii) protection from predators, such as pit vipers (Campbell and Armstrong, 1979; Campbell and Solorzano, 1992), mammals, including bats and foxes (Orr, 1954; Evens, 1988), scorpions (Polis, 1979), birds, such as barn, burrowing, screech and great horned owls (Brown *et al.*, 1986; W.H. Sakai, California, 1988, personal communication; S. Goodell, Utah, 1998, personal communication), parasites, including tachinid flies (Duncan, 1923 (now *Stomatomyia floridensis* (Townsend)); D.B. Weissman, unpublished, by unidentified Tachinidae, on species no. 5); wasps (Dodson and Gwynne, 1984), hairworms (Fig. 3.7; Poinar, 1991) and mermithids (G.O. Poinar, Jr and D.B. Weissman, unpublished); (iv) movement may be easier along the logs than in

Fig. 3.8. Defensive position of adult female Jerusalem cricket from Riverside County, California. In this position, she can bite (note open mandibles) or kick.

the thick vegetation around the logs; and (v) drumming communication may be more efficient using a log as substrate than using saturated ground.

Jerusalem crickets have been collected in ant nests in Nevada and Idaho on several occasions (Clark and Tinkham, 1977; subsequent specimens sent to D.B. Weissman by W.H. Clark), apparently living as commensals.

Defensive Mechanisms

Jerusalem crickets, being large, succulent, potential meals, protect themselves in several ways: (i) they have large mandibles (especially when megacephalic) and easily draw my blood when mishandled. They frequently flip on to their backs (Fig. 3.8) with mandibles agape or can rear up; (ii) both sexes of some species (e.g. no. 4; Duncan,

Fig. 3.9. Adult male from Sonoma County, California, showing long, sharp, rear-leg tibial spines used during kicking. Arrow points to left hook.

Fig. 3.10. Adult male *Stenopelmatus mescaleroensis*, from Chaves County, New Mexico, showing entire regenerated left-side leg (arrow). The leg is shorter, hypopigmented and with poorly developed spines when compared with a normal leg on the right side.

1923; Davis and Smith, 1926; Baker, 1971) stridulate by moving the abdomen past the hind femora or the hind femora past the stationary abdomen, as the abdomen has short spines and the hind femora have raised, roughened areas (abdomino-femoral defence stridulation is common in the Gryllacrididoid families (see Gorochov, Chapter 1, this volume, and detailed discussion in Field, Chapter 15, this volume). Despite their lack of tympana, JCs can probably sense such noises (McVean and Field, 1996); (iii) many species kick and have long, sharp, rear-leg tibial spines (Fig. 3.9); (iv) some individuals run away, while others play dead; (v) species no. 4 and no. 6 both have offensive-smelling anal discharges (from an anal stink gland? Bateman and Toms, 1998; see also

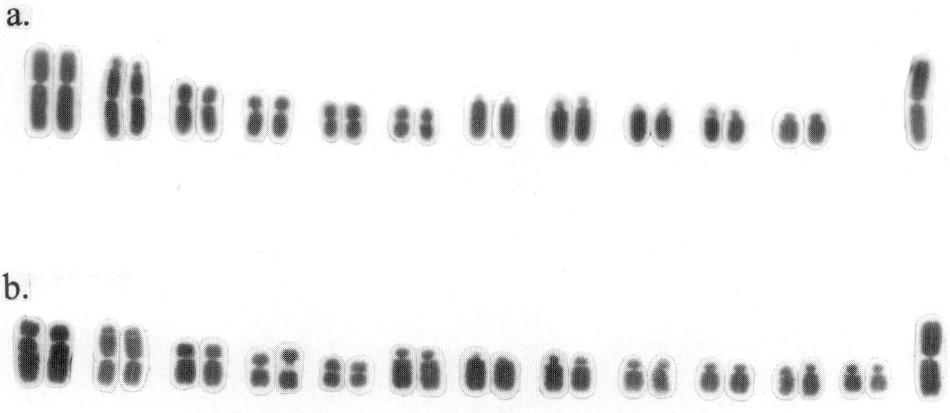

Fig. 3.11. Karyotypes in Jerusalem crickets. (a) Male (from San Bernardino County, California) with 23 chromosomes, the single large X chromosome on the right. (b) Male (from Contra Costa County, California) with 25 chromosomes. Courtesy of N. Ueshima.

Toms, Chapter 4, this volume) which they can shoot up to 10 cm when disturbed; and (vi) they are largely nocturnal (Lutz, 1932), although occasionally crepuscular.

Regeneration

Jerusalem crickets can regenerate legs (Fig. 3.10), which are sometimes lost during moulting. If lost at an early instar, the new leg will 'approximate' a normal one after several moults. These new legs continue to grow with each moult. Similar appendage regeneration is known in the stick insects (Phasmatoptera) and an extinct cricket (Trigonidiidae) (Vickery and Poinar, 1994), as well as the Agraecini katydids (Ingrisch, 1995) and South African king cricket (McDonald and Hanrahan, 1993).

Karyotypes

Jerusalem crickets have a male XO/female XX sex determination system. Males from some 90 populations, representing some 65 species, have been karyotyped with the following results: one JC (species no. 7) has 2n male = 19; three JCs (species nos. 8, 9 and 10) have 2n male = 21; 19 JCs have 2n male = 23; and 43 JCs have 2n male = 25 (see Fig. 3.11 for the latter two idiograms). In each karyotype, the X chromosome is large and metacentric, with varying numbers of metacentric and telocentric autosomes. The mechanism(s) by which these numbers were reduced in time (e.g. Robertsonian fusions) is unclear. The related California anostostomatid genus *Cnemotettix* (see Rentz and Weissman, 1981) has 2n male = 27 for the three species examined (N. Ueshima and D.B. Weissman, unpublished). The above US genera are cytologically similar to Australian anostostomatids and cooloolids (John and Rentz, 1987). Stevens' (1905) count of 2n = 47 for a California JC is clearly in error (John and Rentz, 1987).

Endangered Species

California's present population of 33 million people is predicted to increase to 50 million within 20–30 years. At present levels of habitat destruction and modification, I consider the following taxa to be potentially endangered and deserving of further study: *S. nigrocapitatus*, *S. cahuilaensis*, *S. pictus* and *A. kelsoensis*. Several undescribed JC species will be added to this list with publication of my revisionary studies.

For the Future

How soon after becoming adult do females mature eggs and how is this timing related to mating and oviposition? Do any California species living in areas with a Mediterranean-type climate have an adult reproductive diapause (Weissman and French, 1980; Lightfoot and Weissman, 1991)? Both species no. 1 and species no. 4 first drum several months after becoming adults, a behaviour possibly associated with a period of adult reproductive inactivity.

Is there an egg diapause in some/all species? Is there any parental care?

Acknowledgements

Numerous individuals have helped with the collection of specimens. Most will be thanked in the revisionary monograph but a few deserve particular mention now: G.R. Ballmer, M.S. Caterino, W.H. Clark, F.T. Hovore, V.F. Lee, J.K. Liebherr, D.C. Lightfoot, W.H. Sakai, B.I. Weissman and D.W. Weissman. T.J. Cohn, R.V. Dowell, D.T. Gwynne, B. John, V.F. Lee and D.C. Lightfoot provided invaluable comments on an earlier draft. D. Ubick provided the scanning electron micrograph (SEM).

References

Baker, N.W. (1971) Jerusalem crickets. *Pacific Discovery* 24, 12–13.

Bateman, P.W. and Toms, R.B. (1998) Olfactory intersexual discrimination in an African king cricket (Orthoptera: Mimnermidae). *Journal of Insect Behavior* 11, 159–163.

Brown, B.A., Whitaker, J.O., French, T.W. and Maser, C. (1986) Note on food habits of the screech owl and the burrowing owl of southeastern Oregon. *Great Basin Naturalist* 46, 421–426.

Buckell, E.R. (1930) The Dermaptera and Orthoptera of Vancouver Island, British Columbia. *Proceedings of the Entomological Society of British Columbia* 27, 17–51.

Campbell, J.A. and Armstrong, B.L. (1979) Geographic variation in the Mexican pygmy rattlesnake, *Sistrurus ravus*, with the description of a new subspecies. *Herpetologica* 35, 304–317.

Campbell, J.A. and Solorzano, A. (1992) The distribution, variation, and natural history of the middle American montane pitviper, *Porthidium godmani*. In: Campbell, J.A. and Brodie, E.D., Jr (eds) *Biology of the Pitvipers*. Selva, Tyler, Texas, pp. 223–250.

Clark, W.H. and Tinkham, E.R. (1977) Occurrence of the Jerusalem cricket *Stenopelmatus fuscus* (Orthoptera: Gryllacrididae), in a nest of the Owyhee harvester ant, *Pogonomyrmex owyheei* (Hymenoptera: Formicidae), with taxonomic notes on *Stenopelmatus*. *Journal of the Idaho Academy of Science* 13, 47–52.

Davis, A.C. (1927) Studies of the anatomy and histology of *Stenopelmatus fuscus* Hald. *University of California Publications in Entomology* 4, 159–208.

Davis, A.C. and Smith, L.M. (1926) Notes on the genus *Stenopelmatus* with description of a new species (Orthoptera). *Pan-Pacific Entomologist* 2, 174–180.

Dodson, G.N. and Gwynne, D.T. (1984) A digger wasp preying on a Jerusalem cricket. *Pan-Pacific Entomologist* 60, 297–299.

Duncan, C.D. (1923) Notes on the biology of two species of *Stenopelmatus* (Orth.: Tettigoniidae). *Entomological News* 34, 73–77.

Essig, E.O. (1926) *Insects of Western North America*. Macmillan, New York, 1035 pp.

Evens, J.G. (1988) *The Natural History of the Point Reyes Peninsula*. Point Reyes National Seashore Association, Point Reyes, California, 226 pp.

Gwynne, D.T. (1995) Phylogeny of the Ensifera (Orthoptera): a hypothesis supporting multiple origins of acoustical signalling, complex spermatophores and maternal care in crickets, katydids, and weta. *Journal of Orthoptera Research* 4, 203–218.

Gwynne, D.T. (1997) Evolution of mating in crickets, katydids and wetas. In: Gangwere, S.K., Muralirangan, M.C. and Muralirangan, M. (eds) *The Bionomics of Grasshoppers, Katydids and Their Kin*. CAB International, Madras, India, pp. 281–314.

Hebard, M. (1916) A study of the species of the genus *Stenopelmatus* found in the United States. *Journal of the New York Entomological Society* 24, 70–86.

Hebard, M. (1936) Orthoptera of North Dakota. *North Dakota Experiment Station Bulletin* 284, 1–69.

Hebard, M. (1938) An ecological survey of the Orthoptera of Oklahoma. *Oklahoma Agricultural Experiment Station Technical Bulletin* 5, 1–31.

Henry, C.S. (1985) The proliferation of cryptic species in *Chrysoperla* green lacewings through song divergence. *Florida Entomologist* 68, 18–38.

Ingrisch, S. (1995) The Agraecini, hidden beauties of the tropical forest (Tettigoniidae, Conocephalinae) (a preliminary account on the systematics and biology of the Agraecini of Thailand). *Journal of Orthoptera Research* 4, 74.

John, B. and Rentz, D.C.F. (1987) The chromosomes of four endemic Australian fossorial orthopterans: a study in convergence and homology. *Bulletin of the Sugadaira Montane Research Center* 8, 205–216.

Johns, P.M. (1997) The Gondwanaland weta: family Anostostomatidae (formerly in Stenopelmatidae, Henicidae or Mimnermidae): nomenclatural problems, world checklist, new genera and species. *Journal of Orthoptera Research* 6, 125–138.

Karny, H.H. (1937) *Orthoptera, Family Gryllacrididae, Subfamiliae Omnes*. Genera Insectorum, Fasc. 206, Quatre-Bras, Tervueren, Belgium, 317 pp.

Kevan, D.K.McE. (1982) Orthoptera. In: Parker, S.P. (ed.) *Synopsis and Classification of Living Organisms*, Vol. 2. McGraw-Hill, New York, pp. 352–379.

Lightfoot, D.C. and Weissman, D.B. (1991) Review of the grasshopper genera *Dracotettix* and *Litoscirtus* (Orthoptera: Romaleidae), with a discussion of their origins and life histories. *Proceedings of the California Academy of Sciences* 47, 159–174.

Lutz, F.E. (1932) Experiments with Orthoptera concerning diurnal rhythm. *American Museum Novitates* 550, 1–24.

McDonald, I.R. and Hanrahan, S.A. (1993) Aspects of foraging and behaviour of the king cricket, *Libanasidus vittatus* (Kirby) (Orthoptera: Mimnermidae). *South Africa Journal of Science* 89, 354–358.

McVean, A. and Field, L.H. (1996) Communication by substratum vibration in the New Zealand tree weta, *Hemideina femorata* (Stenopelmatidae: Orthoptera). *Journal of Zoology, London* 239, 101–122.

Orr, R.T. (1954) Natural history of the pallid bat, *Antrozous pallidus* (LeConte). *Proceedings of the California Academy of Sciences* 28, 165–246.

Poinar, G.O., Jr (1991) Hairworm (Nematomorpha: Gordioidea) parasites of New Zealand wetas (Orthoptera: Stenopelmatidae). *Canadian Journal of Zoology* 69, 1592–1599.

Polis, G.A. (1979) Prey and feeding phenology of the desert sand scorpion *Paruroctonus mesaensis* (Scorpionidae: Vaejovidae). *Journal of Zoology, London* 188, 333–346.

Rentz, D.C.F. (1978) New species and records of western Orthoptera. *Pan-Pacific Entomologist* 54, 81–97.

Rentz, D.C.F. and Weissman, D.B. (1981) Faunal affinities, systematics, and bionomics of the Orthoptera of the California Channel Islands. *University of California Publications in Entomology* 94, 1–240.

Smith, L.M. (1929) *Coccospora stenopelmati* gen. nov., sp. nov. A gregarine from *Stenopelmatus* (Orthoptera) from central California. *University of California Publications in Entomology* 33, 57–68.

Stevens, N.M. (1905) Studies in spermatogenesis with especial reference to the 'accessory chromosome'. *Carnegie Institution, Washington, Publications* 36, 1–32.

Tallamy, D.W. and Brown, W.P. (1999) Semelparity and the evolution of maternal care in insects. *Animal Behaviour* 57, 727–730.

Tinkham, E.R. (1965) Studies in Nearctic desert sand dune Orthoptera: a new genus and species of stenopelmatine crickets from the Kelso Dunes with notes on its multi-annual life history and key. Part X. *Great Basin Naturalist* 25, 63–72.

Tinkham, E.R. (1968) Studies in Nearctic desert sand dune Orthoptera. Part XI. A new arenicolous species of *Stenopelmatus* from the Coachella Valley with key and biological notes. *Great Basin Naturalist* 28, 124–131.

Tinkham, E.R. (1970) Studies in Nearctic desert sand dune Orthoptera. Part XII. A remarkable new genus and species of stenopelmatine crickets from the Viscaino Desert, Baja California, Mexico, with key. *Great Basin Naturalist* 30, 173–179.

Tinkham, E.R. (1979) Studies in Nearctic desert sand dune Orthoptera, Part XVI: a new black *Stenopelmatus* from the Mescalero Sands. *Great Basin Naturalist* 39, 226–230.

Tinkham, E.R. and Rentz, D.C. (1969) Notes on the bionomics and distribution of the genus *Stenopelmatus* in central California with the description of a new species (Orthoptera: Gryllacrididae). *Pan-Pacific Entomologist* 45, 4–14.

Vickery, V.R. and Kevan, D.K.McE. (1983) A monograph of the orthopteroid insects of Canada and adjacent regions. *Lyman Entomological Museum Research Laboratory, Memoirs* 13, 1–679.

Vickery, V.R. and Poinar, G.O., Jr (1994) Crickets (Grylloptera: Grylloidea) in Dominican amber. *Canadian Entomologist* 126, 13–22.

Weissman, D.B. and French, E. (1980) Autecology and population structure of *Trimerotropis occidentalis* (Bruner) (Orthoptera: Acrididae: Oedipodinae), a grasshopper with a reproductive dormancy. *Acrida* 9, 145–157.

Williams, S.C. (1968) Scorpion preservation for taxonomic and morphological studies. *Wasmann Journal of Biology* 26, 133–136.

4
South African King Crickets (Anostostomatidae)

Rob B. Toms
Department of General Entomology, Transvaal Museum, Northern Flagship Institution, PO Box 413, Pretoria 0001, South Africa

Introduction

The first king cricket described was the 'monstrous cricket' *Henicus monstrosus* (Herbst), from the Western Cape province of South Africa (Herbst, 1803). Looking back at nearly two centuries of progress in our knowledge of these insects, we can see that early entomologists recognized that king crickets were Orthoptera by lumping them with genera such as *Locusta* and *Gryllus*. In fact, *H. monstrosus* was originally described as *Locusta monstrosa*. Later, it became obvious that king crickets should be placed in different genera, and an excessive number of new genera were described. The early descriptive work invariably used small sample sizes (often one specimen), collected under very difficult conditions (which may have included encounters with various large mammals) and sent to various countries in Europe. The knowledge of the African fauna was strongly affected by these constraints. Given the rudimentary communication of those times, many synonyms were described by biologists, who were not always aware what their colleagues at other institutions and on different continents were doing. Sexual dimorphism and new species descriptions based on single specimens made it difficult to compare species, especially if they were described on the basis of single specimens of the opposite sex. Many nomenclatural problems and synonymies have been sorted out, including recent treatments of the higher taxa (Johns, 1997; Gorochov, Chapter 1, this volume). It is now possible to move forward with a far firmer footing, with most of the basic nomenclatural problems resolved.

Electronic communication, with digitized images of species, should soon be available on CD ROM or the website of the Orthopterists' Society. It should become increasingly easy to get accurate identifications via the internet, although much basic work needs to be done, especially with the African fauna.

In many ways, the African species are similar to their counterparts in other southern continents. All known species are nocturnal omnivores and most adult males have enlarged mandibles. Natural predators of these insects are known to include owls.

Composition of the Fauna

Since 1803, 44 species of African Anostostomatidae in seven valid genera have been described (Table 4.1). Roughly half of these are from South Africa. As with other insects on this continent, the number of species described so far is probably less than half of the total fauna. Several new species and at least one new genus are already represented in collections in South Africa. Also, it is possible that cryptic species exist, where several similar-looking species may be fundamentally different. New techniques, such as DNA

Table 4.1. Valid African genera with numbers of species (from Johns, 1997).

Genus	Species
Borborothus	1 (*opaca*)
Onosandrus	7
Onosandridus	8
Nasidius	12
Henicus	10
Libanasa	4
Libanasidus	2 (*vittatus, impicta*)

analysis, are now starting to be used on New Zealand weta and it is anticipated that these techniques will soon be applied to some of the African fauna. Apart from limited knowledge of *Libanasidus vittatus* and *H. monstrosus*, we know almost nothing about the biology or behaviour of any African species.

Noteworthy species

Borborothus opaca *Brunner*

At present, *Borborothus* is a monotypic genus. This genus is mentioned first because it is primitive, in that the mandibles of the males are not developed and males and females are very similar (apart from their genitalia). The undeveloped mandibles are completely concealed behind the labium (Figs 4.1 and 4.2). These insects are known from George, Jonkersberg and Kogelberg in the Western Cape Province. It is probable that further collecting could lead to the discovery of additional species. The preferred habitat of this genus appears to be in the southern mountainous regions.

Onosandrus *sp.*

Another presumably primitive genus lacking sexually dimorphic head development is *Onosandrus*. Members of this genus are known from indigenous forest in Mpumalanga and Northern Province.

Henicus monstrosus

During the 19th and early 20th century, *H. monstrosus* was probably South Africa's best-known king cricket, and was the only species discussed by Skaife in the 1953 version of *African Insect Life*. This species has now undoubtably lost its position as the best-known species in southern Africa to the 'Parktown prawn', discussed below.

The most remarkable feature of *H. monstrosus* is the face of the males (Figs 4.3 and 4.4), which is unusual even amongst other king crickets. It is thought that the large round face may be used to plug the entrance of their burrows as the males wait for passing females, intruders or prey. The function of the two lateral horns on the face of *H. monstrosus* and other species of *Henicus* is not known. Field observations of these insects are needed. According to Skaife (1953, but not 1979), male *H. monstrosus* make a rasping noise by rubbing their mandibles and maxillae together, and it has been suggested that this gnashing of the teeth constitutes a 'love' song (Skaife, 1953). While many Anostostomatidae have the ability to hear and to produce sounds, the ears of *H. monstrosus* are degenerate and external tympana are wanting (based upon examination of one male in the collection of the Transvaal Museum). Presumably, the same applies to the female, which makes acoustic communication in this species problematic. Lateral abdominal stridulatory pads are also wanting in this species, although scattered lateral abdominal spines are present. These may be indicative of the morphology of ancestors of species that subsequently acquired well-developed lateral abdominal stridulatory pads. Alternatively, they may represent degenerate stridulatory pads. The femora of this species are smooth, but it is not known whether this condition is primitive or advanced. Unfortunately, the stridulatory behaviour of *H. monstrosus* has not been adequately studied nor are recordings of their songs available.However, most species of king crickets produce rather soft sounds, which appear to be important in courtship and defence, but are not known to be important in calling or attracting females. If it is true that *H. monstrosus* sings with its mouth-parts, this would act as another good example of the independent evolution of sound-producing mechanisms, which have evolved independently several times in Anostostomatidae (Field, 1993). The mouth-parts could be used for stridulation but further examination is necessary. Although *H. monstrosus* was described nearly two centuries ago, we still know very little about its biology or behaviour. The genus *Henicus* is distributed along the mountains of the Western and Eastern Cape.

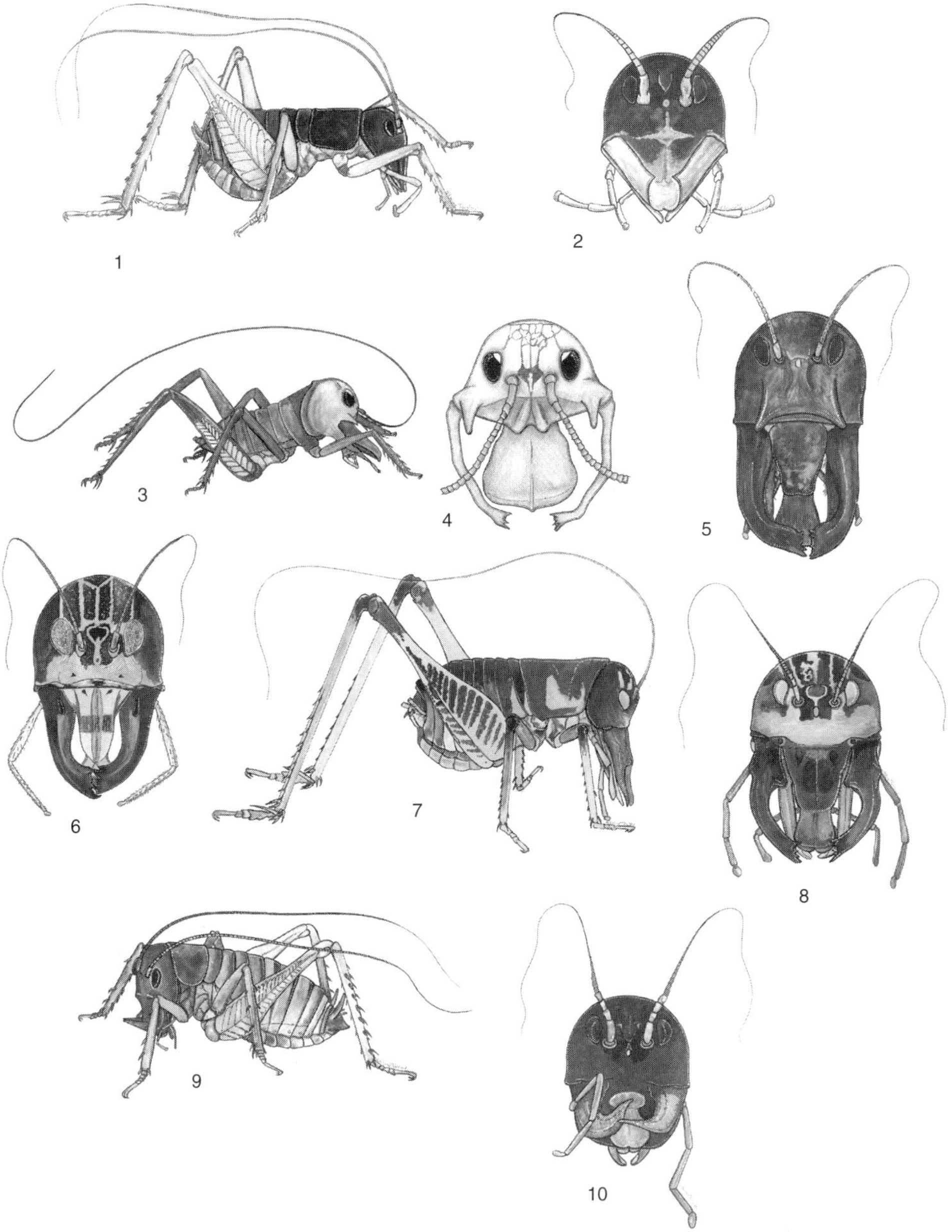

Fig. 4.1. *Borborothus opaca*, lateral view.
Fig. 4.2. *Borborothus opaca*, anterior view.
Fig. 4.3. *Henicus monstrosus*, lateral view.
Fig. 4.4. *Henicus monstrosus*, anterior view.
Fig. 4.5. *Nasidius whellani*, anterior view.
Fig. 4.6. *Libanasa* sp. a, anterior view.
Fig. 4.7. *Libanasa* sp. b, lateral view.
Fig. 4.8. *Libanasa* sp. b, anterior view.
Fig. 4.9. *Libanasidus vittatus*, lateral view.
Fig. 4.10. *Libanasidus vittatus*, anterior view.

Nasidius *sp.*

Twelve species of *Nasidius* have been described, but there are at least another 23 undescribed species (P.M. Johns, personal communication). Males of some species in this genus have well-developed lateral horns on their faces, but other species only have small bumps in the same place. Two *Nasidius* species that possess well-developed horns are *N. whellani* (Chopard) (Fig. 4.5) and *N. longicauda* Karny. The lateral horns on the face of *Nasidius* and *Henicus* species are probably an important shared derived feature, but the function of these horns is unknown (see Field and Deans, Chapter 10, this volume). It is probable that horn development is an advanced feature and that horns developed from small bumps. Members of this genus have been found in open grassland, where they live in burrows.

Libanasa *sp.* (Figs 4.6–4.8)

There are at least six undescribed species of *Libanasa* in South Africa, some of which are only known from single males. It is probable that some new species are only represented by females in the collections; hence more collecting is needed. The taxonomy of these insects should be based on males, because the females are difficult to tell apart. Males of this genus exhibit simple development of the mandibles, which are enlarged, but may have small anterior ridges, which could be precursors of tusks (Fig. 4.6). Other members of this genus may have clearly developed protrusions on their mandibles, suggesting that these could be the precursors of tusks found in *Libanasidus*. This genus has been collected in dry thornveld in areas such as the southern Kruger National Park and Nelspruit. Specimens have also been collected in rotting logs.

Libanasidus vittatus*, the 'Parktown Prawn'*

By far the best-known species in South Africa is *L. vittatus* (Figs 4.9 and 4.10). It is called the Parktown prawn, because they often stumble into swimming-pools in the affluent suburbs of Johannesburg. The natural habitat for this genus is in and around forests in Mpumalanga and Northern Province and probably also in Zimbabwe. During the day, they can be found in burrows or under logs. Specimens have also been collected in gardens in Johannesburg, Randburg and Pretoria. A few articles have been written on the biology these insects, but very little is known about any other species of African king crickets.

The 'Parktown prawn' has captured the imagination of the general public living in Johannesburg and surrounding areas to such an extent that it has stimulated numerous articles in the local and international media (in publications such as *Time* and *The Economist* and on international news broadcasts of the BBC and CNN). They are also popular subjects for school projects. This species already has its own website, and two rampant Parktown prawns have featured on a billboard advertising a local radio station.

One of the largest females collected measures 64 mm from the front of its head to the ovipositor tip (from indigenous forest in the Soutpansberg Mountains in the Northern Province). The ovipositor is 19 mm long and the total length of the insect, including legs and antennae, is 166 mm ($6\frac{1}{2}$ in). South African Jerusalem crickets in the genus *Sia* (formerly *Maxentius*) (Stenopelmatidae; Johns, 1997; see Gorochov, Chapter 1, this volume) may be even larger than the large African king crickets, such as *Libanasidus*. A male Jerusalem cricket may measure up to 250 mm, including legs and antennae (body length = 63 mm; abdominal width = 22 mm), but these insects do not have a strong association with people, so few know that they exist.

Many people dislike or fear the 'prawn', as they frequently wander into houses, where they may even appear inside people's beds at night. One of their least liked attributes is their defence mechanism of jumping when they are molested, often coming to rest on the clothing of the person who disturbed them or a bystander. People have been known to move house to get away from them, and one person slept in her car to avoid close contact with them.

This species has featured in several myths, mostly related to their sudden prominence in Johannesburg in the 1960s. For example, it has been suggested that they are alien invaders or the result of a freak mutation. Fortunately, the historical advantage of museum specimens and the species description by P. Rendall in Barberton (Mpumalanga Province, South Africa) more than 100 years ago, is sufficient proof that this species did not suddenly appear 30 years ago.

The population and notoriety of the 'prawn'

has steadily increased in Johannesburg since 1965 for unknown reasons. In 1985, it was suggested that they could have been introduced into Johannesburg from somewhere near Barberton (Toms, 1985), but a specimen in the Transvaal Museum, collected by George van Son, who was Curator of Entomology at the time, shows that they were already in Pretoria by January 1955. The first museum specimen from Johannesburg was collected by R. du Preez in November 1962. It now seems possible that they were originally present in low numbers and that environmental changes in suburban gardens led to an increase in numbers and range. Apart from urbanization, weather patterns may also have played a role. The increased rainfall over the past three summers led to a decrease in populations, presumably through flooding of burrows and refuges. Unfortunately, we do not have quantitative data associated with changes in population density, but it is possible that this species could act as an environmental indicator. Modern techniques, such as DNA and protein analysis, could indicate whether these insects were introduced into Johannesburg or whether a small local population underwent an increase in range and numbers.

The great interest in these insects has created a superb opportunity for public education through appropriate information presentation, as for example, during the 'Centenary of the Description of *L. vittatus*' in 1999. Even the mild ridicule, from some quarters, of the centenary idea was part of the process of attracting attention. Having raised a public audience, crucial issues, such as the biodiversity crisis, the age of collections and the need to collect and care for specimens, declining funding and even repatriation could be addressed (the holotype of the Parktown prawn is an adult female collected in Barberton and housed in the British Museum of Natural History). This appears to be the first time that the centenary of any insect has been celebrated.

Recent research on the behaviour of *L. vittatus* has revealed that their vile-smelling faeces, ejected in self-defence, also plays a role in intersexual communication and territorial behaviour (Bateman and Toms, 1998a). Males and females are able to discriminate between faeces produced by members of the same and opposite sex and can use these cues to find conspecifics of the opposite sex or avoid confrontation with members of the same sex. Mating behaviour, mate guarding and male–male relative strength have also been studied in these insects (Bateman and Toms, 1998b).

One of the most interesting features of the morphology of *Libanasidus* is the anterior mandibular tusks, which arise as anterior outgrowths of the mandibles in males (Bennet and Toms, 1995) (Figs 4.9 and 4.10). These tusks are similar to those of New Zealand tusked wetas (see comparative discussion of tusks as male secondary sexual characters and Fig. 10.4 in Field and Deans, Chapter 10, this volume). *Libanasidus* and tusked weta both exhibit advanced mandibular development, in contrast to the undeveloped mandibles of *Borborothus* and simple development of mandibles found in genera such as *Nasidius*, where the mandibles are simply elongated and thickened. These mandibles arose from small protrusions on the mandibles of ancestral species, which may have resembled those found in *Libanasa* (Figs 4.6–4.8). Male *L. vittatus* use their mandibles to dig their burrows, by shifting sand with an action rather like that of a bulldozer. They also use their strong mandibles in disputes with other males, by gripping the opponent and throwing him over his back.

Very little is known about the stridulatory mechanisms of African Anostostomatidae. Sound production in *Libanasidus* is similar to that of many New Zealand wetas, which stridulate by rubbing the lateral abdominal stridulatory pads against the modified inner surfaces of the femora (Field, 1993). Stridulation is known to occur during courtship and in self-defence (Bateman and Toms, 1998b). Tibial ears are clearly visible on the front legs, and this is an important derived feature shared with the Gryllidae and Tettigoniidae, but not with Stenopelmatidae (*sensu stricto*). Thus, it is possible that Gryllidae and Tettigoniidae evolved from winged anostostomatid-like ancestors that already had tibial ears (see discussion of phylogeny of these derived characters in Gorochov, Chapter 1, this volume).

Adult insects seem to die at the beginning of winter, and juveniles in various stages of development can be dug up during winter. Both facts suggest that overwintering may occur in the egg stage and as juveniles. Females deposit up to 200 eggs in damp soil by thrusting their long scimitar-shaped ovipositor into the soil repeatedly, apparently laying just one egg each time. The omnivorous diet of these insects is known to include garden pests, such as slugs, snails and cutworms, so they can be regarded as beneficial allies (McDonald and Hanrahan, 1993).

Future Developments

Although certain genera and species are now fairly well represented in collections, there are regions where little or no collecting has been done. During dry weather conditions, king crickets may be difficult to collect and sample sizes may be small, because they tend to hide. Usually, opportunistic collecting is conducted for all Orthoptera in areas where specific goals can be achieved. The roads and infrastructure in South Africa are among the best in Africa, but there may be interference from people in some areas. In the large national parks, one may encounter large mammals, and the nocturnal habits of king crickets make it necessary to work at night. Under these conditions, collecting is generally done close to a vehicle after the general area has been checked with good lights. Some parts of Africa, such as Angola and Mozambique, have been extremely poorly collected because of human activity.

African Anostostomatidae are interesting insects and there is public interest in them. Hopefully, these positive factors can be used to counteract difficult collecting conditions and limited employment and funding opportunities, so that we can learn more about the many species that are poorly known or completely unknown.

Acknowledgements

Thanks are due to P.W. Bateman and L. Field for their help and suggestions. The Transvaal Museum, Northern Flagship Institution, is thanked for financial support. E.H. Taylor is thanked for the figures.

References

Bateman, P. and Toms, R.B. (1998a) Olfactory intersexual discrimination in an African king cricket (Orthoptera: Mimnermidae). *Journal of Insect Behavior* 11(1), 159–163.

Bateman, P. and Toms, R.B. (1998b) Mating, mate guarding and male–male relative strength assessment in an African king cricket (Orthoptera: Mimnermidae). *Transactions of the American Entomological Society* 124(1), 69–75.

Bennet, A. and Toms, R.B. (1995) Sexual dimorphism in the mouthparts of the king crickets *Libanasidus vittatus* (Kirby) (Orthoptera:Mimnermidae). *Annals of the Transvaal Museum.* 36(15), 205−214.

Field, L.H. (1993) Structure and evolution of stridulatory mechanisms in New Zealand wetas (Orthoptera: Stenopelmatidae). *International Journal of Insect Morphology and Embryology* 22, 163–183.

Herbst, I.F.W. (1803) Beschreiβung einiger hochst seltener Heuschrecken. *Neue Schriften der Naturforschender Freunde zu Berlin* 4, 111–120.

Johns, P.M. (1997) The Gondwanaland weta: family Anostostomatidae (formerly in Stenopelmatidae: Henicidae or Mimnermidae): nomenclatural problems, world checklist, new genera and species. *Journal of Orthoptera Research* 6, 125–138.

McDonald, I.R. and Hanrahan, S.A. (1993) Aspects of foraging and behaviour of the king cricket, *Libanasidus vittatus* (Kirby) (Orthoptera: Mimnermidae). *South African Journal of Science* 89, 354–358.

Skaife, S.H. (1953) *African Insect Life.* Longmans, Green, Cape Town, 387 pp.

Skaife, S.H. (1979) *African Insect Life.* C. Struik, Cape Town, 279 pp.

Toms, R.B. (1985) Johannesburg's cricket invader the 'Parkmore prawn'. *African Wildlife* 39, 200−202.

5

Australian King Crickets: Distribution, Habitats and Biology (Orthoptera: Anostostomatidae)

Geoffrey B. Monteith[1] and Laurence H. Field[2]

[1]*Queensland Museum, Brisbane, Queensland 4101, Australia;* [2]*Department of Zoology, University of Canterbury, Private Bag 4800, Christchurch, New Zealand*

Introduction

Australian members of the family Anostostomatidae tend not to occur in towns and cities and thus have not gained the public awareness enjoyed by their often suburban New Zealand and South African relatives. Thus they do not enjoy a popular name comparable to the universally known Maori word, 'weta', used by layman and entomologist alike in New Zealand. The name 'king cricket' was coined for the large Australian species of *Anostostoma* and is now becoming commonly applied to all Australian and African anostostomatids. This practice will be followed here.

The Australian king cricket fauna is rich and unique. It reflects ancient connections to the Gondwanaland supercontinent, while still including several fascinating endemic genera and species. The largest and most bizarre king crickets in the genus *Anostostoma* attracted the attention of early European insect taxonomists, such as Gray (1837) and Burmeister (1838), but most of the remaining fauna remains undescribed and has received little biological study. The only available non-taxonomic literature consists of general comments found in entomological textbooks or popular handbooks under the family name Stenopelmatidae (e.g. McKeown, 1944; CSIRO, 1991; Rentz, 1996). However, substantial collections of Australian material reside in the Queensland Museum, mostly accumulated from comprehensive rainforest collecting over the last 25 years by one of us (GBM). These are accompanied by a wealth of collecting information. These collections have been partly sorted by P.M. Johns (retired, University of Canterbury, Christchurch), whose revised generic view of the Australian fauna was summarized in his recent world synopsis (Johns, 1997). The aim of this chapter is to consolidate the existing unpublished knowledge about this fauna.

Faunal Summary

It is appropriate here to present an overview of the Australian king cricket fauna to form a taxonomic framework for discussions that follow. This section gives a brief account of each genus, including characteristics for recognition, estimated number of species known and distribution within Australia. Synoptic maps of the distribution of all Australian genera are given in Fig. 5.1. This treatment recognizes the Australian fauna as comprising about 60 species in nine genera, of which only 13 species in seven genera are currently described. The distribution and diversity of this fauna are summarized in Table 5.1. The nomenclature of Johns (1997) is followed, and his partial sorting of the Queensland Museum collection, plus much material collected since then, as well as the Australian National Insect Collection in Canberra, is drawn upon for distribution information. Johns places all the Australian genera, except the unplaced genus *Transaevum*, in the subfamily

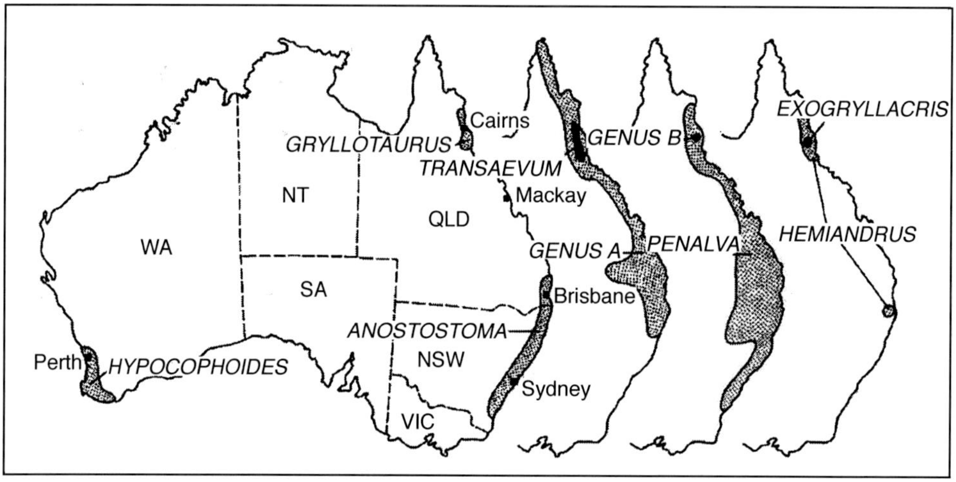

Fig. 5.1. Map of Australia showing geographical extent of each of the nine genera of Australian king crickets (Family Anostostomatidae). WA, Western Australia; NT Northern Territory; SA, South Australia; QLD, Queensland; NSW, New South Wales; VIC, Victoria.

Table 5.1. List of genera of Australian Anostostomatidae showing total species and the estimated numbers of species and genera in the various regions of Australia. Regions shown in the columns are as mapped in Fig. 5.3.

Genus (total species)	S/NSW	N/NSW	SEQ	CQ	WT	CYP	SWA
Anostostoma (6)	1	4	3				
Penalva (16)	1	2	10	2	3		
Genus A (20)		2	6	1	11	1	
Genus B (1)					1		
Hemiandrus (8)			1		7		
Exogryllacris (1)					1		
Gryllotaurus (3)					3		
Transaevum (2)					2		
Hypocophoides (3)							3
Totals							
Species (60)	2	8	20	3	28	1	3
Genera (9)	2	3	4	2	7	1	1

S/NSW, southern New South Wales; N/NSW, northern New South Wales; SEQ, south-eastern Queensland; CQ, central Queensland; WT, wet tropics; CYP, Cape York Peninsula; SWA, south-western Australia.

Anostostomatinae. Two of the genera recognized by Johns are as yet undescribed and are referred to below as Genus A and Genus B.

Genus Anostostoma *Gray 1837 (Anostostomatinae: Anostostomatini)*

This genus includes the 'giant king crickets' in Australia. The genus name *Australostoma* Karny 1931 has been used widely for this group in the past (e.g. Rentz, 1991, 1996), but usage has been stabilized as *Anostostoma* by Johns (1997). The species are all apterous and they differ from all the other Australian genera in having a rough, sandpaper-like dorsal surface to the abdominal segments. They readily stridulate, when disturbed, by rubbing the inner surface of the hind femora against a patch of spines on each side of the base of the abdomen. The ovipositor is greatly shortened in *A. erinaceum* and the new species from Springbrook, but long and robust in other species. Members of the genus occur from Bateman's Bay, in southern New South Wales (NSW), north to about Gympie in southern Queensland. There are five described species, with a small, undescribed one known from Springbrook Plateau on the southern border of Queensland. Most of the species have an enlarged head and jaws in the male, and this applies particularly to the largest species, *A. australasiae*, which can reach 80 mm head/body length with mandibles of 22 mm (Fig. 5.2A). This is the northernmost species and is common in rainforests and nearby wet eucalyptus forests, from Gympie, south to the latitude of Coffs Harbour and Dorrigo in northern NSW. It occurs both in mountains and at quite low elevations, occasionally including the western suburbs of Brisbane City. Similarly, the small, dark *A. erinaceum* sometimes occurs in sandstone remnant habitats within suburban Sydney. *Anostostoma spinosum* is a striking species with black body, bright orange femora and extremely spiny tibiae (Fig. 5.2B). It lacks head enlargement in the male and occurs in high-altitude rainforests on the border between Queensland and NSW. This is the centre of diversity of the genus, with three sympatric species. *Anostostoma opacum* (illustrated in Rentz, 1991) is a small, brown, stocky species that occurs in the mountain rainforests of northern NSW and along the coast in open forests to as far south as near Bateman's Bay.

Described species are as follows:

- *Anostostoma australasiae* Gray, 1837 (type species)
- *Anostostoma erinaceum* (Burmeister, 1839)
- *Anostostoma femoralis* (Walker, 1869)
- *Anostostoma opacum* (Brunner, 1888)
- *Anostostoma spinosum* Karny, 1930

Fig. 5.2. Representative species of the Australian anostostomatid fauna. A. Close-up of head of male of the giant king cricket, *Anostostoma australasiae* (Mt Glorious, SQ. Photo: Queensland Museum). B. Female of the red-legged king cricket, *Anostostoma spinosum*. (Tallebudgera Valley, SQ. Photo: J. Wright, QM). C. Female of the white-kneed king cricket, *Penalva flavocalceata* (Innisfail, NQ. Photo: J. Wright, QM). D. Female of one of the 'big-headed' species of genus A (Mt Finnigan, NQ. Photo: A. O'Toole). E. Male of genus B resting on tree roots growing from the ceiling of a boulder cave on Mt Bartle Frere, NQ (Photo: J. Wright, QM). F. Female of *Hemiandrus* sp. (Mt Finnigan, NQ. Photo: A. O'Toole). G. Female of *Exogryllacris ornata* (Millaa Millaa Falls, NQ. Photo: J. Wright, QM). H. Male of *Gryllotaurus* sp. showing asymmetric horns on face (Mt Sorrow, NQ. Photo: A. O'Toole). I. Side-view of female of *Transaevum* sp. holding a spermatophylax vesicle in mouth-parts after separating from the male (Massey Range, NQ. Photo: H. Janetzki, QM). All photographs reproduced with permission. SQ, southern Queensland; QM, Queensland Museum; NQ, northern Queensland.

Genus Penalva *Walker, 1870 (Anostostomatinae: Anostostomatini)*

Penalva is a large genus of medium to moderately large, wingless species. The species can be recognized by having smooth terga, fore tibiae each with two inner, subapical tibial spines and two tympana, and very narrow margins to the lower edges of the thoracic terga. The ovipositor is always well developed. No sexual dimorphism of the head occurs. This genus has four described species and perhaps 12–15 undescribed species, though Johns (1997) estimates 30 undescribed species. The genus occurs down the east coast of Australia from near Cooktown in north Queensland to Bateman's Bay in southern NSW. Most are from the rainforests of the coastal ranges, but a species group extends into dry habitats in sandstone areas of inland southern Queensland (Carnarvon Range) and NSW (Pilliga Scrub). The centre of diversity is in south-east Queensland, where about ten species occur. The best-known, and largest, species in the genus is the white-kneed king cricket, *Penalva flavocalceata*, with reddish-brown body and striking white patches on the dorsal apices of the femora (Fig. 5.2C). It is widespread at high and low elevations in tropical rainforests of northern Queensland between Cooktown and Townsville (Fig. 5.3) and is often seen at night on national park walking tracks. A nymph of this species is illustrated as *Papuaistus* sp. by Rentz (1996), but that genus is not now regarded as Australian. A *Penalva* species pair from central Queensland (Mt Elliot, Eungella and Conway Range (Fig. 5.3)) is curious in having tiny vestigial tegmina present (Figs 5.4 and 5.5).

Described species are as follows:

- *Penalva (Penalva) lateralis* Walker, 1870 (type species)
- *Penalva (Penalva) flavocalceata* (Karny, 1928)
- *Penalva (Trihoplophora) abnormis* (Brunner, 1888) (type species of subgenus)
- *Penalva (Trihoplophora) uniformis* Karny, 1928

New genus A (Anostostomatinae: Anostostomatini)

This undescribed genus includes small to medium-sized, wingless species (Fig. 5.2D). They have smooth terga, fore tibiae with two tympana and only one subapical spur, a thorax with broad lower margins to the terga and the fastigium between the eyes very narrow. The species show a great diversity of form. Some are large and robust, with well-developed ovipositors, and look very like species of *Penalva*. Others are small, slender species, with greatly reduced ovipositors ('short tails'). Two or three high-mountain species in north Queensland have enlarged heads in the males. A curious species from the Dawes Range, west of Miriam Vale in central Queensland (Fig. 5.3), has hind tibiae in males greatly flattened and with enlarged spines. Genus A includes about 20 species, none of which have been described. Most occur in mountain rainforests along the east coast of Queensland, from the northern tip of Cape York Peninsula, south to Dorrigo in northern New South Wales (Figs 5.1 and 5.3). The only lowland species are the one in Cape York Peninsula and a related, speckled species near Cairns. One isolated species occurs in sandy eucalyptus forest on a high tableland 400 km from the coast in the Carnarvon Range. Centres of diversity for the genus are the tropical mountains between Cooktown and Townsville, with about 12 species, and south-east Queensland, with about six species.

New genus B (Anostostomatinae: Anostostomatini)

This peculiar, undescribed, wingless genus is recognized by having two subapical spurs (one inner, one outer) and one tympanum on the fore tibiae and a narrow fastigium between the eyes (Fig. 5.2E). The hind tibiae are very long, are flattened, have short, close-spaced spines running the whole length and have the apical two pairs of spurs widely spaced. There is no sexual dimorphism. Genus B has a single species known from two populations about 45 km apart in the centre of the wet tropical zone south-west of Cairns in north Queensland (Fig. 5.1). One is on Mt Bartle Frere and the other is on the Walter Hill Range (Fig. 5.3); both are in granite boulder zones in rainforests above 1000 m altitude, where the species occurs in boulder caves.

Genus Hemiandrus *Ander 1938 (Anostostomatinae: Anostostomatini)*

This genus has been known only from New Zealand, but a number of undescribed species from Australia were included by Johns (1997). The Australian *Hemiandrus* species differ from other

Australian genera by having only one subapical spur and one tympanum on the fore tibiae and the meso- and metanota of the thorax greatly reduced in size compared with the pronotum. All species have a uniform appearance and there is no sexual dimorphism. They are small, shiny, wingless, ground species, which have a characteristic 'hump-backed' appearance when foraging (Fig. 5.2F). The ovipositor is always well developed, curved and pointed. *Hemiandrus* has a highly disjunct distribution in Australia, with a compact group of about seven species in the wet, tropical mountains between Cooktown and Ingham in north Queensland, and a single species 1250 km further south on the Lamington and Springbrook Plateaus on the border between Queensland and NSW (Fig. 5.1). This southern species has a pale median band along the dorsum of the abdomen.

Genus Hypocophoides *(Anostostomatinae: Lutosini)*

The only described king cricket from Western Australia was traditionally placed in the South African genus *Onosandrus*. Johns (1997) has transferred the species tentatively to *Hypocophoides*, a genus otherwise known from two Indian species. In Australia, there are two or three undescribed species, which are small, smooth, wingless and ground-burrowing. They all occur in the moist south-west corner of Western Australia, from Albany across to a little north of Perth (Fig. 5.1). They occur in sclerophyll forests and apparently burrow in sandy soil. Specimens from Cape Naturaliste are about 30 mm long, pale in colour with a russet pronotum, and have the general appearance of New Zealand *Hemiandrus*.

The described species is:

- *Hypocophoides lepismoides* (Walker, 1871)

Genus Exogryllacris *Willemse 1962 (Anostostomatinae: Anabropsini)*

This fully winged genus is easily recognized by its very broad, jet-black wings, which are held flat and vertical and which project beyond the end of the body. The head is black, with a white spot in the centre of the face, and the hind legs have conspicuous white patches at the apices of the femora (Fig. 5.2G). No sexual dimorphism is evident. The only known species has a restricted range in mountain rainforests, from the Lamb Range, west of Cairns, south to the Cardwell Range in north Queensland. It has the appearance of a large (55 mm), black katydid and readily flies short distances when disturbed.

The described species is:

- *Exogryllacris ornata* Willemse 1962 (type species)

Genus Gryllotaurus *Karny 1929 (Anostostomatinae: unplaced tribe)*

This fully winged genus includes stout, pale species with wings that wrap around the body and project well past its apex (Fig. 5.2H). The pale pronotum has dark front and rear borders. Males have an enlarged head, with two fixed, curved horns that project forward from the lower margin of the face, just above the enlarged mandibles; the horns are often slightly asymmetric in size and curvature and resemble those of the south African genus *Henicus* (see Fig. 10.4A, Field and Deans, Chapter 10, this volume). *Gryllotaurus* is restricted to mountain rainforests of tropical north Queensland, from Mt Finnigan, near Cooktown, south to the Cardwell Range. There is one described species and perhaps two undescribed species. In life, the species have the appearance of gryllacridids and fly readily. There may be polymorphism in the males, which may complicate the taxonomy.

The described species is:

- *Gryllotaurus bicornis* Karny 1929 (type species)

Genus Transaevum *Johns 1997 (unplaced subfamily)*

This recently described, fully winged genus contains slender, greyish, mottled species, with long narrow wings, which wrap closely around the body (Fig. 5.2I). The hind legs are long, slender and relatively unarmed. There is no sexual dimorphism and the heads of both sexes are small, with no inclination to bite when handled. *Transaevum* may be the most primitive extant member of the Anostostomatidae (P.M. Johns, Brisbane, 1989, personal communication). The genus occurs along streams in mountain rainforests throughout the wet tropical region of north Queensland between Cooktown and Townsville. This is the only one of the three Australian winged genera to extend south to the Paluma Range, just north of Townsville. There is one described species and perhaps a second undescribed one.

The described species is:

- *Transaevum laudatum* Johns 1997 (type species)

Distribution Patterns

Faunal provinces

Australia's faunal provinces reflect both its early Gondwanaland connections and the present geological and climatological conditions that characterize this isolated continent. The king crickets, in turn, show sharply demarcated distributions, determined by Australia's biogeographical history. Much of the fauna is a reflection of endemicity and radiation, resulting from Australia's complete isolation from other land masses between the Mesozoic and the Pliocene (Mackerras, 1973). However, underlying affinities with the wetas and king crickets of New Zealand, southern Africa and India reflect the Gondwana origins of these primitive insects. Included in the Australian fauna are fully winged species, considered to be the most primitive in the world (P.M. Johns, Christchurch, 1998, personal communication).

The three faunal provinces of Australia basically divide the country into two broad coastal regions and the great interior arid region. The Torresian province occupies the entire north, in a band that sweeps eastward around Cape York and down the east coast to the mid-latitudes near Sydney, and includes tropical and subtropical grasslands, woodlands and rainforest. The Bassian province continues the coastal band south and around the continent as far west as near Adelaide, and includes the isolated south-western corner below Perth. The whole of the arid and semi-arid interior region is encompassed by the Eyrean province, in which the hot desert climate extends to the central south coast and most of the west coast (Keast *et al.*, 1959).

Almost the entire king cricket fauna of Australia (95%) resides in the Torresian province, specifically that part of it which comprises a narrow coastal belt, extending about 3000 km, from the tip of Cape York down to the southern border of NSW (Fig. 5.1). In the north, distributions do not extend westward from the monsoon forests of Cape York Peninsula and no king crickets occur in the Northern Territory. This eastern coastal belt is typically no wider than 50–100 km, extending inland from areas of coastal rainfall to the adjacent mountains. The only exceptions are records of genus A from 400 km inland on a high plateau of the Carnarvon Range in central Queensland and of *Penalva* from sandstone areas 300 km inland in both Queensland and NSW. These inland areas contain many other relict forest insect species. The other 5% of the fauna comprises one genus (*Hypocophoides*), found in the western section of the Bassian province in south-western Australia (Fig. 5.1). One of the great enigmas of the distribution of the Anostostomatidae in Australia is their total absence from Tasmania and Victoria, areas where Gondwana representatives of most other insect groups have proliferated. It is presumably neither climate nor habitat factors that cause this absence, because similar temperate/alpine habitats in nearby New Zealand support many weta species.

Faunal transition and diversity patterns along the east coast

The narrow, 3000-km-long, east coastal zone of the Australian continent comprises a chain of wholly or partly isolated mountain/rainforest blocks, which act as a linear archipelago of distributional 'stepping stones' and refuges for moist-adapted faunas. These blocks are separated by several arid, lowland corridors, which form barriers to distribution between the blocks (Fig. 5.3). The pattern of insect distribution along this linear chain was first described by Darlington (1961), for carabid beetles, and more recently by Monteith (1997), for the aradid bugs. The king crickets are seen to follow a similar pattern, with the major centre of their diversity being the Wet Tropics zone between Cooktown and Townsville, with 28 species in seven genera. The next most diverse area is the south-east Queensland zone, which includes the complex mountain systems along the Queensland/NSW border. It has 20 species, but with a much lower generic diversity of only four genera. No other zone has more that eight species in three genera. The following discussion traces the transition of the king cricket fauna of Australia from north to south. The zones referred to, and all the localities mentioned, are shown in Fig. 5.3.

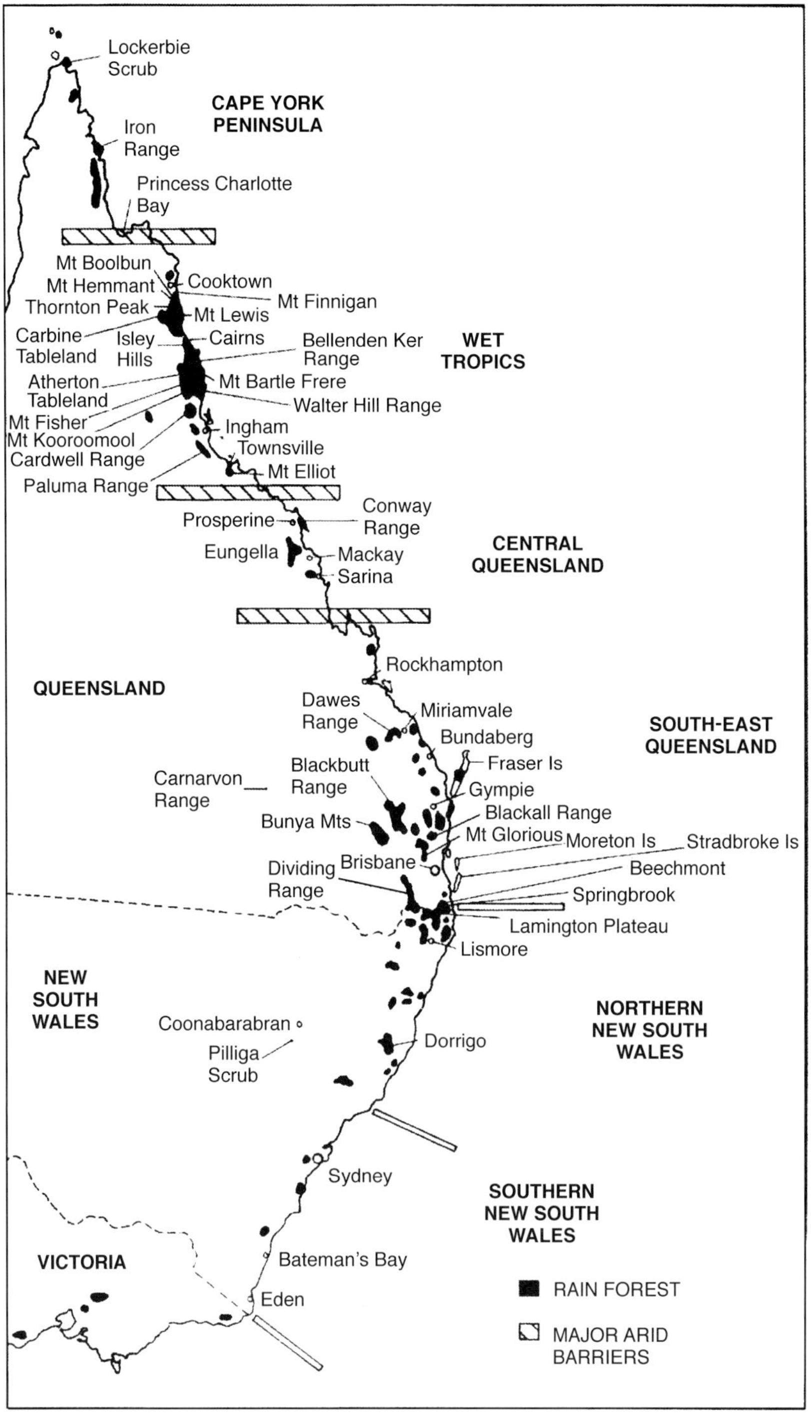

Fig. 5.3. Map of eastern Australia showing distribution of rainforest. Positions of important arid barriers (hatched bars) between the tracts of rainforest in Queensland are shown and the distributional zones used in the text discussion are indicated in capital letters on the right. All localities mentioned in the text are shown.

In the northernmost zone of Cape York Peninsula, at the northern extremity of Australia, the only king cricket is a large, speckled species of genus A. This is found in the lowland, monsoonal rainforests of Lockerbie Scrub and Iron Range. This limited fauna parallels the paucity of Gondwana species in this region for all other groups studied (Kikkawa *et al.*, 1981).

Cape York Peninsula is delineated to the south by an arid corridor at the level of Princess Charlotte Bay. Immediately to the south of this is an extensive complex of rainforested coastal mountains, known as the wet tropics, which includes many massifs that exceed 1000 m and which attract high rainfall. The wet tropics zone has Australia's richest king cricket fauna, in terms of both species and genera. Four of the seven genera of the wet tropics are restricted to the region. These include three genera of fully winged king crickets, namely *Gryllotaurus, Exogryllacris* and *Transaevum*. The fourth endemic, Genus B, is a specialized cavernicolous species, which has evolved in a boulder cave habitat peculiar to granite mountains of considerable antiquity. It shares that habitat with a distinctive cave fauna, including one of its own predators, a giant, primitive spider (*Macrogradungula moonya*), which also has an impeccable Gondwana lineage (Forster *et al.*, 1987). Seven of the eight Australian king crickets that have been attributed to the New Zealand genus *Hemiandrus* also occur in this rich wet tropics zone. Another distinctive element in the wet tropics is a small group of head-dimorphic species of Genus A, all of which occur on mountains at the northern end of the zone, such as Mt Finnigan, Mt Hemmant and Thornton Peak (Fig. 5.3).

Of the wet tropics king crickets, only the large white-kneed king cricket (*P. flavocalceata*) and one species of genus A occur in lowlands. The remaining assemblage of 26 wet tropics species are all found at altitudes exceeding 500 m in mountain systems, such as the Bellenden Ker Range and the Carbine Tableland. Both these ranges are very large massifs, found south and north, respectively, of Cairns in the central wet tropics (Fig. 5.3), and which appear to have the greatest diversity of king crickets in the entire region. The faunas of each are very different, however, and they have almost certainly served as independent, isolated refuges when dry climatic conditions have caused the periodic retreat of rainforests since the Gondwanaland breakup. From these refugial centres of diversity, the king cricket fauna appears to have emigrated to repopulate the adjacent, lower mountains and tablelands when moist conditions returned. South of Townsville lies an arid belt, which separates the wet tropics rainforest system from the Central Queensland rainforest system. This system, centred around Mackay, is focused on the Eungella and Conway mountain ranges and associated coastal rainforest. It extends from Proserpine in the north to as far south as Sarina (Fig. 5.3). This region has a small, characteristic group of three large species (two *Penalva* and one Genus A). Both the *Penalva* spp. have tiny vestigial tegmina, a feature seen nowhere else in the Australian fauna and indicating considerable isolation. One of these species occurs down to about 200 m in the coastal Conway Range near Proserpine. The central Queensland region is peculiar for its apparent lack of a component of small species of *Hemiandrus* and Genus A, which are typical of mountain systems to the north and south. Such species may have been more vulnerable to extinction during past arid phases, which the large, deep-burrowing species have survived.

Another arid belt occurs just north of Rockhampton and, since coastal mountains are absent, so also is wet rainforest absent. The south-east Queensland rainforest system starts from about Miriam Vale and extends down to the northern border of NSW (Fig. 5.3). A new, diverse assemblage of king crickets inhabits this region. Ten species of *Penalva* and six species of genus A occur, often as allopatric representatives in small mountain rainforest systems, such as the Blackbutt, Bunya, Blackall and Dawes Ranges north and west of Brisbane. In the last is found an unusual species of Genus A, in which the male has enormously dilated, paddle-like, hind tibiae with massive, marginal spines. Around the southern parts of the Great Divide, several of the distinctive, tiny, 'short-tailed' species of Genus A occur. In addition, one undescribed species of the disjunct genus *Hemiandrus* is found only in the highest reaches (*c.* 1000 m) of the mountain rainforests of Springbrook and Lamington on the border between Queensland and NSW.

The most spectacular and largest king cricket genus, *Anostostoma*, appears for the first time in the south-east Queensland rainforest system (Fig. 5.3). Three of the five species are found there, the peak diversity for the genus. Crossing the border into northern NSW, only a couple of species of *Penalva* and Genus A persist as far south as about the Dorrigo Plateau. Beyond that, only isolated populations of *Anostostoma* and *Penalva* occur, only very close to the coast, and these finally disappear near Bateman's Bay, shortly before the Victorian border is reached. No king crickets are known from the extensive subalpine habitats of the Australian Alps of southern NSW and Victoria.

Vegetation Affiliation of King Crickets

In eastern Australia, within the range of most Australian king crickets, two quite distinct vegetation types occur: open sclerophyll forest and rainforest. The sclerophyll forests, dominated by *Eucalyptus* and *Acacia* trees, predominate and comprise the typical Australian landscape. Rainforests occur only in very limited areas where rainfall and soil type are favourable. Most king crickets are confined to those circumscribed tracts of rainforest and the greatest diversity is seen in high- altitude areas with high rainfall and heavy cloud. The genera *Gryllotaurus*, *Transaevum*, *Exogryllacris*, Genus B and the Australian *Hemiandrus* are entirely rainforest-restricted, while most of the species of *Penalva*, *Anostostoma* and Genus A are similarly restricted.

The few species that live in the dry, sclerophyll forests appear to be confined to areas with sandy soils, and this is presumably because the hard, compacted soils of other dry forests are not amenable to burrow-making. Examples of these sclerophyll-forest king crickets include the *Penalva* species, which occur in coastal sandy areas of southern Queensland and nearby islands, such as Fraser, Stradbroke and Moreton. Other *Penalva* occur in the sandy open forests of the Pilliga Scrub near Coonabarabran in western NSW. An isolated species of Genus A also occurs in the same habitat on the Carnarvon Range of western Queensland. No rainforest occurs within the range of the Australian species of *Hypocophoides* in the southwestern corner of the continent, and these species are similarly restricted to sandy soils in sclerophyll forests. The only other king cricket found in sclerophyll forests is *A. opacum*, the southernmost species of *Anostostoma*, which occurs in the dense, wet sclerophyll forests of southern NSW.

Behavioural Ecology

No formal behavioural studies have been done on Australian king crickets. The following sections summarize casual field observations on the lifestyle and habits of the species made over many years during collecting trips, primarily by GBM and colleagues at the Queensland Museum. Productive collecting methods have included: long-term, 2-litre, roofed, pitfall traps; flight-intercept traps; digging out of burrows; night-collecting using headlights. The rare winged genera are taken almost solely by night headlighting. Day collecting yields few king crickets, except by burrow digging.

Microhabitats

Within the generalized forest habitat, Australian king crickets all fall into one of four discrete microhabitat preferences: the forest floor, arboreal situations, streamside environments and boulder caves. All species of *Anostostoma*, *Penalva*, *Hemiandrus*, *Hypocophoides* and Genus A live on the forest floor and can often be seen in substantial numbers on warm, wet, summer nights, moving over the leaf litter and engaging in feeding, mating and egg laying. A few species will explore low vegetation and sometimes wander a metre or two up tree-trunks, but the soil surface is their main arena of activity. Most retreat into self-excavated burrows beneath the surface during the daytime. Those for which burrows have not been noted are *Hemiandrus* and *Anostostoma*. *Hemiandrus* species are very small and may merely shelter among the leaf litter during the day. Most *Anostostoma* species, especially *A. spinosum*, are associated with rotten logs and may be found in or under them as daytime retreats.

The two macropterous genera, *Gryllotaurus* and *Exogryllacris*, are essentially arboreal, a habit shared by some of the New Zealand tree and giant wetas. *Gryllotaurus* species resemble gryllacridids in both appearance and behaviour. They move rapidly up tree-trunks and along branches and vines, occasionally making short flights across spaces. Although they lack the broad tarsal pads of gryllacridids, and do not have their ability to cling to twigs and foliage when seized, their bite is equally impressive. *Gryllotaurus* species retreat to nest burrows in logs and stumps in the daytime.

Exogryllacris ornatas frequents the trunks of standing dead trees and logs suspended off the ground. It rests quietly and rarely moves unless disturbed, relying on the camouflage afforded by its enormous dark wings, which shroud most of its body. It does not retreat to a daytime burrow and rarely attempts to bite when handled.

The streamside environment of high-altitude, rocky streams in north Queensland is the habitat of the primitive, but specialized, genus *Transaevum*. At night, large numbers of black nymphs forage on spray-wetted, granite rocks and gravel beds beside running water. When full grown, they ascend sedges and reeds to moult to adults. These leave the rock pavements and can then be found at night on foliage of shrubs up to 2 m high along the stream and in the adjacent rainforest. The slender, mottled adults fly infrequently and do not appear to use daytime retreats.

The only specialized cave-frequenting king cricket in Australia is the sole species of Genus B from high altitudes in tropical Queensland. The caves where it is found are unusual 'boulder caves'. These are formed in ancient, deep-weathered, granite soil profiles where rainforest root systems hold the surface layers intact, but, lower down, runoff from tropical downpours cuts away the weathered soil between giant, residual, subterranean boulders. This process leaves anastomosing chambers, which penetrate to bedrock up to 25 m deep. Genus B occurs in large numbers right to the maximum depth, mostly in the totally dark zone, and has been observed actively foraging there when it is daylight at the surface. Occasional specimens have been taken in boulder crevices and overhangs at the surface, but only adjacent to known subterranean colonies. The species has a specialized herbivorous diet, long, powerful back legs and very long antennae. True cave crickets (Raphidophoridae) occur in the same cave systems but are predatory.

Burrow and retreat construction

Australian king crickets that construct burrows and retreat chambers, to which they regularly return after nocturnal foraging, include species of *Penalva* and Genus A in the soil and species of *Gryllotaurus* in soft wood. The Western Australian genus *Hypocophoides* also apparently lives in soil burrows (Rentz, 1996), but direct observations are not available. The giant king crickets (*Anostostoma* spp.) favour chambers in and under logs, but details are scanty. Figure 5.6 shows the distribution and shape of king cricket burrows in an idealized Australian rainforest.

The best information on soil burrows comes from excavating those of *Penalva* and Genus A species, which may be extremely abundant in favourable sites, such as rainforests with red basalt soils. When the crickets are actively excavating the burrows (e.g. after rain in early summer), their location on the forest floor is obvious from fresh soil push-ups, which occlude the entrances. Once burrows are completed and the push-ups dispersed, the burrow entrances become very inconspicuous. The entrances are maintained at minimum 'squeeze' diameter for the insect, with litter pulled over the entrance after entry. Immediately beneath the entrance, the burrow widens rapidly into a large cylinder, two to three times the entrance diameter and with its walls smoothly plastered by the cricket's mouth-parts. These cryptic burrows are easily located by using a shovel to shave away the top centimetre of soil to reveal the wide-diameter main chambers. The burrow can then be excavated by digging along a flexible probe, such as a green twig, to the terminal chamber. Study casts of the burrows can be made by squeezing plaster of Paris into the entrance and then digging up the cast and scrubbing it clean of soil.

In *Penalva*, the burrows are straight and very wide, up to 4–5 cm diameter in large species, with a terminal section a little wider than the main shaft. They extend vertically downwards for 20–40 cm in flat topography (Fig. 5.6b), but some species prefer steep slopes and earth banks as burrow sites. In these situations, the burrows are horizontal (Fig. 5.6a). The very large species, such as *P. flavocalceata*, and the vestigial-tegmina species from central Queensland favour horizontal burrows. The open-forest species from the Pilliga Scrub forms very deep, wide, elegantly finished burrows in the loose, sandy soil of that region. *Penalva* species appear to use their burrows as simple retreats for both adults and larger nymphs, and we have no evidence of oviposition taking place in them.

Genus A is a large and diverse genus, and burrow details are unknown for many forms, such as the *Penalva*-like, robust species from Cape York Peninsula and the head-dimorphic species from tropical mountains. But at least two species groups of Genus A have been found to build burrows that form subterranean oviposition chambers. Burrows were excavated from a dense colony of a medium-sized Genus A species with a somewhat reduced, blunt-tipped ovipositor at Beechmont, south Queensland. Beneath concealed entrances, the narrow burrows went vertically down for 10–15 cm and then turned horizontally and rose over a hump before descending slightly to become enlarged into a smooth, elongate chamber, which rose again at the end (Fig. 5.6d). This structure would act like a sink trap when flooded and provide an air-locked terminal retreat. Male/female pairs and, in one instance, one male and two females were found in these burrows. On two occasions, single females of tiny, 'short-tailed' species of Genus A have been found in urn-shaped chambers just below the soil surface (Fig. 5.6c). The 'neck' of the urn formed a short passageway to the surface. Both these Genus A burrow types were used both as retreats and as oviposition sites (see below). As stated above, most *Anostostoma* species seem to be associated with large logs. *Anostostoma spinosum* is frequently taken inside hollow logs, and this seems to be its regular retreat type (Fig. 5.6e). The giant *A. australasiae* is sometimes taken in cavities beneath logs (Fig. 5.6f) and there are one or two records of individuals being accidentally found in large, smooth-walled, underground chambers (A. Hiller, personal communication). Considering the large numbers that emerge to forage on rainy nights, it is surprising how rarely these large insects are encountered during daytime searching. Such subterranean chambers with concealed entrances may be the explanation.

On three occasions, *Gryllotaurus* species have been found in neatly made chambers inside rotten wood, always with tight 'squeeze' entrances, and this appears to be typical for the genus. One at Mt Lewis was in a small log on the ground and another at Mt Fisher was in a horizontal, fallen tree-trunk, suspended about 1.5 m above the ground. These were both found in a damaged state. The third find was a group of separate chambers in a large, mossy stump near Mt Kooroomool and these could be carefully examined. The entrances were very inconspicuous and obscured by moss. Behind the entrance, a tubular burrow 1 cm diameter ran horizontally for 3–4 cm and then turned vertically downwards and widened into a chamber about 3 cm diameter and 5–6 cm long. This was extended by a narrow, blind-ended tube about 4 cm long (Fig. 5.6g, h). The inner surface of the chambers was very smooth and well finished. There were about four or five such burrows, all with openings within about 5 cm of each other. One burrow contained an adult male and female; another had a single adult female; the others were smaller and contained half-grown nymphs.

Food

In New Zealand, the large wetas, especially species of *Hemideina* and *Deinacrida*, are major herbivores on green, shrub foliage. In contrast, no Australian king cricket has been seen to feed on foliage, fresh or dead. Most of the species that forage on the forest floor seem to be generalized scavengers and gleaners (Rentz, 1996), taking a great variety of dead and decaying organic matter, such as small berries, dead insects, soft fungi, animal droppings, dead earthworms and so on. A highly effective bait for luring many species is rotten pineapple. This is particularly attractive to *A. australasiae* and *Penalva* species. The former species is troublesome to mammal surveys, since it regularly springs small traps baited with meat, peanut paste and so on. It clearly wanders a considerable distance from its retreat during foraging.

Some species are more specialized. Thus, the arboreal *Exogryllacris* feeds on masses of jelly-like fruiting bodies of fungi growing on the trunks of fallen trees. Both *Transaevum* and genus B feed on black, encrusting, algal/fungal growth on moist rock surfaces, and both spend much time patiently scraping the material from suitable sites. *Transaevum* nymphs browse on rock pavements beside running streams, while an adult was found resting on a palm frond and feeding on a frog dropping filled with insect remains. Genus B feeds on the surface of walls and ceilings of caves, leaving patterns of scraped bare patches where it has fed. It constantly sweeps its long antennae back and forth over the surrounding rock surface, presumably to detect possible predators in the dark.

In addition, observations in the field and in captivity, as well as the presence of large, sharp-toothed mandibles, indicate that some king crickets are also carnivorous. For example, a *P. flavocalceata* has been seen eating a large, freshly killed cicada nymph on the Carbine Tableland. Other *Penalva* individuals have been seen eating earthworms in the forest at night and meal-worms in captivity. They also attack and eat each other if kept together in containers. A large-headed species of Genus A from Mt Hemmant in north Queensland was seen on low foliage consuming a 'semi-slug' (pulmonate mollusc). *Gryllotaurus* has large, sharp mandibles and a quick, roving habit, both observations suggesting a predatory mode of life. Furthermore, *A. australasiae*, when kept in captivity, devours dead cockroaches and mealworms with a peculiar feeding schedule. When food is presented daily, it is ignored for many days until suddenly one day the insect will feed massively on dead insects that are offered. This may reflect the fact that the natural activity cycle in this large king cricket is closely linked to periods of occasional heavy rainfall.

Nocturnal activity patterns

No king crickets are active in the daytime. They tend to emerge soon after sunset to commence foraging and subsequent nocturnal activity. At Mt Glorious, near Brisbane, the large *A. australasiae* may be found soon after dusk on fruit baits laid earlier in the day. Many *Penalva* species emerge somewhat later, about half an hour after full darkness, and within the first 2 h have mated and are seen feeding, ovipositing or sitting still. By 3–3½ h after dark, *Transaevum* males and females have been recorded on streamside rocks and foliage. Generally, all species are still active by midnight.

Weather determines the likelihood of emergence from burrows. During the dry season from April to October, *A. australasiae* does not emerge at all. At this time, *Penalva* seldom emerges at night and, if it does, in smaller numbers than those seen in the wet season. Cold and dry conditions seem to deter surface activity. In the wet season, king crickets emerge if conditions are moist or wet or it is raining. They prefer warm temperatures, and may be found even in heavy rain. In this case, *Penalva* specimens have been observed to shelter on the underside of fern leaves or low foliage.

Sound Production

Sound in *Anostostoma* (observations of *A. australasiae* and *A. spinosum*) is produced by rocking forward and rubbing the abdomen in a vertical movement against the hind legs as they are appressed to the body. All legs remain on the ground, and there is no upraised-leg defence-sound kicking, as seen in New Zealand wetas. The stridulatory behaviour resembles that described for gryllacridids from Western Australia (*Hadrogryllacris* and *Ametrus*: Field and Bailey, 1997) and the New Zealand henicine ground wetas (Field, 1993). Territorial calling, common in New Zealand tree wetas, has not been recorded in Australian king crickets.

The stridulatory structures have been examined in *A. australasiae* and *Penalva* sp. Both species have a femoro-abdominal mechanism typical for the family; however, the structures are of the plesiomorphic type and do not show the specializations seen in the New Zealand anostostomatid fauna (see Field, Chapter 15, this volume). They consist of dense patches of short, sharp pegs on the sides of the first three (*A. australasiae*) or first four (*Penalva*) abdominal tergites, and four to five narrow, chevron-shaped fields of similar pegs on the inner (posterior) surface of the hind femur. Both fields of pegs are engaged when the legs are held close against the body during stridulation. No audible sound production has been noted in *Penalva* species.

Reproduction

Mating

Substantial observations of mating behaviour and nuptial 'feeding' have been made for several species of *Penalva*, for *T. laudatum* and for *A. australasiae*. The most interesting facets to arise are the discovery of courtship behaviour in *Penalva* and *Anostostoma* and the details of spermatophylax production in *Penalva* and *Transaevum*. Courtship, or premating behaviour serving to prepare the female for mating, has never been reported for New Zealand wetas or king crickets (Anostostomatidae). A substrate-drumming behaviour, which apparently serves a courtship role, has been described for Jerusalem crickets (Stenopelmatidae) by Weissman (Chapter 19, this volume).

The most complete observations of mating behaviour were made in December 1992 on the large, undescribed, brachypterous *Penalva* species commonly found on the forest floor at Eungella National Park, west of Mackay. Premating behaviour commences with the male facing the female's rear at an acute angle and repeatedly butting her at the base of the ovipositor. The female moves a few steps in response to each butting and the pair describes several tight circles of stepping and butting. The male then commences the next phase by placing a fore tarsus on the female's ovipositor and vibrating the leg for 3–4 s. The male repeats the application of the tarsus and vibration of the foreleg four to five times, while the female remains stationary. Mating is initiated when the male moves to the front of the stationary female and, while facing away from her, backs under her head. This prompts her to climb forward on to the back of the male until the genitalia of both individuals are in contact, with the female pressing the region at the base of her ovipositor firmly down on to the tip of the male's abdomen. The pair remains stationary for several minutes, during which time two large, transparent, gelatinous spermatophylax vesicles slowly expand from the tip of the male's abdomen (Fig. 5.4a). These reach about 7 mm diameter and are positioned one on each side of the base of the female's ovipositor. When the vesicles have reached full size, the female begins to repeatedly push the tip of her abdomen down against the male's genitalia in several short series of movements lasting about 2 min. The purpose of this action seems to be to force the spermatophore ampulla into her genital opening. Finally, the female breaks away from the male with the whole spermatophylax complex attached to her abdomen. She walks only a few steps before, arching her body, she reaches down between her legs and tears the gelatinous vesicles off with her mandibles (Fig. 5.4b). The small bilobed ampulla remains projecting from her abdomen. The female remains stationary while chewing and palpating the spermatophylax for several minutes, after which the male moves away. The entire premating and mating episode lasted about 9 min. One female of the same species was observed on low foliage manipulating a spermatophylax, rather than on the ground. Postnuptial manipulating of a bilobed spermatophylax was also observed in the white-kneed king cricket, *P. flavocalceata,* at Mt Boolbun, south of Cooktown, in November 1995.

Mating in *Transaevum* always seems to occur on low foliage rather than on the ground. Postmating females, holding a spermatophylax in their jaws, are often found at night on the upper surface of palm fronds and other foliage overhanging creeks (Fig. 5.2I). A peculiar posture accompanies this behaviour. The female stands with legs rigidly extended, body raised off the palm frond and the head extended forward with the spermatophylax

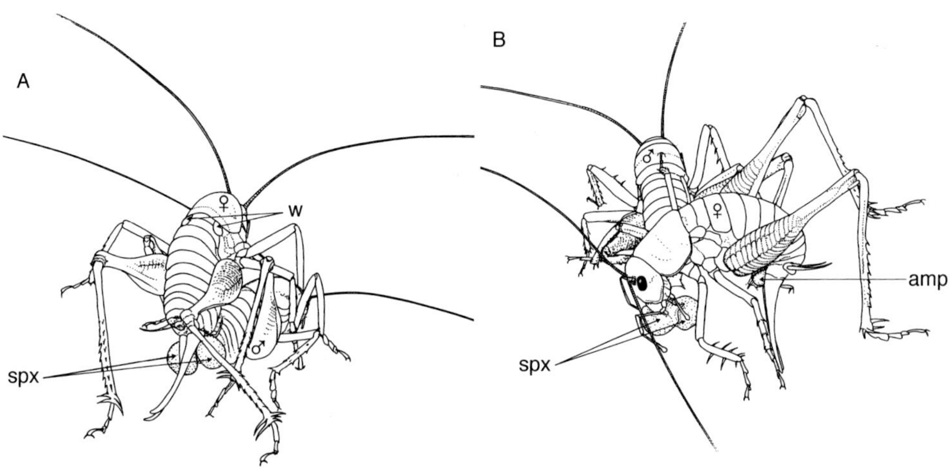

Fig. 5.4. Mating behaviour of *Penalva.* Drawings prepared by Geoff Thompson from photographs by Heather Janetzki of a mating pair of *Penalva* sp. n. at Eungella, central Queensland. A. In mid-copulation, the female rests on the back of the male. The two vesicles of the spermatophylax (spx) being secreted by the male are visible. B. Soon after, the female tears away from the male and reaches under her abdomen to grasp the spermatophylax vesicles in her jaws. The spermatophore ampulla (amp) remains in position at the base of the female's ovipositor. The small vestigial wings (w) (shown in A) of this distinctive species are obvious.

conspicuously displayed. The spermatophylax is continually palpated and rotated in the mouth-parts, but after up to an hour of this activity the vesicle has not been noticeably consumed. This behaviour has been observed a number of times, but its function is unknown. It may be a post-mating signal from the female to other males. No premating behaviour has been observed in *Transaevum*. In one mating observation, the male was suspended by the forelegs on a palm frond and the female faced the opposite direction with her abdomen ventrally apposed to his and joined by the genitalia. This configuration is therefore unlike that observed in *Penalva*. The male underwent abdominal pulsating as the spermatophore, with a single, spherical, gelatinous spermatophylax vesicle about 5 mm diameter, was transferred to the female. When the transfer was complete, the female immediately separated and grasped the vesicle in her mouth-parts.

Fragmentary observations of *A. australasiae* indicate that this species also has some courtship. The male approaches the female, stridulates repeatedly with the femoro-abdominal mechanism and retreats. He again approaches, stridulates and retreats, and may repeat the sequence several more times. Finally, the female backs over the male and copulation commences. The remainder of copulation was not observed and no evidence of spermatophylax production has been seen in *Anostostoma*.

No mating activity has been noted among hundreds of observations of foraging individuals of genus A. Since several members of this genus have been found to oviposit inside their burrows, this may indicate that mating also takes place in the burrows.

Harem formation

No obvious harem formation, as found in the New Zealand tree wetas, has been observed in any of the Australian king crickets. There is one documented example of a male and two females of Genus A sp. (Beechmont, near Brisbane) living in the same burrow with a number of juveniles, with eggs still embedded in the chamber walls. An observation of both sexes occupying each of several adjacent retreats of *Gryllotaurus*, combined with the prominent facial armature and possible polymorphism of the males, indicates that some social activity could occur in this genus.

Oviposition

Only *Penalva* species have been regularly observed ovipositing. Members of this genus all have long, powerful, curved and sharply pointed ovipositors. Females use these to lay eggs directly into open spaces on the forest floor. The abdomen is curved forward between the legs during this operation and the ovipositor is inserted vertically to its full length in the soil (Fig. 5.5).

Several species of Genus A with short ovipositors lay their eggs underground, on the walls of the terminal chamber of the burrow. A female of a small, 'short-tailed' species was dug from an oval soil chamber just beneath the ground surface at Isley Hills, near Cairns, in November 1993. About 20 flattened eggs were deposited on their sides in a row on the wall of the chamber. They were held in position by a coating of thin mud, presumably plastered over them by the female. A similar arrangement of eggs, of a different, larger species with half-length ovipositor, was observed in a chamber excavated at Beechmont, southern Queensland, in May 1997. The eggs were orientated vertically on the wall, side by side in a row. These were also partly plastered over by a layer of mud. Many of these were on the verge of hatching and, when the wall of the burrow was disturbed during digging, numerous first-instar hatchling nymphs began to burst out from the egg cavities. This burrow was occupied by a male and two adult females, so the eggs may have been the progeny of more than one female.

Both species of Genus A mentioned in the previous paragraph produce black, liquid faeces when handled and this material may be what is used to plaster over eggs, to hold them in position, after they are deposited on the burrow walls.

Fig. 5.5. Female of *Penalva* sp. ovipositing in the soil (Eungella, central Queensland. Photo: H. Janetzki).

Fig. 5.6. Profile through the ground layer of an idealized Australian rain forest showing location and configuration of burrows and retreats of king crickets. a, horizontal burrow of large *Penalva* made into a vertical earth bank. b, vertical burrow made in the forest floor by large *Penalva*. c, Small, ovoid burrows made beneath ground surface by 'short-tailed' species of genus A. d, Sinuous burrows with terminal, air-locked chamber made by larger genus A. e, Retreat chamber in the centre of a rotten log used by *Anostostoma spinosum*. f, Retreat chamber in soil beneath log used by *Anostostoma australasiae*. g and h, Concealed burrows gnawed into soft rotten wood of logs and stumps by species of *Gryllotaurus*. Illustration by Geoff Thompson.

Defence Behaviour

The primary defence mechanism of most Australian king crickets is to hide in burrows during the daytime to avoid diurnal visual predators. Only species of *Transaevum* and *Exogryllacris* are not known to use daytime burrows. Secondary mechanisms in response to disturbance by predators are behavioural, and often consist of escape jumping (*Anostostoma, Penalva, Hemiandrus,* Genus A, Genus B), stridulatory behaviour (described below) or flying (*Exogryllacris* and *Gryllotaurus*). The nymphs of *Transaevum*, which feed on wet rocks beside running streams, regularly escape by jumping into the water. They are then carried a few metres downstream before crawling out. In this way, they resemble pygmy grasshoppers (Tetrigidae), which feed in the same microhabitat. While the herbivorous New Zealand wetas produce large, dry, fibrous faecal pellets, these are unknown in the Australian fauna and almost all species produce putrid-smelling, black, liquid faeces, which is used as a defensive deterrent. When species of *Anostostoma, Penalva*, genus A, *Exogryllacris* and *Gryllotaurus* are being handled, it is a common occurrence for the faeces to be suddenly ejected in a strong squirt. The strong smell is very persistent and difficult to wash off.

Sound production and short erratic jumps are made by *A. australasiae* and *A. spinosum* when they are disturbed by turning over a log or rock. The excited insect appears to thrash around while jumping, and simultaneously rubs the legs against the abdomen during this somewhat chaotic and noisy behaviour. Escape jumping is also another secondary defence tactic in these species. However, if further stressed, *A. australasiae* will flip over on its back and lie motionless with all legs spread and mandibles fully gaped. In this position, the ventrum shows the bright transverse yellow and white banding of the abdominal tergites, which may be an aposematic deterrent display (Fig. 5.7). In this immobile display, the king cricket is highly alert and, if touched anywhere on the ventrum, it will rapidly close the legs and mandibles around the offending object and bite it, while simultaneously stridulating loudly. If not touched, the alarmed insect will remain motionless for at least 10 min, while still primed for biting. Similar behaviour has been seen in the smaller

Fig. 5.7. Male of *Anostostoma australasiae* exhibiting the ventral aposematic pattern on the abdomen during its inverted defence display when vigorously attacked (Tallebudgera Valley. Photo: J. Wright, Queensland Museum).

A. spinosum, which has an even more colourful display, with black body, red legs and yellow abdominal bands.

The inverted defence posture of *Anostostoma* is remarkably similar to that described for two species of New Zealand wetas, *Hemideina maori* and *Hemideina ricta*. The existence of this behaviour in two groups that have been geographically separated since the Mesozoic argues for an early evolution of the behaviour in the Anostostomatidae. As discussed by Field and Glasgow (Chapter 16, this volume), experiments with predators show that it is likely to have developed as a successful method of defence against reptiles and against avian predators that do not have long beaks.

Unlike the New Zealand wetas, *A. australasiae* does not show mandible gaping as a threat display in defence. This is surprising, since males have secondary sexual development of extremely elongated mandibles and enlarged heads. However, if handled, this species, like almost all other species of king crickets, will readily bite.

Acknowledgements

Doug Cook has been invaluable in assisting with the fieldwork, often in difficult terrain, during which many of the collections and observations on which the discussion is based were made. The information would have been impossible to present without the provisional taxonomic framework provided by Peter Johns.

References

Burmeister, H. (1938) *Handbuch der Entomologie*. Band 2, Berlin.

CSIRO (1991) *The Insects of Australia: a Textbook for Students and Research Workers*, 2 vols. Melbourne University Press, Melbourne.

Darlington, P.J., Jr (1961) Australian carabid beetles. V. Transition of wet forest faunas from New Guinea to Tasmania. *Psyche* 68, 1–24.

Field, L.H. (1993) Structure and evolution of stridulatory mechanisms in New Zealand wetas (Orthoptera: Stenopelmatidae). *International Journal of Insect Morphology and Embryology* 22, 163–183.

Field, L.H. and Bailey, W.J. (1997) Sound production in primitive Orthoptera from Western Australia: sounds used in defence and social communication in *Ametrus* sp. and *Hadrogryllacris* sp. (Gryllacrididae: Orthoptera). *Journal of Natural History* 31, 1127–1141.

Forster, R.R., Platnick, N.I. and Gray, M.R. (1987) A review of the spider superfamilies Hypochiloidea and Austrochiloidea (Araneae, Araneomorphae). *Bulletin of the American Museum of Natural History* 185, 1–116.

Gray, G.R. (1837) Descriptions of some singularly formed Orthopterous insects. *Charlesworth Magazine of Natural History* n.s. 1, 141–145.

Johns, P.M. (1997) The Gondwanaland weta: family Anostostomatidae (formerly in Stenopelmatidae, Henicidae or Mimnermidae): nomenclatural problems, world checklist, new genera and species. *Journal of Orthoptera Research* 6, 125–138.

Keast, A., Crocker, R.L. and Christian, C.S. (1959) *Biogeography and Ecology in Australia*. Monographiae Biologicae, Junk, The Hague, 640 pp.

Kikkawa, J., Monteith, G.B. and Ingram, G. (1981) Cape York Peninsula: major region of faunal interchange. In: Keast, A. (ed.) *Ecological Biogeography in Australia*. Junk, The Hague, pp. 1697–1742.

McKeown, K.C. (1944) *Australian Insects – an Introductory Handbook*. Royal Zoological Society of New South Wales, Sydney, 303 pp.

Mackerras, I.M. (1973) Composition and distribution of the fauna. In: CSIRO (eds) *Insects of Australia*. Melbourne University Press, Melbourne, pp. 187–203.

Monteith, G.B. (1997) Revision of the Australian flat bugs of the subfamily Mezirinae (Insecta: Hemiptera: Aradidae). *Memoirs of the Queensland Museum* 41, 1–169.

Rentz, D.C.F. (1991) Orthoptera (Grasshoppers, locusts, katydids, crickets). In: CSIRO (eds) *The Insects of Australia*, vol. 1. Melbourne University Press, Melbourne, pp. 369–393.

Rentz, D.C.F. (1996) *Grasshopper Country: the Abundant Orthopteroid Insects of Australia*. University of New South Wales Press, Sydney, 284 pp.

6

The Gryllacrididae: An Overview of the World Fauna with Emphasis on Australian Examples

Roderick J. Hale[1] and David C.F. Rentz[2]

[1]*Department of Zoology, University of Western Australia, Nedlands, Western Australia 6907, Australia;* [2]*Division of Entomology, CSIRO, Canberra, ACT 2601, Australia*

Introduction

In past classifications stenopelmatids and rhaphidophorids were considered as part of the group Gryllacrididae, which was itself placed within the orthopteran section of Locustodea (Tepper, 1892; 1895). Later this group was assigned to the Gryllacrinae, a subfamily of the Tettigoniidae (Tillyard, 1926). However, the Gryllacrididae are here considered a separate family, distinct from both the Rhaphidophoridae and the Stenopelmatidae, as well as the Schizodactylidae or Cooloolidae. Recently, they have been included in their own superfamily known as either the Gryllacridoidea (Key, 1970; John and Rentz, 1987; Rentz, 1991; Nickle and Naskrecki, 1997) or the Stenopelmatoidea (Gorochov, Chapter 1, this volume; Storozhenko, 1997). Some workers now classify the family Gryllacrididae within the superfamily Tettigonioidea (Brown and Gwynne, 1997) while dividing off the wetas and king crickets into a family of their own, the Anastostomatidae (Johns, 1997). Others give the Stenopelmatoidea superfamily status and group them with the Tettigonioidea and Hagloidea within the infraorder Tettigoniidea (Gorochov, Chapter 1, this volume). The discrepancy over the name of the superfamily may have arisen from the fact that both Brunner von Wattenwyl (cited in Rentz and John, 1990) and Karny (1932, 1937) used Stål's 1874 terminology Gryllacrididae for the family name rather than Burmeister's 1838 Stenopelmatidae (Key, 1959; Johns, 1997).

Despite these discrepancies, it appears that the closest relations to the Gryllacrididae are the Stenopelmatidae, which appear to have diverged during the late Cretaceous (Storozhenko, 1997). Gryllacridids bear a number of behavioural as well as morphological similarities to the stenopelmatids (Tepper, 1892; Bland and Rentz, 1991a; Rentz, 1997). The close affinity of the two groups accounts for Gorochov's current placement of the Gryllacrididae as a subfamily of the Stenopelmatidae (thus, Gryllacridinae in Chapter 1, this volume). The first Australian gryllacridid was described in 1842 and by the mid-1920s it was realized that the Gryllacrididae were well represented in Australia by more than 50 species, most of which were believed to belong to the genus *Paragryllacris* (Tillyard, 1926; Rentz and John, 1990). However, biological notes on even common species were extremely rare. For example, the only information known concerning one of the most commonly encountered species, *Paragryllacris combusta* Gerstaecker, was that it lived in holes in trees and between the leaf sheaths of certain plants (Tillyard, 1926). (Later this species was found feeding on emerging cicadas in Sydney; see Rentz and John, 1990.) Since the 1920s the most comprehensive work on the world fauna was conducted by H.H. Karny in 1937, however this work was descriptive and taxonomic in nature, concerned with nomenclature rather than natural history. Since then, there have been a few species descriptions but little else (see Rentz and John, 1990).

It is now known that of the 600 or so species of Gryllacrididae found throughout the world, over 200 species in more than 20 genera are found within Australia (Rentz and John, 1990; Rentz, 1991, 1996, 1997). Despite this very high representation, there is still very little known of the general biology and ecology of these animals (see Rentz, 1996, 1997). There is a lack of information about not only Australian species, but all members of the family worldwide. In fact, research throughout the rest of the world has been limited to observational anecdotes (e.g. Gleason and Johnson, 1985; Shaw *et al.*, 1987; Huff and Coyle, 1992) and it is only in Australia that more thorough investigations into these animals are starting to be made (Morton and Rentz, 1983; Lockwood and Rentz, 1996; Field and Bailey 1997; Hale, 2001). The two most comprehensive works on the Australian gryllacridids (Karny, 1937; Rentz and John, 1990) have dealt mainly with taxonomic descriptions of the numerous genera and species found on the continent. The more recent of these works did include some biological information, and it is the aim of this chapter to summarize such current knowledge and introduce previously unknown aspects of the biology of two gryllacridid species, Genus novo 9 sp. 4 and *Craspedogryllacris* sp. 21, from southwest Western Australia. There are at least 22 species of *Craspedogryllacris* Karny in Australia, most of which are as yet undescribed. Species of *Craspedogryllacris* are long-winged, have characteristic genitalia and usually dark markings on the head and pronotum. Genus nov. 9 comprises particularly robust, wingless species with large heads and a large circular median ocellus. They appear to be related to the more widespread *Ametrus* Brunner. Both genera are common throughout the arid regions of Australia. New information on these two species presented in this chapter is the result of PhD research conducted by R.J. Hale. Genus and species code numbers refer to specimen types held in the Australian National Insect Collection, Canberra.

Systematic Considerations of the Gryllacrididae

The Gryllacrididae are generally robust crickets, often winged although brachypterous and apterous species are known (Tepper, 1892; Rentz and John, 1990; Rentz, 1996). They range in size from less than 1 to 15 cm in length, with the filamentous antennae being generally as long as the body or longer (Rentz, 1996, 1997). All species are thought to be totally nocturnal (Rentz, 1996, 1997). Gryllacridids are characterized by the depressed and soft tarsi with prominent lateral lobes (Fig. 6.1), the presence of pegs on the inner surface of the hind femur that rub against tubercules or hairs on the opposing sides of the abdominal tergites, and the lack of spines or auditory organs on the front tibiae (Tepper, 1892; Kevan, 1982; Rentz and John, 1990; Rentz, 1991, 1996; Field and Bailey, 1997). The simplest superficial distinction from the other major families within the superfamily is that gryllacridids are generally not as long-legged as the cave and camel crickets (Rhaphidophoridae) and are soft-bodied in contrast to the wetas (Anastostomatidae) (Fig. 6.2).

The family has not been divided into subfamilies or tribes. Genera are often large and species illustrate extraordinary development of the genitalia, both in males and females. Unfortunately, this was not properly appreciated by Karny (1937) who based his classifications largely on wing venation, an often unreliable character and subject to considerable variation. Rentz and John (1990) discuss at length the suites of taxonomic characters and their potentiality for use in classifications. In general, it could be said that it was too early in Karny's time to propose a classification of the family. It still may be too early since many tropical regions of the world have been poorly sampled for these orthopterans.

Distribution of the Gryllacrididae

No extensive modern study of the distribution of the family Gryllacrididae has been made. Karny (1929, 1937) presented a treatise on the distribution of the family (*sensu lato*) and although his analyses were based on unreliable characters (Rentz and John, 1990), he did state that Australian gryllacridids showed more relationships with South American (= Gondwanan) forms than those of South-east Asia. Gryllacridids occur throughout the Indian subcontinent, south-western Asia and South Africa (Storozhenko, 1997). There are also gryllacridids in the Americas (mostly South) and many on Pacific and Atlantic

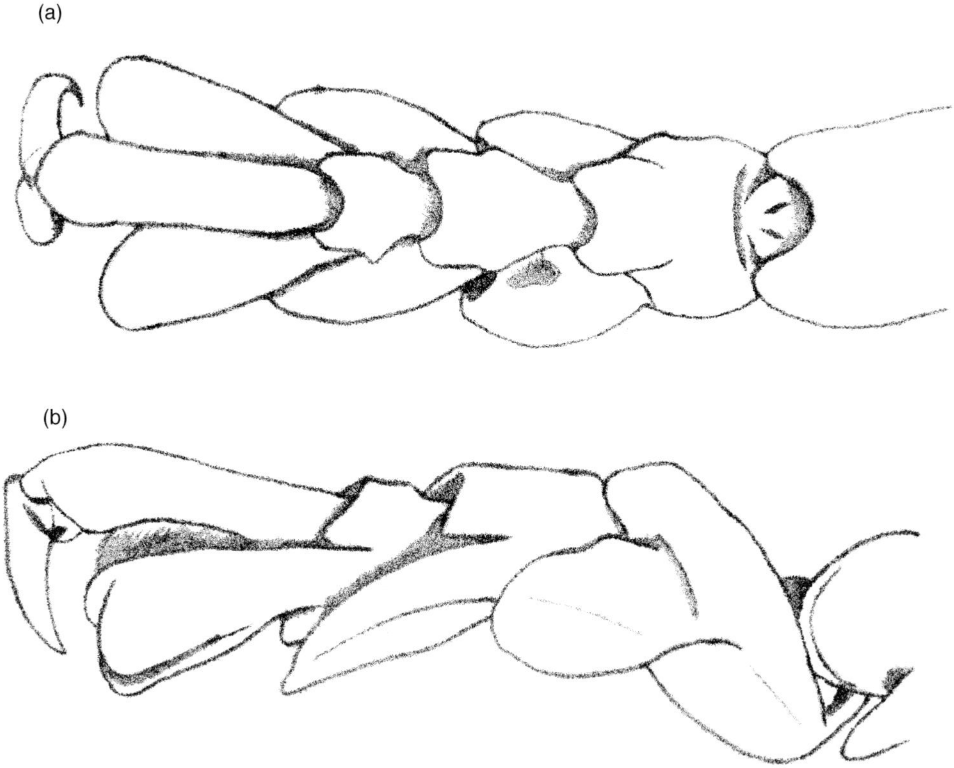

Fig. 6.1. Dorsal (a) and lateral (b) view of a gryllacridid tarsus showing the characteristic dorso-ventral flattening and the prominent lateral lobes on the first and second joints. Redrawn from Rentz (1996) with kind permission from the author.

islands (Tepper, 1892, 1895; Karny, 1929, 1937; Rentz and John, 1990; Rentz, 1991, 1996). Unlike their close relatives, the Stenopelmatidae, only a few genera of Gryllacrididae are known from the northern hemisphere (Bland, 1989; Parmenter *et al.*, 1991; Peck *et al.*, 1992; see Rentz, 1997), and just one species is thought to occur in temperate North America (Kevan, 1982). Recently it has been claimed that Australia has more species of Gryllacrididae per unit area than anywhere else in the world, with about a third of all known species found throughout the continent, from the tropical north-west, through the central arid zone, to the Mediterranean climate of the south-west corner (Rentz, 1997). The distribution of Gryllacrididae within Australia is perhaps better documented than the rest of the world, due mainly to the collecting and rearing efforts of DCFR.

Population density estimates in the literature are rare, but most gryllacridids are thought to occur at low densities (Lockwood and Rentz, 1996). The density of *Bothriogryllacris turris* Rentz was estimated at 500 to 800 inhabited burrows per hectare (Morton and Rentz, 1983). Active searches over three field seasons at Cape Naturaliste in southwest Western Australia, 33°54′S, 115°02′E, were used to calculate the density of a population of *C.* sp. 21. The density of *C.* sp. 21 was estimated at 360 animals ha^{-1}. Yearly estimates of population density were: 1996/97: 333 animals ha^{-1}, 1997/98: 380, and 1998/99: 367. Although this density is lower than that of *B. turris*, because density was estimated by counting active individuals rather than nest sites, 360 animals ha^{-1} is likely to be an underestimate of these silent, highly cryptic, nocturnal orthopterans.

Habitat: Macro- and Microhabitat as Illustrated by the Australian Fauna

Many species of gryllacridid appear to be highly localized and very specific in their habitat (Rentz, 1996). It may be added that there seems to be no defining feature of the macrohabitat that either ensures or precludes the presence of these animals.

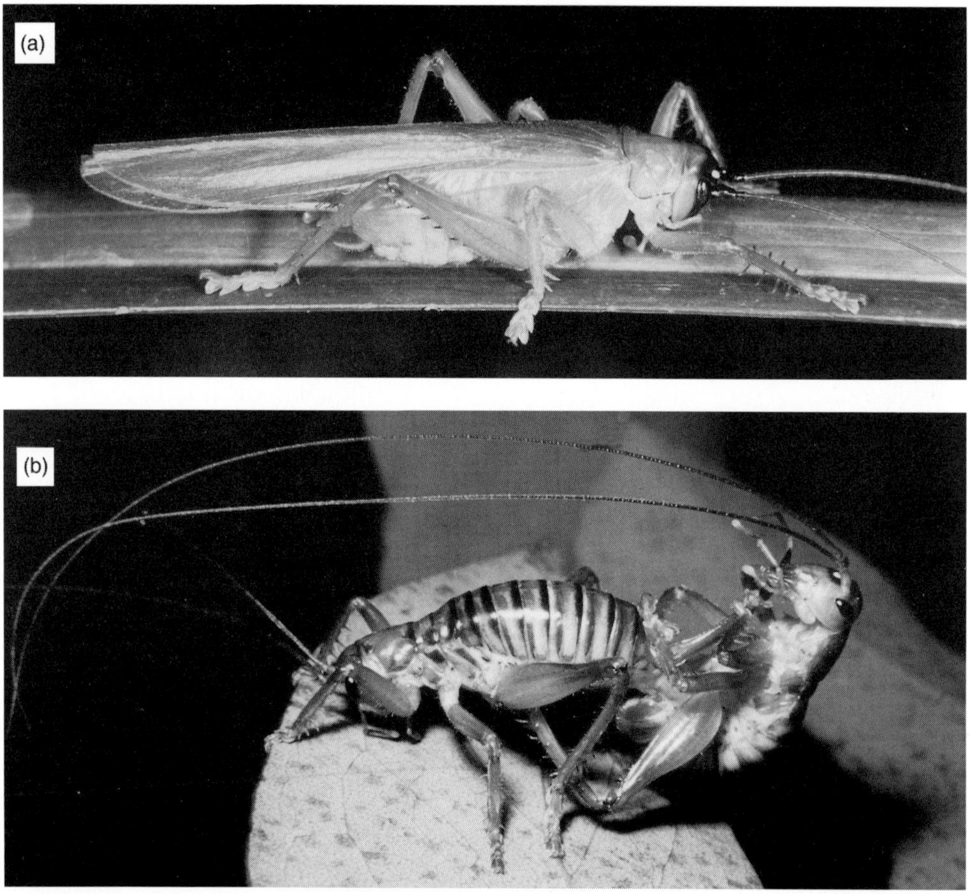

Fig. 6.2. Typical examples of Gryllacrididae: a fully winged species and a small, flightless species, both from Australia. (a) *Nunkeria brochis* Rentz; (b) *Arrolla tibialis* Rentz, mating pair.

Generally, knowledge of the habitats that these orthopterans occupy is scant, although see Rentz and John (1990) for good notes on the species described therein. The extensive range of habitats occupied by gryllacridids can be illustrated with the following examples. *Apotrechus* Brunner von Wattenwyl is a genus that appears to occupy wet and/or coastal forests of south-east Australia, whereas *Cooraboorama* Rentz occurs in grasslands and are known from a very small number of localities in and peripheral to the Australian Capital Territory (ACT). The genera *Pararemus* Ander, *Ametrus* and *Hadrogryllacris* Karny all have representatives that can be found in arid, central Australia in contrast to *Australogryllacris* Karny which is thought to be restricted to rainforest habitats (Tepper, 1892; Rentz, 1996). *Xanthogryllacris* Karny and *Hyalogryllacris* Karny are both widespread genera, occurring in a variety of forest habitats, ranging from rainforest to dry savannah scrub, with species of the former occurring in Papua New Guinea and Indonesia as well as Australia (Rentz and John, 1990; Rentz, 1996). Gryllacridids appear to be active early in succession of fire-swept habitats (see Hansen, 1986; Rentz and John, 1990), although this is likely to be a result of their choice of microhabitat rather than any defining feature of their larger environment. In brief, gryllacridids occur throughout Australia, ranging from the cold temperate island of Tasmania to the warm tropics of far-north Queensland (Rentz and John, 1990; Rentz, 1996, 1997).

Gryllacridids within the same genera can differ widely in the macrohabitats they occupy. This can be demonstrated by examining the environments in which some members of the widespread genus

Arrolla Rentz are found. *Arrolla rotamah* Rentz has been found living in *Acacia* Miller, *Banksia* L. f., broomheath and bracken scrub on coastal dunes in south-eastern Australia. *Arrolla turramurrae* Rentz occupies wet forest habitats, *A. platystyla* Rentz is found in dry sclerophyllous forest of *Eucalyptus rossii* R. Baker and H.G. Smith, *E. mannifera* Mudie and *Acacia*. *Arrolla weiri* Rentz occurs in mixed forest of *Acacia* and *Lomandra* Labill. while *A. lewisi* Rentz is found in the subalpine Kosciusko Range (Rentz and John, 1990; Rentz, 1996).

The microhabitats occupied by gryllacridids seem as diverse and little known as their macro-environment. The common name, formerly used to describe the gryllacridids, emphasized their ability to tie leaves together with silk. However, it is now thought that not all species do this, and the term 'leaf-rolling crickets' may be a misnomer (Rentz, 1996). Although certain species do construct shelters by tying leaves or other materials together, many others spend the daytime encased in burrows or exhibit no 'nest-building' behaviour at all (Tepper, 1892; Morton and Rentz, 1983; Rentz and John, 1990; Rentz, 1996).

Traditional 'leaf-rolling' gryllacridids such as *Apotrechus* species live in leaf litter of the forest understorey where they bind dry leaves, twigs and debris together to form a shelter (Rentz and John, 1990). Such a 'silk-spinning' habit is also displayed by other species of Gryllacrididae such as the 'Pretty raspy cricket', *Xanthogryllacris punctipennis* Walker, which actually cuts leaves into the desired shape before binding them together with silk (R.J. Hale, personal observation). All known stages of 'silk-spinning' species, from nymph to adult, seem capable of producing silk (Rentz, 1996, 1997; Hale, 2001). Those species that do produce silk are thought to secrete it from specialized 'silk glands' near the insect's mouth. The silk is manipulated and attached to surfaces by the long, modified maxillary and labial palps. The labial palps gather the silk strands from the mouth and feed them to the maxillary palps which distribute the silk and fasten it to the various surfaces (Bland and Rentz, 1991b).

A coastal species of gryllacridid, *Arrolla rotamah*, does not bind leaves together but instead lives in subterranean burrows of its own construction within loose sand. Such behaviour is similar to some species of Anastostomatidae (e.g. *Hemiandrus maculifrons* (Johns, 1997) = *Zealandosandrus gracilis* Salmon) which construct nests for themselves, usually as burrows in the ground (Carey, 1971, and Wahid, 1978 – cited in Field and Sandlant, 1983), sometimes with the soil excavated from the burrow distributed in a neat circle around the hole (Tepper, 1895). One of the few West Australian members of the Anastostomatidae, *Hypocophoides lepismoides* Karny (= *Onasandrus* Walker in literature prior to 1997) also appears to display a fossorial, burrow-digging habit (Rentz, 1996; R.J. Hale, personal observation) similar to that of another Australian species, *Anostostoma* Gray, 1837 (= *Australostoma* Karny) *opacum* Brunner 1888 (John and Rentz, 1987; Johns 1997). Henicine stenopelmatids like *Cnemotettix* Caudell are remarkably similar to gryllacridids in both appearance and biology, for example possessing a femoro-abdominal stridulatory apparatus like that of gryllacridids. These crickets occur in coastal central and southern California and adjacent Baja California, Mexico. Rentz and Weissman (1973) related the silk-spinning habit and their burrow construction. The soft-bodied, wingless crickets live in burrows of their own construction, emerging after dark to feed and reproduce. Silk is spun through the mouthparts, exactly as it is in gryllacridids. It is used in a similar fashion both to reinforce loose, sandy burrows and to form tight shelters in which the crickets can remain inactive for days at a time.

Perhaps the most spectacular burrow-building gryllacridid is the Australian arid zone *Bothriogryllacris turris*. This species also builds a silk-lined burrow in the ground, but raises the top of the burrow slightly from ground level. Reasons for this are speculative but may be to reduce the potential for flooding during thunderstorms. Furthermore, the individual seals the burrow with a raised cap or 'trapdoor' composed of sand and soil particles bound together by silk produced by the insect (Morton and Rentz, 1983). Thus, once inside the burrow the individual is protected not only from thunderstorms, but also potential enemies and the heat of the day. When the insects emerge after dark, the door is left open until they return. Another species, *B. brevicauda*, constructs a similar burrow but seals its entrance over with a pebble fixed in place with silk (Morton and Rentz, 1983).

Some species appear to construct nests but are not subterranean in habit. Instead they occupy hollows in living trees (e.g. *Mooracra canobolas* Rentz and *Chauliogryllacris grahami* Rentz) or dead branches (e.g. *Pteropotrechus* spp., *Craspedogryllacris* spp.), using their mandibles to enlarge the existing refuges (Rentz and John, 1990; Rentz, 1996). Making use of and modifying existing refuges in this manner is a behaviour similar to certain New Zealand wetas (*Hemideina crassidens* Blanchard and *H. femorata* Hutton) which occupy chambers within Manuka (*Leptospermum scoparium*) or mixed broadleaf trees. These chambers were initially formed and then vacated by the larvae of cerambycid or curculionid beetles (Field and Sandlant, 1983; Field and Sandlant, Chapter 13, this volume). A number of other gryllacridid species (e.g. *Arrolla turramurrae*, *Paragryllacris combusta, Wirritina brevipes* Ander) have also been observed occupying readily available refuges in various habitats.

Finally, there are gryllacridids (e.g. *Cooraboorama canberrae* Rentz, *Nullanullia maitlia* Rentz) that show no evidence of either burrow excavation or constructing a silk-lined shelter (Rentz and John, 1990). These individuals may be similar to the New Zealand weta, *Deinacrida connectens* Ander, which simply crowds into refuges beneath rocks during the daytime (Field, 1980). Despite these few exceptional species, it would appear from the reports in the literature, either anecdotal (Tepper, 1892; Tillyard, 1926; Bland and Rentz, 1991b; Rentz, 1991) or direct (Morton and Rentz, 1983; Rentz and John, 1990; Rentz, 1996) that nests or burrows are an integral aspect of gryllacridid biology.

The nesting behaviours of *C.* sp. 21 and Gen. nov. 9 sp. 4 were examined to determine whether there was a general preference for individuals to build nests in association with particular vegetation types. The Cape Naturaliste study site is characteristic of low coastal heath of the south-west of Western Australia. Vegetation at this site was surveyed and quantified by sampling a number of 100 m transects and counting the major species present. Ground cover was sparse although patches of leaf litter were present. The heath rarely exceeded 1 m in height, with the major shrub species present being *Spyridium globulosum* Labill. Benth, *Olearia axillaris* DC and *Hakea* spp. Trees were present only in small patches and were dominated by *Eucalyptus gomphocephala* DC, *Acacia cyclops* and *Agonis flexuosa* (Wild.) Sweet. There was no association between the nest sites of *C.* sp. 21 and Gen. nov. 9 sp. 4 and the nearest species of plant to where a nest was located (χ^2 = 2.43, d.f. = 4, P = 0.6567). This lack of association between nest sites and any particular plant species can be explained by the nesting behaviour of the two species. *C.* sp. 21 occupy already existing refuges, and nests are usually located at the base of shrubs, under old bark of tree-trunks or on non-living detritus such as fallen logs, sticks, bark or leaf litter (Table 6.1). This is similar to certain species of stenopelmatids and other gryllacridids which modify already existing natural refuges (Field and Sandlant, 1983; Rentz and John, 1990; Field and Sandlant, Chapter 13, this volume). It would appear that these gryllacridids are therefore not constrained to associate with a resource which may be provided only by a few particular species. Only grass trees (*Xanthorrhoea* spp.) are sometimes found to have nests present within their fronds but the nests are usually located underneath the uppermost green fronds in the denser, dead fronds of the underside of the shrub. Gen. nov. 9 sp. 4 constructs subterranean burrows near the base or within the shelter of shrubs, but there is no preference for any particular plant species (Table 6.1). Like *Bothriogryllacris* spp., the burrows have a single 'trapdoor' entrance, but the burrows extend horizontally rather than vertically under the soil surface (see Morton and Rentz, 1983). The burrows are only slightly longer than the occupying insect. The individual enters the nest head first before turning within the silk-lined burrow and sealing the trapdoor closed after it. The trapdoor is left unsealed but closed while the animal is outside the burrow. The single example of leaf-binding displayed in the field (Table 6.1) was an early instar encased between *Agonis flexuosa* leaves.

Both *C.* sp. 21 and Gen. nov. 9 sp. 4 line the shelter or burrow with silk. Although *C.* sp. 21 tend to occupy existing refuges they also seal the chambers with silk but, rather than a 'trapdoor' as constructed by *Bothriogryllacris turris* or Gen. nov. 9 sp. 4 , the entrance is sealed by simply interlacing silk across the opening.

Both sexes of *C.* sp. 21 and Gen. nov. 9 sp. 4 were found to occupy nests and there was no difference in the incidence of males or females found

Table 6.1. Number of Gen. 9 sp. 4 and *C. sp. 21* found to use each of four types of refuge.

Nest used	Gen. 9 sp. 4	*C. sp. 21*
No nest	0	2
Burrows	23	0
Existing refuges	0	16
Silk-bound leaves	1	1

in nests ($\chi^2 = 0.546$, d.f. = 1, $P = 0.4601$). This observation reinforces remarks in the literature (e.g. Rentz and John, 1990; Lockwood and Rentz, 1996) indicating that both sexes of gryllacridid use nests. Both male and female stenopelmatids use galleries as refuges, however, males defend certain galleries during the night and attempt to establish harems of females within their nest-sites (Field and Sandlant, 1983; see also Field and Sandlant, Chapter 13, this volume).

A few studies have shown that once gryllacridids or stenopelmatids have constructed or modified a nest, the individual appears to display a certain amount of fidelity to that nest (Field, 1980; Field and Sandlant, 1983; Field and Sandlant, Chapter 13, this volume; Rentz and Lockwood, 1996; Hale, 2001). This raises the question of whether *C.* sp. 21 and Gen. nov. 9 sp. 4 individuals use the one nest permanently and if so when do they construct their first nest? If gryllacridids construct a nest at an early age they may continue to use that nest for as long as they can fit in it. Such reoccupation of a nest site by juvenile gryllacridids may indicate the ability of the juvenile to locate its nest after nightly foraging and is discussed at the conclusion of this chapter.

Feeding Biology

Some of the earliest observations established that gryllacridids are carnivorous (Tepper, 1892), and certain American gryllacridids are known to be generalist predators, which feed on caterpillars and other insects (Shaw *et al.*, 1987). It is now established that gryllacridids include carnivorous, herbivorous, florivorous, granivorous and opportunistically omnivorous species, with considerable specificity in food preferences having been noted (Rentz, 1996, 1997). For example, *Arrolla rotamah*, *A. weiri*, *A. tibialis* Rentz and *A. longicauda* Rentz, have been recorded feeding on floral parts of *Acacia*, *Eucalyptus* and introduced composites, but specimens also contained insect parts in their gut contents (Rentz and John, 1990). *Chauliogryllacris grahami* has been observed feeding on floral parts of roadside weed vegetation as has *Nunkeria brochis* Rentz. *Hyalogryllacris* spp. have been observed feeding on floral parts of native bushes and trees (Rentz and John, 1990).

One of the few studies which detailed the diet of a particular gryllacridid concluded that, from the large number of seeds found in the foregut, *Bothriogryllacris turris* is primarily granivorous, with a very high proportion of the seeds eaten being from ephemeral grasses (Morton and Rentz, 1983). Some insect fragments were also identified, with harvester ants *Chelaner* spp., *Pheidole* spp. appearing to be the main prey. Individuals examined in winter were also found to have seeds in their foreguts indicating that, at least this particular species of gryllacridid, feeds to some extent throughout the colder months (Morton and Rentz, 1983).

A study of proventricular structure of seventeen species of gryllacridid suggested that those species with long lateral denticles tended to be predominantly plant feeders rather than carnivores (Bland and Rentz, 1991a). In contrast, specimens with a proventriculus with very short lateral denticles usually had moderate to large amounts of insect fragments inside. Thus it was suggested that the species *Paragryllacris combusta*, *Arrolla lewisi* Rentz, *Ametrus tibialis* Brunner, *Ametrus* sp. 5, *Nullanullia maitlia* and *Hadrogryllacris* sp. 9 are mainly carnivorous in habit (Bland and Rentz, 1991a). Such a finding seems to agree with the biological information that is available, e.g. *Nullanullia maitlia* is thought to be a primarily predaceous species which actively searches vegetation throughout the night (Rentz and John, 1990). However, due to the lack of information on their feeding habits it is not possible to differentiate gryllacridids into strictly carnivorous or herbivorous species. *Arrolla lewisi*, mentioned above as possibly being carnivorous has also been observed feeding on the foliage and fruits of *Bossiaea foliolosa* Cunn. (Rentz, 1996), suggesting that an omnivorous habit is more likely. In the laboratory, all known species are easily reared on a diet of the usual Orthopteran Food Mix (Rentz, 1996) on which normal adults are produced. Finally, there is evidence that some gryllacridids are opportunistically necrophagic, with species of *Ametrus* having been observed feeding on conspecifics squashed on the roadside (Rentz, 1996).

Field observations by R.J. Hale on *C.* sp. 21 and Gen. nov 9 sp. 4 also suggests an omnivorous diet for these gryllacridids. Gen. nov. 9 sp. 4 feed on other insects such as caterpillars, cicadas and katydids (e.g. *Tympanophora* spp., *Kawanaphila* spp.) and often forage around sites that are popular with other insects such as the inflorescences on grass tree (*Xanthorrhoea* spp.) flower spikes. *C.* sp. 21 occasionally eat small snails (*Theba pisana*

Müller) but are predominantly herbivorous, feeding on flowers of native rosemary (*Olearia axillaris*) and peppermint myrtle (*Agonis flexuosa*). Both species feed on the floral parts of shrubs that are in flower. At the start of the summer season, these are predominantly species of *Hibbertia* Andrews and *Melaleuca acerosa* Colla G. Don. Midway through the season they feed on *Hakea* and *Agonis* DC Sweet, and by the end of the season both species feed on the flowers of *Eucalyptus gomphocephala*. The more omnivorous behaviour of Gen. nov. 9 sp. 4 appears similar to that of the genus *Ametrus*, which have mandibles adapted for an omnivorous feeding habit (Bland and Rentz, 1991a; Rentz, 1996). The feeding habits of *C.* sp. 21 appear similar to some other long-winged, species such as *Chauliogryllacris grahami* and *Nunkeria brochis* (Rentz, 1991).

Communication and Mating

Like the stenopelmatids (Davis and Smith, 1926; Field and Sandlant, 1983), gryllacridids appear capable of communicating through substrate vibrations produced by drumming either their tarsi or abdomen on the substrate, causing waves to propagate outwards from the emitter (Rentz and John, 1990; Field and Bailey, 1997). These vibrational waves travel through the substrate, such as wood, leaves, soil or sand on which the animal stands. The insects can detect such vibrational waves via tibial subgenual organs (McVean and Field, 1996). The patterns of drumming carried by the vibrational waves appear to be species-specific and also convey sexual information (Rentz, 1996; Field and Bailey, 1997). A recent examination of the substrate vibration communication systems of two species of Australian gryllacridid (*Hadrogryllacris* sp. unknown and *Ametrus* sp. unknown) revealed that the two sexes duet, with males initiating substrate drumming and the females responding. Such a duetting system may be a mechanism that aids sexually active individuals to locate each other in complex environments (Field and Bailey, 1997). In the field, male *C.* sp. 21 were observed to call from 'perches' such as grass-tree spikes (*Xanthorrhoea* spp.) or thin twigs of shrubs and trees. Males drum on the same twigs used by females when feeding on flowers. Calling males presumably utilize such concentrations of food as a resource area, as in other ensiferans (Brown and Gwynne, 1997). Drumming males were located in close proximity (64 ± 18 cm) to females, as expected because vibrational signals can be transmitted over only a relatively short range (Fuchs, 1976; Markl, 1983; McVean and Field, 1996). Males in the field were observed drumming only on thin twigs and stems, substrates most efficient for transmission of vibrational signals (Michelsen *et al.*, 1982; Stewart and Ziegler, 1984), supporting the conclusions of Rentz (1996) and Field and Bailey (1997) that vibrational communication is used by these Ensifera.

Recent observations on the gryllacridids have established that all known members of this family appear to produce acoustic sound. This has led to the new common name of 'raspy crickets' being proposed for the family (Rentz, 1996). The raspy airborne sound is produced as a defensive display to deter potential predators (Rentz, 1996; Field and Bailey, 1997; R.J. Hale, personal observation) and is not thought to be used in any intraspecific communication. Gryllacridids would appear from their auditory anatomy to have poor sensitivity to airborne sound (Kevan, 1982; Rentz and John, 1990; Rentz, 1991; 1996). Although there have been no physiological studies on raspy crickets, examinations of the auditory systems of wetas (Field *et al.*, 1980; Ball and Field, 1981) indicates that gryllacridids may still be able to detect airborne sound. Stenopelmatids are known to have tibial tympanal organs on one or both sides of the foreleg (Kevan, 1982; Field and Sandlant, 1983) which may explain the greater proportion of taxa showing intraspecific acoustic calling in this family.

Mating has been observed in a number of gryllacridid species in the field. Copulation generally occurs on leaves, twigs or other plant material, with the male curled behind the female in a manner characteristic of this superfamily (Fig. 6.2b) (Rentz and John, 1990; Brown and Gwynne, 1997). Mating lasts for approximately one hour and, when the pairs separate, the female bears a large spermatophylax as would be expected from the long copulation time (Brown and Gwynne, 1997; see Rentz and John, 1990 for specific observations of copulation). Additionally, many gryllacridids appear to be facultatively parthenogenetic with unmated females laying eggs which give rise to progeny that are solely female (Rentz, 1997). Females usually lay eggs directly into the soil but eggs have also been discovered within the burrows of some female gryllacridids. This suggests that there may be some overlap of generations (Rentz and John, 1990). Such phenomena are similar to those found in subsocial burrowing crickets such as *Anurogryllus muticus* (De Geer) (West and Alexander, 1963) and stenopelmatids, which sometimes allow young nymphs within their galleries (Field and Sandlant, 1983; Brown and Gwynne, 1997). However, communal nesting is unlikely to occur within the Gryllacrididae due to their potential carnivorous and possibly cannabalistic feeding habits.

Mating occurs as early as October in both *C.* sp. 21 and Gen. nov. 9 sp. 4 at Cape Naturaliste. The latest recorded mating occurred in early February (for Gen. nov. 9 sp. 4). Copulatory behaviour observed in the field appeared to be exceptionally long-lasting (3.0 ± 1.53 h for Gen. nov. 9 sp. 4, n = 2; 1.55 ± 0.32 h for *C.* sp. 21, n = 3) although the low number of observations does not give a very precise estimate. The copulatory position taken up by mating pairs of these species was typical of the 'male-curled-behind-female' or 'female-above, end-to-end' position found throughout the Stenopelmatoidea, apart from the Rhaphidophoridae (Alexander and Otte, 1967; Kevan, 1982; Brown and Gwynne, 1997). Males maintained such a position by grasping the female's ovipositor with their fore- and sometimes middle-legs (as in Fig. 6.2b). The long duration of mating in these two species may reflect the massive spermatophylax that is passed from male to female and as mentioned above seems characteristic of gryllacridid mating behaviour. Females were often observed to retreat to their burrow with the whole spermatophylax attached and undisturbed. Such an action may be in order to consume the spermatophore within the safety of the nest but may also be a simple response to the imminent onset of daylight. Unlike many weta species, e.g. *Hemideina femorata* (Field and Sandlant, 1983), neither Gen. nov. 9 sp. 4 nor *C.* sp. 21 have been observed copulating within or near to their nest entrance, but seem to mate within vegetation like other gryllacridids (Rentz and John, 1990; R.J. Hale, personal observation).

Females of both Gen. nov. 9 sp. 4 and *C.* sp. 21 oviposit from late November to February. The eggs are deposited by the long needle-like ovipositor, approximately 3 cm into the ground. Eggs were never recovered from within the burrow or nest of these two species, which contrasts with some other species of Gryllacrididae (*Arrolla rotamah* – Rentz and John, 1990; *Pararemus* – Rentz, 1996) and Stenopelmatidae (*Hemiandrus maculifrons* Walker = *Zealandosandrus gracilis* Salmon; *Hemiandrus* sp. unknown – Carey, 1971 and Wahid, 1978 – cited in Field and Sandlant, 1983)). Females of both Gen. nov. 9 sp. 4 and *C.* sp. 21 were observed most often to lay their eggs in soft, penetrable soil or dune sand. Soil is thought to be the ancestral and still the most common oviposition substrate used by orthopterans (Key, 1959; Stauffer and Whitman, 1997). Oviposition behaviour of Gen. nov. 9 sp. 4 and *C.* sp. 21 is similar to that of another gryllacridid, *Bothriogryllacris pinguipes* Rentz, which has been seen to lay its eggs randomly in the gravel-sand bottom of dry stream beds (Morton and Rentz, 1983).

Up to 80 eggs are deposited by female Gen. nov. 9 sp. 4 in a single night (mean 39.40 ± 12.56, n = 36) and females of both species have been observed to oviposit on more than one night. The mean number of eggs laid is fairly consistent with the very few quantitative notes on egg numbers in the literature. Tepper (1895) found 47 and 24 eggs in two females of the species *Pararemus callabonnensis* Karny (= *Eonius callabonnensis* Tepper) and Rentz and John (1990) counted a clutch of around 45 eggs laid by *Arrolla rotamah*. Eggs of Gen. nov. 9 sp. 4 and *C.* sp. 21 hatch during late February and March in the field, when temperatures are generally very warm (25.6 and 24.2°C average daily maximum, respectively). In the laboratory eggs hatched when kept at a constantly high minimum temperature and after complete exposure (soaking) to water. That these two factors are likely to be important for eclosion agrees with knowledge of egg development in the Tettigoniidae, which is temperature and water dependent (Hartley, 1990), and the cave cricket *Hadenoecus subterraneus* Scudder (Rhaphidophoridae) where egg survival is closely correlated to substrate moisture (Griffith, 1992).

First instars of Gen. nov. 9 sp. 4 and *C.* sp. 21 are small but moult frequently during the ensuing months so that by May they have increased four-fold in size. Juveniles overwinter and reach maturity by the following summer. Adults were found during the late spring and summer months at Cape Naturaliste with peak numbers recorded in January for Gen. nov. 9 sp. 4, and November for *C.* sp. 21 (Fig. 6.3). The lifespan of these species may be shorter than that of several species of stenopelmatid which are known to live for several years. Field observations of *C.* sp. 21 suggest that the life span of this species may be only a single year or slightly longer. Mature animals observed towards the end of the season (March/April) are often infested with large numbers of a small mites on the animal's joints near the thorax and under the wingfolds (Fig. 6.4). Not only the average number of mites but also the number of individuals with infestations was much lower (only 21 cases of infestation out of 137 individuals examined in October–December 1996/97) at the start of the season than by the end (Fig. 6.4). Mite infestation is known to occur also in the wetas of New Zealand (Sharell, 1971) and other ensiferans (Tindale, 1928). No adult gryllacridids of either species were found in the field from May through to September. This life history pattern is similar to a Queensland gryllacridid *Bothriogryllacris turris*, of which both adults and nymphs were found in October, but by January all individuals observed were adult. Gravid females were found from October through to February, and although the precise period of hatching was not determined, only young nymphs were found from June to August (Morton and Rentz, 1983).

Nesting Behaviour of Gryllacridids: Site Recognition and Fidelity

Although biological information on the Gryllacrididae is scarce, the nest-building behaviour of these orthopterans has been noted or at least suggested, in nearly every relevant publication (Tepper, 1892; Tillyard, 1926; Kevan, 1982; Morton and Rentz, 1983; Rentz and John, 1990; Lockwood and Rentz, 1996; Rentz, 1996). This disproportionate reporting suggests that nest building is an important component of gryllacridid biology. Nests are used by animals for a variety of reasons, but the most likely explanations for why gryllacridids occupy burrows or other shelters are predator avoidance or shelter from the environment. The arid-inhabiting gryllacridid *Bothriogryllacris turris* Rentz is thought to construct its burrows to avoid predators or to conserve moisture within its arid environment (Morton and Rentz, 1983). In support of the latter idea, there is known to be a link between the environmental temperature and the depth to which some insects dig their burrows (Weaving, 1989). Furthermore, burrow-dwelling insects are known to vary their foraging activity in relation to the prevailing environmental conditions, adopting a more sedentary habit when conditions are not favourable. This allows them to retreat to their nearby burrow at dawn and hence avoid desiccation (Rasa, 1994). Such a strategy may explain burrow construction in arid-dwelling gryllacridids, where suitable natural shelter may be scarce.

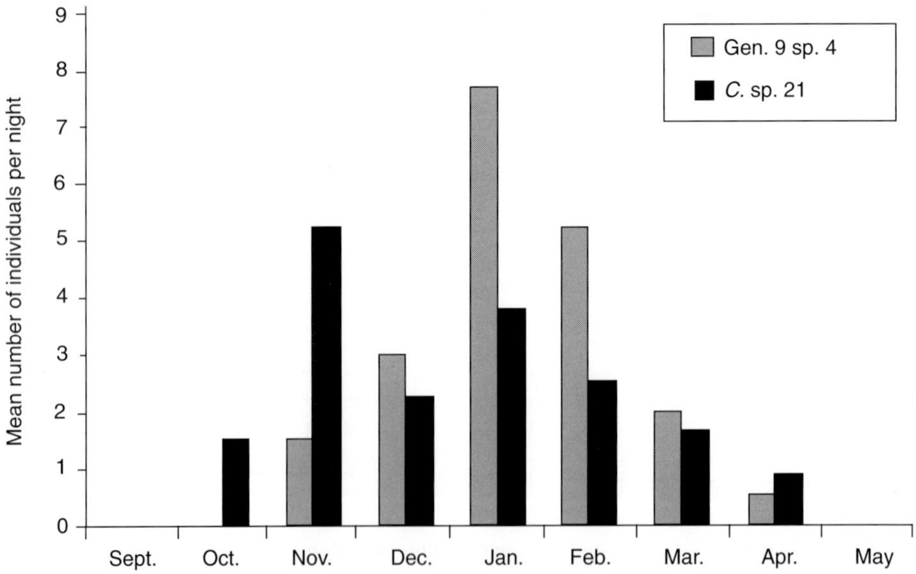

Fig. 6.3. Numbers of Gen. nov. 9 sp. 4 and *C.* sp. 21 sighted over four seasons (1995, 1996, 1997, 1998) at Cape Naturaliste, Western Australia. These values include juveniles as well as adults. The actual numbers of both species found in a month was standardized by dividing the number of individuals caught by the number of nights spent searching for them.

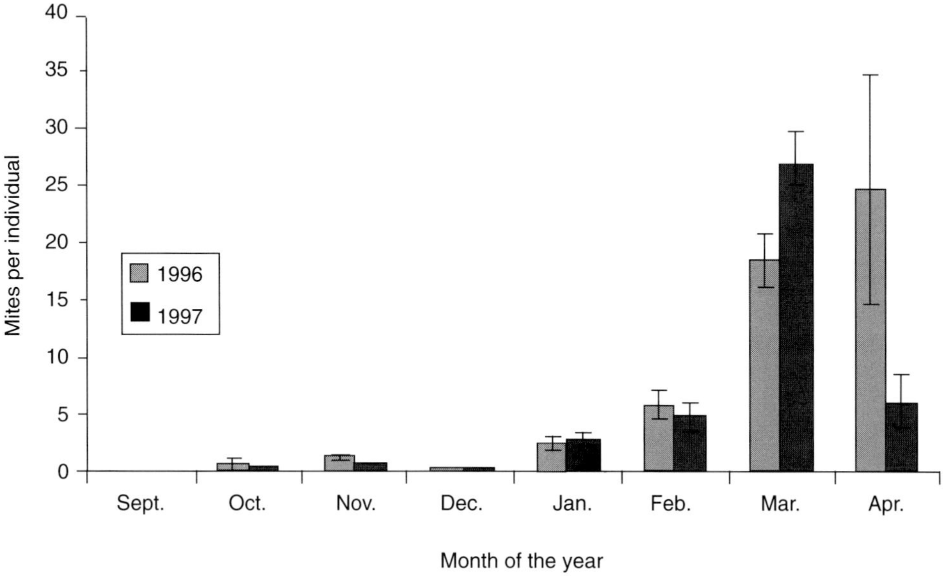

Fig. 6.4. Average number of mites per individual infesting *C.* sp. 21 at Cape Naturaliste. Error bars are the standard errors of the mean. Total number of individuals checked for mites during the entire season was 79 for 1996 and 116 in 1997.

However, nesting behaviour is not displayed only by those species that dwell in Australia's arid interior. The construction of nests by temperate species may not be only to create a favourable microclimate, but also to provide a retreat from predators. There have been no studies that have specifically looked at the predators of the Gryllacrididae although anecdotal references in the literature, combined with similar observations on stenopelmatids, indicate that there are a variety of potential vertebrate and invertebrate predators. A study in North America found that the Jerusalem cricket *Stenopelmatus fuscus* Haldeman was one of the primary prey items taken by burrowing owls *Athene cunicularia* (Gleason and Johnson, 1985). Tuataras, *Sphenodon punctatus*, are known to be important predators on the New Zealand tree weta *Hemideina crassicruris* (Moller, 1985). In Australia, potential vertebrate predators may include similar reptiles such as varanid lizards, as well as birds and insectivorous mammals. One such mammal, Leadbeater's possum *Gymnobelidus leadbeateri* McCoy, is known to prey on the forest raspy cricket *Apotrechus unicolor* Brunner (Rentz, 1996). Known invertebrate predators of gryllacridids are lampshade spiders *Hypochilus sheari* and *H. coylei* (Huff and Coyle, 1992), megalomorph spiders (*Migas* spp.) (Field and Sandlant, Chapter 13, this volume), scorpions (R.J. Hale, personal observation), and the digger wasps *Sphex ichneumoneus* Linnaeus and *S. vestitus* Smith (Brockmann, 1985; Rentz and John, 1990).

Construction of a burrow or nest, whether it be to provide refuge from predators or the environment, is likely to be a costly investment and building a new nest after nightly foraging would require both time and energy. The cost of inhabiting a nest would be markedly decreased if an individual displayed site fidelity and reused the same nest. In order to reuse its nest, an individual gryllacridid would need to be capable of recognizing its own nest site, and it was recently demonstrated in the laboratory that certain species of gryllacridid can recognize their own nest from those of their conspecifics (Lockwood and Rentz, 1996; Hale, 2001). The ability of *Hyalogryllacris* sp. 14 to recognize its nest was shown to be mediated by an 'autorecognition' chemical compound, or combination of compounds, which are associated with the nest (Lockwood and Rentz, 1996). The compound was shown to be volatile and may possibly aid the insect in locating its nest after nightly foraging (Lockwood and Rentz, 1996). Despite the consequent failure in the laboratory to demonstrate long-distance homing behaviour based on chemical cues, it is clear that gryllacridids show long-term nest-site fidelity in the field (Hale, 2001; unpublished data).

The two species of gryllacridid studied at Cape Naturaliste emerged from their nests approximately 1–1.5 hours after standardized sunset (ϕ = 20.54, r = 0.99, P < 0.01). There was no difference in the mean emergence times of the two species (Watson's F-test: F = 1.39, P = 0.25, d.f. = 32). The lag from sunset to emergence seems at first glance to be excessively long (Fig. 6.5a). Sunset is defined as the instant in the evening under ideal meteorological conditions when the upper edge of the sun's disc is coincident with an ideal horizon (Anon., 1999). From this definition it becomes clear that it is not necessarily dark at this time, in fact it is not until the end of evening nautical twilight, when the centre of the sun is at a depression angle of 12° below an ideal horizon, that it becomes dark for normal practical purposes (Anon., 1999). Nautical twilight at this location ends at 20.35 for a standardized 19.30 sunset (Anon., 1999). Thus, when compared to the emergence times of the two species of gryllacridid studied, it can be seen that both species emerge shortly after the onset of full darkness. This result confirms reports that gryllacridids appear to be nocturnal (Rentz, 1996, 1997) and establishes that at least these two species appear fully nocturnal and display little crepuscular activity.

Once out from their nests, Gen. nov. 9 sp. 4 spent the majority of the night feeding and moving around their habitat (Fig. 6.6a). Movements over short distances within flowering shrubs on which the insect was feeding were categorized as feeding behaviour and included as 'foraging' rather than 'moving' (Fig. 6.6a). The movement category only contained movements over greater 0.5 m or those that did not coincide with foraging/feeding behaviour. That these gryllacridids spend over one-third of the night travelling around their immediate environment reinforces the fact that they are highly active, mobile insects (Rentz and John, 1990).

'Grooming' was characteristically performed when the individual emerged from its nest. Like all gryllacridids, Gen. nov. 9 sp. 4 have extremely long antennae which they run through their maxilla, using their palps to clean. The second most prominent form of grooming involves them cleaning their tarsi in a similar manner. 'Other' activities (Fig. 6.6a) undertaken during the night were scored when individuals remained stationary without feeding or grooming. It also included encounters with conspecifics which were relatively rare and always ended in immediate withdrawal by both parties, unless one was a courting male. If mating ensued, it took up a significant proportion of the insect's night (for example, one particular pair of Gen. nov. 9 sp. 4 spent 40% of their night coupled together).

A large proportion of nights are spent foraging and feeding although an activity which had a major impact on females' time budget was oviposition. Gravid females spent about one-third of their night laying eggs at the expense of both travelling and feeding (Fig. 6.6b). The amount of time spent grooming by gravid females also increased and reflects the intense cleaning of the ovipositor undertaken between bouts of oviposition.

Both species return to their nest or shelter (Table 6.1) at 04.17 ± 00.06 h (Fig. 6.5b). There was no difference in the mean return times for Gen. nov. 9 sp. 4 and *C.* sp. 21 (Watson's F-test: F = 1.56, P = 0.22, d.f. = 23). Upon reaching the nest-site individuals commonly spend some time examining the area immediately around the nest. Antennal activity is high and the insect often palpates the burrow entrance or the vegetation immediately around the burrow. Two incidental observations in the field support the idea of these insects making use of a chemical recognition compound (see Lockwood and Rentz, 1996; Hale, 2001). Firstly, individuals could often be seen inspecting and palpating the flagging tape that was used to mark the positions of individual nest-sites. Due to the difficulty in detecting nest entrances, it was necessary to place the tape as close as possible to the entrance and was it consequently often encountered by the returning gryllacridid. Secondly, a female gryllacridid was seen to pass close by an unoccupied nest entrance that was not her own. The individual spent some time examining the area around the nest and went so far as to lift the trapdoor and insert her head. However, she withdrew in less than 30 s without fully entering and moved away from the foreign nest, eventually returning to her own burrow that was over 1 m away.

The time the two species of gryllacridid returned to their nest was standardized to an 05.30 sunrise. Thus, they return to the nest 1.25 hours prior to sunrise. Although again, this may seem to unnecessarily shorten the insects' active period, it emphasizes the non-crepuscular habit of these

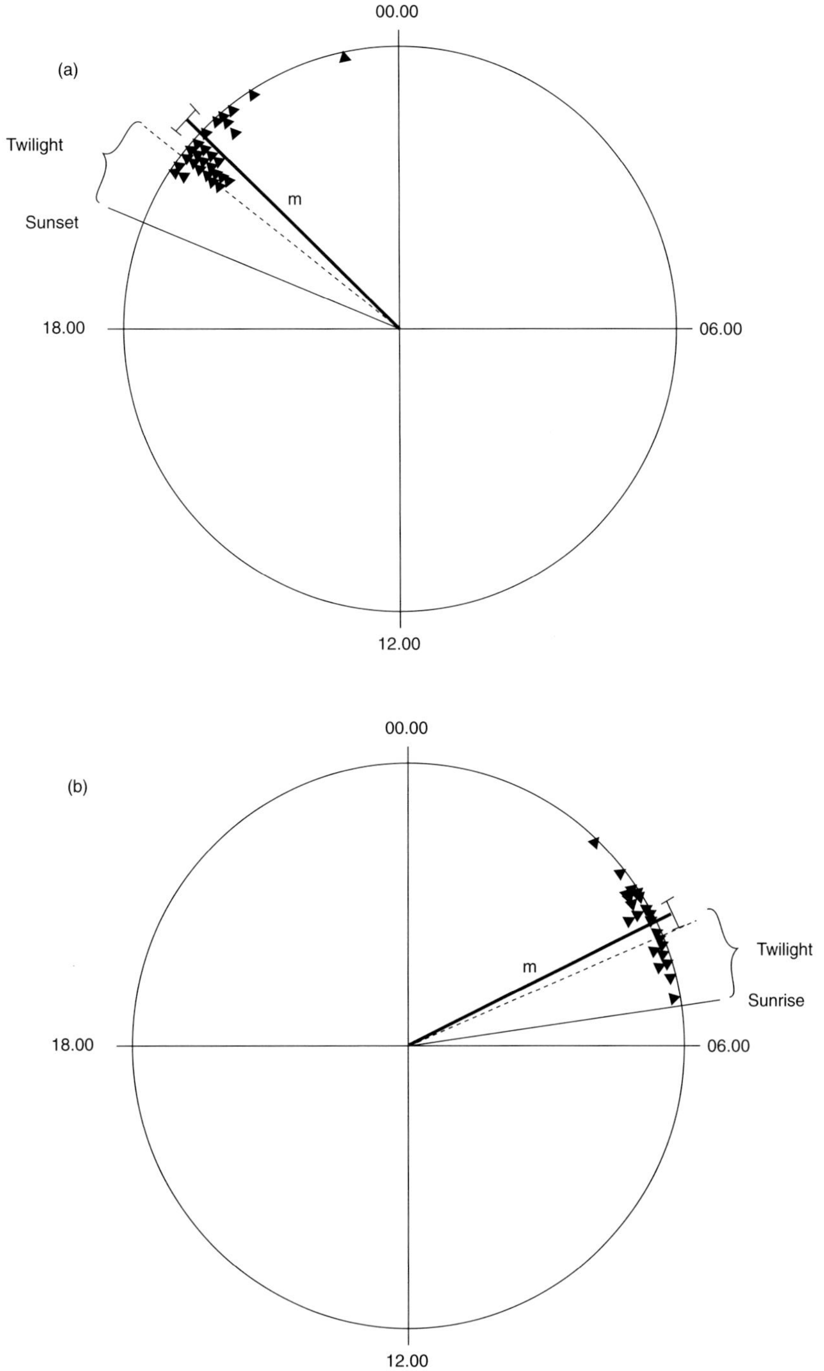

Fig. 6.5. (a) Emergence (*n* = 33) and (b) return (*n* = 25) times for adult gryllacridids (both Gen. nov. 9 sp. 4 and *C.* sp. 21 pooled) at Cape Naturaliste. Time is in the 24-h format, hence 00.00 is midnight. Triangles on the circumference of the circle are independent data points. The line labelled 'm' is the mean vector, the error bars are 95% confidence intervals for the mean. All data points were standardized to a 19.30 sunset (a) and 05:30 sunrise (b). Dashed lines indicate the end (or beginning) of standardized twilight.

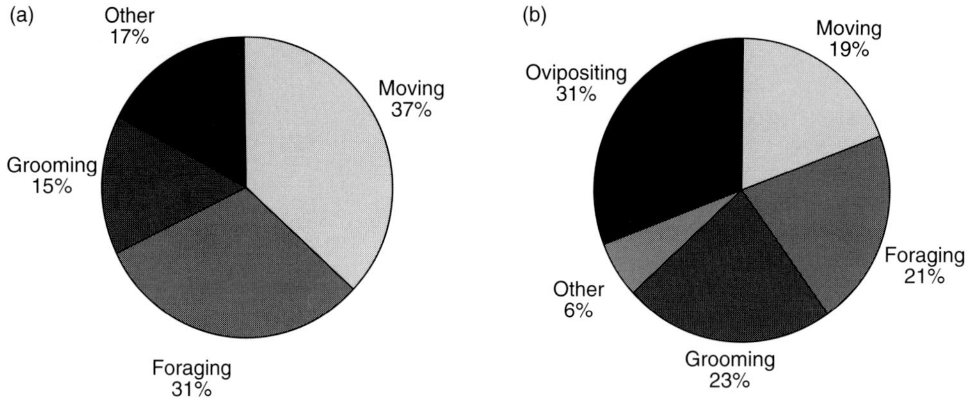

Fig. 6.6. Activity time budgets for Gen. nov. 9 sp. 4 for (a) males and non-ovipositing females and (b) ovipositing females. Adult Gen. nov. 9 sp. 4 individuals are, on average, active for 8 h per night during summer.

Orthoptera, as the sky begins to lighten with the onset of morning nautical twilight, which begins at 04.28 when sunrise is 05.30 (Anon., 1999). The gryllacridids are likely to be able to sense the change in light intensity (Menzel, 1981), and return to their own nests as twilight commences. To enter the nest Gen. nov. 9 sp. 4 individuals invert themselves and use their forelegs to lift the trapdoor and enter the burrow which is only slightly longer than the insect. These insects must however be able to turn within the burrow as, soon after entering the nest, they can be seen sealing the trapdoor shut with silk from their mouth.

It is becoming apparent that gryllacridids, like their weta relatives, display strong nest-site fidelity with individuals reusing the same nest for considerable lengths of time. This raises important questions as to how these ensiferans return to such a specific location night after night. When in close proximity to nest-sites, it seems that gryllacridids are capable of recognizing their own nest from conspecifics via an 'individual-specific' chemical marker (Lockwood and Rentz, 1996; Hale, 2001). Individual nest recognition by gryllacridids mirrors similar abilities displayed by other insects, especially the Hymenoptera (see Wehner, 1992). The orientational and homing abilities of gryllacridids appear as advanced as those of any orthopteran (Baker, 1980; Beugnon and Campan, 1989; Greenfield *et al.*, 1989) and this rather poorly understood family of ensiferans may rival the much-studied hymenopterans in their nesting, site-recognition and navigational capabilities.

Acknowledgements

The authors would like to thank Dr W. Bailey, Dr M. Cooper and Dr P. Cranston for comments on the manuscript. Dr J. Lockwood's comments on an earlier version of this manuscript are also appreciated. Previously unpublished data presented here by R.J. Hale are the result of research at the University of Western Australia whilst funded by an Australian Postgraduate Award. Gen. nov. 9 sp. 4 and *C.* sp. 21 individuals were collected and studied under Department of Conservation and Land Management (Western Australia) permits SF001697 and SF002021.

References

Alexander, R.D. and Otte, D. (1967) The evolution of genitalia and mating behaviour in crickets (Gryllidae) and other Orthoptera. *Miscellaneous Publications of the University of Michigan Museum of Zoology* 133, 5–62.

Anon. (1999) Australian Surveying and Land Information Group. www.auslig.gov.au/geodesy/astro/sunrise.htm

Baker, R. (1980) *The Mystery of Migration.* Macdonald Futura Books, London and Sydney.

Ball, E.E. and Field, L.H. (1981) Structure of the auditory system of the weta *Hemideina crassidens* (Blanchard, 1851) (Orthoptera, Ensifera, Gryllacridoidea, Stenopelmatidae). I. Morphology and Histology. *Cell and Tissue Research* 217, 321–343.

Beugnon, G and Campan, R. (1989) Homing in the field cricket *Gryllus campestris*. *Journal of Insect Behavior* 2, 187–198.

Bland, R.G. (1989) An annotated list of the Orthoptera of Beaver Island, Lake Michigan, USA. *Great Lakes Entomologist* 22, 39–44.

Bland, R.G. and Rentz, D.C.F. (1991) Studies in Australian Gryllacrididae: the proventriculus as a taxonomic character. *Invertebrate Taxonomy* 5, 443–455.

Bland, R.G. and Rentz, D.C.F. (1991b) External morphology and abundance of mouthpart sensilla in Australian Gryllacrididae, Stenopelmatidae, and Tettigoniidae. *Journal of Morphology* 207, 315–325.

Brockmann, H. J. (1985) Provisioning behavior of the great golden digger wasp *Sphex ichneumoneus* (Sphecidae). *Journal of the Kansas Entomological Society* 58, 631–655.

Brown, W.D. and Gwynne, D.T. (1997) Evolution of mating in crickets, katydids and wetas (Ensifera). In: Gangwere, S.K., Muralirangan, M.C. and Muralirangan, M. (eds) *The Bionomics of Grasshoppers, Katydids and Their Kin.* CAB International, Wallingford, pp. 281–314.

Davis, A.C. and Smith, L.M. (1926) Notes on the genus *Stenopelmatus* with description of a new species (Orthoptera). *Pan-Pacific Entomologist* 2, 174–181.

Field, L.H. (1980) Observations on the biology of *Deinacrida connectens* (Orthoptera: Stenopelmatidae), an alpine weta. *New Zealand Journal of Zoology* 7, 211–220.

Field, L.H. and Bailey, W.J. (1997) Sound production in primitive Orthoptera from Western Australia: sounds used in defence and social communication in *Ametrus* sp. and *Hadrogryllacris* sp. (Gryllacrididae: Orthoptera). *Journal of Natural History* 31, 1127–1141.

Field, L.H., Hill, K.G. and Ball, E.E. (1980) Physiological and biophysical properties of the auditory system of the New Zealand weta *Hemideina crassidens* (Blanchard, 1851) (Ensifera: Stenopelmatidae). *Journal of Comparative Physiology A* 141, 31–37.

Field, L.H. and Sandlant, G.R. (1983) Aggression and mating behavior in the Stenopelmatidae (Orthoptera, Ensifera), with reference to New Zealand wetas. In: Gwynne, D.T. and Morris, G.K. (eds) *Orthopteran Mating Systems. Sexual Competition in a Diverse Group of Insects.* Westview Press, Boulder, Colorado, pp. 120–146.

Fuchs, S. (1976) The response to vibrations of the substrate and reactions to the specific drumming in colonies of carpenter ants (Camponotus, Formicidae, Hymenoptera). *Behavioral Ecology and Sociobiology* 1, 155–184.

Gleason, R.S. and Johnson, D.R. (1985) Factors influencing nesting success of burrowing owls *Athene cunicularia* in southeastern Idaho, USA. *Great Basin Naturalist* 45, 81–84.

Greenfield, M.D., Alkaslassy, E., Wang, G. and Shelly, T.E. (1989) Long-term memory in territorial grasshoppers. *Experientia* 45, 775–777.

Griffith, D.M. (1992) The effects of substrate moisture on survival of adult cave beetles *Neaphaenops tellkampfi* and cave cricket eggs *Hadenoecus subterraneus* in a sandy deep cave site. *National Speleological Society Bulletin* 53, 98–103.

Hale, R.J. (2001) Nest utilisation and recognition by juvenile gryllacridids (Orthoptera: Gryllacrididae). *Australian Journal of Zoology* (in press).

Hansen, J.D. (1986) Comparison of insects from burned and unburned areas after a range fire. *Great Basin Naturalist* 46, 721–727.

Hartley, J.C. (1990) Egg biology of the Tettigoniidae. In: Bailey, W.J. and Rentz, D.C.F. (eds) *The Tettigoniidae: Biology, Systematics and Evolution.* House Press, Bathurst, pp. 41–70.

Huff, R.P. and Coyle, F.A. (1992) Systematics of *Hypochilus sheari* and *Hypochilus colei* two southern Appalachian lampshade spiders (Aranae: Hypochilidae). *Journal of Arachnology* 20, 40–46.

John, B. and Rentz, D.C.F. (1987) The chromosomes of four endemic Australian fossorial orthopterans: a study in convergence and homology. *Bulletin of the Sugadaira Montane Research Centre* 8, 205–216.

Johns, P.M. (1997) The Gondwanaland weta: family Anostostomatidae (formerly in Stenopelmatidae, Henicidae or Mimnermidae): nomenclatural problems, world checklist, new genera and species. *Journal of Orthoptera Research* 6, 125–138.

Karny, H.H. (1929) On the geographical distribution of the Pacific gryllacridids. *Fourth Pacific Science Congress, Batavia (Bandoeng), Java. Archiv für Zoologie* 20A(13), 1–86.

Karny, H.H. (1937) Orthoptera fam. Gryllacrididae, subfamiliae Omne. *Genera Insectorum Fascicule* 206, 1–317.

Kevan, D.K.McE. (1982) Orthoptera. In: Parker, S.P. (ed.) *Synopsis and Classification of Living Organisms*, Vol. 2. McGraw Hill, New York, pp. 252–379.

Key, K.H.L. (1959) The ecology and biogeography of Australian grasshoppers and locusts. In: Keast, A.L., Crocker, R.L. and Christian, C.S. (eds) *Biogeography and Ecology in Australia.* Dr W. Junk, The Hague, pp. 192–210.

Key, K.H.L. (1970) Orthoptera. In: *The Insects of Australia.* CSIRO, Melbourne University Press, Melbourne, pp. 323–347.

Key, K.H.L. (1989) Gryllacridoidea Stål, 1874 (Insecta, Orthoptera): proposed precedence over Stenopelmatoidea Burmeister, 1838. *Bulletin of Zoological Nomenclature* 46, 25–27.

Lockwood, J.A. and Rentz, D.C.F. (1996) Nest construction and recognition in a gryllacridid: the discovery of pheromonally mediated autorecognition in an insect. *Australian Journal of Zoology* 44, 129–141.

Markl, H. (1983) Vibrational communication. In: Huber, F. and Markl, H. (eds) *Neuroethology and Behavioral Physiology*. Springer-Verlag, Berlin, Heidelberg, pp. 332–353.

McVean, A. and Field, L.H. (1996) Communication by substratum vibration in the New Zealand tree weta, *Hemideina femorata* (Stenopelmatidae: Orthoptera). *Journal of Zoology* (London) 239, 101–122.

Menzel, R. (1981) Spectral sensitivity and colour vision in invertebrates. In: Autrum, H. (ed.) *Handbook of Sensory Physiology, Vol. VII/6A. Invertebrate Photoreceptors*. Springer-Verlag, Berlin, New York, pp. 503–580.

Michelsen, A., Fink, F., Gogala, M. and Traue, D. (1982) Plants as transmission channels for insect vibrational songs. *Behavioural Ecology and Sociobiology* 11, 269–282.

Moller, H. (1985) Tree wetas *Hemideina crassicuris* (Orthoptera: Stenopelmatidae) of Stephens Island, Cook Strait. *New Zealand Journal of Zoology* 12, 55–70.

Morton, S.R. and Rentz, D.C.F. (1983) Ecology and taxonomy of fossorial carnivorous gryllacridids (Orthoptera: Gryllacrididae) from arid central Australia. *Australian Journal of Zoology* 31, 557–579.

Nickle, D.A. and Nasrecki, P.A. (1997) Recent developments in the systematics of Tettigoniidae and Gryllidae. In: Gangwere, S.K., Muralirangan, M.C. and Muralirangan, M. (eds) *The Bionomics of Grasshoppers, Katydids and their Kin*. CAB International, Wallingford, UK, pp. 41 – 58.

Parmenter, R.R., Macmahon, J.A. and Gilbert, C.A.B. (1991) Early successional patterns of arthropod recolonization on reclaimed Wyoming strip mines USA: the grasshoppers Orthoptera, Acrididae and allied faunas (Orthoptera: Gryllacrididae, Tettigoniidae). *Environmental Entomology* 20, 135–142.

Peck, S.B., Walker, T.J. and Capinera, J.L. (1992) Distributional review of the Orthoptera of Florida. *Florida Entomologist* 75, 329–342.

Rasa, O.A.E. (1994) Behavioural adaptations to moisture as an environmental constraint in a nocturnal burrow-inhabiting Kalahari detritivore *Parastizopus armaticeps* Peringuey (Coleoptera: Tenebrionidae). *Koedoe* 37, 57–66.

Rentz, D. (1991) Orthoptera. In: Naumann, I.D. (ed.) *The Insects of Australia: a Textbook for Students and Research Workers*, 2nd edn, Vol. I. CSIRO, Australia, Melbourne University Press, pp. 369–393.

Rentz, D.C.F. (1996) *Grasshopper Country: The Abundant Orthopteroid Insects of Australia*. CSIRO, University of New South Wales Press.

Rentz, D.C.F. (1997) The world's most unusual gryllacridid (Orthoptera: Gryllacrididae). *Journal of Orthoptera Research* 6, 57–68.

Rentz, D.C.F. and John, B. (1990) Studies in Australian Gryllacrididae: Taxonomy, biology, ecology and cytology. *Invertebrate Taxonomy* 3, 1053–1210.

Rentz, D.C. and Weissman, D.B. (1973). The origins and affinities of the Orthoptera of the Channel Islands and mainland California. Part 1. The genus *Cnemotettix*. *Proceedings of the Academy of Natural Sciences of Philadelphia* 125, 89–120.

Sharell, R. (1971) *New Zealand Insects and Their Story*. Collins, Auckland.

Shaw, P.B., Owens, J.C., Huddleston, E.W. and Richman, D.B. (1987) Role of arthropod predators in mortality of early instars of the range caterpillar *Hemileuca oliviae* (Lepidoptera: Saturniidae). *Environmental Entomology* 16, 814–820.

Stauffer, T.W. and Whitman, D.W. (1997) Grasshopper oviposition. In: Gangwere, S.K., Muralirangan, M.C. and Muralirangan, M. (eds) *The Bionomics of Grasshoppers, Katydids and Their Kin*. CAB International, Wallingford, pp. 231–280.

Stewart, K.W. and Ziegler, D.D. (1984) The use of larval morphology and drumming in Plecoptera systematics, and further studies of drumming behavior. *Annals de Limnologie* 20, 105–114.

Storozhenko, S.Y. (1997) Fossil history and phylogeny of orthopteroid insects. In: Gangwere, S.K., Muralirangan, M.C. and Muralirangan, M. (eds) *The Bionomics of Grasshoppers, Katydids and Their Kin*. CAB International, Wallingford, pp. 59–82.

Tepper, J.G.O. (1892) The Gryllacrididae and Stenopelmatidae of Australia and Polynesia. *Transactions of the Royal Society of South Australia* 15, 137–178.

Tepper, J.G.O. (1895) Descriptions of new or little-known species of Blattariae, Gryllacrididae and Stenopelmatidae collected at Lake Callabonna, S. A. *Transactions of the Royal Society of South Australia* 19, 19–24.

Tillyard, R.J. (1926) *The Insects of Australia and New Zealand*. Angus and Robertson, Sydney.

Tindale, N.B. (1928) Australasian mole-crickets of the family Gryllotalpidae (Orthoptera). *Records of the South Australian Museum* 4, 1–42.

Weaving, A.J.S. (1989) Habitat selection and nest construction behavior in some afrotropical species of *Ammophila* (Hymenoptera: Sphecidae). *Journal of Natural History* 23, 847–872.

Wehner, R. (1992) Arthropods. In: Papi, F. (ed.) *Animal Homing*. Chapman & Hall, London.

West, M.J. and Alexander, R.D. (1963) Sub-social behavior in a burrowing cricket *Anurogryllus muticus* (De Geer) Orthoptera: Gryllidae. *The Ohio Journal of Science* 63, 19–24.

7
The Evolutionary History of Tree Weta: A Genetic Approach

Mary Morgan-Richards, Tania King* and Steve Trewick
Zoology Department, University of Otago, PO Box 56, Dunedin, New Zealand

Introduction

New Zealand tree weta (*Hemideina*) have been the biological material for a number of genetic studies. The genus contains few species, all of which are endemic to New Zealand. Currently, seven species are recognized within this group (Ramsay and Bigelow, 1978; Morgan-Richards, 1995a), and this small species number has the advantage that the group as a whole has been studied in more depth than many weta genera. Another aid to the study of tree weta is that they are widely distributed and, in many regions of New Zealand, are very abundant. Studies have concentrated on describing genetic variation within and among species and striving to understand the processes that produce this variation. The *Hemideina* species that have been most intensively examined are *H. maori* (the alpine tree weta), *H. thoracica* (the Auckland tree weta) and *H. crassidens* (the Wellington tree weta). Investigations have described chromosome variation, hybrid zones, gene flow and genetic differentiation within these species, using molecular genetic techniques. These investigations shed some light on the recent evolutionary history of tree weta species and the genetic relationships among populations within species. Here, we review the distribution of the New Zealand tree weta, their genetic structure and the extent of hybridization. In addition, we report on genetic studies of the single giant weta species that is sufficiently abundant and widespread to warrant population genetic studies (*Deinacrida connectens*).

Distribution Patterns

Of the seven described *Hemideina* species, *H. thoracica* and *H. trewicki* occur north of Cook Strait, and *H. ricta*, *H. maori*, *H. femorata* and *H. broughi* occur to the south. *Hemideina crassidens* is the only species with a distribution that spans the Strait. For the most part, these species exist in local allopatry or parapatry with other members of the genus (Trewick and Morgan-Richards, 1995). This holds even where the current distribution of neighbouring species is a mosaic; local sites remain exclusively the domain of one or other species. In most instances, a single recognized species is found in a particular area. Tree weta are found on most of the offshore and lacustrine islands that were connected to the main islands during periods of lowered sea level associated with recent glaciations. Maps of the species distributions are presented in Chapter 2 (Gibbs, this volume). However, a detailed knowledge of the fine-scale patterns of distribution of species is an essential component of genetic studies. The distribution of the northern species (*H. thoracica*, *H. trewicki*, *H. crassidens*) has been explored here in some detail (Trewick and Morgan-Richards, 1995). As a result of this work, it is possible to make the important

*Present address: School of Biology, University of Leeds, Leeds LS2 9JT, UK.

distinction between local allopatry and geographical sympatry.

Hemideina thoracica (Auckland tree weta) is found from Levin on the Manawatu coast northwards to Cape Reinga. On Mount Taranaki the species is replaced by *H. crassidens* above about 650 m a.s.l., *H. thoracica* occurring at the highest altitude on the northern slopes. A narrow sympatry zone exists above 650 m a.s.l. Within this area, members of the two species share roost cavities and males and female adults of the two species have been observed cohabiting. Although this behaviour clearly provides the opportunity for hybridization between the two species, none has been documented. A similar pattern exists on Mount Ruapehu, where the two species meet at about 950 m a.s.l. The current disjunct distribution of *H. crassidens* can be explained by a southward range expansion of *H. thoracica* after the last ice age, replacing *H. crassidens* from lowland central North Island but leaving isolated pockets of *H. crassidens* at high altitude. The interaction between these two species may involve competitive exclusion.

Although *H. crassidens* spans Cook Strait, there are no consistent genetic differences that distinguish populations across this barrier (Morgan-Richards *et al.*, 1995). A separate morphological study, however, found a significant difference between stridulatory ridge counts of the north and south populations (for morphometric analysis, see Field and Bigelow, Chapter 9, this volume). A subspecies, *H. crassidens crassicruris*, is restricted to Stephens Island in Cook Strait, where it occurs in large numbers (Moller, 1985). Although not genetically differentiated (26 allozyme loci), *H. c. crassicruris* is, on average, slightly larger than *H. c. crassidens*, and has a subtle but distinctive pigmentation (illustrated in Figs 9.7 and 9.8, Field and Bigelow, Chapter 9, this volume; Morgan-Richards *et al.*, 1995). Its geographical isolation should ensure continued reproductive isolation from other populations of *H. crassidens*, but it is not considered sufficiently divergent to justify full species status.

Hemideina trewicki (Hawkes Bay tree weta) is restricted to southern and central Hawkes Bay, where it is locally sympatric with *H. thoracica* in the northern parts of its range. Individuals of both species are found in the same trees and even side by side in the same roost holes. The species is apparently parapatric, with *H. crassidens* to the south of Hawkes Bay near Dannevirke.

Hemideina ricta (Banks Peninsula tree weta) is restricted to the eastern tip of Banks Peninsula on the Canterbury coast. Because of its limited distribution, this species is legally protected. It is sympatric with *H. femorata* but, although *H. ricta* has been found from 40 to 806 m a.s.l., there is evidence of near-complete altitude separation of the two species, with most *H. ricta* found above 400 m (Townsend *et al.*, 1997). This may be the result of competition with *H. femorata* for roost holes at lower altitudes ($<$ 400 m a.s.l.) and of different climate tolerance.

Hemideina femorata (Canterbury weta) is found in the east of the South Island, north to the Clarence River Valley, throughout Canterbury and south to Fairlie (Ramsay and Bigelow, 1978). The distribution extends along the east coast from Christchurch to Kaikoura and as far west as Nelson, where this species is geographically sympatric with *H. crassidens* (G. Ramsay, Auckland, 1992, personal communication).

Hemideina maori (alpine 'tree' weta) is found throughout the central South Island, but its range is not known to overlap with that of any other tree weta. Considerable size and colour variation has been documented among populations of this species.

Deinacrida connectens (alpine scree weta) is found in high elevations, where it shelters among rocks. It lives on most mountain ranges of the South Island above ~1200 m a.s.l., where it is subjected to freezing conditions during winter months.

Genetic Differentiation Between Species

Six of the species of tree weta that are currently recognized were described from morphological characteristics that distinguished them. Subsequent studies have confirmed that five of these are genetically distinct. The sixth species (*H. ricta*) will be discussed in more detail later. In addition, genetic studies have identified another species (*H. trewicki*) that was not previously recognized because of its parapatry with and morphological similarity to *H. crassidens* (Morgan-Richards, 1995a; Morgan-Richards *et al.*, 1995).

There exists the opportunity to quantify the extent of interbreeding where the distribution of weta includes areas of overlap between taxa. Behavioural observations of captive individuals have been made and allow us to say whether copulations occur, but to quantify the extent of successful matings, one needs to examine genetic exchange between the taxa. Evidence of copulation between described species has been accumulated, suggesting weak or non-existence prezygotic recognition systems, but how does this relate to actual gene exchange? That is, how strong/well-established are *in vivo* barriers to gene exchange between described tree weta species?

Hybridization vs. Mating

From captive observations, most adult tree weta seem ready to mate with any other adult tree weta regardless of the species. Sharing tree cavities, attempts to mate and the production of hybrid offspring have been documented between most of the tree weta species that are in contact in nature. Of the five parapatric species combinations studied in captivity, four attempted to mate across species:

1. *H. crassidens* × *H. thoracica*
2. *H. crassidens* × *H. trewicki*
3. *H. thoracica* × *H. trewicki*
4. *H. femorata* × *H. ricta*.

In contrast, an *H. crassidens* male placed in a container with an adult female *H. femorata* did not attempt to mate. This species pair is the only naturally occurring combination of southern and northern species of tree weta.

1. Visually, the Wellington tree weta (*H. crassidens*) is easily distinguished from the Auckland tree weta (*H. thoracica*) by the banded abdomen and dark pronotum of *H. crassidens* compared with the uniform brown of *H. thoracica* and its pale pronotum with black markings (see Figs 9.7 and 9.8, Field and Bigelow, Chapter 9, this volume). Genetic studies have confirmed the distinctiveness of these two species. On the other hand, the weta themselves seem to have more difficulty telling the species apart and, where their ranges overlap, adults are known to share galleries (Trewick and Morgan-Richards, 1995). When kept, *H. crassidens* and *H. thoracica* females that mate with conspecifics laid fertile eggs. Also, heterospecific pairs will mate in captivity, but so far have not laid eggs. No morphological or genetic evidence of gene flow between these two species has been found (Morgan-Richards *et al*., 1995). In allozyme studies, parapatric populations of the two species on Mount Taranaki showed fixed differences at five loci, as well as fixed chromosome differences.

2. The Hawkes Bay weta (*H. trewicki*) looks almost identical to *H. crassidens*; however, allozyme data show clear separation of these two species (Morgan-Richards *et al*., 1995). Like us, these two species have difficulty telling each other apart and, when held in captivity, they will mate and produce hybrids. Whether their hybrids are fertile is not known but initial observations suggest that hybrids are also produced naturally in a contact zone in the Wairarapa region (M. Morgan-Richards, unpublished data). Unfortunately, it is difficult to study the interaction of these two species, because their morphological similarity means that to determine what species one has found (or if it is a hybrid) requires killing that animal and studying its chromosomes and other genetic traits. These two species differ in both the number of chromosomes they have and the shape and size of some of their chromosomes (karyotype). Such chromosome differences are likely to result in hybrids that are partly or completely sterile. This is because the chromosome differences cause disruptions to the normal pairing process of homologous chromosomes during the first division of meiosis. Disruption of this cell division leads to the production of gametes (eggs and sperm) with additional or missing chromosomal material, with either case yielding gametes that are unlikely to produce viable weta. Hybrid weta may produce low numbers of viable gametes, substantially reducing the fertility of hybrids. In an initial survey of tree weta from the Wairarapa region, two allozyme loci were found to distinguish the two species (Morgan-Richards, 1995a), indicating that these species are maintaining separate gene pools. However, the sharing of alleles at other loci, such as *Pgd*-1 'fast', may be due to gene flow. A detailed study with larger samples from adjacent populations in Hawkes Bay and Wairarapa is needed to determine if viable hybrids are leading to the exchange of alleles between these two species.

3. *Hemideina thoracica* and *H. trewicki* can be clearly differentiated, using both morphology and genetics; however, these two species have been found sharing roost galleries at a number of sites in Hawkes Bay. At Mohi Bush Reserve in Hawkes Bay, a putative first-generation (F_1) hybrid between *H. thoracica* and *H. trewicki* was collected. This animal had a chromosome complement with elements from both species and was heterozygous

for two diagnostic allozyme loci. Although three allozyme loci differentiate these two species where they are sympatric, one is sex-linked. *Pgm*-1 is carried on the X chromosomes and therefore males have a single copy (are hemizygous) (Hebbert, 1984; Morgan-Richards and Townsend, 1995). The male F_1 hybrid had the *Pgm*-1 allele characteristic of *H. thoracica* and therefore was probably the product of a mating between a female *H. thoracica* and a male *H. crassidens*. Examination of meiosis in this hybrid suggests that such hybrids are probably infertile. The chromosome differences between these two species means that the production of chromosomally balanced, functional gametes (eggs and sperm) is very unlikely. In addition, the lack of genetic introgression between the two parental species suggests that such hybridization is rare and/or F_1 hybrids are infertile (Morgan-Richards, 1995a).

4. *Hemideina femorata* and *H. ricta* mate in captivity (Field, 1993a) and hybrids between these two species have been collected in the field (Morgan-Richards and Townsend, 1995). An area of overlap occurs east of Akaroa Harbour on Banks Peninsula. Weta collected from this region showed that there are 11 allozyme loci that distinguish the two species. Two individuals were identified as first-generation hybrids, as they were heterozygous at all 11 loci. One hybrid was male and therefore hemizygous at the sex-linked locus *Pgm*-1 (in this case it was the allele characteristic of *H. ricta*, indicating that this species was probably the mother and *H. femorata* the father). No genetic introgression was detected in either parental species using these 11 nuclear markers. This may be because hybridization is rare, although these two hybrids were from a sample of only 14 animals. It is more likely that no introgression was detected due to infertility of such hybrids between this species pair. No difference in karyotype has been detected between these two species (Morgan-Richards, 1995b), although the allozyme data show that genetically they are well differentiated (Morgan-Richards and Townsend, 1995).

Genetic Structure as Evidence of Past Fragmentation of Weta Species

Most evidence from tree weta indicates that genetic isolation is the result of postmating barriers to gene exchange, not from behavioural isolating mechanisms. Postmating isolating mechanisms maintain strong levels of genetic differentiation (structuring) among species, even in sympatry. How well do these barriers function at lower taxonomic levels? Where species are subdivided into morphological types or chromosomal races, is there any evidence of gene exchange among these variant subpopulations?

In the weta species that have been examined using molecular techniques, extensive structure has been found. The level and distribution of genetic variation and its partitioning into populations may derive from stochastic and deterministic forces of random genetic drift, mutation, migration and natural selection. Natural selection can favour adaptations to local environments and can ultimately lead to genetic differentiation of local populations. The movement of individuals (migration) and gametes (gene flow) among populations opposes this differentiation (Slatkin, 1985).

The existence of genetic subdivision and genetic structure within species can reflect patterns of past geographical subdivisions. Geophysical and climatic changes, such as sea-level fluctuations, volcanic activity, mountain building and shifting forest cover result in habitat fragmentation. During the last 3 million years, New Zealand's biota has been exposed to glacial periods with dramatic changes in forest cover and the position and extent of the alpine zone and shorelines. In addition, the southern alps have arisen during this time and North Island volcanic activity has continued. Such forces have the potential to cause the fragmentation of species ranges, as well as reuniting populations of terrestrial flightless animals. Whilst New Zealand's geophysical history has presumably been responsible for the loss of many populations and species (extinction), it must also have led to additional biodiversity through differentiation of populations.

Examples of Genetic Structuring within Taxa

Chromosome variation

Cytogenetic studies have found that *H. crassidens* and *H. thoracica* are subdivided into races that have different numbers of chromosomes (Morgan-Richards, 1995b, 1997). Table 7.1 describes variation in chromosome number within and among *Hemideina* species. The chromosome variation exhibited by these weta species is a result of chromosome mutations, such as fissions, fusions, translocations and inversions. In most cases, every individual within a population has the same karyotype, but this may differ from the karyotype of neighbouring populations. The origin of chromosomal races is generally thought to be a result of small populations of related individuals recolonizing areas or being geographically isolated from other populations long enough to fix new mutations in the population. In all but a few exceptional cases, these karyotype changes appear to be selectively neutral (Marchant and Shaw, 1993). Chromosome races have been observed and studied in many Orthoptera, as well as a wide range of mammals, insects and plants. Although factors that influence the rate of chromosome change in species are not well understood, it is likely that those factors that increase genetic differentiation, such as high levels of inbreeding, geographical isolation and a high mutation rate, will also increase the rate of chromosome evolution.

The consequence of reproduction between weta from different chromosomal races has only been determined in a few cases. Between some chromosomal races of *H. thoracica*, adults will mate and produce viable offspring. These offspring may have low fertility, due to the difficulties of chromosome pairing during meiosis (part of the process of producing sperm and eggs).

The Wellington tree weta

There are two chromosome forms of *H. crassidens*, one on the west coast of the South Island between Haast and the Buller River and a second found in northern South Island and the North Island. The northern chromosomal race has 15 chromosomes in the male and 16 in the female (Morgan-Richards, 2000). As is commonly seen in Orthoptera, weta have an XX/XO system of sex determination where females have two sex chromosomes and males have one. In the south, *H. crassidens* have 19(XO) and 20(XX) chromosomes. The increase in chromosome number probably resulted from fission (breakage) of two large metacentric chromosomes to form four medium-sized acrocentric chromosomes (Robertsonian translocation). Such chromosome mutations are common between races of many plant and animal species (John and Lewis, 1966; White, 1973). In breeding experiments between the 19 and the 15 chromosomal races of *H. crassidens*, F_1 hybrids were successfully reared to adulthood. The chromosomes of the F_1 hybrids, a mixture of the two parental types, aligned to form trivalents during meiosis. It is likely that they would produce gametes with a balanced set of chromosomes and potentially these hybrids may suffer little loss of fertility. However, in nature all individuals within almost all populations of *H. crassidens* have the same karyotype, with variation found only between populations. It is of interest to note that the current division of this species' range by Cook Strait does not mark the change in chromosome number. Some earlier geographical isolation of north and south may be responsible for the differentiation of the two races.

The Auckland tree weta

Collection and study of *H. thoracica* from throughout its range revealed that this morphologically uniform species is subdivided into at least eight chromosomal races (Morgan-Richards, 1997; Fig. 7.1). Most populations are characterized by a single distinctive karyotype, but, in many cases,

Table 7.1. Chromosome variation in New Zealand tree weta (*Hemideina*).

Species	Number of chromosomal races	Number of chromosomes ($2n$ = XO)
H. thoracica	8	11–24
H. crassidens	2	15–19
H. trewicki	1	17
H. femorata	1	25
H. ricta	1	25
H. maori	1	25
H. broughi	1	25

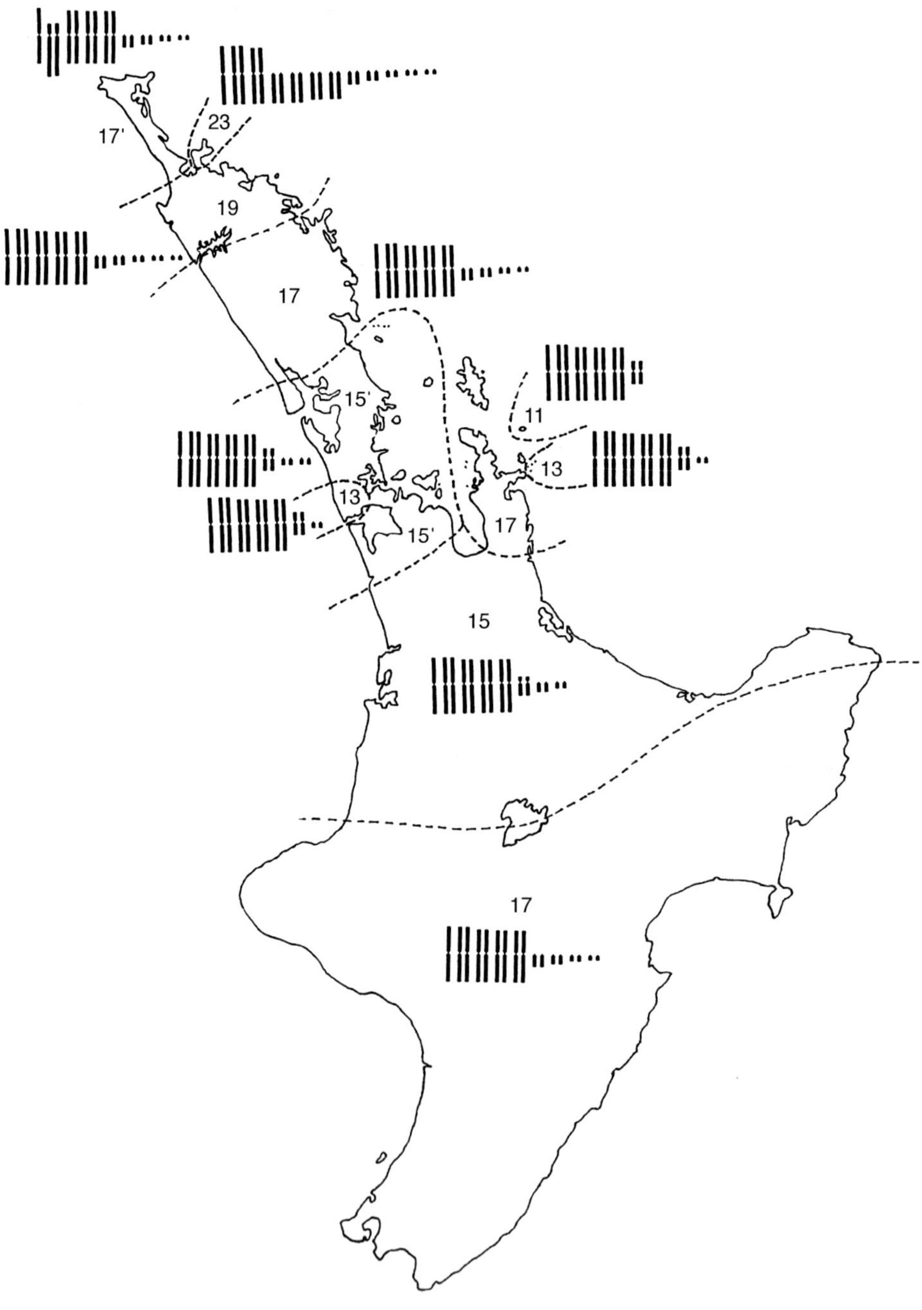

Fig. 7.1. The distribution of chromosome races of the Auckland tree weta (*Hemideina thoracica*) in the North Island, New Zealand.

this differs from neighbouring populations by chromosome fusions, fissions, inversions and addition or loss of chromosome material. Diploid numbers range from 11 to 24. Weta in the far north have 17 chromosomes, as shown on the map. They are replaced by weta with 19 chromosomes near Kaitaia, and this karyotype extends about 50 km to Puketi Forest. South of there, weta with 17 chromosomes are found, but their karyotype differs from the far north 17 in that one pair of autosomes is metacentric rather than submetacentric. Thus, the whole of the North Island can be divided into regions, each distinguished by distinctive karyotypes of *H. thoracica*. This extends to some offshore islands, which differ from the nearest sampled mainland population. For example, weta from Coromandel peninsula have the common 17 karyotype, but weta from Double Island have 13 and those from Cuvier Island have 11. As neither of these islands has been isolated from the mainland for more than 20,000 years (Stevens, 1980), this is evidence that chromosome evolution is probably fairly rapid in this species. Island populations are often small and geographically isolated, leading to increased inbreeding and increased probability of the fixation of new mutations. Of the eight chromosome forms observed within *H. thoracica*, six are found north of Auckland. It is possible that the division of Northland into many small islands during the Pliocene resulted in the genetic fragmentation and diversity of this species in the north of its range (Morgan-Richards, 1997).

The chromosomal variation within *H. thoracica* is not (in most cases) concordant with population structure observed using other nuclear markers. This lack of concordance is indicative of the reticulate nature of intraspecific evolution, where gene flow between populations prevents sustained genetic divergence of 'races'.

The alpine scree weta

The only giant weta species that is widespread (*D. connectens*) has also been shown to be highly variable for its karyotype (set of chromosomes) (Morgan-Richards and Gibbs, 1996). Diploid chromosome numbers range from 17 (XO) to 22 (XX). The identified karyotypes are found in populations of scree weta with discrete geographical ranges. In contrast to the two *Hemideina* species discussed above, at least four populations of *D. connectens* are karyotypically polymorphic.

Protein variation

Allozyme studies of weta have found structure that is broadly concordant with geography in *H. crassidens*, *H. thoracica*, *H. maori* and *D. connectens* (Morgan-Richards *et al.*, 1995; Morgan-Richards and Gibbs, 1996; Morgan-Richards, 1997). In these four species, a number of populations were sampled (14, 37, eight and 16, respectively). A similar amount of genetic differentiation was found among populations within each species (conspecific Nei's D ranged from 0 to 0.14). The geographical distance covered in each study was also similar. An association of genetic distance with geographical distance was found in *D. connectens* (Morgan-Richards and Gibbs, 1996) – evidence of isolation by distance. Isolation by distance does not, however, rule out the possibility that chromosome variation limits gene flow in these species.

Mitochondrial DNA variation

Studies of variation in mitochondrial DNA (mtDNA) have also found structure that is concordant with past (and present) geographical isolation (King, 1997; T.M. King and G.P. Wallis, in preparation). The Rock and Pillar range in Central Otago is home to two very distinctive colour forms of the alpine tree weta, *H. maori*. In the north of the range the typical form is yellow, with black stripes across the dorsal surface of the abdomen and plain creamy yellow on the ventral surface. To the south, the striped form is replaced by a less common melanic form. Throughout the range, morphologically intermediate individuals are found that exhibit varying degrees of speckling on the yellow areas, forming a continuum between the two 'parental' taxa. Molecular genetic analyses have revealed significant differentiation of mtDNA between the striped and dark forms, although there is little evidence of differentiation of nuclear markers (isozymes, chromosomes, randomly amplified polymorphic DNA (RAPDs), microsatellites (King *et al.*, 1996, 1998)). Mitochondrial differentiation, coupled with a clinal distribution of body colour, strongly suggests that the Rock and Pillar range portrays a hybrid zone. Most hybrid zones are believed to result from secondary contact of differentiated forms. This explanation is consistent with the likely mechanism of speciation within the genus as a whole. One can test this hypothesis by comparing phylogenetic relationships among a number of populations of this species. If, instead, the hybrid zone arose *in situ* (i.e. through primary contact), one would expect the mtDNA of the two colour forms on the Rock and Pillar range to cluster as sister taxa on a phylogenetic tree. Weta from locations through the South Island were collected. *Hemideina ricta* (Banks Peninsula) and *H. femorata* (Kaikoura and Banks Peninsula) were chosen as outgroups. Mitochondrial DNA, consisting of approximately 605 base pairs, was sequenced from all individuals. Figure 7.2 shows the results of a cladistic analysis of the data (King, 1997; T.M. King and G.P. Wallis, in preparation).

The resulting tree suggests that the two Rock and Pillar populations are not sister taxa. Rock and Pillar striped weta (Y) are more closely related to other Central Otago populations (Dunstan Range, Crescent and Harwich Islands) than they are to the dark form. This demonstrates that whatever event led to the two colour forms meeting on the Rock and Pillars was preceded by isolation of the two forms. It is also interesting to note that *H. ricta*, the Banks Peninsula weta, clusters with a population of *H. maori* and not outside the tree, as

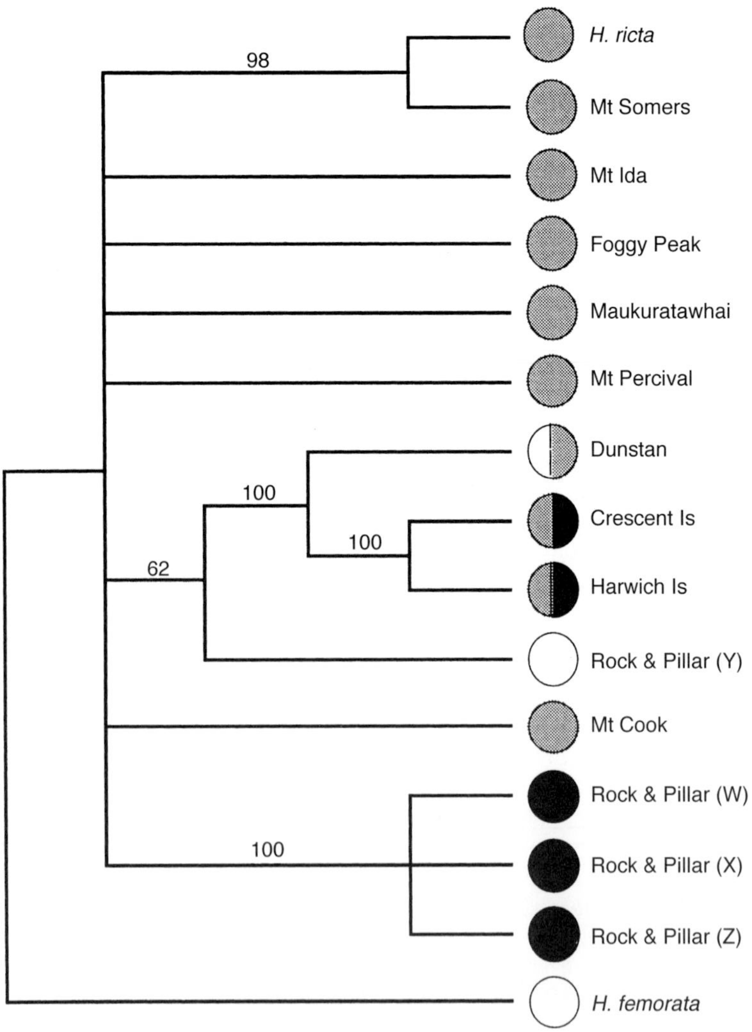

Fig. 7.2. Body colours of weta in each population of *H. maori*, *H. ricta* and the outgroup, *H. femorata*, mapped on to a weighted, bootstrap tree. Colours on circles represent black, yellow (striped) and intermediate categories, indicated by black, white and grey shading, respectively. In the Rock and Pillar range populations 'Y' represents a yellow haplotype, while 'W, X and Z' are black haplotypes. Numbers along branches are bootstrap values. (From King, 1997.)

expected. This could be the result of introgression of *H. maori* mtDNA into *H. ricta*, although the two species are currently allopatric. Alternatively, *H. ricta* could be a species recently derived from *H. maori* and of questionable taxonomic status. Comparisons between the two species of allozymes, chromosomes, coloration, stridulatory structures and behaviour (Field, 1982, 1993b; Morgan-Richards, 1995b; Morgan-Richards and Townsend, 1995; see also Field and Bigelow, Chapter 9, this volume) all support the latter explanation.

The most significant recent geological event in the South Island was the Kaikoura Orogeny (5–2 million years ago (mya)), which gave rise to many present-day South Island mountain ranges. To determine if the timing of splitting events seen on the tree are consistent with this event, a molecular clock can be applied to the mtDNA sequence data. Brower (1994) took cases involving arthropod groups for which accurate geological dates were available in order to extrapolate a formula for calculating the relationship between mtDNA change and time. Using this formula, we estimated that the Rock and Pillar striped lineage split from the black lineage between 2.13 and 1.83 mya, which is consistent with the timing of the Kaikoura Orogeny. The present-day Rock and Pillar striped lineages apparently did not split from other Central Otago populations until 1.11–0.89 mya. This result endorses the view that geological events played a significant role in shaping the genetic structure of the tree weta species we see today.

Similarly, evolution of the alpine-adapted giant weta *D. connectens* and its population fragmentation can be correlated with the emergence of the mountain ranges it inhabits in the South Island (Trewick *et al.*, 2000). Individuals sampled from mountain-top populations were found to share similar or identical mtDNA haplotypes, and genetic difference among these populations was correlated with the geographical distance separating them. Although most mountain-top populations carried unique haplotypes, scree weta found on neighbouring mountains within any one range had closely related haplotypes. The most striking situation is that between populations in the Nelson and Marlborough region. Although separated by only ~60 km, these populations are strongly differenetiated, indicating an absence of gene flow between these regions for an extended period. The highest genetic distances among populations indicate that the deepest well-defined lineages originated some 4 million years ago. This suggests that populations were initially isolated from one another as the mountains of South Island emerged during the Pliocene. Climate cycling and, notably, altitudinal shifts in the alpine zone will probably have provided a mechanism for the continued isolation and, to a limited extent, migration of scree weta between mountains during the Pleistocene. It is likely that, under current conditions, scree weta populations are now isolated from one another since they are not found below the alpine zone.

Barriers to Gene Flow

Most animal species do not consist of a single, panmictic population. Populations can become separated by geographical barriers or their distributions can be discontinuous as a result of ecological patchiness. Alternatively, gene flow may be partially impeded as a result of divergence produced during past (temporary) separations. Whether mixis is prevented by obvious physical barriers or more subtle mechanisms associated with genetic differentiation, both will result in genetically divergent populations.

In weta, the flow of genes between measurably genetically and/or morphologically differentiated forms is a demonstration of their conspecific status. However, although differentiated forms may interbreed, there often remains sufficient genetic differentiation/distinction between populations that were isolated in the past to prevent complete fusion in the present. Examination of present-day patterns allows us to reconstruct past subdivisions.

The alpine tree weta

The alpine weta, *H. maori*, provides us with an opportunity to quantify the extent of gene flow within a species. The northern, yellow and black striped form abruptly switches to a southern, melanic form within an area of less than 1 km. Examination of mtDNA using restriction analyses of a 2.3 kilobase (kb) fragment of the genes comprising cytochrome oxidase subunit 2 (COII) and part of cytochrome oxidase subunit 3 (COIII) revealed six composite haplotypes on the mountain range (Fig. 7.3). Three were associated with the striped form (designated U, V and Y) and three

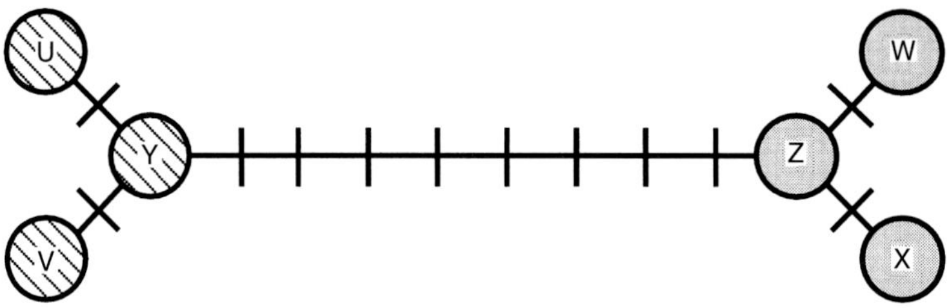

Fig. 7.3. Network showing the relationships between each composite mitochondrial DNA haplotype. The network was derived from a strict consensus tree of three shortest trees, generated using the branch and bound option of PAUP (RI = 0.857, CI = 0.8). Bars along each branch represent the gain or loss of a restriction site. Stippled (W, X, Z), black; crosshatched (U, V, Y), striped. (Adapted from King, 1997.) PAUP, Phylogenetic Analysis Using Parsimony; RI, retention index; CI, consistency index.

with the dark form (W, X and Z) (T.M. King and G.P. Wallis, in preparation). Most variation described separates the striped from the dark variant, with minimal differences present within each lineage.

The extent of gene flow can vary among genes, due to their means of transmission (maternal versus biparental) and the selective value that the gene or linked genes confer. mtDNA is typically inherited through the female line, meaning that it can only reflect patterns of movement of females. To get a broader picture of the movements of all animals, one needs to use biparentally inherited markers. Our experiments failed to find any evidence for fixed or even significant frequency differences in any nuclear markers (King *et al.*, 1996, 1998). However, one type of marker – microsatellites – showed high levels of variation and, despite being unable to distinguish the colour variants, it does allow us to quantify relative levels of variation at different hierarchical levels in the population.

As well as investigating structure between colour forms, we can also look for yet finer-scale differentiation. On the Rock and Pillar range, the landscape is dominated by large outcrops of schist, termed tors. These outcrops comprise layers in which weathering has produced cracks, which provide a habitat for weta. Weta are only found between slabs of schist, and not in the intervening vegetation. This provides a convenient, testable means to subdivide the population by tors.

A hierarchical analysis with categories of body colour and rock tor was used in order to quantify the extent of genetic structuring at each level (T.M. King and G.P. Wallis, in preparation). Data were collected along a transect that crossed the transition zone between striped and dark weta. Up to ten adult weta were collected at rock tors along the transect. They were photographed against a standard background to allow colour categorization and a small amount of antenna was removed for DNA analysis. Each weta was characterized for its mitochondrial haplotype and for two microsatellite loci. Analysis of molecular variance (AMOVA) (Excoffier *et al.*, 1992) was performed, using WINAMOVA version 1.5 (Excoffier, 1995), with body colour as the top division. We excluded intermediates in order to avoid the possible confounding effects of introgression on differentiation between the two parental body colours. Tors were defined as a second division. Mitochondrial and nuclear markers were analysed separately. Results of the AMOVA are shown in Table 7.2.

As expected, mtDNA showed that the vast majority of overall variation (69%) was due to differentiation between the colour types. Variation between sites within colours was lower but still statistically significant.

The results for microsatellites were somewhat different. When weta were differentiated on the basis of body colour, there was no evidence of genetic differentiation at either microsatellite locus. However, when mtDNA was used to categorize them instead of body colour, there was significant variation at one locus between colours and among sites within colours. Variation among sites overall was also statistically significant and accounted for most of the variation seen. In general, there was little evidence that populations were subdivided with respect to nuclear DNA.

There is evidently a dichotomy between maternally and biparentally inherited markers, which can be explained in several ways. The pattern may be historical. Prior to the secondary contact event,

Table 7.2. Summary of AMOVA results. Variance components for each hierarchical level are given. Significance estimates of each statistic are derived from 100 permutations.

	Variation among colours	Variation among sites within colours	Variation among sites within total
mtDNA	69%*** ($P < 0.001$)	11%** ($P = 0.004$)	20%*** ($P < 0.001$)
Microsatellites			
Grouped by body colour			
Locus 1	−1% ($P = 0.499$)	5% ($P = 0.198$)	96% ($P = 0.151$)
Locus 2	−1% ($P = 0.567$)	−1% ($P = 0.511$)	102% ($P = 0.574$)
Grouped by mt lineage			
Locus 1	5%* ($P = 0.02$)	5%* ($P = 0.049$)	90%* ($P = 0.004$)
Locus 2	−1% ($P = 0.932$)	2% ($P = 0.287$)	99% ($P = 0.309$)
Grouped by body colour and mt lineage			
Locus 1	−2% ($P = 0.555$)	−1% ($P = 0.462$)	103% ($P = 0.504$)
Locus 2	17% ($P = 0.052$)	−9% ($P = 0.920$)	92% ($P = 0.411$)

the mtDNA of each colour variant may have differentiated but not their nuclear DNA. Selection on nuclear markers may be stronger than on mtDNA, resulting in little change in the nuclear genome. The effective population size (N_e) of the mitochondrial genome, is only 1/4 that of the nuclear genome, as a result of it being haploid and maternally inherited. If a small, founding population gave rise to one or other of the present-day colour variants, then this would have a more pronounced effect on mtDNA variation than on nuclear DNA. Mutational processes may also exert differential influence on the evolution of genetic markers. Microsatellites are believed to mutate in a stepwise fashion, creating new alleles that are close in size (smaller or larger) to existing alleles. This would mean a greater likelihood of convergent mutation in microsatellites than is found in mtDNA. Weta social structure may also play a role in producing the dominance of structuring in mtDNA over nuclear DNA. Asymmetry in mating preference or success, resulting in F_1 hybrids backcrossing solely with parentals that are the same as their female parent, will homogenize nuclear DNA but retain structuring in mtDNA. Alternatively, it is possible that the majority of migration among tors is by males, who would pass on their nuclear DNA but not mtDNA. As single males defend territories and potentially harems of adult females, there may be an excess of males, who are forced to disperse in search of territories and/or mates.

The Auckland tree weta

Within another weta species, the flow of alleles between two populations has been studied. In the central North Island, two *H. thoracica* populations meet. In this case, there is no morphological feature that distinguishes the two populations, but chromosome number readily identifies the two groups. These two populations cannot have existed beside LakeTaupo prior to 2000 years BP, as volcanic activity covered the central North Island with ash, burying the forest.

A contact zone between these two chromosomal races of the Auckland tree weta has been examined to determine the extent of gene flow among populations. South of Lake Taupo, *H. thoracica* has 17 chromosomes, compared with 15 chromosomes immediately north of the lake. Weta were studied for nuclear and mitochondrial markers from six locations along a 24 km stretch on the eastern shore of Lake Taupo, where the two chromosomal races make contact (Fig. 7.4.). Adjacent to Taupo airport, weta were collected that were heterozygous for the two chromosome complements. Two allozyme loci (*Icd-2* and *Pdg*) have alleles that are characteristic of either the northern or southern populations and a pair of mitochondrial haplotypes show this same geographical pattern. Weta collected from within the contact zone had a mixture of the southern and northern nuclear markers, showing that hybrids between the two chromosomal races are fertile and allow the exchange of alleles between the races. However, only the northern mtDNA was found within the contact zone between the two chromosomal races. Because Lake Taupo is a volcanic crater lake, we know that the contact between these two chromosomal races cannot have existed for more than 2000 years. For these two chromosome races to have created a hybrid zone, these weta would have had to have moved 80–100 km per generation. However, the fact that samples within the contact zone were monomorphic for their mitochondrial haplotype but polymorphic for both allozyme and chromosome markers suggests that males may disperse further than females in this species. (Morgan-Richards *et al.*, 2000).

Summary Remarks

The seven species of *Hemideina* are mostly allopatric. Where two tree weta species come into

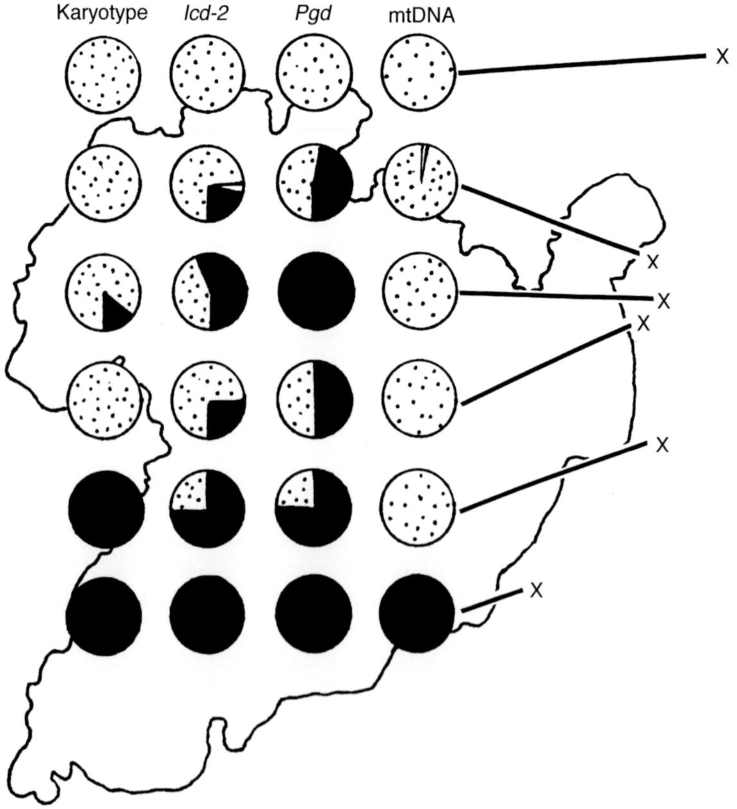

Fig. 7.4. A hybrid zone between two chromosome races of the Auckland tree weta (*Hemideina thoracica*) at Lake Taupo. Weta samples taken from six sites on the shore of a volcanic crater lake show a change in frequency of karyotype, alleles at two allozyme loci (*Icd-2* and *Pgd*) and mitochondrial DNA haplotype.

contact at the edge of their range, lack of species-specific recognition leads them to share cavities and attempt to mate. However, these matings generally result in infertile, if any, offspring, thus maintaining the genetic independence of the species. This lack of mate recognition suggests that speciation has, in general, been allopatric in this genus and only long geographical isolation of populations could have given rise to the species differentiation seen today.

At least three tree weta species are subdivided into colour and chromosome morphs (or races). Where weta populations have differentiated to some extent, mating leads to fertile offspring and gene flow between 'races'. Recent mountain building and climate changes have apparently not isolated tree weta populations long enough to result in the formation of new species but have produced intriguing patterns of genetic variation. Mountain emergence is, however, implicated in the evolution and population structuring of the alpine scree weta, *D. connectens*. This genetic structure provides evidence of the effect of volcanic activity, shoreline changes and mountain building, and suggests new ideas about the dispersal and mating behaviour of tree weta.

Acknowledgements

We thank G. Gibbs, C. Daugherty, R. Hitchmough and G. Wallis for assistance during the course of these studies and for comments on this chapter. The Department of Conservation provided the authority to collect weta from the Conservation Estate and many private landowners allowed access to their properties for the purposes of weta collection. Thanks to the University of Otago, Victoria University of Wellington, the Royal Forest and Bird Protection Society and the Marsden Fund (PVT-601) and Foundation for Research Science and Technology (UOO704) for funding.

References

Brower, A.V.Z. (1994) Rapid morphological radiation and convergence among races of the butterfly *Heliconius erato* inferred from patterns of mitochondrial DNA evolution. *Proceedings of the National Academy of Sciences USA* 91, 6491–6495.

Excoffier, L. (1995) *Analysis of Molecular Variance*, version 1.5. Geneva.

Excoffier, L., Smouse, P.E. and Quattro, J.M. (1992) Analysis of molecular variance inferred from metric distances among DNA haplotypes: application to human mitochondrial DNA restriction data. *Genetics* 131, 479–491.

Field, L.H. (1982) Stridulatory structures and acoustic spectra of seven species of New Zealand wetas (Orthoptera: Stenopelmatidae). *International Journal of Insect Morphology and Entomology* 11, 39–51.

Field, L.H. (1993a) Observations on stridulatory, agonistic, and mating behaviour of *Hemideina ricta* (Stenopelmatidae: Orthoptera), the rare Banks Peninsula weta. *New Zealand Entomologist* 16, 68–74.

Field, L.H. (1993b) Structure and evolution of stridulatory mechanisms in New Zealand wetas (Orthoptera: Stenopelmatidae). *International Journal of Insect Morphology and Embryology* 22, 163–183.

Hebbert, D.R. (1984) Dosage compensation of the sex-linked enzyme phosphoglucomutase in the Orthoptera. *Heredity* 53, 361–369.

John, B. and Lewis, K.R. (1966) Chromosome variability and geographic distribution in insects. *Science* 152, 711–721.

King, T.M. (1997) Geographic distribution, systematics and molecular genetic analyses of colour morphs of the alpine weta (*Hemideina maori*), with specific reference to a hybrid zone. Unpublished PhD thesis. University of Otago, Dunedin, New Zealand.

King, T.M., Wallis, G.P., Hamilton, S.A. and Fraser, J.R. (1996) Identification of a hybrid zone between distinctive colour variants of the alpine weta *Hemideina maori* (Orthoptera: Stenopelmatidae) on the Rock and Pillar range, southern New Zealand. *Molecular Ecology* 5, 583–587.

King, T.M., Hanotte, O., Burke, T. and Wallis, G.P. (1998) Characterisation of four microsatellite loci in tree weta (Orthoptera: Stenopelmatidae): their potential application to the study of *Hemideina*. *Molecular Ecology* 7, 663–664.

Marchant, A.D. and Shaw, D.D. (1993) Contrasting patterns of geographical variation shown by mtDNA and karyotype organisation in two subspecies of *Caledia captiva* (Orthoptera). *Molecular Biology and Evolution* 10, 855–872.

Moller, H. (1985) Tree wetas (*Hemideina crassicruris*) (Orthoptera: Stenopelmatidae) of Stephens Island, Cook Strait. *New Zealand Journal of Zoology* 12, 55–69.

Morgan-Richards, M. (1995a) A new species of tree weta in the North Island of New Zealand (Orthoptera: Stenopelmatidae: *Hemideina*). *New Zealand Entomologist* 18, 15–23.

Morgan-Richards, M. (1995b) Weta karyotypes: the systematic significance of their variation. Unpublished PhD thesis, Victoria University of Wellington, New Zealand.

Morgan-Richards, M. (1997) Intraspecific karyotype variation is not concordant with allozyme variation in the Auckland tree weta of New Zealand, *Hemideina thoracica* (Orthoptera: Stenopelmatidae). *Biological Journal of the Linnean Society* 60, 423–442.

Morgan-Richards, M. (2000) Robertsonian translocations and B chromosomes in the Wellington tree weta, *Hemideina crassidens* (Orthoptera: Anostostomatidae). *Hereditas* 132, 49–54.

Morgan-Richards, M. and Gibbs, G.W. (1996) Colour, allozyme and karyotype variation show little concordance in the New Zealand giant scree weta *Deinacrida connectens* (Orthoptera: Stenopelmatidae). *Hereditas* 125, 265–276.

Morgan-Richards, M. and Townsend, J.A. (1995) Hybridisation of tree weta on Banks Peninsula, New Zealand, and colour polymorphism within *Hemideina ricta* (Orthoptera: Stenopelmatidae). *New Zealand Journal of Zoology* 22, 393–399.

Morgan-Richards, M., Daugherty, C.H. and Gibbs, G.W. (1995) Specific status of the tree weta from Stephens Island, Mt Arthur and Mt Holdsworth. *Journal of the Royal Society of New Zealand* 25, 301–312.

Morgan-Richards, M., Trewick, S.A. and Wallis, G.P. (2000) Characterization of a hybrid zone between two chromosomal races of the weta *Hemideina thoracica* following a recent volcanic eruption. *Heredity* (in press).

Ramsay, G.W. and Bigelow, R.S. (1978) New Zealand wetas of the genus *Hemideina*. *The Weta (News Bulletin Entomological Society NZ)* 1, 32–34.

Slatkin, M. (1985) Gene flow in natural populations. *Annual Review of Ecology and Systematics* 16, 393–430.

Stevens, G.R. (1980) *New Zealand Adrift: the Theory of Continental Drift in a New Zealand Setting*. A.H. and A.W. Reed, Wellington, 442 pp.

Townsend, J.A., Brown, B., Stringer, I.A.N. and Potter, M.A. (1997) Distribution, habitat and conservation status of *Hemideina ricta* and *H. femorata* on Banks Peninsula, New Zealand. *New Zealand Journal of Ecology* 21, 43–50.

Trewick, S.A. and Morgan-Richards, M. (1995) On the distribution of tree weta in the North Island, New Zealand. *Journal of the Royal Society of New Zealand* 25, 1–9.

Trewick, S.A., Morgan-Richards, M. and Wallis, G.P. (2000) Phylogeographic pattern correlates with Pliocene mountain-building in the alpine scree weta (Orthoptera, Anostostomatidae). *Molecular Ecology* 9, 657–666.

White, M.J.D. (1973) *Animal Cytology and Evolution*, 3rd edn. Cambridge University Press, Cambridge, 961 pp.

Part II
Morphology and Anatomy

8
Morphology and Anatomy of New Zealand Wetas

Barry O'Brien[1] and Laurence H. Field[2]

[1]*Department of Biological Sciences, University of Waikato, Private Bag 3105, Hamilton, New Zealand;*

[2]*Department of Zoology, University of Canterbury, Private Bag 4800, Christchurch, New Zealand*

Introduction

Morphology and anatomy provide a foundation upon which most other biological studies of a group may proceed, and yet there is no comprehensive account of the morphology and internal anatomy of the anostostomatid fauna of any region. The New Zealand Deinacridinae (tree wetas and giant wetas) have been described selectively and without synthesis, while very little exists for the Anostostomatinae (ground wetas). A pioneering study of the tree weta *Hemideina crassidens* by Maskell (1927) provides much careful description, but the illustrations are limiting and there is no account of the musculature. The skeleton and musculature of the head of the giant weta *Deinacrida rugosa* and a developmental account of the female terminalia of the same species have been meticulously presented by Ramsay (1955, 1965). The developmental morphology of male and female external genitalia and internal reproductive organ anatomy are reviewed (for an anostostomatine species) by Stringer and Cary (Chapter 21, this volume). Neurobiological studies have added detailed accounts of specific aspects of neuromuscular anatomy and sensory structures (e.g. O'Brien, 1984; Field, Chapter 22, this volume), but there has been little recent attention to the digestive, circulatory or excretory organ systems.

Much of the above material is difficult to obtain, if published, or is lodged in theses and has not been published. Furthermore, our knowledge is incomplete in several areas, which could easily be investigated by those with access to endemic anostostomatids. Therefore, one of the aims of this chapter is to encourage studies of morphology and anatomy of these interesting insects by indicating where knowledge is deficient.

We present a description of the musculoskeletal anatomy, followed by an account of the various organ systems. Wherever possible, this has been based on the tree weta *Hemideina*, with further material from the closely related giant weta *Deinacrida*. In condensing the material, we have focused attention on those features which are novel or distinctive, leaving the typically orthopteran character of the wetas to be conveyed by the illustrations. While the description of the head capsule condenses the source material substantially, the account of the leg and thoracic musculature is presented in full detail, because it has not appeared in print elsewhere. Moreover, the apparent bias towards the musculoskeletal system and the relatively cursory treatment of several of the organ systems simply reflect the state of our current knowledge. A histological and anatomical description of sensory and reproductive systems in the Chilean cratomiliine *Cratomelus* is given by Angulo (Chapter 11, this volume).

The only weta for which we have a more or less complete description of the morphology and anatomy is the large cave weta, *Macropathus filifer*, examined in detail by Richards (1955). A member of the stenopelmatoid family Rhaphidophoridae, this relative of the tree weta has many similarities, although it is not sexually dimorphic and there is no mechanism for sound production or for hearing. Comparisons of this animal with the deinacridine wetas are made in several of the organ system descriptions. All abbreviations used in the figures are defined at the end of the chapter text, rather than in the figure legends.

The Head

The repugnant appearance of the mature male tree weta stems mainly from its disproportionately elongated head, which may exceed 40% of the total body length in some species. It is substantially enlarged and more heavily tanned than in either the females or the preceding male instars. In *Hemideina*, the differences between the sexes stem largely from changes in proportion appearing mainly at the final moult, the head of the male being both taller and broader, without an equivalent change in the third axis. Early-instar males closely resemble females in head appearance, the structure of the head being essentially the same in the two sexes. This secondary sexual dimorphism, which is not found in the closely related giant wetas (*Deinacrida* spp.), is more fully addressed by Field and Deans (Chapter 10, this volume).

The cervical region allows considerable freedom of movement and the head may be held in various positions. *Hemideina* may rest with the head, thorax and abdomen all closely applied to the substrate, in which case the long axis of the head is parallel to the rest of the body and the mandibles are pointed forwards. This resembles the primitive prognathous condition but is a secondary consequence of the increased head size. An active male weta holds the head tilted obliquely forward, as in Fig. 8.1-1. Females and early instars tend to hold the head more nearly vertical, as do the related *Deinacrida* species. For ease of comparison with other orthopterans, the head is described here as if the long axis is vertical, with the mandibles and other mouth-parts carried ventrally and the occipital opening to the neck region (occipital foramen, Fig. 8.1-2) on the posterior surface.

The head skeleton and mouth-part descriptions draw heavily on Ramsay's (1955) detailed account of the giant weta *D. rugosa*. To support later chapters on sexual dimorphism, sense organs and neuromuscular coordination, the mandibles have been described largely from studies of *Hemideina maori*, together with additional notes on the tentorium and head capsule (O'Brien, 1984).

The following description refers to Fig. 8.1-3. Viewed from the front, the head capsule is rounded above, broad and with a flattened frontal region. Much of the epicranium in *Hemideina* shows markings representing the origin of the mandibular adductor muscles. In most genera, the parietal sutures (pts) are present as broad grooves. A fine median line, the coronal suture (cos), passes down from the vertex almost to the level of the antennae, where it divides the median ridge, or fastigium (fast), near the ends of which are the lateral ocelli (lo). Below the fastigium is the median ocellus (mo). All three are minute, pale, circular areas. In *Macropathus*, only the median ocellus is present. The compound eyes appear frontally directed, but in *Hemideina* the head is tilted upward during activity, extending the field of view above. A distinctive white patch (wp) is found on the inner dorsal margin of each eye, the cornea in this region not being clearly differentiated into ommatidial facets. This region is polarization-sensitive in other orthopterans (Nilsson *et al.*, 1987) and its prominence in the nocturnal tree wetas is unexplained.

The ability to bite strongly shows in the development of sulci and in heavy tanning of parts of the head capsule. The anterior tentorial pits (atp) are clearly evident, but the epistomal sulcus, which runs between them and delineates the clypeus and frons in many orthopterans, is not evident externally. Instead, the cuticle in this region is extensively thickened and heavily tanned. A raised subocular ridge (fgr) runs from the eye to the anterior mandibular articulation. This region, the lower gena and subgena (sug) are heavily sclerotized and the deep subgenal sulcus or pleurostomal suture (pss) indicates a well-developed internal strengthening ridge, which is continuous with the anterior tentorial arms (Fig. 8.2-5, 6).

The tentorium is the internal skeletal component of the head capsule (Fig. 8.2-5, 6). It is a well-developed X-shaped structure, bracing the capsule laterally and along the frontoposterior

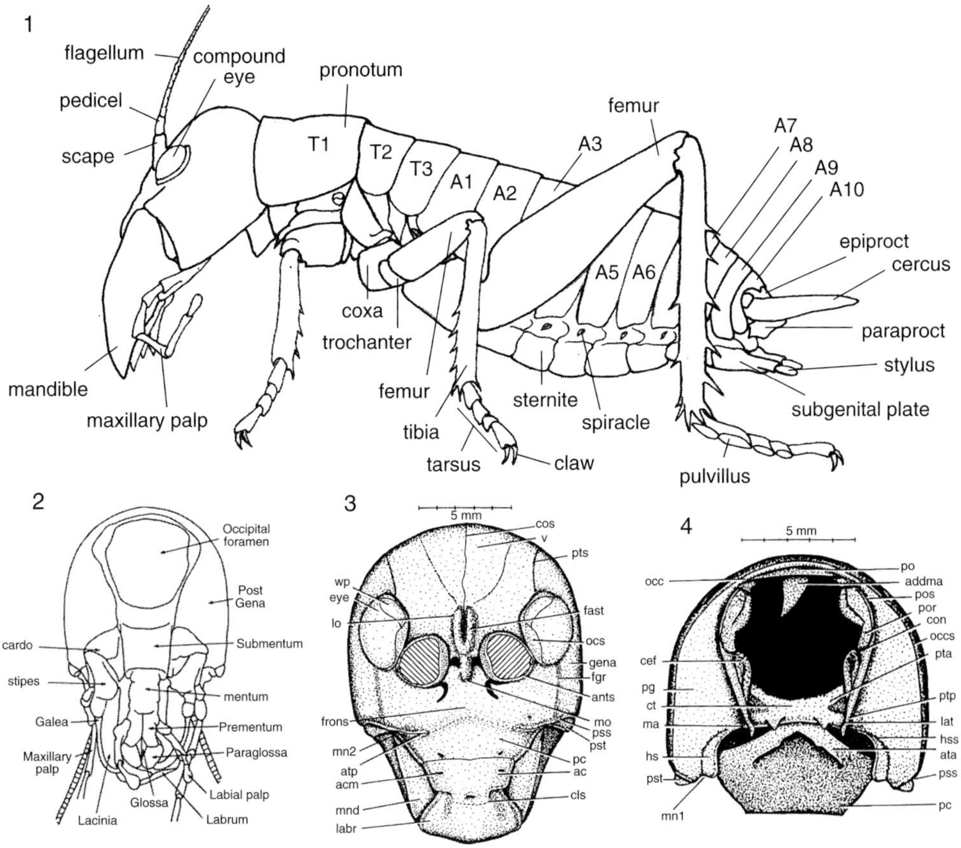

Fig. 8.1. 1, Lateral view of an adult male New Zealand tree weta, *Hemideina maori* (after Smith, 1979). 2, Posterior view of the head of an adult male *Hemideina maori* (from O'Brien, 1984). 3, Anterior view of the head of adult male giant weta *Deinacrida rugosa* (from Ramsay, 1955). 4, Posterior view of the head capsule of *Deinacrida rugosa* (from Ramsay, 1955). All abbreviations are given at end of chapter.

axis, as well as providing attachment sites for muscles associated with several of the mouthparts, the antennae, the pharynx and the cervical region (see Fig. 8.12-70). The large anterior arms run continuously into the subgenal ridge to form strong internal bracing in the region of the postgena, mandibular articulations and frons, particularly in male tree wetas (Fig. 8.2-6). Externally the tentorial pits (atp) are found in the pleurostomal suture on the frontoclypeal region (Fig. 8.1-3). The anterior arms give rise to the slender dorsal arms (dta), which connect loosely with the hypodermis between the antennal and ocular sutures (see Fig. 8.4-18). The shorter posterior arms (ptm) arise medially from the corporotentorium (ct) and are continuous with the postoccipital ridge (por), bracing the head capsule along the ventral margin of the occipital foramen (Fig. 8.1-4). The posterior tentorial pits (ptp), marking the invagination of the posterior tentorial arms, are continuous with the post-occipital suture (pos, Fig. 8.1-4).

The frons is normally bounded below by the epistomal suture, but this is not evident except in the lateral regions, where the anterior tentorial pits are clearly evident. Thus the clypeus (consisting of the anteclypeus (ac) and postclypeus (pc)) is not clearly demarcated externally from the frons. In *Hemideina*, this region is thickened and heavily tanned, particularly in the males, as it comes under considerable laterally directed tension during biting. The anteclypeus is demarcated from the labrum by the clypeolabral suture (cls, Fig. 8.1-3), and is a flexible, membranous structure, which can be completely withdrawn under

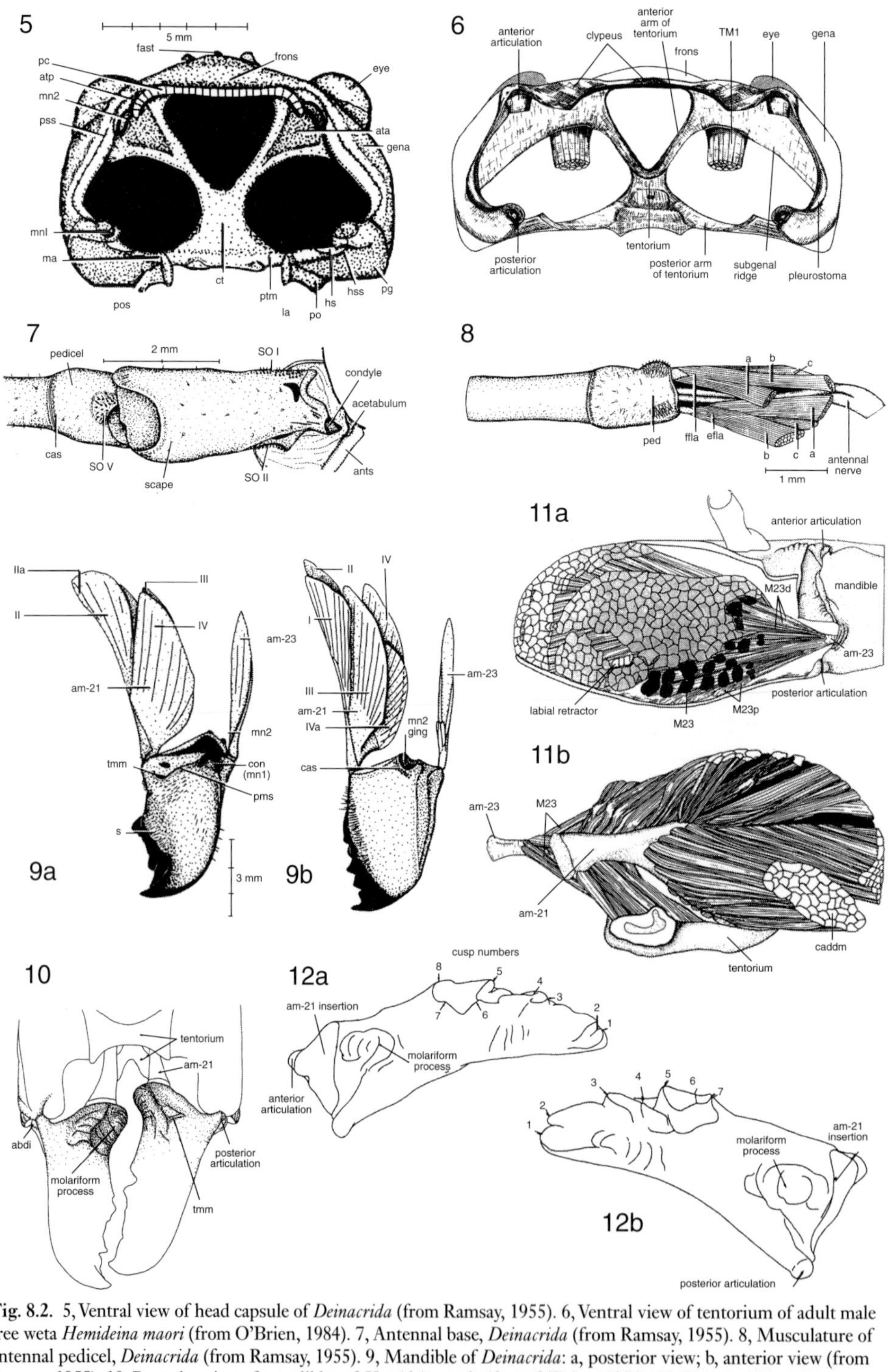

Fig. 8.2. 5, Ventral view of head capsule of *Deinacrida* (from Ramsay, 1955). 6, Ventral view of tentorium of adult male tree weta *Hemideina maori* (from O'Brien, 1984). 7, Antennal base, *Deinacrida* (from Ramsay, 1955). 8, Musculature of antennal pedicel, *Deinacrida* (from Ramsay, 1955). 9, Mandible of *Deinacrida*: a, posterior view; b, anterior view (from Ramsay, 1955). 10, Posterior view of mandibles of *Hemideina* male (from O'Brien, 1984). 11, Tergal adductor and abductor of mandible, *Hemideina*: a, lateral view; b, medial view (from O'Brien, 1984). 12, Mandibular cusp patterns of *Hemideina maori*: a, left mandible; b, right mandible (from O'Brien, 1984). All abbreviations are given at end of chapter.

the rigid postclypeus when the labrum is retracted.

A posterior view of the head capsule (Fig. 8.1-2, 4) shows a flat postgena (pg), bounded laterally by the occipital suture (occs), which demarcates the rear of the head capsule from the collar-like occiput (occ) surrounding the large opening to the neck (occipital foramen, Fig. 8.1-2)). The inner margin is demarcated by the postoccipital suture, which contains the posterior tentorial pits. A condyle (con, ball-and-socket mechanism) and a facet (cef) on the postoccipital ridge articulate with the cervical sclerites. More ventrally, the post-occiput bears the labial articulation (lat). The acetabulum of the primary mandibular articulation is formed from the ventral margin of the post-gena, where it meets the hypostoma (hs). An asymmetrical apodeme (addma), serving as an origin for muscle bundles from both left and right mandibular adductors is found on the dorsal post-occiput.

The antennae

The antennae are situated at the inner margin of the compound eyes (Fig. 8.1-3). In *Deinacrida* these are a little longer than the body, but in the tree and ground wetas they may be much longer. On the head capsule is a chitinous ring, the antennal sclerite (Fig. 8.2-3, ants), which has a single point of articulation with the basal joint or scape of the antenna (Fig. 8.2-7). There is no antennifer or pivot process on the sclerite, which, instead, has a concave acetabulum, which receives a condyle borne on an articulating process (antartis) produced from the base of the scape. Distally, the scape articulates at two points with the pedicel (ped), one a defined condyle on the pedicel and the other a looser connection with a ridge on the scape. The two segments move essentially in two planes at right angles, with the scape moving vertically and the pedicel moving largely horizontally. The pedicel gives rise to the elongate flagellum of many annuli. Two morphological types of sensilla occur on the flagellum, trichoid and placoid, both being much more abundant distally (described by Angulo, Chapter 11, this volume). The flagellum is readily broken, but regeneration will take place. No sexual dimorphism occurs in the antennae, as has been reported in other allied groups.

Three defined patches of trichoid sensilla are found on the proximal portion of the scape with a further two on the pedicel (Fig. 8.2-7, SO I, II, V; not all patches are shown). These contact the turgid conjunctival membranes when the joints are moved. A ring of campaniform sensilla (cas, defined in Field, Chapter 22, this volume) circles the distal rim of the pedicel.

Two sets of muscles move the antenna. The three moving the scape arise on the tentorium, and within the scape a pair of antagonists controls the pedicel. A single antennal depressor (see Fig. 8.4-18, dpab) arises in two places on the tentorium and inserts on the outer ventral angle of the scape. The two antennal levators (ola, ila) arise from the anterior and dorsal arms of the tentorium, respectively. The action of these three muscles restricts movement of the scape essentially to the vertical plane. Flexion and extension is achieved at the pedicel by a pair of muscles from within the scape (Fig. 8.2-8, efla, ffla). Each of these is divided into three, thereby increasing the muscle mass available to move the elongate flagellum.

Mouth-parts

A posterior view of the mouth-parts *in situ* is shown for the head of *Hemideina thoracica* (Fig. 8.1-2), while detailed descriptions of isolated mouth-parts are given subsequently.

The labrum

The labrum (labr) is carried by the anteclypeus (Fig. 8.1-3). It is ovate and the distal margin is notched in the midline. Its undersurface has a median groove, the epigusta (epg), on either side of which is a thick clothing of hairs, most of which are directed towards the groove (Fig. 8.3-15). The undersurfaces of the labrum and clypeus constitute the epipharynx, which bears a diversity of sensilla. It carries a transverse sclerite, produced on either side into a small chitinous framework, the tormae (toa, tob), which carry the posterior retractor muscles of the labrum (Fig. 8.3-15).

Mandibles

The large mandibles can gape widely and bite strongly in all tree and giant wetas. The two articulations effectively confine movement to a single plane, although the posterior articulation may dislocate on complete closure during powerful biting. The mandibular base is approximately triangular,

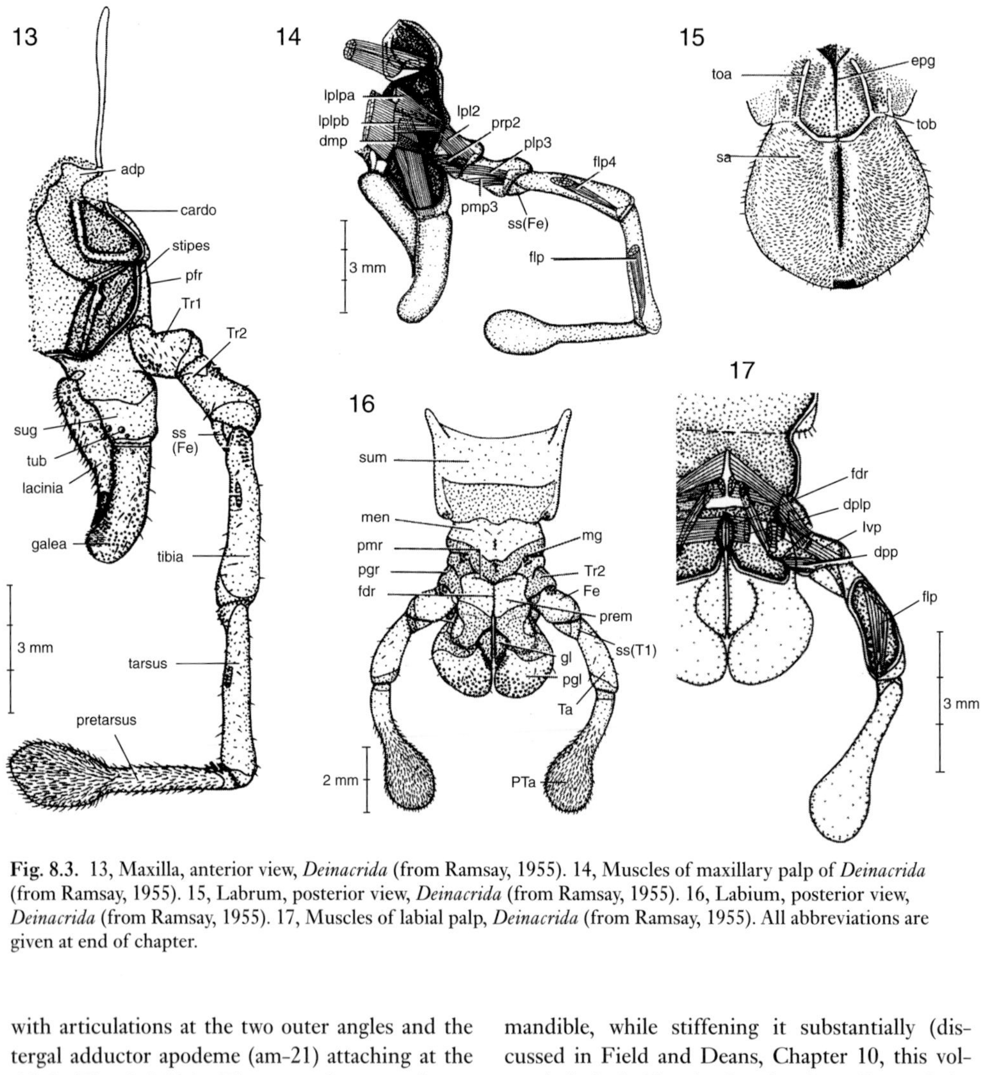

Fig. 8.3. 13, Maxilla, anterior view, *Deinacrida* (from Ramsay, 1955). 14, Muscles of maxillary palp of *Deinacrida* (from Ramsay, 1955). 15, Labrum, posterior view, *Deinacrida* (from Ramsay, 1955). 16, Labium, posterior view, *Deinacrida* (from Ramsay, 1955). 17, Muscles of labial palp, *Deinacrida* (from Ramsay, 1955). All abbreviations are given at end of chapter.

with articulations at the two outer angles and the tergal adductor apodeme (am-21) attaching at the third (Fig. 8.2-12a). The posterior or primary articulation (Fig. 8.2-6, 12c; Fig. 8.2-9a, mn1) is a spheroidal condyle matching a concave acetabulum on the posterolateral margin of the head capsule. The anterior articulation (Fig. 8.2-6) is a solidly reinforced U-shaped collar or ginglymus (ging), bearing more loosely on a thickened pillar on the clypeus (Fig. 8.2-9b, mn2), which then overlaps the articulation, further restricting upward movement. The single abductor (opener muscle) apodeme (am-23) inserts on a raised portion of the lateral margin between the two articulations (Fig. 8.2-9b, 10).

In male *Hemideina*, a carina on the anterolateral margin increases the apparent width of the mandible, while stiffening it substantially (discussed in Field and Deans, Chapter 10, this volume). It is lacking in females, juveniles and the related *Deinacrida*, suggesting that its primary function may be either for display or for strengthening the elongated mandibles. The mandible tapers and curves sharply inward at the tip. The medial surface of the distal half of the mandible is a heavily sclerotized cusp region (Fig. 8.2-9a, 9b).

The longer left mandible always overlaps the right and has a slightly different cusp pattern. The interlocking cusps produce an efficient shearing mechanism at the distal cusps and a crushing action at the more proximal cusps (Fig. 8.2-12a). Males and females have similar cusp patterns. In the most distal region, illustrated for *H. maori* in Fig. 8.2-12a, 12b, a distal apical tooth is followed by an elongate ridge, with a sharp cutting edge maintained by abrasion against its opposite. This process, thigosis, is shown in Fig. 10.8 of Field and Deans (Chapter 10, this volume). Cusp 3 is crescentic, with extended trailing ridges, while cusp 4 is another elongate ridge. Some variation may be found in this region, perhaps through physical damage. On the left mandible, the proximal cusps 5–8 are low points, connected by lower ridges to form a quadrate crushing region, which matches a similar three-cusp configuration (cusps 5–7) on the right mandible. Cusp 5 on the right locates in the depression surrounded by cusps 5 to 8 on the left mandible, and cusp 7 on the left locates in the depression surrounded by the cusps 5 to 7 on the right.

A large sclerotized protuberance, the molariform process, is found on the posterior surface near the insertion of the adductor apodeme (Fig. 8.2-10). This unique feature of male tree wetas appears to strengthen the mandible and probably has a display function, as it shows no signs of abrasion, nor can the two processes connect to play any role in mastication (a strengthening function is implicated, as discussed by Field and Deans, Chapter 10, this volume). On the posterior surface, there is a pigmented indentation (tmm), from which a postmandibular suture (pms) runs out towards the posterior articulation (Fig. 8.2-9a, 10). The pigmented region marks the insertion of the tentorial adductors and the associated muscle receptor organ (vmro), described by Field (Chapter 22, this volume).

External sensory structures, such as campaniform sensilla (Fig. 8.2-9b, cas) and numerous hair sensilla of various lengths (Fig. 8.2-9a, s), are described in Field (Chapter 22, this volume).

Mandibular musculature

The two features of particular interest in the muscles of the weta mandibles are the proliferation of the main adductor muscle and the retention and apparent proliferation of the mandibular muscles associated with the tentorium. These are developed more extensively in *Hemideina* than in any other orthopteran described.

Four distinct muscle groups are associated with the mandible. They have been numbered in accordance with Matsuda (1965). The principal adductor is the massive tergal adductor (M21, muscle 9 of Snodgrass (1928)), which fills most of the volume of the head capsule above the level of the antennae, fibres arising from the vertex, gena, parietal and occipital regions, as well as the postocciput and posterior arms of the tentorium (Fig. 8.2-11a, 11b). An asymmetrical apodeme protruding from the postocciput serves as an origin for fibres from both left and right mandibular adductors (addma, Fig. 8.2-4; caddma, Fig. 8.2-11b). As it is relatively small and firmly anchored to the epicranium, it does not constitute a mechanical coupling between the two mandibles. A multilobed apodeme (am-21) receives the insertions of all these fibres, which converge from widely disparate angles (Fig. 8.2-9, 11). Between the fibre insertion and its attachment to the mandible, the apodeme is stout, rigid and heavily tanned. It fans out before attaching obliquely to the mandible margin by a narrow zone of untanned flexible cuticle.

The sole abductor of the mandible (M23) lies in the angle between the gena and the occiput, with origins on both. All these fibres insert on a thin flexible apodeme (am-23), which attaches to a protuberance on the lateral basal margin of the mandible (abdi, Fig. 8.2-10). A small fan-shaped group of fibres arises more anteriorly on the gena and inserts on to a short branch of the apodeme close to its insertion (M23, Fig. 8.2-11).

M26 lies within the cavity of the mandible (see Fig. 22.13B in Field, Chapter 22, this volume). From the hypopharynx, a long flexible apodeme passes through a sling of tissue and angles abruptly into the body of the mandible, giving rise to a loose conical mass of fibres that insert on to the lateral wall of the mandible. Although derived from a primary sternal adductor and termed a hypopharyngeal adductor by many morphologists (Snodgrass, 1928; Matsuda, 1965), it is active during the threat display when the mandibles are abducted, and it assists retraction of the hypopharynx (O'Brien, 1984). It is active during feeding and may contribute to opening of the mouth. This primitive muscle is absent from many orthopterans.

The tentoro-mandibular muscles

A group of small muscles arises on the anterior arm of the tentorium and inserts on two different parts of the basal region of the mandible. They are homologous with a group of primary sternal adductors, which are present as small remnant muscles in several primitive insect orders and in some, but not all, other orthopterans. They correspond to muscle 25 in Matsuda's nomenclature. The various components are here named TM1, TM2a and 2b and vmro, the ventral muscle receptor organ (see Fig. 22.13A in Field, Chapter 22, this volume). TM1 arises on the tentorium (Fig. 8.2-6) and inserts on a well-defined indentation of the posterior face of the mandible (tmm). The much smaller TM2a arises from an apodeme immediately adjacent to TM1, bends over the raised insertion of TM1 and inserts just distal to it. At its proximal end, TM2a divides and sends an even finer branch, TM2b, to the anterior face of the mandible. A mechanoreceptor cell body lies alongside this minute muscle and has dendrites in series with its insertion (dmro, Fig. 22.13B in Field, Chapter 22, this volume). A muscle receptor organ (vmro) arises and inserts immediately beside TM2a but spirals round it approximately 180°. The functioning of this complex stretch receptor is discussed in Field (Chapter 22, this volume). With a cross-sectional area of less than 1 mm, TM1 can make only a minor contribution to the total adductor force and the branches of TM2 seem insignificant in mechanical terms, but their associated sensory structures suggest that they play a significant role in mandibular functioning (see Field, Chapter 22, this volume). These muscles are found in *Deinacrida* (except for *2b*) but apparently not in *Macropathus.*

The presence of tentoro-mandibular adductors and the hypopharyngeal adductor is generally considered to be a primitive condition intermediate between that found in apterygotes and Ephemeroptera and the more advanced condition reported from most mandibulate insects, where only the tergal adductor and abductor are present (Matsuda, 1965). TM adductors are found among the Blattaria, Mantodea and Phasmida and in some other orders, although not in all species. They are considered to be absent from most holometabolous insects, although Matsuda (1965) refers to their being found in tabanid flies. Honomichl (1976, 1978) has reported these muscles from larval and adult beetles of different families. In each case, they have associated sensory structures which appear homologous to those found in wetas (described in Field, Chapter 22, this volume). Their presence or absence may thus depend on functional requirements and, while their primitive origins are not in dispute, they should not simply be regarded as vestigial structures of little functional significance.

As the mechanics of the mandibles would seem to be broadly similar amongst the Orthoptera, the control systems presumably face similar requirements. The association of several sense organs with the tentoro-mandibular musculature raises the question of how control is achieved in those species that apparently lack these muscles.

The maxillae

The maxillae are of typical orthopteran form, but distinguished by elongate palpi, which is a familial character (Fig. 8.3-13). In the cave weta, *Macropathus*, these reach unusual lengths. Proximally, the triangular cardo articulates with the head capsule via a single condyle. A flexible membrane forms a hinged joint with the stipes, from which a palpifer (pfr) bears the palp. The stipes also bears two terminal lobes, the lacinia and galea. The lacinia bears two firm distal teeth and one flexible tooth, as well as numerous trichoid sensilla and two groups of campaniform sensilla. Along the anterior surface is a row of six or seven large, thin-walled, conical tubercles (tub), similar in appearance to Brunner's organ in the hind femora of acridids (Fig. 8.3-13). Further tubercles of this type are found on the subgalea (sug) and the blunt-tipped galea, which also bears numerous trichoid and basiconic sensilla, probably serving as contact chemoreceptors. The palpus consists of five complete segments and one subsegment (ss), between the second and third segments, which receives the insertion of a muscle not described from other orthopterans. Ramsay argues that this subsegment is homologous to the femur, disagreeing here with Snodgrass's interpretation of the subsequent segment being the femur. The expanded terminal portion of the palp has a dense covering of trichoid and basiconic sensilla.

The musculature of the maxillae is in general similar to that of other Orthoptera, although there are two additional muscles in the palp which have not been described from other species (prp2, pmp3, Fig. 8.3-14). The first of these is a promotor of the second palp segment; it arises ventrally in the basal palp segment and inserts medially on the anterior proximal rim of the second segment. The second (pmp3) promotes the third palp segment, arising from from the floor of the second palp segment and inserting mainly on to the subsegment described above. The presence of these may reflect the unusual length of the palps, which are very active when the animal is feeding. The more distal segments are moved by flexors (flp, flp4).

In *Hemideina* the lacinia and galea are elongated in the adult males. The enlargement of the mandibles has shifted the cusp region further from the mouth, and the elongation of the distal elements of the maxillae probably results from the altered requirements of food manipulation, rather than from involvement in display, although this has not been tested.

The labium

The submentum (sum), mentum (men), prementum (prem) and a ligula, composed of paired paraglossae (pgl) and glossae (gl), are all present (Fig. 8.3-16). The three proximal regions are separated by membranous areas, allowing the mentum and prementum to be retracted beneath the proximal regions when the labium is raised by the labial retractor (Fig. 8.2-10). The prementum is grooved medially, revealing the origin of the labium as fused second maxillae, and laterally, demarcating the palpiger (pgr). The median groove (fundarima, fdr) connects with an anterior premental sclerotization (pmr), from which a number of palp and ligula muscles arise. The glossae and paraglossae are attached to the distal membrane of the prementum. The glossae are small, straight and pointed and each bears a number of short, stout, trichoid sensilla on its anterior and mesial surfaces. The paraglossae are large, blunt, curved lobes, almost twice as long as the glossae, which they overtop laterally. Each is mainly membranous but contains four sclerotized areas. There are two groups of trichoid sensilla, one mesial and one distal. Basiconic and campaniform sensilla are scattered over both surfaces. In *Hemideina* males, the labium is elongated, mainly through enlargement of the submentum and mentum.

Each palpus consists of three segments, the third being swollen distally and thickly clothed with trichoid sensilla, and moved by a flexor muscle (flp, Fig. 8.3-17). In addition to this general orthopteroid organization, there is a subsegment (ss/Ti) between the first and second segments. In this feature, it resembles the maxillary palps. As with the maxillary palp, Ramsay (1955) interprets the associated musculature as indicating that this was once a complete segment, in this case homologous to the tibia. Similar evidence suggests that the palpiger was also once a complete segment and not simply a subdivision of the prementum. In this interpretation, all six primitive segments of the palp can still be traced in *Deinacrida* and the palpiger is homologous with the proximal trochanter of the maxillary palp, rather than with the palpifer of the maxilla.

Primitive features of the labial musculature of *Deinacrida* include the presence of several muscles not found in acridids by Albrecht (1953) and Snodgrass (1928) and some not present in *Grylloblatta* (Walker, 1931). A promotor of the palpiger (dplp) has a different function from that described by the above workers, who found it to insert on the proximal palp segment, and a promotor muscle of the middle palp segment (dpp) has not been described before (Fig. 8.4-18). The division of the labial palp levator muscle suggests that a more refined movement is possible. The possibility that retention of an apparently primitive feature might also involve more recent specialization is a recurring theme in the interpretation of weta musculature.

Hypopharynx

This is a fleshy, tongue-shaped lobe (Fig. 8.4-20) forming the floor of the preoral cavity, or cibarium. A median transverse groove (gr) divides it into a proximal basipharynx (bph), continuous laterally with the mentum, and a free distal portion (dph). Prominent bands of trichoid sensilla are located laterally (bts, Fig. 8.4-20, 21). Anteriorly, the distal lobe has a paired band of fine trichoid sensilla (tb, Fig. 8.4-20) and a large area of similar sensilla covering the ventral tip (td). Laterally, the wall of the distal portion contains two sclerotizations, the salivia (sal, Fig. 8.4-21).

The musculature of the hypopharynx follows the usual orthopteroid pattern. A hypopharyngeal retractor (hypr, Fig. 8.4-21) arises on the postocciput and inserts on the basal sclerite (salivia). Paired muscles within the lobe dilate the salivary orifice. A median muscle arising on the tentorial body dilates the oral aperture. Two other muscles associated with the hypopharynx also probably alter the mouth aperture; the so-called hypopharyngeal adductor (M26) which arises on an apodeme (am-26) and inserts within the mandible (Fig. 8.4-20, 21) and the retractor of the mouth angle (rao), which arises on the frons and inserts on the arm of the pharyngeal sclerite (prg). The

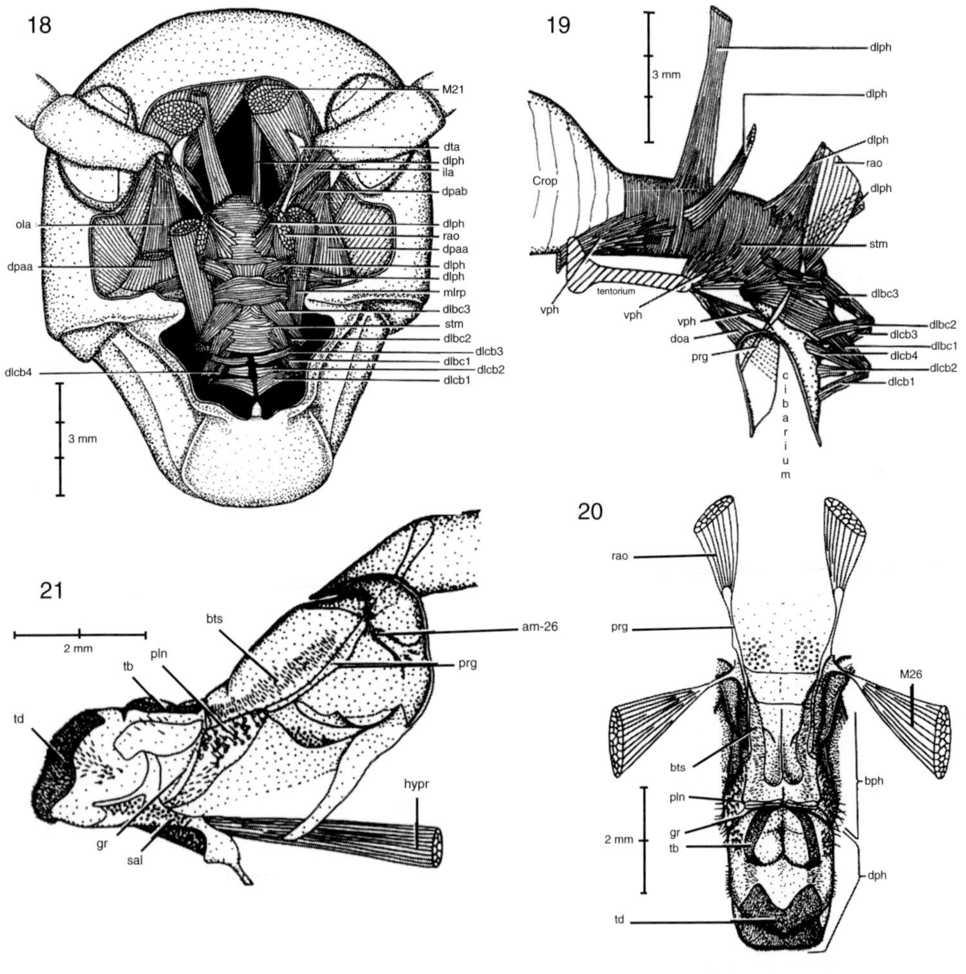

Fig. 8.4. 18, Anterior dissection of head of *Deinacrida* (anterior labral retractor muscles removed) showing antennal and stomodaeal muscle anatomy (after Ramsay, 1955). 19, Lateral view of stomodaeal musculature of *Deinacrida* (after Ramsay, 1955). 20, Frontal view of hypopharynx, *Deinacrida* (from Ramsay, 1955). 21, Lateral view of hypopharynx, *Deinacrida* (from Ramsay, 1955). All abbreviations are given at end of chapter.

hypopharyngeal adductor is also a retractor of the hypopharynx, rather than an adductor of the mandible (see under mandibular musculature).

Muscles of the stomodaeum

The stomodaeum (stm) is moved by numerous dilators arising on the head capsule and tentorium as well as by the circular and longitudinal layers ensheathing the crop and pharynx (Fig. 8.4-18, 19). The intrinsic musculature is dealt with in the section on the alimentary canal. The multiple pairs of dilators are designated according to the regions of the stomodaeum on which they operate, as follows: oral aperture (doa), the buccal cavity (dlbc1,2,3), the pharynx (dlph) and the cibarium (dlcb1–4) (Fig. 8.4-18, 19). With the notable exception of the cibarial dilators, the stomodaeal musculature is very similar to that described for *Dissosteira* (Snodgrass, 1928). *Deinacrida* has four pairs of cibarial dilators, two of these not being found in other orthopteroid insects (Fig. 8.4-19, dlcb1–4). As three of these pairs (dlcb1,2,4) arise under or close to the anteclypeal macula (acm, Fig. 8.1-3), the second and fourth pairs may be secondary divisions of the first, as in the thysanuran

Thermobia, where this muscle is represented by nine bundles (Chaudonneret, in Matsuda, 1965). This may be a consequence of the large size of the weta, which may exceed 90 mm in length, excluding the ovipositor, or perhaps its unusual mode of feeding. *Deinacrida* ingests large portions of leaf material, which is both bitten by the mandibles and torn by the maxillae and labium extending ventrally while grasping material held by the mandibles. After feeding voraciously for perhaps 15 min, the insect retires and, following a quiescent period, it begins to masticate food apparently regurgitated from the pharynx.

Cervix

The neck region is largely membranous and the head has considerable freedom of movement. *Hemideina thoracica* has two pairs of lateral cervical sclerites (lcs, Fig. 8.5-23). The anterior and smaller of the pair articulates with the postoccipital ridge by a condyle (con, Fig. 8.1-4). The posterior and larger sclerite is separated from the anterior by a transverse invagination or apodeme. In the ventral surface of the neck is a pair of faintly chitinized small circular patches, the ventral cervical sclerites (vcs).

Thorax

It is useful to review the basic anatomy of the thorax in order to have a grounding for understanding the positions of origins for the various coxal and trochanteral muscles of the leg. Each of the three thoracic segments (prothorax, mesothorax and metathorax) is composed of two plates: the dorsal tergite and the ventral sternite. These are joined by the membranous body wall, in which are embedded a number of chitinous plates (sclerites), more or less continuously fused and of evolutionarily complex origin. Together they constitute the pleuron. The coxa (Cx), or basal segment of each leg, is attached to the junction of each pleuron and sternite (Fig. 8.5-22, 23).

The tergite (also called the notum) is composed of three sclerites: a large anterior sclerite, the scutum, which is fused to a narrower posterior sclerite, the scutellum (see Fig. 9.7 in Field and Bigelow, Chapter 9, this volume). The prescutum forms a broad anterior and lateral region, fused to the scutum and scutellum and demarcated by a prominent suture.

The pleuron is composed of two major sclerites fused along the prominent pleural suture (ps), which runs diagonally from the anterior dorsal corner of the pleuron to the ventral posterior

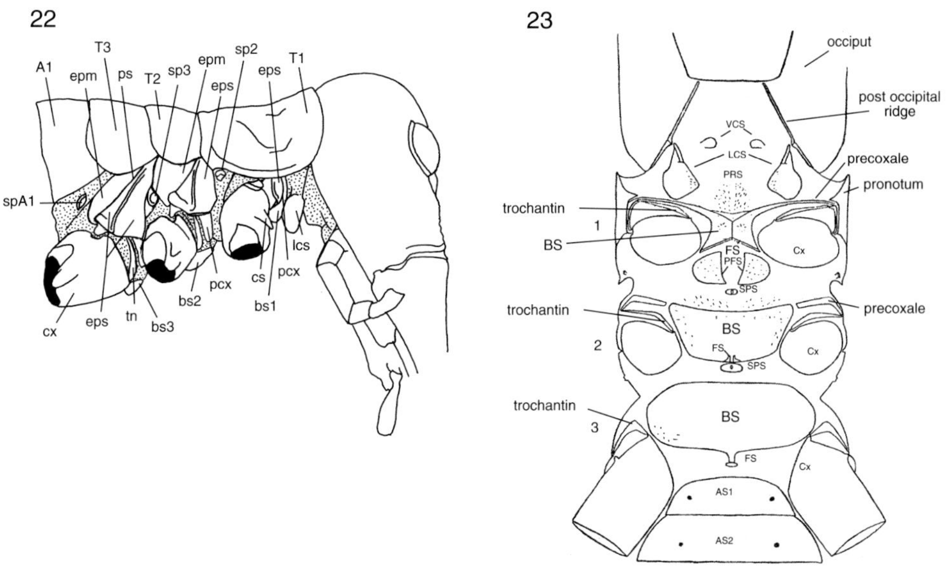

Fig. 8.5. 22, Lateral view showing external thoracic anatomy of male tree weta, *Hemideina crassidens*. Stippled areas indicate pleural and joint articulation membranes. 23, Ventral view of cervical and thoracic regions of *Hemideina thoracica* (after Maskell, 1927). All abbreviations are given at end of chapter.

pleural corner. Internally, this suture is marked by the prominent pleural ridge (pr), which ends ventrally in the coxal process, an elaboration forming the condyle of the thoracocoxal joint (see Fig. 8.7-37). The anterior pleural sclerite is called the episternite (eps), while the posterior sclerite is called the epimeron (epm, Figs 8.5-22 and 8.8-37). Ventrally, the pleural ridge forms a foot-shaped projection, the pleural arm (pa, see Fig. 8.7-37), which extends into the body cavity over the coxal aperture and above the coxal process (c). It is joined to the sternite by strong, very short pleurosternal muscles (PS1, 2, see Fig. 8.7-32, 33). This structure reinforces the hollow thoracic box by providing a triangular brace across the two ventral corners.

The sternite is divided into two transverse plates, an anterior, inflated basisternite (BS) (described as the verasternite by Maskell (1927)) and a narrower posterior furcasternite (FS), which spans the two coxae. Internally, the furcasternite gives rise to the furca (sternal apophysis), which is a paired, branched structure projecting dorsally into the body cavity and laterally toward the coxal aperture.

Prothorax

The pronotum is very broad and overlaps the insertions of the legs, the cervix in front and the anterior margin of the mesonotum. The first spiracle (sp2) has migrated forward from the mesothorax to be found just below its lateral margin. In *Hemideina*, a small portion of the episternite is visible below the lateral margin of the pronotum, to which it is very closely apposed, but it extends dorsally, beneath the pronotum, for some distance and is about one-third the length of the pronotum. The thickened ventral margin articulates with the trochantin (tn, Figs 8.5-23 and 8.7-37) (a small scimitar-shaped sclerite articulating with the anterior margin of the coxa) and the thickened posterior margin is the entopleural ridge or pleural suture (ps, Fig. 8.5-22). The epimeron is absent or reduced to a minute part hidden below the pronotum.

In the prothorax, the sternite is more complex than in the other thoracic segments (Fig. 8.5-23). A small faintly chitinized area in the cervix is probably the prosternite (prs). Behind this is the basisternite (bs), which is produced into long arms directed forward and laterally, forming the precoxal bridges or precoxales (pcx), which connect the basisternite and episternite. The furcasternite is fused with the basisternite. There is no postcoxale. The postfurcasternite (PFS) is divided into two by an untanned region, behind which is the small oval spinasternite (SPS), bearing a single median spina.

Mesothorax

The mesonotum is less broad than the pronotum. It overlaps the anterior portion of the metanotum. The mesopleuron consists of an anterior episternum and a posterior epimeron, joined at the pleural suture, which projects inwards as the pleural ridge or entopleuron. Its ventral end articulates with the coxa via the enlarged coxal process (c, see Fig. 8.7-37). The slender trochantin articulates with a small process on the anterior margin of the coxa, while the other end abuts the base of the episternum. The second thoracic spiracle is situated below and slightly behind the epimeron (Fig. 8.5-22).

Metathorax

The metanotum is overlapped by the mesonotum. Both episternite and epimeron are present and separated by the pleural suture (ps). As in the mesothorax, the pleural suture runs obliquely backwards to the pleural coxal process. The trochantin articulates with a process on the anterior margin of the coxa and extends to the base of the episternum. The sternite consists of a large basisternite, with the small furcasternite fused to the posterior margin. There is no separate spina, unlike the thorax in the rhaphidophorid *Macropathus*, which is apparently unique in this feature (Richards, 1955). Internally, the furcasternum gives rise to the furca (sternal apophysis), which is a paired, branched structure projecting dorsally into the body cavity and laterally toward the coxal aperture. Each side of the sternal apophysis (mtsa) has a vertical stem which gives rise to a horizontal arm (ar, Figs 8.6-30 and 8.7-37). The arm divides into an anterior and posterior thumb (ath, pth, respectively) and, at its distal end, produces a flat blade, called the apophysis wing (w), which projects directly beneath the pleural arm (Fig. 8.6-29, 31). Each stem is joined by a horizontal bridge, the cut end of which is seen in Fig. 8.7-37.

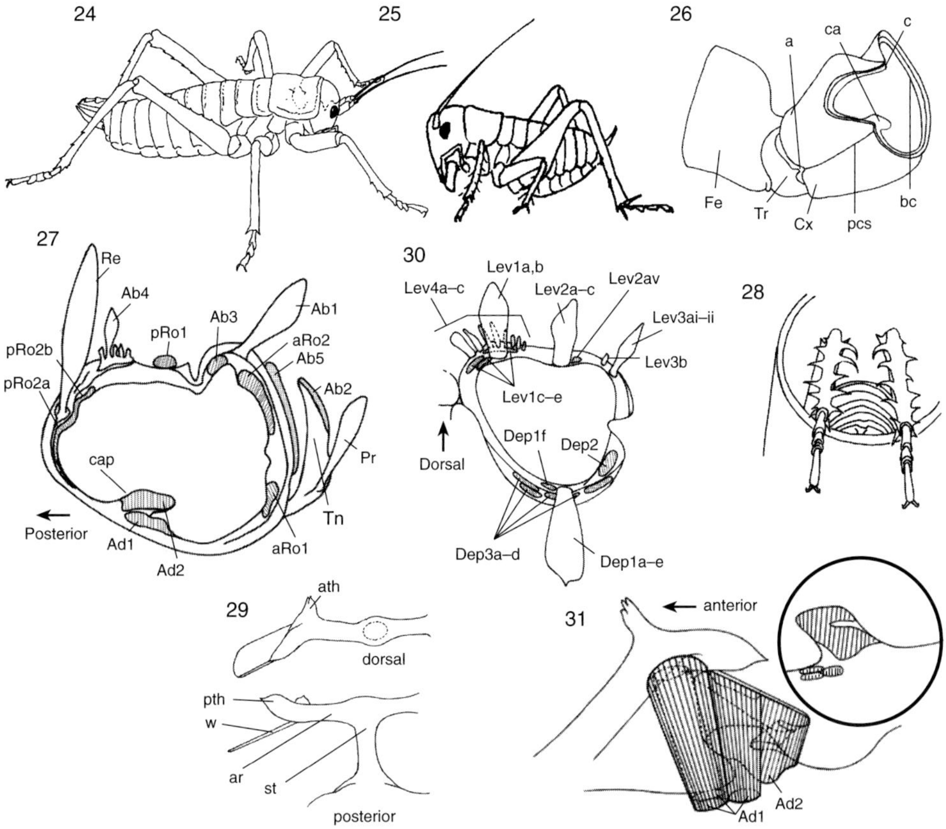

Fig. 8.6. Details of weta leg morphology and anatomy. 24, *Deinacrida connectens*, an alpine giant weta (Deinacrideinae), showing the slender shape of the hind leg femur. 25, *Hemiandrus maculifrons*, a henicine (anostostimatine) weta with an enlarged hind femur for jumping. 26, Posterior view of left metacoxa and trochanter. 27, Proximal rim of right metacoxa showing apodemes and muscle attachment sites (hatched) on basicoxite and trochantin. Proximal orientation shown by arrow. 28, Posterior view of *Deinacrida rugosa*, showing recurved proximal spines on hind tibiae. 29, Metasternal apophysis, dorsal (upper) and posterior (lower) views. 30, Proximal rim of trochanter showing apodemes and muscle attachment sites (hatched). Dorsal orientation indicated by arrow. 31, Attachment sites of the coxal adductor muscles (four components) from coxal apophysis to metasternal apophysis, showing origins on underside of wing. Inset: proximal view of insertion sites (hatched) on coxal apophysis. All abbreviations are given at end of chapter.

External Anatomy of the Legs

The legs are attached to the three thoracic segments, and are accordingly named the prothoracic legs (forelegs), the mesothoracic legs (midlegs) and the metathoracic legs (hind legs). The leg consists of six segments, the coxa, the trochanter, the femur, the tibia, the tarsus and the pretarsus (Fig. 8.1-1). Particular leg segments may be distinguished by the pro-, meso- or meta- prefix. The following account will deal with the metathoracic leg of the tree weta, *Hemideina femorata,* as a representative example. It should be borne in mind that the anostostomatine and deinacridine wetas are all secondarily apterous, and hence lack the differences seen between the pro- and the meso-, metathoracic legs in pterygote orthopterans. In these, direct flight muscles (basalar and subalar) connect the wing bases in the meso- and metathorax to the coxae of the respective two pairs of legs (Matsuda, 1963).

The coxa is a short, truncated, conical segment attached to the body at the intersection of the pleuron and the sternum. On its anterior side,

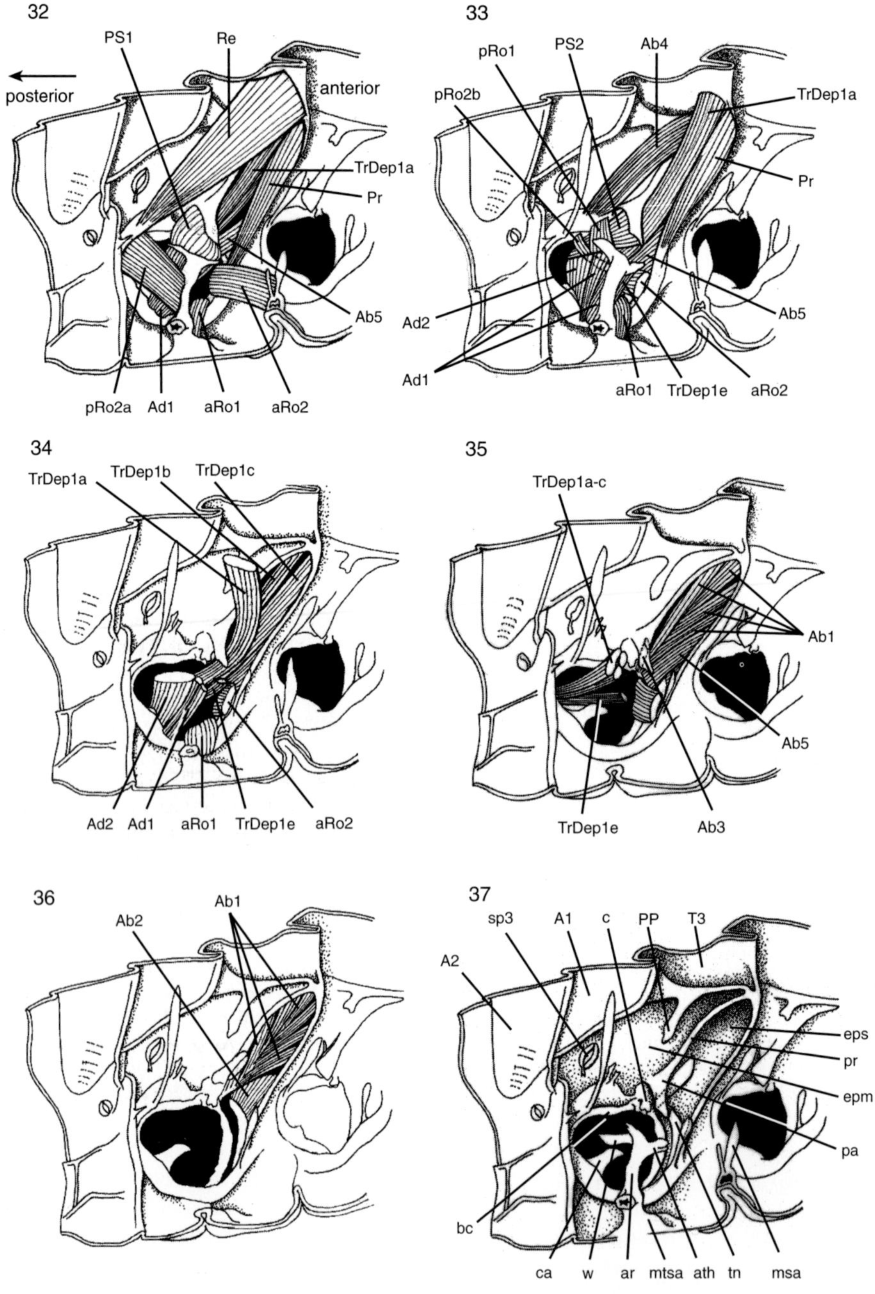

Fig. 8.7. Extrinsic coxal muscles and lateral aspect of left metathorax, showing progressive removal of overlying muscles. 32, Intact anatomy, posterior on left, anterior on right. Note internal impression of stridulatory ridges on second abdominal tergite, robust size of coxal remotor muscle (Re) and unusual origin of second anterior rotator (aRo2) on mesothoracic sternal apophysis. 33, Removal of rotators and remotor reveals muscles originating on metasternal apophysis, and unusual fourth abductor (Ab4) which often inserts on an apodeme at both ends. 34, Tripartite trochanteral depressors (Dep1a–c) and coxal adductors (Ad1–2) revealed by removal of apophysis. 35, Overlying coxal abductors (Ab1, 5) deeply embedded in episternite. Trochanteral depressor components clearly seen. 36, Superficial coxal abductor (Ab2) now seen in episternite. 37, Skeletal elements of meso- and metathorax. All abbreviations are given at end of chapter.

closely linked but within the membranous articulation to the body, is a narrow plate, the trochantin. This is thought to be the evolutionarily derived remnant of the subcoxa, and retains a loosely linked connection to the coxa, which represents a primitive anterior pleuro-coxal articulation (Snodgrass, 1927). Proximally the coxa is rimmed by a narrow internal circular shelf, known as the basicoxite (bc, Fig. 26). This provides an internal surface to which many of the coxal muscles attach, and it is demarcated externally by a fine suture (the basicostal suture). However, the coxal rim is very different from that of the pterygote orthopterans, in which the basicoxite is considerably widened into a plate-like surface (meron) to allow for the attachment of the basalar and subalar

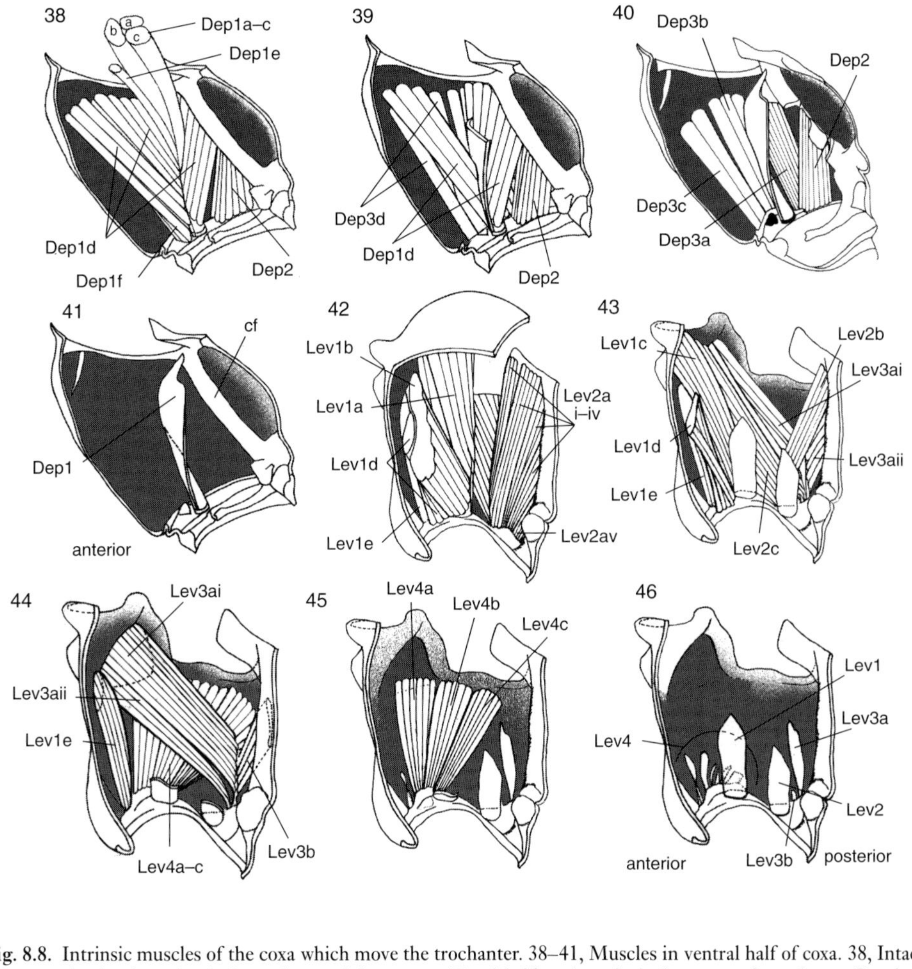

Fig. 8.8. Intrinsic muscles of the coxa which move the trochanter. 38–41, Muscles in ventral half of coxa. 38, Intact anatomy, showing large intrinsic trochanteral depressors (Dep1d, 1f) and extrinsic depressors from thorax (Dep1a–c, 1e). 39, Trochanteral apodeme cut to show underlying trochanteral depressors (Dep2, 3d) inserting below trochanteral rim. 40, Complete removal of first depressor bundles shows remaining superficial components of third trochanteral depressor (Dep3c). 41, Skeletal elements of ventral half of coxa, including intact depressor apodeme. 42–46, Muscles in dorsal half of coxa. 42, Intact dissection showing first and second trochanteral levators (Lev1a–e, Lev2ai–v). Ventral section of coxa included to show origins of first levator components (Lev1a, b, d, e), which cross midline to insert on dorsal half of coxa. 43, Underlying components of first, second and third levators revealed by removal of Lev1a,b, 2a. 44, All components attached to first and second levator apodemes removed. Third levator (Lev3a, 3b) and apodeme insertions clearly seen. 45, Fan-shaped fourth levator (Lev4a–c) against coxa wall. 46, Skeletal elements of dorsal half of coxa. Note partially circled apodeme cluster of fourth levator (Lev4). All abbreviations are given at end of chapter.

direct flight muscles (Snodgrass, 1927). Along the posterior face, another major (posterior) suture (pcs) runs from the proximal rim, at a point marked internally by a thumb-like projection, the coxal apophysis (ca), to the posterior articulation with the trochanter. A broad inner fold (cf, Fig. 8.8-41) along this suture provides surface area for intrinsic muscle origins within the coxa. The coxa articulates with the thorax by means of a single condyle (con, Fig. 8.6-26). This monocondylic joint allows an enormously versatile range of movement of the leg about its articulation, while the remaining joints in the leg are restricted to a single plane of movement by dicondylic (two-condyle) joints. The trochanter has an extreme range of levation against the coxa, as seen when the legs are raised in the defence display. This occurs by virtue of the deeply concave dorsal coxal margin, which allows the trochanter to move well in against the coxa as it is raised. The coxa itself is activated by massive muscles within the thorax, to provide the leg with strength and proximal stability about the monocondylic joint, while the more distal dicondylic joints restrict movement to one plane. They provide the leg with articulation strength at the expense of muscle bulk.

The metatrochanter is a greatly shortened segment, which makes a dicondylic connection to the coxa (a), but which is fused to the more distal femur (Fig. 8.6-26). Hence there is no movement of the trochanterofemoral joint. The pro- and mesotrochanters are much longer cylindrical segments, as is also the case in the locust and cricket.

The hind femur is stoutly constructed but not enlarged proximally as much as in jumping orthopterans, such as the cricket and locust (Fig. 8.6-24 for *Deinacrida*; Fig. 9.2 (in Field and Bigelow, Chapter 9, this volume) for *Hemideina*). It is about 1.5 times the length of the pro- and mesofemora in tree wetas. External longitudinal reinforcing ridges, as seen in the locust, are absent in all wetas, as well as in the cricket (Thakare, 1972b). Distally, the metafemur forms a heavy dicondylic articulation with the tibia, which itself is very strong, thick and partially flattened dorsally. A special feature described for the femoro-tibial joint – not observed in any other orthopteran – is a pair of lubricating pads on the inner contact surfaces of the distal femur flanges, the genicular lobes, which overlap and bear against the proximal surface of the tibia. These flanges apply lateral pressure to the sides of the tibia, which, owing to its angled surfaces, is automatically flexed to an angle of about 70° when the muscles of the femur relax (Hoyle and Field, 1983).

The robust nature of the hind tibia is due, in part, to the heavy armament of spines projecting from the antero- and posterodorsal aspect of the length of the tibia (Fig. 9.1A, in Field and Bigelow, Chapter 9, this volume). Such strong hind tibiae are not found in ground wetas or in most other ensiferan or caeliferan Orthoptera. This reinforcement may relate to the major dependency of tree wetas on the hind tibiae as defence weapons, even after being grasped by a predator (e.g. Fig. 16.5, in Field and Glasgow, Chapter 16, this volume). Tibial spines in New Zealand anostostomatids range from the strong, very sharp structures in tree wetas and giant wetas to short, delicate spines, which are more numerous on the thin anostostomatine tibiae. A highly effective defence adaptation is seen in the strong, recurved tibial spines near the femoro-tibial joint of the hind leg in the giant wetas, *D. rugosa* and *Deinacrida parva* (Fig. 8.6-28). The insects close the tibiae against anything poking close to the abdomen and, when withdrawn, the recurved spines rake the intruding object.

The meso- and protibiae are subcylindrical and slightly appressed anteroposteriorly. Each protibia bears two ears of equal size, the tympanal organs (described in Field, Chapter 22, this volume). Tympanal organs are not found in the Anostostomatinae, even though femoro-abdominal stridulatory structures are present.

The tarsus is composed of three cylindrical segments, the most proximal of which has two pulvilli (pads, Fig. 8.1-1) beneath, while more distal ones have a single pulvillus, as does the pretarsus. The claws are strong, and especially enlarged in the arboreal giant wetas.

Musculature of the Hind Leg

The muscles of the hind leg of *H. femorata* will be compared with descriptions of hind leg muscles of the related pterygote orthopterans, the locust (*Locusta* (Albrecht, 1953)) and the cricket (*Gryllus bimaculatus* (Thakare, 1972a)). This comparison is of interest because it allows an examination of the differences between winged species and the wingless tree wetas. It is especially interesting because it brings to light undescribed differences, which appear to be related to retained features of the primitive ancestry of wetas. Typically, only the foreleg is described in comparative works, since the mid- and hind legs are modified to receive bifunctional muscles from the wings. The following comparison with the apterous tree weta brings to light: (i) the presence of more muscles than seen in the other orthopterans; and (ii) several curious differences in modifications of muscles and apodemes, presumably associated with the loss of wings, with retained ancestral conditions and/or with adaptations for the defence posture. All references to muscle numbers in the locust and cricket refer to the two papers cited above, which present hind leg muscle descriptions. Extrinsic muscles are those which insert on to the proximal rim of a segment but have origins in the next proximal segment (or in the thorax, in the case of the coxa and trochanter). Intrinsic muscles are those arising within a segment and operating the next more distal segment.

Coxa

Extrinsic muscles

The immediate impression gained when viewing the internal thoracic wall is the marked difference from the thorax of pterygote orthopterans, due to the lack of the powerful direct and indirect flight muscles. There are no phragmata bearing dorsolongitudinal indirect flight muscles, nor are there the prominent vertical columns of basalar/subalar direct flight muscles arising from a modified coxal rim (meron, described below). Instead, the most noticeable muscles are the coxal remotor, promotor and rotators seen in Fig. 8.7-32 (Re, Pr, Ro). These are all extrinsic muscles, with origins in the thorax and insertions on to the coxa.

For convenience in understanding the complex organization of extrinsic coxothoracic muscles, they are divided according to their sites of origin: tergite, sternite and pleuron. This scheme is especially important in trying to attach names and possible homologies for the muscles in the tree weta.

The proximal rim of the tree weta metacoxa bears the insertions of 14 extrinsic muscles, which have origins in the thorax (Fig. 8.6-27). Four muscles insert by means of apodemes and the remaining ten muscles insert directly on to the coxal rim, arthrodial membrane and trochantin. Three of the apodemes (Pr, Re and Ab1, Fig. 8.6-27) are large, blade-like structures, typical for other orthopterans, but the fourth (Ab4) apparently has not been described in other orthopterans. It consists of a linear cluster of one small blade and four tiny apodemes mounted on the posterodorsal coxal rim. A similar dispersion into many tiny apodemes is also found for the tree weta trochanteral levators (Figs 8.6-30 and 8.8-46). These may have implications for interpreting primitive characters in tree wetas, compared with the more modern orthopterans.

COXAL MUSCLES ORIGINATING FROM THE TERGITE. The promotor (Pr) is a single muscle arising on the anterior edge of the metanotum and inserting on to a spatulate apodeme on the ventral end of the trochantin (Figs 8.6-27 and 8.7-32, 33). In the cricket, it comprises two muscles, one stout and one thin (149, 149a), which insert via two apodemes on to the trochantin in a similar location to that in the tree weta. In the locust, the promotor is a single muscle (118) inserting on to the anterior angle of the coxa.

The massive remotor (Re) is the largest muscle inserting on to the coxa. Its origin is on the posterolateral corner of the metanotum, near the junction with the metapleuron, and it inserts on to the dorsal posterior rim of the coxa by a broad apodeme (Figs 8.6-27 and 8.7-32). In contrast, two remotors are found in both the locust (119, 120) and the cricket (150, 150a), one of which is very robust, while the other is thin. The insertion locations on the coxa are very similar in all three insects.

COXAL MUSCLES ORIGINATING FROM THE STERNUM. The anterior rotators (aRo1, 2) comprise two muscles inserting directly on the anterior basicoxite surface and originating on the sternal apophyses (Figs 8.6-27 and 8.7-32–34). The first anterior rotator (aRo1) inserts below the midpoint of the anterior rim and joins the ventromedial side of the stem of the metasternal apophysis (Fig. 8.7-34), much as seen in the cricket (151) and in the locust (121). However, the second anterior rotator (aRo2) inserts above the anterior coxal rim midpoint and extends anteriorly to the lateral arm of the mesosternal apophysis. This additional muscle has not been described for the locust or for the cricket, nor has a metathoracic coxal muscle been described attaching to the mesosternum.

Three posterior rotators are found in the tree weta (pRo1, pRo2a, pRo2b). As with the anterior rotators, they are identified by their origins on the metasternal apophysis. All three insert without apodemes on to the basicoxite. The first posterior rotator (pRo1) has its origin on the anterior thumb and edge of the wing of the apophysis, and it inserts on the dorsal side of the basicoxite, next to the joint condyle (Figs 8.6-27 and 8.7-32). The second posterior rotator has two components. The largest, pRo2a, arises along the posterior face of the apophysis arm and inserts on to the basicoxite, ventral to the remotor apodeme (Figs 8.6-27 and 8.7-32). The small second posterior rotator, pRo2b, arises beside pRo2a on the apophysis arm and inserts on the basicoxite, just dorsal to the remotor apodeme (Figs 8.6-27 and 8.7-33). Three posterior rotators are found in the locust hind leg (122, 123, 124) and two are found in the cricket (152, 153). All originate on the sternal apophysis, as in the weta, and insert variously from the dorsal to the posterior edges of the coxal rim.

In contrast to the single adductor found in the locust and cricket, the tree weta has two adductors, which combine to give greater bulk than the equivalent muscle in the former insects. The increase in bulk arises from the first adductor (Ad1), which comprises three closely inserted muscles on the ventroposterior coxal rim (Figs. 8.6-27, 31 (and inset) and 8.7-33, 34). These originate on the ventral surface of the sternal apophysis wing and the corner where the wing joins the apophysis arm (Fig. 8.6-31). A unique modification of the coxal rim, the thumb-like coxal apophysis (cap, Figs 8.6-27 and 8.7-37), accommodates the insertion of the second adductor (Ad2). This is a short, fan-shaped muscle, which also originates on the wing of the sternal apophysis (Figs 8.6-27, 31 and 8.7-33, 34).

COXAL MUSCLES ORIGINATING FROM THE PLEURON. Five abductors have been found in the tree weta, compared with only two each in the locust and cricket. The first abductor (Ab1) in the weta appears homologous to the large abductors inserted via apodemes in the other two groups (126 in the locust, 154 in the cricket). It originates as a fan-shaped muscle, beneath the pleural ridge, over much of the dorsal face of the episternum (Fig. 8.7-35, 36) and inserts via a large apodeme (Fig. 8.6-27) on the dorsal posterior rim of the coxa. Originating beneath it and more ventrally on the episternum is the second abductor (Ab2, Fig. 8.7-36). This sheet muscle narrows to insert on to the dorsal rim of the trochantin (Fig. 8.6-27), and appears homologous to the second abductor of the cricket (155), which also inserts on the trochantin. The third abductor (Ab3) is a very small muscle, originating on the inner corner of the pleural arm and inserting on to the dorsal rim of the coxa, between the joint condyle and the apodeme of Ab1 (Figs 8.6-27 and 8.7-35). It has no homologue in the cricket or in the locust. The fourth abductor (Ab4, Fig. 8.7-33) is unusual, because it originates on a pleural process, an apodeme-like projection extending ventrally from the dorsal ridge of the epimeron (pp, Fig. 8.7-37). This muscle, therefore, appears to be attached at both ends by an apodeme and, as such, is completely atypical in the arthropod skeletomuscular scheme. In some preparations (as illustrated), the origin also extends dorsally on to the anterior dorsal corner of the epimeron, posterior to the origin of the trochanteral depressor TrDep1 (Fig. 8.7-37). The dorsal 'apodeme' is not described for the thoracic skeleton of the locust or for the cricket (e.g. Thakare, 1972b). The fifth abductor (Ab5) is a broad sheet muscle originating on the episternite, ventral to the origins of Ab1, and inserting on the coxal rim outside the insertion of aRo2 (Figs 8.6-27 and 8.7-35). No equivalents to this muscle have been described in the cricket or locust.

The first trochanteral depressor is made up of six separate muscles. Four are extrinsic, with origins in the thorax, while two are intrinsic to the coxa. Dep1abc, is a long cylindrical muscle divided into three parallel bundles, which extend proximally from the trochanter, through the coxa and dorsally under the pleural arm and on to the pleural wall. Their origins are on the dorsal anterior corner of the episternum (Fig. 8.7-32–34). These bundles are inserted on to the ventral rim of the trochanter by the large depressor apodeme (Dep1a–e, Figs 8.6-30 and 8.8-38, 41). This tripartite muscle is also found in the locust and the cricket. The other extrinsic trochanteral depressor (Dep1e) is a slender cylindrical muscle with its origin on the ventral surface of the anterior thumb of the metasternal apophysis (ath, Fig. 8.6-29). It inserts on to the anterior side of the depressor apodeme (Fig. 8.8-38), and is reported in neither the locust nor the cricket.

Intrinsic muscles of the coxa

The bulk of the muscular mass contained within the coxa provides power to raise (levate) or lower (depress) the trochanter, via the dicondylic coxo-trochanteral joint. Of the intrinsic first trochanteral depressor components, Dep1d is a massive muscle, forming a broad fan, with origins along the proximal ventral coxal wall, below the basicoxite, and extending on to the ventral side of the coxal fold. Its two halves insert on to the anterior and posterior surfaces of the depressor apodeme (Fig. 8.8-38, 39). The remaining first intrinsic depressor, Dep1f, is a narrow muscle, which broadens near its insertion (on to a brown-pigmented spot) on the ventral trochanteral rim, anterior to the depressor apodeme. It narrows and its fibres twist toward the origin, anterior to that of Dep1d (Fig. 8.8-38).

The second intrinsic depressor, Dep2, is a thick compact mass of about 20 fibres with a broad origin on, and ventrally adjacent to, the coxal fold. It inserts on to the ventral trochanter rim, posterior to the depressor apodeme (Figs 8.6-30 and 8.8-38–40).

The third intrinsic depressor, Dep3a–d, is a thick sheet of fibres, divided into four components and lying superficially against the ventral coxal wall, with a ventral origin near the basicoxite. They insert on to a fold of the arthrodial membrane along the ventral margin of the coxo-trochanteral joint (Figs 8.6-30 and 8.8-39–40).

None of these accessory depressor muscles, which lack apodemes, has been reported for the locust. However, in the cricket, two pairs are found inserted on to the trochanteral rim, on the anterior and posterior sides of the depressor apodeme. Only in the tree weta have additional muscles been found – those inserting on the arthrodial membrane.

A similar proliferation of trochanteral levators, accompanied by an increased number of apodemes, highlights the difference between the tree weta and the other two orthopterans. The first trochanteral levator consists of five muscles: two large muscles (Lev1a,b) attached to the large anterior apodeme (Fig. 8.8-46) and three smaller muscles (Lev1c,d,e) inserting on to the proximal trochanteral rim (Fig. 8.6-30). Lev1a is a large cylindrical muscle, originating just proximal to the origin of Dep1d on the ventral coxal surface; hence, it crosses the coxal midline to its dorsal insertion on the Lev1 apodeme (Fig. 8.8-42). Lev1b similarly originates on the midventral coxal surface and crosses the coxal midline to insert on the Lev1 apodeme (Fig. 8.8-42). Of the three trochanteral levators without apodemes, Lev1c is the largest. It originates on the proximal anterior wall of the coxa and inserts on the dorsal trochanteral rim, dorsal to the base of the Lev1 apodeme. Lev1d is a narrow bundle of five to six fibres which originate adjacent to Lev1c on the midventral coxal surface and twist as they insert on the trochanteral rim, anterior to the base of the TrLev1 apodeme. Another slender muscle, Lev1e, originates beside Lev1d and inserts anterior to the latter on the trochanteral rim (Fig. 8.8-42–44).

Trochanteral levator 2 is comprised of seven separate muscle bundles inserting on to the Lev2 apodeme and one inserting on to the trochanteral rim (Figs 8.6-30 and 8.8-46). The largest is a fan-shaped muscle, Lev2ai–iv, with four bundles originating on the underside (distal) of the coxal apophysis and dorsal surface of the coxal fold. All insert on the base and ventral surface of the Lev2 apodeme. A fifth, shorter bundle, originating distally on the coxal fold, Lev2av, inserts on the trochanteral rim beside the Lev2 apodeme (Figs 8.6-30 and 8.8-42). Trochanteral levator 2b also originates distally, on the underside of the coxal apophysis and basicoxite, and inserts on to the proximal end of the Lev2 apodeme (Fig. 8.8-43). Finally, a long, stout muscle, Lev2ci–ii (separate components not illustrated), originates on the proximal anterior wall of the coxa, in a deep evagination of the anterodorsal corner, and crosses the anteroposterior midline to insert on to the dorsal surface of the Lev2 apodeme (Fig. 8.8-43). This muscle is composed of two slightly flattened bundles of fibres.

The third trochanteral levator, Lev3a–b, comprises three bundles attached to a third levator apodeme. They form a bulky muscle, which is split on to the anterior and posterior sides of the coxa. Lev3ai originates in the proximal anterior coxal evagination, crosses the anteroposterior midline and inserts on to the anterior face of the Lev3 apodeme. A parallel narrow bundle of fibres, Lev3aii, splits from Lev3ai and inserts separately on to apodeme Lev3. The smaller and shorter Lev3b originates on the dorsal surface of the coxal fold and inserts on to the posterior face of the Lev3 apodeme (Figs 8.6-30 and 8.8-44).

The fourth trochanteral levator, Lev4a–c, is a tripartite fan-shaped muscle, with a broad superficial origin just distal to the basicoxite rim. The components insert on to a series of apodemes, graded in length, on the anterior rim of the trochanter (Figs 8.6-30 and 8.8-45, 46). The apodemes vary in number in different dissections, but range from four to nine fine cuticular extensions. They are similar to the apodeme cluster of the fourth coxal abductor (Fig. 8.6-27). Neither cluster of apodemes has been reported for other orthopterans.

Trochanter

With the fusion of the trochanter to the femur, a loss of muscles associated with the trochantero-femur joint has occurred. There is only one muscle bundle originating in the trochanter, Flc, an extrinsic bundle from the tibial flexor, described below.

Femur

The tibia attaches by the dicondylic femur–tibia joint and sends two long, blade-like apodemes into the femur. As in virtually all orthopterans, two major muscles occupy nearly the whole of femur, the tibial levator and depressor, while a third diminutive muscle forms part of a complex that depresses the tarsal claws. The few differences found in the tree weta are described below. The tibial levator is described as an extensor muscle, Exta–c, conforming to terminology commonly used by physiologists (Burrows, 1996). It is the largest muscle in the leg, with an origin covering approximately 60% of the anterior, dorsal half of the femur surface, commencing from the proximal end. The major part of the muscle, Exta, inserts on to the proximal 70% of the anterior face of the extensor apodeme (Ext, Fig. 8.9-49) in two rows, each of which is composed of fibre bundle doublets for the proximal half and singlets for the remaining distal half (Fig. 8.9-47). The origins of these bundles correspond to dark patches on the exterior of the femoral cuticle. Extb consists of a series of fibres with origins along the length of the anterior dorsal face of the femur and inserting on almost the full length of the extensor apodeme (Fig. 8.9-48). A further component of the extensor is Extc, a flat pair of fibre bundles originating on the proximal femur margin, adjacent to the trochanter, and inserting on to the end of the extensor apodeme (Fig. 8.9-48). Unlike the saltatorial anostostomatine wetas and the locust and cricket, the Ext bundles do not originate in a chevron pattern along the anterior femur surface. This is typical for the whole of the Deinacrideinae.

The accessory extensors originate on the distal femur. AccExta is a compact series of short bundles originating on the anterior femur surface along a line of external dark markings. It inserts on the distal end of the extensor apodeme near its attachment to the tibia. Its mate, AccExtb, originates on an equivalent position on the posterior femur surface (Fig. 8.9-47, 48).

The tibial depressor complex is called the tibial flexor, in accordance with the above physiological terminology. It has two major muscle masses, with origins along the posterior (Fla) and ventral (Flb) femur wall (while the tibial extensor occupies the anterior wall). Insertions of these bundles occupy most of the length of the flexor apodeme (Fig. 8.9-49). A third, separate bundle, Flc, arises from the trochanter and inserts on to the proximal tip of the flexor apodeme (Fig. 8.9-48).

As seen with the extensor, two small accessory flexors are found at the distal end of the femur. The anterior, AccFla, originates beneath a distinct line of dark dots on the exterior of the distal anterior face of the femur. The posterior, AccFlb, forms a single row of bundles along the posterior femur, without an external marker.

All of the intrinsic femoral muscles resemble the plan of the locust and cricket, except that the cricket apparently has only one accessory flexor muscle.

Tarsus

Two muscles levate and depress the tarsus through the dicondylic tibiotarsal joint. The tarsal levator, TaLev, is a small, fan-like muscle of some 12 bundles, originating distally on the anterior medial side of the tibia, from the level of spine 3 distally to the terminal spine (5). In some species, the origins are marked externally by a row of dark dots on the tibia. The fibres insert along the full length of the tarsal levator apodeme, on its anterior side (Fig. 8.9-50, 51).

The depressor of the tarsus, TaDep, is a long, pinnate muscle, originating along at least 75% of the length of the tibia's ventral surface. It inserts along the full length of the depressor apodeme (Fig. 8.9-50, 51). The above two muscles do not differ from those in the locust and cricket.

The depressor of the tarsal claws is the retractor unguis, RetUnga–c, a tripartite muscle distributed in the femur and tibia along an apodeme

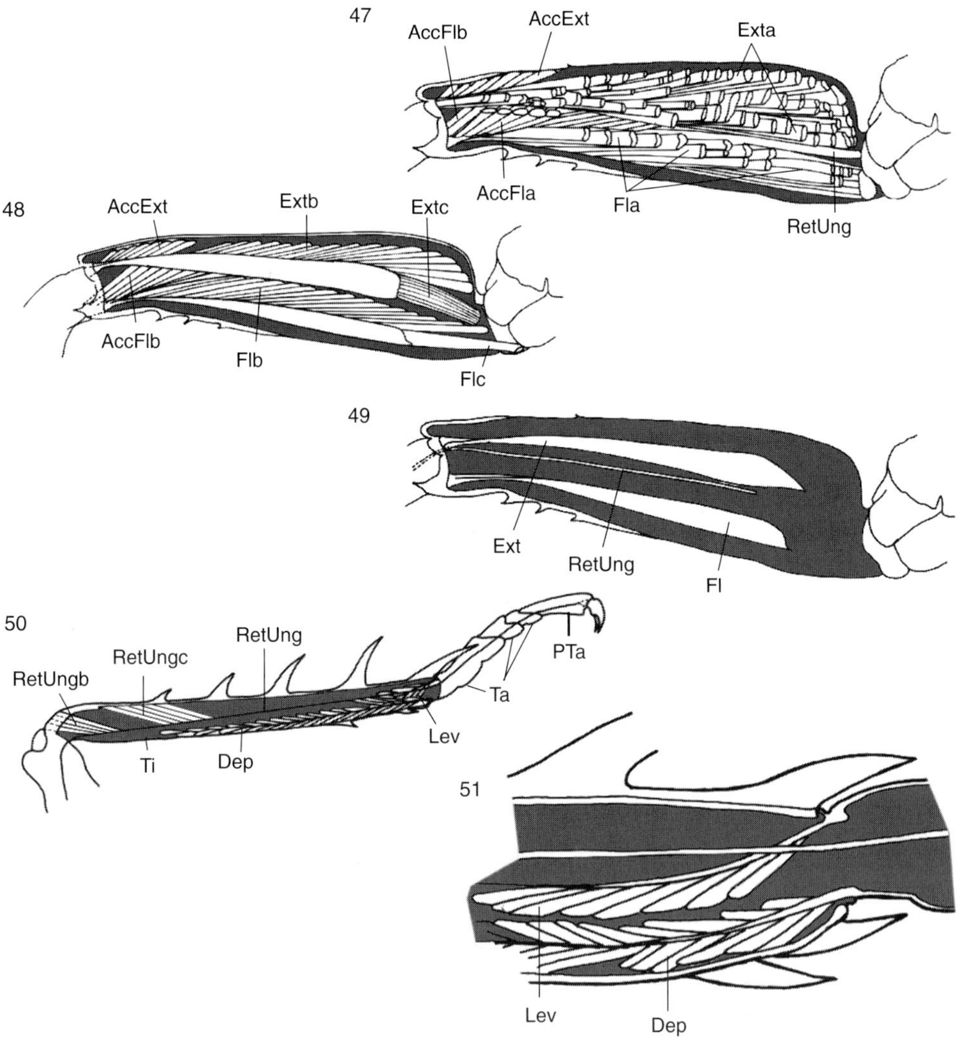

Fig. 8.9. Femoral and tibial muscles. 47–49, View of left femoral muscles from posterior side. 47, Intact anatomy. 48, Posterior bundles removed from tibial extensor and flexor, revealing accessory flexor and extensor bundles (AccFlb, AccExt). Note Flc bundle with origin in trochanter, and Extc insertion on to end of extensor apodeme. Retractor unguis muscle removed. 49, Skeletal elements of femur, including thin apodeme of retractor unguis (RetUng) (but excluding fine apodeme of femoral chordotonal organ; see Fig. 22.11, Field, Chapter 22, this volume). 50, View of posterior side of tibia, showing second and third components of the retractor unguis (RetUngb, c) as well as the tarsal levator and depressor. Retractor unguis apodeme extends from claws in pretarsus (Pta) through tibia and into femur. 51, Detail of tarsal levator and depressor, viewing posterior side of tibia. Retractor unguis apodeme passes above muscles. All abbreviations are given at end of chapter.

(RetUng, Fig. 8.9-49) which extends from the unguitractor plate (attached to the tarsal claws) to the retractor unguis head in the femur. The largest section is a slender, cylindrical muscle, RetUnga, originating just distal to the trochanterofemoral joint and extending almost half the length of the femur before inserting on to its thread-like apodeme (Fig. 8.9-47). Two other components are in the tibia. RetUngb originates as four to five bundles on the dorsal face at the extreme proximal end of the tibia, immediately before it bends and becomes elongate. These bundles insert on the RetUng apodeme posterior to the bend. RetUngc is a row of ten to 12 fibres along the posterior face of the tibia, just distal to the preceding fibres, and extends to below spine 2. The fibres insert midway along the retractor unguis apodeme (Fig. 8.9-50). At the extreme distal end of the tibia, the large, semi-opaque tibiotarsal chordotonal organ originates on the anterior tibial wall, as well as the tarsal levator and depressor apodemes, and tapers proximally to insert on the RetUng apodeme. This may easily be confused with a muscle associated with the apodeme. The preceding retractor unguis muscle components are found in the locust and cricket, with little or no change in anatomy.

Summary of Leg Muscle Comparisons

Table 8.1 indicates the differences found in the leg muscles of the three representative orthopterans. In addition, a partial list is available for the brachypterous grasshopper *Barytettix*, from a study by Arbas (1983). Unfortunately, an anatomical study of a related apterous stenopelmatoid orthopteran, the cave weta *Macropathus* (Rhaphidophoridae), lacks sufficient detail of leg muscles to be included below (Richards, 1955).

It is clear that the greatest variation amongst the orthopterans in Table 8.1 occurs in the proximal muscles operating the coxa and trochanter. Evidently, the distal leg musculature evolved to achieve an optimal functional anatomy, which has remained stable in the face of the diversification and isolation of New Zealand wetas. The differences found in the tree weta are likely to be due to: (i) loss of flight and consequent change or loss of musculature; and (ii) specific adaptation for use of the hind legs in the defence posture.

Consequences of flightlessness

In interpreting the anatomical consequences of flight loss, it is first useful to review hypotheses for evolutionary modifications for flight in insects. The evolution of a pterothorax (wing-bearing meso- and metathorax) included not only the expansion of notal plates into wings and associated hinge structures, but also the appearance of specialized pleural sclerites, e.g. basalar and subalar, and of internal modifications, such as the phragmata (the dorsally positioned transverse septa to which the dorsolongitudinal muscles attach). Extant muscles utilized for movement of the coxa and trochanter became important for moving the tergite and, directly or indirectly, the wings. To accommodate the large basalar and subalar flight muscle insertions on the coxa, the basicoxite became greatly expanded into the meron. In acridids, one of the basalars was changed in attachment from coxa to sternite to become a pure flying muscle, as did the third basalar, which became detached from the trochanter. Pleural muscles and tergosternal muscles were newly evolved as pure (dorsoventral) flight muscles (Tiegs, 1955).

Table 8.1. Comparison of numbers of individual muscle components in the hind leg of apterous (tree weta), brachypterous (*Barytettix*) and pterygote (locust, cricket) orthopterans. The number of apodemes associated with each muscle is indicated in parentheses. 'Cluster' refers to a tightly spaced series of four to six tiny apodemes on the segment rim.

	No. of muscle components (apodemes)			
Muscle	Tree weta	*Barytettix*	Cricket	Locust
Coxal promotor	1 (1)	1 (1)	2 (1)	1 (1)
Coxal remotor	1 (1)	2 (2)	2 (2)	2 (2)
Anterior coxal rotator	2 (0)	1 (0)	1 (0)	1 (0)
Posterior coxal rotator	3 (0)	3 (0)	3 (0)	3 (0)
Coxal abductor	5 (1+cluster)	2 (2)	2 (1)	2 (1)
Coxal adductor	2 (0)	1 (0)	1 (0)	1 (0)
Trochanteral depressor	11 (1)	–	9 (1)	5 (1)
Trochanteral levator	14 (4+cluster)	–	7 (2)	2 (0)
Tibial extensor	3 (1)	–	3 (1)	3 (1)
Tibial flexor	3 (1)	–	3 (1)	3 (1)
Tarsal levator	1 (1)	–	1 (1)	1 (1)
Tarsal depressor	1 (1)	–	1 (1)	1 (1)
Retractor unguis	3 (1)	–	2 (1)	3 (1)

Against this background, brachypterous (short-winged) and apterous (lacking wings) orthopterans underwent reduction or loss of muscles, rather than pressing them into new uses. The following generalizations may be made, as exemplified by *Barytettix*. Muscles used to drive the wings were usually lost. The dorsolongitudinal muscles and associated phragmata disappeared, as did the tergosternal muscles (both being indirect flight muscles, which distort the thoracic box to produce indirect movement of the wings). The pleuroalars and first basalar were also lost. Tiegs (1955) noted that in an Australian acridid (*Monistria*) and a tettigoniid (*Acridopeza*), both of which have brachypterous females, all of the muscles recognized in fully winged males (e.g. the pterygote plan) were present, but those specifically associated with the vestigial wings and with the flight function, were reduced to mere wisps. Thus apterous orthopterans, such as the tree weta, would be expected to have retained only those muscles important for operating the legs.

Accordingly, in *Hemideina*, many of the modifications for flight are lost. There are no pleural sclerites associated with wing function and no tergal phragmata. Dorsolongitudinal muscles and tergosternal muscles are lacking. The direct flight muscles, basalar and subalar, are lacking, along with any meron expansion of the basicoxite. However, two pleuro-coxal muscles are found, which could represent transfer of function from a pterygote ancestral line back to the coxa, and which might explain the increased number of weta coxal abductors (Table 8.1). These are abductors Ab4 and Ab5, neither of which occurs in the cricket or in the locust as abductors. However, in *Barytettix*, Arbas (1983) described a basalar (128) and subalar (129), which originated on the most dorsal margin of the episternum and epimeron, respectively, and inserted on to the coxal rim in anterolateral and dorsal positions, respectively. These have similar attachment sites to Ab5 (origin on episternum, insertion on anterior outer coxal rim) and Ab4 (origin on dorsal epimeron pleural process, insertion on apodeme cluster of dorsoposterior coxal rim).

Adaptation of musculature for defence display

A tree weta is able to rotate the metathoracic femur forward to bring the hind leg above the head at an angle of up to 140° to the body axis. This extraordinary posture, with the tibiae almost touching the back of the head, is used in the defence display (see Figs 16.1 and 16.6 in Field and Glasgow, Chapter 16, this volume). During provocation, the legs are brought down rapidly from the display position and pressed against the stridulatory ridges on the abdomen to produce a warning sound. Such extreme levation, as well as the requirement to press the leg against the abdomen, are in concordance with the observed increase in numbers of trochanteral levator and coxal abductor muscle components (Table 8.1). Furthermore, the coxal remotor is the most massive of the coxal and trochanteral muscles. Therefore, one of the reasons for the large muscle complement in the hind leg of the tree weta appears to be a special enhancement of musculature as an adaptation for an unusual defence behaviour.

Abdomen

In *Hemideina* the abdomen is not heavily sclerotized, particularly in the sternal regions. It tends to be distinctly shorter and narrower in the male, although it may become distended in threat displays.

The tergites cover the dorsal and lateral aspects of the abdomen. These are darkly tanned on the anterior portion and on the posterior margin, giving a banded appearance (illustrated in Fig. 9.8, Field and Bigelow, Chapter 9, this volume). They are connected to the sternites by membranous pleural regions, in which there are small, faintly chitinized pleurites on segments 3 to 8, and also the spiracles. The latter are found on the first and second thoracic and the first eight abdominal segments. The file for the stridulatory apparatus is borne on the lateral margins of the second tergite (see Fig. 15.1, Field, Chapter 15, this volume).

In the abdomen of the male (Fig. 8.1-1) there are ten distinct tergites, the tenth bearing a large epiproct (ept), resulting from fusion with segment 11. The epiproct and paired paraprocts surround the anus (AN). Between the tenth tergite and the paraproct of each side is a cercus (CRC, Fig. 8.11-66), consisting of a small basipodite and a long cigar-shaped segment bearing many sensory setae (described in detail in Field, Chapter 22, this volume). There are nine sternites, the ninth forming the hypandrium (HYP), which forms the lower part of the genital chamber. It is large and bears a pair of styli on its posterior margin (Fig. 8.1-1; see also Fig. 21.5 in Stringer and Cary, Chapter 21, this volume).

The female also has ten tergites and paired cerci, but only eight sternites, the eighth forming the subgenital plate (SGP). The ovipositor is long and tubular with a slight upward curve (see Fig. 21.7, Stringer and Cary, Chapter 21, this volume). It is composed of three pairs of valvulae: a ventral pair, a dorsal pair, which overlaps the ventral pair, and an inner pair (VP, DP, IP, respectively, Fig. 8.11-66), which is the smallest and is enclosed by the other two pairs. Between the bases of the dorsal valvulae is a median sclerite, which bears an apodeme for the attachment of the principal muscles concerned with moving the ovipositor. Ventrally, each member is connected to the other by a transverse bar, to which the inner valvulae are also connected. The inner pair is connected by a dorsal transverse hoop. Their bases are connected by a ventral transverse bar, to which the dorsal valvulae are also connected (Fig. 8.11-66). The bases of the dorsal and inner pairs are connected together for some distance. The female also has an epiproct, well-developed paraprocts (ppt) and slightly shorter cerci (Fig. 8.11-66; see also Fig. 21.7, Stringer and Cary, Chapter 21, this volume).

In the male, the sternite of the ninth abdominal segment is enlarged, forming the hypandrium, or subgenital plate. The penis (pn) lies in a cavity bounded by the hypandrium below, the paraprocts above and unsclerotized cuticle surrounding these (Fig. 8.11-67). It is a single structure, wide, flattened and spinulose. Its grooved dorsal surface makes it appear three-lobed, the central lobe containing the ejaculatory duct, with the thicker lateral lobes forming the side walls. There is a wide, transverse opening, bounded by dorsal and ventral lips, at the end of the penis.

A pair of middorsal hooks, known as falci, occurs on tergite 10 of males. These are especially prominent and darkly tanned in *Deinacrida* spp., but are found also found in *Hemideina* and *Hemiandrus*. They recurve forwards (hooks, Fig. 21.4, Stringer and Cary, Chapter 21, this volume), and are used by the male to secure attachment to the female during copulation. Experimental demonstration of their essential function was made by ablation of one or both falci in *Hemideina*, after which it became difficult or impossible for the male to secure a connection to the female's genitalia (described in Field and Jarman, Chapter 17, this volume). Gibbs (1999) illustrated a series of falci for newly described *Deinacrida* spp. in New Zealand.

No description of the abdominal musculature of *Hemideina* has been made, nor is there an adequately illustrated account of the male genitalia.

In the rhaphidophorid *Macropathus*, Richards (1955) has described the abdominal and genital musculature and illustrated the terminalia of both sexes in exacting detail.

The Alimentary Canal

The recent interest in the maintenance and culture of various wetas, and consequent interest in their diet, has not been matched by comparative work on either the anatomy of the gut or the physiology of digestion. Much of what follows is based on the early work of Maskell (1927) on the mainly herbivorous *H. thoracica* and the detailed description of the large, cave dwelling rhaphidophorid *Macropathus* (Richards, 1955).

The alimentary canal is clearly differentiated into the usual three regions of fore-, mid- and hind-gut. In *Hemideina* it is nearly twice the length of the body, while in *Macropathus* it is proportionately shorter (Fig. 8.10-52, 53).

The foregut

The foregut, or proctodaeum, consists of the buccal cavity, oesophagus, crop and proventriculus or gizzard, all of which have a cuticular lining. After leaving the head, the narrow oesophagus broadens into the crop (CR), which fills much of the thorax and part of the abdomen (Fig. 8.10-53). Lining the lumen is a chitinous intima (CH, Fig. 8.10-57), secreted by the epithelium (EP) immediately below it. A basement membrane separates this from a well-developed layer of longitudinal muscle, surrounded by a layer of circular muscle. The oesophagus shows little differentiation, but sections of the crop show indentations, some of which have serrated margins, forming small teeth. Towards its posterior end, the crop narrows and the intima is thrown into six folds, which form the junction of the crop with the proventriculus.

The proventriculus (pro, Fig. 8.10-52) is the most complex structure of the gut. Within it, the

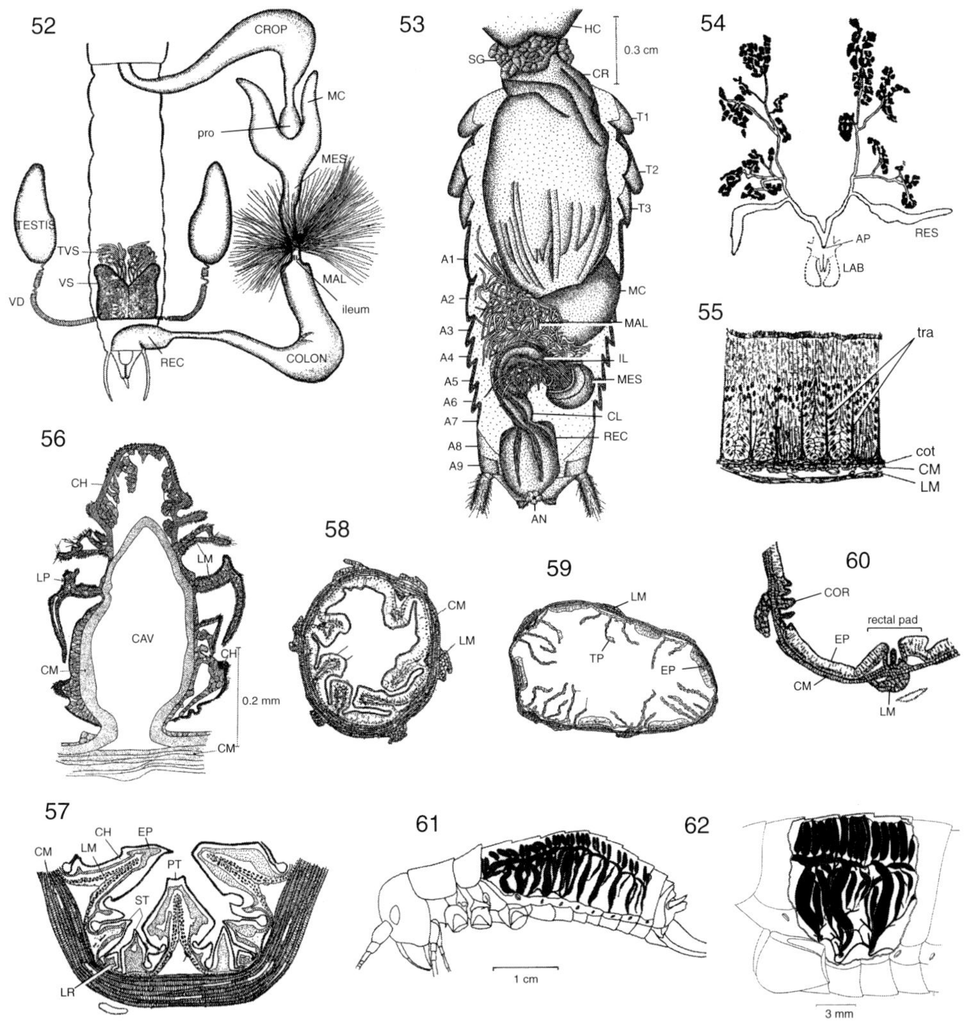

Fig. 8.10. 52, Male reproductive system and alimentary canal reflected from body of *Hemideina* (from Maskell, 1927). 53, Alimentary canal of *Macropathus* shown intact in the body (from Richards, 1955). 54, Salivary glands of *Hemideina* (from Maskell, 1927). 55, Longitudinal section of mesenteric wall of *Hemideina* (from Maskell, 1927). 56, Transverse section of primary tooth of proventriculus of *Macropathus* (from Richards, 1955). 57, Transverse section of wall of proventriculus of *Hemideina* (from Maskell, 1927). 58, Transverse section of ileum of *Hemideina* (from Maskell, 1927). 59, Transverse section of colon of *Hemideina* (from Maskell, 1927). 60, Portion of transverse section of rectum of *Hemideina* (from Maskell, 1927). 61, Tracheolar air sacs underlying the cuticle of *Hemideina* (from Field, 1978). 62, Detail of the air sacs, which underlie the stridulatory file, in the first several abdominal segments (from Field, 1978). All abbreviations are given at end of chapter.

six longitudinal folds developed in the crop become large, with each fold bearing a row of at least 20 large, triangular, primary teeth (PT, Fig. 8.10-57), which project into the lumen. Between these rows are two rows of smaller or secondary teeth (ST, Fig. 8.10-57), separated by a single small chitinous ridge (LR), which is not divided into teeth. The cuticle is thickened and heavily tanned on these ridges and teeth but is flexible between them. Many fine hairs are found on the thickened regions. The proventriculus contains the same tissue layers as the crop. Below the epithelium that produces the cuticle is a layer of longitudinal muscle (LM), which is most extensively developed within each primary tooth. Surrounding all of this is a greatly thickened layer of circular muscle (CM), described as striated in *Hemideina* but as unstriated in *Macropathus*. As visceral muscle is striated in other insects, an ultrastructural approach is needed to shed light on this apparent anomaly.

At the posterior end of the proventriculus the six folds are continued on without the teeth, projecting a short distance into the mesenteron (MES, Fig. 8.10-52). The dorsal and ventral folds each bear a flap valve, the cardiac valve. These overlap and can form a partition between the mesenteron and gizzard, restricting communication between the mesenteron and gizzard to grooves in the other four folds. The pattern, described here from *Hemideina*, varies slightly in *Macropathus*, particularly in the structure of the primary teeth of the proventriculus, but similar structures are present. This elaborate array of structures serves as both a filter and a valve, controlling the transfer of food from the crop to the midgut, while allowing the passage of digestive fluid forward from the midgut to enable enzymatic digestion to take place in the non-secretory crop. The brown fluid occasionally regurgitated during defensive biting is thought to originate in the midgut.

Both Maskell and Richards use the term gizzard for the proventriculus, implying a grinding function that has not been clearly established for either *Hemideina* or *Macropathus*. The two insects differ in the structure of the primary teeth, those of *Macropathus* being hollow and lined with circular muscle and having elaborate lateral projections (LP) which seem more suited to filtration than to trituration (Fig. 8.10-56).

The midgut

The midgut or mesenteron (MES) is produced into two large forwardly directed caeca (MC), which lie dorsal and ventral to the proventriculus (Fig. 8.10-52). In this, the digestive region of the gut, the arrangement of the tissues differs from that of the crop. A delicate connective tissue layer (cot) bounds the muscle layers (Fig. 8.10-55). In *Hemideina*, the longitudinal muscle is external to the circular, as is generally found in the insect mesenteron, although Richards (1955) describes a third, innermost layer of smooth longitudinal muscle in *Macropathus*. Inside the muscle layer is connective tissue, from which arise trabeculae, which project into the epithelium at regular intervals (tra, Fig. 8.10-55), dividing it into a honeycomb-like array of four- or five-sided pits. At the bottom of each pit is a nest of small cells, which constantly divide to produce a succession of cells moving up the walls to form a coherent layer at the mouth of the pits. The tall columnar cells have regular microvilli forming a striated border adjacent to the lumen. There is no cuticular lining to the lumen. In a cross-section of the midgut of *Hemideina*, numerous concentric peritrophic membranes may be seen between the food contents and the epithelium, formed apparently by the delamination of the inner margin of the epithelial cells. Masses of cast-off cells may be seen free in the lumen between the peritrophic membrane and the epithelial layer.

The Malpighian tubules (MAL) are found at the junction of the mesenteron and hind-gut, where they converge on ureters opening into the ileocaecal groove (Fig. 8.10-52). They are described more fully later in this chapter.

The hind-gut

The hind-gut or proctodaeum is marked off from the mesenteron by a distinct ileocaecal groove and is differentiated into an ileum, colon and rectum (IL, CL, REC).

The short ileum is lined with a thin layer of cuticle. The epithelium of columnar cells is often thrown into folds but is capable of being extended (Fig. 8.10-58). Outside the layer of circular muscle, there are six longitudinal bands (LM), from which fibres may cross the circular layer and run in the epithelial folds, effectively forming a third muscle layer.

The colon is long, bent back on itself and about twice the diameter of the ileum. There are six longitudinal bands of columnar epithelium (EP), alternating with six longitudinal areas bearing elongate tag-like processes (TP), which spread out amongst the contents of the colon (Fig. 8.10-59). Each consists of a double layer of small epithelial cells below a thin overlying cuticle. Below the epithelium is a delicate connective-tissue layer and then a band of circular muscle fibres, within which appear a few longitudinal bundles. Outside this are the six bundles of longitudinal muscle, as in the ileum, now disposed between the regions of columnar epithelium.

A slight constriction separates the colon from the rectum. The areas that in the colon bore the tag-like processes persist as narrow corrugated areas (COR) with thickened cuticle. Between these are the six longitudinal bands of taller epithelial cells, bearing a thinner cuticula. These bands are more extensive than in the colon and are now known as the rectal pads (Fig. 8.10-60). The organization of the muscle layers persists, much as found in the colon. The longitudinal muscle groups are attached to the body wall between the ninth and tenth abdominal tergites. In the anal region, the rectal pad structure is lost, the wall is thrown into many deep folds and the layer of circular muscle is much thickened.

In *Macropathus*, Richards (1955) details a 'gland organ' of unknown function along the margins of the pads, although nothing in the figures or description suggests a glandular function. A similar structure has been described from *Stenopelmatus* (Davis, 1927). The role of the rectal pads in water balance has not been studied in these insects.

Salivary glands

There is one pair of diffuse salivary glands (SG) lying in the thorax below and beside the crop. Each gland consists of a number of acini or lobules. The ducts from these progressively converge until the two main ducts on each side unite, forming a single duct, which opens to the exterior immediately below the hypopharynx and above the labium. The main duct on each side has a salivary reservoir (RES, Fig. 8.10-54). Richards (1955) described 'typical glandular tissue' from the glands in *Macropathus*, but no more detailed histological account is available.

Circulatory System

The dorsal blood vessel or heart extends in *Hemideina* from the posterior end of the abdomen into the prothorax and is continued as the aorta into the head. It is suspended from the dorsal body wall by fine elastic filaments. There are eleven ampullae, with paired incurrent ostia (valves). The enlarged appearance of the vessel at these points gives it a chambered appearance, but the vessel is not partitioned. The ostia are laterally placed at about the centre of the ampullae, except on the smaller eleventh ampulla, where they are terminal and situated dorsally. An aortic valve, consisting of about 12 annular fibres, separates the heart from the aorta, just anterior to the first chamber. The aorta then continues forward before bending ventrally to open between the paired corpora cardiaca in the head.

The rhaphidophorid *Macropathus* also has 11 ampullae, this being one of the primitive features typical of Orthoptera (Nutting, 1951).

Hemideina has both a dorsal and a ventral diaphragm, partitioning the haemocoele into pericardial, visceral and perineural sinuses, although the ventral diaphragm is restricted to the thorax only. The ventral diaphragm is present in most Orthoptera, though it is often incomplete. It is absent in *Macropathus*, there being no separate perineural sinus. The dorsal diaphragm is a thin, transparent, nucleated membrane, interlaced with numerous fine fibres and perforated by numerous apertures. A light brown mass of nephrocytes is found within the pericardial sinus.

On either side of the heart in each segment, a pair of alary muscles fans out over the surface of the dorsal diaphragm. In *Hemideina*, as in *Macropathus* and in crickets, there are 11 pairs, corresponding to the number of chambers and ostia. A dense mass of fat body is found between the alary muscles of *Macropathus*.

The abundant haemocytes have been described only superficially. Maskell reported two principal types from *Hemideina*, one spherical, with granular contents, and one spindle-shaped, with clear cytoplasm. These are difficult to relate to current systems of haemocyte classification. The ionic composition of the haemolymph is described in Neufeld and Leader (Chapter 25, this volume).

Excretory System

The principal excretory structures in insects are the Malpighian tubules (MAL), which empty into the alimentary canal at the junction of the mid- and hind-guts (Fig. 8.10-52, 53). There are about 200 tubules in *H. thoracica*, arranged in six groups. Each group opens into one of the six ampullae or ureters distributed equidistantly around the gut. The elongate, blindly ending tubules lie freely in the abdominal cavity, although a few penetrate the diaphragm to lie partly in the pericardial sinus. As in most orthopterans, the tubules in *Hemideina* exhibit a serpentine motion, although neither Maskell (1927) nor Richards described muscles from these structures. There are more than 250 tubules in *Macropathus* and these collect into paired ureters. A delicate peritoneal membrane surrounds a more defined basement membrane, which supports the single layer of large epithelial cells surrounding the lumen, with a brush border being described from *Macropathus*. No more detailed histological or ultrastructural data are available.

Nephrocytes or pericardial cells are present as a mass of light-brown tissue on either side of the heart within the pericardial sinus and also associated with the ventral diaphragm in the thorax. The binucleate cells are loosely packed and often arranged in a linear series (Maskell, 1927). These cells are generally considered to take up from the haemolymph high-molecular-weight chemicals, which the Malpighian tubules may be incapable of excreting. Neither their ultrastructure nor their physiology has been studied in wetas.

Urate cells storing granular products of uric acid were found in the fat body of *Macropathus* (Richards, 1955).

The role of the rectal pads in the reabsorption of water and salts has been documented in many insects. While the pads are well developed in *Hemideina* and *Macropathus*, their physiology and ultrastructure are not known.

Fat Body

The fat body is a white, opaque tissue, consisting of a loose meshwork of strands invested in delicate connective tissue. It is a cellular structure, but the cell walls have often broken down. It occurs throughout the body and is extensively developed in the ventral abdominal region in both *Hemideina* and *Macropathus*. Urate cells, containing little fat but many granules of uric acid derivatives, have been described from *Macropathus* (Richards, 1955). Imms (1931) says that insects that live in caves have an extra large fat body, as is found in *Macropathus*.

The Respiratory System

Hemideina has spiracles derived from the second and third thoracic segments and a further eight abdominal spiracles, thus showing the primitive holopneustic arrangement (Figs 8.1-1 and 8.5-23). A distinctive feature of the respiratory system of *Hemideina* is the large number of dilated tracheae it contains, particularly in the first three segments of the abdominal region (Fig. 8.10-61, 62). Many of these pass through very narrow tracheal tubes at each end, and may be thought of as tracheal sacs. Air sacs in many insects are thin-walled, with the supporting taenidia absent or poorly developed, but here the taenidia are well developed. Together they produce an almost continuous air chamber, lying immediately below the hypodermis, and are clearly associated with sound production. Possible functions are to expand the abdomen to aid contact between the stridulatory elements and/or to act as an acoustic resonator. Related species of similar size, such as *Hemiandrus* (Anostostomatinae) and *Macropathus* (Rhaphidophoridae), which lack both stridulatory and auditory structures, have a smaller number of narrower, sparsely distributed air sacs (Field, 1978). The general arrangement of the tracheal system of *Hemideina* has not been examined, apart from other special features associated with the auditory system (described in Field, Chapter 15, this volume) A detailed account of the pattern found in *Macropathus* is given in Richards (1955).

The Reproductive System of the Male

The testes are two white, pear-shaped bodies lying in the abdomen on either side of the gut (Figs 8.10-52 and 8.11-63). They comprise numerous elongated follicles (fol), each ending with a short efferent duct. The follicles are at the periphery of the testis, with their efferent ducts leading into the central region before uniting to form a single vas deferens (VD), which leaves the testis at its broader end. Each follicle is surrounded by a thin peritoneal layer. Together with a thin layer of fat body, this serves to bind the follicles into a testis.

Groups of spermatogonia are produced at the blind end of the follicle, each group surrounded by a thin sheath or cyst, also containing nurse cells. Different stages of spermatogenesis occur as this spermatocyst passes down the follicle, until near the efferent duct spermatids and spermatozoa are found. At this stage, the spermatozoa are free, but, at about the junction with the vasa deferentia, the spermatozoa become aggregated into sperm bundles, each comprising all the sperm previously enclosed in a spermatocyst.

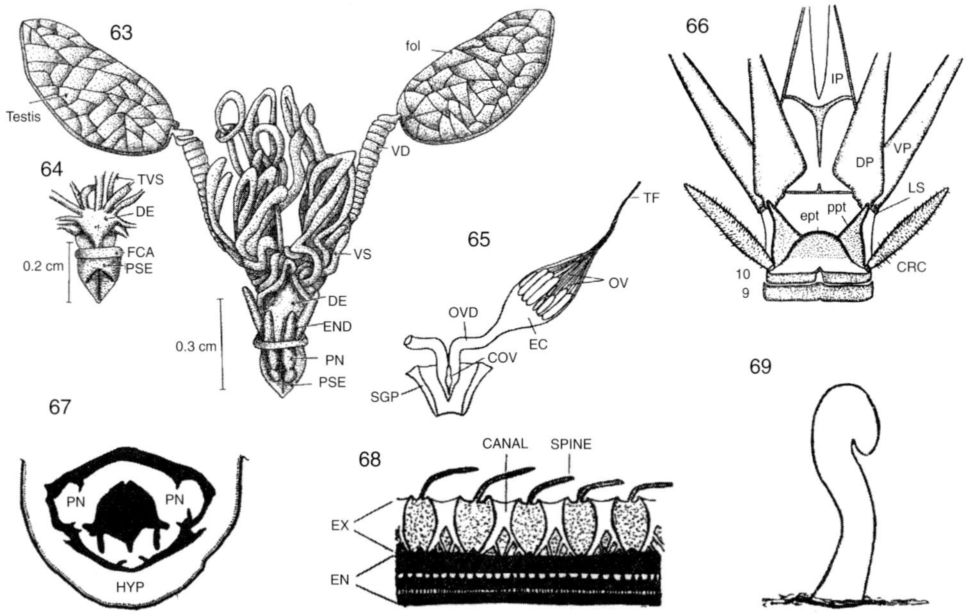

Fig. 8.11. 63, Male reproductive system of *Macropathus* (from Richards, 1955). 64, Ventral view of lower portion of male reproductive system (from Richards, 1955). 65, Female reproductive system of *Hemideina* (from Maskell, 1927). 66, Dorsal view of the ovipositor of *Hemideina* with the valves separated (from Maskell, 1927). 67, Transverse section of penis and adjoining regions of *Hemideina* (from Maskell, 1927). 68, Longitudinal section of chorion of egg still in ovariole, *Hemideina* (from Maskell, 1927). 69, Transverse section of spermatheca of *Hemideina* (from Maskell, 1927). All abbreviations are given at end of chapter.

In *Macropathus*, mature spermatozoa in free bundles are found scattered throughout the whole of the sperm tube and are not confined to the end nearest the efferent duct. Sometimes the bundles are solitary but more frequently several are linked together.

The vasa deferentia are a pair of very closely coiled tubes, which lead from the testes to the seminal vesicles (VS). The wall of the vas consists of a peritoneal coat, with an epithelium lining the lumen. The paired vesicles each have a small cavity. Their surface is covered by very numerous accessory glands or tubules of two types. A bunch of larger tubules (TVS) opens into the ventral surface of the vesicles, while the numerous very small tubules cover the rest of the surface of the vesicles. In *Hemideina*, the tubules produce secretions but do not contain sperm. Richards' description of the vesicles of *Macropathus* suggests that they consist of the tubules, which act as the sperm storage organs. In *Hemideina*, the vesicles unite just before opening into the ejaculatory duct, whereas, in *Macropathus*, the vesicles open by several tubules into the single, large, median ejaculatory duct (DE, Fig. 8.11-64), suggesting that in the latter case some of the tubules may be developed from the duct and be ectodermal in origin. This thick-walled duct is lined with cuticle and has a well-defined layer of muscle. At this level, a number of sperm bundles are aggregated into a spermatophore, a viscid mass formed from the secretions of the accessory glands and tubules.

The description given above for *Hemideina* is similar to that for many orthopterans, but in specific detail it varies from the rhaphidophorid *Macropathus* in a number of areas, including the structure of the penis, the seminal vesicles and accessory tubules, the spermatophore and the path of the spermodesms within the follicle.

The Female Reproductive System

In *Hemideina*, this system consists of a pair of ovaries, a pair of oviducts, leading into a median oviduct, and an unpaired spermatheca (Fig. 8.11-65; see also Fig. 21.9, Stringer and Cary, Chapter 21, this volume).

The ovaries in a mature female are large, pear-shaped bodies, situated in the abdomen at either side of the gut. Each consists of about 40 ovarioles (OV) or egg tubes. What appears to be the lower third of the ovary is really a calyx (EC), or much dilated upper end of the oviduct, which serves as a pouch and may hold 18 to 20 eggs at once. The ovarioles are panoistic, lacking specialized nutritive cells. Each ovariole has a terminal filament, a slender, thread-like prolongation of the peritoneal layer. All the filaments of one ovary are bound together to form a common thread, the suspensory ligament, which inserts into the pericardial diaphragm. The germarium, the region below the terminal filament (TF), appears as a mass of nuclei. From this region are differentiated the ova and the cells of the follicular epithelium. The main portion of the ovariole is the vitellarium, which contains a linear series of ova, which increase in size as they approach the egg calyx. Each ovum is surrounded by its own coat of follicular epithelium and consists of a mass of yolk, with a large nucleus or germinal vesicle and a nucleolus or germinal spot. In most insects, including *Macropathus*, at its proximal end each ovariole narrows to a fine duct, the pedicel, which connects to the oviduct (OVD). Maskell (1927) makes no mention of these in *Hemideina*.

The oviduct begins as an expanded calyx and then continues as a short narrow tube, before uniting with the other oviduct to form a short median oviduct (COV), which opens by a slit-like aperture upon the upper face of the subgenital plate (SGP), which connects to the seventh sternite (Fig. 8.11-65). In a gravid *Macropathus* female, the most distal eggs have matured and severed the connection of the pedicels to the calyx of the oviduct, most of the eggs lying loosely in the body cavity on either side of the hind-gut, from where they pass to the calyx of the oviduct. The ovarioles are bound together by a thin layer of connective tissue, within which the follicular epithelium grows inwards in such a way that each ovum is enclosed in a complete follicle or sac. Besides its nutritive function, this epithelium secretes the egg shell or chorion, which is well developed by the time the ovum reaches the calyx. The chorion is brown in colour and tough and on its outer surface shows a hexagonal pattern of broad, spinose ridges surrounding cavities or canals, from which finer branching canals spread out deeper into the chorion (Fig. 8.11-68). Below the exochorion (EX) are three dense parallel layers of chitin, forming the endochorion (EN). The micropylar area is situated at the anterior pole of the egg. The egg is elongate, oval and about 5–6 mm long. Detailed descriptions of egg structure and surface features are provided in this volume by Angulo (Chapter 11) and Stringer (Chapter 20).

The structure formed from the joining of the paired oviducts has been termed a vagina in both *Hemideina* and *Macropathus*. As Richards believed that this structure is not penetrated during copulation, it is probably better described as a common oviduct (COV), terminating in a gonopore, which opens into the genital chamber.

The spermatheca in *Hemideina* is a median blind tube, whose distal end is bent upon itself (Fig. 8.11-69). It opens into the genital chamber on a triangular plate just above the opening of the vagina. The chitinous lining is pierced by fine canals, thought to carry glandular secretions from the lower of the two cell layers beneath. A basement membrane separates these from a band of muscle, containing fibres orientated in various directions. The spermatheca of *Macropathus* is a median blind tube, which branches into two unequal tubes halfway along its length. This is generally considered to be a more primitive condition. Many orthopterans have an accessory gland opening into the genital chamber, but this has not been reported in either *Hemideina* or *Macropathus*.

The Nervous System

The brain (BR) is situated anterior to the oesophagus (Fig. 8.12-70). From its lower ventral margin, the paired circumoesophageal connectives (COC) pass around the oesophagus and tentorial plate to connect with the suboesophageal ganglion in the posterior part of the head. This, in turn, gives rise to long paired connectives, which pass back below the tentorial plate to connect with the first thoracic ganglion of the ventral nerve cord (VNC).

From the upper portion of the brain (the protocerebrum, PL, Fig. 8.12-73, 74) arise three small ocellar nerves (LON, MON), one to each ocellus. These are also present in *Macropathus*, despite its having only one functional ocellus. On each side, a short, robust optic nerve (ON) connects the protocerebrum to the optic lobe (OL), which produces seven or eight short tracts to the eye. Below the protocerebrum lies the deutocerebrum (DL), from

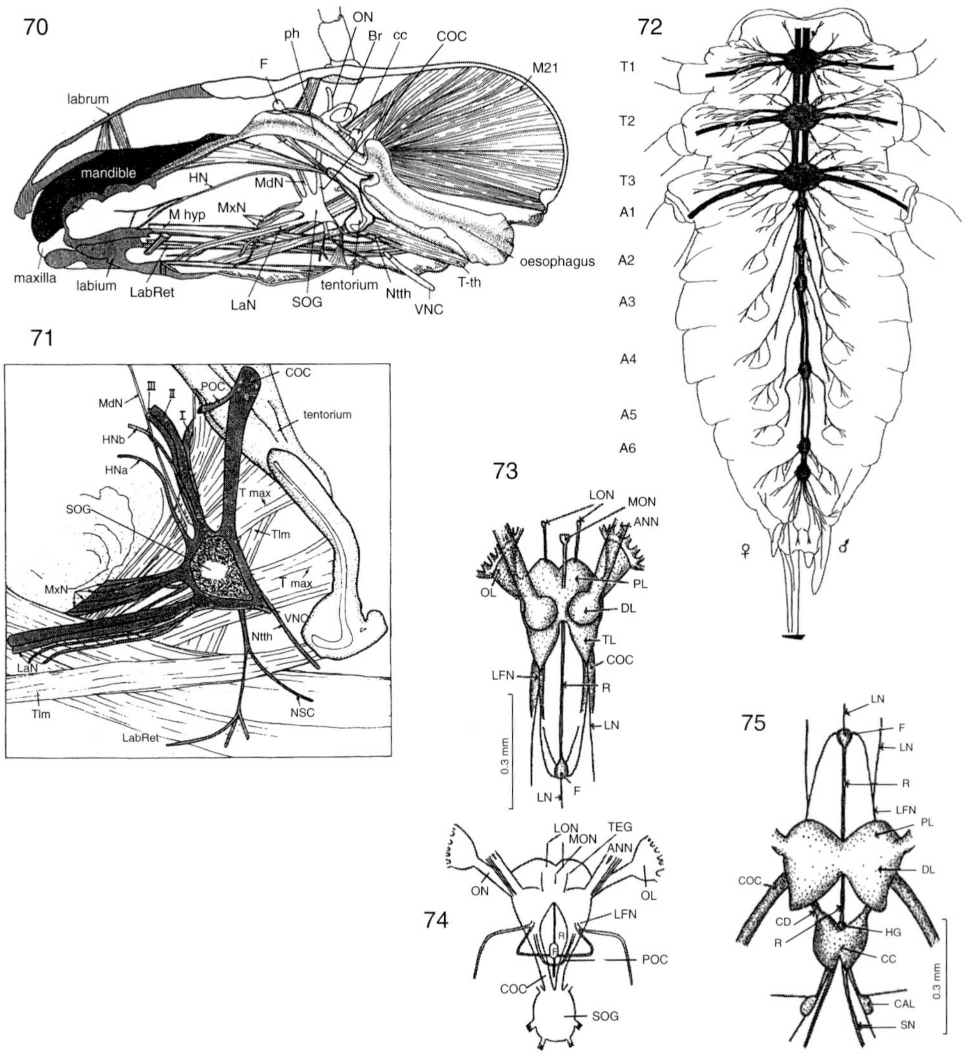

Fig. 8.12. 70, Parasagittal section of the head of *Hemideina* showing location of central nervous system (brain, corpus cardiacum, suboesophageal ganglion and associated nerves) (from O'Brien, 1984). 71, Suboesophageal ganglion of *Hemideina*, lateral view (from O'Brien, 1984). 72, Ventral nerve cord of *Hemideina maori*. Thoracic and abdominal segments are labelled T1-T3 and A1-A6 (remainder unlabelled), respectively, to show fusion of ganglia and resulting non-correspondence of ganglia with abdominal segments (after Smith, 1979). Note nerves from the third thoracic ganglion innervate A1 segment, and nerves from the first separated abdominal ganglion (second and third fused abdominal ganglia) innervate A2 and A3 segments. 73, Frontal view of brain, *Macropathus* (after Richards, 1955). 74, Brain and suboesophageal ganglion of *Hemideina*, anterior view (from Maskell, 1927). 75, Posterior view of brain showing sympathetic system, *Macropathus* (after Richards, 1955). All abbreviations are given at end of chapter.

which arise the antennary nerves (ANN) and the fine tegumentary nerves (TEG). The third division of the brain is the tritocerebrum (TL), which is less distinct externally than the other regions. It consists of two separate lobes connected by the post-oesophageal commissure (POC). From it also pass the labrofrontal nerves (LFN), which innervate the labrum, as well as connecting to the frontal ganglion (F, Fig. 8.12-74). The circumoesophageal connectives pass from the posterior portion of the tritocerebrum to the suboesophageal ganglion, lying below the oesophagus (SOG, Fig.

8.12-70). This ganglion innervates all the mouthparts and their associated musculature, except for the labrum (Figs 8.12-70, 71). From it arise three large trunks to each mandible (MdN I, II, III), two fine nerves to the hypopharynx (HNa,b) as well as large trunks to the maxillae (MxN) and to the labium (LaN). It also innervates the tentorothoracic musculature (Ntth) and has sensory branches to the neck region (NSC). Finally, it gives off paired connectives that pass through the occipital foramen to the thorax, giving rise to the ventral nerve cord (Fig. 8.12-70).

The ventral nerve cord consists of three thoracic and six abdominal ganglia, connected by paired longitudinal connectives into a chain extending the whole length of the body (Fig. 8.12-72). From the suboesophageal to the third thoracic ganglion, between any two adjacent ganglia is an extra pair of finer nerves, lying outside the stouter main connectives. Maskell (1927) was probably in error in interpreting these as being additional connectives. It is more likely that they represent a fusion of two nerves, one from each ganglion. A similar nerve in the mesothorax of *Locusta* carries both sensory and motor elements, supplying several regions of the thorax (Campbell, 1961).

Several pairs of nerves arise from each thoracic ganglion to supply the muscles of the corresponding segment, with the metathoracic ganglion also innervating part of the first abdominal segment (since the first abdominal ganglion has fused with the third thoracic ganglion in *Hemideina*) (Fig. 8.12-70). Each ganglion gives rise to a pair of large nerves (nerve V), which innervate the legs. There are six abdominal ganglia. Two pairs of nerves arise from the first ganglion to innervate the second and third abdominal segments, indicating that the the second and third abdominal ganglia are fused. A single pair arises from each of the next four ganglia and three pairs from the sixth ganglion, which is markedly larger than the others, being a composite of the fused ganglia of the posterior segments 8–11. It gives off nerves to the cerci and muscles, to the ovipositor, ovaries and spermatheca in the female and to the seminal vesicles in the male. The abdominal musculature is innervated from all these ganglia, as well as from the metathoracic ganglion.

In *Macropathus*, the pattern is similar, although there is one extra pair of nerves from each of the first five ganglia and the first abdominal ganglion has moved into the thorax and, in some specimens, appears close to fusion with the metathoracic ganglion (Richards, 1955).

There are stomatogastric and ventral sympathetic systems. In the former (Fig. 8.12-75), a small frontal ganglion (F) lies on the oesophagus, giving off four nerves: a frontal nerve to the labrum (LN), laterally a pair of nerves to the tritocerebrum and a median recurrent nerve (R), which runs back along the oesophagus to a hypocerebral ganglion (HG). This is connected to the paired corpora cardiaca (CC), which also connect to the brain. The hypocerebral ganglion gives off a pair of nerves that innervate the crop before producing the ingluvial ganglia, which also give off nerves to the crop and gizzard. On each side, a corpus allatum (CAL) lies behind the corpus cardiacum and is connected to it by a fine nerve. In *Macropathus*, the corpora cardiaca have fused in the midline.

The ventral sympathetic (or 'median') system lies just above the ventral nerve cord. It leaves the suboesophageal ganglion as a fine median cord, initially doubled, each part having a small ganglion, from which two nerves arise. From the third thoracic ganglion, each ventral nerve-cord ganglion gives off a short fine nerve to a small ganglion, which, in turn, gives off a pair of nerves, there being eight pairs in all, with a median connective from the metathoracic to the sixth abdominal ganglion. Branches of these nerves supply the abdominal spiracles. Thoracic ganglia 1 and 2 supply the first and second thoracic spiracles, while ganglion 3 supplies the first abdominal spiracle. Maskell (1927) illustrates two pairs of sympathetic nerves between the third thoracic and the first abdominal ganglion, none between the second and third abdominal ganglia and two pairs between the third and fourth, suggesting that the sympathetic system is not following the pattern of condensation shown by the ventral nerve cord. The abdominal sympathetic system is anatomically more elaborate, with three fine nerves paralleling the ventral cord from the suboesophageal to the sixth abdominal ganglion.

Discussion

Retention or modification of function of primitive characters

While the New Zealand representatives of the Stenopelmatoidea (Anostostomatidae and Rhaphidophoridae) are in many ways typical orthopterans, they possess a number of features peculiar to the group. A common theme that recurs in the anatomy of the wetas is that of primitive characters. Some characters are generally shared amongst orthopterans and reflect the primitive origins of the Orthoptera, but others reveal the primitive position of wetas within the order. Examples of the former include a multichambered heart with an aorta and a ventral nerve cord with three thoracic and six distinct ganglia. Characters in the latter group are most abundantly seen expressed in the numbers and organization of muscles. An exception is found in the structure of the labial palps, which have an extra segment, which is thought to represent the primitive origins of these appendages, based upon the segmental plan of the walking legs.

Close examination of the musculature repeatedly reveals the primitive position of anostostomatids within the Orthoptera, but these features seldom appear simply as vestigial structures with little function. In the mouth-parts, some additional labial muscles are not found in acridids or other orthopterans (e.g. the promotor of the middle palp segment), while others have different functions (e.g. the palpiger promotor). In the stomodaeum, the increased number of cibarial dilators is similar to that seen in thysanurans and, while suggesting a primitive feature, may well serve the unusual feeding mode of masticating food regurgitated from the pharynx (described earlier for the giant weta *D. rugosa*). As a further example, the undoubtedly primitive tentorial adductor muscle of the mandibles has clearly evolved into an unusually sophisticated sensory function (muscle receptor organ, also described in Field, Chapter 22, this volume).

The leg muscles are strikingly different when compared with those of locusts and crickets. The acridid orthopterans (represented by the locust) are, in many respects, considered to be highly evolved compared with the stenopelmatoid orthoptera, while the crickets have an ancient lineage but show many modern features. Snodgrass (1927), Tiegs (1955) and Matsuda (1963) have all pointed out the trend in the reduction of the number of leg muscles during the evolution of higher insects. This trend is accompanied by a reduced number of apodemes and the condensing of muscles on to the remaining apodemes, especially in the coxa and trochanter. The tree weta has eight apodemes on the coxa and trochanteral rims, two of which comprise a cluster of four to six small apodemes (Fig. 8.6-27). The cricket has seven and the locust has five (Table 8.1). We suggest that the large number of apodemes, as well as the presence of clusters of small apodemes rather than single larger ones, are primitive characters which have been retained in the tree wetas, and which serve as evidence for the more primitive status of the Anostostomatidae. Additional support for this hypothesis comes from the greater number of muscles in the legs of the tree weta compared with the muscles of the other groups. The argument applies regardless of whether muscle components, or insertion sites, are counted on the two segments. Especially cogent, for example, is the number of muscles without apodemes on the trochanteral rim in the weta. Here, of 25 muscle components operating the trochanter, nine insert directly on the rim. In the cricket, of 16 components, four insert on the trochanter rim, while, in the locust, three of the seven components insert on the trochanter rim (Table 8.1). Thus a trend is seen in progressive reduction in number of muscles, as well as fusion of muscle components. The large number of unfused components with independent insertions on the trochanter in the weta suggests the most primitive condition.

Lack of anatomical knowledge

Certain aspects of the anatomy have simply been neglected, notably the exoskeleton and the musculature of the abdomen. Richards (1955) described the abdominal musculature of *Macropathus*, but there is no equivalent account for other anostostomatids. Ramsay (1965) has given a detailed description of the ovipositor in *Deinacrida*, comparing the origin of the constituent elements with patterns found in other orthopterans, but there is no well-illustrated description of the male terminalia of any species.

The internal organs are poorly known and would repay careful histological examination. Similarly, the reproductive system has only been studied superficially in the New Zealand species, leaving Angulo's descriptions of cratomeline anatomy and histology as the major work in this area (Chapter 11, this volume). The alimentary canal is typically orthopteran, with the complex folding and thickening of the proventriculus that becomes most elaborately developed in the tettigoniids; but there has been little comparative work on the gut, despite the known variation in diet amongst carnivorous and herbivorous anostostomatids of New Zealand, Australia and Africa. For example, comparison of proventricular teeth could assist in species differentiation, especially in the highly similar anostostomatine species. Histologically, the alimentary tract is known only superficially, and anomalies exist in early work, which suggests that smooth muscle exists in some layers (unexpected in arthropods). Examination at the ultrastructural level is required. Little is known of the rectal pads in any species, and their role in the regulation of water content has not been established.

Only certain limited aspects of the central nervous system have been studied in any depth (e.g. mapping of motor neurons and dorsal unpaired medial neurons, central projection of tympanal (hearing) sensory neurons), despite attention given to the physiology of proprioception and motor control. Unusual sensilla on the mouth-parts, discoveries of undescribed tympanal structures and a new form of stretch receptor organ at the tibiotarsal joint in the leg all suggest that the sensory system may have peculiarities not so far reported in other orthopterans. As with many other structures, the morphological descriptions of what is known indicate the need for investigation at the ultrastructural level.

Abbreviations Used in Figures

a – coxo-trochanteral articulation, *A1-10* – abdominal segments 1 to 10, *Ab1-5* – apodemes of coxal abductor muscles, *abdi* – insertion of mandibular abductor, *ac* – ante clypeus, *AccExt* – accessory extensor of tibia, *AccFla, b* – posterior and anterior accessory flexors of tibia, *acm* – anteclypeal macula, *Ad1-2* – first and second coxal adductor, *addma* – common apodeme for mandibular adductors, *adp* – arm of cardo, *am-21* – apodeme of tergal adductor of mandible, *am-23* – apodeme of mandibular abductor, *am-26* – apodeme of hypopharyngeal retractor ('hypopharyngeal adductor of mandible'), *AN* – anus, *ANN* – antennary nerve, *ants* – antennal sclerite, *AP* – opening of salivary ducts, *ar* – arm of metasternal apophysis, *aRo1–2* – first and second anterior rotators of coxa, *AS1,2* – first and second abdominal sternites, *ata* – interior tentorial arm, *ath* – anterior thumb of metasternal apophysis, *atp* – anterior tentorial pits

bc – basicoxite, *Beps* – basal margin of the prothoracic episternite, *bph* – basipharynx, *Br* – brain, *BS1-3* – basisternite of thoracic segments, *bts* – band of trichoid sensilla

c – coxal process, *ca* – coxal apophysis, *caddm* – portion of adductor muscle 23 sharing a common apodeme, *CAL* – corpus allatum, *cap* – coxal apophysis, *cas* – campaniform sensilla, *CAV* – cavity, *CC* – corpus cardiacum, *CD* – connective to deutocerebrum, *cef* – cervical facet, *cf* – coxal fold, *CH* – chitinous layer, *CL* – colon, *cls* – clypeolabral suture, *CM* – circular muscle, *COC* – circumoesophageal connective, *con* – condyle, *COR* – corrugated areas, *cos* – coronal suture, *cot* – connective tissue, *COV* – common oviduct, *CR* – crop, *CRC* – cercus, *cs* – coxal spine, *ct* – corporotentorium, *Cx* – coxa

de – ejaculatory duct, *Dep1a–e* – first trochanteral depressor (or apodeme) for components *a–e, Dep1f* – first trochanteral depressor (or insertion site) component *f, Dep2* – second trochanteral depressor (or insertion site), *Dep3a–d* – third trochanteral depressor (or insertion sites), components *a–c*, *dl* – deutocerebrum, *dlbc1,2,3* – dilators of the buccal cavity; *dlcb1,2,3,4* – dilators of the cibarium; *dlph* – dilator of pharynx, *dmro* – dorsal mandibular muscle receptor organ, *doa* – dilator of oral aperture, *dp* – dorsal pair, *dpab* – depressor of antenna (parts *a,b*), *dph* – distal hypopharynx, *dplp* – depressor of maxillary palp, *dpp* – depressor of middle palp segment, *dta* – dorsal arm of tentorium

EC – calyx of ovary, *efla* – extensor of antennal flagellum, *EN* – endochorion, *END* – endapophysis, *EP* – epithelium, *epg* – epigusta, *epm* – epimeron, *eps* – episternite, *ept* – epiproct, *EX* – exochorion, *Ext* – tibial extensor apodeme, *Exta–c* – tibial extensor components a–c

F – frontal ganglion, *fast* – fastigium, *FCA* – chitinized arch connecting rami, *fdr* – fundarima, *Fe* – femur, *ffla* – flexor of antennal flagellum, *fgr* – fronto-genal (sub-ocular ridge), *Fl* – tibial flexor apodeme, *Fla, b* – posterior and anterior tibial flexor, *flp* – flexor of distal palp segment, *flp4* – flexor of maxillary palp segment 4, *fol* – follicle, *FS* – furcasternite

ging – gingylmus, *gl* – glossa, *gr* – median transverse groove

HC – head capsule, *HG* – hypocerebral ganglion, *HNa,b* – nerves to hypopharynx, *hs* – hypostoma, *hss* – hypostomal suture, *HYP* – hypandrium, *hypr* – hypopharyngeal retractor

IL – ileum, *ila* – inner levator of antenna, *IP* – inner pair of valvulae

lab – labium, *labr* – labrum, *LaN* – labial nerves, *lat* – labial articulation, *lcs* – lateral cervical sclerite, *Lev1a,b* – trochanteral levator (or apodeme) for components *a,b, Lev1c–e* – trochanteral levator (or apodemes), components *c–e, Lev2a–c* – second trochanteral levator (or apodeme), components *a–c*, *Lev2av* – second trochanteral levator (or apodeme), component *av*, *Lev3ai–ii* – third trochanteral levator (or apodeme), components *ai–ii*, *Lev3b* – third trochanteral levator (or apodeme), component b, *Lev4a–c* – fourth trochanteral levator (or apodeme complex), components *a–c*, *lfn* – labrofrontal nerve, *lin* – fused lingulae, *lm* – longitudinal muscle, *ln* – labral nerve, *lo* – lateral ocellus, *lon* – lateral ocellar nerve, *lp* – lateral projection, *lplpa,b* – levator of maxillary palp, *lpl2* – levator of second palp segment, *lr* – long chitinous ridge, *lvp* – levator of middle palp segment

M21 – tergal adductor of mandible, *M23,d,p* – abductor of the mandible with dorsal and posterior bundles, *M-26* – hypopharyngeal retractor ('hypopharyngeal adductor of mandible'), *ma* – maxillary articulation, *MAL* – malpighian tubules, *MC* – mesenteric caecum, *MdN I,II,III* – mandibular nerves, *men* – mentum, *MES* – mesenteron, *mg* – insertion ridge of mental retractor, *M hyp* – hypopharyngeal retractor muscles, *mlrp* – posterior retractor of labrum, *mnl* – primary mandibular articulation (posterior articulation), *mn2* – anterior mandibular articulation, *mnd* – mandible, *mo* – median ocellus, *MON* – median ocellar nerve, – *mtsa* – metathoracic sternal apophysis, *MxN* – maxillary nerves

NSC – sensory nerve to cervical region, *Ntth* – nerve to tentoro-thoracic musculature

occ – occiput, *occs* – occipital suture, *ocs* – ocular sclerite, *OL* – optic lobe, *ola* – outer levator of antenna, *ON* – optic nerve, *OV* – ovarioles, *OVD* – oviduct

pa – pleural arm, *pc* – post clypeus, *pcs* – posterior suture, *pcx* – precoxale, *ped* – pedicel, *pfr* – palpifer, *PFS* – postfurcasternite, *pg* – postgena, *pgl* – paraglossa, *pgr* – palpiger, *ph* – pharynx, *PL* – protocerebrum, *pln*- paralingula, *plp3* – productor of the 3rd palp segment, *pmp3* – promotor of the third palp segment, *pmr* – premental ridge, *pms* – postmandibular suture, *PN* – penis, *po* – postocciput, *POC* – post-oesophageal commissure, *por* – postoccipital ridge, *pos* – post occipital suture, *pp* – pleural process, *pplp* – promotor of palpiger, *ppt* – paraproct, *Pr* – coxal promotor apodeme, *pr* – pleural ridge, *prem* – prementum, *prg* – pharyngea sclerite, *pro* – proventriculus, *pRo1* – first posterior rotator, *pRo2a–b* – second posterior rotator, components *a* and *b*, *prp2* – promotor of second palp segment, *PRS* – prosternite, *ps* – pleural suture, *PS1* – first pleuro-sternal muscle, *PS2* – second pleurosternal muscle, *PSE* – pseudosternite, *pss* – pleurostomal suture, *pst* – pleurostoma, *PT* – primary tooth, *PTa* – pretarsus, *pth* – posterior thumb of sternal apophysis, *ptm* – posterior tentorial arms, *ptp* – posterior tentorial pit, *pts* – parietal suture

R – recurrent nerve, *rao* – retractor of the mouth angle, *Re* – coxal remotor apodeme, *REC* rectum, *RES* – reservoir, *RetUng* – retractor unguis apodeme, *RetUnga–c* – retractor unguis components, *Ro* – coxal rotator

s – sensilla, *sa* – sensory area, *sal* – saliva, *sg* – salivary gland, *SGP* – subgenital plate, *SN* – stomatogastric nerve, *SOI–V* – groups of trichoid sensilla, *SOG* – sub-oesophageal ganglion, *sp2–3* – spiracles of thoracic segments, *spA1* – spiracle of first abdominal segment, *SPS* – spinasternite, *ss* – subsegment, *ST* – secondary tooth, *stm* – stomodaeum, *sug* – subgalea, *sum* – submentum

T1–3 – thoracic segments 1 to 3, *Ta* – tarsus *TaDep* – tarsal depressor, *TaLev* – tarsal levator, *tb/td* – trichoid sensilla, *TEG* – tegumentary nerve, *TF* – terminal filament, *Ti* – tibia, *TL* – tritocerebrum, *Tlm* – tentorolabial muscles, *TM1,2a,2b* – tentoro-mandibular adductor muscles, *Tmax* – tentoromaxillary muscles, *tmm* – tentoro-mandibular macula, *tn* – trochantin, *toa,b* – arms of torma, *TP* – tag-like processes, *Trl,2,* – trochanter, parts 1 and 2, *tra* – trabeculae, *TrDep1abc* – trochanteral depressor 1, with three components, *T-th* – tentorothoracic muscles, *tub* – sensory tubercles, *TVS* – large tubules of seminal vesicles

v – vertex, *VCS* – ventral cervical sclerites, *VD* – vas deferens, *vmro* – ventral mandibular muscle receptor organ, *VNC* – ventral nerve cord, *VP* – ventral pair of valvulae, *vph* – ventral dilators of pharynx, *VS* – seminal vesicles covered by smaller tubules

w – wing of sternal apophysis, *wp* – white patch

References

Albrecht, F.O. (1953) *The Anatomy of the Migratory Locust.* University of London, Athlone Press, 118 pp.

Arbas, E.A. (1983) Thoracic morphology of a flightless Mexican grasshopper, *Barytettix psolus*: comparison with the locust *Schistocerca gregaria*. *Journal of Morphology* 176, 141–153.

Burrows, M. (1996) *The Neurobiology of an Insect Brain*. Oxford University Press, Oxford, 682 pp.

Campbell, J.I. (1961) The anatomy of the nervous system of the mesothorax of *Locusta migratoria migratorioides* R & F. *Proceedings of the Zoological Society, London* 137, 403–432.

Davis, A.C. (1927) Studies on the anatomy and histology of *Stenopelmatus fuscus* Hald. *University of California Publications in Entomology* 4, 159–208.

Field, L.H. (1978) The stridulatory apparatus of New Zealand wetas in the genus *Hemideina* (Insecta: Orthoptera: Stenopelmatidae). *Journal of the Royal Society of New Zealand* 8(4), 359–375.

Gibbs, G.W. (1999) Four new species of giant weta, Deinacrida (Orthoptera: Anostostomatidae: Deinacridinae) from New Zealand. *Journal of the Royal Society of New Zealand* 29(4), 307–324.

Honomichl, K. (1976) Feinstruktur eines Muskelreceptors im Kopf von *Dermestes maculatus* De Geer (Insecta, Coleoptera). *Zoomorphologie* 85, 59–71.

Honomichl, K. (1978) Feinstruktur zweier Propriocep-toren im Kopf vom *Oryzaephilus surinamensis* (L.) (Insecta, Coleoptera). *Zoomorphologie* 90, 213–226.

Hoyle, G. and Field, L.H. (1983) Defense posture and leg-position learning in a primitive insect utilize catch-like tension. *Journal of Neurobiology* 14, 285–298.

Imms, A.D. (1931) *Social Behaviour in Insects*. Methuen, London.

Maskell, F.G. (1927) The anatomy of *Hemideina thoracica*. *Transactions and Proceedings of the New Zealand Institute* 57, 637–670.

Matsuda, R. (1963) Evolution of the thoracic musculature in insects. *University of Kansas Science Bulletin* 44(11), 509–534.

Matsuda, R. (1965) *Morphology and Evolution of the Insect Head*. Memoir of the American Entomological Institute No. 4. Ann Arbor, Michigan.

Nilsson, D., Labhart, T. and Eichendorf, A. (1987) Photoreceptor design and optical properties affecting polarization sensitivity in ants and crickets. *Journal of Comparative Physiology A* 161, 645–658.

Nutting, W.L. (1951) A comparative anatomical study of the heart and accessory structures of the orthop-teroid insects. *Journal of Morphology* 89, 501–597.

O'Brien, B. (1984) Mandibular movements and their con-trol in the weta, *Hemideina maori* (Orthoptera : Ensifera: Stenopelmatidae). PhD thesis, University of Canterbury, Christchurch, New Zealand.

Ramsay, G.W. (1955) The exoskeleton and musculature of the head, and the life-cycle of *Deinacrida rugosa* Buller, 1870. MSc thesis, Victoria University, Wellington, New Zealand.

Ramsay, G.W. (1965) Development of the ovipositor of *Deinacrida rugosa* Buller (Orthoptera: Gryllacridoidea: Henicidae) and a brief review of the ontogeny and homology of the ovipositor with par-ticular reference to the Orthoptera. *Proceedings of the Royal Entomological Society of London (A)* 40, 41–60.

Richards, A.R. (1955) Anatomy and morphology of the cave-orthopteran *Macropathus filifer* Walker 1869. *Transactions of the Royal Society of New Zealand* 83, 405–452.

Smith, P.E. (1979) Behavioural and reflexive analysis of a defence response in *Hemideina maori* (Orthoptera: Stenopelmatidae). BSc (Hons) thesis, University of Canterbury, Christchurch, New Zealand.

Snodgrass, R.E. (1927) *Morphology and Mechanism of the Insect Thorax*. Smithsonian Miscellaneous Collections 80, Smithsonian Institution, Washington, DC, 108 pp.

Snodgrass, R.E. (1928) *Morphology and Evolution of the Insect Head and its Appendages*. Smithsonian Miscellaneous Collections 81, Smithsonian Institution, Washington, DC, 158 pp.

Thakare, V.K. (1972a) Studies on the skeleto-muscular mechanism of legs in the Indian field cricket *Gryllus bimaculatus* de Greer (Gryllidae, Orthoptera). *Zoologisches Anzeiger, Leipzig* 188, 372–399.

Thakare, V.K. (1972b) Studies on the skeleto-muscular mechanism of the metathorax in the Indian field cricket *Gryllus bimaculatus* de Greer (Gryllidae: Orthoptera: Insecta). *Journal of Animal Morphology and Physiology* 19, 111–122.

Tiegs, O.W. (1955) The flight muscles of insects – their anatomy and histology; with some observations on the structure of striated muscle in general. *Philosophical Transactions of the Royal Society, London* 238, 221–359.

Walker, E.M. (1931) On the anatomy of *Grylloblatta cam-podeiformis* Walker. 1. Exoskeleton and musculature of the head. *Annals of the Entomological Society of America* 24, 519–536.

9
Morphometric Analysis of *Hemideina* spp. in New Zealand

Laurence H. Field and Robert. S. Bigelow†
Department of Zoology, University of Canterbury, Private Bag 4800, Christchurch, New Zealand

Introduction

In this chapter, we assess the reliability of leg spination and stridulatory ridge counts as morphological characters for distinguishing species of New Zealand tree wetas in the genus *Hemideina*. Additionally, coloration pattern is considered for usefulness as a taxonomic character. The classification of spination patterns (chaetotaxy) has been used widely in insect taxonomy, but its application to the legs of orthopteroid insects is relatively new. Nevertheless, evidence from developmental studies of insects shows that leg spine and bristle patterns are reliably consistent within species. For example, Tanaka and Kitamura (1992) found that the spur patterns on the tibia and femur of the German cockroach (*Blatella germanica*) were basically similar, but were modified differently to form a distinct pattern for each pair of legs. Thus leg spur pattern could be used as one of the characters to recognize the different instars. Furthermore, the distinct pattern of each leg remained unchanged throughout postembryonic development. Some variation in patterns occurred, but it was confined to certain spurs and did not detract from the general reliability of the character.

Initial descriptions of several species of *Hemideina* and subsequent identification keys have incorporated leg chaetotaxy to distinguish species within the genus (Salmon, 1950, 1955). Spination patterns are also used in species diagnoses of other wetas, e.g. *Deinacrida* (giant wetas) and *Hemiandrus* (Salmon, 1955; Gibbs, 1999). However, in routine use of keys, one of us (RSB) encountered individuals of *Hemideina maori* with spination patterns which did not conform to the described diagnostic patterns. Moller (1985) also found Salmon's diagnostic spination characters for *Hemideina crassicruris* (now a subspecies of *Hemideina crassidens* (Morgan-Richards *et al.*, 1995)) to vary amongst individuals, and a study of spination in *Deinacrida connectens* confirmed significant amounts of intraspecific variation (Morgan-Richards and Gibbs, 1996). These observations thus raise the question of the reliability of chaetotaxy for taxonomic purposes in wetas, especially in the genus *Hemideina*.

Although stridulatory ridge count has not been utilized for taxonomic diagnoses in wetas, it has been considered for use as a taxonomic character in crickets, tettigoniids and grasshoppers (e.g. Walker and Carlysle, 1975; Morris and Walker, 1976; Pitkin, 1976; den Hollander and Barrientos, 1994). From an earlier numerical analysis of ridge counts, it was clear to us that this character varied amongst species of *Hemideina* (Field, 1978), although the extent of intraspecific variation was

† Bob Bigelow passed away on 4 January 2000, leaving behind a legacy of orthopteran research and many appreciative students.

not established. We therefore included it in the present numerical analysis.

Intraspecific variation in *Hemideina* was analysed in large data sets (up to n = 695 for one species) for *H. crassidens, H. femorata, H. maori, H. ricta,* and *H. thoracica*, as well as small data sets for *H. broughi* and the newly described cryptic species, *H. trewicki* (Ramsay and Bigelow, 1978; Morgan-Richards, 1995). The numbers and positions of spurs and spines on the femora and tibiae of legs II and III and counts of abdominal stridulatory ridges (except for *H. broughi,* which lacks stridulatory ridges (Field, 1978)) are compared and discussed. Notes on characteristic coloration patterns are included, inasmuch as this has recently been subjected to genetic analysis (see Morgan-Richards *et al.*, Chapter 7, this volume).

Methods

Alcohol-preserved specimens from all major museum collections in New Zealand, as well as privately collected specimens of the authors, were inspected and notes were recorded on the presence or absence of articulated and non-articulated spines of the middle and hind femora and tibiae. These included the non-articulated prolateral and retrolateral apical spines, which protrude from the flattened distal apices of the hind femur (Fig. 9.1A), and the retrolateral spine on the distal end of the mid-femur. A single articulated spine, retrolateral on the dorsal aspect of the middle tibia (arrow, Fig. 9.1B), was examined. This spine is located one-third back from the tibial apex. Occasionally, an extra spine was found just distal to this one. The number of stridulatory ridges on each side of the second abdominal tergite and the sex and maturity of most specimens were also recorded. Substantial sample sizes were used for *H. femorata, H. thoracica, H. crassidens, H. maori* and *H. ricta*, while conclusions on *H. trewicki* and *H. broughi* are limited by small numbers of specimens. Total sample sizes differed somewhat between the spine analysis and the ridge-count analysis.

Results

Leg spination modal patterns

The frequencies of occurrence of these spines are shown in Table 9.1 for the seven species studied. To clarify the major trends in the table, cell frequencies with percentage occurrence of < 5% have been left blank (hence percentages in each

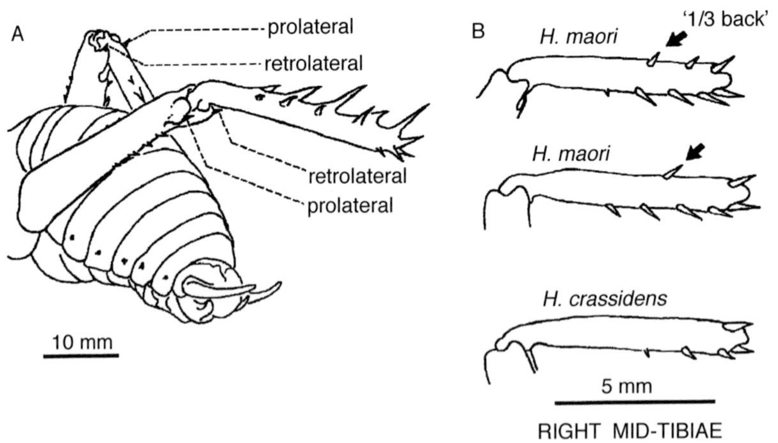

Fig. 9.1. Illustrations of spines used in the chaetotaxonomic study of *Hemideina* tree wetas. A. Hind femur non-articulated apical spines. The prolateral apical spine is on the lateral (anterior if leg is rotated 90° to body axis) face of femur, while the retrolateral apical spine is medial (posterior if leg rotated as above). B. Articulated '1/3 back' spine on dorsal aspect of mid-femur (solid arrows) in two specimens of *H. maori*, but lacking in *H. crassidens* (bottom). The upper specimen of *H. maori* shows the 'extra' articulated tibial spine which occasionally appeared on *H. maori* and rarely (< 5%) on *H. ricta*.

Table 9.1. Patterns of distribution for articulated and non-articulated spines on the mid- and hind legs of all seven species of *Hemideina*. In the 13 patterns (numbered in left-hand column), solid symbols indicate the presence of a spine. Percentages of individuals having different patterns are indicated in columns beneath each species name (values < 5% are omitted).

	HIND LEGS				MIDDLE LEGS												
	Hind femur apical spines				Femur retro apical spine		Tibia articulated dorsal spines				Percentage specimens with pattern indicated						
	Right		Left				Right		Left								
	Prolat	Retrolat	Prolat	Retrolat	Right	Left	1/3 back	Extras	1/3 back	Extras	$n = 9$ *broughi*	$n = 14$ *trewicki*	$n = 162$ *femorata*	$n = 212$ *thoracica*	$n = 695$ *crassidens*	$n = 261$ *maori*	$n = 81$ *ricta*
1	◆	◆	◆	◆	◆	◆	◆		◆		67	79	27	11			
2	◆	◆	◆	◆			◆		◆		11		27				
3	◆	◆	◆	◆	◆		◆		◆				9				
4	◆	◆	◆	◆		◆	◆		◆		11		6				
5	◆	◆			◆	◆	◆		◆			7					
6	◆	◆	◆	◆	◆	◆						7		59	61		
7	◆	◆	◆	◆										10	21		
8	◆	◆	◆				◆		◆		11						
9	◆		◆				◆	◆	◆	◆						6	
10	◆		◆				◆		◆							46	61
11	◆						◆		◆								6
12			◆				◆		◆							8	
13							◆		◆							11	12

column do not total 100%). The first point clearly shown by the data is that no one pattern of spination (i.e. combination of hind femur apical spines, middle femur retrolateral apical spines and middle tibia dorsal spines) is unique for any weta species. Therefore, at best, it is only possible to discuss modal patterns to distinguish the different species. A modal pattern is that which is found in the greatest percentage of specimens for a species. Accordingly, the species fell into three groups, in each of which two species share the same modal pattern, plus a fourth group containing only *H. femorata*, with a peculiar chaetotaxic condition. Thus, *H. broughi* and *H. trewicki* shared modal pattern 1 (67% and 79% of specimens, respectively; Table 9.1), having the greatest number of spines of all *Hemideina* species patterns. *Hemideina femorata* differed from all other species by having a peculiar bimodal distribution, in which two patterns were equally split within 54% of the individuals while two additional patterns were shown by an additional 15% of individuals. In the next group, *H. thoracica* and *H. crassidens* shared modal pattern 6, in which the mid-tibial '1/3 back' spine was missing. Finally, *H. maori* and *H. ricta* shared modal pattern 10, in which all retrolateral apical spines on mid- and hind femora were missing. The amount of variation in additional minor patterns was not consistent among species. *Hemideina thoracica* and *H. crassidens* showed the least variation: about 60% of individuals had the modal pattern and 21% shared one or two closely related spination patterns. In *H. maori* and *H. ricta*, 46% of individuals for each species shared the modal pattern, but an additional 25–35% shared three or six (respectively) additional spination patterns, i.e. there was much more variation in spination in these two species.

Variability in individual spine expression

Some spines appeared to be quite reliably expressed within species, while others were less stable and had erratic expression amongst individuals. The most stable spine was the articulated '1/3 back' mid-tibial spine. It was reliably found in all individuals of all species, except for *H. thoracica* and *H. crassidens*, in which it was consistently lacking. Therefore, lack of this spine allows the latter two species to be separated from all other *Hemideina* species. Occasionally (5% occurrence), an extra mid-tibial articulated spine (Fig. 9.1B) appeared in *H. maori* (and in two *H. ricta* specimens), thus contradicting Salmon's key, which used the single spine as a unique character to identify this species. This extra spine never occurred in the other species.

Both hind femur apical spines (prolateral, retrolateral) showed greater variability in expression. They always occurred in *H. thoracica, H. crassidens* and *H. femorata*. In *H. broughi* and *H. trewicki*, these spines were sometimes absent. Similarly, only the prolateral apical spine occurred in *H. maori* and *H. ricta*, but it was absent in 19% and 13% of individuals, respectively. Its bilateral mate, the retrolateral apical spine, was never found in these two species, thus setting them apart from the other species. In general, the hind femur apical spines were not consistent characters in *H. broughi, H. trewicki, H. maori* and *H. ricta*.

The middle femur retrolateral apical spine showed the greatest variation in occurrence. The most unusual case was for *H. femorata*, where it was equally present (27%) or absent (27%) on both mid-legs, and thus it formed two modal patterns (1 and 2) equally represented in the majority of the specimens. In the remainder of the specimens of *H. femorata,* and in *H. broughi*, either the left or right leg lacked this spine. A similar situation occurred for *H. thoracica* and *H. crassidens,* where the spine was either present or absent on both legs, in a 6 : 1 or 3 : 1 ratio, respectively. Finally, the spine reliably never occurred in the *H. maori* and *H. ricta* pair of species; however, its presence or absence would not be a unique diagnostic character for any of the *Hemideina* species.

Two hybrid individuals have been found which represent crosses between *H. femorata* and *H. ricta* (Morgan-Richards and Townsend, 1995). Their spination is more characteristic of *H. femorata* than of *H. ricta*. One hybrid had pattern 1 (Table 9.1) and exactly matched *H. femorata*. The other showed the same pattern but it lacked the prolateral hind femur apical spine.

Leg morphology

In addition to leg spination differences between *Hemideina* spp., differences were also found in hind leg morphology (Fig. 9.2). This is apparent from the relative length and thickness of the femur and tibia. The two leg segments are thinnest in *H. thoracica* (Fig. 9.2A), intermediate in *H. crassidens* and *H. femorata* (Fig. 9.2B, C, respectively) and

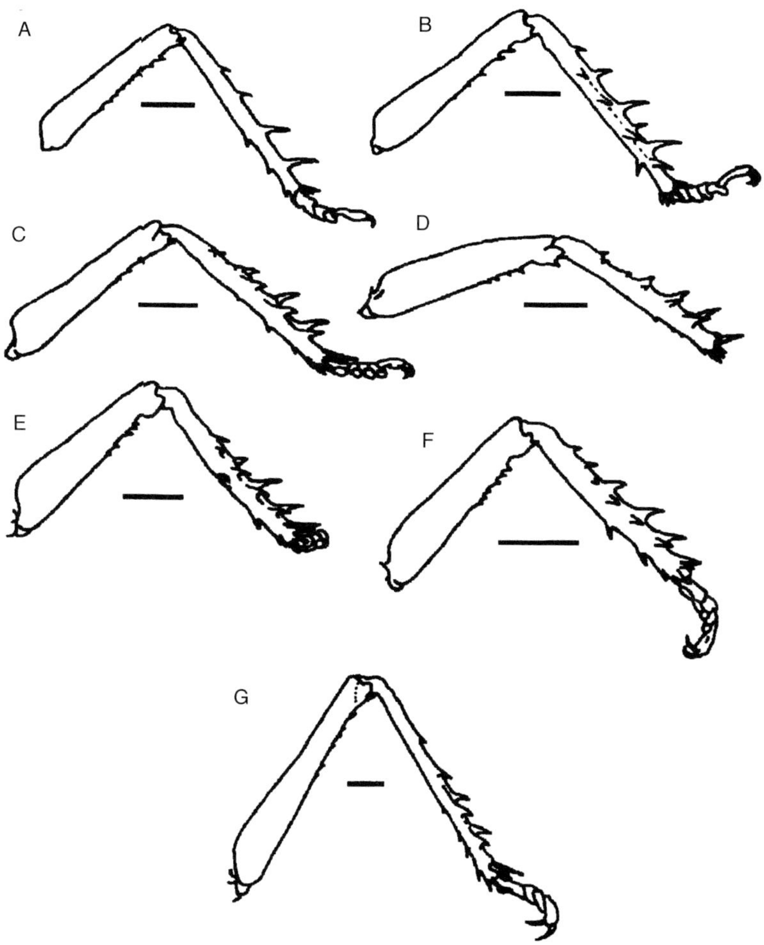

Fig. 9.2. Hind leg morphology of all *Hemideina* species, showing relative widths of femur and tibia, as well as differences in non-articulated tibial spine morphology. A. *H. thoracica*. B. *H. crassidens*. C. *H. femorata*. D. *H. ricta*. E. *H. maori* (formerly *H. alterna*, from Rock and Pillar Range, Otago, New Zealand). F. *H. maori* (from Porter's Pass, Canterbury, New Zealand). G. *H. broughi*. Note relatively thin, long tibiae in *H. thoracica, H. crassidens*, and *H. femorata*, shorter, thicker tibiae in *H. maori* and *H. ricta*, and extremely long thin tibia in *H. broughi*. Scale bar (A–G) = 5 mm.

thickest in *H. maori* (including the Rock and Pillar Range form) and *H. ricta* (Fig. 9.2D, E, respectively). For comparison, we show the leg of *H. broughi* (doubtfully of the genus *Hemideina* (P.M. Johns, Christchurch, 1985, personal communication)), which is relatively longer and narrower than any of the above species (Fig. 9.2F).

Stridulatory ridges

A row of heavily tanned cuticular ridges occurs on the ventral, anterior margin of abdominal tergite II in all *Hemideina* spp. except *H. broughi* (Field, 1978) (Fig. 9.3A). These occur in both sexes and, together with an array of cuticular pegs on the posterior face of the hind femur (Field, 1993), they form the femoro-abdominal stridulatory apparatus used in inter- and intraspecific communication (see Field, Chapter 15, this volume).

Morphological variation

Variation occurs in the morphology of stridulatory ridges, as well as in the number of ridges between

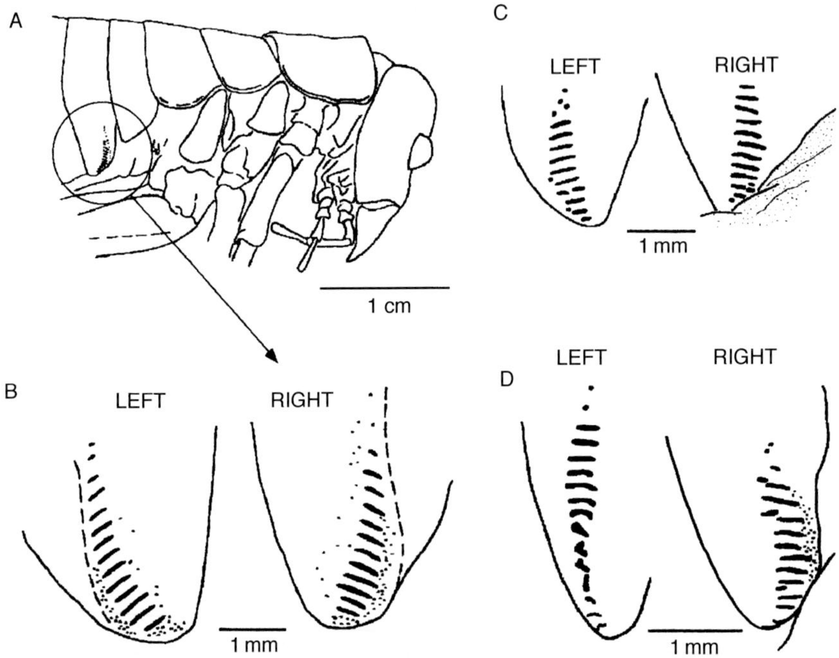

Fig. 9.3. Location, morphology and variation of the stridulatory ridges on the second abdominal tergite of tree wetas, using *H. maori* for examples. A. Position and orientation of file on right side of a female. B. Enlarged view of left and right file from same individual, showing with 14 ridges on the left side and 12 ridges on the right. C. Morphological variation in ridges of the left and right sides of a male. D. Extreme variation in ridge morphology of males from different locations in the South Island of New Zealand.

sides within individuals and number of ridges among individuals within species. For example, the left row of ridges in Fig. 9.3B (*H. maori*) contains 14 ridges, while the right row of the same individual contains only 12. Morphological differences between the right and left sides of a single individual (e.g. Fig. 9.3C), as well as between individuals (Fig. 9.3D), may be very marked.

Interspecific variation

The distributions of the ridge counts (sum of left plus right sides) for the different species fell into two distinct groups: one with species having relatively few ridges (*H. femorata, H. thoracica, H. crassidens* and *H. trewicki*) and one with species having many ridges (*H. maori* and *H. ricta*) (Fig. 9.4). These broad groupings are shown along the top of Fig. 9.4 as means (± SD) of pooled locality samples for each species. Although apparently distinct ridge-count distributions are suggested for each species, the actual pooled frequency distributions for the species within a group overlapped considerably (not shown), thus preventing unique classification of species by this character.

Intraspecific variation

Variation within species is indicated by the regional samples, shown in the lower half of Fig. 9.4. The samples are arranged in geographical order to enhance trends. They should be compared with overall distributions for all *Hemideina* species (Fig. 9.5). The results clarify the basis for the generally large spread seen in pooled species distributions, and showed several different geographical trends. For example, *H. thoracica* had a highly significant clinal decrease in ridge count from north (Northland (ND)) to south (Taupo (TO)) in the North Island (although the Coromandel (CL) population was not different from that of the adjacent ND region) (Table 9.2, Fig. 9.5). *Hemideina crassidens* had a disjunct pattern related to the separation of the North Island

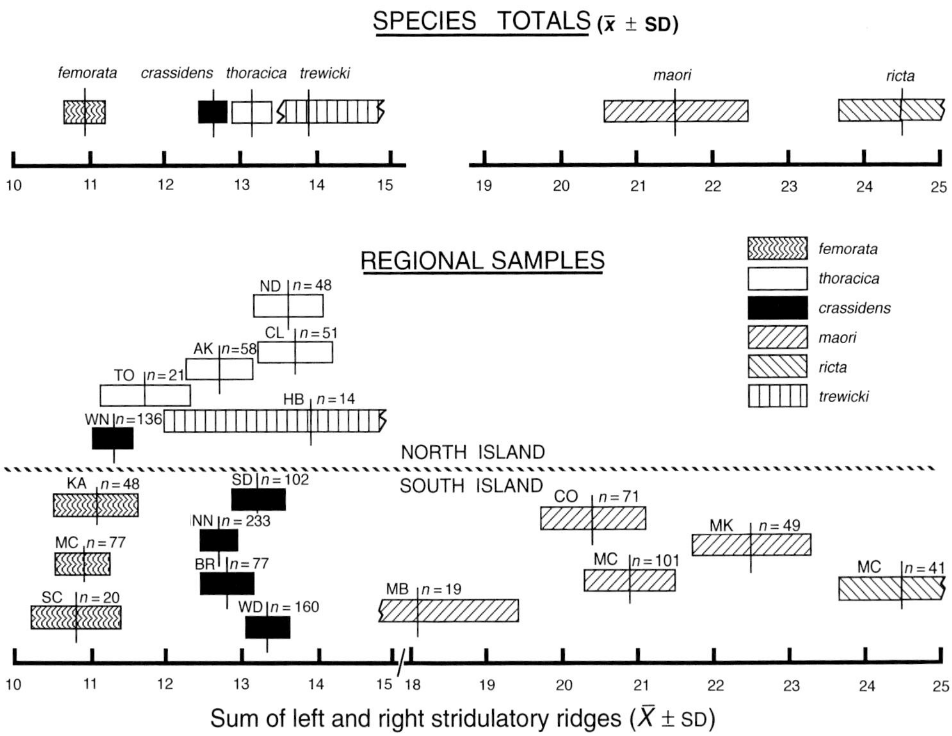

Fig. 9.4. Summary of summed left and right stridulatory ridge counts shown as mean values (± SD) for *Hemideina* species. Top: Distribution of pooled sample means from all collection localities for each species, showing two groupings, one with species having low ridge counts (*H. femorata, H. crassidens, H. thoracica* and *H. trewicki*), and the other with species having high ridge counts (*H. maori* and *H. ricta*). Bottom: Regional breakdown of means (± SD) for each species. Dashed line indicates division between North Island and South Island localities. Two-letter locality codes for New Zealand follow Crosby *et al.* (1976): AK, Auckland; BR, Buller; CL, Coromandel; CO, Central Otago; HB, Hawkes Bay; KA, Kaikoura; MB, Marlborough; MC, Mid Canterbury; MK, Mackenzie; ND, Northland; NN, Nelson; SC, South Canterbury; SD, Marlborough Sounds; TO, Taupo; WD, Westland; WN, Wellington. *n*, Sample size from each locality.

population (Wellington, WN) from the adjacent South Island populations by the Cook Strait (see Gibbs, Chapter 2, this volume, for distribution maps). The dramatic difference between the low ridge count of *H. crassidens* samples from the North Island WN sample and the higher ridge count from the nearest South Island population (Marlborough Sounds, SD) was highly significant. The populations along the western side of the South Island were also significantly different. In contrast, *H. femorata* (along the eastern side of the South Island) had a uniformly low ridge count with no regional differences (Table 9.2). The alpine *H. maori* had the largest variation in the genus. There is a clinal variation along the Southern Alps (from Marlborough region (MB) through Canterbury (MC) to the McKenzie region (MK), including Mount Cook), in which each of the samples is highly significantly different from its nearest neighbour. The disparate population in Central Otago (CO) is isolated on the eastern Rock and Pillar Range, and was formerly considered to be a separate species until analysed genetically (see Morgan-Richards *et al.*, Chapter 7, this volume). The occurrence of *H. ricta* on Banks Peninsula is too restricted to make regional comparisons.

Statistical analyses of intraspecific variation within regional samples (Table 9.2) showed that the coefficient of variation was usually low (1.8–5.5) except in the cases of small sample size (*H. trewicki, H. maori*). However, there was an unusually large coefficient of variation in ridge count for *H. ricta* (11.0), when compared with this

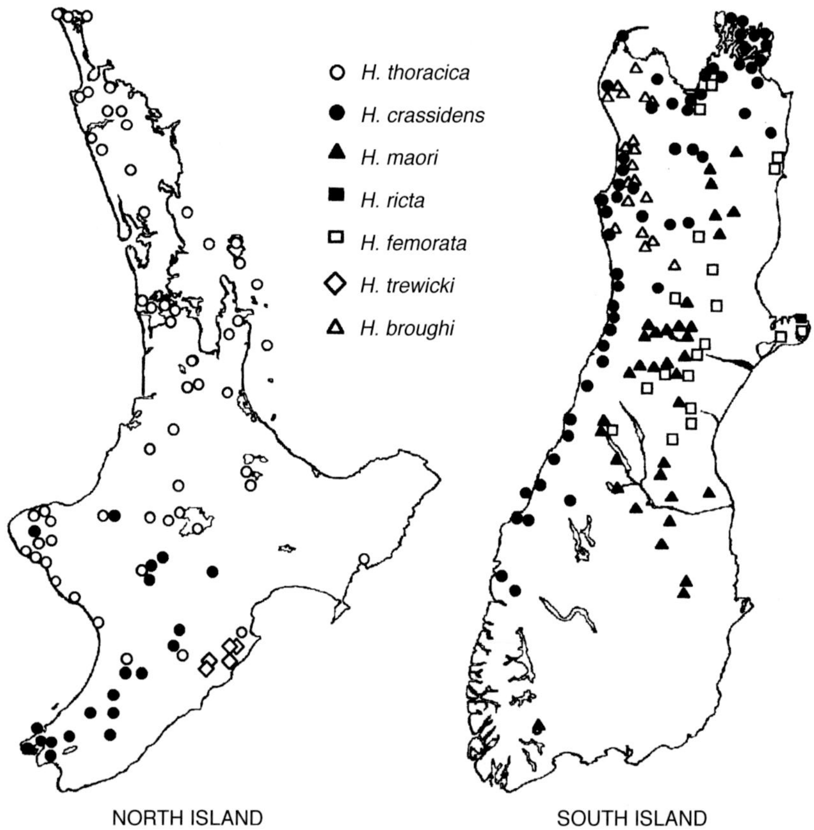

Fig. 9.5. Collection localities for all *Hemideina* species examined in this study. Note distinct lack of species in the south of the South Island. The East Cape of the North Island was not covered in this study but is known to be inhabited by *H. thoracica* (Morgan-Richards, 1997).

trait in other *Hemideina* species or with intensively studied traits, such as mammalian skeletal variation (usual values: 5–6 (Simpson *et al.*, 1960)).

Sexual differences

Males had a higher mean of total ridge count compared with females for those species with a large sample size (Fig. 9.6). This was the case for *H. femorata, H. thoracica, H. crassidens* and *H. maori.* The differences were highly significant using the *t* test (Table 9.3).

Body colour pattern

The pronotum and the abdominal tergites are distinctively coloured in most species of tree wetas. Usually a dark pattern of brown to black markings occurs over a general body ground colour on these structures. The variation in patterns of dark markings, rather than the ground colour, appeared to be useful for distinguishing the species. We have examined small samples from different localities within the ranges of each species to determine if the patterns were consistent. The results suggest that these patterns could serve as useful diagnostic characters.

Pronotum

Three of the species (plus one subspecies) have pronota with a dark ground colour (dark brown to black) and markings that are slightly darker. The other four species have a pale (straw-coloured) background, upon which dark brown to black markings are prominently displayed. In all species, the markings are associated with the sutures and grooves of the three major pronotum plates: the

Table 9.2. Statistics of distributions of summed (left and right sides) stridulatory ridge counts for each *Hemideina* species based upon collection localities. Results of *t* tests for differences between means are shown on the right. Each *t* value compares the mean in the same line with the mean in the line below.

Species	Location	Sample size	Mean no. ridges	SD	CV	*t* value	*P*
femorata	KA	48	11.1	0.6	5.0	1.93	NS
	MC	77	10.9	0.3	2.9	0.30	NS
	SC	20	10.9	0.6	5.5		
thoracica	TO	21	11.7	0.6	5.1	7.94	< 0.001
	AK	58	12.7	0.5	3.5	10.05	< 0.001
	ND	48	13.6	0.5	3.5	1.02	NS
	CL	51	13.7	0.5	3.6		
crassidens	WN	136	11.3	0.3	2.2	48.02	< 0.001
	SD	102	13.2	0.4	2.7	14.95	< 0.001
	NN	233	12.7	0.2	1.8	2.24	< 0.02
	BR	77	12.8	0.4	2.9	11.54	< 0.001
	WD	160	13.3	0.3	2.0		
trewicki	HB*	14	13.9	2.0	14.4		
maori	MB	19	18.1	1.3	7.2	14.82	< 0.001
	MC	101	20.9	0.6	2.9	13.68	< 0.001
	MK	49	22.5	0.8	3.6	15.23	< 0.001
	CO	71	20.4	0.7	3.4		
ricta	BP*	49	24.5	2.7	11.0		

CV, coefficient of variation; NS, no significant difference; *P*, probability of no difference between sample means; SD, standard deviation; *t* value, Student's *t* value. Two-letter locality codes for New Zealand follow Crosby, Dugdale and Watt (1976): AK, Auckland; BP*, Banks Peninsula; BR, Buller; CL, Coromandel; CO, Central Otago; HB, Hawkes Bay; KA, Kaikoura; MB, Marlborough; MC, Mid Canterbury; MK, Mackenzie; ND, Northland; NN, Nelson; SC, South Canterbury; SD, Marlborough Sounds; TO, Taupo; WD, Westland; WN, Wellington. Two species are restricted to a single locality (*) and could not be tested.

prescutum, the scutum and the scutellum (Fig. 9.7A). Either the prominent groove (arrow, Fig. 9.7A) in the prescutum is set off by a wide tapered band (*H. femorata*, *H. thoracica* and *H. trewicki*, Fig. 9.7G, H and I, respectively) or the adjacent anteromedial raised region is more or less marked as an oval patch (*H. crassidens*, *H. c. crassicruris*, *H. ricta* and *H. broughi*, Fig. 9.7B, C, E and F, respectively). In *H. maori*, the patch forms a distinct black oval on a light ground (Fig. 9.7D). Other sutures are marked by narrower dark irregular lines and dark spots. Distinctive differences include a dark anteroposterior band along the lateral scutum margins in *H. femorata* (Fig. 9.7G), black posterolateral marks in *H. thoracica* and *H. trewicki*, but paler brown in *H. femorata* (Fig. 9.7H and I vs. Fig. 9.7G, respectively). The entire pronotal margin is characteristically black in *H. maori* and *H. femorata*, although the lateral margins are brown in the latter species. In *H. ricta*, the entire margin is darker than the ground colour. The anterior and posterior margin is black in *H. thoracica* and *H. trewicki*. The posterior margin is darker than the ground colour in *H. crassidens*, *H. c. crassicruris* and *H. broughi*.

Abdominal tergites

The ground colour of abdominal tergites varies considerably around what seems typical for each species (for early descriptions, see Salmon, 1950). However, we found that the patterns and shapes of dark bands on the anterior and/or posterior dorsal tergite margins were consistent within each species and hence may be useful for diagnostic purposes (Fig. 9.8). Four species and one subspecies have a

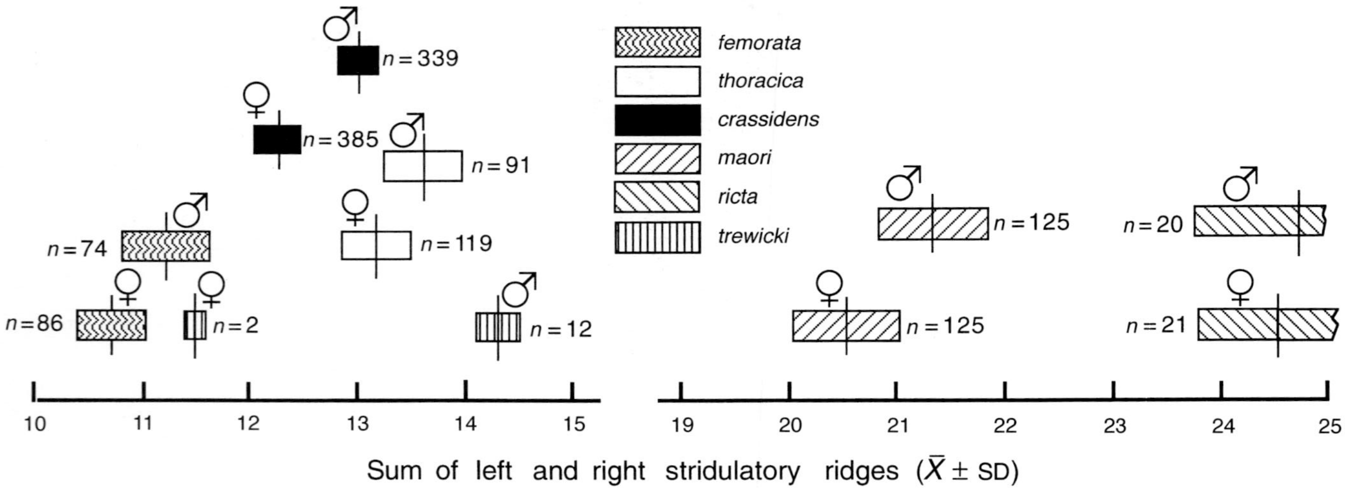

Fig. 9.6. Variation and differences in left and right summed stridulatory ridge counts between males and females for all *Hemideina* species. Data presented as mean ± SD; *n*, sample size. Note consistently larger number of ridges in males compared with that in females.

Table 9.3. Statistics for summed (left and right sides) stridulatory ridge counts for male and female *Hemideina* species. Results of *t* tests for differences between means are shown on the right. Each *t* value compares the mean in the same line with the mean in the line below.

Species	Sex	Sample size	Mean no. ridges	SD	CV	*t* value	*P*
femorata	M	74	11.2	0.4	3.4	9.03	< 0.001
	F	86	10.7	0.3	3.0		
thoracica	M	91	13.6	0.4	2.7	9.01	< 0.001
	F	119	13.2	0.3	2.4		
crassidens	M	339	13.0	0.2	1.5	52.33	< 0.001
	F	385	12.3	0.2	1.5		
trewicki	M	11	14.2	2.0	14.4	1.18	NS
	F	3	12.7	1.7	13.3		
maori	M	125	21.3	0.5	2.5	12.08	< 0.001
	F	125	20.5	0.5	2.4		
ricta	M	20	24.7	1.0	4.0	0.73	NS
	F	21	24.5	0.8	3.1		

CV, coefficient of variation; NS, no significant difference; *P*, probability of no difference between sample means; SD, standard deviation; *t* value, Student's *t* value.

broad black or black/brown transverse band on the anterior margin (*H. crassidens, H. c. crassicruris, H. trewicki, H. femorata* and *H. maori*). In *H. crassidens, H. trewicki* and *H. femorata*, the anterior band widens at the dorsal midline, forming a triangular enlargement of dark colour, which projects posteriorly (Fig. 9.8). In *H. crassidens, H. c. crassicruris* and *H. trewicki*, this medial projection of dark colour may reach the posterior tergite margin; thus these species have very similar banding patterns, but the *crassidens–trewicki* distinction is easily made by comparison of pronotum colour pattern (Fig. 9.7B, C, I; Morgan-Richards, 1995). These two species differ from *H. femorata* by possessing a narrow black band along the posterior tergite margin, whereas in *H. femorata* the same band is usually only slightly darker than the tergite ground colour; rarely, there is a darker posterior margin (compare lower vs. upper illustration for *H. femorata*, Fig. 9.8). The triangular enlargement of the anterior band is absent in *H. maori*. The band is considerably wider than that of the previous species, of even width and always black. Even in the melanistic race of *H. maori* (see Morgan-Richards *et al.*, Chapter 7, this volume), the black band may be seen against the very dark brownish-black ground colour (compare lower vs. upper illustration for *H. maori*, Fig. 9.8). Also, there is no posterior marginal band in this species. Although one published identification key (Meads, 1990) has utilized the presence of tiny black spots in the pale areas between dark bands as a distinguishing character, we find that such spots are lacking in some populations and are an unreliable character. The prominent dark transverse bands are absent in *H. thoracica, H. broughi* and *H. ricta*. Instead, a pale posterior marginal band, narrowly edged with black, is found on *H. thoracica* tergites, while the posterior band in *H. ricta* and *H. broughi* is slightly darker than the tergite ground colour (Fig. 9.8). Neither of the latter two species has an anterior marginal band.

The overall abdominal banding pattern shows a progression amongst the species from (a) alternate dark and pale transverse bands (*H. maori*), to (b) alternate dark and pale bands with midline expansion of the dark bands, which themselves remain separated (*H. femorata*), to (c) dark and pale bands with midline merging of the dark bands of sequential segments (*H. crassidens* and *H. trewicki*), to (d) almost even abdominal coloration with slightly darker posterior bands (*H. ricta* and *H. broughi*), to (e) even abdominal coloration with lighter posterior bands (*H. thoracica*).

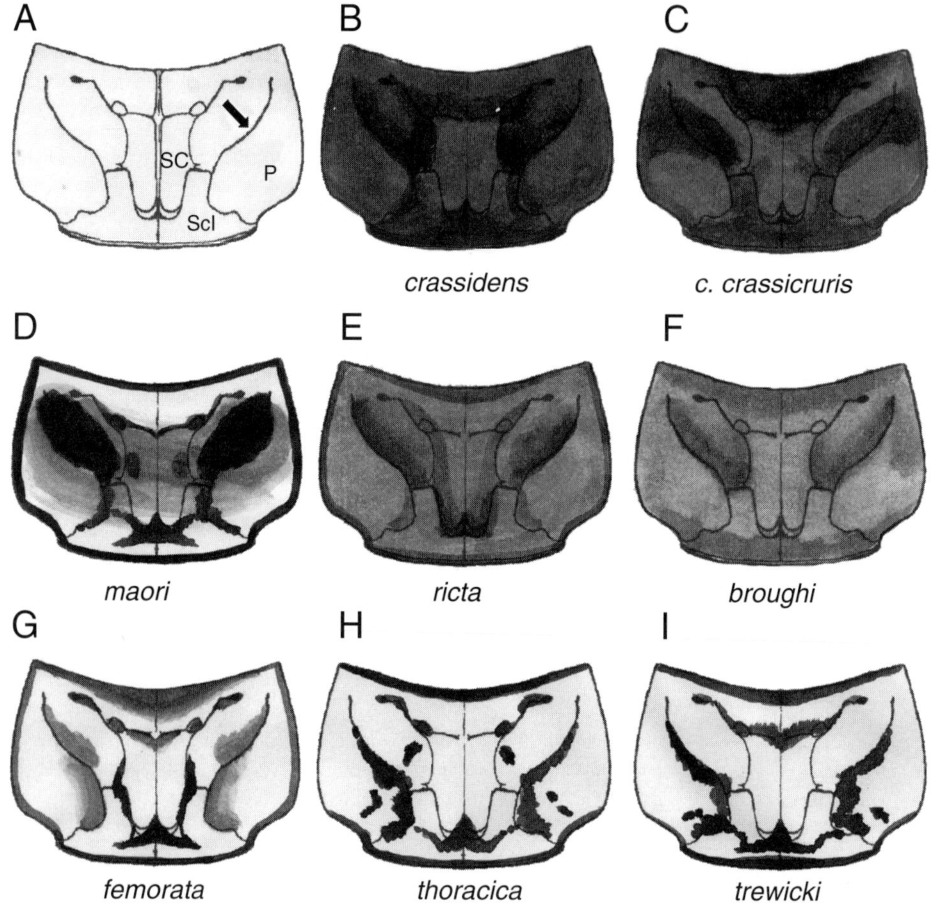

Fig. 9.7. Coloration patterns for the pronotum in *Hemideina* species. A. Structure of generalized pronotum, based upon fusion of scutum (SC), scutellum (Scl) and prescutum (P). Arrow marks prominent groove in prescutum. B–I. Dorsal views of pronota for species indicated beneath each illustration.

Discussion

Few studies have analysed the intraspecific variation of diagonostic taxonomic characters in very large samples of orthopteran species. Certainly in the descriptions of *Hemideina* spp., authors have prepared diagnoses on the basis of relatively few, even single, specimens (e.g. *H. brevaculea* (Salmon, 1950)). The advantage of analysing a large number of specimens lies in determining the stability of a character or assemblage of characters within a species as a whole or within populations of a species. The current interest in analysing intraspecific population variance in genetic characters (see Morgan-Richards *et al.*, Chapter 7, this volume), as well as behavioural characters, such as acoustic call patterns (see Field, Chapter 15, this volume), highlights the caution with which diagnostic taxonomic characters should be selected.

Our results clearly show that neither leg spination nor stridulatory ridge count serves as a reliable diagnostic character by which species of *Hemideina* may be separated. Although some species show relatively little variance about the modal expression of a character (e.g. *H. crassidens*, Table 9.1), none was uniquely defined by any single leg spine pattern or number of stridulatory ridges. Therefore we suggest that the currently available keys for determination of *Hemideina* species are unreliable and much in need of revision.

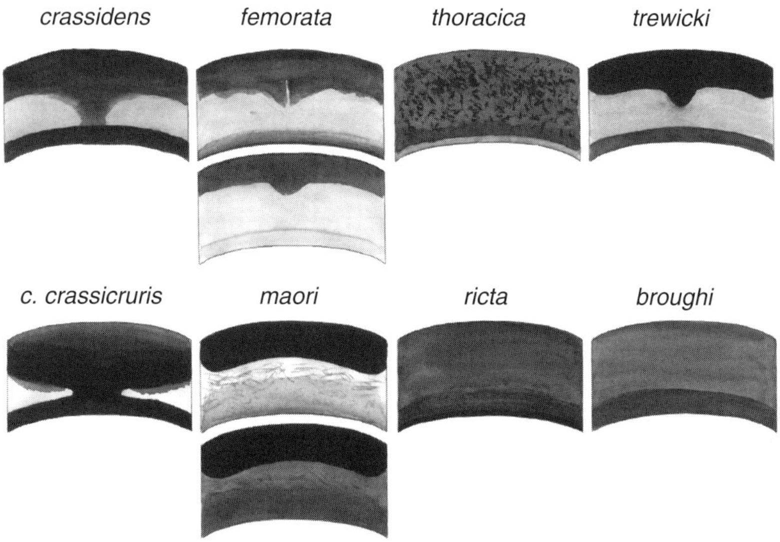

Fig. 9.8. Dorsal views of coloration patterns of abdominal tergites of *Hemideina* spp. Anterior margins are upwards. Variants are shown for *H. femorata* (with and without dark posterior marginal band) and *H. maori* (usual and melanistic forms).

Previous studies (Morgan-Richards *et al.*, 1995) have shown that two species, *H. maori* and *H. ricta*, have colour variants that could confound the discrimination of these species. Furthermore, mitochondrial DNA (mtDNA) evidence shows strong similarities between *H. maori* and *H. ricta* and questions the validity of a full species distinction (see Morgan-Richards *et al.*, Chapter 7, this volume). However, we suggest that the banding pattern on abdominal tergites, rather than depth or hue of abdominal coloration, as well as the dark/light pattern of the pronotum, could serve as useful practical characters for species separation in the field (Fig. 9.8).

Suites of characters taken together are probably the most useful way to separate the species on morphological grounds. When leg spination, stridulatory ridge count, pattern of pronotum and abdominal banding are all considered, differences become more apparent.

In contrast to their appropriateness as taxonomic characters, leg spination and stridulatory ridge count appear to be very useful cladistic characters. In cladistic analyses, where the aim is to gain insight into possible historical (phylogenetic) relationships between and within populations and species, these two characters could be expected to show interesting trends.

For example, *H. crassidens* shows a marked change in stridulatory ridge count in the North Island population (WN) compared with that of all the South Island populations (Fig. 9.4). This difference seems most likely to have resulted from the present postglaciation fragmentation of this species between the two islands. At the time of the last (Otiran) glaciation in the late Pleistocene, about 20,000 years ago, a major land bridge linked the two islands with lowland well below the snowline, due to lowered ocean levels (Fleming, 1962; Stevens, 1980). This condition must have allowed a contiguous range of *H. crassidens* from the North Island south along the west coast of the South Island, the latter of which was apparently glacier-free along a coastal belt. However, a simple explanation of subsequent fragmentation is confused by the finding that there are two chromosome races of *H. crassidens*, one of which extends along the west coast of the South Island up to the Buller region, and the second of which occupies the range on both the north and south sides of the Cook Strait (Morgan-Richards *et al.*, Chapter 7, this volume). As will be seen below for other South Island species, cooling and warming during this last Pleistocene glaciation could well have imposed such additional compounding fragmentation of populations. The interesting unanswered question is whether the ancestral population had a lesser number of ridges or a greater number of ridges, or, rephrased, did intraglacial invasion occur from south to north, or vice versa? Attempts at answering this may gain leverage from considering ridge phenotypes and respective distributions of closely related species of *Hemideina* (M. Morgan-Richards, Dunedin, 1999, personal communication).

Another interesting trend is the apparent cline in stridulatory ridge count in *H. thoracica* populations from Taupo to Northland (excluding the Coromandel population) (Fig. 9.4). Recent chromosome analysis of races of *H. thoracica* has shown that nine karyotypes exist across the North Island, but, unlike the stridulatory ridges, the change in chromosome number does not show clinal variation. In fact, the chromosome races have rather sharp boundaries and their disjunct distributions appear to correlate more with isolation events caused by the formation of islands in a Pliocene archipelago (Morgan-Richards and Wallis, 1998) than with any environmental graded change (a more likely factor to cause a cline). There are no obvious morphological character differences associated with the chromosome races; hence it is not clear whether changes in chromosome number could influence external morphology. At least, it is known that changes in chromosome number do not present barriers to gene flow among populations (Morgan-Richards, 1997).

The ridge counts for the alpine *H. maori* show a possible cline along the Southern Alps of the South Island, with disjunct populations associated with isolated eastern mountain ranges (Fig. 9.4). King (1997) has used evidence from mtDNA analysis that the major mountain-building episodes of the Kaikoura Orogeny have led to fragmentation and genetic isolation of alpine populations of *H. maori* (see also Morgan-Richards *et al.*, Chapter 7, this volume).

In contrast, the lowland *H. femorata* has apparently not undergone fragmentation, at least based upon ridge-count data. Reconstruction maps show that Pleistocene glaciers extend only partially eastward from the mid-island Southern Alps chain, leaving a broad outwash plain along the entire eastern extent of the South Island (Fleming, 1962). This region could have served as a stable habitat for *H. femorata*, which may have maintained gene flow throughout its range.

Acknowledgements

The authors would like to thank the following institutions and people for the generous loan of material: the Otago Museum, the National Museum, the Mt Albert Research Centre, P.M. Johns and M. Morgan-Richards. Our thanks to Graeme Ramsay for discussions about chaetotaxy, its application to morphometrics and relevant references.

References

Crosby, T.K., Dugdale, J.S. and Watt, J.C. (1976) Recording specimen localities in New Zealand: an arbitrary system of areas and codes defined. *New Zealand Journal of Zoology* 3, 69.

den Hollander, J. and Barrientos, L. (1974) Acoustic and morphometric differences between allopatric populations of *Pterophylla beltrani* (Orthoptera: Tettigoniidae: Pseudophyllinae). *Journal of Orthoptera Research* 2, 29–34.

Field, L.H. (1978) The stridulatory apparatus of New Zealand wetas in the genus *Hemideina* (Insecta: Orthoptera: Stenopelmatidae). *Journal of the Royal Society of New Zealand* 8, 359–375.

Field, L.H. (1993) Structure and evolution of stridulatory mechanisms in New Zealand wetas (Orthoptera: Stenopelmatidae). *International Journal of Insect Morphology and Embryology* 22, 163–183.

Fleming, C.A. (1962) New Zealand biogeography: a paleontologist's approach. *Tuatata* 10, 53–108.

Gibbs, G.W. (1999) Four new species of giant weta, *Deinacrida* (Orthoptera: Anostostomatidae: Deinacrideinae) from New Zeanand. *Journal of the Royal Society of New Zealand* 29, 307–324.

King, T.M. (1997) Geographic distribution, systematics and molecular genetic analyses of colour morphs of the alpine weta (*Hemideina maori*), with specific reference to a hybrid zone. Unpublished PhD thesis, University of Otago, Dunedin, New Zealand.

Meads, M. (1990) *The Weta Book: a Guide to the Identification of Wetas.* DSIR Land Resources, Lower Hutt, New Zealand, 36 pp.

Moller, H. (1985) Tree wetas (*Hemideina crassicruris*) (Orthoptera: Stenopelmatidae) of Stephens Island, Cook Strait. *New Zealand Journal of Zoology* 12, 55–69.

Morgan-Richards, M. (1997) Intraspecific karyotype variation is not concordant with allozyme variation in the Auckland tree weta of New Zealand, *Hemideina thoracica* (Orthoptera: Stenopelmatidae). *Biological Journal of the Linnean Society* 60, 423–442.

Morgan-Richards, M. and Gibbs, G.W. (1996) Colour, allozyme and karyotype variation show little concordance in the New Zealand Giant Scree Weta *Deinacrida connectens* (Orthoptera: Stenopelmatidae). *Hereditas* 125, 265–276.

Morgan-Richards, M. and Townsend, J.A. (1995) Hybridisation of tree weta on Banks Peninsula, New Zealand, and colour polymorphism within *Hemideina ricta* (Orthoptera: Stenopelmatidae). *New Zealand Journal of Zoology* 22, 393–399.

Morgan-Richards, M. and Wallis, G. (1998) Genetic structure of the Auckland tree weta and the Pliocene archipelago of Northland. *Geogenes* 98 (March), 55–58.

Morgan-Richards, M., Daugherty, C.H. and Gibbs, G.W. (1995) Taxonomic status of tree weta from Stephens Island, Mt Holdsworth and Mt Arthur, based on allozyme variation. *Journal of the Royal Society of New Zealand* 25, 301–312.

Morris, G.K. and Walker, T.J. (1976) Calling songs of *Orchelimum* meadow katydids (Tettigoniidae). I. Mechanism, terminology and geographic distribution. *Canadian Entomologist* 108, 785–800.

Pitkin, L.M. (1976) A comparative study of the stridulatory files of the British Gomphocerinae (Orthoptera: Acrididae). *Journal of Natural History* 10, 17–28.

Ramsay, G.W. and Bigelow, R.S. (1978) New Zealand wetas of the genus *Hemideina*. *The Weta* 1, 32–34.

Salmon, J.T. (1950) A revision of the New Zealand wetas: Anostostominae (Orthoptera: Stenopelmatidae). *Dominion Museum Records, Entomology* 1, 121–177.

Salmon, J.T. (1955) A key to the tree and ground wetas of New Zealand. *Tuatara* 6, 19–23.

Simpson, G.G., Roe, A. and Lewontin, R.C. (1960) *Quantitative Zoology*, rev. edn. Harcourt Brace and Company, New York, 440 pp.

Stevens, G.R. (1980) *New Zealand Adrift: the Theory of Continental Drift in a New Zealand Setting.* A.H. and A.W. Reed, Wellington, 442 pp.

Tanaka, A. and Kitamura, S. (1992) Spur patterns on the tibia and femur of the German cockroach (Blattodea; Blatellidae). *Annals of the Entomological Society of America* 85, 767–774.

Walker, T.J. and Carlysle, T.C. (1975) Stridulatory file teeth in crickets: taxonomic and acoustic implications (Orthoptera: Gryllidae). *International Journal of Insect Morphology and Embryology* 4, 151–158.

10
Sexual Selection and Secondary Sexual Characters of Wetas and King Crickets

Laurence H. Field[1] and Neil A. Deans[2]
[1]*Department of Zoology, University of Canterbury, Private Bag 4800, Christchurch, New Zealand;* [2]*Fish and Game Council, Nelson, New Zealand*

Introduction

Males in many genera of the Anostostomatidae show secondary sexual characters in the form of enlarged heads (megacephaly), enlargement of mandibles and additional head capsule or mandibular weapons, such as elongated spines and tusks. In New Zealand, these distinctive features attracted the attention of early naturalists and taxonomists (Hutton, 1897; Hudson, 1920; Karny, 1937; Beier, 1955), and were sometimes used to illustrate Darwin's principle of sexual selection (Hudson, 1920). The proportions reached by some secondary sexual structures are bizarre and spectacular, often giving the insects a fearsome appearance. Ethnic names reflect the repulsion response of humans to these creatures. For example, the Maori name for the tree weta, *taipo*, means 'devil who comes by night', while the largest giant weta is called *wetapunga*, named after the God of Ugly and Deformed Things. No less striking are the king crickets, such as *Henicus monstrosus* from South Africa and *Anostostoma australasiae* from Australia, and even some of the remarkably distorted Caribbean species, such as *Licodea cerberus* from Cuba.

Although some secondary sexual structures have been illustrated in the literature, there are no published descriptions of their behavioural function or detailed analyses of structure. We present in this chapter a review and new descriptions of the morphological variety of such structures with a worldwide treatment, new data on details of allometric growth in New Zealand tree wetas, structural analyses of mandibles and behavioural applications of male mandibular weapons related to sexual selection.

Morphological Variety of Secondary Sexual Characters

Four kinds of male secondary sexual characters are found in the Anostostomatidae. These include: (i) development of tusks or knobs projecting anteriorad from the mandibles; (ii) enlarged and sometimes asymmetrically distorted crania; (iii) elongated and/or bowed mandibles, often reinforced; and (iv) spines or lobes projecting from the gena or frons.

Tusks

Tusks projecting from the male mandibles are found in *Libanasa* and *Libanasidus* from Africa and *Motuweta*, *Anisouris* and the undescribed 'Raukamara' tusked weta from New Zealand. They vary from the spectacular elephantine tusks seen in *Motuweta isolata* to small knobs on the proximal anterior surface of *Libanasa* (Fig. 10.1C–D, E). All tusks taper at the tip to a sharp or blunt point, and are either slightly downturned (*Libanasa capicola*, 'Raukamara' tusked weta (Fig. 10.1A, B, F)), weakly upturned (*M. isolata*,

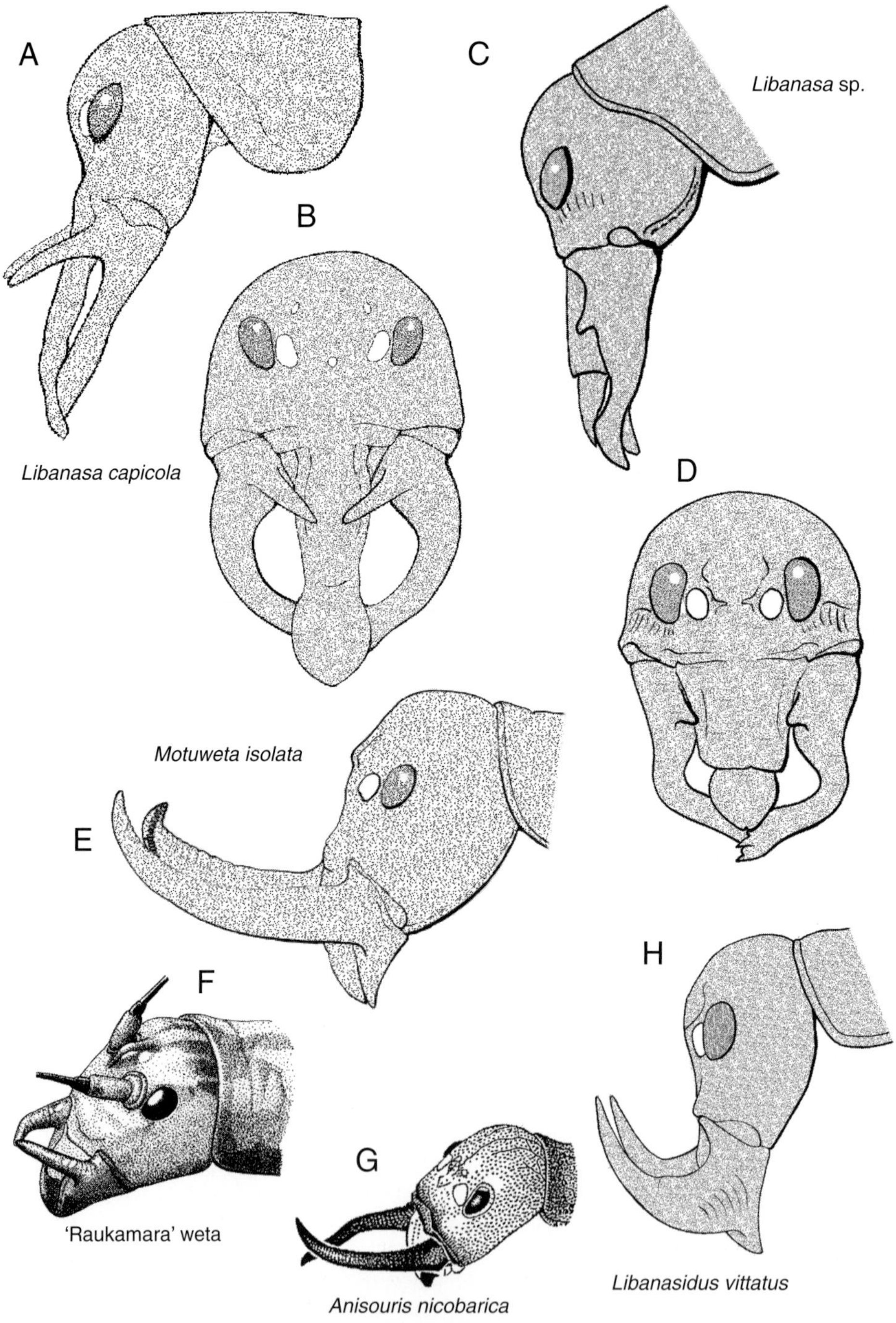

Fig. 10.1. Tusks as secondary sexual characters of male anostostomatids. All tusks project from the anterior aspect of the mandibles. A, B. *Libanasa capicola* male, South Africa. C, D. *Libanasa* sp. male, South Africa. E. *Motuweta isolata* male, lateral view, New Zealand. F. Undescribed New Zealand 'Raukamara' weta, male. G. *Anisouris nicobarica* male, H. *Libanasidus vittatus* male, South Africa. A, B: courtesy of D. Otte, with permission. C, D: specimen courtesy of D. Rentz. G, courtesy of D. Helmore, with permission.

Anisouris nicobarica, Fig. 10.1E, G) or strongly upturned (*Libanasidus vittatus*: Fig. 10.1H). In *M. isolata*, the tusks are robust and strong at the base, while those of *L. vittatus* are shorter, much more compact and extremely thick at the base. These contrast with the long, slender tusks seen in *A. nicobarica* and the shorter slender tusks of the 'Raukamara' weta. A tendency is more or less expressed for tusks to point medially or to cross at the tips; in the latter case, the left overlaps with the right. Within a single species, tusks do not necessarily cross in all males.

In the two New Zealand long-tusked species (*M. isolata, A. nicobarica*), the crossed structures bear stridulatory tubercles or ridges on apposable surfaces near the tips, thus providing insight into the function of the strongly crossed condition in these particular species. Because the tusks are borne by the mandibles, sound is produced by simultaneous rubbing of the tusk surfaces as the mandibles are sharply opened (see Field, Chapter 15, this volume; Field, 1993). The wetas must be able to differentially control mandibular movements for eating versus those for sound production. No stridulatory structures are associated with the tusks of African species. The tusks of *L. vittatus* either cross slightly or meet at the sharp tips (Fig. 10.1H). Mandibular protrusions of *Libanasa* sp. (Fig. 10.1C, D) project from the same proximal location as tusks in other genera, but they are attenuated to short conical structures pointing forwards and would seem to have little use in male agonistic interactions.

The longest tusks are those of *M. isolata*, which reach up to 44% of the insect's head and body length. The holotype male, with a head and body length of 61 mm, has tusks measuring 26 mm from base to tip (Johns, 1997). Another mature male had 24 mm tusks with a body length of 55 mm (Meads, 1990).

The development of mandibular tusks appears to be an allometric process related to sexual maturity and male body length. Male nymphs of *L. vittatus* lack tusks but have small knobs, which presumably are tusk precursors. Only the last two male instars (considered to be adults: 40–60 mm in length) have tusks, and they show proportionately longer tusks with increasing body length (Bennett and Toms, 1995). Similarly, *M. isolata* males with smaller body size have shorter and less curved tusks (e.g. the paratype has a body length of 42 mm and tusk length of 9 mm), even though both males were sexually mature as judged by other traits (Johns, 1997). Both instances may also represent examples of the early male maturation phenomenon described for *Hemideina* spp. in New Zealand (see Field and Jarman, Chapter 17, this volume). In all cases, detailed histological and behavioural studies of male sexual maturity conducted in parallel with the development of secondary sexual characters would be of great use in elucidating the question of alternative mating strategies in males with large and small body forms or secondary sexual characters.

Enlarged and asymmetrical crania (male megacephaly)

Almost all males of anostostomatid species have greatly enlarged head capsules compared with those of females (e.g. Figs 10.2D–E and 10.3D–E, F, G). The enlarged cranial condition, termed megacephaly, has been well described for New Zealand tree wetas (*Hemideina*). In some species, the mandibles reach proportions that border on being cumbersome as the animals walk and feed. For example, *H. crassidens* has races or subspecies on some offshore islands (Morgan-Richards *et al.*, 1995) where adult male body size is greater than that of mainland populations, and the resulting megacephaly is extreme. In the subspecies *H. crassidens crassicruris*, on Stephens Island in the Marlborough Sounds, the maximum recorded male body length (including head) was 86 mm, of which the head length was 37 mm (43% of total body length) (see Figs 10.5A and 10.6A–C). Another population of large *H. crassidens* occurs on Open Bay Islands, off the west coast of the South Island. Here an 80 mm male was collected with a head length of 31 mm (39% of total body length). In a sample of 18 adult Stephen Island males in the collections of the Museum of New Zealand (Wellington), the average head length was 40% of the total body length (L. Field, unpublished data). Such males often seem to trip over the tips of their mandibles as they walk along tree branches. The fascinating question of whether offshore islands promote gigantism in tree wetas remains to be investigated.

In many genera, the male head becomes proportionately wider than its length, giving a width/length (W/L) ratio greater than 1.0 (Table 10.1). The extreme is reached in *H. monstrosus*, with a ratio of 1.66, the result of which is a very

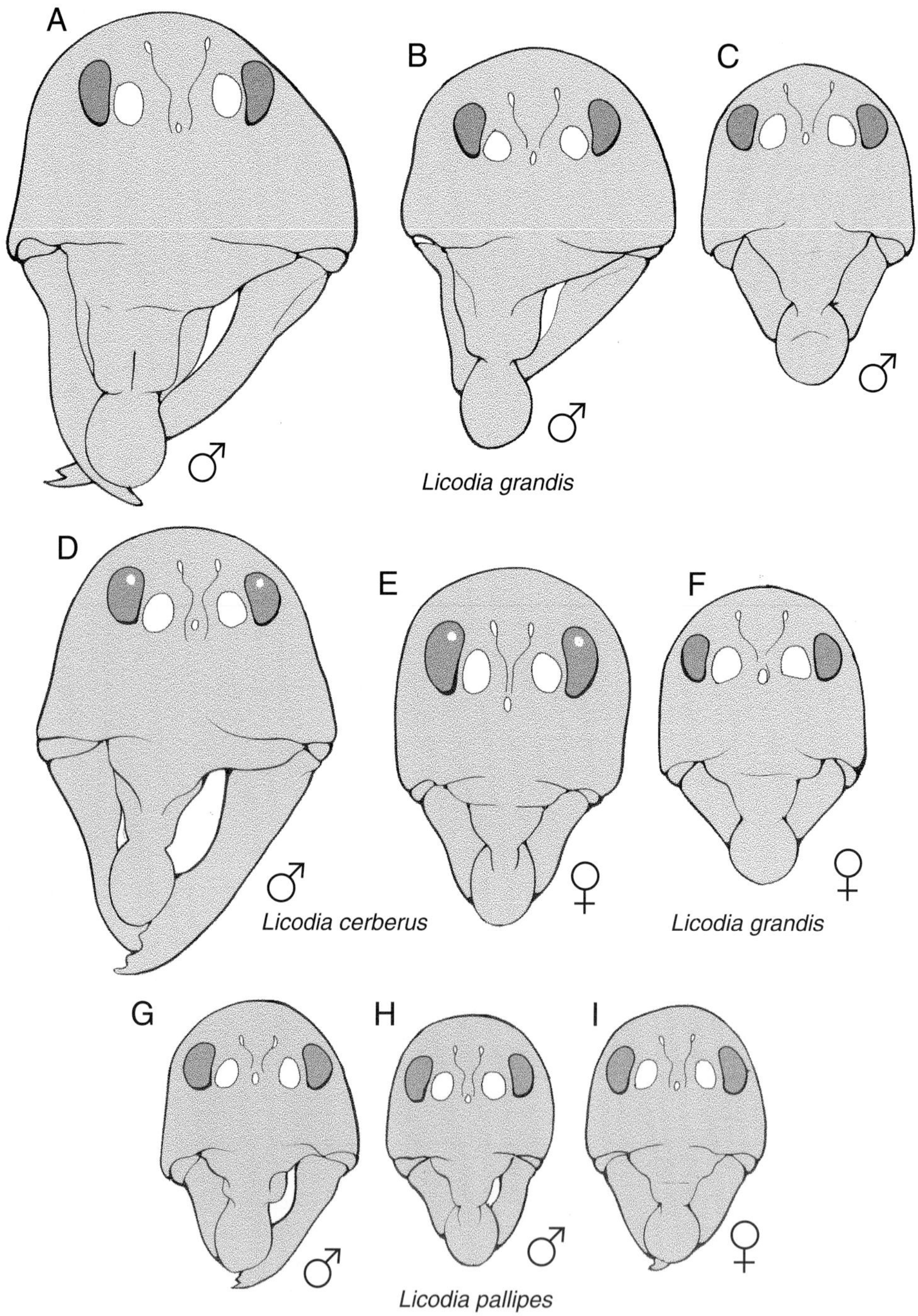

Fig. 10.2. Examples of megacephaly and asymmetrical crania in Anostostomatidae. A–C, F. *Licodia grandis*, Haiti. Size series of males showing asymmetry in two larger instars (A, B), but with symmetrical cranium in smaller male (C). F, female, with symmetrical cranium. D, E. *Licodia cerberus*, Cuba. Male and female, respectively. Note asymmetry in male only. G–I. *Licodia pallipes*, Hispaniola. Males of two instars have asymmetrical crania, but female (I) cranium is symmetrical. A–I, courtesy of D. Otte, with permission.

wide separation of the mandibular articulations (Fig. 10.4A; also discussed in Toms, Chapter 4, this volume). In some of the high-ratio species, where the male cranium is some 30–60% wider than long, there is an accompanying increase in mandibular length, as discussed below. On the

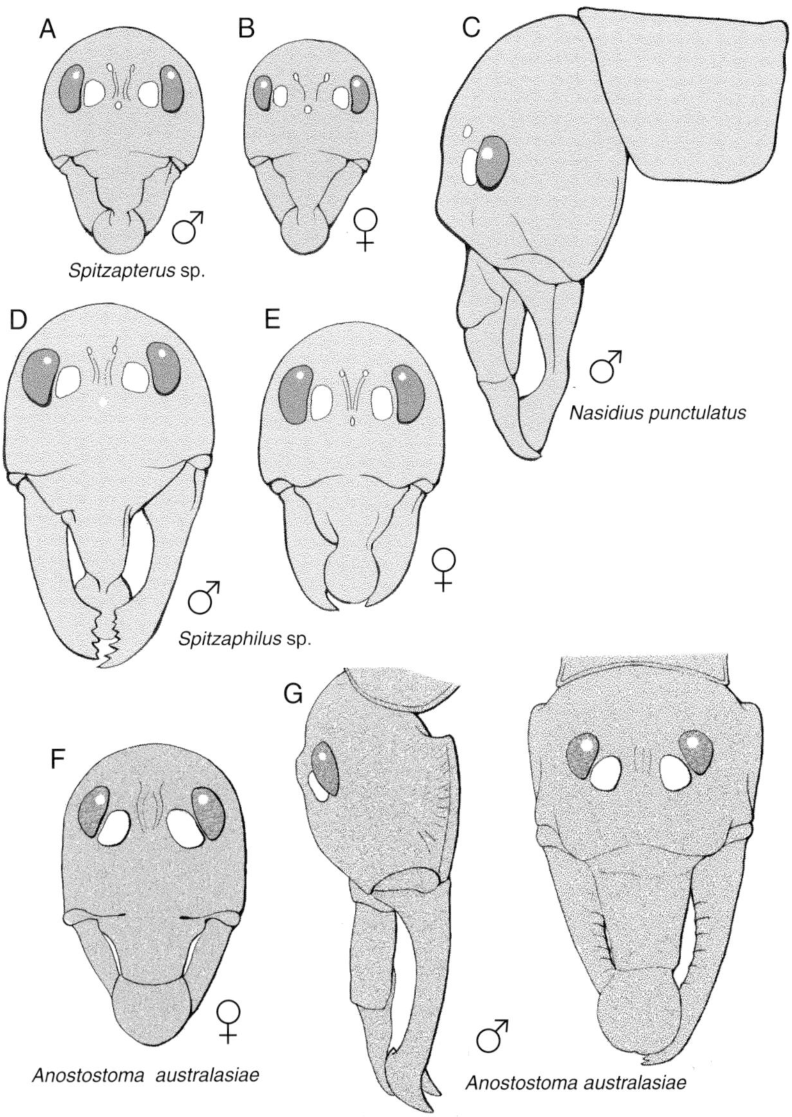

Fig. 10.3. Enlarged mandibles as secondary sexual characters of Anostostomatidae. A, B. *Spitzapterus* (= *Spitzaphilus*?) sp., Madagascar. Male (A) has only slightly larger mandibles and megacephalic cranium, compared with female (B). C. *Nasidius punctulatus* male, South Africa. Mandibles bowed, do not occlude except at the spatulate tips, which are armed with pointed cusps. D, E. *Spitzaphilus* sp., Madagascar. Male mandibles greatly elongated, not sharply bowed laterally (compare with *Libanasa* spp., Fig. 10.1A, C), and only occlude at tips. F. Female *Anostostoma australasiae*, Australia showing elongation of mandibles to moderate degree, with only distal occlusion. G, H. *Anostostoma australasiae* male, Australia. Greatly elongated mandibles only occlude at tips, which are spatulate and armed with pointed cusps. A–E, courtesy of D. Otte, with permission. G, H, specimens kindly provided by G. Monteith.

Table 10.1. Ratios of head width (W) to head length (L) in selected male and female species of Anostostomatidae from Australia, Africa, New Zealand and the Caribbean region. Note that the male crania are wider than those of females (higher ratio), except for the elongated crania of the New Zealand wetas (*Hemideina*).

Genus	Species	Male W/L	Female W/L
Anostostoma	*australasiae*	1.35	1.08
Hemideina	*crassidens*	0.91	1.00
Henicus	*monstrosus*	1.66	–
Henicus	*whellani*	1.21	–
Libanasa	*capicola*	1.58	–
Licodia	*cerberus*	1.48	1.05
Licodia	*grandis*	1.58	1.14
Licodia	*pallipes*	1.30	1.21
Motuweta	*isolata*	0.80	–
Spitzaphilus	sp.	1.27	1.19
Spitzapterus	sp.	1.26	1.16

other hand, male tree wetas of New Zealand (*Hemideina*) have elongated heads, leading to a ratio of less than 1.00.

The most bizarre form of megacephaly is found in the species of *Licodia* from islands in the Caribbean. Adult males have asymmetrical crania, in which the left side is greatly distended laterally (Fig. 10.2A–G). As a result, the left mandibular articulation is displaced laterally, far from the midline, and the left side of the clypeus is distended proportionately to the left. The mandibles are highly asymmetrical, as discussed in the next section. The fastigium, left eye, lateral ocellus and antenna are also displaced somewhat to the left. The right side is not distorted. Curiously, the postclypeus and labrum are also not distorted, and they cover the right mandible in normal fashion. Owing to the restricted nature of the distended cranium, the midline is grossly curved to the left in the dorsal and medial region but then swings to the right in the ventral half of the head.

The distended cranium appears to occur in the final and penultimate male instar of *Licodia* (Fig. 10.2A, B) but, as with the tusked genera, not in earlier male instars (Fig. 10.2C). Females show no asymmetrical cranial features in the specimens studied.

Based upon study of male New Zealand tree wetas, megacephaly appears to have evolved concomitantly with enlarged mandibles, as a way of housing the enhanced mass of the mandibular adductor muscles. With increases in mandible length, an increasingly unfavourable lever ratio requires ever more powerful mandibular adductors in order to develop sufficient force to make the mandibles functional. If the mandibles are used in battles as gripping devices, the muscle mass, and hence cranial volume, would be controlled by the same selective forces driving enhancement of the mandibles as secondary sexual characters. Reasons for the evolution of widely separated mandibular articulations are not so clear. Behavioural observations are required first, to learn how the mandibles are deployed in order to gain insight into the advantages of separating the articulations. One possibility is that use of the mandibles to outgrip the opponent could lead to selection for wider articulation separation. This would be predicted, for example, in *H. monstrosus* (Fig. 10.4A).

Other selective forces may (also) have led to megacephaly for entirely different reasons. The enlarged head, and especially the greater width, would provide a larger frontal outline of the male head. This could serve as a visual cue signalling superior fighting ability during male–male encounters, or it could serve as a cue to females in the selection of the fittest males. Similar enhancement of male head outline has occurred in a number of animal species, such as the African lion.

Mandibles

While the mandibles of male anostostomatids show the usual orthopteran prognathous condition, they are often more massive and more elongated than those of females. The difference may be only slight (*Spitzapterus*, Fig. 10.3A, B) or it can

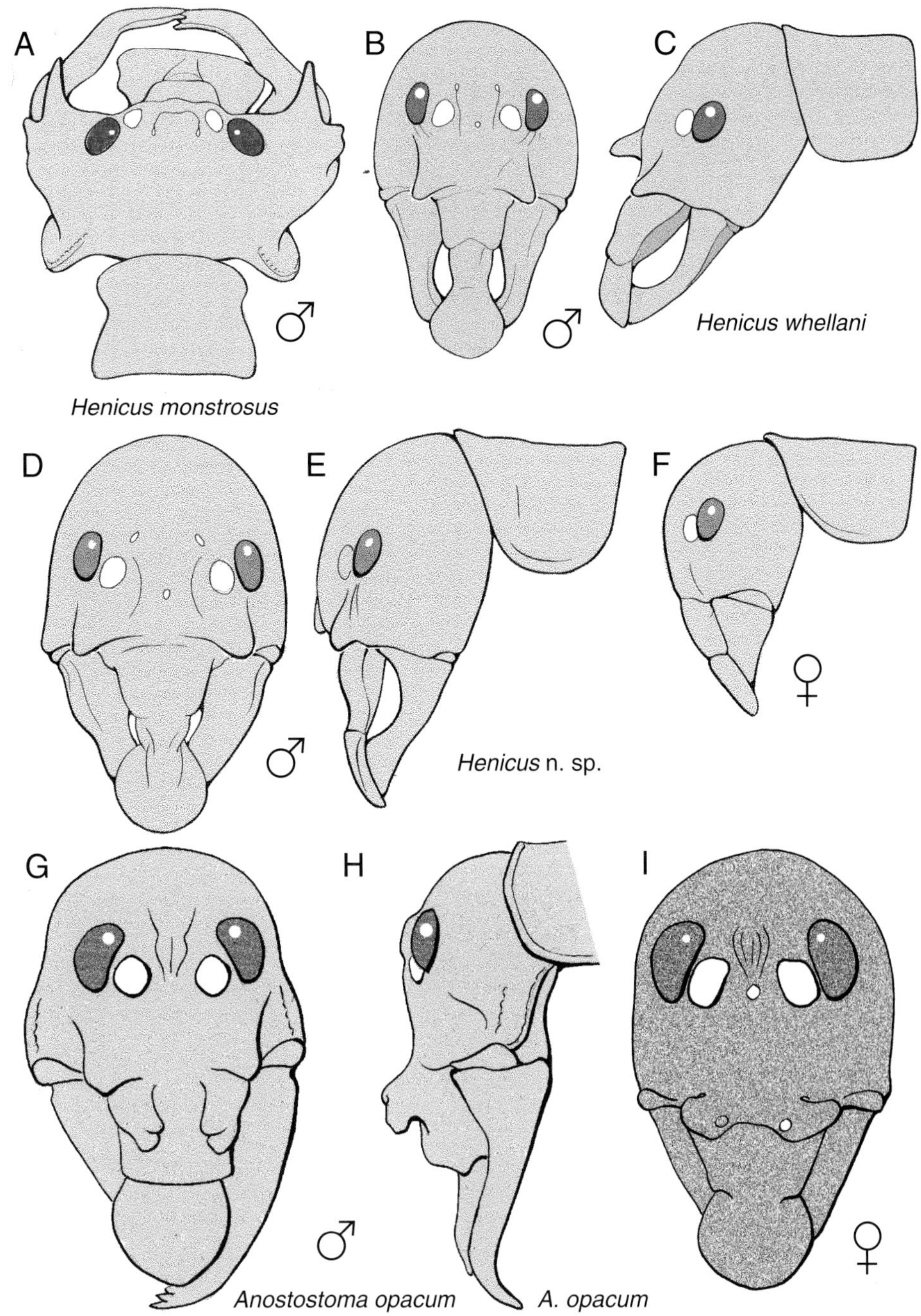

Fig. 10.4. Examples of armament on frons, clypeus and gena of male Anostostomatidae. A. *Henicus monstrosus,* South Africa. Note anterior genal spines and flared shield-like plate developed on posterior of gena. B, C. *Henicus whellani,* Zimbabwe. Blunt spines project from frons. D–F. *Henicus* sp., Malawi. Male (D, E) has blunt protrusions on frons, while female (F) lacks any enhanced cranial characters. G–I. *Anostostoma opacum*, Australia. Male (G, H) has bilobed protrusions projecting from clypeus, which are lacking in female (I). All males have elongated mandibles which only occlude at tips, and which are armed with sharp pointed cusps. A–F, courtesy of D. Otte, with permission.

reach extreme scales, as seen in *Spitzaphilus* sp., *H. crassidens* and *A. australasiae* (Figs 10.3D, 10.6A–C and 10.3F–G, respectively). The morphology varies from elongated, relatively straight mandibles (*Spitzaphilus* sp., *A. australasiae*) to the highly bowed structures seen in *Libanasa* and *Nasidius* (Figs 10.1C and 10.3C). In nearly all but the New Zealand species, the mandibles of males do not have continuous occlusion along their length, but instead are tapered structures which meet only at the tips, which become more or less spatulate and are armed with a set of sharply pointed cusps. Such cusp patterns suggest that most anastostomatids are predatory and use the mandibles as grasping and tearing structures. Except for the tusked wetas and some Henicinae, most of the New Zealand wetas, on the other hand, are herbivorous and have occlusion with incisor and molar cusps along at least the distal 50% of the length (described below, and in Field, Chapter 23, this volume).

In anostostomatids, male mandibles usually comprise more than half the total head length. In Table 10.2, the ratio of mandible length to cranial length (from vertex to epistomial suture) is given for a variety of genera.

Many interesting questions arise about the use and functional limitations of the mandibles in the extreme examples listed above. What closing forces can be developed? Are the mandibles used only for ritualized display during male–male combat, or are they designed to grip with considerable force, as in *L. vittatus* and *Hemideina maori*? What mechanisms limit the evolution of even greater lengths? Unfortunately, many of the species are nocturnal and difficult to collect. The large *A. australasiae* is perhaps the most likely to yield information, as it may be collected reasonably easily in the Queensland rainy season (described in Monteith and Field, Chapter 5, this volume).

Elaboration of frons, clypeus and gena

Tusks and spines are found not only on the mandibles, but also on the frons and genae. These are prominent in species of *Henicus*. In the examples shown in Fig. 10.4, the frons bears a pair of blunt tusks protruding from the ventrolateral corners of *H. whellani* and *Henicus* sp. (Fig. 10.4B–E). In the Australian *Anostostoma opacum*, two bilobed, bulbous protrusions project from the dorsolateral corners of the clypeus, which itself is raised into a strong, hemicylindrical structure (Fig. 10.4G, H). In *H. monstrosus*, a stout spine projects anteriorly from each ventrolateral corner of the genae (Fig. 10.4A).

A further embellishment of cranial enlargement is seen in some genera: expansion of the lateral gena or postgena into flanges or flared plates. The former is best seen in *A. australasiae* (Fig. 10.3H), but is also present in *A. opacum* (Fig. 10.4G, H), *Nasidius punctilatus* (Fig. 10.3C) and *Libanasa* sp. (Fig. 10.1C, D). In the most elaborate

Table 10.2. Mandible morphometry of Anostostomatidae from Australia, Africa, New Zealand and the Caribbean region. ML, mandible length from outer articulation edge to tip; CL, cranial length from vertex to epistomial suture. Length asymmetry is a ratio of left (L) to right (R) mandible length.

Genus	Species	ML/CL	Length asymmetry
Anostostoma	*australasiae*	1.41	L > R: 1.09
Nasidus	*punctulatus*	0.67	–
Licodia	*grandis*	1.46	L > R: 1.38
Licodia	*cerberus*	1.36	L > R: 1.42
Licodia	*pallipes*	1.08	L > R: 1.59
Spitzapterus	sp.	0.63	L > R: 1.04
Spitzaphilus	sp.	1.43	L > R: 1.15
Libanasa	*capicola*	1.40	L > R: 1.07
Libanasa	sp.	1.28	L > R: 1.09
Henicus	sp.	0.73	L > R: 1.02
Henicus	*monstrosus*	1.25	L > R: 1.04
Henicus	*whellani*	0.81	L > R: 1.05
Motuweta	*isolata*	0.34	–
Hemideina	*crassidens*	0.94	L > R: 1.16

case (*H. monstrosus*, Fig. 10.4A), a large vertically flared plate extends outward along the full length of each postgena. The function is unknown for these structures. It is possible that they are cranial reinforcing devices to counteract the enormous stresses developed by the mandibular muscles. The unusually wide separation of the mandibles in *H. monstrosus* may, in turn, impose high cranial stresses during strong biting, which may be resisted by the very large postgenal flared plates. Toms (Chapter 4, this volume) suggests that the head may be used to block the burrow entrance in this species.

Comparisons with Other Orthopterans

Secondary sexual structures similar to those discussed for the Anostostomatidae have evolved convergently in other Orthoptera. In the American subfamily Listroscelidae, family Tettigoniidae, males have developed large tusks on the mandibles (*Listroscelis*) as well as long hooked mandibles (*Carliella*) (A. Gorochov, St Petersburg, 1999, personal communication). In the Pleminiinae (South America), males of *Dicranostomus monocerus* and *Dicranostomus nitidus* have extremely long, thin tusks, which protrude anteriorly from the mandibles and curve medially near the tips (Beier, 1962), as seen in *Motuweta*. Moreover, a medial large spine projects anteriorly from the fastigium, in unicorn fashion. In the same subfamily, *Gnathoclita* males have strong, elongated mandibles, which meet only at the tips, similar to those of *Spitzaphilus*. The genus *Disceratus* from Ecuador has short tusk-like spines projecting forwards from the clypeus, as well as elongated mandibles, broadened at the tips into sharp pointed cusps, very much like those of *A. australasiae*. Megacephaly is prominent in *Gnathoclita* and *Disceratus*.

The Gryllidae also show similar secondary sexual characters in males. For example, various processes occur on the frons, scape and genae in *Scobia*, *Loxoblemmus*, *Coiblemmus*, *Conoblemmus* and *Homaloblemmus* and elongated, reinforced mandibles are found in *Velarifictorus horridus*, *Velarifictorus sulcifrons* and *Velarifictorus aspersus*, as well as *Scapsipedus saussurei* (A. Gorochov, St Petersburg, 1999, personal communication).

Fluctuating Asymmetry in Secondary Sexual Characters

In some animal species, the loss of bilateral symmetry in secondary sexual characters used as weaponry negatively affects male success in securing females (e.g. Møller, 1992). Because the tusks in the African species, *L. vittatus*, are used in male–male competition for females, Bateman (2000) quantified a number of tusk measurements to determine whether male success is affected by such asymmetry. Males were matched in terms of head and body size, and tusk length and angle to clypeus were recorded for every male. Fights were instigated by placing cotton wool with female faeces in the arena, and a trial ended when one male managed to throw the opponent or one male decamped. The results indicated that males with greater symmetry in tusks were more successful in combat than those with asymmetry. The most critical index of tusk asymmetry was the angle of the tusk to the clypeus, rather than tusk length. Presumably males with longer tusks are at an advantage over those with shorter tusks, but in the above study tusk length was relatively equal, since this index varies with head size, which, in turn, was controlled in the experiments.

Further studies of the effects of fluctuating asymmetry, the random deviation from perfect bilateral symmetry in a morphological trait, are needed to highlight ways in which secondary sexual characters in anostostomatids determine male mating success. While the tusks of *L. vittatus* are used to grip the cranium to throw the opponent, those of *M. isolata* are used as levers to dislodge and lift the opponent off the substrate. Asymmetry may disadvantage the latter in different ways or may have no effect at all on mating success. The highly asymmetrical mandibles of *Licodia* may be an interesting case where greater asymmetry confers an advantage rather than a disadvantage to males, since asymmetry has clearly been selected for as a secondary sexual character.

Mandibular Functional Morphology in New Zealand Tree Wetas (*Hemideina*)

Of at least 37 recognized species of wetas in New Zealand, only six species in *Hemideina* show

Fig. 10.5. Mandibular differences in New Zealand tree wetas. A. Enormous head of *Hemideina crassidens crassicruris*, which comprised 43% of the insect's total body length (Stephens Island, New Zealand). Mandibles are straight and long. B. Lateral view of adult male *Hemideina femorata*, which has shorter, down-curved and heavily reinforced mandibles (South Island, New Zealand).

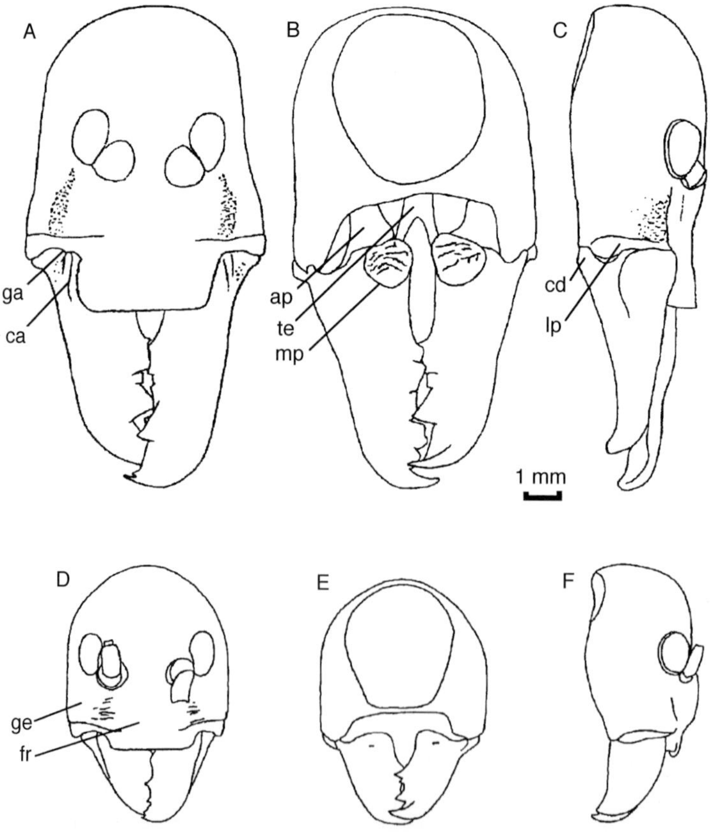

Fig. 10.6. '*Hemideina crassidens* form' of mandible in New Zealand tree wetas, shown in male (A–C) and female (D–F) *H. crassidens.* The form is characterized by straight, narrow and greatly elongated mandibles with small carina (ca), large molariform process (mp) and medially curved distal tips (B, C). ap, apodeme of mandibular adductor muscle; cd, condyle of mandibular articulation; fr, frons; ga, ginglymoid articulation; ge, gena; lp, lateral pleurostoma; te, tentorium.

pronounced adult male secondary sexual characters. We have examined the morphology and functional properties of mandibles with respect to allometric growth, structural integrity and behaviour in most of these species. In general, all tree wetas are herbivorous, and, in contrast to many of the predatory anostostomatids in Australia and Africa, which have multiple, sharp terminal cusps, the *Hemideina* wetas have shearing and masticating cusps.

Sexual dimorphism in mandible and head capsule

The mandibles of adult tree wetas consist of two regions: an outer incisor region and a proximal molar region (see O'Brien and Field, Chapter 8, this volume, for detailed description). Both regions have three teeth (cusps); those of the incisor region are arranged serially, while the cusps of the molar region are arranged in a triangle and undergo a complementary interlocking when the mandibles are closed (see Fig. 10.8A). As in all Orthoptera, the left mandible is always longer and overlaps the right in the incisor region. The result is an efficient shearing mechanism for cutting by the incisors, while the molar cusps do not overlap, and they act as crushing mechanisms for mastication. However, males have mandibles that can reach three times the length of female mandibles, and which show greater morphological variation amongst *Hemideina* species than do those of females (Figs 10.6, 10.7). In general, the male mandible is more massive and can have well-

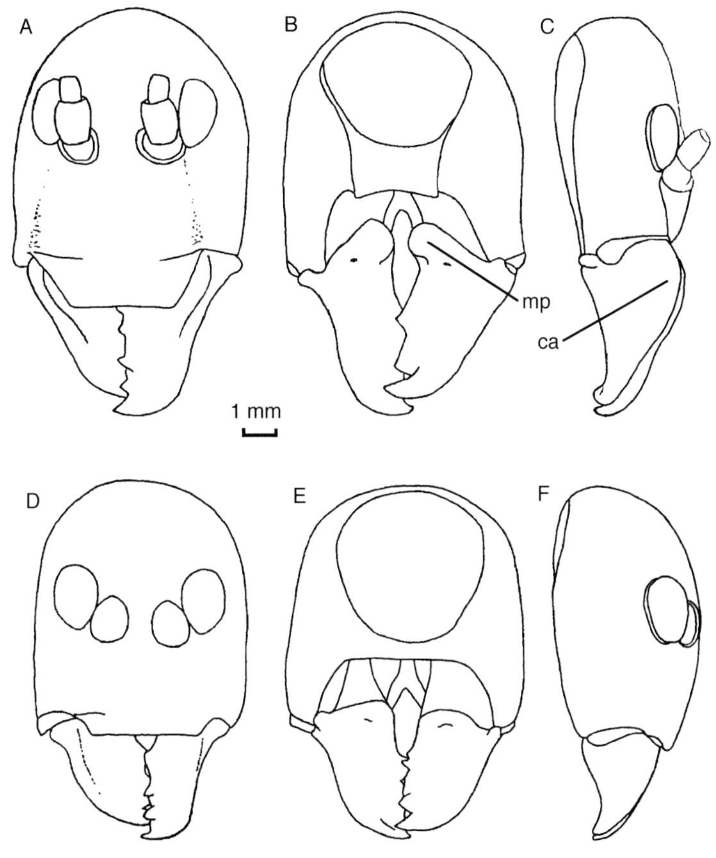

Fig. 10.7. '*Hemideina femorata* form' of mandibular morphology in New Zealand tree wetas. The morphology is shorter and broader, with a massive frontolateral reinforcing carina (ca), small molariform process (mp) and down-curved tips in the male (A–C). These characters are lacking in the female (D–F).

developed reinforcing features, such as a frontolateral carina (Fig. 10.5B). In addition to an increase in head capsule width and volume (to allow for increased mass of mandibular adductor (closer) muscles and larger mandibular articulations), males show elongation of the gena, frons and clypeus, which results in an elongated head capsule. Especially in *H. crassidens* and *H. thoracica*, the head capsule width is increased at the level of the genae, compared with female head capsules (Fig. 10.6A, D). Also, the frontal and lateral regions of the male head capsule are heavily reinforced with ridges and sulci and have heavily tanned cuticle, coloured dark brown to black.

Mandibular forms

The mandibles of *Hemideina* species fall into two morphological categories. Those of *H. crassidens, H. thoracica* and *H. trewicki* are in one category, while those of *H. femorata, H. maori and H. ricta* are in the other. The '*H. crassidens* form' comprises a greatly lengthened, rather narrow mandible with a small carina (ca), a large ventral molariform process (mp) and medially curved distal tips (Fig. 10.6). The '*H. femorata* form' is not as elongated, and is very broad, with a massive carina, a small molariform process and posteriorly curved distal tips (Fig. 10.7A, C). The differences in forms are related to structural properties governing maximum force capabilities in biting and the degree to which the mandibles gape apart. These attributes, in turn, relate to success in agonistic interactions (see 'Mandibular Weapons and Sexual Selection', below) and ability to defend against predators.

Thigosis

In the herbivorous New Zealand tree wetas, the shearing function is aided by a process of self-sharpening of the terminal cusps. The dicondylic articulation of the mandible to the head capsule consists of an anterior ginglymoid articulation (ga, Fig. 10.6A) and a posterior condyle (cd, Fig. 10.6C). There is sufficient lateral elasticity in the ginglymoid articulation and lateral pleurostoma (lp, Fig. 10.6C) to allow a mandible to slightly rotate axially about the posterior condyle during closure. The result is that the left mandible tip rides up and over the right mandible tip as closure is completed. The lower surface of the left mandible then scrapes against the upper surface of the right mandible. An examination of the medial edges of the incisor regions of male mandibles showed that a cutting edge is maintained by abrasion of one side of each mandible (Fig. 10.8B, C), a process known as thigosis in mammals (Every, 1972). Thigosis is especially evident on the elongated incisor teeth of males, and it most probably allows efficient shearing of leaves during feeding in a manner similar to that of scissors. Thigosis is also found to a lesser extent on the molar teeth, where it allows efficient mastication through good interlocking of teeth.

Allometric growth

Estimates of allometric growth rate (the non-linear increase in length) of male mandibles was made by comparing the length of the hind femur with mandible length in a large series of male and female wetas in the following species: *H. crassidens, H. thoracica, H. femorata, H. ricta* and *H. maori*. For comparison with a species which does not show allometric growth, data from *Deinacrida connectens* were included.

The pronounced difference in growth rate between male and female mandibles appears to commence at about the eighth instar in tree wetas (which have a maximum of ten instars). This was confirmed in one previous study of *H. crassidens*, where the divergence between male and female growth trends occurred at the eighth instar, males having mean head lengths of 11.5 mm and mean femur lengths of 14 mm (Spencer, 1995). Growth curves for males and females of *Hemideina* species in the present study (Fig. 10.9A–F) best fitted the exponential equation, $y = ax^b$, where a represents the y intercept and b is the exponent characterizing the extent of allometry. This indicates how different mandibular growth rate is from that of the hind femur as the animal matures. Estimates of how well the calculated curves fit the data are given by correlation coefficients (r^2). Specifically, these determine how much of the mandible length variation is predicted by variation in femur length, where 1.0 indicates that there is a perfect correlation between the two variates. The r^2 values for *Hemideina* ranged from 0.72 to 0.98, with the exception of *H. femorata* females ($r^2 = 0.53$) and *D. connectens* males ($r^2 = 0.46$).

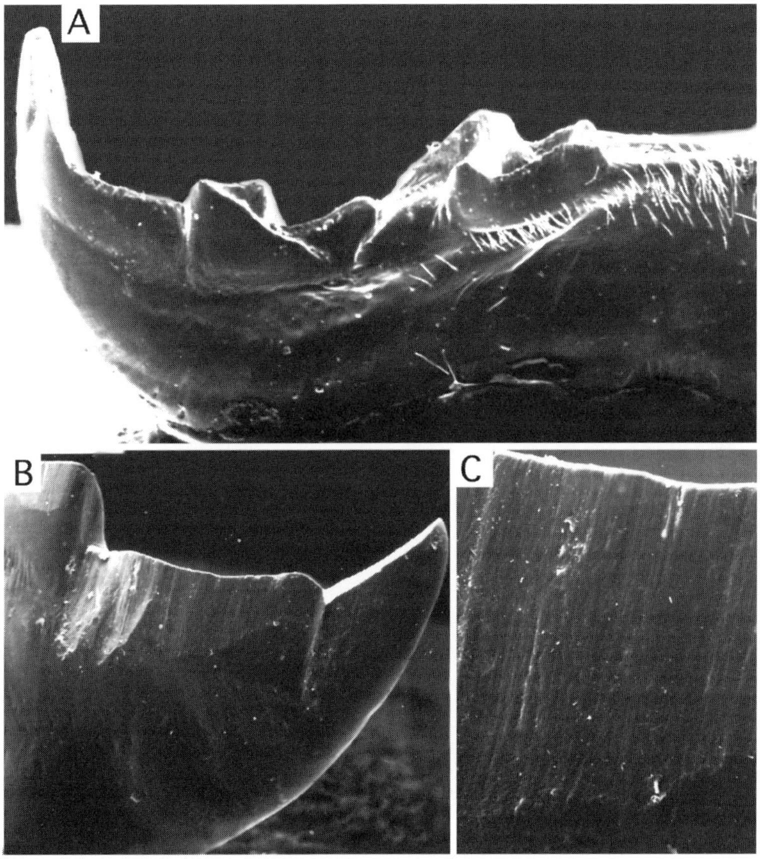

Fig. 10.8. Scanning electron micrographs of the mandible of *H. femorata*. A. Cusps comprise distal incisors for shearing leaves, and proximal molars for crushing. B. The overlapping incisors of both mandibles abrade each other, thus maintaining a sharp cutting edge. The process is known as thigosis. C. Close-up of the effect of thigosis on an incisor.

Values for b are given in Fig. 10.9. If $b > 1$, positive allometry occurs; if $b = 1$, then the relationship is isometric and mandible growth rate is linearly related to femur growth rate; if $b < 1$, negative allometry exists. Females of *H. crassidens, H. thoracica* and *H. femorata* are either isometric or slightly negatively allometric, while females of *H. ricta* are slightly positively allometric and those of *H. maori* are strongly allometric. The latter is unusual because the degree of male and female allometry is similar (1.62 and 1.65, respectively). The male mandibles are still larger, since the a value for males is greater, and the male curve lies above that for females.

Male allometry is strongly positive for all species of *Hemideina*, where b values ranged from 1.65 to 1.98 (Fig. 10.9). There are no trends in b value which differentiate the *crassidens* mandibular form from the *femorata* form. In *H. crassidens*, the data are sufficient to show that the early instars have an isometric growth rate similar to that of females, until a femur length of about 12 mm (Fig. 10.9A). Thereafter the allometric growth rate increases very rapidly, giving a higher b value (2.08) than that for the whole curve ($b = 1.98$). The same isometric growth in juveniles is suggested by the data for *H. ricta* (Fig. 10.9E), although the sample size is small.

The longest mandibles are found in *H. crassidens* (up to about 20 mm), which also has the greatest body size. In *H. crassidens* and *H. maori,* isolated populations (either races or subspecies (see Morgan-Richards *et al.*, Chapter 7, this volume)) are found with much larger body size than

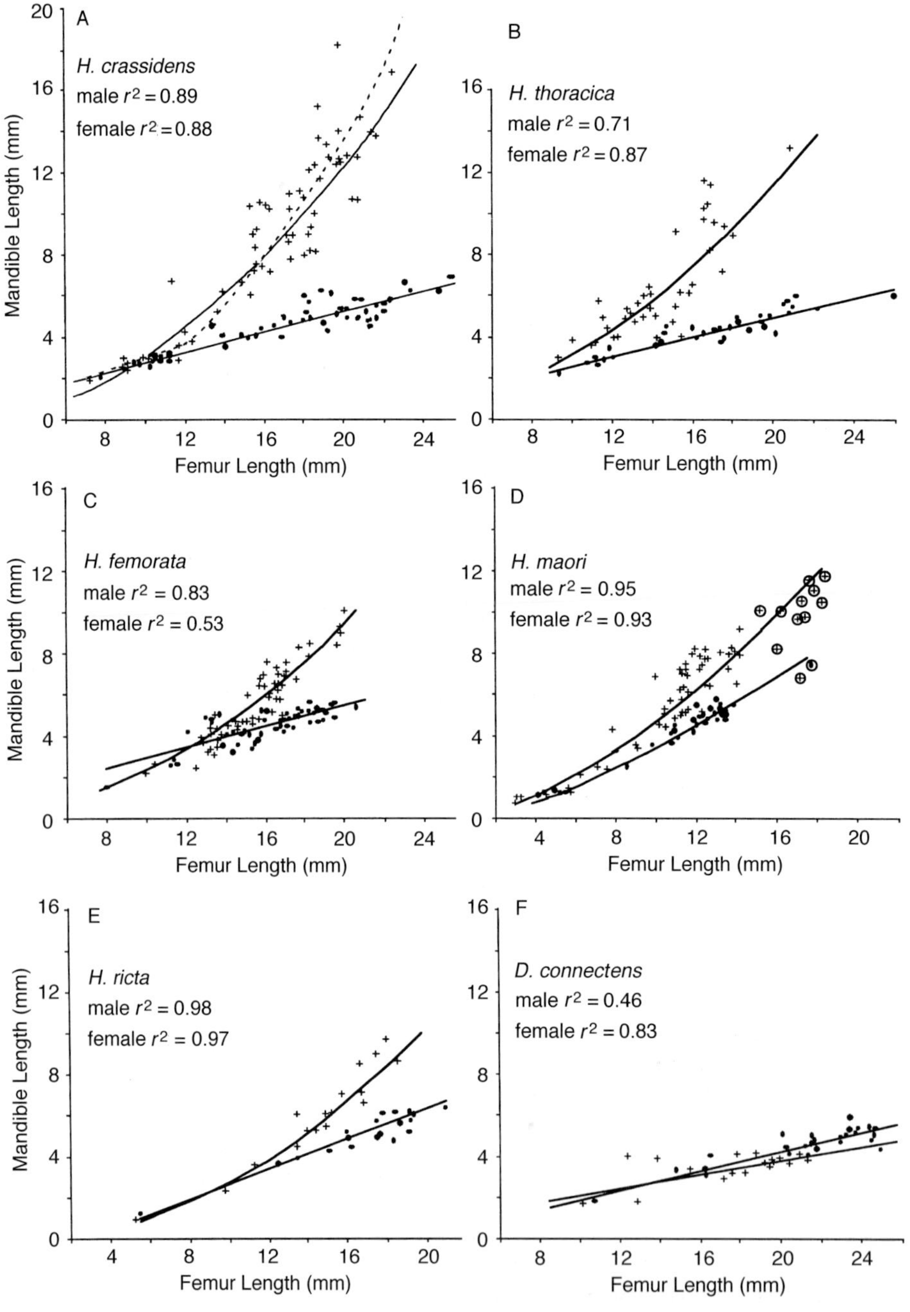

Fig. 10.9. Allometric growth of tree weta mandibles (*Hemideina* spp.) as plotted against femur length. Males (+), females (•). Solid lines give fitted curves from the equation $y = ax^b$, where a and b are constants for males and females of each species: A. *H. crassidens*, males: $a = 0.33$, $b = 1.97$, females: $a = 0.38$, $b = 0.87$; B. *H. thoracica*, males: $a = 0.03$, $b = 1.98$, females: $a = 0.26$, $b = 0.97$; C. *H. femorata*, males: $a = 0.02$, $b = 2.0$, females: $a = 0.39$, $b = 0.88$; D. *H. maori*, males: $a = 0.10$, $b = 1.65$, females; $a = 0.08$, $b = 1.62$; E. *H. ricta*, males: $a = 0.04$, $b = 1.25$, females: $a = 0.15$, $b = 1.25$; F. *Deinacrida connectens*, males: $a = 0.42$, $b = 0.77$, females: $a = 0.23$, $b = 0.99$. The value of b gives the extent of allometric difference in growth rate of mandibles compared with that of the femur.

normal. An example of the resulting increase (20–25%) in mandible size for these populations is indicated by the circled data points for males of the alpine weta, *H. maori,* in Fig. 10.9D.

Variation in mandible size is much greater in males than in females, as seen in the increased spread of points in Fig. 10.9A–E. This is undoubtedly due to the phenomenon of early maturation seen in *Hemideina* males, where a terminal moult can be reached in the eighth and ninth instars and the male is apparently sexually mature even though it has a small body size compared with tenth (final)-instar males (see Stringer and Cary, Chapter 21, this volume).

The lack of positive allometry in the giant weta species is exemplified by the data for *D. connectens* (Fig 10.9F), where mandible length increases isometrically. There is no significant difference in mandible length between the sexes (*F* test, d.f. = 44), although hind femora were significantly larger ($P < 0.001$) in adult females compared with adult males.

Functional anatomy associated with large forces

The increase in mandibular length coincides with a great increase in head capsule size and, presumably, adductor muscle volume. This suggests that male wetas have developed the ability to create large biting forces, and that the mandibles are not just elongated structures with a bite weakened by a large lever ratio. If the former case is correct, it might be expected that the mandible structure is modified to withstand large forces.

We examined this by confirming a sexual difference first in volumes of adductor muscles and then in biting force. Muscle volume was estimated by measuring water displacement, using a calibrated burette. Biting force was measured by allowing wetas to bite against two parallel steel rods (2 mm dia.) coated with plastic, one of which was attached to a linear force transducer (Grass FT-10, linear response 0.0–10.0 kg). When the mandibles squeezed the rods toward each other, force was recorded on a Sanborne paper chart recorder. Forces could be recorded for both tip and molar regions in males, but only for tip regions in females.

Next, mandibular structural modifications were studied by: (i) sectioning mandibles, embedded in araldite, with a diamond saw to study cuticular modifications; (ii) scratching mandibles against a series of minerals of known hardness according to the Moh's scale to determine tooth hardness; and (iii) loading samples of head capsule cuticle and whole mandibles to determine stress/compression resistance. The last tests were made by embedding the half-head capsule, of both dried and recently dead specimens, with muscles removed, in araldite, holding in a rigid clamp and slowly moving against a force transducer (Grass FT-10) in the plane of biting until breakage.

Muscle volume

Adductor muscle volumes were significantly different (*t* test, $P < 0.001$) between male and female adult wetas of two species. The mean volume for male *H. crassidens* was almost 3.5 times that for female adults ($\bar{X} = 0.15$ cm^3 vs. $\bar{X} = 0.043$ cm^3, respectively; $n = 6$ for each sex). In *H. femorata*, the mean male muscle volume was four times that for females ($\bar{X} = 0.14$ cm^3 vs. $\bar{X} = 0.035$ cm^3, respectively; $n = 6$ for males, $n = 4$ for females). Thus larger male head capsules have proportionately larger volumes of adductor muscles, which should lead to greater biting force in these individuals.

Biting force

The maximum mandibular biting force (in Newtons, where 1 N = 1 kg m s^{-2}) was measured for adult females ($n = 8$) and subadult and adult males ($n = 10$) of *H. crassidens*. On average, males could exert greater maximum force than females (Fig. 10.10). The mean female bite force at the mandibular tips was 6.1 N, while male mean bite force at the tips was 8.7 N and this increased to 11.7 N in the molar region. Some of the females could achieve bites as hard as the tip forces achieved by males, but the male molar forces reached much greater levels. Although females have smaller head capsules and adductor muscles, they must have very similar muscle lever ratios to those of males to be able to achieve such forces.

In a much more detailed study of mandibular function, O'Brien (1984) measured biting forces in the large race (Rock and Pillar Mountains, Otago, New Zealand) of *H. maori.* He found that the males could typically achieve forces around 11.8 N near the tip, and that the maximum force achieved near the tip by a single mandible was 18.6 N. To place such biting forces into perspective, 18.6 N is

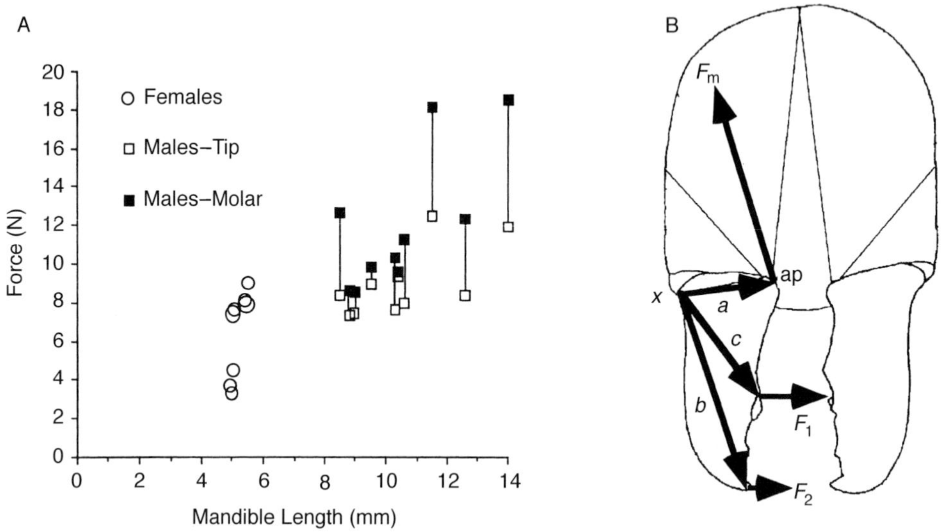

Fig. 10.10. A. Maximum biting forces recorded from adult and subadult male and adult female *H. crassidens*. The smaller female mandibular size meant that females could bite only as hard as some males at their mandibular tips, but that males could achieve much larger biting forces in their molar regions. Measurements of individual tip and molar forces are linked by a vertical line for each male. B. Diagram of mechanical advantages achieved by the adductor muscle on bite force at the incisor and molar regions. The adductor pulls with force F_m at its point of attachment (ap) on lever arm *a*, about the fulcrum (*x*). The force developed in the incisor region is given by $F_1 = F_m(a/b)$, while that developed in the molar region is given by $F_2 = F_m(a/c)$.

equivalent to a force of 629 kg cm^{-2} delivered along the 46 μm wide cutting edge of the incisors, or a tension of 4.7 kg being applied to the adductor muscle apodeme, which closes a single mandible. The area of the adductor apodeme insertion on to the mandible is enlarged to spread the force per unit area of coupling to the mandible. Nevertheless, for the largest *H. maori* mandible studied (length = 12 mm), the estimated tension at the apodeme coupling would reach about 43 kg cm^{-2}, for an attachment area of 10.9 mm^2.

O'Brien (1984) showed that male wetas are able to create such large forces by the combination of massive, highly pinnate adductor muscles, which nearly fill the head capsule, and a large mechanical advantage gained from wide displacement of the apodeme attachment site from the hinge line. The anatomical arrangement of a wide head capsule, with wide mandibles articulated laterally, allows the apodeme insertions on the mandibles to be displaced as much as 4 mm medial to the articulation of each mandible (Fig. 10.10B). This gives a mechanical advantage of 1 : 0.38 for forces developed at the mandible tip and 1 : 0.63 for forces in the molar region (*a*/*b* and *a*/*c*, respectively, Fig. 10.10B).

Structural modifications

Two mandibular features stand out as possible modifications in the male mandible to increase structural integrity against large forces. These include the carina and the molariform process. The carina (Fig. 10.6A) is a prominent ridge commencing near the anterior articulation and extending along the frontolateral margin to a different extent in the various *Hemideina* species. The molariform process is a more or less heavily sclerotized proximal protuberance on the ventromedial side of the mandible (Fig. 10.6B).

In the *H. crassidens* form of mandible the carina is short, while it is higher and much longer in the *H. femorata* form. In both cases, the carina would seem to reinforce an area undergoing compression during biting. Carina length may correlate with the difference in force developed in the two mandible forms. The former has less mechanical advantage (e.g. approximately 1 : 0.31 from Fig.

10.6B), due to longer mandibles, and cannot develop such high forces as in the latter form. Behavioural studies (Sandlant, 1981) showed that, in the *H. crassidens* form, the elongated morphology is used to gain dominance status in ritualized agonistic fights, wherein each male tries to encompass the other's mandibles (see Fig. 10.13). In this form, there appears to have been an evolutionary drive to increase mandibular length. In contrast, the *H. femorata* mandible form seems to be adapted for withstanding high biting forces. Certainly, the most massive carinae occur in *H. maori,* which achieves the highest biting force.

The molariform process is a secondary sexual character that is peculiar to adult male *Hemideina* mandibles, but which has an unknown function. Its name implies a molar function, but the processes do not contact each other when the mandibles are closed, nor are they in a position to be useful for food manipulation, since they lie behind the mouth. Furthermore, they show no sign of wear or abrasion from mastication. There is no evidence of use during mating or during fighting with other males (O'Brien, 1984). They may structurally reinforce the basal mandible region where the adductor apodeme attaches, based upon evidence of extreme cuticular tanning given below. At this position, a high tensile force would develop during a strong bite and the extended surface area may help to dissipate this force. The largest and most protruding molariform processes occur in the *H. crassidens* form, while those of the *H. femorata* form protrude less and are somewhat less massive.

Cuticular tanning

The patterns of heaviest tanning, together with transverse sections at different levels of the mandibles, are shown for two species of *Hemideina* (Fig. 10.11). The cuticle is darkest on the cusps along the full length of the medial side and along the lateral side (e.g. Fig. 10.11A). These are the two regions which receive maximum tensile (stretch) and compression forces, respectively. The carinae, molariform processes and articulation area are also equally heavily tanned (e.g. Fig. 10.11B). All these regions are black in *H. crassidens* and *H. thoracica* and black-brown in *H. maori*. *Hemideina ricta* and *H. femorata* have dark brown tanning in these regions.

Transverse sections of cuticle clearly confirm that tanning in the medial and lateral mandible sides, the carina and molariform process and the tip of the mandible extends through the full depth of the cuticle in all species. However, species differences occur in the depth of tanning and thickness of cuticle. In *H. maori* the cuticle is relatively thin throughout the mandible, while it is thickest in *H. femorata*. Also, in *H. maori* most of the mandible circumference has a thin layer of very dark tanning, while in *H. crassidens* the thin layer is seen only on the dorsal and ventral surfaces in some regions (dark line in sections, Fig. 10.11A). Only the proximal region near the articulation has the thin layer in *H. femorata* (Fig. 10.11B).

Cuticle hardness

Presumably, the combination of thicker cuticle and darker tanning leads to stronger cuticular structure. To determine whether this is the case, hardness of cuticle was tested against various minerals in the Moh's scale (Table 10.3).

Preserved specimens were not affected by the alcohol, since they had the same results as freshly dead material. The tips of the mandibles were the hardest part of the mandible in all species, while the molars were slightly softer in species with the *H. femorata* form of mandible compared with those with the *H. crassidens* form. *Deinacrida connectens*, which does not engage in male agonistic battles involving mandibular jousting (Field, 1980), had relatively softer mandibular tips and molar cusps. The head capsule cuticle was softer than the mandibular cusps in the three species tested. It therefore appears that darkness of colour indicates the hardness of cuticle in the mandibles and head capsule and that darker areas are likely to be stronger than lighter areas.

Fracture studies

To test the above surmise, mandibles and pieces of head capsule cuticle were subjected to breakage while measuring stress imposed transversely to the plane of the sample. For a given thicknesses of head capsule cuticle (of equal area), darker pieces had a higher breaking strength than lighter pieces. In Fig. 10.12A, the filled symbols represent the darkest cuticle, partially filled symbols are from medium cuticle and open symbols are from the lightest cuticle. Also, as cuticle became thicker, its breaking strength increased approximately as a

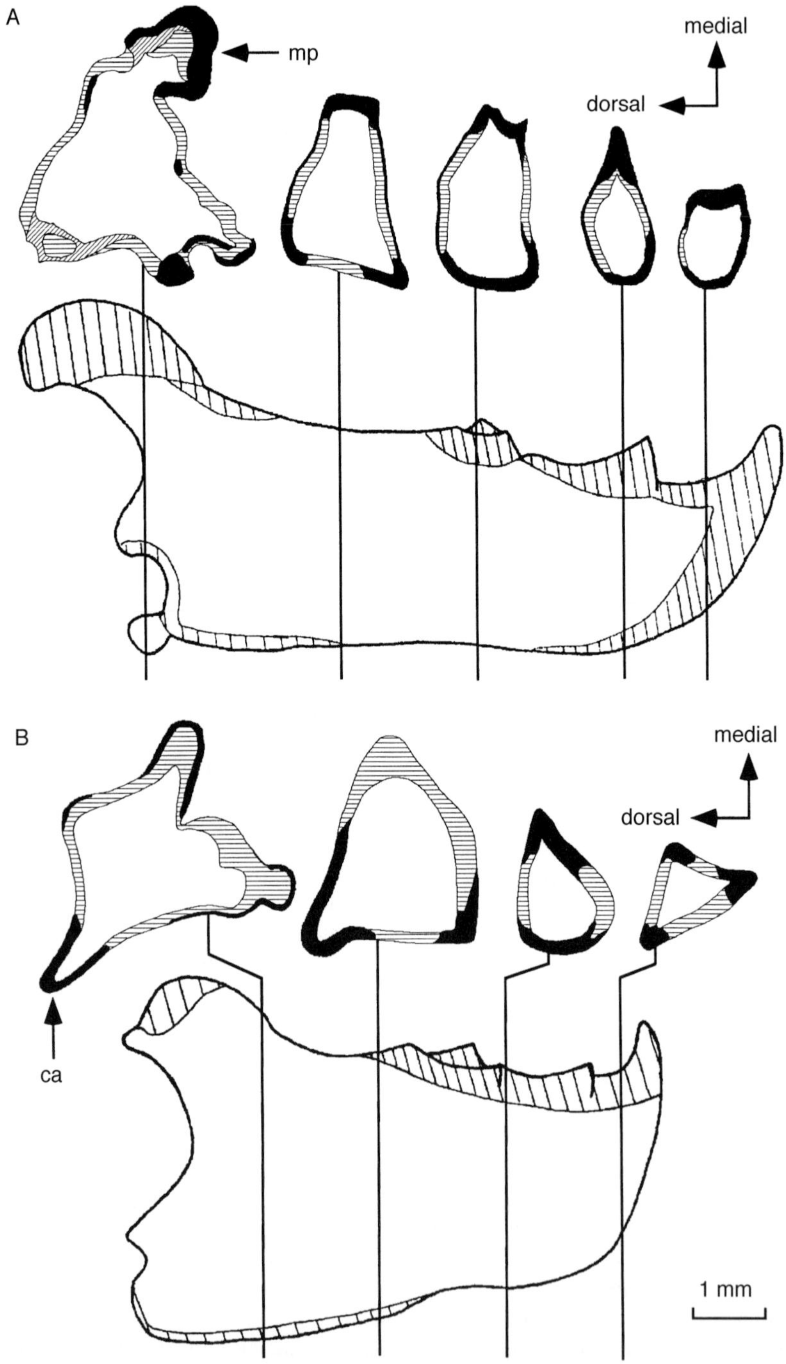

Fig. 10.11. Areas of darkest tanning and depth of tanning in transverse sections of mandibles from two species of *Hemideina*. The sections are shown above dorsal views of isolated mandibles. Hatched areas indicate regions which are black in living wetas (presumably most highly tanned). In the sections, these areas of cuticle are indicated in black, while lighter regions were proportionately more brown. A. *H. crassidens*. B. *H. femorata*. ca, Carina; mp, molar process.

Table 10.3. Demonstration of hardness of parts of mandibles and head capsule. The darkest cuticle from mandibular tips and molar teeth and lightly tanned cuticle from the head capsule were scratched against minerals of known hardness according to the Moh's scale (Pough, 1957). Each mineral can scratch the one beneath it on the scale as follows: 1: talc, 2: gypsum, 2–2.5: muscovite, 2.5–3: galena, 3: calcite, 3–3.5: barite, 3.5–4: sphalerite, 4: fluorite (harder levels not relevant).

Species	Male: tip, molars	Female: tip, molars	Light cuticle
H. crassidens	3+	3	2–2.5
H. thoracica	3+	3	–
H. femorata	3	2.5	2–2.5
H. ricta	3	2–2.5	–
H. maori	3	2.5	2–2.5
D. connectens	2.5	2.5	–

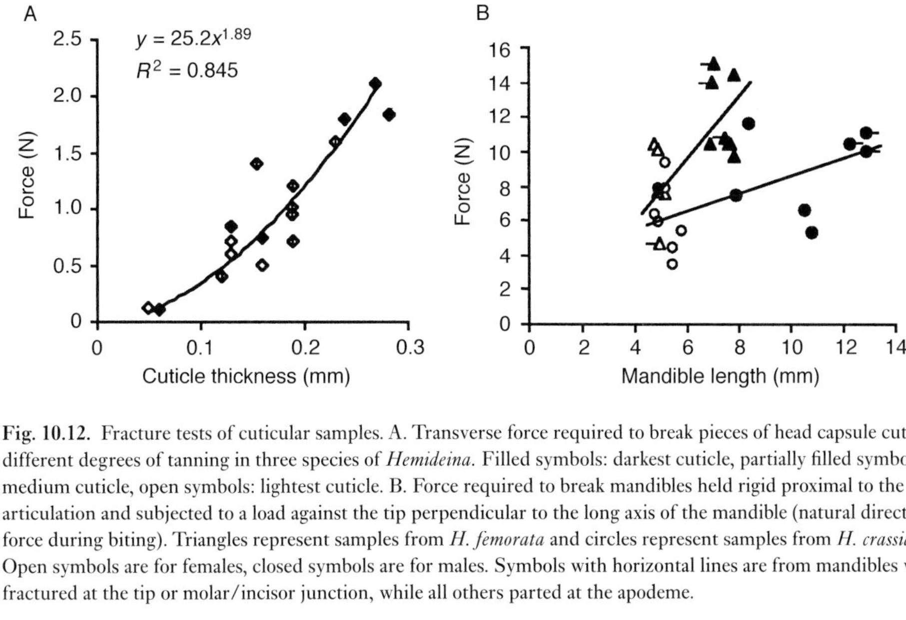

Fig. 10.12. Fracture tests of cuticular samples. A. Transverse force required to break pieces of head capsule cuticle of different degrees of tanning in three species of *Hemideina*. Filled symbols: darkest cuticle, partially filled symbols: medium cuticle, open symbols: lightest cuticle. B. Force required to break mandibles held rigid proximal to the articulation and subjected to a load against the tip perpendicular to the long axis of the mandible (natural direction of force during biting). Triangles represent samples from *H. femorata* and circles represent samples from *H. crassidens*. Open symbols are for females, closed symbols are for males. Symbols with horizontal lines are from mandibles which fractured at the tip or molar/incisor junction, while all others parted at the apodeme.

power function (Fig. 10.12A). Both results help to interpret the increased darkness observed in the cranium, gena and frons of adult male head capsules and the dark tanning of adult male mandibles. In *H. maori*, O'Brien (1984) observed that the head capsule visibly deformed during strong biting, thus attesting to the high stresses developed by the massive adductor muscles.

Two species (*H. crassidens* and *H. femorata*, representing the two forms of mandible morphology) were tested for maximum breaking strength. Most of the tested mandibles parted at the junction between the apodeme and the mandibular base. This result may be due to the dried condition of some of the test material embedded in araldite. The relationship between mandible length and breaking strength (Fig. 10.12B) was not consistent for the two species. In *H. femorata*, which has a large flared carina, the male mandibles had a higher breaking strength than those of the females (filled versus open triangles, respectively). Presumably, the carina allowed for greater reinforcement of the larger male mandibles. However, in *H. crassidens*, which has a small carina and a proportionately longer male mandible, there was a wide variation in breaking strength, with the longer mandibles breaking at the same or lower imposed force than the shorter ones (filled circles, Fig. 10.12B). The female mandibles had similar breaking strengths to those of female *H. femorata* of the same size. The symbols marked with a horizontal line in Fig. 10.12B represent mandibles that parted with tensile fractures either at the tip or in the region between the incisor and molar cusps.

These results confirm the role of the carina as a reinforcing structure, since the *H. femorata* form had greater strength per unit length than mandibles of the *H. crassidens* form. Moreover, it appears that the *H. crassidens* form is not likely to be used for developing high tensile forces during biting, but instead serves more as a weapon for maximizing the gape during ritualized male agonistic battles (Sandlant, 1981; see also Field, Chapter 18, this volume).

Mandibular Weapons and Sexual Selection

The original concept of sexual selection arose from Darwin's observations (1859) that male animals often had elaborate characters that seemed to persist in contradiction to the forces of natural selection. He proposed that sexual selection is 'a struggle between individuals of one sex, generally the males, for possession of the other sex. The result is not death to the unsuccessful competitor, but few or no offspring.' In many animals, sexual selection operates through male secondary features which serve to attract females. But, in the context of the present review, sexual selection will favour male features of weaponry which lead to success over rivals in monopolizing or attracting females. Thus, the elaborate secondary sexual characters described above for wetas and king crickets are thought to have evolved through sexual selection. There is a wealth of possibilities for the study of sexual selection in these insects, especially in Africa, Australia and New Zealand, but to date behavioural knowledge is available only for New Zealand tree wetas and the South African *L. vittatus*.

Hudson (1920) first proposed that sexually dimorphic megacephaly in New Zealand tree wetas was the indirect result of male competition for tree galleries. The enlarged mandibles would allow victorious males to defend groups ('harems') of females within the galleries, and presumably to have greater reproductive success through such monopolization. A later behavioural study confirmed this idea (Sandlant, 1981) and led to a description of the tree weta mating system as a defence resource polygyny (Field, Chapter 18, this volume).

Aggressive and agonistic use of mandibular weapons

Use of mandibles by Hemideina

Studies of male agonistic behaviour in *Hemideina*, in laboratory colonies and in the natural environment, have shown that the mandibles are used both in threat displays and in combat involving wrestling and jousting (Sandlant, 1981; Deans, 1982; Moller, 1985; R. Ewers and G. Cowley, Christchurch, 1998, personal communication; see also Field, Chapter 18, this volume). In encounters where actual contact does not occur, dominant males give stridulatory sounds and gape the mandibles toward an opponent, with the clypeus and hypopharynx retracted (Fig. 10.13A). This combined display can have a signal value, as it is sometimes sufficient to terminate the encounter and the loser retreats. However, if the encounter escalates, the males initially grapple while in close contact, and then make lunging attacks at each other with open mandibles (Fig. 10.13B–D). Both males usually maintain the mandibular gape during this phase. The mandibles are used either to batter the opponent in an attempt to dislodge it, or to outgrasp the opponent's head or gaped mandibles. Occasionally, the latter manoeuvre allowed the lunging weta to grasp and throw the opponent off the tree-trunk.

The prime event which usually signalled victory for one opponent was a lunge in which the loser's mandibles were outgrasped by those of the winner (Fig. 10.13C–D). This was almost invariably determined by which male had the greatest mandible length and hence the largest mandible gape. In our studies of *H. crassidens* the gape was about 70° to 80°, while O'Brien (1984) recorded gape angles of 75° in *H. maori* males.

In addition, the lunge was sometimes accompanied by brief bites to the opponent's head, but no damage to the head capsule was ever observed in *H. crassidens*, *H. maori* and *H. ricta*. Although several cases of head puncture were seen in *H. femorata* (Sandlant, 1981), the above combat behaviour rarely seemed to involve severe injury to combatants for tree wetas in general. This observation strongly suggests that the use of the mandibles in agonistic fights is ritualized and that signals other than injury or death determine the outcomes of battles.

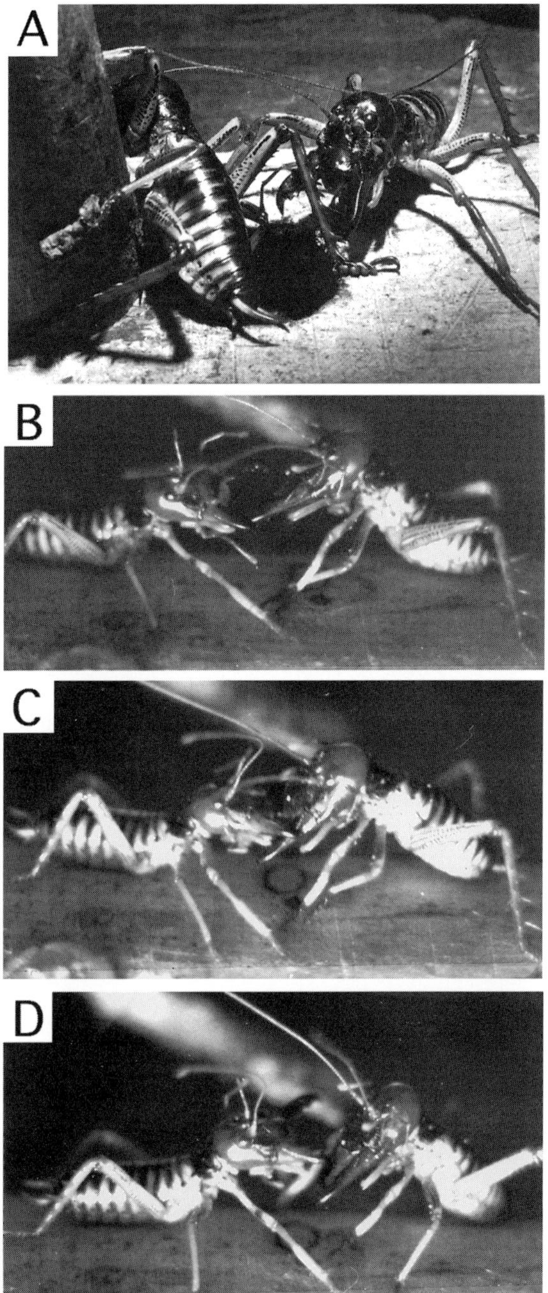

Fig. 10.13. Deployment of mandibular weapons. A. Threat display of mandible gaping by a large intruder male which had just pulled the resident male out of the gallery entrance. B–D. Ritualized use of gaped mandibles to outgrasp the opponent in a battle between two male *H. crassidens*. Photographs were taken 0.5 s apart. B. Both males joust with heads and bodies raised and mandibles gaped. C. Left-hand male lunges forward in an attempt to grasp mandibles of larger right-hand opponent, which recoiled upward and backward, (D) thereby foiling the grasping attempt. The unsuccessful male maintained its gape (D), but the right-hand male immediately reciprocated with a lunge and outgrasped the left-hand male's mandibles, which caused it to lose and decamp.

Use of mandibular tusks

The largest, recently described New Zealand tusked weta, *M. isolata* (Johns, 1997), also engages in male fighting behaviour. Males of this rare, endangered species guard the entrances of large burrows in the soil and apparently compete for burrows, rather than directly for females. Aggressive behaviour can be artificially increased by limiting hiding or burrowing space (M. McIntyre, Wellington, 1992, personal communication).

Males fight by approaching head on and interlocking each other's heads with the tusks (Fig. 10.14). In the only available video sequence of this species, it was found that the left tusk of one male was engaged over the opponent's right and the right tusk was engaged under the opponent's left. Thus each male could push the opponent as well as lift the opponent by levering upwards beneath the head. There is no attempt by the males to use the tusks as pinching devices (Meads, 1990). A battle consisted of pushing and circling, while remaining interlocked, and attempting to flip the opponent upwards. Males also attempted to pin the opponent against obstacles on the substrate. In one case, the loser was lifted momentarily upward and clear of the substrate. This act may have decided the battle, for just afterwards the loser disengaged and decamped. Often the resident male (in a laboratory terrarium) won the battle over a burrow if allowed to occupy a burrow for at least 2 h (M. McIntyre, Wellington, 1992, personal communication).

Raukamara tusked weta

While aggressive behaviour has not been investigated in this secretive species, several instances of violent defence of burrows by males have been observed in captive wetas. In addition, at least one captive female died apparently from puncture wounds inflicted by a male with tusks, suggesting that the straight, sharply pointed tusks of this species are used directly as weapons, rather than as jousting or grappling structures (M. McIntyre, Wellington, 1992, personal communication).

Agonistic male fights in Libanasidus vittatus

This South African tusked king cricket uses the tusks in mate-guarding fights, where recently mated males defend the female vigorously against intruder males (Bateman and Toms, 1998). The tusks are used to grasp the opponent's head during attempts to overturn him and thus are apparently used as pincers, in contrast to the use of tusks by *M. isolata* as levers. No rotary overturning attempts have been documented for the latter.

Fig. 10.14. Head-to-head clamping with tusks in battling males of *Motuweta isolata*. The elephantine tusks project forward from the rather reduced mandibles, which are not themselves used in fighting. Both males stayed locked in this mutual pushing and circling behaviour until one disengaged and decamped after having been lifted off the substrate by the other. Photograph by B. Robertson, Victoria University.

Evidence for Mechanisms of Sexual Selection

Although the presence of enlarged mandibles and associated weaponry and of male agonistic competitions suggests that sexual selection occurs, questions of how this could happen remain unanswered. Presumably, there are reproductive returns for males that can monopolize either females or areas that attract females, and thereby obtain greater mating success. In tree wetas, there has been no analysis of the supposed female preference for larger secondary sexual characters in males; however, females congregate in retreats for which males compete and this could provide a mechanism for the enhanced mating success of dominant males (Hudson, 1920; Gwynne and Jamieson, 1998).

Male competition for retreats that attract females

Only one study has approached the question of how selection for large secondary sexual characters could operate in tree wetas. This was an analysis of the relationship between head size and male success in acquiring mates in *H. maori*, a large alpine species that inhabits cavities beneath rocks (Gwynne and Jamieson, 1998). During the summer breeding season, single adult males usually live with several adult females, as well as earlier nymphal stages of both sexes, in the cavities. The wetas emerge at night to forage and oviposit. Mating can occur within and around the rock cavities, and a dominant male will readily mate with new females introduced to a cavity (Gwynne and Jamieson, 1998). Thus, as in other tree weta species, adult males appear to maintain harems of females within the rock cavities, for which they compete agonistically. This defence resource polygynous mating system is discussed by Field and Jarman (Chapter 17, this volume).

Gwynne and Jamieson (1998) found up to seven females with single adult males occupying single cavities during the alpine breeding season. For 43 such groups, they found a significant correlation between harem size, male head width and male left mandible length. The conclusion was that males with larger secondary sexual characters won the opportunity for greater mating success than that available to smaller males. Unlike lucanid beetles, where males with larger weapons win fights to occupy oviposition sites and thus gain greater mating success (Eberhard, 1982), *H. maori* (and presumably other tree weta species) males monopolize retreat sites that attract females. Those with larger weapons have more females in their harem and an enhanced opportunity to mate with them as they return to the retreat cavity; this presumably leads to greater reproductive success.

Female mate choice versus male mating success

It remains to be proved whether actual paternity success results from the above scenario for any tree weta species. A complicating factor is that, in *H. crassidens*, *H. femorata* and *H. ricta*, females appear to be choosy about partners and they often disengage from mating attempts by males (Jarman, 1982; Spencer, 1995; L.H. Field and N.A. Deans, personal observation). For example, Spencer (1995) found that 68% of mating attempts (n = 95) in *H. crassidens* were broken off by the female. The crucial experimental questions arising from these observations involve determining whether females assess male fitness based upon size of male secondary sexual characters, body weight or some other criterion.

In addition, there is evidence that some male tree wetas (but not *H. maori* males) mature at an earlier instar (eighth or ninth) while other males and all females mature at the tenth instar. The earlier-maturing males appear to adopt a sneak and wander tactic for mating with females outside the galleries, rather than fighting for gallery possession to access females (Spencer, 1995). In this case, the selective advantage accruing to large-weaponed males that possess galleries would be diminished by the sneaking males outside the galleries. The runaway selection process would either be opposed or be split into two optimal processes, leading to selection for large and small male characters simultaneously. The result would be a stable system of alternative mating strategies with satellite males and dominant males.

However, the direction of evolution of such a system of alternative mating strategies would be determined by female selection biases in choosing males. It cannot always be assumed that females choose to mate with the largest or strongest males. Kaneshiro (1993) has proposed that females can create a stable polymorphic mating system, in which females show different preferences for males of different phenotypes. Clearly, the alternative male phenotype in tree wetas, as well as in New Zealand and African tusked wetas with and without tusks, would be an interesting case in which to test this model.

The alternative mating strategy system would also result in sperm competition between the two groups of males (Sandlant, 1981). One function of guarding a gallery might be to limit the opportunity for gallery-inseminated females to emerge and make themselves available for mating (and subsequent displacement of sperm) by sneaking males. Studies are needed on the mating success of males of different size classes which use different mating strategies.

Sexual Selection versus Natural Selection, and Gigantism

In various models of sexual selection, and especially that espoused by Fisher (1930), it is assumed that the forces of natural selection will counterbalance the runaway process of sexual selection. While female preference for enlarged secondary sexual characters will drive enlargement of those characters, the environment will select against extreme phenotypes which exceed the optimum for survival. An excellent example of the counterbalance of these two evolutionary forces is afforded by the description of the gallery ecology of *H. femorata*, the tree weta that commonly lives in manuka and kanuka trees (*Leptospermum scoparium* and *Kunzia ericoides*, respectively) in the South Island of New Zealand (see Field and Sandlant, Chapter 13, this volume). The wetas occupy galleries made exclusively by the larvae of one species of long-horned wood-boring beetle, the manuka beetle *Ochrocydus huttoni* (Cerambycidae). Thus the narrow size distribution of gallery entrances available to wetas in these trees is determined by the sizes of beetles emerging after eclosion. There is a rather uniform maximum size of gallery entrance available to male wetas as they mature and, once their head width exceeds this dimension, they exclude themselves from the protection offered by the gallery against diurnal predators. Data on *H. femorata* head width and gallery entrance diameter show that large males live in galleries with entrance diameters nearly the same size as the head width. It follows that those males whose head widths exceed available entrance diameters have a high probability of being removed from the population by natural selection from avian and mammalian predators. Indirect evidence of this is the repeated observation that very large males are not found elsewhere in the forest habitats (L.H. Field and N.A. Deans, personal observation). The conclusion is that the allometric growth of head capsule and mandibles is limited by the size of the host beetle species that provides the gallery refuges for the wetas, and sexual selection is counterbalanced by natural selection.

Gigantism and release from the stabilizing role of natural selection

What happens when the distribution of gallery entrance diameters is no longer a constraint on maximum development of head size? Will sexual selection allow runaway enlargement of secondary sexual characters in tree wetas if the limiting selection force of gallery diameter is relaxed? The answer appears to be yes, although natural selection will ultimately control the sexual selection process at some new level of enhancement. An example is seen in the offshore island (Stephens Island and Open Bay Islands) races of tree wetas (*H. crassidens*), which apparently show gigantism. In these populations, the large body size of males is associated with very large head and mandible length, as discussed earlier. At the same time, a variety of diurnal retreats for these wetas are found in trees with large openings where branches have broken and rotted back or trunks are split and excavated (Moller, 1985). There are large entrance openings and males can easily occupy the galleries and accumulate harems of females. Thus it appears that the response to a release from size restrictions imposed by gallery dimensions is the selection of larger secondary sexual characters in the males. The same process probably also applies to mainland populations where gallery dimensions are large. However, new forces come into play from natural selection, which limit the Fisherian process. Larger galleries and entrance diameters allow rat, mouse and weasel predators access to the wetas, and the function of the gallery as a diurnal refuge from predators must decrease in proportion to the increase in gallery entrance diameter.

The differential selection model of sexual selection could eliminate the stabilization role of natural selection

Kaneshiro (1993) discussed a model in which differential mating propensities in females within an interbreeding population could lead to a balanced mating system where natural selection need not be imposed as a stabilizing mechanism of secondary sexual character development. By having a range of mating types segregating in both sexes and a strong genetic correlation between male mating success and female discrimination, sexual selection itself could act as a stabilizing force in maintaining a normal distribution of mating types, including those with large secondary sexual characters, in subsequent generations. This effect could apply to the mating system of tree wetas if females show different mating responses to the satellite and tenth-instar males.

References

Bateman, P.W. (2000) The influence of weapon asymmetry on male–male competition success in a sexually dimorphic insect, the African king cricket *Libanasidus vittatus* (Orthoptera: Anostostomatidae). *Journal of Insect Behaviour* 13, 157–163.

Bateman, P.W. and Toms, R.B. (1998) Mating, mate guarding and male–male relative strength assessment in an African king cricket (Orthoptera: Mimnermidae). *Transactions of the American Entomological Society* 124, 69–75.

Beier, M. (1962) Tettigoniidae (Pseudophyllinae II). *Das Tierreich, Berlin* 74, 1–468.

Bennett, A. and Toms, R.B. (1995) Sexual dimorphism in the mouthparts of the king cricket *Libanasidus vittatus* (Kirby) (Orthoptera: Mimnermidae). *Annals of the Transvaal Museum* 36, 205–214.

Darwin, C. (1859) *The Origin of Species.* Modern Library, New York.

Deans, N.A. (1982) The functional morphology of the mandibles in the genus *Hemideina*. Honours thesis, University of Canterbury, New Zealand.

Eberhard, W.G. (1982) Beetle horn dimorphism: making the best of a bad lot. *American Naturalist* 119, 420–426.

Every, R.G. (1972) *A New Terminology for Mammalian Teeth Founded on the Phenomenon of Thigosis.* Pegasus Press, Christchurch, 64 pp.

Field, L.H. (1980) Observations on the biology of *Deinacrida connectens* (Orthoptera: Stenopelmatidae), an alpine weta. *New Zealand Journal of Zoology* 7, 211–220.

Field, L.H. (1993) Structure and evolution of stridulatory mechanisms in New Zealand wetas (Orthoptera: Stenopelmatidae). *International Journal of Insect Morphology and Embryology* 22, 163–183.

Fisher, R.A. (1930) *The Genetical Theory of Natural Selection.* Clarendon Press, Oxford, 272 pp.

Gwynne, D.T. and Jamieson, I. (1998) Sexual selection and sexual dimorphism in a harem-defending insect, the alpine weta (*Hemideina maori*, Orthoptera: Stenopelmatidae). *Ethology, Ecology and Evolution* 10, 393–402.

Hudson, G.V. (1920) On some examples of New Zealand insects illustrating the Darwinian principle of natural selection. *Transactions and Proceedings of the New Zealand Institute* 52, 431–438.

Hutton, F.W. (1897) The Stenopelmatidae of New Zealand. *Transactions and Proceedings of the New Zealand Institute* 29, 208–242.

Jarman, T.H. (1982) Mating behaviour and its releasers in *Hemideina crassidens* (Orthoptera: Stenopelmatidae). MSc thesis, University of Canterbury, Christchurch, New Zealand.

Johns, P.M. (1997) The Gondwanaland wetas: family Anostostomatidae (formerly in Stenopelmatidae, Henicidae or Mimnermidae): nomenclatural problems, world checklist, new genera and species. *Journal of Orthoptera Research* 6, 125–138.

Kaneshiro, K.Y. (1993) Habitat-related variation and evolution by sexual selection. In: Kim, K.C. and McPheron, B.A. (eds) *Evolution of Insect Pests/Patterns of Variation.* Wiley and Sons, New York, pp. 89–101.

Karny, H.H. (1937) Orthoptera. Family Gryllacrididae. Subfamily Omnes. *Genera Insectorum de Wytsman* 206, 1–317.

Meads, M.J. (1990) *The Weta Book: a Guide to Identification of Wetas.* DSIR Land Resources, Lower Hutt, New Zealand, 36 pp.

Møller, A.P. (1992) Patterns of fluctuating asymmetry in weapons: evidence for reliable signalling of quality in beetle horns and bird spurs. *Proceedings of the Royal Society of London B* 248, 199–206.

Moller, H. (1985) Tree wetas (*Hemideina crassicruris*) (Orthoptera: Stenopelmatidae) of Stephens Island, Cook Strait. *New Zealand Entomologist* 18, 15–24.

Morgan-Richards, M., Daugherty, C.H. and Gibbs, G.W. (1995) Taxonomic status of tree weta from Stephens Island, Mt. Arthur and Mt. Holdsworth, based upon allozyme variation. *Journal of the Royal Society of New Zealand* 25, 301–312.

O'Brien, B. (1984) Mandibular movements and their control in the weta *Hemideina maori* (Orthoptera:

Ensifera: Stenopelmatidae). PhD thesis, University of Canterbury, Christchurch, New Zealand.

Pough, F.H. (1957) *A Field Guide to Rocks and Minerals*, 2nd edn. Houghton Mifflin, Boston, 349 pp.

Sandlant, G.R. (1981) Aggressive behaviour of the Canterbury weta *Hemideina femorata* (Orthoptera: Stenopelmatidae): its adaptive significance in resource allocation. MSc thesis, University of Canterbury, New Zealand.

Spencer, A.M. (1995) Sexual maturity in the male tree weta *Hemideina crassidens* (Orthoptera: Stenopelmatidae). MSc thesis, Victoria University of Wellington, Wellington, New Zealand.

11

Anatomy, Development and Behaviour of the Chilean Red Cricket, *Cratomelus armatus* Bl.

Andrés O. Angulo

Departamento de Zoologia, Facultad Ciencias Naturales y Oceanográficas, Universidad de Concepción, Casilla 4040, Correo 3, Concepción, Chile

Introduction

In 1851, Blanchard described the genus *Cratomelus* based upon *C. armatus* Bl., the type species of the genus (Fig. 11.1-1, 2). According to Borror (1898), the absence of stridulating behaviour, the front and middle tibiae with articulated ventral spines and the smooth front wings almost without tegmen place it in Gryllacrididae (subfamily Stenopelmatinae). Subsequently, *Cratomelus* was placed in the Stenopelmatidae (superfamily Gryllacridoidea) (CSIRO, 1991). Finally, Johns (1997) and Gorochov (see Chapter 1, this volume) placed this genus and species in the family Anostostomatidae Blanchard, 1997, and in the subfamily Cratomelinae Brunner, 1888.

The original description says that 'it lives in Concepción and has a certain resemblance to several crickets'. Since then, collections have extended the distribution from Chiloe, Aysén, in the south, to the north of Santiago. The width of the distribution extends from the coast, eastward to the pre-Andean zone. Within this region, *C. armatus* is found burrowing in moist soil or forest floors up to 1000 m a.s.l.

It is of great interest that the southern populations, especially those of Chiloe, are darker than those of the north. Finally, being an omnivorous animal, it has the ability to survive in limited food environments. However, its distribution is sensitive to the relative humidity, and therefore it is nocturnal or crepuscular.

External Anatomy of Sense Organs

Compound eyes

The eyes have a suboval form, 2 mm in length by 1.5 mm maximal width. The surface is divided into numerous facets of ommatidia, which are tightly juxtaposed and form sinuous contours, giving a general subhexagonal shape. Some setae are irregularly inserted (Fig. 11.1-3, 4, 5).

Antennal anatomy

The number and types of sensilla vary along the antenna according to form and size (Fig. 11.2-9, 10), although in the scape (or first antennal segment, sc) they are scarce, except for a few very small trichoid sensilla (Benoit, 1975). In the pedicel (second antennal segment, ped) close to the articulation with the scape, the Johnston's organ (jo) is found. Although the trichoid sensilla are sparse, they increase distally towards the articulation with the flagellum (fla). Along the flagellum the number of trichoid sensilla varies greatly, from very few in the first segments, increasing rapidly in the middle area, where the density is great, and remaining dense until the apical region of the antenna, where the density diminishes slightly. Starting a quarter of the antennal length from the base, a small quantity of campaniform sensilla appear, and they increase up to three-quarters of the antennal length, where their density is very

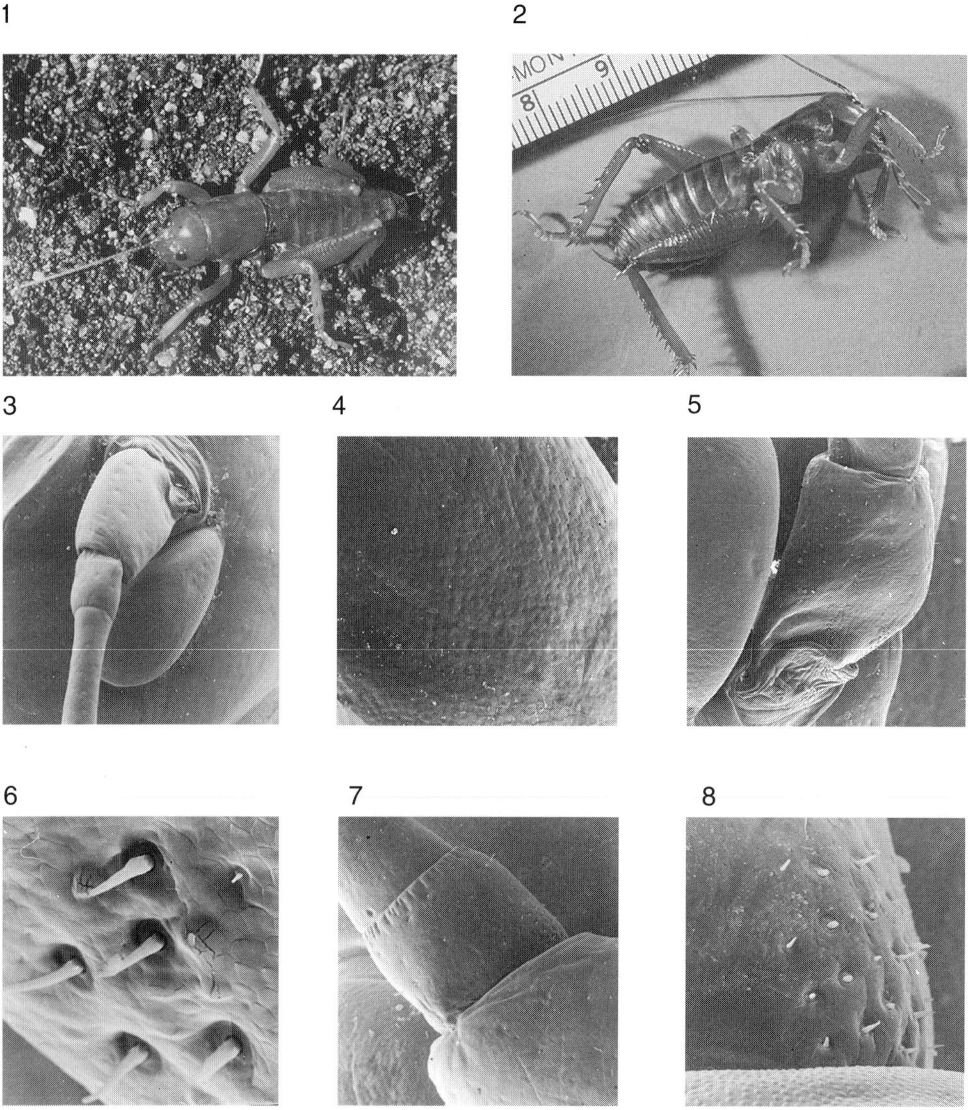

Fig. 11.1. 1, Male *Cratomelus armatus.* 2, Female *Cratomelus armatus.* 3, Scanning electron micrograph (SEM) of panoramic view of the compound eye (20×). 4, SEM showing surface of facets of compound eye (80×). 5, SEM showing ommatidia and setae of the compound eyes (800×). 6, Sensillum trichoidea on the antenna (800×). 7, Base of pedicel (60×). 8, Base of pedicel: Johnston's organ (150×).

great (five to six in each segment). The number then decreases and disappears in the last antennal segments. Occasionally, from the middle part of the antenna to the end, large chaetica sensilla are observed, one every one to two segments. Along the antenna variable quantities of very small basiconic sensilla also exist; these are very difficult to observe and count.

Antennal hair-plate morphology

An antennal hair-plate organ (jo) appears as a series of 17 to 25 transparent protuberances of vesicular aspect, closely grouped in the proximal region of the pedicel (Figs 11.1-8 and 11.2-10). This resembles the larger of several hair-plate organs found on New Zealand giant wetas in the

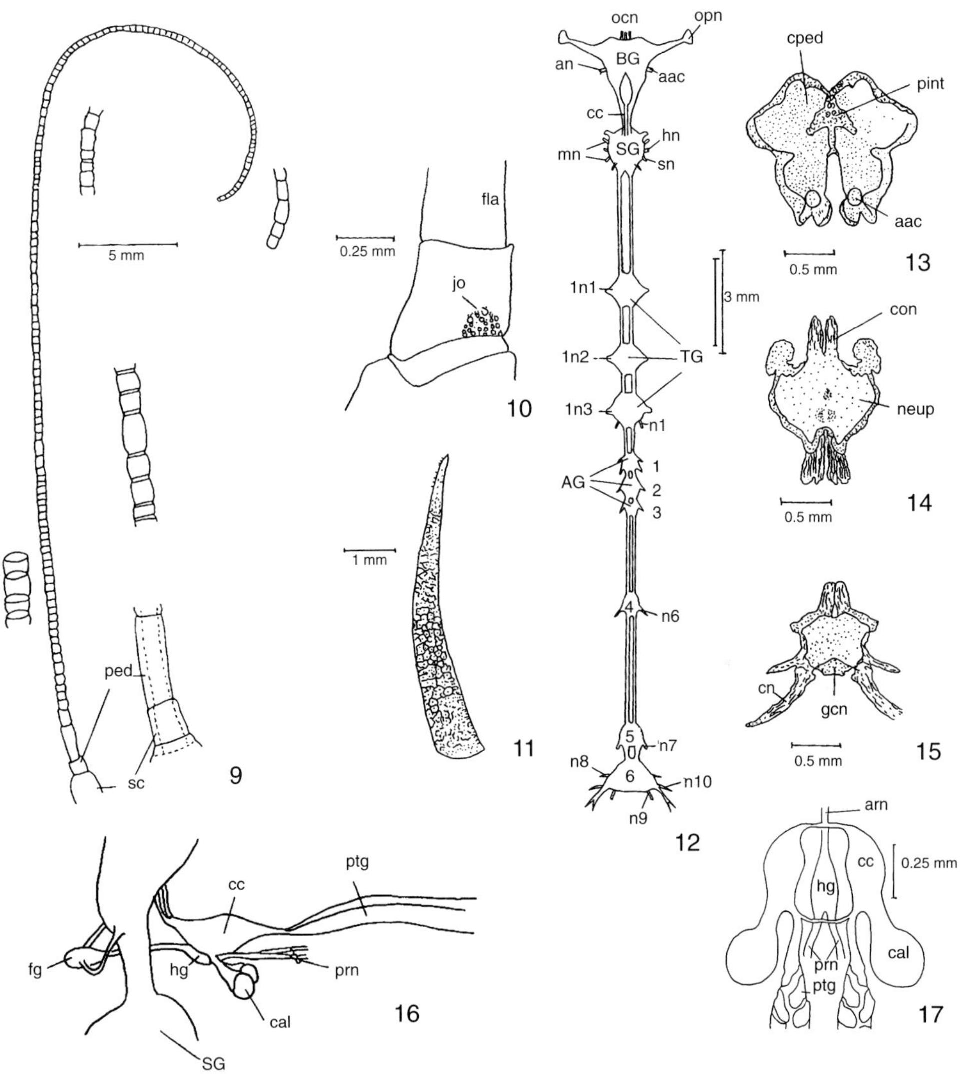

Fig. 11.2. 9, General view of the antenna. 10, External aspect of Johnston's organ. 11, General view of the cercus. 12, Ganglion chain of central nervous system. Note semicondensation of first three abdominal ganglia. 13, Supraoesophageal ganglion sagittomedial view. 14, Thoracic ganglion, sagittomedial view. 15, Caudal ganglion sagittomedial view. 16, Lateral view of stomatogastric ganglion and retrocerebral complex. 17, Dorsal view of stomatogastric and retrocerebral complex. See text for abbreviations.

same antennal location (described in O'Brien and Field, Chapter 8, this volume) and undoubtedly serves to indicate the position of the antenna as it moves against the arthrodial membrane of the scape.

Cercal anatomy

The proximal 80% of the cercus is covered with numerous trichoid sensilla, appearing as very long filiform hairs decorated at the base by regular petaloid structures (Fig. 11.2-11). Towards the end of the cercus the number of filiform hairs decreases and disappears. Uniformly, throughout the cercus, shorter trichoid sensilla, similar to those observed in the antenna, and basiconic sensilla, larger than the antennal ones, are found. This is similar to the cercal sensilla array found in the New Zealand tree weta, *Hemideina* (see Fig. 22.9 in Field, Chapter 22, this volume).

Sensillum types

Simple trichoid sensillum

This type of sensillum is considered to be one of the most primitive sense organs of the insects. Externally, it consists of one thin hair, which varies between 25 and 100 µm, which is inserted in a protuberance of the cuticle. Internally, it is in contact with a group of three to five cells, at least one of which is a sensory neuron. This type of sensillum is the most usual one, since it is found throughout the whole antenna and cercus.

Campaniform sensillum

This sensillum appears externally as a transparent dome less than 1 µm thick, with a more or less circular form. The maximum width varies between 18 and 30 µm. The innervation of this sensillum type is formed by a single and extended sensory neuron, surrounded by one to three joined cells. The dendrite of the sensory cell penetrates to the underside of the dome. Smaller campaniform sensilla are found between 25% and 50% of the antenna length, and larger ones are situated from 75% of the antenna to the end.

Chaetica sensillum

Externally, this sensillum is very similar to the trichoid sensillum, but it is essentially different in size. The chaetica sensillum is stronger than the hair of the trichoid sensillum, and its basal articulation is larger (Fig. 11.1-7). The chaetica is innervated by one neuron, accompanied by three to four cells, which surround it. The size of these sensilla from the base to the end of the hair is 90–100 µm.

Basiconic sensillum

Externally, this sensillum forms a conic nipple approximately 7 µm in height in the antennae and 11 µm in the cerci. The internal morphology was not clearly observed, but two cells that connected with the underlying nerve were observed.

Large trichoid sensillum (filiform sensillum)

These are exclusively sensilla of the cerci. Single sensory hairs reach 250–600 µm in length. They are articulated in a convexity of the cuticle with regular petaloid decoration. The length of the hair is greater in the middle part of the cercus and diminishes towards the extremes. Internally, the hair is associated with two to four neurons surrounded by four, and in some cases five, cells.

Remarks

Studies carried out by Angulo (1969), Guzmán *et al.* (1970a, b) and Delpin (1972), on *C. armatus* Bl., suggest a primitive status of the Anostostomatidae in relation to other ensiferans (Gryllidae) and to caeliferans (Acrididae). For example, the great quantity of the trichoid sensilla on the ‘red cricket’ suggests a primitive condition, since their structure is considered to be an old character in the evolution of the arthropod groups (Snodgrass, 1925).

The well-developed Johnston’s organ, which is typically sensitive to the movement and vibration of the antennae by air, together with the brachypterous wing condition, may reflect earlier phylogenetic affinities to Gryllidae and Acrididae – that is, intermediate flying forms.

Internal Anatomy

Nervous system

Morphological description of the ganglion chain

The ganglionic chain consists of 11 ganglia (Fig. 11.2-12), including the two cephalic ganglia: supra-oesophageal (BG, brain) and the sub-oesophageal (SG), three thoracic (TG, pro, meso and meta) and six abdominal (AG). Each ganglion is connected with the adjacent one by means of two nerve cords or connectives. The cerebral supra-oesophageal ganglion, in its general aspect, is typical for insects. The protocerebrum in a frontal view is bilobed, and laterally it gives rise to the optic regions, which are composed of the optic lobe – approximately half the width of the cerebral ganglion – and the optical cup with the optical nerves (opn) – one-quarter the ganglion width. In a posterior view, the medial region (pars intercerebralis) gives rise to the two ocellar nerves (ocn), one of which is bifurcated and innervates the two superior ocelli (Guzmán *et al.*, 1974).

In its medial region (deutocerebrum), the ganglion gives rise to the antennal complex (one antennal nerve (an), and three accessory antennal nerves (aac). Posteriorly, at the boundary of the tritocerebrum and in the deutocerebrum, the two occipital nerves arise. The widely bilobed posterior region represents the two tritocerebra, which give rise laterally to the labral and frontal part and the circumoesophageal connective (cc).

The suboesophageal ganglion gives rise laterally to the main mandibular nerve (mn), and medially to the secondary and tertiary mandibular nerves. The following then arise, moving ventrally: the hypopharyngeal nerve (hn), the salivary nerve (sn), the cervical nerve and finally the connective to the prothoracic ganglion.

The thoracic ganglia all have similar innervation to the legs (ln1–3, Fig. 11.2-12). The third thoracic ganglion is larger than the first and second, since it is fused with the first abdominal ganglion. The second thoracic ganglion is closer to the third than to the first one.

In the abdominal cord, each of the six abdominal ganglia gives rise to nerves that innervate the corresponding abdominal segments, except the first abdominal, which innervates the second and third abdominal segments. Also, the caudal or sixth abdominal ganglion innervates the eighth, ninth and tenth abdominal segments. The arrangement of the six abdominal ganglia along the ventral cord is irregular, since the first three are located anteriorly, the fourth has an intermediate position and the fifth is close to the sixth, which is approximately twice as great in size as the fifth. The sixth gives rise to the cercal nerves (cn, Fig. 11.2-15).

The larger size of the third thoracic ganglion and its innervation of the first abdominal segment indicate that it has fused with the true first abdominal ganglion. Similarly, the true second and third abdominal ganglia have fused, as shown by the innervation to the second and third abdominal segments. In the New Zealand tree weta, the same fusion and innervation pattern exists (see O'Brien and Field, Chapter 8, and Fig. 23.13 in Field, Chapter 23, this volume).

Histological Description of Ganglia

A longitudinal section of a ganglion (Fig. 11.2-13) shows the neurilemma, which covers the whole ganglion, the connectives and the nerves that originate from it. This neurilemma (or neural plate or 'neural laminella') (Smith, 1968) is formed by an amorphous matrix of connective tissue and collagen fibres. Immediately beneath the neurilemma and close to it, one or more layers of neuroglia cells are observed: the perineurium (Smith, 1968). More centrally, the cortex is formed by neuron cell bodies and interspersed glial cells.

According to their function and morphology, these cells can be grouped as follows:

1. The association neurons (interneurons): small in size with a spherical nucleus, chromatin forming irregular granules deeply stained and hardly distinguishable reduced cytoplasm. One type of these neurons has been named globular cells (Du Porte, 1961).
2. Motor neurons: these are greater in size, with oval or spherical nuclei, and chromatin forming two or more granules with less density than the interneurons, and a more distinguishable cytoplasm.
3. Secretory neurons or giant cells of Hartweek (Demerec, 1950). They have a great size, an ovoid nucleus, diluted chromatin and a deeply stained nucleolus. The cytoplasm is abundant and dense, with a great quantity of secretion granules.
4. Neuroglia cells: small in size, polymorphic, with elliptical or fusiform nitid nuclei. The cytoplasm is so scarce that sometimes it is observed as thin membranes that originate from the different regions of the nucleus.

Other types of accessory elements observed in this area are the tracheole and the channels. Both structures are also found at the border of the cortex and neuropile. The channels are spaces delimited by a thin sheet of fusiform cells.

In the middle of the ganglion is found the neuropile, which constitutes the greater part of the ganglion mass. It is surrounded by the cortex, except in the area where the connectives and nerves begin. It is formed by the axons and their ramifications and dendritic preocesses. Microscopically, the neuropile is observed as a net of fibres transversely and longitudinally cut. Among these fibres secretory granules occur. In a ganglion, the neuropile may be 'structured' (centres or glomeruli) or unstructured' (Maynard, 1962).

The connectives are externally covered by a continuation of the neurilemma. In the inner part, the axons of the neurons, the fusiform nuclei of the glia cells and secretion granules are found (the latter being similar to those of the secretory neurons and of the neuropile of the ganglia).

Histological description of the supraoesophageal ganglion

The ganglion, externally surrounded by the neurilemma, is formed by proto-, deuto- and tritocerebral regions which are common to all insects.

Protocerebrum

The most characteristic aspect of this ganglion region is the neuropile, where the following structures are found:

1. The fungiform bodies (corpora pedunculata) (cped, Fig. 11.2-13): These are formed by a large calyx and a short peduncle. This is wider in its distal region, where more coloured areas, which could correspond to a greater condensation of fibres, are observed. Laterally to the fungiform bodies, the optical tract is observed, continuing toward the posterior extreme to join the neuropile.
2. Protocerebral bridge (Fig. 11.2-13): This is divided and is observed as two fasciculi immediately below the pars intercerebralis (pint).
3. The central body. This is located in the posterior part. The radial arrangement of its fibres gives it a characteristic fan-like aspect. Surrounding the central body and even between their fibres, neuroglia cells were observed.
4. The ventral bodies: they are two small fibre masses situated behind and below the central body.

In the protocerebrum cortex, we find four kinds of cells previously described. The globular cells (interneurons) are found in great numbers between the anterior region of the calyx and the neurilemma. Among them, axons that originate from the calyx towards the ocellar nerves are observed.

The greatest number of neurosecretory cells is in the pars intercerebralis (Fig. 11.2-13). They are also widespread between the motor neurons, which are placed basolaterally to the fungiform bodies.

According to the general description of a ganglion, the glia cells are found below the neurilemma, widespread among the different types of neurons and forming the perineurium (Guzmán *et al.*, 1974).

Deutocerebrum

The neuropile is formed by two circular antennal centres (aac, Fig. 11.2-13), each comprising approximately ten glomeruli. From this antennal centre and laterally to it, the antennal nerve begins. The commissural deutocerebral track, which joins the two antennal centres, originates here. The occipital centre is below the deutocerebral commissure; from this centre, the occipital nerves originate. In the deutocerebral cortex, neurosecretory cells and association neurons are also found.

Tritocerebrum

The subcircular neuropile consists of a dense mass of fibres. In the posterior region, approximately four glomeruli are seen, joined to the occipital centre by fibre tracts.

From the tritocerebral centre two nerves originate: the commissural and the labrafrontal. The circumoesophageal connective, which joins the supraoesophageal ganglion with the suboesophageal, arises from the tritocererbrum as a thick trunk of fibres. In this area of the ganglion, as in the rest of the cortex, the neuroglia and the motor, secretory and association neurons are located.

Suboesophageal ganglion

The neurilemma, cortex and different kinds of neurons and glia cells forming the perineurium are similar to those of other ganglia. The neuropile is simpler than that of the supraoesophageal, since it is not differentiated into glomeruli. Only some fibre condensations are observed, especially in the areas where the connectives and nerves begin. In addition, the same spaces, longitudinal and transverse fibres and secretory granules that are in the former ganglion are observed.

Thoracic ganglia

Tracheoles are present immediately below the neurilemma and some cross the cortex, reaching the neuropile. The neurilemma is followed by the perineurium, which presents the same arrangement as in the former ganglia. The cortex surrounds the neuropile (neup) and, in the first thoracic ganglion, shows areas of greater concentration of neural, motor and secretory bodies and a dorsal median and two ventrolateral areas separated by the median ventral fissure. In the ventrolateral area of the second thoracic ganglion is found a cluster of secretory neurons. At the nerve roots the cortex is reduced to only a thin cover of glia cells (Fig. 11.2-12, 14).

The neuropile of the thoracic ganglia is very similar to that of the suboesophageal ganglion, without glomerular differentiation, but with fibre tracts and secretion granules. In the second thoracic ganglion, the neuropile contains medial fibre tracts arranged longitudinally and transversely in section. In the third thoracic ganglion, the neuropile is not so condensed as in the second, and the spaces are greater in size (Guzmán *et al.*, 1974).

Abdominal ganglia

These are smaller than the thoracic ganglia, but the neurilemma, perineurium and tracheoles are similar in form and arrangement. The cortex is directed laterally (Fig. 11.1-7) with association and secretory neurons. Its neuropile also shows large spaces, fibres and granules.

In a transverse section of the connectives that join the first to the second abdominal ganglion, five giant fibres are observed in the first and about seven in the second, some of which have small fusiform nuclei. In the rest of the connectives, widespread among the fibres, some larger nuclei with granular chromatin are observed.

Caudal ganglion (sixth abdominal)

This is larger than the previous five abdominal ganglia, but the neurilemma and perineurium are similar. The cortex is formed by neurons of different sizes, including the large cell bodies of the giant neurons (gcn) located in the dorsal region (Fig. 11.2-15). The neuropile is similar to the previous ones, with giant fibres clearly apparent both in transverse and longitudinal sections.

Stomatogastric and retrocerebral system: a macroscopic description

Neurosecretory and retrocerebral system

In lateral view, two nerves arise from the neurosecretory protocerebral centres and enter the corpus cardiaca (cc, Fig. 11.2-16). Each of the latter, subfusiform structures is enlarged apically and gives rise to a lateral posterior nerve, which connects to its respective corpus allatum (cal), of spherical form. The enlargement is called the calyx and has a subcircular form. The nerves to the oblong prothoracic glands (ptg) arise from the apical extreme of the corpora cardiaca, and occupy approximately the length of the thorax. This system is dorsal to the pharynx. In a dorsal view, the general structure has a notably depressed aspect.

Stomatogastric nervous system

This lies above the oesophagus. Anteroventrally to the supraoesophageal ganglion is the subtriangular frontal ganglion (fg); the connectives that originate from the respective anterior tritocerebral hemispheres enter the basal angles of this ganglion (Fig. 11.2-17). The hypocerebral ganglion (hg) is found between the corpora cardiaca towards the ventral region, joined to each one by short, wide bridges of the same tissue as that of the corpora cardiaca. From its anterior extreme, the anterior recurrent nerve (arn) originates and, from the posterior, two posterior recurrent nerves (prn) extend to their respective oesophageal ganglia.

Stomatogastric and retrocerebral system: histological description

Prothoracic gland

This is formed by multiple acini and covered by a prolonged capsule of connective tissue surrounding each acinus separately. The acini are composed of secretory cells in two states of activity. Some have lightly granular cytoplasm with few vacuoles and a large, lightly stained nucleus (in some of which a nucleolus is observed). Others have little cytoplasm, large vacuoles and small, compact, deeply stained nuclei. The acini communicate with each other by means of a net of tubules formed by a single layer of cylindrical epithelial cells. These tubules join together forming a single duct that goes toward the corpora cardiaca (Guzmán *et al.*, 1974).

Corpora cardiaca

The corpus cardiacum is covered by a layer of connective tissue with fusiform nuclei laterally and spherical nuclei in the dorsal and ventral regions. Each corpus cardiacum is composed of two kinds of cells: neurosecretory and chromophil. The neurosecretory cells are larger, with small, dense nuclei and highly granular cytoplasm. The cellular form is variable; in some nuclei, a nucleolus is clearly observed. The chromophil cells may have different forms, but the cytoplasm is lightly granular and the nucleus is variable from polygonal to ovoid. The size of these cells varies but is not greater than that of the neurosecretory cells. In the central region of the corpora cardiaca are found fibres orientated dorsoventrally. Among them, irregular spaces exist.

Corpora allata

These are formed by epithelial, columnar, pseudostratified cells with a basal membrane (external). Some caliciform cells are widespread among the epithelial cells and, towards the interior, amorphous cells with diffused cellular borders and conspicuous large nuclei of granular chromatin are observed. In this central mass of cells, relatively large intercellular spaces and little connective tissue are observed. In the calyx region are nerve fibres, which intermix with the amorphous cells of the central mass.

Hypocerebral ganglion

This is surrounded by an extension of the connective tissue that surrounds the corpora cardiaca. Below it and in the region where the ganglion does not communicate with the corpora cardiaca, one or more layers of subcylindrical cells are observed. The neuropile is reduced to axons, which give rise to the posterior and anterior recurrent nerves. The cortex is formed by neurons of regular size (Guzmán *et al.*, 1974).

Reproductive System

Testes

Histology of the immature testis

In juveniles, the testes (t) are two almost transparent, ovoid masses, partially covered by adipose tissue, which stains with Sudan III (Fig. 11.3-18). The surrounding adipose tissue is formed by large cells, some with sporadic vacuoles and others with homogeneous cytoplasm. The nuclei are large, with granular chromatin, and are easily observed in the cells that have vacuoles. Their location is always parietal. This adipose tissue (ad) also surrounds each tubule and is sometimes reduced to a sheet of planar cells (Guzmán *et al.*, 1970a).

In histological section, stained with haematoxylin–eosin, transverse and longitudinal planes of seminiferous tubules (sf) are observed. A planar epithelium, which seems to invaginate towards the interior, divides the tubule and its external part into lobuli. In the base of these lobuli, a triangular cell of greater size than the rest is observed and, in the division itself, fusiform nuclei are seen. In each lobule, cells whose nuclei are always in the same phase of the meiotic process are observed. This facilitates the delimitation of the lobuli (in agreement with De Wilde (1974) about Acrididae). Each tubule ends in an efferent duct, whose wall is formed externally by the prolongation of the epithelium that covers the tubules. Towards the lumen, cylindrical cells with ovoid nuclei are observed. Adipose and unstructured tissue occurs amongst the tubules.

The efferent duct in transverse section is externally surrounded by two or more layers of fusiform cells. Towards the lumen, the tissue is structured by several layers of cells with diffuse contours and by nuclei with granular chromatin. This lumen is circular and contains an amorphous, eosin-positive substance. No spermatozoa were observed in sections, thus confirming the immature state of the tissue.

Histology of the mature testis

Macroscopically, there is no difference from the immature testis, except a greater opacity. Histologically, adipose tissue surrounds the whole testis in the same form as in the immature state, but it is more abundant than that observed in the latter.

Internally, seminiferous tubules are seen in transverse, oblique and longitudinal planes. The internal structure of the tubule is similar to that observed in the immature phase.

The vas efferens is also similar to that of the immature state. In this case it has a spiral tubular structure, externally surrounded by one or more sheets of planar epithelial cells with ovoid nuclei. Towards the lumen, it has one or more layers of

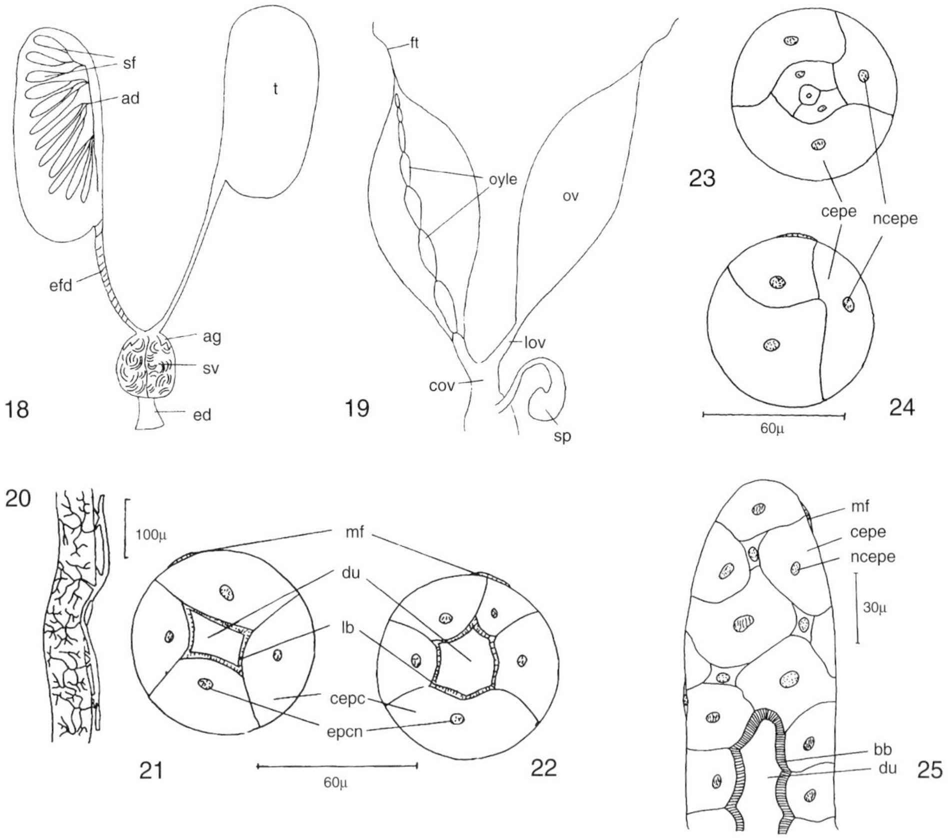

Fig. 11.3. 18, Male reproductive system. 19, Female reproductive system. 20, Tracheation of the Malpighian tubule. 21 and 22, Transverse view of the proximal region of the Malpighian tubule. 23 and 24, Transverse view of the apical region of the Malpighian tubule. 25, Oblique view of the Malpighian tubule. See text for abbreviations.

cylindrical cells with contours that are not well defined and with spherical nuclei. Clusters of spermatozoa can be detected (Guzmán *et al.*, 1970a).

Seminal vesicle

Histology of immature seminal vesicle

As previously mentioned, this spherical, transparent organ is located in the medial inferior part of the abdomen. At a low magnification (315×), a saculiform structure of thick walls is observed and, in a more advanced state, still immature, this structure presents invaginations, bridges and remaining cavities, which in the future could be the tubules of the mature vesicle (sv). Two zones are distinguished:

1. External zone. Two or three sheets of fusiform cells are concentrically located and become more cylindrical towards the lumen. The nuclei are spherical or oblong, with compact or granular chromatin.
2. Internal zone. This is a zone of cells of diffused contours with large nuclei and granular chromatin. The cells are arranged in a perpendicular form to those of the first zone. In the layers that confine the lumen, the cells become cylindrical and most of them have large, compact nuclei.

With greater development, deep invaginations of the wall of this zone form internal bridges. In this arrangement, the internal zone of the wall is located between both sides of the external zone.

Histology of the mature seminal vesicle

Macroscopically, the seminal vesicle is covered by a transparent membrane, through which the parallel tubules are seen. In the histological section, it can be seen that each tubule ends in a vas efferens (Du Porte, 1964), all of which join to form the vas deferens. Each vesicle is surrounded by a membrane of fusiform, planar cells. Inside the vesicle, transverse and oblique sections of tubules are observed. These were classified in two groups: those that present spermatozoa and those that do not. The wall of the tubules with spermatozoa is formed by an external layer of fusiform cells with long nuclei and another internal layer formed by cylindrical cells with spherical nuclei. In the lumen, there is an eosin-positive substance, in which clusters formed by spermatic heads are seen. The tails of the spermatozoa were not observed in the section. The same histology was observed in the tubules without spermatozoa. Among the tubules, widespread cells are seen. Some are spherical and others fusiform.

In medial transverse section of the seminal vesicle, all the tubules end in a common cavity and have a digitiform appearance in their borders. Striated muscle fibres are around this lumen and among the tubules. Within the lumen, heads of sperm form parcels in an amorphous eosin-positive substance.

A structure corresponding to the ejaculatory sack proximally and to the penis distally is observed. Its surface is covered with scaly transparent structures of unknown composition, which become wider and blunt towards the body cavity. When stained with acetic lactic orcein, each structure appears to have a circular extended base and a sharpened tip. In the dissection, the ejaculatory sack forms a tubular structure through which the ejaculatory duct (ed) passes. The muscular tissue that forms both the ejaculatory sack and the penis is composed of clusters of striated fibres pleuridirectionally arranged. The ejaculatory duct is composed of a sheet of planar cells with poorly staining nuclei. In its lumen, an eosin-positive amorphous substance, like that observed in the tubules, is seen. Spermatozoa were preferentially located towards the walls and at the base of the duct (Guzmán *et al.*, 1970a)

Frotis of spermatozoa

A frotis is made with spermatozoa from the spermatheca. The spermatozoa were 500 μm long, with heads approximately 25 μm long (Giemsa stain (Davenport, 1964)). In a physiological solution for insects (Patton, 1963), active flagellar movements were confirmed.

Ovaries

Histology of the immature ovary

In the immature state the transparent ovaries are formed by groups of fusiform ovarioles of the panoistic type (Fig. 11.3-19). In longitudinal section, the immature ovary (ov), and each ovariole (oyle), is covered by a layer of planar epithelial cells. Proximally, each ovariole ends in a pedicellum, which opens towards the oviduct (cov) (Du Porte, 1964). At its distal extreme, a terminal filament (ft) is formed by cubical cells with relatively large nuclei. The germarium, with cells of greater size, lies adjacent. These cells correspond to the closely packed oocytes, with nuclei that stain poorly with haematoxylin. In the vitellarium, oocytes undergoing maturation increase in size, including nucleus and cytoplasm, maintaining Hertwig's relation. The nucleolus gradually disperses and reaches the maximum in the pedicellum. Each oocyte is surrounded by a sheet of planar follicular cells. Proximally, each pedicellum is made of cylindrical cells with basal nuclei, covered externally by a layer of planar cells, which ensheathe all ovarioles. The calyx is formed by an external layer of fusiform cells with spherical nuclei and by an internal layer of cylindrical cells with a basal nucleus. The lateral oviduct (lov) is histologically similar to the calyx (Guzmán *et al.*, 1970a).

Histology of the mature ovary

The mature ovaries contain yellowish ovarioles, which are larger than the immature ones. Each ovary is covered by a transparent membrane. The sack-like spermatheca duct (sp) ends in the bursa copulatrix.

In longitudinal section, oocytes of different sizes were observed. Using the techniques previously described, it was not possible to visualize the whole cytoplasm; only dispersed vitelline plates were seen (De Wilde, 1974). Each ovariole has one external cellular sheet similar to that described for the immature ovariole. Within the ovarioles, linearly arranged oocytes gradually increase in size as they approach the pedicellum. The larger ones have an external layer of cylindrical cells; these possess a large nucleus, granular chromatin and a compact spherical nucleolus. The cytoplasm contains vitelline placoid structures. Nuclei did not stain, even using Feulgen stain. In the smaller oocytes, the external layer has planar cells and its cytoplasm is characterized by having a protein vitellum. In the pedicellum region near the oocytes, transverse fusiform cells and highly vacuolated amorphous cells were observed.

The duct of the spermatheca ends in the bursa copulatrix. The spermatheca has an ovoid morphology and the duct is sinuous. In the histological section, the spermatheca has an external layer of planar cells, followed by two or more layers of striated muscular fibres, and towards the lumen it exhibits several layers of cylindrical cells, which become cubical towards the lumen (Guzmán *et al.*, 1970b).

Excretory System

The Malpighian tubules, which are part of the insect excretory system (in conjunction with the proctodeal and rectal epithelium), are found between the mesenteron and the proctodeum, freely projecting into the haemocoele. The primitive number of tubules is considered to be six and they are always multiples of two.

Morphology and cellular structure of the Malpighian tubules

The Malpighian tubules in *C. armatus* Bl. are arranged in six groups, or clusters, in symmetrical form, at the junction of the mesenteron and the proctodeum. Each cluster has 26 tubules, which meet in an inconspicuous channel, which communicates with the lumen of the digestive tube. Each tubule is provided with a tracheolar ramification that branches along the full tubule length (Fig. 11.3-20). The length of the tubules is 8 to 12 mm, depending on the development state of the individual. Isolated tubules in physiological saline (Meisenheimer's) show especially active movements (Delpin, 1972).

Transverse section in the proximal region of the tubule

Striated muscular fibres (mf) and respective nuclei (and some tracheolar ramifications) are found in the wall of the tubule (Fig. 11.3-21, 22). On the inner surface, a basal membrane (lb) supports large cubical, epithelial cells (cepc, four to five cells in each section) with nuclei displaced towards the lumen (du). The nuclei (epcn) are compact and sometimes found in initial mitotic states. Large chromosomal bodies are observed (Demerec, 1950). The cytoplasm of these cubical epithelial cells is highly granular, as demonstrated with Tionina stain. The granular aspect is probably due to crystals of urates and other catabolites (Berkaloff, 1960). The luminal surface of the cubical epithelial cells has a brush border (bb, Fig. 11.3-25) with a length of approximately one-quarter to one-third of the cell's height (Delpin, 1972).

Transverse section at the apex of the tubule

Depending on the section, approximately three to nine cells, which correspond to the blind extremity of the tubule, are observed. From the exterior, striated muscle fibres, basal membrane (clearly seen with fast green stain) and cubical epithelial cells are distinguished (cepe, Fig. 11.3-23, 24). The muscle fibre arrangement is helicoid. They show the same aspect and form in oblique, longitudinal and transverse sections. These striated muscle fibres are accompanied by tracheolar ramifications, which ensure adequate oxygenation. The cuboidal epithelial cells have nuclei (ncepe) in initial phases of mitosis and a granular cytoplasm. In oblique sections in the distal zone and close to the middle of the tubule, small cells have a substellar form and correspond to the cubical epithelial cells, where the section included the nucleated region (Fig. 11.3-25) (Delpin, 1972).

Eggs: Morphology and Chorion

The eggs measure approximately 3–4 mm in length. They are fusiform, yellowish white and of irregular surface. To obtain more details of the chorion, they were observed by scanning electron microscopy (SEM).

The egg surface presents a complex structure, externally sculptured by a net of cells. Most of them are hexagonal, with less frequent pentagonal cells inserted between them (Fig. 11.4-26, 27). With greater magnification the borders of the raised cells are observed, giving a hexagonal and

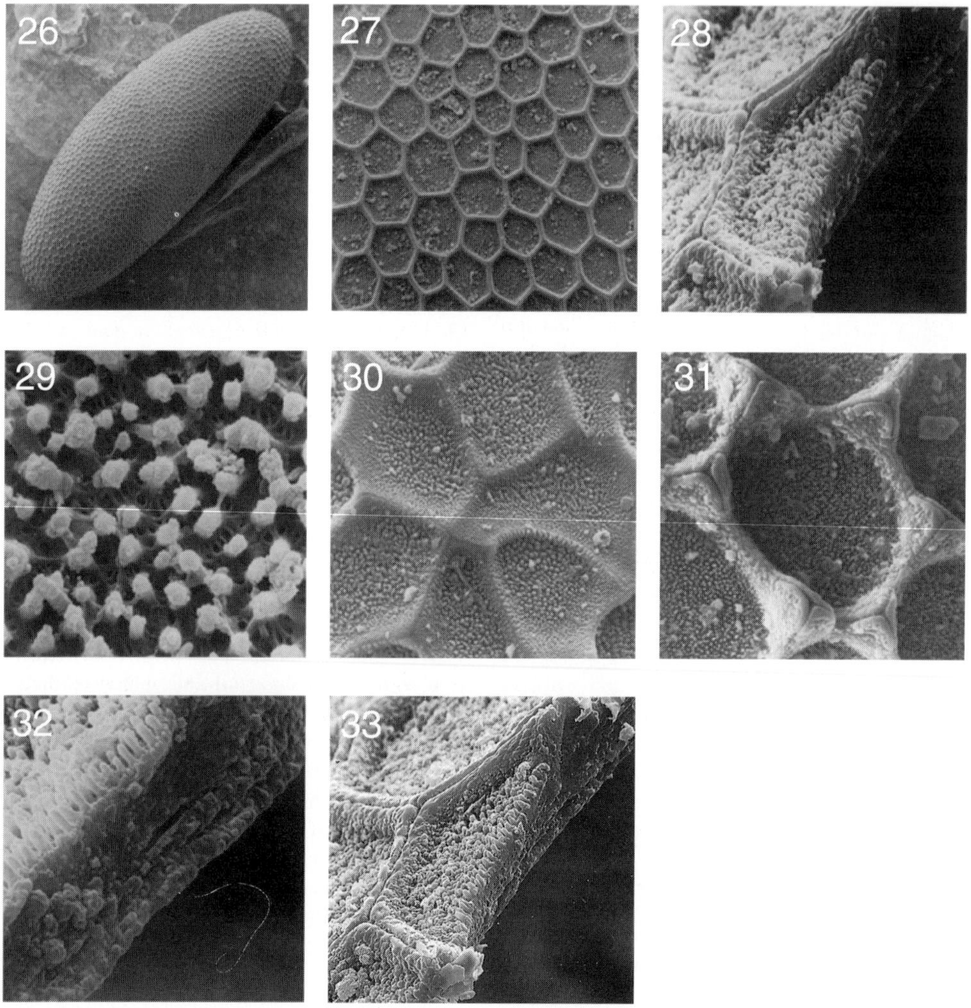

Fig. 11.4. 26, Panoramic view of the egg (SEM). 27, External sculpture of egg (SEM). 28, Magnified view of of the egg surface (SEM). 29, Inner surface of egg chorion (SEM). 30, Micropilar cells (SEM). 31, Vegetative pole cells (SEM). 32 and 33, Fractured edge of chorion showing layering and sculptured inner surface.

pentagonal relief aspect (Fig. 11.4-28). The interior of the cells has a trabecular appearance with airspaces and superficial and functional air-pyles (Fig. 11.4-29). Towards the animal pole of the egg, which corresponds to the micropilar region, there is a rosette-like terminal structure formed by five to six pentagonal and hexagonal cells. Towards this region, the suture lines in relief become less distinguishable (Fig. 11.4-30). The vegetative pole does not have any special structure, except that the terminal cells have slightly more elevated borders than the rest of the surface (Fig. 11.4-31).

By breaking the chorion, a clear division between the exochorion and endochorion is seen. The exochorion has a trabecular aspect, while the endochorion is more complete (Fig. 11.4-32).

Oviposition

In the laboratory, gravid females initially search for an appropriate place for laying, often in the corners of the rearing cages. After approximately 1 or 2 weeks, they lay anywhere on the sand in a disorderly way. This may be the result of a restricted habitat. Oviposition occurs in humid sand at 3–4 cm depth.

Maintenance and Rearing

Maintenance

About 40 individuals were collected in humid shady places in Concepción. In the laboratory, they were placed in wooden cages, covered by metallic mesh lids for good ventilation. The cages contained sand previously sterilized at 80°C. The environment was controlled for temperature (6–8°C) and relative humidity (80–90%). In order to maintain this humidity, water was sprinkled once a week.

Each cage had two Petri dishes (3.5 cm dia.), one for food and the other for water. Cage maintenance was done once a week, which was sufficient to maintain the food in good condition. The amount of sand was adjusted to maintain a homogeneous humidity and optimum ventilation.

Food requirements

Solid food

The 'red cricket' is omnivorous and it can be fed with concentrated food pellets for dogs, cats, etc. and pieces of apples and the like. It is also possible to feed it with oats or ground wheat, partially augmenting the diet with cheese.

It is important to monitor protein content, and especially to provide high quantities of protein for females prior to egg production. Pellets, milk powder, oats and Nestum (Table 11.1) are used in the laboratory. In the feeding dishes, the following combinations, all in equal quantities, are presented for a weekly diet: (i) pellets, oats, milk powder and Nestum; (ii) oats, Nestum and milk powder; and (iii) pellets and oats.

The food is provided in relation to the quantity consumed, based upon an average of three individuals in 1 week. The greatest preference was shown for the pellets (91.6% acceptance), followed by milk powder (45.8% acceptance) and Quaker Oats and Nestum (37.5% acceptance).

With the above diet, maintenance and care is easy and bioassays are more efficient. Covering with a metallic net is also important, since it helps to avoid excessive condensation of water inside the habitat, as occurs in closed plastic cages. Moreover, it also precludes the need for antimycotic additives, such as the ascorbic acid used by Angulo and Weigert (1978).

Humidity requirements

In the natural environment where the 'red cricket' is found, the relative humidity oscillates between 80 and 90%. It seems that, due to its primitive condition, this animal has a deficient water retention mechanism, which incapacitates it from living in the arid, dry fields of the north of Chile.

Rearing

Based upon their behavioural characteristics, it is necessary to take some precautions in 'red cricket' rearing, as mentioned in the following paragraphs (Aracena and Angulo, 1989).

A common rearing cage

Four crickets were placed per plastic or wooden cage of 32 cm × 21 cm × 10 cm, filled up to one-third of its height with sand and pine-needle

Table 11.1. Nutritional contents (percentage) of each food used. Data obtained directly from the food boxes corresponding to each commercial product: Pellet Can-Can, Molinera El Globo, Nestum Nestlé, Milk Powder Calo and Quaker Oats.

	Pellets	Milk	Oats	Nestum
Fat	7.0	26.0	8.2	5.3
Proteins	23.0	27.2	12.5	10.5
Carbohydrates		37.3	64.0	71.2
Fibre	5.0		2.0	4.4
Humidity		3.5	13.3	6.0
Others		60		2.6

humus, and supplied with the required food and humidity. Fresh food, such as fruits and cheese, was not used, as in Angulo and Weigert (1978), since it readily rots in the humid environment of the rearing cage. As the humidity requirement is so high, it is convenient to add periodically to the substrate a 1.25% w/v sorbic acid solution (its action is antimycotic) until approximately the relative humidity previously mentioned is reached.

Oviposition cage

Another plastic cage, half filled with sand, is set into the cage already described. In it are placed a small Petri dish with water and two cardboard cylinders to allow hiding. For separating the eggs, the sand is periodically inspected, without disturbing the crickets.

Egg incubation

Recently oviposited eggs are arranged in a tray with sand and pine humus. Once the stratum is well saturated with sorbic acid solution (1.25% w/v), they are maintained at 25–27°C. It is necessary to periodically control the relative humidity. After approximately 30 days, eclosion occurs.

Ecdysis (moulting)

In ecdysis, the exoskeleton opens dorsally in the medial line of the pronotum and then continues to the meso- and metanotum (ecdysis line), allowing the emergence of the animal with its new exoskeleton. The process lasts 2–3 h. When the nymph emerges, it preferentially seeks a dark place while its white–rose-coloured exoskeleton becomes hardened and pigmented. This occurs in approximately 24 h. During this time, it acquires a dark purple colour or other colour characteristic of its nymphal stage. However, the tegmen is initially orange and then becomes dark. It is interesting to observe that, when isolated from other crickets, the animal eats its exuvium.

Growth and Development

Histological technique for incubated eggs

The following histological technique has been developed to analyse the different states of embryonic development of the Orthoptera, using eggs of *C. armatus*, after 21 days of incubation (25–27°C), as an example (Angulo, 1969).

In acridid eggs, besides the two sheets of maternal covering, there is an internal wax-like layer associated with the vitelline membrane and external exochorion (Dave, 1965). In *Acheta* (Gryllidae), the water absorption period commences with the disintegration of the maternal epicuticle, and therefore it can be concluded that it is this layer which prevents the water absorption (Dave, 1965). The elimination of this wax-like layer, necessary for good staining of the embryonic states, has been attained using a prefixation technique, added to the classical histological process for this material (von Borstel and Linsley, 1959). The eggs are put in a capsule containing sodium hypochlorate (3% w/v) for 5 min and then washed copiously with distilled water to eliminate the excess sodium hypochlorate. Afterwards, they are immersed in Kahle solution (von Borstel and Linsley, 1959) and punctured to permit penetration of this fixative.

Description of the instars

Instar I

The colour of the body is whitish, the pronotum semicircular and 2.5 mm in length. The intersegmental membranes are visible between each abdominal tergite, particularly between the narrower sides. The antennae have a length of 11 mm and are 2.6 times the body length.

Instar II

The colour of the body and pronotum is as above. The antennae are approximately 25 mm long and are 2.5 times the body length.

Instar III

The colour is dark brown and somewhat reddish, with only the hind legs brownish red. The suboval pronotum is 3.5–4.0 mm long. The antennae are approximately 26 mm long and are 2.0 times the body length.

Instar IV

The colour is brownish red; the suboval pronotum is 5.0–6.0 mm long. The antennae are 28 mm long and are approximately 1.9 times the body length. The wing buds in the respective lateral depressions of the meso- and metanotum are seen externally through the transparent tegumen.

Instar V

The colour is dark brownish red; the suboval pronotum is 6.0–7.0 mm long. The antennae are 32 mm long and are approximately 1.8 times the body length. Externally, each wing bud has lengthened to exceed the length of its respective segment.

Instar VI

The general colour is dark reddish; the suboval pronotum is 8.0–9.0 mm long. The antennae are 34 mm long and are approximately 1.36 times the body length. The wings exceed the corresponding segment in length, but the anterior wings do not cover the posterior ones.

Adult

The colour is dark reddish to dark; the suboval pronotum is 9.0–10.0 mm long. The antennae are 43 mm long and are approximately 1.2 times the body length. The first pair of wings (tegmina) completely covers the second pair. Only the last third of the first abdominal segment can be seen.

Growth and metabolic features of developmental stages

The progression through the six nymphal stages of 'red cricket' growth, as well as the adult state, are plotted according to pronotum length in Fig. 11.5-34. The outstanding features are the following:

1. *Individual variation.* Individual variation in pronotum length only becomes evident from stage III. Above this stage, the range of individual variation remains constant for upper and lower limits. In stage III, the variation range is 0.75 mm; in stages IV, V, VI and the adult, the range is 1.0 mm.
2. *Absence of class intervals or measure ranges.* In the special physiological stages (E1–E4), metabolism and energy would be primarily invested in other essential processes necessary for postembryonic development. The growth rate would be slower, until it becomes structurally necessary for the next moult. In E1 and E2, organogenesis processes are activated by the reproductive apparatus. In E3 the processes initiating wing development are observed and in E4 sexual maturation begins. When the pronotum length is over 8.0 mm, the 'red crickets' are completely mature (Guzmán *et al.*, 1970a, b).

Outstanding features of nymphal instars

The most noticeable features in the instars of the 'red cricket' are the pronotum, the wings, the antennae and the body colour.

Pronotum

This is the most stable external morphological feature, since it is not affected by the physiological stages previously mentioned. Its distension apparently becomes increasingly difficult due to its solidity (demonstrated by the plateau in the upper curve of Fig. 11.5-35). The pronotum shape is more circular or subcircular in the early stages I and II. In the subsequent stages and in the adult, its shape tends to be oval and suboval, due to a greater growth in length rather than in the animal's width.

Wings

The wing development becomes noticeable in instar IV. In the depressions of the lateral borders of the meso- and metanotum, wing buds are seen, which in stage V are externally visible. The length of these wing buds does not exceed that of the respective segment, as seen in stage VI, where the mesothoracic wings are especially longer (protegmina), and finally, in the adult state, only the posterior third of the first abdominal segment is visible.

Antennae

The relation between the length of the antennae and the length of the 'red cricket' body is an acceptable criterion for each stage, but it has two disadvantages: on the one hand, they have a very extensible abdomen, especially the egg-replete females, and, on the other hand, the antennae are easily damaged, leading to errors in classification.

In comparing the growth of the antennae with that of the pronotum (Fig. 11.5-35) it is observed that antennal growth is affected by physiological

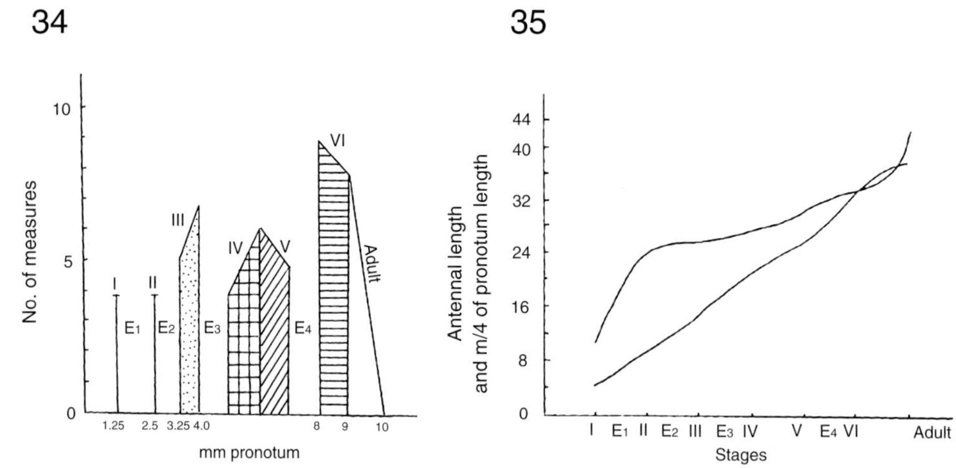

Fig. 11.5. 34, Graph of pronotum length versus relative number of measures of different stages of development. 35. Graph of the relation of pronotum length and antennal length, versus different ontogenetic stages of development (instars). Upper curve plots pronotum length (in mm/4 of y-axis scale); lower curve, antennal length.

stages EII and EIII, as seen in the most depressed region in the curve of the antennal growth.

Colour

In some specimens examined, the colour described for each stage was constant with respect to the first two instars. In the rest of the stages, as in the adult one, there was a tendency to melanism and genuine darkening of the reddish colour. Consequently, some specimens were completely dark in their general colour. However, avoiding the consideration that the pigmentation increases with senescence, the isolated geographical population stocks at Chiloé (42° 30′ S, 73° 55′ W), Mocha Island (38° 22′ S, 73° 56′ W) and others present this phenomenon.

Behaviour

The following observations on the behaviour of the 'red cricket' are useful for understanding the problems that appear in the laboratory during rearing and maintenance.

Cannibalism

In laboratory rearing, moderate cannibalism only occurred in later instars. Thus, the nymphs of stage VI and the adults have a great tendency to cannibalism, presumably due to the physiological demands of sexual maturation and egg development in females. It is optimal to have only four crickets (two males and two females) in each rearing cage, since overcrowding leads to aggression and partial or general injuries. These, in turn, increase the possibility of moderate cannibalism.

Moreover, if the diet supplied is limited or of low protein content, cannibalism is easily induced (Angulo and Weigert, 1978).

Building of galleries

The crickets start investigating the substrate and tend to make their excavations at the corners of the rearing cages. Once they have found a place, they push the head into the soil, make a rapid up–down movement similar to that of a shovel and, with the two pairs of hind legs, clear the entrance with rapid kicking. They continue opening this tunnel with the head, excavating 4–5 cm, and then excavate horizontally. Deeper sand is vigorously cleared with the last pair of legs. This constitutes one of the most notable displays.

A mixture of plaster was used to obtain internal moulds of the galleries. Two stereotyped shapes are constructed: those with a blind bottom (Fig. 11.6-36) and others with double entrances (U-shaped galleries, Fig. 11.6-37, 38). The blind-bottom galleries built at the cage corners have a sand promontory at the entrance, which has a 'trench' appearance. Those built out on the open surface have smooth, flat entrances. The blind-

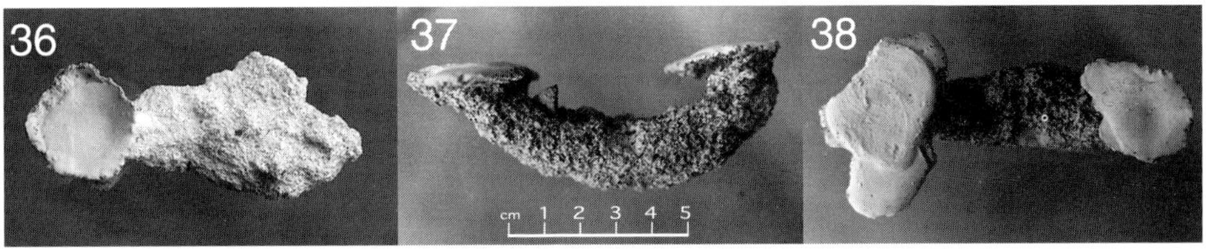

Fig. 11.6. 36, View from substrate surface of casting of blind-end gallery. 37, Side-view of casting of U-shaped gallery. 38, Surface view of casting of U-shaped gallery.

bottom galleries vary from 5 to 7 cm in depth, with an entrance diameter of 1.5–2 cm. The U-shaped galleries are 7–8 cm deep, with 3–7 cm between each opening. A third kind of gallery is that built under the food dish, about 2 cm in depth and with a diameter of 4 cm at the entrance.

Jumping capacity

In the early instars, the 'red cricket' is capable of jumping very high (50 to 60 times its body length). However, in the following instars, especially in the last two and as an adult, it seldom jumps, and then only weakly. This is due presumably to the structure and weight of its body, which, in the early instars is lighter and easily impelled by the hind legs. Jumping clearly serves as an escape mode from predators in the early instars.

Copulation

Copulation lasts from 45 to 60 min, and occurs in relatively dark and safe places. The female mounts the male, which remains quiet. In the rearing cages it is frequently observed.

Aggression

Adult *Cratomelus* of the same sex are highly aggressive if housed together; often the interactions end in mutual elimination. For example, when four mature males and one immature male were put in a cage (stage V according to Angulo and Weigert, 1978), a dead mature male was found 6 days later and the cage was in great disorder. Two days later another pair of mature males died and finally an adult and immature individual remained. This was not common when equal numbers of each sex were housed together, as indicated in Table 11.2.

A similar tendency occurs with the females. Upon placing six females in a cage, after 20 days only one survived. It seems that living together is easier when the individuals are of different instars, since adult individuals, independent of their sex, can live with nymphs without great difficulty (Aracena, 1988; Aracena and Angulo, 1989). A summary of survival times (in days) for various numbers of individuals (of both sexes) housed together is presented in Table 11.2.

Table 11.2. Duration in days after which animals died when housed in groups with numbers of each sex indicated (♥ = lived more than 20 days; # = dead male; + = dead female, – = no data).

Number of males	Number of females			
	1	2	3	4
1	♥20	+12	–	–
2	#2	♥20	–	–
3	#4	–	♥20	–
4	–	–	#3	+1

Generally, in most animals, intraspecific fighting seldom ends in mortal aggression since less aggressive attitudes are usually adopted, such as submission behaviour (Vaz-Ferreira, 1984). The interactions of *C. armatus* appeared to be narrowly focused on establishing a basic unit that results in a couple; the result is a female and a male that share the same territory without fighting and without competition from others of either sex. Moreover, this facilitates copulation and subsequent female oviposition.

References

Angulo, A.O. (1969) Técnica histolùgica para huevos incubados de Orthoptera. *Bol. Soc. Biol. Concepción (Chile)* 41, 207.

Angulo, A.O. and Weigert, G. (1978) Estados ninfales, etología y crianza de *Cratomelus armatus* Blanchard (grillo rojo) (Orthoptera: Gryllocrididae). *Bol. Soc. Biol. Concepción (Chile)* 51, 41–49.

Aracena, C.V. (1988) Conducta e interacción social de 'grillo rojo' *Cratomelus armatus* Blanchard, 1852 (Orthoptera: Gryllacrididae). Thesis, University of Conceptión, Chile, 42 pp.

Aracena, C.V. and Angulo, A.O. (1989) Técnicas y datos para la crianza de grillo rojo *Cratomelus armatus* (Bl., 1852). *Comun. Mus. Reg. Concepción* 3, 61–64.

Benoit, I.L. (1975) Sensilos antenales y cercales de *Cratomelus armatus* Blanchard (Orthoptera: Gryllacrididae). *Bol. Soc. Biol. Concepción (Chile)* 49, 33–38.

Berkaloff, M.A. (1960) Contribution à l'étude des tubes de Malpighi et de l'excrétion chez les insectes. Observation au microscopie électronique. *Ann. Sci. Nat., Paris (Sér. 12). Zool.* 2, 869–947.

Blanchard, E. (1851) Ortopteros. In: Gay, C. (ed.) *Historia Física y Política de Chile*, Vol. VI, *Zool.* Paris, France, pp. 37–39.

Borror, D.J., Triplehorn, C.A. and Johnson, N.F. (1989) *An Introduction to the Study of Insects*. 6th edn. Saunders College Publishers, New York, 825 pp.

CSIRO (1991) *The Insects of Australia*, Vol. 1. Cornell University Press, Ithaca, New York.

Dave, K.G. (1965) *Reproduction in the Insects*. Univ. Reviews in Biology, Oliver and Boyd, Edinburgh and London, pp. 23–25.

Davenport, H.A. (1964) *Histological and Histochemical Techniques*. W.B. Saunders, New York.

Delpin, M.E. (1972) Análisis estructural de los túbulos de Malpighi en *Cratomelus armatus* (Bl.) (Orthoptera: Gryllacridiidae). *Bol. Soc. Biol. Concepción (Chile)* 44, 1161–1167.

Demerec, M. (1950) *Biology of Drosophila*. Wiley and Sons, New York, 632 pp.

De Wilde, J. (1974) *Physiology of Insects*. Academic Press, New York, pp. 10–58.

Du Porte, E.M. (1961) *Manual of Insect Morphology*. Reinhold, New York, pp. 200–214.

Du Porte, E.M. (1964) *Manual of Insect Morphology*. Reinhold, New York, Chapman and Hall, London, pp. 155–184.

Guzmán, E., Angulo, A. and Delpin, M. (1970a) Análisis estructural del aparato reproductor masculino de Gryllacridiidae. *Bol. Soc. Biol. Concepción (Chile)* 42, 137–151.

Guzmán, E., Angulo, A. and Delpin, M. (1970b) Análisis estructural en ovario de Gryllacridiidae. *Bol. Soc. Biol. Concepción (Chile)* 42, 167–175.

Guzmán, E., Angulo, A.O. and Delpin, M. (1974) Histomorfologia del sistema nervioso y neurosecretor de *Cratomelus armatus* Bl. (Orthoptera: Gryllacridiidae). *Bol. Soc. Biol. Conceptión, Chile* 48, 231–242.

Johns, P.M. (1997) The Gondwanaland weta: family Anostostomatidae (formerly in Stenopelmatidae, Henicidae or Mimnermidae): nomenclatural problems, world checklist, new genera and species. *Journal of Orthoptera Research* 6, 125–138.

Maynard, D.M. (1962) Organization of neuropil. *American Zoologist* 1, 79–96.

Patton, R.L. (1963) *Introductory Insect Physiology*. W.C. Saunders, New York.

Smith, D.S. (1968) *Insects Cells, Their Structure and Function*. Oliver and Boyd, Edinburgh.

Snodgrass, R.E. (1925) *Principles of Insect Morphology*. McGraw Hill, New York, pp. 464–509.

Vaz-Ferreira, XX (1984) Etologia: estudio biológico del compartamiento animal. *Ser. Biología* No. 29. OEA, 150 pp.

von Borstel, R.C. and Linsley, D.L. (1959) Insect embryo chromosome techniques. *Stain Technology* 34, 23–26.

Part III
Ecology

12
The Ecology of Some Large Weta Species in New Zealand

Mary McIntyre
*School of Biological Sciences and Island Biology Research Programme,***
Victoria University, Box 600, Wellington, New Zealand

Introduction

Some New Zealand weta are unusually large and bulky. The name 'giant weta' is normally reserved for members of the genus *Deinacrida* (Anostomatidae: Deinacridinae), which includes the largest species. These reach up to about 45 g in weight for field-collected specimens, while the common tree weta, *Hemideina crassidens*, as a point of comparison, typically range from 4 to 8 g on mainland New Zealand. So-called 'giant' forms also occur in the tusked weta (Anostomatidae: Anostostomatinae) and cave weta (Rhaphidophoridae). This account provides an evolutionary perspective for their large body size and reviews information on the ecology of the Cook Strait giant weta (*Deinacrida rugosa*), wetapunga (*Deinacrida heteracantha*) and the giant tusked weta (*Motuweta isolata*) as case-studies (M. McIntyre, unpublished data unless otherwise attributed) of large weta. Information on the ecology of the Poor Knights weta (*Deinacrida fallai*) and Mahoenui weta (*Deinacrida mahoenui*) is also briefly reviewed.

Giantism and Island Isolation

The trend to giantism is a feature of isolated island faunas. The islands of New Zealand have been isolated geologically since the mid-Cretaceous period, and terrestrial biota carried adrift at the time of separation included ancestors of weta. Large size is seen in several New Zealand animal groups. Most notably, these include flightless birds, for example giant moa (Dinornithidae; extinct *c.* AD 1600), takahe (Rallidae) and kakapo (Psittacidae), a putative[1] giant gecko (Reptilia: Gekkonidae) and large land snails (Mollusca: Rhytidae) and other invertebrates. The weta are the largest and best known of the insects.

New Zealand is also unique in having no native land mammals, apart from three bat species, one of which is now believed to be extinct. Large size has evolved independently in several taxa. For weta, the scarcity of ground-living predatory endotherms with nocturnal habits and a well-developed olfactory sense has probably been the most significant factor.

* Contribution No. 10 from the Island Biology Research Programme, Victoria University of Wellington.
[1] Known from a mislabelled skin and accounts by Maori.

The absence of mammalian predators and competitors has also allowed the evolution in New Zealand of taxa that serve to some degree as ecological homologues of mammals that occur elsewhere (reviewed in Daugherty *et al.*, 1993[2]). In this regard, the giant weta are sometimes referred to as 'invertebrate mice'. Their nocturnal and more or less omnivorous habits, combined with large size, plus the fact that adults, even of the more arboreal species, may crawl about on the ground at night, make this an appealing, popular description to highlight a combination of characteristics more typically found in small mammals elsewhere.

The size of giant weta is best appreciated by body weight (Table 12.1), as this distinguishes the largest species from the more common tree weta more clearly than dimensions of hard body parts or body length. The two largest species (*D. heteracantha* and *D. fallai*) both reach weights more than double that of a house mouse occurring on mainland New Zealand (Murphy and Pickard, 1995).

Large body size in animals is often associated with ability to exploit high volumes of low-grade diet. While all weta are to some extent omnivorous, the largest species are primarily vegetarian. It seems that the absence of predatory mammals has made it possible for some New Zealand species to adopt this way of life, despite high costs in terms of mobility and concealment. The carnivorous *M. isolata* is an exception.

New Zealand also has some endemic animal species inhabiting alpine zones that belong to taxa not known elsewhere for this – notably parrots (Psittacidae), lizards (Scinciidae), cicadas (Hemiptera) and some grasshoppers and weta (Orthoptera). These have diversified into alpine habitats as the opportunity arose during an extended period of mountain building, brought about by tectonic activity in the Miocene/Pliocene period. Some of the *Deinacrida* species have also adopted alpine habitats, but they are not as large as the lowland ones.

Current Status

The largest weta species now survive, with one exception, only on small offshore islands, which lack most or all introduced mammalian predators and especially rats, or in mountain habitats above the tree line, where these are scarce (reviewed in Gibbs, Chapter 2, this volume).

Rats (kiore, *Rattus exulans*) were brought to New Zealand by Polynesian settlers 1000 years ago, possibly earlier (Holdaway, 1997), while a suite of other night-hunting predators, most notably Norway rats (*Rattus norvegicus*), ship rats (*Rattus rattus*), feral cats and mustelids (*Mustela erminea*, *Mustela furo* and *Mustela nivalis*), were introduced by European settlers in the past 200 years.

Gibbs (1998a) considers the large body size of weta, combined with a lack of appropriate defensive or escape behaviour and the distinctive odours of all weta in this family, to be the major predisposing factors to mammal predation and thus the decline of the species. The more ground-living species are especially vulnerable. Added to this is the fact that, although lowland giant weta may be active at night in temperatures down to 3–4°C, they face predators which maintain a high body temperature and have increased energy demand in these conditions. Furthermore, most indigenous predators, mainly birds and reptiles (reviewed in Gibbs, 1998a), are diurnal or at least rely extensively on visual, possibly also tactile, rather than olfactory senses to detect their prey.

Islands thus have dual significance with regard to large-bodied weta. While large size has evolved over geological time on the isolated archipelago of New Zealand, small offshore islands, formed mostly by rising sea levels in the post-Pleistocene era, now provide a refuge for some of the larger species, which can no longer survive in their former range.

The lowland giant weta include the better-known species. While one species (*D. fallai*) is considered endemic to the Poor Knights Islands, the others have all probably suffered severe range restriction as a result of both predation and habitat loss. These survive as relict populations on an offshore island that is relatively little exposed to human influence (*M. isolata*), an island with some introduced mammals (*D. heteracantha*), old farmland with exotic vegetation on offshore islands (*D. rugosa)* and mainland habitat remnants (*D. mahoenui*).

[2] Table 4 in this article erroneously names *D. rugosa*. This should be *D. heteracantha*. Body length, excluding legs, may reach 6–7 cm.

Table 12.1. Adult body size in lowland giant weta. The 'Raukumara' tusked weta is included as a point of comparison for *Motuweta isolata*.

	Females				Males			
	n	Wt (g)	Rht. lgth (mm)	B. lgth* (mm)	*n*	Wt (g)	Rht. lgth (mm)	B. lgth (mm)
'Giant' weta:								
Deinacrida rugosa†	97	20.5 ± 2.9 (26.6)	31.6 ± 2.0	54–67	43	9.5 ± 1.5 (13.5)	27.1 ± 2.1	42–49
D. heteracantha†‡	14	32.2 ± 6.3 (41.0)	50.5 ± 2.4	60–73	9	7.4 ± 1.4 (20.0)	47.4 ± 4.2	52–57
D. mahoenui§	7	14.1 ± 0.9 (15.0)	34.1 ± 0.6	n/a	3	8.9 ± 0.5 (11.3)	33.5 ± 1.0	n/a
Tusked weta:								
M. isolata†	58	18.7 ± 2.6 (25.0)	36.7 ± 5.7	50–70	51	14.6 ± 3.6 (22.8)	32.5 ± 2.0	46–65
'Raukumara' sp.‖	5	(3.4–3.5)¶	23.5 ± 2.0	30–36	5	3.3 ± 0.4 (3.7)	21.7 ± 1.9	29–36

* Dorsal midline of head → tip of abdomen (abdomen extensible but provides visual 'size' impression).
† Unpublished field data.
‡ Data from Meads and Ballance (1990); Meads and Notman (1993); Gibbs and McIntyre (1997).
§ Data from Richards (1994), P. Barrett, Wellington Zoo, personal communication.
‖ Data from McIntyre (1998a), P. Burge, Victoria University, unpublished field data.
¶ Two measurements only.
Wt, weight, mean ± SD (max); Rht. lgth, length of right hind tibia, B. lgth – body length; n/a, not available.

Giant Weta (*Deinacrida* spp.)

The *Deinacrida* species have a distinctive broad body with a rounded head and short mandibles (see Figs 2.7–2.12, Gibbs, Chapter 2, this volume). Females are conspicuously larger than males (Table 12.1), and it is the female size that is usually quoted for advocacy purposes. While female body weight in the larger species is impressive, it varies with development of the egg mass. Captive females that are unmated or not receiving suitable triggers for oviposition (as often seems to happen in captivity) will accumulate eggs and become (sometimes markedly) heavier than the maximum recorded in the field.

Some lowland species typically take refuge during the day in vegetation, while others sit on the soil underneath vegetation or other objects, although they all make nocturnal visits on to the ground, at least to oviposit. They are best sampled by searching with a spotlight at night, when they are active. They are easy to capture and handle, but, depending on the rarity of the species and the type of habitat, may be difficult to find or study.

The initial reaction of giant weta to a spotlight beam is usually to freeze. This is probably an effective response to disturbance by a potential predator which relies largely on visual cues. At night, they can be readily picked up, examined and returned to the site of capture with minimal disturbance. It is a remarkable experience to hold a large insect in the hand (Fig. 12.1) and obtain various measures with little need to subdue it. While this is often interpreted as docility, it is more likely to be a well-honed response to disturbance or novelty in an environment where, until the introduction of night-hunting mammals, this was appropriate.

While night searching is limited to accessible sites, the use of radiotelemetry has greatly extended appreciation of weta activity and use of habitat (Richards, 1994; M. McIntyre, 1992, unpublished data). Richards (1994) also used luminous 'beta lights' attached to weta as night markers to spot them on vegetation. The use of a night-vision scope has also made it possible to observe without the intrusion of artificial light, where access within range (about 5 m) is possible.

A mix of field data and captive observations indicate that there is a 3-year lifespan for active stages, with typically nine instars in *D. rugosa* (Ramsay, 1955; McIntyre, 1992, 1993), *D. fallai* (Richards, 1973) and *D. mahoenui* (Sherley and Hayes, 1993) and ten in *D. heteracantha* (Richards, 1973). The winter period seems to be a significant marker in this regard. There are two winters spent as a juvenile and a third as an adult. Most oviposition probably occurs through the autumn and winter. This situation is seen in field survey data as three distinct size (= year) cohorts, with the size range in juvenile cohorts depending on the time of year when measurements are made. There are probably most life stages present at all times, with overlapping adult generations. Nevertheless, as

Fig. 12.1. Female *D. fallai* on Tawhiti Rahi (Poor Knights Islands) (photo M. McIntyre).

might be expected, there are fewer weta seen over the winter and these tend to be adults. For *D. rugosa*, at least, juveniles are most conspicuous in the later summer and autumn and hard to find in winter. It seems that this may be not only because they are smaller and harder to see, but also because they are less vagile than adults and considerably less active in winter.

Cook Strait Giant Weta, *D. rugosa*

(see Fig. 2.10, Gibbs, Chapter 2, this volume)

This is a medium-sized ground-living species with a bulky body, relatively short legs (Table 12.1) and solitary habits. Ramsay (1955) described the morphology and growth stages. The species survives as remnant populations on three small offshore islands – Mana Is., 217 ha; Stephens Is., 150 ha; Middle Trio Is. (and two adjacent islets), 13 ha (see Fig. 2.9, Gibbs, Chapter 2, this volume) – in the Cook Strait area. These islands are naturally free of rats and the larger two have been deforested for farming.

On Mana Island, *D. rugosa* occurs in old pasture and coastal scrub. The island was farmed for over 100 years until 1987, when it acquired conservation status. It is now the subject of an ambitious programme to restore ecological communities typical of the coastal forest once characteristic of the adjacent mainland area (Department of Conservation, 1999).

Although the island is fortuitously without rats, it has had mice (*Mus musculus*), which were probably subject to periodic irruptions when the island was farmed (Pickard, 1984). Livestock were removed in 1986 and the mice eradicated in the winter of 1989. This was followed by a marked increase in nocturnal counts of giant weta (Fig. 12.2) seen on the ground or climbing on shrubs in the abandoned pasture (Fig. 12.3a). The corresponding, artificially high, initial densities (Table 12.2) reflect a temporary situation with few predators of weta on the island.

The gradual decline after an initial explosion in numbers (Fig. 12.2, Table 12.2) is expected as the island ecosystem becomes more diverse with the planting of forest and recovery and planned reintroductions of other native bird and reptile species. These include some predators of weta (Department of Conservation, 1999). Relatively high densities of weta in the meantime facilitate ecological investigation – both as a baseline for ecological management and as a contribution to knowledge of the large weta species.

Nocturnal activity is influenced by weather, in particular the ambient temperature (Fig. 12.4-9). There are notably more adults seen on still warm nights, although this is relative to time of year, recent weather and, in the medium term,

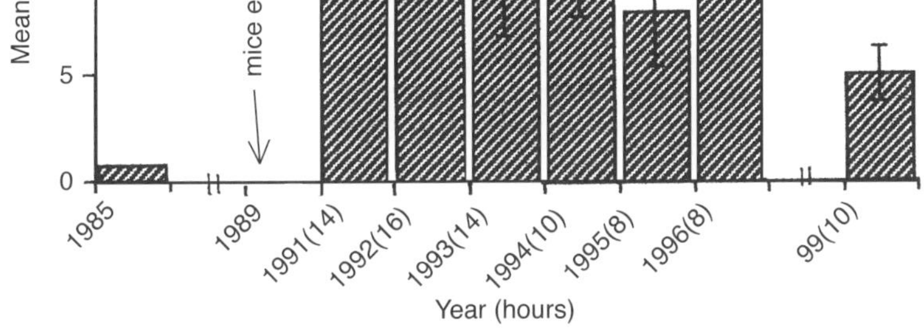

Fig. 12.2. Nocturnal capture rates of *D. rugosa* > 1.5 g, following the eradication of mice on Mana Island; data 1991–1999 from standardized 2-hour nocturnal counts over 3–7 nights in late February–early March.

Fig. 12.3. *D. rugosa* habitat in old farmland on Mana Is.; note mosaic of planted and naturally seeding shrubs coming though pasture grass; researchers estimate weta dispersal on the previous night (photo B. Robertson). b, *D. rugosa* female with radiotransmitter (photo B. Robertson, Victoria University). c, Faecal pellet from adult female *D. rugosa* (photo B. Robertson, Victoria University). d, Stephens Island; note deforested state (photo B. Robertson, Victoria University). e, Forest habitat of *D. heteracantha* on Little Barrier Island extends from sea level to 700 m; weta were recorded in the foreground scrub by Richards (1973) but are no longer found in this habitat (photo M. McIntyre). f, A large female *D. heteracantha* cryptic in tree fern foliage during the day (photo M. McIntyre).

population changes. In windy conditions, the weta tend to move on to low vegetation but stay inside the foliage, while moonlight has no demonstrated effect. Most weta are seen within about 2 h after dusk, which probably corresponds to the peak of their circadian period. Activity drops off notably after this on cooler nights.

Adults are frequently seen in male–female pairs on the ground. The smaller male invariably trails close behind (within antenna length) the female. Occasionally there are two or three males. One of these may eventually put one or both forelegs on the abdomen of the female. On vegetation there is often a male nearby and below the female. The orientation of males is distinctive and probably relates to olfactory cues.

Table 12.2. Density estimates for *D. rugosa* in the Weta Valley, Mana Island, following the eradication of mice. Estimates are derived from mark–recapture data obtained in nocturnal counts using the 'CAPTURE' procedure (USGS Patuxent Wildlife Research Centre, 1996).

Year	Survey nights	Weta* count	Est. no. ± SE	95% CI	Biomass[2] (kg ha^{-1})
1991	7	126	231 ± 31.5	185–313	42
1996	4	68	148 ± 28.5	108–225	27
1999	5	38	88 ± 13.5	67–122	16

* Weta > 1.5 g.
† Based on mean cohort body weight for each of the 3-year classes.

Whether more adults are seen at night on the ground or on vegetation seems to be associated with pairing behaviour. This also varies with weather and the time of year. In warm, still, autumn weather, males appear on the ground conspicuously before females and the trailing behaviour becomes quickly evident. On other occasions, feeding is the dominant activity, most weta are seen on vegetation and these are mostly females or juveniles.

The trailing may be prolonged and is often aborted. There may also be a delay or inhibition of mating, as most copulation occurs during the day. Copulating pairs located in the daytime are well hidden beneath vegetation and, as also noted by Ramsay (1955), there are usually several spent spermatophores on the ground. Male and female stay together until the next night at least, when the male decamps. If there is a change to cool or wet weather, they may not part for several days. There is thus, by insect standards, a fairly prolonged and variable period of association, although there is no overt courtship.

Information on habitat use and dispersal was obtained by fitting some adult females[3] found at night with miniaturized radio transmitters (Fig. 12.3b). Associated males were identified with colour marks. During the day, weta take refuge on the ground sitting in a dry site on the soil

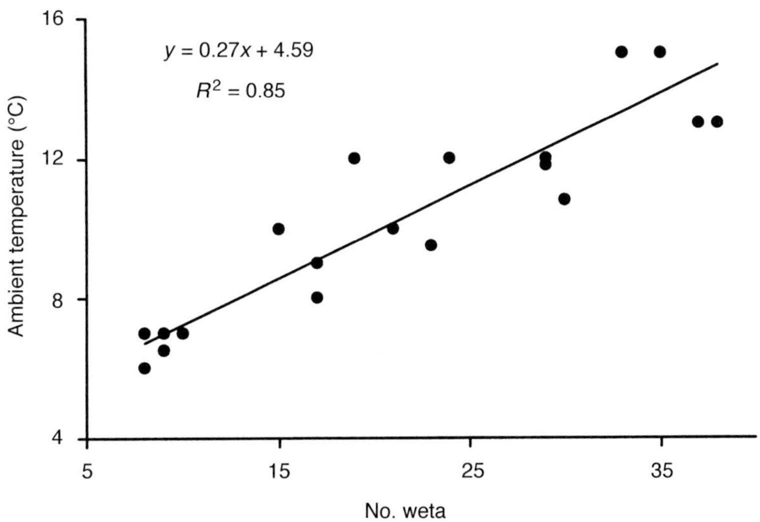

Fig. 12.4. Nocturnal counts of *D. rugosa* 1991–1993 in relation to ambient temperature; data from standardized 2-hour counts (n = 20 nights).

[3] The transmitters add about 1.4 g to weta weight. They are too bulky for the smaller males. A battery life of up to 27 days allowed for intensive short-term surveillance.

underneath vegetation or other objects. The rank pasture grass on Mana Is. provides well for this. Ramsay (1955), observing captive weta, notes that they are highly thigmotactic at this time. In the field, they tend to position themselves beside the base of a grass clump or piece of wood or in the small depressions produced by soil creep on sloping ground. Their positioning gives the impression that the weta is relatively protected from trampling as weight from above is diffused. This may have been significant with regard to the heavy flightless birds that characterized the prehuman biota of New Zealand. Suitable dense cover at ground level appears to be a prime habitat requirement. They are nevertheless unprotected from animals such as rats, and would be defenceless in a rat encounter.

The thigmotaxis disappears after dark and weta emerge from under vegetation on to tracks through the rank grass or climb on shrubs (including small trees, 3–5 m high, in reafforestation plots of canopy). They are invariably back on the ground by day. The distance between day refuges of individual weta provides a measure of overnight dispersal (McIntyre, 1993). The pattern for females seems to be that they stay within a small area (perhaps a metre radius) over several days and sometimes up to a week, and then migrate (maximum recorded 56 m) overnight to a new site and repeat the pattern. This seems to be associated with mating followed by oviposition and possibly a quiescent period, and then another migration, which is likely to attract a new male. The timing might be prolonged by inactivity in cool or wet weather and is probably much extended over winter and in the later reproductive life of the female. Once females leave a refuge site they do not return. Reports of captive adult weta repeatedly returning to the same day site to the extent of sitting with the same orientation (Ramsay, 1955; M. Meads, Wellington, 1998, personal communication) may be a product of confinement.

It appears that free-ranging adult females are quite vagile, attracting and mating with a number of males over several months and so dispersing both their genetic interests and their eggs. In contrast, those juveniles of a size amenable to repeat observations (> 1.5 g, instar 6+, according to Ramsay (1955)) tend to be sedentary. They migrate vertically on to vegetation at night, especially where the grass grows through the lower foliage of the shrub, but are rarely seen on larger trees, where they must climb up the trunk and then out on to branches, or moving along the ground (McIntyre, 1993).

There is a particular association with the shrub species *Ozothamnus leptophylla* (tauhinu or native cottonwood). This plant persists around cliffs and coastal margins but is eventually overtaken by other vegetation elsewhere. On Mana it grows through old pasture as an early successional species and weta densities are highest where this occurs (Fig. 12.3a). Weta are found on these shrubs at night throughout the year and yet are rarely seen on plantings of other early–mid-successional native shrub and small tree species. They feed on the new shoots and flowers of tauhinu and are also attracted to flowers of native flax (*Phormium* spp. and spear-grass (*Aciphylla squarrosa*). They also feed on clover, especially the flowers, and other 'improved' pasture grasses and introduced herbs.

The faecal pellets resemble those of a rat (Fig. 12.3c). However, they have a distinctive weta odour when fresh and sometimes also have longitudinal stripes, possibly produced by a rectal imprint or secretion. They comprise mostly dry fibrous material, which is very persistent. A mature female produces three to four pellets per day, each about 2 cm long and up to 10% of body weight when fresh. Pellets accumulate in the relatively dry environment under tauhinu bushes. They are also found singly along what appear to be walkways underneath the grass sward. Beauchamp (1990) investigated the possibility of using pellet counts to indicate abundance, but concluded that variable persistence makes this difficult – and those deposited in drier sites can probably long outlast the life of a weta! He also recorded giant weta pellets around 60% of the coastal margin and in a range of open habitat types. Subsequent searches for weta signs in a gully with the only patch of older mid-successional second-growth forest on the island have not been successful.

On Mana Island, it is possible to see four kinds of weta *(D. rugosa, H. crassidens, Hemiandrus bilobatus* (Anostostomatidae*)* and *Gymnoplectron edwardsi* (Rhaphidophoridae)) on a small shrub at night. This reflects the state of biotic communities in transition from farm to forest, with low numbers of species but temporarily high numbers of a few of these, including weta and small lizard species (McIntyre, 2000). This situation will change as the forest regenerates and there is a greater range of indigenous predators of weta present, including tuatara (Department of Conservation, 1999). Lower densities and stronger distinctions in habitat use by the different weta species are predicted. Information on habitat use by giant weta, in particular the association with tauhinu, supports the suggestion of Ramsay (1955) and observations of Meads and Notman (1992) of translocated weta on Maud Island that *D. rugosa* is a species of coastal and shrubland habitats and the forest edge. It will probably be excluded from much of its present habitat on the island as forest regenerates, but is expected to persist around forest margins and on exposed scrublands and coastal cliffs. Continued monitoring will be important in this regard.

On Stephens Island (Fig. 12.3d), *D. rugosa* coexists with tuatara (Reptilia; *Sphenodon punctatus*), which have probably been a major predator of weta through much of their evolutionary history. The ecology of the island is unusual in that, although the vegetation is in a very degraded state, rich nutrients of marine origin are deposited by vast numbers of sea birds, which visit the island to breed from July to late December. Here tuatara densities reach artificially high levels, with a maximum of 2000 ha^{-1} in two small areas of remnant forest (Newman, 1982). Tree weta, *Hemideina crassicruris*[4], also reach estimated densities of 5300 ± 2200 ha^{-1} (Moller, 1985) in the same forest, although these are arboreal and mostly out of range of tuatara, which typically do not climb. The status of *D. rugosa* living on the ground within this regime has not been assessed, but densities are clearly much lower than on Mana Is. Tuatara will eventually be reintroduced to Mana Is. (Department of Conservation, 1999).

There are also tuatara surviving in coastal forest on Middle Trio Island. Meads and Notman (1992) note that only two *D. rugosa* have been seen there in the past 45 years. J. Marris (Lincoln University, 2000, personal communication) collected a specimen from each of the rocky islets of North Trio and South Trio and made a brief unproductive search of likely habitats on the main island of Middle Trio in 1995.

Populations of *D. rugosa* have been successfully established on Maud Is. (309 ha; Meads and Moller, 1978; Meads and Notman, 1992), off the northern South Island, and Somes Is. (32 ha; McIntyre, 1998a) in Wellington Harbour. The former has had stoats but no rats, while ship rats were eradicated from Somes Is. in 1990. Both islands were previously farmed but are now reserved for conservation, and both have probably had giant weta in the past. Somes Is. and Mana Is. are open to visits by the public.

Wetapunga, *D. heteracantha*; Poor Knights Giant Weta, *D. fallai*; and Mahoenui Giant Weta, *D. mahoenui*

These three species spend the day above ground in vegetation. They all have longer legs in relation to a less bulky body (Table 12.1) than *D. rugosa*, and this seems to be associated with the more arboreal habits. They survive in very different situations ecologically.

The largest giant weta is *D. heteracantha* (Table 12.1; see also Fig. 2.7, Gibbs, Chapter 2, this volume). The Maori name 'wetapunga' refers to the embodiment of ugliness, and is formally reserved for this species (Scott and Emberson, 1999). This species survives only on Little Barrier Island, 3083 ha (Fig. 2.6, Gibbs, Chapter 2, this volume), although it has previously been more widespread (see Gibbs, Chapter 2, this volume). It occurs in habitat ranging from second-growth forest on the lower slopes of the island to mid-level tall kauri forest and undisturbed cloud forest above this up to 720 m (Fig. 12.3e). The island also has kiore, which are widespread and mostly active on the ground, and had feral cats until they were eradicated in 1980.

It appears that arboreal habits enable this species to coexist with the kiore. However, the fact that the weta are often in tall canopy, combined with the rugged nature of the island, makes it difficult to assess the state of the population – which is of particular interest, especially since cats were removed. Occasional sightings of weta or findings of the large dry pellets by conservation workers suggest that the weta are widely dispersed but occur in local patches and densities are low. Meads and Notman (1993) cite an encounter rate of one weta per 8.8 person-hours' searching time overall and one faecal pellet per 3 person-hours during daytime. The weta are also no longer found in low coastal shrubs (Fig. 12.3e) where they were

[4] This name has reverted to *H. crassidens* (Morgan-Richards *et al.*, 1995).

described by Richards (1973) as common (Meads and Ballance, 1990; Meads and Notman, 1993; Gibbs and McIntyre, 1997), and this is cause for concern. Present ecological knowledge is based on the small numbers reviewed in Gibbs and McIntyre (1997) and a further study, using radiotelemetry and a night-vision scope, in May–June 1997 (autumn–early winter) (M. McIntyre, unpublished data).

As for *D. rugosa*, there is probably a range of growth stages present at any particular time but with a seasonal pattern, at least in those seen from ground level. Gibbs and McIntyre (1997) found large juveniles and a few small adults (i.e. post-moult body weights) in November–December (spring). By May (autumn), they found heavier adults (including a copulating pair) and larger numbers of mid-sized juveniles, which, according to measurements of captive weta (Richards, 1973), would become adult by the next year. Meads and Notman (1993) found some large adults in October, which had presumably overwintered as adults, as well as several subadult and a few small instars.

The adults often perch in foliage during the day, sometimes in ungainly postures, which they maintain throughout the daylight period, aligned against some piece of vegetation in a way that breaks the visual outline of the body (Fig. 12.3f). Like *D. rugosa* on the ground, they are quite vagile, moving about at night in the intermeshing canopy, remaining inactive in cooler weather and not returning to a day site once they have left. This is also consistent with the wide spread of occasional weta pellets found on the ground. Males appear to be more mobile and range more widely than females, often apparently without descending to lower levels for periods of perhaps 14–16 days and then unexpectedly being found on the ground. Several subadult weta, monitored by Gibbs and McIntyre (1997), in contrast, emerged from their hiding places early in the evening and sat for varying periods nearby before migrating vertically into foliage, returning to shelter sites in bark, tree crevices or epiphytes before daylight. Individual variations in this pattern apparently relate to the moulting cycle.

After looking inconclusively for clear evidence to link nocturnal weta sightings and weather or dark nights, it was quite a surprise to find females on the ground ovipositing in dark sections of the lunar cycle – just as the giant tusked weta use this period to emerge from burrows in the soil (see Fig. 12.7). Predictions of 'good' times to find weta on the ground in the period following full moon, made on the basis of those successful for tusked weta, were met for two female wetapunga monitored intensively in advance by radiotelemetry. One of these spent up to 2 h on the ground, while there was also a male on low foliage nearby. The weta are clearly very vulnerable to rat predation at this time, as well as predation by the indigenous brown kiwi (Aves) and, in past times, probably also tuatara. It will be particularly valuable to monitor and review the population response to the proposed eradication of kiore from Little Barrier Island (Department of Conservation, 1994).

I speculate that the adult weta come to ground more often than previously supposed, and will tend to do so at the darkest times of the moon cycle, especially the period after full moon, using the ground as a route between trees and, through the late autumn and winter, in association with reproductive activity, when they also spend more time on the ground. Females have also been occasionally observed ovipositing on the ground at this time of year in the middle of the day at high altitude on the island (anonymous, entries in bunkhouse logbook).

The sister species *D. fallai* (Fig. 12.1) is found in moderate numbers in largely undisturbed pohutakawa (*Metrosiderso excelsa*) forest on the Poor Knights Islands (Fig. 2.6, Gibbs, Chapter 2, this volume), where it coexists with moderate–low densities of tuatara. These islands are more distant from the New Zealand mainland than other islands with giant weta and there is no evidence of wider distribution in the past. Like wetapunga, *D. fallai* spend the day above ground, hiding in the loose bark of pohutakawa or perching in foliage. Barrett (cited in Brown, 1995) notes that they may shift to more suitable sites as the day progresses. On the island, it is easy to see weta on pohutakawa trunks at night. The biology of this species is known mainly from captive observations of Richards (1973) and, more recently, from a behavioural study of captive weta in an outdoor enclosure of about 9 m^2 at the Wellington Zoo (Brown, 1995).

Brown (1995) found that, as for wetapunga, the adults moved about much more at night than juveniles, in captivity often repeatedly covering the perimeter of their enclosure, and that the males devote a large proportion of the night searching for females. Age, sex and body weight are key variables affecting activity patterns of individual weta, and the heavier adults are more likely to be involved in reproductive activity. He also found that individuals shifted between resting sites about every 10 days and, in captivity at least, they also form short-term and often mixed-sex resting associations during the day – although this may be a product of the captive environment. Such information is difficult to obtain in the field and is of particular value if it can be complemented with field data.

The *Deinacrida* species surviving on mainland New Zealand tend to be the smaller large-bodied species (Table 12.1). At one of two sites in the central North Island, the Mahoenui giant weta, *D. mahoenui*, survives in the dense prickly foliage of gorse (*Ulex europaeus*) (Sherley and Hayes, 1993). This is an exotic weed, which invades disturbed sites as a primary colonizer, although it is eventually overtaken by native species, provided there is a seed source available. There is an unusual challenge for the conservation managers in declaring the site a reserve, since they must now also maintain this early successional status! The gorse apparently provides protection from predators, shelter and food, making it an ideal habitat for giant weta (Sherley and Hayes, 1993). With a canopy height of less than 2 m, it also makes them relatively accessible to study – at least by researchers with suitable protective clothing.

Sherley and Hayes (1993) examined seasonal phenology and habitat use in gorse. They found more weta on steeper slopes with some open areas with livestock present and in less dense mid-aged to older bushes. I suggest, with reference to the behaviour of adult *D. rugosa,* that it may be important, at least for adults, to have a mosaic of open and more enclosed vegetation, which facilitates dispersal and mate-finding behaviour. Stronge *et al.* (1997) concluded that goat browsing, which stimulates dense regrowth of gorse, might not be an important influence on weta survival. It is nevertheless likely that goat browsing helps to limit the re-establishment of other shrub and tree species, which would eventually grow through and replace the gorse.

Richards (1994) examined activity patterns in captive and free-ranging weta. She found that they spend the day perching deep inside bushes, usually at branch junctions, and migrate vertically at night. Barrett (cited in Brown, 1995) also notes that they often sit on the tops of gorse bushes in the morning and migrate deeper into the foliage as the day progresses. On summer nights, they have a bimodal pattern of activity, though they may feed throughout the period of darkness, while in winter there is relatively little activity, a peak in the early part of the night and no activity recorded in the last third of the dark period. They also form daytime resting associations, as observed in captive *D. fallai* (Brown, 1995).

Richards (1994) also suggests that most weta remain within a small area of up to 3 m^3 for much of their lives and often return repeatedly to rest in the same site. She does not distinguish adults and juveniles in this regard, but notes that adults make occasional movements to new sites, sometimes across pasture and mostly in late summer. This latter point seems consistent with information on dispersal of adult *D. rugosa* and *D. heteracantha* and the impression that juveniles are more sedentary. Perhaps the enclosed nature of the gorse habitat and shelter it affords in this regard is conducive to more sedentary habits in this species.

It is implicit in the field data of Sherley and Hayes (1993) and Richards (I. Stringer, Massey University, 1994, personal communication) that weta densities are relatively high in gorse. By implication, this is associated with exclusion (by the dense prickly foliage down to ground level) of mammal predators, in particular rats, possums, mustelids and hedgehogs. There are very few weta persisting at a second site in remnant forest about 20 km distant (Sherley and Hayes, 1993).

Giant Tusked Weta *Motuweta isolata*

The tusked weta (Anostostomatinae) have a body type and burrowing habits characteristic of ground weta (*Hemiandrus* spp.) with which they may coexist, but are easily distinguished by the tympanum on the forelegs (absent in ground weta) and tusks in males (see Fig. 2.15, Gibbs, Chapter 2, this volume). The tusk becomes evident from about the fourth instar as a 'knob' on the front of the mandible. The adult females are similar in size to the largest males or only slightly heavier, with a long ovipositor and no tusks. This is in contrast to the *Deinacrida* species, in which the female is considerably larger than the male (Table 12.1). Tusked weta also lack the robust cuticle and heavily spined middle and hind legs of the *Deinacrida* species. As for ground weta generally, they are predators and scavengers on other invertebrates and seem to take relatively little plant material.

As in *Deinacrida* spp., there appears to be a span for active stages of *M. isolata* which extends over three winters in the field (unpublished data). In captivity they can be reared to the adult stage in a mean of about 18 months (min. 15.5; C. Winks, personal communication; Winks and Ramsay, 1998), with two winters spent in the juvenile stages. Preliminary field data (McIntyre, 1998b) indicate a similar period for the smaller Raukumara tusked weta (Table 12.1). This species is found along stream banks in mainland forests of the eastern North Island (Fig. 2.13, Gibbs, Chapter 2, this volume).

Motuweta isolata survives, as far as is known, only in coastal forest on Middle Island (13 ha; Fig. 12.5a) in the Mercury group (see Fig. 2.13, Gibbs, Chapter 2, this volume). This is uninhabited and naturally free of introduced mammal pests. As on Stephens Island, the soil is enriched by sea birds, but the forest, in contrast, is relatively undisturbed (Atkinson, 1964). There is an unusually rich community of invertebrates and reptiles, including moderate densities of tuatara. This is in contrast to the highly modified islands, such as Mana Is. and Stephens Is. which have fewer species but artificially high densities of some.

The soil is extensively undermined by burrowing sea birds and tuatara, friable and prone to dryness. There is little or no ground cover of vegetation or leaf litter and there is no fresh water on the island. Ecological information has been obtained mainly by scanning for weta at night, with some limited use of radiotelemetry to monitor adults of both sexes. Steep cliffs restrict access to parts of the island and the need to limit human impact on the fragile and unstable environment is a prime consideration in undertaking research or wildlife surveys.

The weta dig shallow chambers in the soil (Fig. 12.5b) and seal the entrance with a plug of soil, apparently cemented with saliva. Empty weta burrows have a characteristic round exit hole (Fig. 12.5c), with an ovoid chamber that fits closely to the body of the weta, so indicating the size of the occupant, at least for juveniles and solitary adults. These are usually obliterated by the sea birds within a short time.

Fig. 12.5. a, Middle Island (Mercury group), the sole location for *M. isolata* (photo M. McIntyre). b, *M. isolata* exit hole in ground swept bare by burrowing sea birds (photo M. McIntyre). c, *M. isolata* female located by radiotelemetry in cut-away underground chamber; note chamber constructed under tree root, seeds of canopy tree on bare soil (photo M. McIntyre). d, Large and medium-sized adult male *M. isolata* (the largest and smallest males have not been seen together at the same time of year; photo B. Robertson, Victoria University).

There is an element of chance in finding weta out of their burrows at night, and there has been much searching without success. Generally, adults are more predictable than juveniles and are prominent in the early–mid-part of the year. Searches undertaken over several nights when weta are active typically encounter representatives of two size cohorts – these are adult or near-adult, separated by one–two instars from mid-sized or smaller weta (> 1 g) – depending on the time of year (Fig. 12.6-14). A third cohort, consistent with the 3-year life cycle, comprises small instars. The smallest instars (<< 1.0 g) spend most of their time in the soil (Winks and Ramsay, 1998) and are seldom seen in the field.

There is a seasonal pattern, as follows (Fig. 12.6). In early January (midsummer), large near-adult juveniles (typically ~8–12 g) are prominent. A few of the males are as large as or larger than some adults (up to 16 g), but have the juvenile tusk form (non-overlapping, symmetrical and smooth). There are also representatives of a cohort that will become adult 1 year hence. These weigh about 4–5 g and will overwinter one or two instars advanced from this. By late February–March (autumn), many of these larger juveniles have become adult, while representatives of the smaller-sized cohort are also at least one instar advanced. The largest individuals have been found at this time of year: males 23 g, females 25 g at time of capture (and up to 27–28 g in captivity). By May–June (early winter), mostly adults are seen at night and occasionally mid-sized juveniles, which will probably have an extended period over winter before moulting next. Males range widely and are less likely to return to a site once having left, while females seem to use one or a few burrows within an area extending for perhaps 10–20 m. Females oviposit on the surface, often using small depressions in the soil. They often seem to be paired with small–medium-sized males, but I suggest that these are roving males, which, by staying close, will find opportunities to mate. Copulation takes place in underground chambers and the pairs may associate for several days. The habits of the largest males and the role of the tusks in male interactions in this regard are unknown.

By October–November (spring), small overwintering juveniles (1–2 g) within the size range of ground weta (*Hemiandrus pallitarsus*) become more active at night. These seem to stay in or near the loose, dry, leaf litter, where they are quite cryptic.

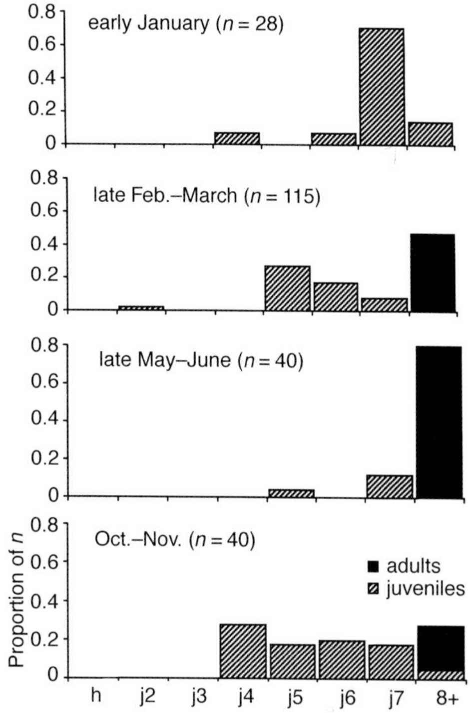

Fig. 12.6. The seasonal occurrence of *M. isolata* according to body size (juveniles) or adult status (no size distinction) 1991–1994. h, Hatchling (very small instars); j, juvenile size groupings indicated by proportionality of hard parts.

The few adults encountered at this time of year seem to be near-spent adults. Females typically had low body weights for their size, a flaccid abdomen, damaged or missing appendages and other injuries. The smallest males were seen at this time, some similarly damaged, possibly by tuatara.

The largest males are similar in size to females (refer to Table 12.1), although their bulk is made up by an enlarged head capsule and tusks, whereas in females the body is more robust. However, males can differ over one and possibly two instars in size at maturity (field range 8.6–23 g; Fig. 12.5d). This suggests that there are strategies for access to females in which body size and timing of maturation may be key factors. There seems to be a similar situation in the Raukumara tusked weta (McIntyre, 1998a). Furthermore, there is a significant difference in tusk length between the adult males seen in February–March (mean tusk length 19.0 mm ± 5.2 mm, n = 26, with the largest ones seen only at this time) and those seen at other times (mean 13.3 mm ± 2.6 mm, $n = 20$; $P_{(t)}$, 0.05). Tusks are used in jousting contests (see Fig. 10.14, Field and Deans, Chapter 10, this volume), at least by the largest males, but the ecological setting for this and the role in the reproductive biology of the species are not yet known. Observations of captive weta indicate that residency seems to be a prime requirement for such a dispute, from which one contestant will eventually withdraw. There are also stridulatory pegs (see Fig. 10.1, Field and Deans, Chapter 10, this volume) on the tusks. To date, the tusk stridulation has been noted only as a shrill alarm call. In larger males, the pegs are often worn flat on the upper surface. Jousting males also make hissing sounds – possibly via the spiracles.

The large juvenile males, as seen in December, presumably become the largest adult males seen in March. It seems that these delay maturation, perhaps persisting for a third winter as juveniles. Whether this is genetically fixed or subject to some maternal or environmental influence that delays maturity has implications for captive rearing as a conservation method. The duration of egg incubation and time of year when eggs hatch might be important in this regard. Winks and Ramsay (1998) found a minimum egg incubation period of 9 weeks, but a typical period of 6–8 months in captive weta. Several small clues also suggest that the largest males might be relatively short-lived as adults and not survive through the winter.

The effects of external conditions are added to seasonal factors in their influence on nocturnal activity. The influence of ambient light and moon phase (Fig. 12.7) is most notable. Most weta emerge from the soil only in the dark spells before moonrise or after moonset. It seems that there has to be at least 80 min of dark for weta to appear and the highest capture rates have been obtained when there has been about 2½ h of darkness. I suggest that the shorter dark periods compress nocturnal activity and there are not necessarily more weta seen over the time of the new moon, when all or most of the night is dark.

These times can be predicted for research and survey purposes according to the lunar cycle and do not appear to be conspicuously overridden by cloud cover. Thus weta do not necessarily emerge in the early part of the evening in accordance with circadian entrainment, but their activity may be delayed to the later part of the scotophase or suppressed entirely around the time of full moon. Perhaps there is a superimposed lunar periodicity.

The lunar effect is believed to be driven by tuatara, which are probably major predators of the larger instars and adults. Tuatara territories include weta habitat and weta may even construct their chambers inside the outer part of occupied tuatara burrows. Tuatara bask in the day but most of their activity is crepuscular and nocturnal. They are primarily visual hunters that sit motionless in wait for potential prey and require some movement to engage their attention. The sparseness of ground-cover vegetation makes for prime tuatara habitat with unimpeded vision and a need for coexisting weta to adopt suitably evasive behaviour. Likely predators of the smaller instars also include several species of nocturnal lizards and a giant centipede, which may reach up to about 20 cm in length. The last of these provides a good reason for the weta to seal its burrow.

The effects of weather are added to this. Weta are more likely to be seen in conditions that are still, mild and damp. In windy conditions, they are often found in more sheltered parts of the island and likely areas can be predicted according to wind direction, provided other conditions are also suitable.

It seems that these weta will sit out unfavourable periods of moonlight or dry or cool weather in their burrows. One adult monitored through a period of full moon did not emerge for 8 days (McIntyre, 1991). Juveniles have long inactive periods, which seem to be associated with moulting. Winks and Ramsay (1998) found that captive juveniles spend periods from 2 weeks to 2 months in the soil without feeding while moulting, while McIntyre (1998a) found a similar situation in 'Raukumara' sp.

Weta out at night are very active, unlike the *Deinacrida* species. While the smaller instars tend to sit low on the ground and freeze when disturbed, initially at least, the larger stages can leap unpredictably up to metre when disturbed and may also tumble on landing. This can be confusing for researchers – and is probably likewise for a predator. Weta viewed without the disturbance of a night light are very mobile, continually engaging unusually long palps and antennae. Most are seen on the ground, but adults are occasionally observed on tree-trunks and have been detected (by radiotelemetry) above in foliage.

The weta are not abundant and, although large, can be surprisingly cryptic. Over a 4-year period,

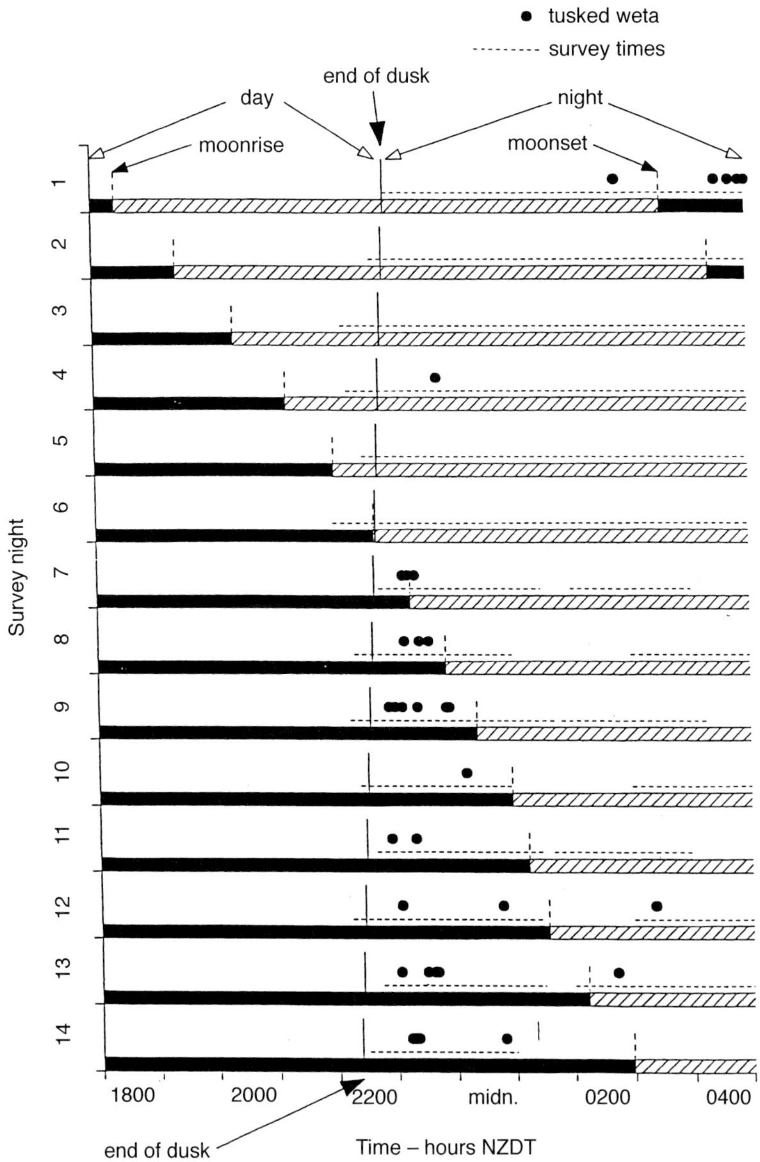

Fig. 12.7. The typical occurrence of late instar and adult *M. isolata* with regard to time of moonrise and moonset over a 2-week period in January 1993. Solid bars, moon below horizon; cross-hatched bar, moon above horizon at sea.

239 weta were found in 118 nights of searching, and > 90% of these were seen at times of night with no moon. Estimates from a 14-day survey session undertaken in autumn, when nocturnal activity is greatest, indicate sampling from a pool of about 120 active mid-sized juvenile and adult weta (95% CI, 95–178) (M. McIntyre, unpublished data). This is equivalent to about 415 ha^{-1}. However, they are very localized on the island and most were found within three small patches of habitat totalling about 0.2 ha.

Several features, reviewed in McIntyre (1998a), suggest that tusked weta might be limited by availability of moisture in the environment at drier times of the year. The localized habitat patches are well shaded by forest canopy and seem to be in locations where the soil retains some moisture over the typically dry summer and autumn period. By May–June, when adults, particularly males, range more widely, the sun is low in the sky and moisture is no concern.

The Mercury islands are volcanic remnants which have been isolated from mainland New Zealand for about 8500 years. The weta would undoubtedly have once been on the nearby islands and was probably also more widespread before the arrival of rats. Comparisons with the smaller Raukumara tusked weta, which inhabits stream banks in mainland forest (Gibbs, 1998b), suggest that it is perhaps marooned in a marginal environment (McIntyre, 1998a). It will be possible to evaluate the relative influence of predation and environment in hindsight with weta translocated to islands with few or no tuatara and access to damper sites. C. Winks (Landcare Research, Auckland, 1999, personal communication) is developing methods for captive rearing and I. Stringer (Massey University, 1999, personal communication) is developing methodology and protocols for translocation of captive reared weta and monitoring their establishment on nearby islands from which kiore have been eradicated.

Concluding Comments

The various situations in which the large lowland weta species survive highlight the special character of each species as well as some general characteristics of large weta. All of these species are represented by a single or small number of remnant populations. Since these populations are either on offshore islands or 'islands' of otherwise suitable habitat and usually small area, they are isolated from each other and thus subject to the ecological and genetic concerns which beset small populations and possibly also those which affect declining populations, as distinguished by Caughley (1994). Conservation management (Department of Conservation, 1998; see also Sherley, Chapter 26, this volume) is addressing the situation with habitat restoration and recent and planned translocations, mainly from captive reared stocks. It nevertheless faces the challenge of protecting an ancient biota in a substantially altered modern environment, in which the relatively small areas involved are always vulnerable to perturbation. Thus monitoring is essential and intervention may sometimes be necessary.

This scenario also has some very fortunate aspects. Seeding new populations by translocation to suitable island sites (Department of Conservation, 1998) is providing opportunities to examine the population ecology and behaviour of translocated weta in reponse to 'empty' weta habitat, different environmental parameters and fewer or different predators. It thus has the potential to provide fundamental insights into species ecology, as well as achieving practical objectives. Ecological restoration, as on Mana Is. (Department of Conservation, 1998; McIntyre, 2000) and Somes Is. (Department of Conservation, 2000; McIntyre, 1998b) also offers the opportunity to predict and test the ecological implications for weta and other biota of putting communities together from existing and introduced components, rather than the more conventional approach of analysing existing assemblages of species to seek information on the function of communities.

There is also the opportunity to obtain in hindsight some appreciation of the relative influence of environment and predation in maintaining weta populations. Manipulations such as the removal of mice from Mana Island and the planned removal of kiore from Little Barrier Island would not ordinarily be available as experimental variables. Likewise, the colonization of gorse by Mahoenui weta is elucidating both the influence of mammal predation by excluding it and the ability of the species to adapt its behaviour and ecology to opportunities provided by the new environment – in this case, organized by the weta itself!

If island restoration projects are even moderately successful in raising weta numbers and security, there is also the potential, in conjunction with ongoing ecosystem restoration programmes, of setting up rigorous experimental procedures, with controls, that can answer specific questions about weta or the dynamics of communities with weta, which would not be ethically or physically possible with fragile natural populations. As suggested by Diamond (1990), there are opportunities to learn much about species in undertaking to protect them. In the case of large weta, these are also icon species, which attract public attention, and are thus also flagship species for the other less spectacular New Zealand invertebrates.

Acknowledgements

The fieldwork on which much of this review is based was made possible by the support of Victoria University and the Department of Conservation and owes much to the stimulus and input from George Gibbs. Funding assistance has been provided by National Geographic (USA), the Department of Conservation, Victoria University, the World Wide Fund for Nature, New Zealand, and the Forest and Bird Protection Society. The Department of Conservation provided permits and logistic support, while conservation staff and volunteers have assisted with many unsociable hours' scanning for rare weta at night.

References

Atkinson, I. (1964) The flora, vegetation, and soils of Middle and Green Islands, Mercury Group. *New Zealand Journal of Botany* 2(4), 385–402.

Beauchamp, A. (1990) *Status of the Giant Weta on Mana Island.* Report to Department of Conservation, Wellington Conservancy, 4 pp.

Brown, J. (1995) Behaviour of captive Poor Knights giant weta (*Deinacrida fallai*). BSc Hons thesis, Victoria University, 100 pp.

Caughley, G. (1994) Directions in conservation biology. *Journal of Animal Ecology* 63, 215–544.

Daugherty, C., Gibbs, G., and Hitchmough, R. (1993) Mega-island or micro-continent? New Zealand and its fauna. *Trends in Ecology and Evolution* 8, 437–442.

Department of Conservation (1994) A strategy for the management of kiore (*Rattus exulans*) on New Zealand Islands. Unpublished draft document, Department of Conservation, New Zealand, 21 pp.

Department of Conservation (1998) *Threatened Weta Recovery Plan.* Threatened Species Recovery Plan No. 25, Department of Conservation, New Zealand, 45 pp.

Department of Conservation (1999) *Mana Island Ecological Restoration Plan*, Department of Conservation, New Zealand, 149 pp.

Department of Conservation (2000) *Matiu/Somes Island – A Plan for Conservation Management.* Department of Conservation, 156 pp.

Diamond, J. (1990) Learning from saving species. *Nature* 343, 211–212.

Gibbs, G. (1998a) Why are some weta (Orthoptera: Stenopelmatidae) vulnerable yet others are common? *Journal of Insect Conservation* 2 (3–4), 161–166.

Gibbs, G. (1998b) *Raukumara Tusked Weta: Discovery, Ecology and Management Implications.* Conservation Science Advisory Notes 218, Department of Conservation, New Zealand, 17 pp.

Gibbs, G. and McIntyre, M. (1997) *Abundance and Future Options for Wetapunga on Little Barrier Island.* Science for Conservation 48, Department of Conservation, New Zealand, 24 pp.

Holdaway, R. (1997) Arrival of rats in New Zealand. *Nature* 384, 225–226.

McIntyre, M. (1991) *Middle Mercury Island Tusked Weta Project.* Interim report to Department of Conservation, Waikato Conservancy, April, 3 pp.

McIntyre, M. (1992) *Dispersal and Preliminary Population Estimates of the Giant Weta,* Deinacrida rugosa, *Following the Eradication of Mice from Mana Island.* Report to the Department of Conservation. Department of Conservation, New Zealand, 9 pp.

McIntyre. M. (1993) *A Baseline Survey of the Giant Weta,* Deinacrida rugosa, *Following the Eradication of Mice from Mana Island.* Report to the Department of Conservation, Wellington Conservancy, New Zealand, 15 pp.

McIntyre, M. (1998a) *Raukumara Tusked Weta: Field and Captive Observations.* Conservation Science Advisory Notes 219, Wellington Conservancy, New Zealand, 35 pp.

McIntyre, M. (1998b) *Monitoring the Release of Giant Weta,* Deinacrida rugosa, *on Matiu-Somes Is.* Report to Department of Conservation, Wellington Conservancy, New Zealand, 10 pp.

McIntyre, M. (2000) *Monitoring Some Interim Effects of Ecological Change on Mana Island.* Report to Department of Conservation, Wellington Conservancy, New Zealand, 21 pp.

Meads, M. and Ballance, A. (1990) *Report on a Visit to Little Barrier Island.* Ecology Division Report No. 27, Department of Scientific and Industrial Research, New Zealand, 9 pp.

Meads, M. and Moller, H. (1978) *Introduction of the Giant Weta* (Deinacrida rugosa) *to Maud Island and Observations of Tree Wetas, Paryphantids and Other Invertebrates.* DSIR Ecology Division Report No. 4/15/9, Department of Scientific and Industrial Research, New Zealand, 14 pp.

Meads, M. and Notman, P. (1992) *Resurvey for Giant Wetas* (Deinacrida rugosa) *Released on Maud Island, Marlborough Sounds.* DSIR Land Resources (New Zealand) Technical Record No. 90, 34 pp.

Meads, M. and Notman, P. (1993) *Giant Weta* (Deinacrida heteracantha) *Survey of Little Barrier Island, October 1992.* Landcare Research Contract Report No. 393, 18 pp.

Moller, H. (1985) Tree weta (*Hemideina crassicruris*) (Orthoptera: Stenopelmatidae) of Stephens Island Cook Strait. *New Zealand Journal of Zoology* 12, 55–69.

Morgan-Richards, M., Daugherty, C. and Gibbs, G. (1995). Taxonomic status of tree weta from

Stephens Island, Mt Holdsworth and Mt Arthur, based on allozyme variation. *Journal of the Royal Society of New Zealand* 25(2), 301–312.

Murphy, E. and Pickard, J. (1995) House mouse. In: King, C.M. (ed.) *The Handbook of New Zealand Mammals.* Oxford University Press, Oxford, pp. 225–242.

Newman, D. (1982) Tuatara, *Sphenodon punctatus*, and burrows, Stephens Island. In: Newman, D. (ed.) *New Zealand Herpetology. Proceedings of a Symposium Held at Victoria University of Wellington, New Zealand, January 1980.* Occasional Publication No. 2, New Zealand Wildlife Service, Wellington, pp. 213–221.

Pickard, R. (1984) The population ecology of the house mouse, *Mus musculus*, on Mana Island. MSc thesis, Victoria University, 233 pp.

Ramsay, G. (1955) The exoskeleton, musculature of the head and the life-cycle of *Deinacrida rugosa* Buller 1870. MSc thesis, Victoria University College.

Richards, A. (1973) A comparative study of the biology of the giant wetas *Deinacrida heteracantha* and *D. fallai* (Orthoptera: Henicidae) from New Zealand. *Journal of Zoology, London* 169, 195–236.

Richards, G. (1994) Ecology and behaviour of the Mahoenui giant weta, *Deinacrida* nov. sp. MSc thesis, Massey University, 184 pp.

Scott, E. and Emberson, R. (1999) *Handbook of New Zealand Insect Names. Common and Scientific Names for Insects and Allied Organisms.* Bulletin 12, Entomological Society of New Zealand.

Sherley, G. and Hayes, L. (1993) The conservation of a giant weta (*Deinacrida* sp. Orthoptera: Stenopelmatidae) at Mahoenui, King Country: habitat use and other aspects of its ecology. *New Zealand Entomologist* 16, 5–68.

Stronge, D., Fordham, R. and Minot, E. (1997) The foraging ecology of feral goats, *Capra hircus,* in the Mahoenui giant weta reserve, southern King Country, New Zealand. *New Zealand Journal of Ecology* 21(1), 81–88.

USGS Patuxent Wildlife Research Centre (1996) ftp://ftp.im.nbs.gov/pub/software/capture

Winks, C. and Ramsay, G. (1998) *Captive Rearing of the Middle Island Tusked Weta.* Landcare Research Contract Report LC9899/10, 24 pp.

13
The Gallery-related Ecology of New Zealand Tree Wetas, *Hemideina femorata* and *Hemideina crassidens* (Orthoptera, Anostostomatidae)

Laurence H. Field and Graham R. Sandlant
Department of Zoology, University of Canterbury, Private Bag 4800, Christchurch, New Zealand

Introduction

New Zealand tree wetas (*Hemideina* spp.) are large, flightless orthopterans which commonly live in vacated tunnels made by other insects in branches and trunks of trees (Miller, 1984; Field, 1978). Because wetas are nocturnal insects, it is presumed that they occupy the tunnels in the daytime to escape predation by visual predators, such as birds (supporting evidence reviewed by Field (1978) and Field and Rind (1992)). However, ecological and behavioural studies have shown that the wetas also centre much of their nocturnal activity around the tunnels, and often individuals occupy specific tunnels for extended periods of time (Sandlant, 1981; Moller, 1985; Ordish, 1992). Furthermore, tree wetas enlarge and groom the tunnels (present report). Therefore, we designate the tunnels as galleries, implying that they serve as relatively stable homes from which the nightly activity cycle of wetas is based.

Besides potentially serving as protection from predators, galleries represent an important resource which is central to the reproductive investment strategy of male tree wetas. Males aggressively compete for galleries in order to acquire and retain one or more females (in harems) for mating and to hold the females in isolation from other reproductively competitive males (Sandlant, 1981; Field and Sandlant, 1983; see also Field, Chapter 18, this volume).

Tree wetas have never been reported to excavate their own galleries *de novo*, and all evidence suggests that they opportunistically acquire empty galleries which are available in the habitat. In support of this, there seems to be little preference by wetas for specific tree species. The common South Island weta, *Hemideina femorata*, has been reported to inhabit a wide variety of trees, including kanuka (*Kunzia ericoides*), lacebark (*Hoheria angustafolia*), tree fuschia (*Fuschia excorticata*), mahoe (*Melicytus ramiflorus*), beech (*Notofagus solandri*) and broadleaf (*Grisolinea littoralis*) (Little, 1981; Sandlant, 1981). Similarly, trees inhabited by the parapatric *Hemideina crassidens* include mahoe (*M. ramiflorus*), manuka (*Leptospermum scoparium*), putaputaweta (*Carpodetus serratus*), seven-finger (*Schleffara digitata*), kohekohe (*Dysoxylum spectabile*) and cabbage tree (*Cordyline australis*) (Asher, 1977; Moller, 1985; Sandlant, 1981; L.H. Field and G.R. Sandlant, personal observation). Furthermore, tree wetas sometimes make use of cavities in other plants besides trees. *Hemideina crassidens* has been found in kiekie (*Freycinetia banksii*), coastal veronica (*Hebe elliptica*), New Zealand flax (*Phormium tenax*), gorse

(*Ulex europeaus*), the hollow rhizomes of bracken fern (*Pteridium aquilinum*) and the dried, hollow stems of tutu (*Coriaria arborea*), karaka (*Coprosma robusta*) and tree fern fronds (*Dicksonia squarrosa*) (Asher, 1977; Meads and Moller, 1978; Sandlant, 1981; P.M. Johns and W.C. Clark, Christchurch, 1985, personal observations). If tree wetas normally acquire galleries opportunistically, it follows that the weta population density must be more or less limited by the total gallery resource provided by the habitat. Furthermore, the head capsule width of any weta must limit the proportion of the total resource that the insect can utilize, since head capsule width limits the minimum entrance diameter of galleries that the insect can enter. Little is known about the gallery resource available to *H. femorata* and to *H. crassidens*, or how fully this resource is utilized.

Galleries would also appear to represent an ideal buffer against environmental changes in humidity, since the tree wood is moist and wetas are particularly susceptible to desiccation (L.H. Field and G.R. Sandlant, personal observation). Nothing is known about the changes of humidity in weta galleries, but during our collecting efforts we have noticed that galleries which are very dry or very damp (or wet) seldom contain tree wetas. Humidity requirements for anostostomatids and stenopelmatids are discussed by Angulo (Chapter 11, this volume) and Weissman (Chapter 3, this volume).

This study explores the above ecological aspects, gallery availability and utilization by tree wetas, and attempts to determine whether limitations are imposed upon weta populations by the gallery resource of forest habitats.

Materials and Methods

Study sites in the South Island of New Zealand included kanuka forests in the foothills near Kaikoura, Canterbury Province, and a mixed broad-leaved/podocarp forest in the vicinity of Westport, Westland Province.

Seven 100 m straight transects (1.5 m wide) were run through the study areas; the total number of trees, the number of galleries per tree and the distance of each tree along the transect were recorded for calculation of gallery distribution and abundance. Although some kanuka trees form several trunks from a common base close to the ground, each trunk was counted as a 'tree' in this study. The rationale was that a weta searching for galleries at night would have to inspect each trunk separately, effectively treating each as a separate tree. The data for vertical distributions of galleries and entrance diameters were gathered by taking the respective measurements for all galleries encountered while thoroughly searching all trees in large sections of each forest, rather than restricting the search to transects. Foliage density estimates of these search areas were made by photographing clear silhouettes of groves of trees (including a 1 m scale bar) in the respective forests, making xerox enlargements of the photographs and taking readings of the relative percentage of light transmitted through equivalent 1 m^2 areas of the xeroxed images, using a calibrated selenium photocell and a standard light source.

To study gallery occupancy, trees intended for agricultural clearing were destructively sampled by sectioning with a chain-saw and splitting the sections containing galleries. Wetas were collected and maintained for behavioural experiments, while other occupants were preserved for later identification. Internal gallery dimensions were measured.

Behavioural studies were carried out in enclosures (containing tree branches with galleries, food and water) in the laboratory, where wetas were maintained on a 12 h light : 12h dark reversed light cycle at 15°C, and fed native tree leaves *ad libitum*. Agonistic interactions of males and general nocturnal behaviour of males and females

Fig. 13.1. (*opposite*) Formation and characteristics of galleries excavated in kanuka trees by the cerambycid beetle *Ochrocydus huttoni*. a, Longitudinal section of kanuka tree showing gallery through heartwood. Well-packed strips of frass, chewed from the tunnel lining by the beetle larva, grade upwards into finer chewed material and finally into a powder-like lining of the pupal chamber (above the level shown). b, A newly chewed escape hole from the long-horned beetle's pupal chamber has a sharply defined margin (above), while an old gallery (below) has a swollen margin, which has been kept open though nibbling by wetas as secondary residents. c, Entrances to eight galleries in a kanuka tree occupied by *Hemideina femorata*. Scale = 10 cm. d, Larva of *O. huttoni* in pupal chamber. Note thin partition separating the two central galleries in this heavy infestation. Scale for 1, 2, 4 = 1 cm.

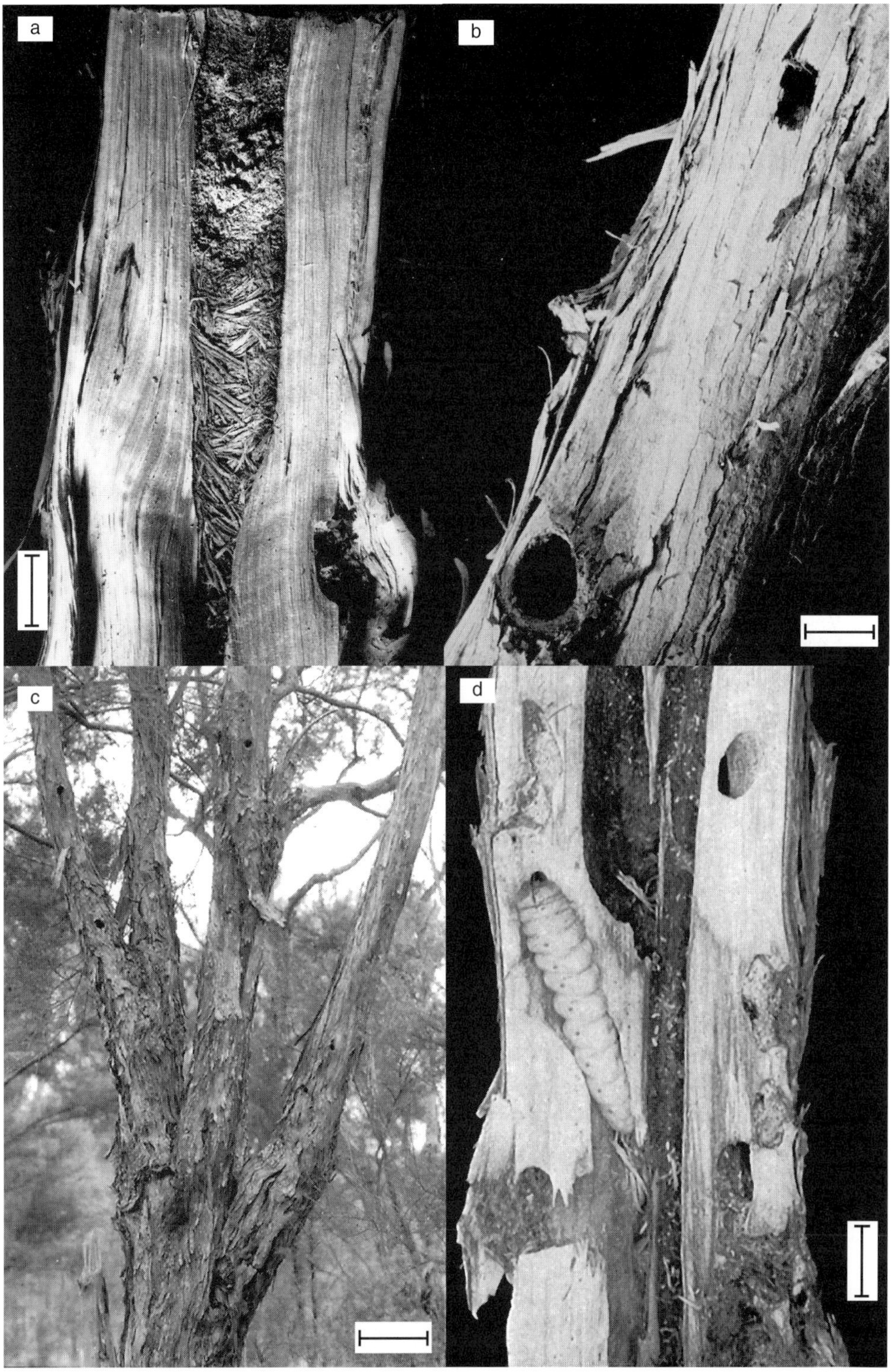
a
b
c
d

centred about galleries were observed under 15 W red light illumination.

Results

Formation and structure of galleries

Most data were collected from the kanuka forests inhabited by *H. femorata* in Canterbury. Galleries were almost exclusively made by the kanuka longhorn beetle, *Ochrocydus huttoni* Pascoe (Cerambycidae). Beetle larvae hatch out of eggs laid in crevices of bark on mature (usually) trees and burrow directly into the heartwood, where vertical tunnels are then excavated. As a larva grows, the tunnels are enlarged to a diameter of approximately 12 mm and to a maximum length of 800 mm (minimum = 50 mm, median = 158 mm, n = 113). They are oval in cross-section. Chewed wood dust is ejected through small lateral holes, and is especially noticeable in piles at the base of infected trees in summer (January and February). A single tree may have numerous parallel tunnels, each from a single larva, but they invariably remain separated (even if only by a thin partition). Prior to pupation, a larva excavates an ovoid, elongate pupal chamber at the top of its tunnel and begins to strip long shreds of wood from the tunnel wall. These are packed tightly in the tunnel (often for a length exceeding 100 mm), forming a plug below the pupal chamber. The upper section of the packed material grades into finer strips, then into particles and finally the remains of the emptied gut contents are deposited as a fine flour lining the pupal chamber (Fig. 13.1a, d). The larva then forms a pupa and eventually undergoes eclosion into the adult beetle, which bores its way out of the chamber laterally leaving a cleanly incised hole about 10–13 mm in diameter. Trees then show prominent holes in the main trunks and branches,

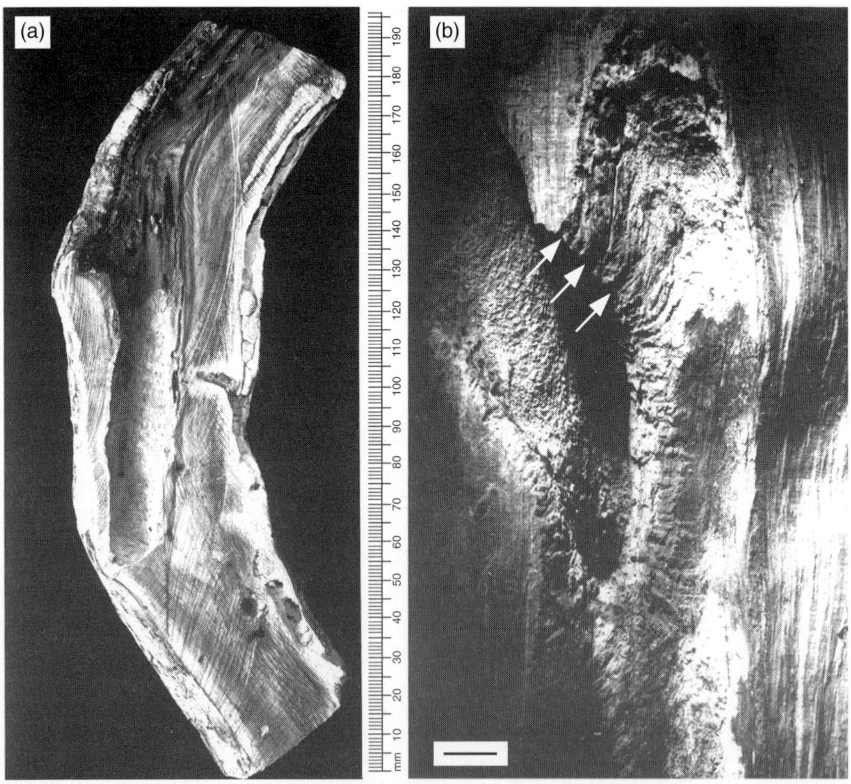

Fig. 13.2. Evidence for enlarging galleries by weta nibbling. a, Natural rotting of heartwood in this sawn branch of *Melicytus* led to a hollow (arrow), which was elongated by picking out bits of the soft wood (below dashed line). b, View from interior of split kanuka trunk showing nibbled margin of opening between two parallel chambers to enlarge the gallery. Scale for a = mm (vertical ruler); for b = 1 cm.

which are available to wetas and which will last the lifetime of the tree (Fig. 13.1a).

The appearance of holes left by newly emerged beetles differs markedly from those subsequently occupied by tree wetas, as the latter are swollen and enlarged, due to regular nibbling of the tree cambium around the margin by wetas (Fig. 13.1b). In the absence of such grooming, we have shown that kanuka bark regeneration occludes experimentally bored holes within $1\frac{1}{2}$ years. Once wetas have occupied cerambycid tunnels, they nibble at accessible surfaces to enlarge and lengthen the interior, as well as to open passages joining neighbouring tunnels (Fig. 13.2a, b). The incision marks left by weta mandibles result from coarse pinch-and-pull movements and differ markedly from the fine semicircular paths of sharply cut transverse rows left by the cerambycid larva mandibles.

Gallery Distribution and Abundance

Galleries in the areas surveyed had a clumped distribution. This may be seen in Fig. 13.3A, B and C, which shows the frequency of trees and galleries along three of the transects. Often galleries were clustered in a single kanuka tree, or even on a single stem (kanuka produces multiple trunk-like stems from one base).

In a detailed analysis of the transects (Table 13.1A), the tree density varied from 0.29 to 1.51

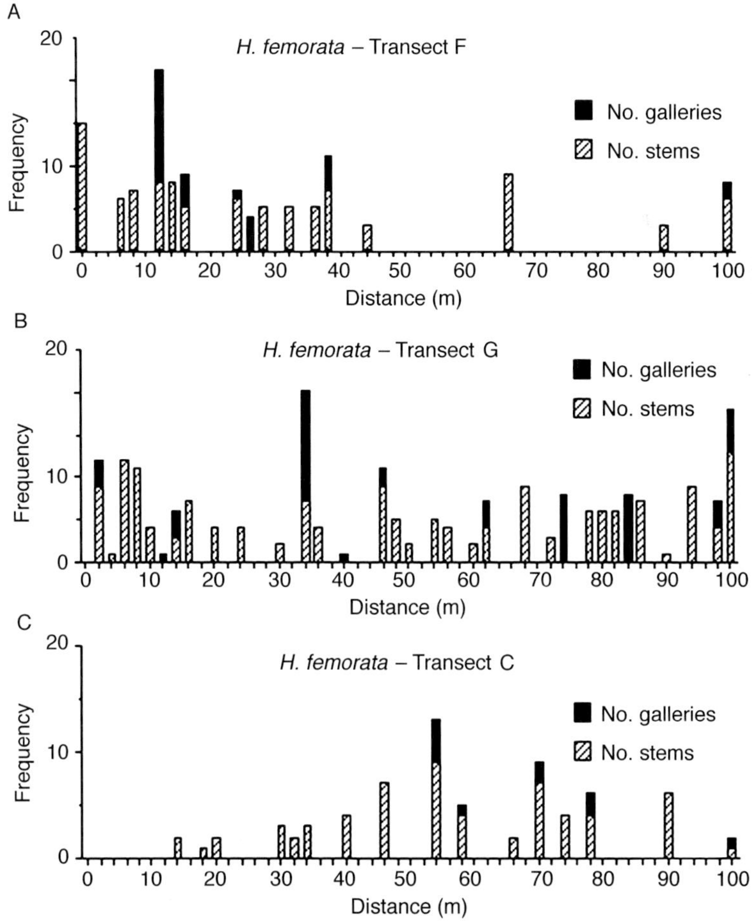

Fig. 13.3. Histograms showing distribution of tree-trunks (hatched bars) and clumped distribution of galleries (solid bars) along two transects through the inner forest (A, B) and one along the forest edge (C). Occasional trees had large numbers of galleries. Fewer galleries were found on edge transects.

Table 13.1. Results of transects (1–7) through forest habitats of *Hemideina crassidens* and *Hemideina femorata*. Two transects (3, 4) extended along the forest edge, while the remainder cut through the inner forest. A. The mean density of trees is expressed as tree-trunks m^{-2}, since kanuka trees can grow with more than one trunk emerging from the base. B. The within-tree gallery densities are also mean values calculated from the totals of galleries divided by the total trees with galleries. C. The gallery frequency in the forest is given as the probability (in %) that a tree in each transect will contained a gallery. A mean estimate of gallery frequency in the forest edge vs. the inner forest is given at the bottom of the table.

	Transect						
	H. crassidens		*H. femorata*				
	1 Inner	2 Inner	3 Edge	4 Edge	5 Inner	6 Inner	7 Inner
A Density of trees							
Total trees	226	86	29	95	64	107	152
Total area	150	90	100	150	150	150	150
Trees/m^2	1.51	0.96	0.29	0.63	0.43	0.71	1.01
B Within-tree gallery density							
Trees with galleries	23	6	1	4	5	10	13
Total galleries	46	8	1	7	10	28	51
Galleries per tree	2	1.3	1	1.75	2	2.8	3.9
C Gallery frequency in forest							
Total trees	226	86	29	95	64	107	152
Trees with galleries	23	6	1	4	5	10	13
P (tree has gallery) (%)	10.20	6.90	3.40	4.20	7.80	9.30	8.60

Mean inner = 8.6%
Mean edge = 3.8%

trees m^{-2}, with greater average density in the west coast forest. Out of 312 trees examined in the west coast broad-leaved forest, 29 contained galleries, whereas 33 trees out of 447 examined in the eastern kanuka forest contained galleries. Of the trees which contained galleries, the mean density varied from 1 to 3.9 galleries per tree (Table 13.1B).

Transects 3 and 4, which were taken along the edge of the kanuka forest, had lower gallery densities than those from within the forest; the mean number of galleries per tree was 1.38 for the edge, compared with 2.4 for trees within the forest (all transects pooled). This may be related to the observation that kanuka trees on the forest edge were small, with dense branching extending close to the ground, whereas the trees of the inner forest had tall, relatively bare trunks with only a crown of top branches. The latter may allow easier access to flying female beetles searching for oviposition sites. The distribution of gallery frequency within trees (Fig. 13.4) showed that most trees had three or fewer galleries. Although the maximum number of galleries per tree in the transect data was 13, greater numbers were occasionally encountered: the maximum in the study was 32 galleries in a single tree. The combined data of Table 13.1A and B allowed estimates of the probability that a tree would contain at least one gallery, for each of the transects (Table 13.1C). Probability values ranged from 3.4% to 10.2%. Thus, at best, one of every 11 trees, on average, contained a gallery (within the forest); at worst, one of every 29 trees contained a gallery (on the forest edge).

Since wetas must walk between trees, or along conjoined canopies, to explore for galleries, it was of interest to calculate the average distance between gallery-containing trees. Mean distances (± SD) ranged from 5.2 ± 6.5 m to 20.6 ± 16.7 m, and the greatest transect distance between two such trees was 62 m. To put this into perspective, adult wetas have a body length of about 35 to 40 mm.

The vertical distribution of galleries was compared with the foliage density in kanuka and broad-leaved trees to assess the amount of vertical exploration a weta might face when seeking a new gallery once it encounters a tree-trunk at night. In

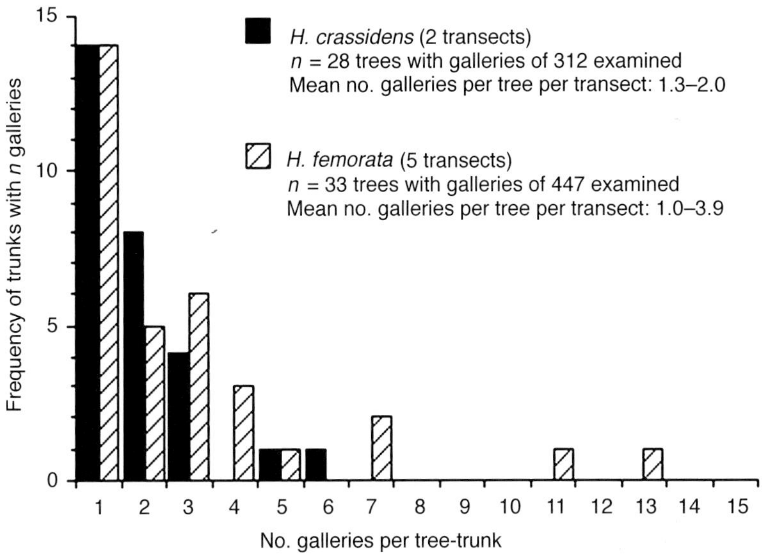

Fig. 13.4. Distribution of number of galleries within individual tree-trunks from transects through *H. crassidens* habitat (solid bars) and *H. femorata* habitat (hatched bars). Sample size (*n*) and ranges of mean values are shown in keys. Although the maximum number of galleries was 13 in the transect data, other trees outside the transects (e.g. Fig. 13.8) had greater numbers of galleries (maximum = 32).

both forest habitats, the peak gallery density was between 1.5 and 2.5 m above the forest floor (horizontal bars, Fig. 13.5). This was well below the peak foliage density, which occurred in the crowns of trees, which reached heights of 6.5 m (kanuka) and 11 m (broad-leaved trees).

Gallery entrance diameter

All *Hemideina* species show megacephaly in the adult males, and hence the diameter of gallery entrances might be a restricting factor in gallery accessibility to males. To assess this, both weta habitats were widely sampled for holes that could serve as weta galleries, and their diameters were measured (Fig. 13.6). The distribution of entrance diameters in the *H. femorata* kanuka habitat (hatched bars) was unimodal, with a mean diameter of 11.3 mm (± 2.3 mm, SD). In the *H. crassidens* broad-leaved habitat (solid bars), the distribution of entrance diameters contained four peaks, with small diameters of 3–8 mm having the greatest abundance.

To obtain an indication of whether gallery entrance diameter restricts weta access, head capsule widths and respective entrance diameters of the occupied galleries were obtained for a restricted data set (*n* = 60 for males, 50 for females) for *H. femorata* (Fig. 13.7). The diagonal line on each graph indicates equality for the two measurements. Females always occupied galleries with entrances larger than their head capsule widths, as shown by the distribution of points clustering well below the line of equality (Fig. 13.7A). However, the distribution of male head capsule widths was closer to the line of equality, and a number of males had head capsules almost the same diameter as the gallery entrance (Fig. 13.7B). The greater sample size of entrance diameters (*n* = 110) obtained from the data in Fig. 13.6 in the *H. femorata* habitat is superimposed upon the data set of Fig. 13.7B to show that the sample obtained in that figure was not biased, and that there actually are no larger galleries available to males in the kanuka forests.

Occupation of galleries

Many other invertebrate species occupied the tree galleries examined in this study (Table 13.2). The most abundant inhabitants were tree wetas (62% in the kanuka forest, 41.6% in the broad-leaved forest), followed by cave wetas and spiders at much lower frequencies (e.g. 5.6% and 10.6% for spiders in kanuka and broad-leaved forests, respectively). Other arthropods included native cockroaches,

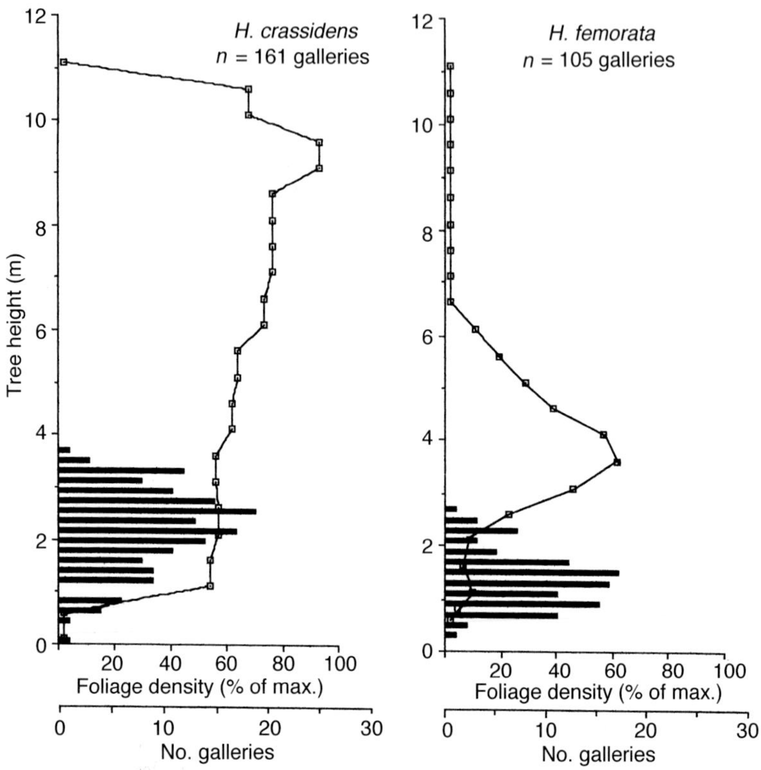

Fig. 13.5. Distribution of foliage density and number of galleries, versus vertical height of trees in the two weta habitats. For both species, gallery distribution peaked within 3–4 m above the ground, well below the foliage crown. □–□, Foliage density; horizontal bars, gallery abundance.

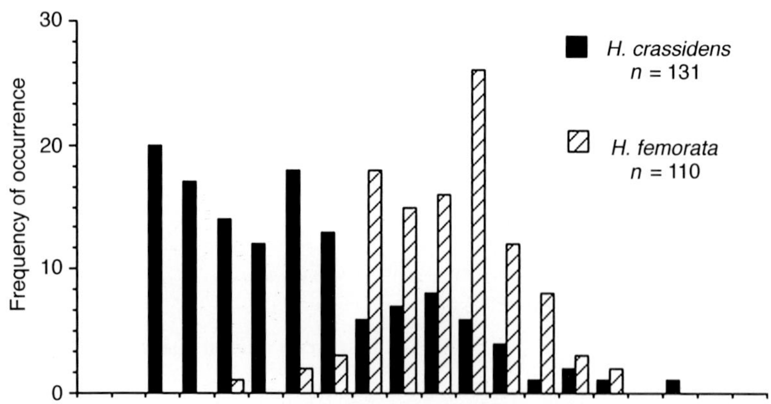

Fig. 13.6. Distribution of gallery entrance diameter in all galleries assessed in the two weta habitats, regardless of occupancy or non-occupancy. The broad-leaved trees in the *H. crassidens* habitats had a preponderance of small entrances and at least four peaks in the distribution (solid bars), suggesting that several species of wood-borers make the galleries. The roughly unimodal distribution (hatched bars) in the kanuka trees of the *H. femorata* habitat strongly suggests that only one species (the kanuka long-horned beetle) makes the galleries. *n*, Sample size.

beetle larvae, slaters (terrestrial isopods) and a centipede. Native slugs also occupied galleries. An example of the actual distribution of occupants is shown for a kanuka tree from Canterbury, which contained 25 galleries (Fig. 13.8). This tree had an unusually large numbers of galleries and was not typical of the gallery-containing trees in the forests; however, such trees allowed optimal data gathering. The accompanying list of occupants allows several insights to be gained. First, wetas coinhabit galleries with other small arthropod species (beetles, woodroaches, slaters, cave wetas and small spiders). Secondly, wetas avoided galleries with spiders, especially the large megalomorph genus *Migas*, which can prey upon tree wetas (L.H. Field and G.R. Sandlant, personal observation). If a gallery contained two entrances, one of which was occupied by a *Migas*, wetas would use the other for access to the gallery. Wetas also avoided slugs in galleries. Thirdly, a surprisingly large number of galleries remain unoccupied (Table 13.2: 18.4% of kanuka trees, 42.5% of broad-leaved trees).

Many accounts (Field and Sandlant, 1983; Miller, 1984) have suggested that tree weta males

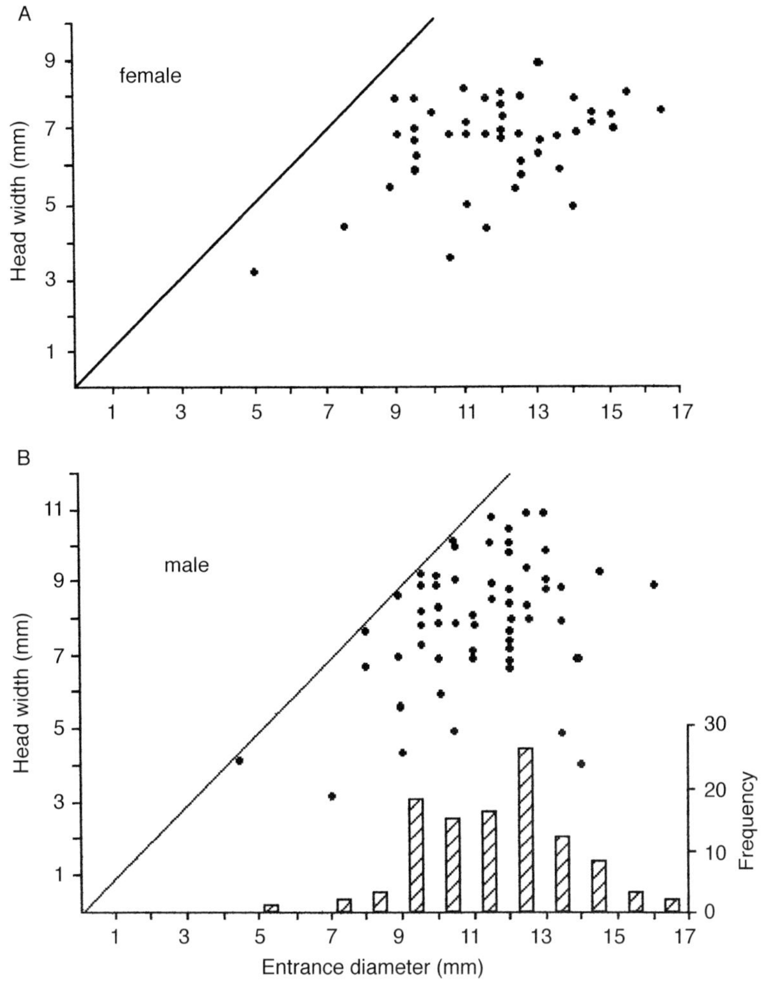

Fig. 13.7. Plots of head capsule width versus entrance diameter of the gallery in which each measured weta lived (*H. femorata*). While females always lived in galleries with ample entrance sizes (A), males showed cases (B) where head capsule size is only slightly smaller than entrance diameter (i.e. falls just below the line of equality for both parameters). This suggests that head capsule size is a limiting factor in gallery availability for males. The distribution of all measured entrance diameters in the *H. femorata* habitat (from Fig. 13.6) is given in B for comparison.

Table 13.2. Composition of coinhabiting animals in galleries of forest habitats of *H. femorata.* and *H. crassidens.* For each weta habitat, the columns give the number of galleries and percentage of the whole sample occupied by each animal species. *n,* Total sample size in each habitat.

	H. femorata (n = 266)		*H. crassidens* (n = 113)	
Category	Galleries	%	Galleries	%
Tree wetas	165	62	47	41.6
Cave wetas	18	6.8	6	5.3
Spiders	15	5.6	12	10.6
Slaters	11	4.1		
Cockroaches	8	3		
Native slugs	3	1.1		
Beetle larvae	3	1.1		
Centipede	1	0.4		
Empty	49	18.4	48	42.5

maintain harems of females, but published numerical data are scant. Therefore the distribution of weta occupation density was also analysed in this study. Data were pooled from several different collection efforts over several years. Entire trees were dissected to exhaustively count the number of wetas per gallery. For *H. femorata*, 53.6% of the occupied galleries contained groups composed of one male plus one or more females (Table 13.3). For *H. crassidens,* the level of group occupancy was 20.8%. Thus a large percentage of occupied galleries for both species contained only a single weta. Histograms of the distribution of number of wetas per gallery (Fig. 13.9) showed that, of the groups, single male–female pairs were most common, but that groups of up to four females with one male were found. The degree to which a male may be able to accumulate a harem of females is likely to be affected by the population sex ratio of males to females. The sex ratios from this study (bottom of Table 13.3) indicate a larger proportion of females to males in *H. crassidens*, compared with that in *H. femorata*.

Discussion

Are weta populations limited by the gallery resource?

Earlier studies have raised the suggestion that gallery availability in the weta habitat may limit the weta population size (Field and Sandlant, 1983; Moller, 1985). In principle, as long as there is severe competition by wetas for gallery resource, the number of galleries of a given size in a habitat should determine the weta population density, size, age structure and even sex ratio (Moller, 1985). The present study attempted to determine whether any evidence exists to support this contention.

Total number of galleries as a limiting factor

The total number of galleries per unit area of forest should represent the maximum refuge space available to a weta population, assuming that wetas do not move into other less suitable refuges in the absence of galleries. Although not specifically

Table 13.3. Incidence of occupation by groups of one male plus one or more females per gallery for both tree weta species. The sex ratios also comprise adults only.

Species	Total galleries	No. adult male/female groups	%
H. femorata	166	89	53.6
H. crassidens	76	14	18.4
Sex ratios:	*H. femorata*	53% males : 47% females	
	H. crassidens	40% males : 60% females	

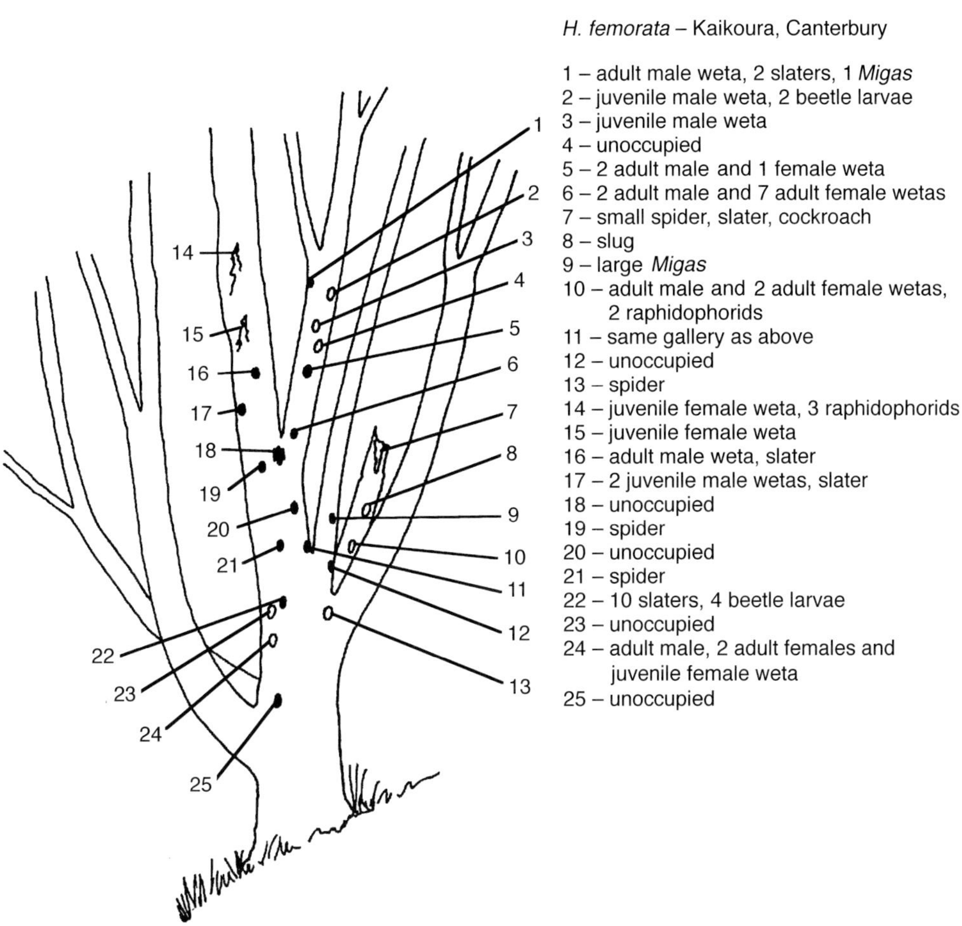

Fig. 13.8. A sketch of a kanuka tree containing an unusually large number of galleries, and a list of all invertebrate occupants found in the galleries. The wetas were *H. femorata*, slaters are terrestrial isopods, raphidophorids are cave wetas and *Migas* is a large megalomorph spider, which can prey upon tree and cave wetas. Note tendency of tree wetas to cohabit with other arthropods, but not with the slug or *Migas*.

tested, our collecting experience suggests that wetas will occupy any hollow refuge in the absence of galleries, but that this is an uncommon scenario in the natural habitat. Tree galleries appear to offer superior protection from predators, while less robust makeshift refuges, e.g. under leaves or in hollow logs on the forest floor, are more likely to be accessible to predators. Based upon these arguments, it seems reasonable to consider that gallery abundance could be a limiting factor for weta populations.

Thus it was surprising to find relatively high percentages of unoccupied galleries in both habitats studied (e.g. Fig. 13.8, Table 13.2). The conclusion must be that neither population of wetas was large enough to utilize the full gallery resource available. The unoccupied galleries did not comprise any extreme segment of the distribution of gallery sizes; instead they were apparently of acceptable sizes for wetas.

One reason for the lack of full occupation of galleries may be that predation pressure in the forest habitats is high enough to reduce the weta population densities to levels that preclude gallery availability as a limiting factor.

Another explanation is that wetas show a preference for living in groups (Fig. 13.9). Although total gallery number did not appear to limit population density, the accompanying assumption of severe competition for galleries is not necessarily

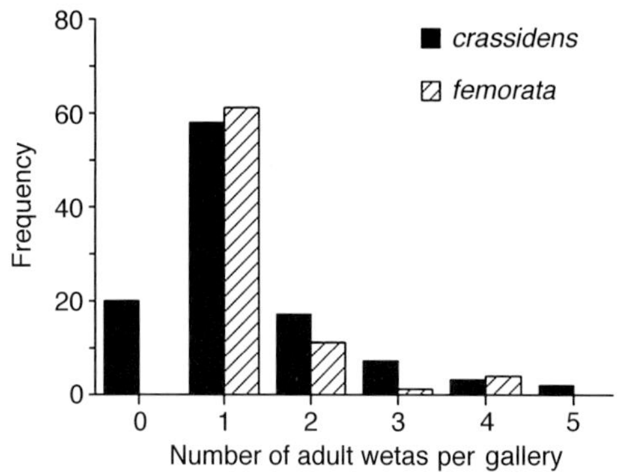

Fig. 13.9. Distribution of the number of adult wetas per gallery from destructive sampling studies of trees, summarized in Table 13.3. Although most galleries contained single wetas, both species showed a tendency for males to live with harems of females (galleries with more than one weta always contained just one male in this study).

ruled out. There could still be severe local competition within single trees, since behavioural studies have shown that males compete for galleries that contain females (Field and Sandlant, 1983). This is a resource-defence polygyny mating system (Emlen and Oring, 1977), which can readily lead to locally high levels of male competition, because each tree constitutes an island of clumped gallery resource. Under high predation pressure, a male's risk of competing for the gallery resource locally may be considerably less than the risk of descending to the ground and attempting to search for another tree devoid of competing males.

In the resource-defence polygyny mating system of wetas, the resource competed for is the combination of females plus galleries, rather than only galleries (Field and Sandlant, 1983; see also Field and Jarman, Chapter 17, this volume). With this view of the resource, it becomes clear that, if males selectively accrue females in galleries, the local tree weta population could become clumped in some galleries, while others remain unoccupied.

If any galleries in living trees are allowed to remain unoccupied, weta populations risk permanent loss of such galleries from the total gallery resource. This is because gallery entrances of living trees become occluded by cambium growth in the absence of active nibbling of the opening by wetas. In kanuka trees, occlusion can be completed within 18 months (L.H. Field and G.R. Sandlant, personal observation). Therefore it is likely that empty galleries observed in this study had been occupied relatively recently, which, in turn, implies that wetas change galleries. Possibly the unoccupied galleries serve as intermittent refuges for single (satellite?) males until they can gain access to galleries with a group of females. Long-term monitoring of such galleries would give insight into the role they serve for weta populations, as well as the degree of fidelity of individual wetas for certain galleries.

Effect of gallery entrance diameter as a limiting factor

The diameter of a weta's head capsule sets a lower limit on the range of galleries that the insect can utilize as a refuge. In this way, the range of gallery entrance diameters available in the habitat could control the population composition of wetas. The effect should be most severe for adult males, since they show development of cranial megacephaly as a secondary sexual character. In the three final instars (out of ten), in which males show sexual maturity (Spencer, 1995), there is an allometric increase in cranial diameter (see Field and Deans, Chapter 10, this volume). If a habitat provides a high proportion of large entrance diameters within the range available, a correspondingly greater proportion of tenth-instar males should be supported in the population. Conversely, the population should be selectively limited by a narrow range of entrance diameters, such as that found in the kanuka forests, where a single host beetle provides the galleries used by wetas. This principle holds in general. For example, on Stephens Island, in Cook Strait, New Zealand, large refuges are available in the form of holes in tree-trunks where branches have broken, hollow trees and large cracks in trunks. Male head sizes are larger than those from nearby islands or from mainland populations (Moller, 1985). A similarly marked increase in male weta head size was found on Open Bay Island, New Zealand, where large gallery entrances are available (L.H. Field and G.R. Sandlant, personal observation).

Effect of gallery depth as a limiting factor

Galleries of greater depth should house more wetas, and therefore depth could act as a regulating factor on the social composition and age distribution of a weta population, but probably not on the population density (Moller, 1985). A correlation analysis of the number of occupants vs. depth of galleries in kanuka forests showed that the above concept holds true (Sandlant, 1981; Field and Sandlant, 1983). In Sandlant's (1981) initial study of kanuka galleries, the mean depth was only 158 mm (although the maximum was 900 mm, and elsewhere *O. huttoni* galleries are reported to exceed 1000 mm (Hosking, 1978)). In the kanuka forests, where short galleries abound, a high proportion of single or paired occupancy is expected and clumping of wetas in harems will be rare. Given the male tendency to sequester females, a high rate of gallery contesting and exchange is also expected, since few galleries would prove suitable. On the other hand, an abundance of deeper galleries would result in a situation where more males with harems remain in residence, since the male tendency is then to defend galleries and attempt to block female egress (Field and Sandlant, 1983).

Clumped gallery distribution

The clumped distribution of galleries in both habitats probably reflects a bias in female wood-boring (gallery-producing) insects to select only certain trees for oviposition. At least two factors may be operating to produce selectivity. First, trees normally differ in water content, depth of bark, hardness of wood and amount of volatile by-products, as well as physical parameters, such as trunk diameter, density of branching and location in the forest. Also, unhealthy trees which have suffered from physical damage or attack by fungi may display a unique set of physical/chemical characters, which set them apart from their conspecific neighbours. It is conceivable that ovipositing beetles make decisions about optimal substrate sites based upon direct or indirect detection of variation in such characters. For example, the scolytid beetle *Platypus* sp. is attracted to damaged *Nothophagus* trees by released terpenes (G. Hosking, 1994, Rotorua, personal communication). Thus, the clumping of galleries could simply reflect a high infestation of wood borers in unhealthy or damaged trees within a forest. Secondly, some wood-boring coleopterans produce aggregation pheromones which cause enhanced infestation of trees that have received an initial beetle invasion. This occurs in the Scolytidae and Cerambycidae (G. Hosking, 1994, Rotorua, personal communication), and could result eventually in a clumped distribution of galleries if it occurred in New Zealand insect species whose galleries are utilized by wetas. Nothing is known about pheromones in *O. huttoni*, but the larger New Zealand cerambycid, *Prionoplus reticularis*, appears to produce an aggregation pheromone (Edwards, 1961).

Another scenario for clumping requires investigation. It is possible that the spatial distribution of ongoing gallery formation by wood-borers is uniform in forests, and that the clumped distribution of weta galleries is the result of the behavioural preference of wetas to aggregate. Wetas could maintain open entrances in local trees of the aggregation by nibbling, while more distant trees would heal over newly formed galleries in the absence of uniform exploitation by wetas.

Effect of gallery distribution on searching by wetas

The data from this study showed that, at best, a weta searching for a tree with at least one gallery can expect success in one out of 11 trees, if it is wandering within the forest. If it wanders near the forest margin, the hit rate drops to one out of every 29 trees. Given the best scenario of 1.5 tree-trunks m^{-2} (Table 13.1), and given that the weta samples along a 1-metre-wide linear transect, it is possible to estimate that every eleventh tree would occur 7 m apart. Since it is searching in the dark, it is unlikely that the weta could aim straight for each tree in its transect, but, allowing this benefit, it would then have to examine each trunk encountered. The area of trunk required to search on each tree is given by $A = \pi dh$, where d = diameter and h = height of trunk. Assuming an average trunk diameter of 15 cm, and a gallery occurrence within the first 3 m above the ground (Fig. 13.9B), the search area for one trunk is approximately 1.4 m^2. For 11 trees this amounts to 15.6 m^2. To appreciate the enormity of such a search on the trees alone, a 40 mm adult weta occupies about 20 cm^2 of substrate, and it would have to cover 7800 times this area on the 11 tree-trunks, on average, to locate a gallery. It must be appreciated that wetas are flightless and are therefore required to descend to the ground after examining each tree.

If a weta were required to execute the above search by tactile means alone, the chances of success would seem remote within one nightly period. Any time spent on the ground at night would put the insect at peril from predators, which abound in New Zealand (rats, feral cats, wekas (large rails), stoats, moreporks (owls) and hedgehogs); additional poorly orientated time between trees increases the peril. Ecological data suggest that wetas are averse to moving on the forest floor. Of nearly 400 wetas observed at night, Moller (1985) found only 1% on the ground. Over an 11-day period, the median distance moved of 64 marked wetas (*H. crassidens* on Stephens Island, New Zealand) was 1.0 m. Even for such movements, the search period is limited to the night, due to intense pressure from diurnal, mostly avian, predators. Even if wetas moved between trees through interjoined canopies, a much greater search time is entailed in finding the trunk of a new tree and in exploring its entire trunk area.

The thrust of the above is to make a strong case for concluding that wetas must use some form of chemical cue(s) to aid in their search for new galleries. Guiding volatile cues could emanate from trees that have been damaged by a gallery cut through the sapwood and cambium, in the same way that New Zealand beech trees release terpines, which attract scolytid beetles, when damaged (G. Hosking, 1994, Rotorua, personal communication). The gallery could either release natural volatiles from the tree or volatiles remaining from the cerambycid occupant; the latter would dissipate with time and probably would not be an important factor for old galleries. Additionally, empty galleries previously occupied by wetas may well release one or more volatiles remaining from the wetas, since there is no evidence that wetas utilize pheromones (see Field and Jarman, Chapter 17, this volume). In any case, recent work has shown that odour plumes can descend along a tree-trunk as heavy volatiles mix and remain in a boundary layer next to the bark (R. Carde, Honolulu, 1997, personal communication). A vertical odour plume that marks a tree and then spreads in decreasing exponential fashion away from the base of the tree would provide a highly efficient means for the weta to find the tree using olfactory senses.

An additional tactic in finding trees with weta-occupied galleries could be to wait on the base of tree-trunks and to listen for acoustic signals from calling resident males. If the calling sound were coincident with vibrations detected on the tree-trunk, the searching weta could confirm the presence of at least one gallery in the tree (see Field, Chapter 15, this volume).

Counterbalance of natural selection and sexual selection for male characters

In classical selection theory, first put forth by Fisher (1930), the forces of natural selection counterbalance the runaway process where sexually selected characters develop into extreme features that reduce the survivability of individuals that inherit them (Kaneshiro and Giddings, 1987). The present data appear to represent a clear case in which this process happens in male *Hemideina* tree wetas. Males undergo an allometric increase in cranial width in growing to the final (tenth) instar. As they reach the extreme limit of gallery sizes available, they will be unable to occupy a diminishing resource until they can no longer obtain refuge from diurnal predators, resulting in their elimination from the population. In wetas, those males which develop the largest head capsules and longest mandibles are, in general, most dominant and have the largest harems of females (Sandlant, 1981; Moller, 1985). Although no study has established whether, in fact, the largest males are most reproductively fit, it appears that there is sexual selection for the enhanced secondary sexual characters of final-instar males. Hence the environment acts to limit the extreme growth of these characters, while the characters themselves appear to confer a reproductive advantage on the males and the system becomes counterbalanced.

Despite the strong case for sexual selection driving enlarged features in tenth-instar males, there is a complicating additional feature in the reproductive system of tree wetas. It appears that the eighth- and ninth-instar males also reach sexual maturity (although histological studies have not proved this), inasmuch as they vie for mating access to females and show the same mating behaviour as tenth-instar males (Spencer, 1995). Once mature, the males do not undergo further moults. These males are able to occupy galleries with small entrances, and hence enjoy a size refuge wherein they can sequester females and escape agonistic eviction by large males. They also wander much more than the tenth-instar males and attempt many opportunistic matings away from galleries. The mating success of these satellite males is reduced (eighth-instar males = 22%, tenth-instar males = 44%; Spencer, 1995). Nevertheless their influence may place an additional limit on any selective drive for extreme secondary sexual characters.

There is an interesting variation in selection pressure acting against the large males. In the single-host gallery-provider system (kanuka longhorn beetle), the unimodal distribution of male weta head widths is superimposed on a normal distribution of entrance diameters. The weta modal value is about 2 mm smaller than the modal entrance diameter (Fig. 13.7B), but there is reasonable congruence of the distributions. In such a system, a high selection pressure forces the weta distribution to conform to that of the hole diameters made by the beetle, since the latter is the sole provider. No such selection pressure occurs in the multiple host gallery-provider system, such as that found in the broad-leaved forest, and male head widths should be correspondingly larger. This is suggested by the large mean head width (11.6 mm) measured for tenth-instar males from a North Island population (Spencer, 1995). On the other hand, as gallery entrance size increases and larger, more dominant males are supported, selection pressure against these males increases from another source: small mammalian predators, such as rats and weasels, which can invade the large galleries. Thus, there is still a counterbalance of natural selection acting against the sexual selection that drives the enlargement of cranium and mandibles in males.

References

Asher, G.W. (1977) Ecological aspects of the common tree weta (*Hemideina thoracica*) in native vegetation. Unpublished study, Zoology Department, Victoria University of Wellington, Wellington, New Zealand.

Edwards, J.S. (1961) The ecology and behaviour of the adult *Prionoplus reticularis* (Coleoptera: Cerambycidae) with general remarks on the reproduction in the Cerambycidae. *Quarterly Journal of Microscopical Science* 102, 519–529.

Emlen, S.T. and Oring, L.M. (1977) Ecology, sexual selection and the evolution of mating systems. *Science* 197, 215–223.

Field, L.H. (1978) The stridulatory apparatus of New Zealand wetas in the genus *Hemideina* (Insecta: Orthoptera: Stenopelmatidae). *Journal of the Royal Society of New Zealand* 8, 359–375.

Field, L.H. (1993) Structure and evolution of stridulatory mechanisms in New Zealand wetas (Orthoptera: Stenopelmatidae). *International Journal of Insect Morphology and Embryology* 22, 163–183.

Field, L.H. and Rind, F.C. (1992) Stridulatory behaviour in a New Zealand weta, *Hemideina crassidens*. *Journal of Zoology, London* 228, 371–394.

Field, L.H. and Sandlant, G.R. (1983) Aggression and mating behaviour in the Stenopelmatidae (Orthoptera, Ensifera), with reference to New Zealand wetas. In: Gwynne, D.T. and Morris, G.K. (eds) *Orthopteran Mating Systems – Sexual Competition in a Diverse Group of Insects.* Westview Press, Boulder, Colorado, pp. 120–146.

Fisher, R.A. (1930) *The Genetical Theory of Natural Selection.* Clarendon Press, Oxford.

Hosking, G.P. (1978) Ochrocydus huttoni *Pascoe (Coleoptera: Cerambycidae).* Forest and Timber Insects in New Zealand No. 30, Forest Research Institute, New Zealand Forest Service, Rotorua, 4 pp.

Kaneshiro, K.Y. and Giddings, L.V. (1987) The significance of asymmetrical sexual isolation and the formation of new species. In: Hecht, M.K., Wallace, B. and Prance, G.T. (eds) *Evolutionary Biology.* Plenum, New York, pp. 29–43.

Little, G.A. (1981) Food consumption and utilisation in two species of weta (*Hemideina femorata* and *H. maori*: Stenopelmatidae). Unpublished study, Department of Zoology, University of Canterbury, Christchurch, New Zealand.

Meads, M.J. and Moller, H. (1978) Introduction of giant wetas (*Deinacrida rugosa*) to Maud Island and observations of tree wetas, paryphantids and other invertebrates. Unpublished report, File No. 4/15/9, New Zealand DSIR Ecology Division, Lower Hutt, 21 pp.

Miller, D. (1984) *Common Insects in New Zealand.* A.H. and A.W. Reed, Wellington, New Zealand, 139 pp.

Moller, H. (1985) Tree wetas (*Hemideina crassicruris*) (Orthoptera: Stenopelmatidae) of Stephens Island, Cook Strait. *New Zealand Journal of Zoology* 12, 55–69.

Ordish, R.G. (1992) Aggregation and communication of the Wellington weta *H. crassidens* (Blanchard) (Orthoptera: Stenopelmatidae). *New Zealand Entomologist* 15, 1–8.

Sandlant, G.R. (1981) Aggressive behaviour of the Canterbury weta *Hemideina femorata* (Orthoptera: Stenopelmatidae): its adaptive significance in resource allocation. MSc thesis, University of Canterbury, New Zealand.

Spencer, A.M. (1995) Sexual maturity in the male tree weta *Hemideina crassidens* (Orthoptera: Stenopelmatidae). MSc thesis, Victoria University of Wellington, Wellington, New Zealand.

14
Parasites of Anostostomatid Insects

David A. Wharton, Robert Poulin and Claudine L. Tyrrell
Department of Zoology, University of Otago, PO Box 56, Dunedin, New Zealand

Introduction

Insects are parasitized by a wide range of organisms, including: bacteria, rickettsia, viruses, fungi, protozoa, nematodes, nematomorphs, platyhelminths (Tanada and Kaya, 1993) and other arthropods. There are few reports of parasites infecting Anostostomatidae and all are from New Zealand (Table 14.1). Published records are confined to fungi, protozoa and animal parasites (nematodes, nematomorphs and mites) and this review will focus on these groups. Details are provided on the life cycle of the parasites and on their effects on the host, as well as on other aspects of their biology, when additional information is available.

Fungi

Entomogenous fungi are transmitted by spores, which are either passed from host to host during physical contact or dispersed by the wind, or both (Madelin, 1966). Spores adhere to the insect cuticle; from there, the fungus penetrates the exoskeleton of the host and grows in its haemocoele. *Cordyceps kirkii* has been reported infecting *Deinacrida rugosa* from Stephens Island, New Zealand, with stromata growing from the intersegmental membranes, knee joints, bases of antennae, eyes and femur (Cunningham, 1923). Members of the genus *Cordyceps* typically release toxins, which sooner or later induce host death (Evans, 1988). Prior to their death, many *Cordyceps*-infected insects climb vegetation to die in exposed positions. This phenomenon has been termed 'summit disease', and it is believed to facilitate wind dispersal of fungal spores (Evans, 1988; Samson *et al.*, 1988).

Gregarines

Dale (1967) reported unidentified gregarines from the intestine of the weta *Hemideina thoracica*. This is the only record of gregarines in anostostomatids that could be found in the literature. These protozoans, however, are common parasites of insects, including the order Orthoptera, and they are likely to infect many other species in the family Stenopelmatidae.

Insects acquire gregarines through ingesting spores containing sporozoites of the parasites. Sporozoites develop into trophozoites, the feeding stages of gregarines, which attach to the epithelium of the host's intestine. Several hundred trophozoites can often be found in the same host. Most species of gregarines are unusual among protozoans in exhibiting only sexual reproduction. Trophozoites eventually become sexual forms, called gamonts. Two gamonts unite to form a cyst, in which gamete production and syngamy take place. Each newly formed zygote then undergoes sporogony to generate sporozoites, enclosed

Table 14.1. Parasite records in anostostomatid insects.

Parasite group	Parasite species	Host species	Geographical occurrence	Location in host	Prevalence	Intensity (mean no. per host)	Reference
Fungi	*Cordyceps kirkii*	*Deinacrida rugosa*	Stephen Island, NZ	Throughout body	?	?	Cunningham, 1923
Protozoa	Gregarines	*Hemideina thoracica*	Throughout NZ?	Intestine	?	?	Dale, 1967
Nematoda	*Wetanema hula*	*Hemideina thoracica*	Throughout NZ?	Hind-gut	43%	5.4	Dale, 1967
	Wetanema sp.	*Hemideina maori*	Rock and Pillar range, NZ	Hind-gut	82.4%	21.2	Tyrrell *et al.*, 1994
	Wetanema sp.	*Hemideina femorata*	Kaikoura,NZ	Hind-gut	97.5%	37.8	Tyrrell, 1991
	Cephalobellus sp.	*Hemideina thoracica*	Throughout NZ?	Hind-gut	?	?	Dale, 1967
Nematomorpha	*Gordius dimorphus*	*Hemiandrus* spp. *Zealandosandrus* spp.	Throughout NZ?	Body cavity	~6%	Usually one	Poinar, 1991a R. Poulin (unpublished)
	Gordionus diblastus	*Hemideina* spp. *Hemiandrus furcifer*	Throughout NZ?	Body cavity	?	Usually one	Poinar, 1991a
	Euchordodes nigromaculatus	*Deinacrida* spp.	South Island, NZ	Body cavity	?	Usually one	Poinar, 1991a
Acarina	?	*Hemideina thoracica*	?	External surface	?	?	Maskell, 1926

NZ, New Zealand.

within a spore. Spores pass out in the host's faeces to await ingestion by another host and begin the cycle anew.

Gregarine infections are not lethal but they do deplete the host's nutrient reserves. In other orthopterans, gregarine infections result in retarded growth (Harry, 1970; Zuk, 1987a) and suppressed reproduction (Zuk, 1987b; Simmons, 1993). These effects are particularly severe when hosts are faced with dietary restrictions, and gregarines can therefore be an indirect source of insect mortality during periods of food shortage.

Nematomorphs

Three described species of hair-worms are known to parasitize four genera of New Zealand weta (Poinar, 1991a). These same worms also infect members of other orthopteran families in New Zealand (Acrididae: Poulin, 1995; Rhaphidophoridae: Thomas *et al.*, 1998a) as well as other insect taxa (Zervos, 1989). They also infect stenopelmatids in western North America (Weissman, cited by Poinar, 1991a).

Hair-worms are large parasites, reaching 20–30 cm in length. The species infecting weta have complex life cycles, requiring two different hosts for completion (Poinar, 1991b). Adult worms live freely in fresh water, where they mate and lay eggs. Larvae hatched from these eggs burrow into or are ingested by the aquatic juvenile stages of insects, such as mayflies or caddis-flies, in which they encyst. There they remain dormant during their first host's metamorphosis and emergence from water up until the time of its death. Omnivorous insects like weta feed on dead insects and accidentally ingest hair-worm cysts. Inside the weta host, hair-worm larvae leave their cyst, burrow through the gut wall and enter the body cavity, where they will develop over several months toward adulthood. Typically, a single mature worm emerges from each infected weta, suggesting either that infection rates are very low or that competition occurs among young worms inside the weta, such that a single parasite survives to develop. The parasite feeds first on the fat bodies of its host before consuming internal organs. By the time it is ready to leave its host, it usually occupies the entire abdomen of the weta.

Because adult worms must reach fresh water to mate and lay eggs, they would benefit if their hosts were to visit any water body when the worms are ready to emerge. The literature contains many anecdotal reports of insects parasitized by hair-worms jumping into water with the parasite emerging from the host immediately following immersion (Poinar, 1991b; Schmidt-Rhaesa, 1997). This has led many authors to suggest that hair-worms have the uncanny ability to trigger a craving for water in their host. To this date, however, there is no solid evidence for hair-worm-induced hydrophilia in weta or other insects.

Hair-worms have very severe effects on their hosts. First, they are a cause of developmental instability in the insect host. For instance, cave weta (family Rhaphidophoridae) harbouring hair-worms display significantly greater asymmetry in hind femur length than uninfected conspecifics (Thomas *et al.*, 1998b). Secondly, parasitized hosts are functionally castrated. Ovaries do not develop or, if they do, they are consumed by the parasite. Thirdly, the parasite almost invariably kills its host when emerging from it at the end of its development. This makes hair-worms parasitoids rather than true parasites.

The prevalence of hair-worms is usually low, i.e. less than 10% in well-sampled host populations (Schmidt-Rhaesa, 1997). The only estimate available for a stenopelmatid host is of 6% in a population of the ground weta *Zealandosandrus gracilis* from an alpine region of the South Island of New Zealand (R. Poulin, unpublished data). Thus they are unlikely to be a major source of mortality in weta populations, despite being highly pathogenic.

Parasitic Nematodes

Mermithid nematodes

Mermithid nematodes have a similar life cycle to that of nematomorphs, with a parasitic larval stage in the haemocoele of insects and a free-living adult. They infect a wide range of insects, including orthopterans (Poinar, 1975). There have been no published reports of mermithids infecting weta, although they are probable hosts. Poinar (1991a) mentions that mermithids and nematomorphs are often confused in museum collections and he recorded no mermithids in the specimens he examined.

Steinernematid and heterorhabditid nematodes

These families of nematodes have attracted a lot of interest as potential biological control agents of insect pests. They carry symbiotic bacteria, which, when released into the haemocoele of the host, cause septicaemia and death. They infect a wide range of insects and have been recorded from orthopterans (Poinar, 1990). They have not been reported as infecting anostostomatids, although it seems likely that they would do so. It is difficult to observe natural populations of this nematode within its host, as the host is rapidly killed after infection. Host susceptibility is usually established experimentally.

Thelastomatid nematodes

Thelastomatid nematodes are members of the order Oxyurida and have been described from the hind-guts of millipedes and of four orders of insects: Coleoptera, Blattodea, Diptera and Orthoptera. Oxyurids are restricted to the hind-gut of their hosts and nearly all feed on bacteria and are associated with hosts that utilize the hind-gut as a site of bacterial fermentation (Adamson, 1994).

Wetanema hula has been recorded from the hind-gut of *H. thoracica* (Dale, 1967). *Wetanema* spp. are also present in the hind-guts of *Hemideina maori* (Tyrrell *et al.*, 1994) and *Hemideina femorata* (Tyrrell, 1991). These nematodes are morphologically similar to *W. hula*, but their species identity has not been confirmed. However, size differences between *Wetanema* spp. from *H. femorata, H. maori* and *H. thoracica* may indicate that they are different species (Table 14.2). Male and female *Wetanema* sp. from *H. maori* are significantly larger than those from *H. femorata* (t test, $P < 0.001$; Tyrrell, 1991). Dale (1967) also reports the presence of male nematodes of the genus *Cephalobellus* in the intestine of *H. thoracica*.

The intensity of infection (the number of nematodes per host) of *H. maori* with *Wetanema* sp. increases with host size, although the correlation is not significant (Fig. 14.1). Infection prevalence (the proportion of hosts infected) in large weta (tibia length > 15 mm) was 75%, whereas prevalence in small weta (tibia length < 15 mm) was only 30%. The highest intensity of infection recorded was 184 nematodes from one weta (Tyrrell *et al.*, 1994). As the nematodes can be as large as 5.5 mm in length and are confined to a small space (the hind-gut), such high worm burdens may represent a stress to their host, although thelastomatids are generally considered to be benign (Poinar, 1975).

In a survey undertaken in 1991 (Tyrrell, 1991; Tyrrell *et al.*, 1994), infection prevalence (taking only large weta) was high throughout the year (Fig. 14.2). Mean intensity varied from a high of 51 nematodes per weta in February to a low of six nematodes per weta in March. There was, however, no clear seasonal pattern to infections. The proportions of different nematode stages varied throughout the year (Fig. 14.3), although again there was no clear seasonal pattern. All nematode stages were found within the weta throughout the year.

Infections consisting of female nematodes only were observed much more frequently than male-only infections in *H. maori*, whereas male-only infections were not observed in *H. femorata* and about a third of infections consisted of females only (Table 14.3). For *H. maori* infected with adult nematodes, 26.7% contained all female nematodes and 13.3% all males and 60% were mixed. For infected *H. femorata*, 40.5% contained all female nematodes, 59.5% were mixed and there were no all-male infections. After dissection, females from these female-only infections were observed to have

Table 14.2. *Wetanema* spp. reported from *Hemideina* spp.

Host sp.	Nematode sp.	Female size (mm)	Male size (mm)
Hemideina femorata	*Wetanema* sp.	4.06 ± 0.06*	1.36 ± 0.03
Hemideina maori	*Wetanema* sp.	5.52 ± 0.23	2.43 ± 0.08
Hemideina thoracica	*Wetanema hula*	6.38	1.70

* Mean ± 1 SE, standard errors not stated for *W. hula*. (Data from Tyrrell, 1991 (*H. femorata, H. maori*) and Dale, 1967 (*H. thoracica*).)

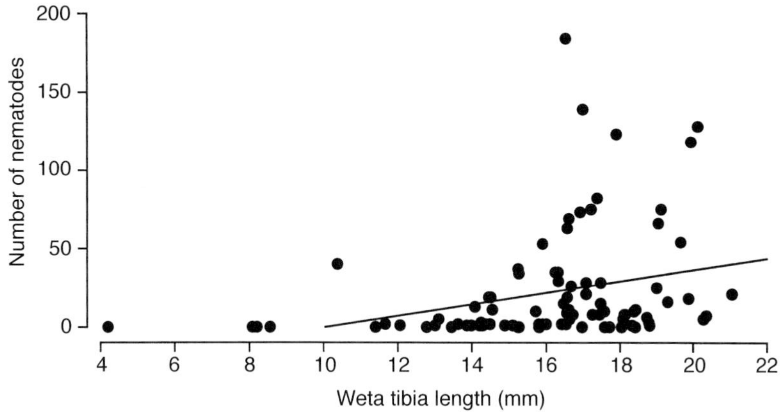

Fig. 14.1. Intensity of infection of *Wetanema* with the size of *Hemideina maori*, as indicated by tibia length.

sperm stored in their reproductive tracts in a seminal receptacle. This suggests that males had been present but had been lost or removed from the gut.

Nematode sex ratios observed were also heavily biased towards females, with a mean ratio of female : male worms in infrapopulations (parasites within individual weta) of 4.5 in mixed infections of *H. maori* and of 6.5 in *H. femorata* (Table 14.3). There was, however, considerable variation in infrapopulations observed within weta in mixed infections, with males outnumbering females in some infections but with females outnumbering males by as much as 46 : 1 in other infections. For the suprapopulation (the total nematode population in weta examined), the female : male ratio was similar in nematodes from single sex infections and those from mixed infections in *H. maori* (Table 14.3). The ratio in the overall suprapopulation (mixed and single-sex infections) was also similar (2.1). This may indicate that the sex ratio is not maintained by interaction between the sexes but by some other mechanism. A genetic mechanism that ensures that there are twice as many females as males is one possibility, but the evidence that male worms are lost from female-only

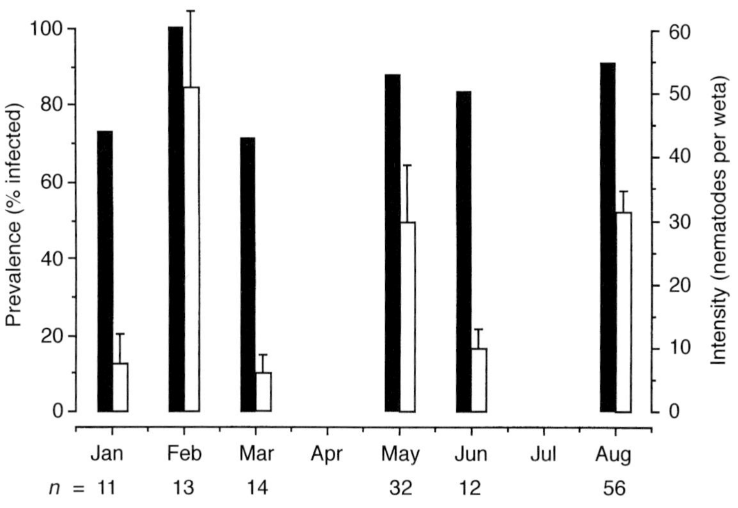

Fig. 14.2. Changes in intensity (□) and prevalence (■) of *Wetanema* in *Hemideina maori* with sample date. Vertical lines represent the standard error of the mean (*n* = the number of weta in each sample) (from Tyrrell *et al.*, 1994).

infections suggests that the ratio is maintained by differential mortality of male and female nematodes. In the spatially and resource-limited environment that the hind-gut of a weta represents, it may be advantageous for there to be more females than males, particularly if males are removed after mating. This would leave the resource available for egg production by females, using stored sperm for fertilization.

Unusual population structures have been

Table 14.3. Sex ratios and composition of infections of *Wetanema* spp. in *Hemideina* spp. (data from Tyrrell, 1991). Infections consisting of larvae only not included.

Wetanema spp. *infection*	*Hemideina maori*	*Hemideina femorata*
% all female nematodes	21.3	34.1
% all male nematodes	10.6	0
% mixed infection	47.9	50
% uninfected	17	4.5
Mean number of males from all-male infections (range)	3.3 (1–20)	–*
Mean number of males from mixed infections (range)	7.9 (1–55)	9.4 (1–53)
% of single males from all-male infections	60%	–*
% of single males from mixed infections	28.9%	18.2%
Mean number of females from all-female infections (range)	3.9 (1–16)	11.4 (1–33)
Mean number of females from mixed infections (range)	16.4 (1–127)	24 (1–83)
Average female : male nematodes in mixed infections (± SE, range)	4.5 ± 1.3 (0.4–46)	6.5 ± 1.8 (0.3–36)
Overall female : male nematodes in mixed infections	2.1	2.6
Overall female : male nematodes in single sex infections	2.3	–*

*No all-male infections were observed.

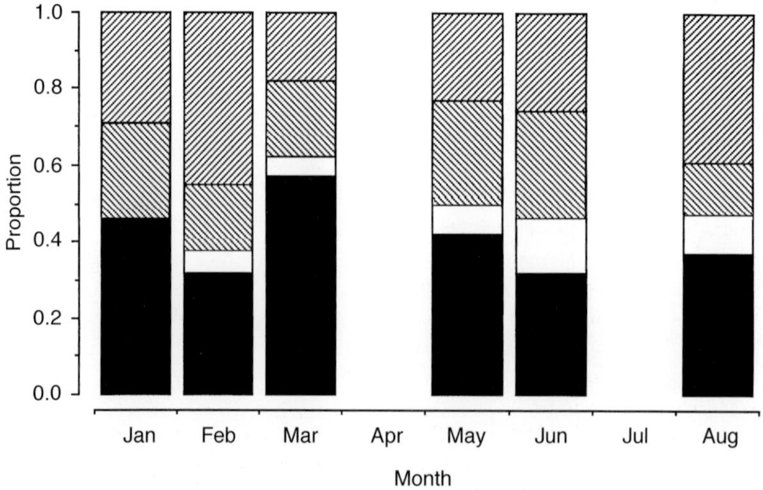

Fig. 14.3. Changes in the proportions of adult female (■), immature female (□), male (▧) and juvenile (▨) *Wetanema* in *Hemideina maori* with sample date (from Tyrrell *et al.*, 1994).

reported in thelastomatid nematodes infecting other arthropods. In some cockroaches, the predominant infrapopulation structure is one male and one or more females. This is the case for *Blatticola monandros* infecting *Parellipsidion pachycercum*, *Protellus dixoni* infecting *Drymaplaneta variegata* and for *Thelastoma bulhoesi* and *Leidynema appendiculata* infecting *Periplaneta americana* (Zervos, 1988a, b; Adamson and Noble, 1992). Infrapopulations of *Thelastoma periplaneticola* and *Hammerschmidtiella diesingi* infecting *P. americana*, however, consisted of several males and females (Adamson and Noble, 1992). Zervos (1988a) lists a number of examples of thelastomatid infections consisting of only one male or where males are outnumbered by females. Single male infections did not predominate in *Wetanema* sp. infections in *H. maori* or *H. femorata* (Table 14.3), but the sex ratios observed do indicate that the population of male nematodes is regulated by some mechanism. Oxyurids exhibit haplodiploidy, in which males are haploid and originate from unfertilized eggs and females are diploid and arise from fertilized eggs. This may favour the production of biased sex ratios (Adamson, 1989).

Hemideina maori inhabits the alpine zone of the South Island of New Zealand. It is regularly exposed to temperatures between −5 and −9°C and occasionally to as low as −11°C (Ramlov *et al.*, 1992; Sinclair, 1997). It is freezing-tolerant, surviving ice formation in its haemolymph and other extracellular compartments (Ramlov *et al.*, 1992; Sinclair and Wharton, 1997). All stages of *Wetanema* sp. are found within their host throughout the year, suggesting that the nematode can survive periods during which the host is frozen in the field (Tyrrell *et al.*, 1994). In laboratory studies, live nematodes were recovered from weta that had been frozen to temperatures as low as −61°C. In experiments where weta were frozen to three temperature bands (−10 to −19°C, −20 to −29°C and below −30°C), there was no significant decrease in nematode survival with temperature (Tyrrell *et al.*, 1994). The lower lethal temperature of *H. maori* is about −11°C (Ramlov *et al.*, 1992). The nematode is thus able to survive much lower temperatures than that survived by its host. The assessment of survival in a parasitic nematode is, however, difficult. The parasite is dependent upon its host for survival and will die after its host dies or after removal from the host. This will impose a stress in addition to the freezing stress on the nematode.

Freezing at a high subzero temperature is advantageous to a freezing-tolerant animal (Duman *et al.*, 1991). If the nematode acted as an ice-nucleating agent, it could assist the survival of its host by ensuring freezing at a high temperature. However, *Wetanema* sp. does not appear to be able to act in this way, since there was no correlation between the intensity of the nematode infection and the supercooling point of the weta and no significant differences between the supercooling points of infected and uninfected weta (Tyrrell *et al.*, 1994).

The freezing of *Wetanema* sp. was observed directly on a microscope cold stage (Tyrrell *et al.*, 1994). The nematode froze by ice nucleating from the external medium, with freezing starting at the posterior end of the body and travelling forwards. Nematodes observed to freeze on the cold stage survived for 6 h or more. The nematode is thus freezing-tolerant and mirrors the cold-tolerance strategy of its host; suggesting the coevolution of cold-tolerance strategies of the parasite and its host. The physiology of freeze tolerance by *H. maori* is discussed by Neufeld and Leader (Chapter 25, this volume).

Although the life cycle and transmission of *Wetanema* sp. has not been studied, it is likely to have the typical thelastomatid pattern, consisting of a direct life cycle in which the nematode develops to an infective larva within the egg. Whether the embryo or larva within the egg can survive low temperatures outside the host is not known, but the increase in the proportion and intensity of juveniles in late winter may indicate the acquisition of new infections (Tyrrell *et al.*, 1994). The presence of an eggshell in nematodes can prevent inoculative freezing and allow the egg to supercool and avoid freezing in the presence of external ice (Wharton, 1995).

Acarina

The presence of mites on the external surface of weta has been reported by Maskell (1926) and in several general accounts of weta biology. The relationship between mites and insects is often a phoresis, in which the mite uses the host for transport. A report of mites with their mouth-parts permanently embedded in the neck membranes of weta (Ramsay, 1975), however, suggests an ectoparasitic relationship. The identification, prevalence, intensity and host range of these mites are unknown.

Conclusions

Insect parasites are usually studied because of their potential as biological control agents of pest species. The lack of information on parasites of Anostostomatidae, especially outside New Zealand, is partly a reflection of their low importance as agricultural pests or as vectors of human or animal diseases.

Parasites could, however, be important for the conservation biology of anostostomatids. Several species of New Zealand weta are either endangered or protected, and the potential impact of parasites on their fragmented populations is unclear. In general, the role of parasitism in insect conservation is ignored by most researchers (e.g. Gaston *et al.*, 1993). Lethal parasites, such as nematomorphs and mermithid nematodes, although usually not very prevalent, could still cause significant losses in weta populations. One important aim for future investigations will be to quantify the influence of parasites on the population biology of threatened anostostomatids and to use the information in the design of conservation strategies.

References

Adamson, M.L. (1989) Evolutionary biology of the Oxyurida (Nematoda): biofacies of a haplodiploid taxon. *Advances in Parasitology* 28, 175–228.

Adamson, M.L. (1994) Evolutionary pattern in the life histories of Oxyurida. *International Journal for Parasitology* 24, 1167–1177.

Adamson, M.L. and Noble, S. (1992) Structure of the pinworm (Oxyurida: Nematoda) guild in the hindgut of the American cockroach, *Periplaneta americana*. *Parasitology* 104, 497–507.

Cunningham, G.H. (1923) A singular *Cordyceps* from Stephen Island, New Zealand. *Transactions of the British Mycological Society* 8, 72–75.

Dale, P.S. (1967) *Wetanema hula* n. gen. et sp., a nematode from the weta *Hemideina thoracica* (White). *New Zealand Journal of Science* 10, 402–406.

Duman, J.G., Xu, L., Neven, L.G., Thursman, D. and Wu, D.W. (1991) Hemolymph proteins involved in insect subzero-temperature tolerance: ice nucleators and antifreeze proteins. In: Lee, R.E. and Denlinger, D.L. (eds) *Insects at Low Temperatures*. Chapman and Hall, New York and London, pp. 94–127.

Evans, H.C. (1988) Coevolution of entomogenous fungi and their insect hosts. In: Pirozynski, K.A. and Hawksworth, D.L. (eds) *Coevolution of Fungi with Plants and Animals*. Academic Press, London, pp. 149–171.

Gaston, K.J., New, T.R. and Samways, M.J. (1993) *Perspectives on Insect Conservation*. Intercept, Andover, UK.

Harry, O.G. (1970) Gregarines: their effect on the growth of the desert locust (*Schistocerca gregaria*). *Nature* 225, 964–966.

Madelin, M.F. (1966) Fungal parasites of insects. *Annual Review of Entomology* 11, 423–448.

Maskell, F.G. (1926) The anatomy of *Hemideina thoracica*. *Transactions of the New Zealand Institute* 57, 637–670.

Poinar, G.O., Jr (1975) *Entomogenous Nematodes*. E.J. Brill, Leiden.

Poinar, G.O., Jr (1990) Biology and taxonomy. In: Gaugler, R. and Kaya, H.K. (eds) *Entomopathogenic Nematodes in Biological Control*. CRC Press, Boca Raton, pp. 23–62.

Poinar, G.O., Jr (1991a) Hairworm (Nematomorpha: Gordioidea) parasites of New Zealand wetas (Orthoptera: Stenopelmatidae). *Canadian Journal of Zoology* 69, 1592–1599.

Poinar, G.O., Jr (1991b) Nematoda and Nematomorpha. In: Thorp, J.H. and Covich, A.P. (eds) *Ecology and Classification of North American Freshwater Invertebrates*. Academic Press, New York, pp. 249–283.

Poulin, R. (1995) Hairworms (Nematomorpha: Gordioidea) infecting New Zealand short-horned grasshoppers (Orthoptera: Acrididae). *Journal of Parasitology* 81, 121–122.

Ramlov, H., Bedford, J. and Leader, J. (1992) Freezing tolerance of the New Zealand alpine weta, *Hemideina maori* Hutton (Orthoptera; Stenopelmatidae). *Journal of Thermal Biology* 17, 51–54.

Ramsay, G.W. (1975) Mites and ticks. In: Knox, R. (ed.) *New Zealand's Nature Heritage*, Vol. 5. Hamlyns, Hong Kong, pp. 1790–1796.

Samson, R.A., Evans, H.C. and Latgé, J.-P. (1988) *Atlas of Entomopathogenous Fungi*. Springer-Verlag, Berlin.

Schmidt-Rhaesa, A. (1997) *Freshwater Fauna of Central Europe*, Vol. 4/4: *Nematomorpha*. Gustav Fischer Verlag, Stuttgart.

Simmons, L.W. (1993) Some constraints on reproduction for male bushcrickets, *Requena verticalis* (Orthoptera: Tettigoniidae): diet, size and parasite load. *Behavioral Ecology and Sociobiology* 32, 135–140.

Sinclair, B.J. (1997) Seasonal variation in freezing tolerance of the New Zealand alpine cockroach *Celatoblatta quinquemaculata*. *Ecological Entomology* 22, 462–467.

Sinclair, B.J. and Wharton, D.A. (1997) Avoidance of intracellular freezing by the freezing-tolerant New Zealand alpine weta *Hemideina maori* (Orthoptera: Stenopelmatidae). *Journal of Insect Physiology* 43, 621–625.

Tanada, Y. and Kaya, H.K. (1993) *Insect Pathology*. Academic Press, San Diego and London.

Thomas, F., Ward, D.F. and Poulin, R. (1998a) Fluctuating asymmetry in an ornamental trait in the cave weta *Pleioplectron simplex* Hutton (Orthoptera: Rhaphidophoridae): no role for parasites. *Canadian Journal of Zoology* 76, 931–935

Thomas, F., Ward, D.F. and Poulin, R. (1998b) Fluctuating asymmetry in an insect host: a big role for big parasites? *Ecology Letters* 1, 112–117.

Tyrrell, C. (1991) Freezing survival of *Wetanema* sp. (Nematoda: Thelostomatidae) and their hosts. BSc Hons thesis, University of Otago, Dunedin, New Zealand.

Tyrrell, C., Wharton, D.A., Ramlov, H. and Moller, H. (1994) Cold tolerance of an endoparasitic nematode within a freezing-tolerant orthopteran host. *Parasitology* 109, 367–372.

Wharton, D.A. (1995) Cold tolerance strategies in nematodes. *Biological Reviews* 70, 161–185.

Zervos, S. (1988a) Population dynamics of a thelastomatid nematode of cockroaches. *Parasitology* 96, 353–368.

Zervos, S. (1988b) Evidence for population self-regulation, reproductive competition and arrhenotoky in a thelastomatid nematode of cockroaches. *Parasitology* 96, 369–379.

Zervos, S. (1989) Stadial and seasonal occurrence of gregarines and nematomorphs in two New Zealand cockroaches. *New Zealand Journal of Zoology* 16, 143–146.

Zuk, M. (1987a) The effects of gregarine parasites on longevity, weight loss, fecundity and developmental time in the field crickets *Gryllus veletis* and *G. pennsylvanicus*. *Ecological Entomology* 12, 349–354.

Zuk, M. (1987b) The effects of gregarine parasites, body size, and time of day on spermatophore production and sexual selection in field crickets. *Behavioral Ecology and Sociobiology* 21, 65–72.

Part IV
Behaviour

15
Stridulatory Mechanisms and Associated Behaviour in New Zealand Wetas

Laurence H. Field
Department of Zoology, University of Canterbury, Private Bag 4800, Christchurch, New Zealand

Introduction

Sound production in insects is accomplished either by stridulation – the frictional rubbing of cuticular parts – or by strigilation – the setting of drum-like cuticular membranes into oscillatory motion through rapid muscular contraction (Haskell, 1961). Such sounds are most commonly used as defence or alarm signals, but, especially in the Orthoptera, sound patterns are used for social communication in mating, territorial and agonistic behavioural contexts (Alexander, 1967). Generally, insect defensive sounds have broad-band frequency characteristics and are not temporally patterned. In contrast, social signals, which are most highly developed in the ensiferan Orthoptera, often show distinctive patterns and species-specific frequency characteristics. These are generated by stridulation of highly modified wing tegmina in the relatively modern ensiferan families, such as the crickets (Gryllidae) and bush crickets (Tettigoniidae) (Bailey, 1991). In the primitive ensiferan families (e.g. Anostostomatidae, Gryllacrididae, Haglidae), wings are usually lacking, and stridulation is accomplished by rubbing other body parts together, such as legs and abdomen (Field and Bailey, 1997). The stridulatory mechanisms and sound patterns in these groups probably reflect the origins of acoustic communication in Orthoptera, particularly since many of the primitive genera lack ears and appear to use sound only for defence, as a plesiomorphic (primitive) condition (Field, 1993a).

New Zealand wetas are frequently heard at night, as they make scratchy patterned calls in the forest. When disturbed from their daytime refuges in tree galleries or ground burrows, they display spectacular leg-kicking defence behaviours, accompanied by sound bursts (see Field and Glasgow, Chapter 16, this volume). This chapter reviews the variety of known stridulatory mechanisms, sound patterns and inherent messages, as well as behavioural roles of stridulation. Most information is from the tree wetas (*Hemideina*), inasmuch as they show the greatest elaboration in social signalling. In addition to airborne sound production, it has recently been discovered that substrate vibration is used as a mode of communication in tree wetas and ground wetas (subfamily Henicinae), and the stage is set for future research in this area. A description of communication by abdominal substrate drumming in North American Jerusalem crickets (Stenopelmatinae) is given in Weissman (Chapter 19, this volume).

Stridulatory Apparatus

The ubiquitous stridulatory structures, first described for New Zealand wetas (Graber, 1874) and subsequently found in all other Anostostomatidae, conform to an abdomino-femoral mechanism (Field, 1993a). This consists of patches of spines, pegs or a file of ridges on each side of one or more abdominal segments (tergites), which are rubbed by apposing pegs on the inner (medial) surface of each hind femur (Fig. 15.1). One to seven tergites may bear stridulatory structures, but, if one, it is invariably tergite II, that closest to the base of the hind leg. In *Hemideina* and *Deinacrida*, the spines, ridges or pegs of the tergite always face into the direction of movement of pegs on the leg, so that the apposing apices of the structures repeatedly catch and release as the leg sweeps past the abdomen in an arc movement. To ensure engagement, individual pegs in a field are usually orientated along an arc that is similar to the arc of travel of the femur across the tergite (Field, 1978, 1982; McVean, 1986).

Similar structures occur in the stridulatory mechanisms of both sexes for all species of New Zealand wetas examined. Stridulatory pegs are lacking on the hind femora of early *Hemideina* instars (probably up to the fourth instar – data for *H. maori*), although abdominal file ridges occur right through the instars, and the number of ridges appears to remain constant during maturation (Field, 1978).

In a study of 35 species (seven genera) of New Zealand wetas, Field (1993a) described three additional stridulatory mechanisms besides the abdomino-femoral one. These occur in one or two species each, and will be described in the appropriate sections below. The mechanisms include mandibulo-mandibular (tusked wetas), pleuro-coxal (*Hemideina broughi*) and tergo-tergal (*Deinacrida rugosa, Deinacrida parva*) structures.

Enlarged lateral air sacs occur beneath the file ridges in the abdominal tergites of *Hemideina* spp. and are thought to act as reverberatory chambers for enhancing stridulatory sounds (Field, 1978).

Subfamily Henicinae (ground wetas and tusked wetas)

Hemiandrus

All species of *Hemiandrus* in New Zealand lack tibial tympanal organs and are presumably deaf to airborne sound. The 16 species have a femoro-abdominal mechanism usually consisting of patches of tiny spines on at least three abdominal tergites and small rows or patches of pointed pegs on the femur, as represented in Fig. 15.1A and B. The general pattern in *Hemiandrus* of many tiny spines/pegs on several tergites appears to be the plesiomorphic condition for the Anostostomatidae,

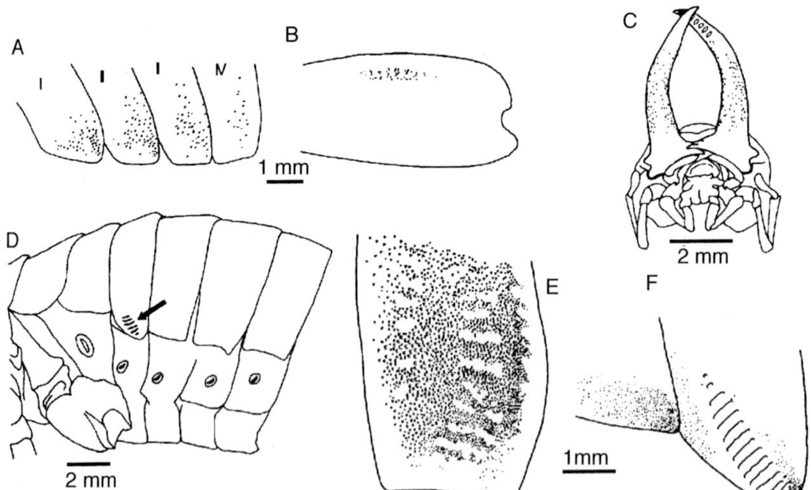

Fig. 15.1. Stridulatory structures in henicine and deinacridine wetas of New Zealand. A–B. Femoro-abdominal pegs on the abdominal tergites and inner surface of the femur characterize the mechanism found in *Hemiandrus* spp. (here shown for *H. maculifrons*). C. In New Zealand tusked wetas, tubercles on apposing surfaces of the tusks (here shown for *Anisouris nicobarensis* (= *H. monstrosus*)) engage when the tusks are opened, and produce sound. D. Stridulatory ridges (arrow) on the second abdominal tergite of *Hemideina* spp. are rubbed by the inner surface of the femur. E–F. Enlarged views of the femoral component (E) and tergal ridges (F) of *Hemideina maori*. (A–C from Field, 1993a; D–F from Field, 1978; all with permission.)

inasmuch as it is found in 90% of the species worldwide and in the earliest fossil anostostomatid representatives from the Queensland Triassic, as well as in other families of the Stenopelmatoidea (Riek, 1970; Field, 1993a; P.M. Johns, New Zealand, 1991, personal communication; see also Gorochov, Chapter 1, this volume).

Three groups of species are recognized in *Hemiandrus*, based upon the density of pegs on the tergite component (Table 15.1). Ten of the species are undescribed, but their stridulatory structures have been illustrated, together with tabulated collection data, by Field (1993a).

In the 'sparse' group, the femoral component is very reduced or lacking; in the latter case (the first two species in Table 15.1), sound production may not be possible. The tergal component is also reduced to several tens of pegs on tergites I–III plus a few on the remaining tergites in *Hemiandrus fiordensis*. In the 'medium' group, pegs are concentrated on tergites I–III in *Hemiandrus* sp. B and evenly distributed over the tergites in the other species. The unusual femoral component of *Hemiandrus* sp. C consists of sharp spines. The 'dense' group contains the greatest variety of component morphology. The tergal component ranges from tiny rounded knobs (*Hemiandrus* spp. D and G) to conical pegs (*Hemiandrus similis, Hemiandrus anomalus, Hemiandrus* spp. E, F, H and I) to sharp, elongate spines (*Hemiandrus maculifrons* (Fig. 15.2A). The femoral component consists of sharp conical pegs orientated distally and arranged in transverse bands (Figs 15.1B and 15.2B) (Field, 1993a).

Tusked wetas

Two species of henicines have elongated tusks projecting anteriorly from the ventrally orientated mandibles. Although the tusks bear similar stridulatory structures, the two species differ considerably in their femoro-abdominal stridulatory mechanisms (Field, 1993a). *Anisouris nicobarensis* (= *Hemiandrus monstrosus*) is a diminutive species which has a rudimentary file on abdominal tergite II, composed of many ridges arranged into two to three columns, while the femoral component consists of blunt pegs similar to those of other *Hemiandrus* species. The tusks curve medially toward the tips and cross slightly (Fig. 15.1C). The apposing surfaces bear oval tubercles, which are rubbed together. The giant tusked weta, *Motuweta isolata*, has long tapered tusks, which curve and cross (left over right) distally when the mandibles are closed. Distal annular ridges form into heavy tubercles on the contacting surfaces of the tusks. These are rubbed against each other as the weta opens the overlapping mandibles. In the femoro-abdominal mechanism, the tergal structures consist of broad, blunt pegs on tergites I–IV, which are rubbed by numerous, small, pointed femoral pegs (see Fig. 10.1E, G in Field and Deans, Chapter 10, this volume).

Table 15.1. Stridulatory structure characteristics of the three groups of *Hemiandrus* species. The first two species of the 'sparse' group may have microscopic structures on the femur not visible at the magnifications used (100×). (Alphabetical species designations after Field, 1993a.)

Group	Species	Tergites with pegs/spines	Tergite	Femur
Sparse	*H. subantarcticus*	I II III	Pegs	Lacking
	H. fiordensis	I II III IV V VI	Pegs	Sharp pegs
	H. sp. A	I II III	Pegs	Pegs
Medium	*H. gracilis*	I II III IV V	Pegs	Pegs
	H. sp. B	I II III IV V VI	Pegs	Pegs
	H. sp. C	I II III IV V	Pegs	Spines
Dense	*H. similis*	I II III IV	Sharp pegs	Pegs
	H. anomalus	I II III IV	Spines	Spines
	H. maculifrons	I II III IV	Spines	Pegs
	H. sp. D	I II III IV	Pegs	Knobs
	H. sp. F	I II III IV	Spines	Knobs
	H. sp. G	I II III IV V VI	Knobs	Pegs
	H. sp. H	I II III IV V	Pegs	Pegs

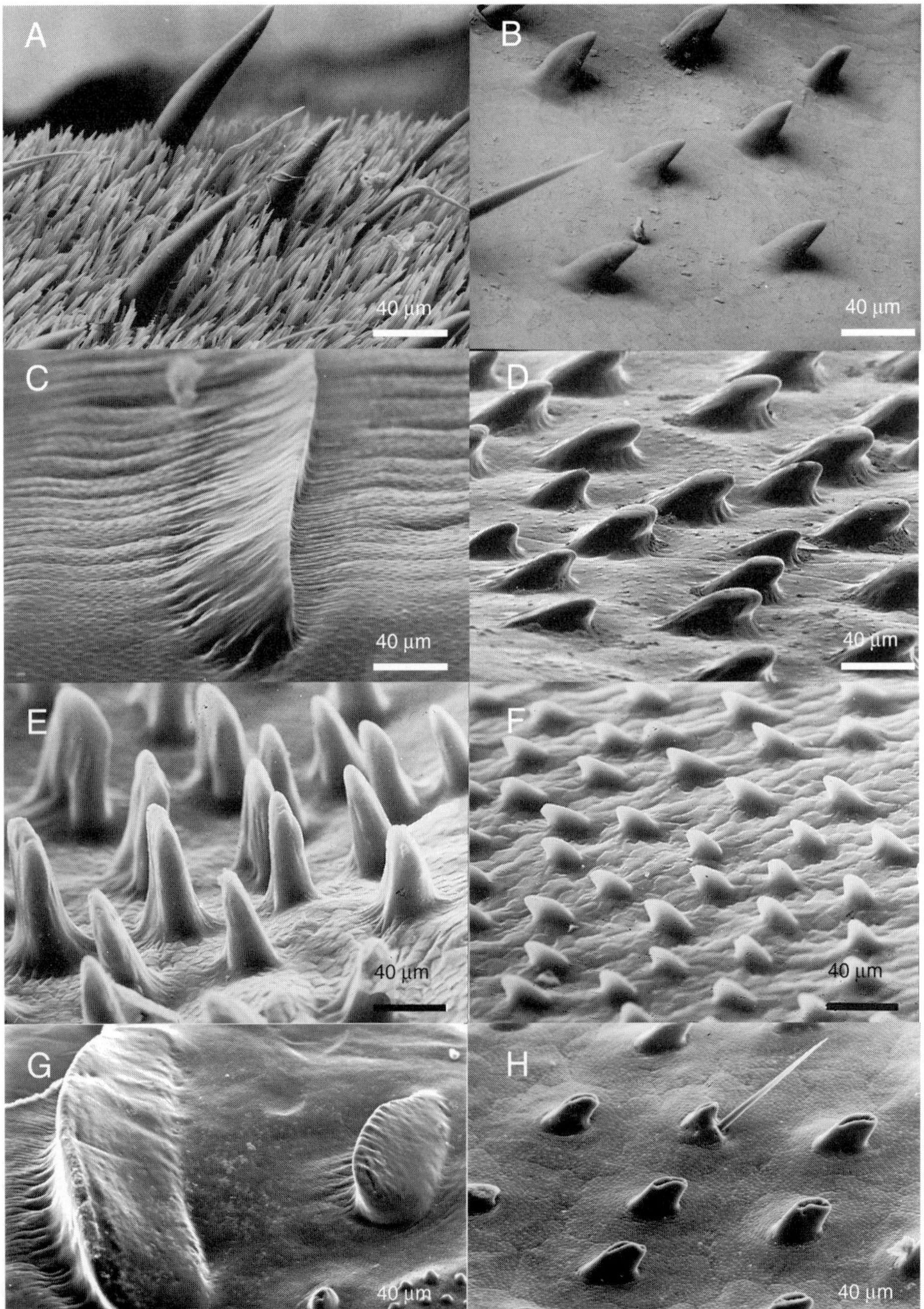

Fig. 15.2. Scanning electron microscope views of stridulatory components of New Zealand wetas. A–B. Tergal spines (A) are rubbed laterally by femoral spines (B) in *Hemiandrus maculifrons*. C–D. File ridge (C) of *Hemideina crassidens* faces directly into the leading apices of the femoral pegs (D). E–F. Flat, slat-like pegs sit erect on the tergites of *D. connectens* (E) and are rubbed laterally by tiny keel-like pegs on the femur (F). G–H. Massive crescentic tergal ridges of *D. rugosa* (G) are engaged by molar-like femoral pegs (H), which sweep laterally across the ridge edge. C–H from Field (1982) with permission.

Subfamily Deinacridinae (tree and giant wetas)

Stridulatory structures have been described for 18 species (including four undescribed ones) in *Hemideina* and *Deinacrida*. The femoro-abdominal structures are restricted to the first one or two tergites of the abdomen, with a trend toward formation of tergal file ridges, while the femoral component shows a diversity of peg shapes, which become the largest and most elaborate in the giant wetas (*Deinacrida*). All deinacridine wetas have large, paired tympanal organs (ears) on the front tibiae, and, in *Hemideina* at least, have good hearing at frequencies up to 5 kHz (Field *et al.*, 1980).

Hemideina

Eight species are recognized, of which *H. broughi* probably belongs to a separate genus (undescribed), based upon many morphological differences from other *Hemideina* (G. Ramsay, Auckland, 1997, personal communication). The other seven species contain a femoro-abdominal mechanism only, characterized by a single file on tergite II (Figs 15.1D, F and 15.2C), and patches of blunt pegs on the femur (Figs 15.1E and 15.2D). The sharp margin of each file ridge projects forward (Fig. 15.2C), such that it engages the femoral pegs as they are brought to bear on the tergite. The rounded apex of each femoral peg is canted in the direction of travel across the file ridges, thus assisting in catching on to the ridges (Fig. 15.2D) (Field, 1978).

Hemideina species fall into two distinct groups, based upon the number of file ridges (Field, 1993a). In the first group (*H. maori, H. ricta*), ten to 14 ridges occur in a long, narrow file extending from the ventral posterior corner of tergite II, and the femoral pegs are very small (Fig. 15.1E, F). In the second group (*H. crassidens, H. femorata, H. thoracica, H. trewicki*), the file contains three to eight ridges and is shorter and wider than in the first group. Also the file ridges are higher and more massive. The femoral pegs are much more massive and larger than those in the first group (Field, 1982).

Hemideina broughi femoro-abdominal structures differ markedly from the above description of femoro-abdominal mechanisms. It also has a pleuro-coxal stridulatory mechanism. In the former, hundreds of very tiny elongate spines (10–20 μm) occur on abdominal tergites II and III, apposed by many equally tiny spinose pegs on the femur; no file ridge is present. The pleuro-coxal mechanism consists of patches of larger spines on pleura I, II and III, apposed by a band of similar coarse spines on the dorsal surface of the hind coxa. No recorded sounds have ever been ascribed to this component, but it probably contributes to the high frequency characteristics (extending into the ultrasonic range) known for *H. broughi* (L.H. Field, personal observation).

Deinacrida

The 11 known species show the greatest variety of stridulatory structures found in the family, and probably the most elaborate structures found in the Orthoptera. The femoro-abdominal mechanisms fall into four species groups (Field, 1993a). The structural trend progresses from enhancements of the plesiomorphic theme (large numbers of simple pegs) in the first group, to the development of a file resembling that of *Hemideina* in the second group, followed by reduction of the file ultimately to a single massive ridge on tergite II (third and fourth group). In addition, a tergo-tergal mechanism occurs in the third and fourth groups.

In the first group (*Deinacrida connectens, Deinacrida carinata, Deinacrida tibiata*), the abdominal component consists of a dense patch of thin blade- or slat-like pegs mounted perpendicularly to the surface of tergites I and II (Figs 15.2E and 15.3A). These are apposed by numerous minute, keel-like pegs, with a sharp apex, on the femur (Figs 15.2F and 15.3B). The flat sides of the abdominal pegs are parallel and orientated perpendicularly to the movement of the femoral pegs (Field, 1982).

The second group comprises three undescribed alpine species of *Deinacrida*. The group is characterized by the presence of a single file on tergite II, which is very similar to the file in *Hemideina* (Fig. 15.3C). There is a large variation in the number of file ridges. The femoral component consists of numerous short spinose pegs which are somewhat larger than those of the first group (Fig. 15.3D).

The third group includes the largest of the giant wetas (*Deinacrida heteracantha, Deinacrida fallai*) and an undescribed smaller species (*Deinacrida* 'Mahoenui'). The file and pegs of this group show the most elaborate morphology seen

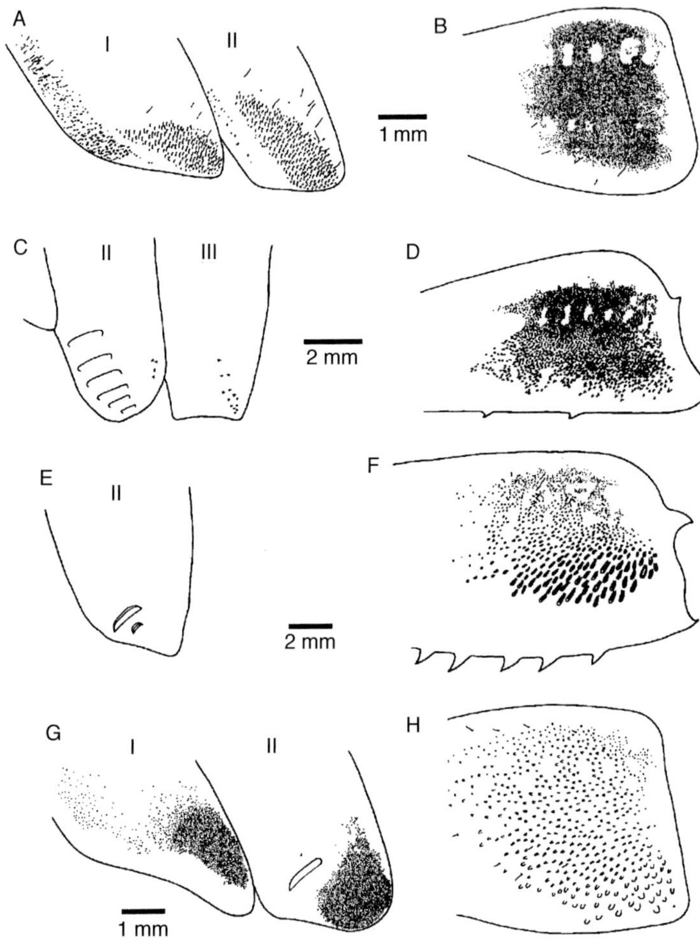

Fig. 15.3. Variety of femoro-abdominal structures found in the four groups of *Deinacrida* spp. of New Zealand. A–B. *D. connectens* group, in which a dense patch of slat-like pegs on abdominal tergites I and II (A) are rubbed by minute flat pegs on the femur (B). C–D. Alpine group (three undescribed species) having a file on the second abdominal tergite (C) and many very tiny pegs on the femur (D). E–F. The *D. heteracantha* group with massive crescentic ridges on the first tergite (E) and elongate sculptured pegs on the femur arranged radially from the joint axis. G–H. *D. rugosa* group with a single massive ridge on tergite II and a field of tiny spines on tergites I and II (G). The femoral component consists of sculptured pegs (H). (All from Field, 1993a, with permission.)

in any orthopteran stridulatory organ. The tergal component has become a pair of massive crescentic ridges (one in *D. fallai*) on tergite II, the dorsal one of which is higher and longer than the ventral one (Fig. 15.3E). The leading (dorsal) edge of each ridge is elaborated into two lips separated by a groove. The femoral pegs are large, raised, elongate structures, with a double-lipped top formed by a medial groove. These 'pegs' are arrayed radially on the femur, such that they scrape laterally across the largest tergal ridge (Fig. 15.3F). Thus each engagement of peg to ridge should involve up to four strike/release events (Field, 1982).

The fourth species group (*D. rugosa* and *D. parva*) has both femoro-abdominal and tergo-tergal stridulatory mechanisms. In the femoro-abdominal mechanism, the tergal file is reduced to a single, massive, unembellished ridge on tergite II, below which is a patch of tiny spinose pegs (Figs 15.2G and 15.3G). Sometimes a second smaller ridge is present. The femoral pegs resemble miniature molar teeth that have been compressed laterally, with a medial groove (Figs 15.2H and 15.3H). The femoral pegs can make two strike/release events with each pass across the tergal ridge (Field, 1982).

The tergo-tergal mechanism in the third and fourth groups appears to be unique in the Orthoptera, although it occurs in Hymenoptera and Lepidoptera (Dumortier, 1963). The mechanism consists of a large pair of bilaterally arranged patches of short, curved spines on the dorsal anterior margins of several abdominal tergites. These are rubbed by patches of blunt articulated pegs on the underside of the posterior margin of each overlapping tergite (Fig. 15.4A). In *D. rugosa* the second, third, fourth and fifth tergites possess this mechanism (Fig. 15.4B), while in *D. heteracantha* the third, fourth, fifth and sixth tergites are modified. In *D. fallai* only the second and third tergites are involved (Ramsay, 1953; Richards, 1972).

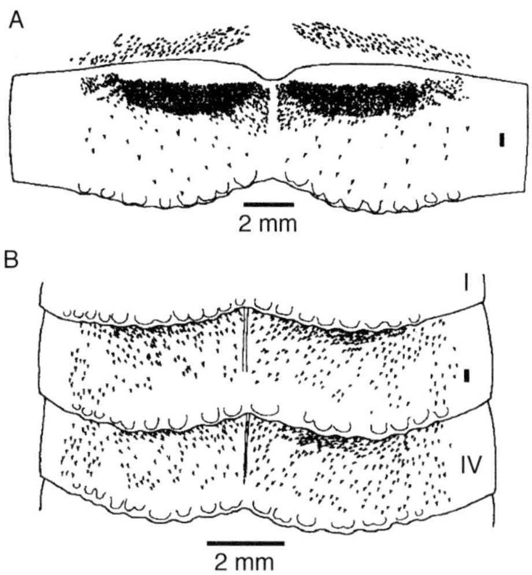

Fig. 15.4. Tergo-tergal stridulatory mechanism found on the dorsal aspect of tergites II–IV in the *D. rugosa* group. Densely arrayed spines on the anterior margin of a tergite (dark field in A) are overlapped by thick articulated sensillar pegs on the underside of the next anterior tergite (shown reflected forward, light field in A). Telescopic extension of the abdomen causes structures to engage. B. Dorsal view of tergites (*D. parva*) shows thickened posterior margins of tergites underlain by anterior fields of spines. From Field (1993a) with permission.

Evolution and diversity of stridulatory structures in New Zealand wetas

Two general observations of anostostomatid species worldwide suggest that stridulatory structures originated as defence mechanisms, in which the simple spine/peg patch dominated the femoro-abdominal theme. First, the majority of the species lack ears and presumably cannot hear the sounds they produce. Secondly, most of the species have patches of tiny spines/pegs on several or all of the abdominal tergites, and simple spine/peg fields on the hind femur; this condition is also seen in fossil relatives of anostostomatids. Thus this femoro-abdominal format is taken as the plesiomorphic condition, which evolved as a predator defence mechanism. This conforms to the principle that when sounds of insects serve primarily for defence, they are often broad-band, unpatterned and thus similar amongst many species. In this way, the species share the common task of educating predators, a process similar to that seen in visual mimicry (Edmunds, 1974; Bailey, 1991).

The striking feature of New Zealand Anostostomatidae is the remarkable elaboration of cuticular structures and diversification from the plesiomorphic theme. Although most of the *Hemiandrus* species retain the plesiomorphic structures (with variations in spine morphology), they are also deaf and thus conform to the general family state. The great embellishment has occurred in *Hemideina* and *Deinacrida*, in which the femoro-abdominal structures, the stridulatory tusks, tergo-tergal spines and articulated sensilla and pleuro-coxal spines all represent apomorphic mechanisms.

The diversity of New Zealand weta structures may be due to the long isolation of the island country. Although they show Gondwana origins, the weta species are now all endemic to New Zealand. They appear to have developed the full gamut of stridulatory mechanisms from the plesiomorphic theme, while isolated from other ancestral stocks. In association with this development, wetas also show enhanced secondary sexual megacephaly in *Hemideina* males and highly sensitive tympanal organs in *Hemideina* and *Deinacrida*. The ears in *Hemideina* are specifically tuned to the peak frequency of conspecific sounds (Field *et al.*, 1980). Taken together, the above processes suggest an evolutionary transition in sound production, in the New Zealand anostostomatid species, from a defensive role against predators to a more expanded role of defence plus intraspecific social communication (discussed below). Once social communication arose, selection pressure must have existed to evolve isolating mechanisms between species during the early stages of radiation in New Zealand. The diversity of stridulatory structures may well reflect the results of that pressure on weta speciation. In *Deinacrida* (groups three and four) and in *Hemideina*, the major changes have been toward larger and more massive stridulatory structures, culminating in the heavy, single- and double-lipped ridges and pegs of *D. rugosa, D. parva, D. heteracantha* and *D. fallai*. Loud penetrating sounds are produced by the stridulatory structures of these giant wetas and of the *Hemideina* species (Field, 1982). Therefore, evolution of more robust stridulatory structures in wetas has led to the ability to produce louder sounds in association with social communication. In the case of tree wetas, the enhanced amplitude of sound signals is advantageous for calling between individuals separated in different trees amongst a population (Field and Rind, 1992). Only anecdotal evidence exists for intraspecific communication in *Deinacrida* (two species: Field, 1980a; Sherley and Hayes, 1993), and it remains to be determined whether the loud sound amplitude is advantageous for long-distance communication in those species.

It is possible to construct an evolutionary scheme based upon similarity of stridulatory structures (Field, 1993a). In general, the species groups described above represent those species most closely related genetically, as shown by allozyme studies (see Morgan-Richards *et al.*, Chapter 7, this volume). In Fig. 15.5, a simplified diagram shows that, from the ancestral plesiomorphic morphology seen in the henicids, two lines led to the extant tusked wetas and to the deinacridine genera. Of the latter, *H. broughi* shows affinities to both *Hemideina* and *Deinacrida* species. The *Deinacrida* groups show a sequence from the near-plesiomorphic *D. connectens* morphology to the well-formed files of the 'alpine wetas' group, to file

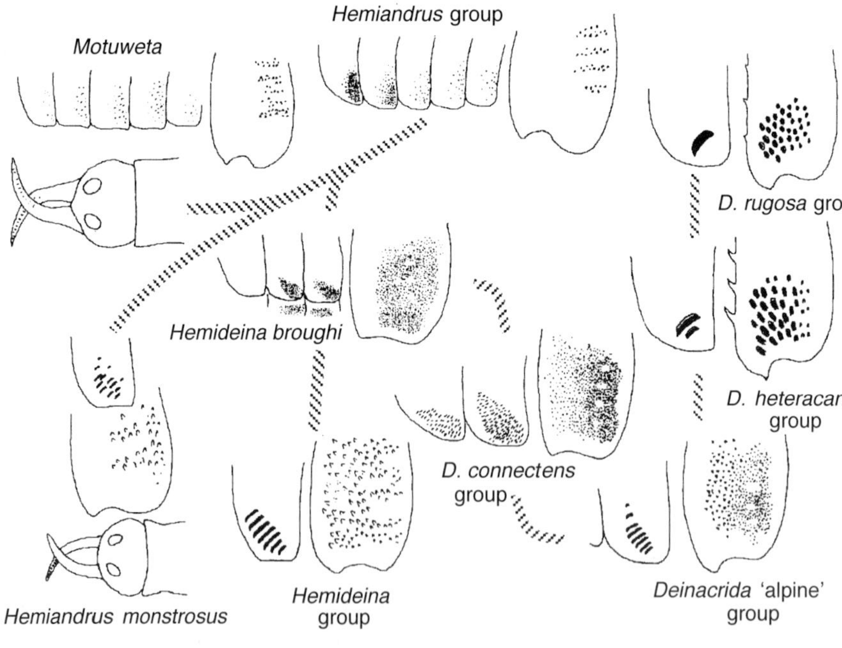

Fig. 15.5. Proposed scheme for the evolution of New Zealand weta stridulatory mechanisms. Diagram simplified to omit ancestral origins of these modern groups. The plesiomorphic condition is represented by the *Hemiandrus* mechanism of simple patches of fine spines or pegs. This gave rise to the tusked weta group (in which one species developed a crude abdominal file) and to *Hemideina broughi*, which lacks a tergal file. The latter gave rise to the remaining *Hemideina* group and the *Deinacrida* groups. The *Hemideina* species developed tergal files, and intraspecific communication. The *Deinacrida* groups developed a tergal file, which became reduced to one or two massive ridges and elaborate femoral pegs. Little, if any, social communication has been demonstrated for these groups. From Field (1993a) with permission. Note: *H. monstrosus* = *Anisouris nicobarensis*.

reduction and femoral peg embellishment in the *D. heteracantha* and *D. rugosa* groups.

In the evolution of New Zealand wetas, the stridulatory file appears to have risen independently three times: in *A. nicobarensis* (= *H. monstrosus*), in *Hemideina* species and in three of the *Deinacrida* groups. The allozyme data indicate that the tusked wetas have little genetic affinity with the other two groups, which themselves are genetically well separated (Morgan-Richards, 1997).

The tusked condition has arisen in southern African species of *Libanasidus* and *Henicus*, as well as in New Zealand wetas, but only the New Zealand species show stridulatory organs on the tusks (Field, 1993a; Bennett and Toms, 1995). In all known species, the male bears tusks rather than the female. Presumably these secondary sexual characters have developed through sexual selection pressure, conferring a considerable reproductive advantage upon the males. The situation is complicated by the existence of small, sexually mature males that do not bear tusks (Bennett and Toms, 1995). In this case, selection for tuskless males may have supported an alternative mating strategy, much as envisioned for the small adult satellite males in *H. crassidens* of New Zealand (see Field and Deans, Chapter 10 and Stringer, Chapter 20, this volume).

Stridulatory Sounds

While the crickets, bush crickets and even gomphocerine grasshoppers have evolved elaborate communication systems with acoustic signals, the more primitive roots of orthopteran signalling systems appear to be reflected in the less sophisticated, but extremely interesting, anostostomatids. Only *Hemideina* species have been examined in detail, inasmuch as they exhibit more social behaviour than their deinacridine relatives in New Zealand. No studies have been made of acoustic behaviour in the latter group. Furthermore, opportunities abound for acoustic studies of the behavioural ecology of king crickets in Australia and Africa and the Jerusalem crickets in North America. Even in cases where species lack tympanal ears, there is now a good basis for examining the potential role of substrate vibratory communication, as demonstrated for wetas later in this chapter and for the closely related Gryllacrididae of Australia (Field and Bailey, 1997; see also Hale and Rentz, Chapter 6, this volume).

Sound structure and frequency characteristics

The components of the femoro-abdominal mechanism of New Zealand wetas are universally arranged so that sound is produced when the femoral pegs move downward relative to the abdominal ridges (or pegs). This happens either when the raised, extended, hind leg rotates posteriorly and downward to engage the femur against the abdomen (sound on the down-stroke, as in defence behaviour) or when the abdomen moves upward against the femur while all six legs are on the substrate (sound on the up-stroke, as in aggression, mating and calling behaviour of tree wetas, discussed below). Each stroke produces a sound syllable, composed of many damped pulses, which arise from individual releases of pegs against ridge(s) (Field, 1978, 1982; McVean, 1986). Tree wetas (*Hemideina*) produce a variety of sound patterns, in which groups of syllables are arranged into echemes (Fig. 15.6), and different echeme patterns are produced in a number of behavioural contexts, described below (Field and Rind, 1992).

The frequency spectra of all weta sounds are broad-band (non-resonant), owing to the frictional nature of their sound-producing mechanisms. Spectra from defence sounds (or aggression sounds of *H. thoracica*) have been reported for five species of *Hemideina* and three species of *Deinacrida* (Field, 1982; McVean, 1986). For *Hemideina* sounds, the spectra had peak energies at 2–5 kHz and rapidly tapered off at higher frequencies, with little ultrasonic component (Fig. 15.7A). Broad overlap of peaks occurred amongst species; hence frequency is not likely to be used in heterospecific acoustic discrimination by tree wetas. The peak frequency of stridulatory sounds made by *H. crassidens* (2.5 kHz) matches the peak hearing sensitivity in that species (Field *et al.*, 1980).

Frequency spectra of sounds from *Deinacrida* species extend over a higher range and have peak frequencies extending well into the ultrasonic band (Fig. 15.7B). Spectra for *D. heteracantha* and *D. connectens* have double peaks (20 and 50 kHz and 7 and 35 kHz, respectively), while the spectrum for *D. rugosa* has a broad peak from 10 to 20 kHz). No simple relationship exists between the elaborate and varied morphologies of stridulatory structures and spectral frequencies for the above species, although the larger structures emit more intense sound than the small ones (Field, 1982).

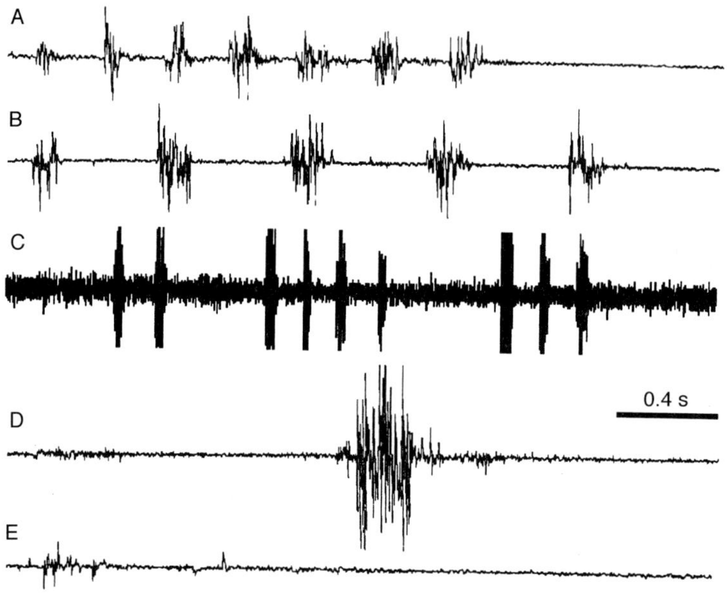

Fig. 15.6. Sound pattern repertoire produced by *Hemideina crassidens*. A. One echeme of the aggression pattern, containing seven syllables. B. One echeme of the mating pattern, containing five syllables produced at a slower pace. C. Three echemes (of two, four and three syllables) of the calling pattern. D. A single syllable of the defence sound during a leg kick. E. The disturbance sound showing its ill-defined and sporadic nature.

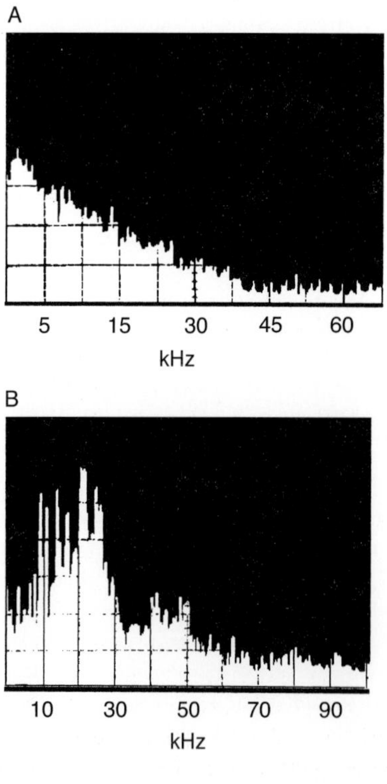

Fig. 15.7. Linear magnitude frequency spectra of defence sounds produced by *Hemideina crassidens* (A) and *Deinacrida heteracantha* (B). Vertical axis in arbitrary voltage units. From Field (1982) with permission.

Sound production and substrate drumming in henicine wetas

In the large tusked weta (*M. isolata*, from Middle Mercury Island, New Zealand), loud single syllables of sound are produced when the tusks are abruptly opened (Fig. 15.8D). This has been likened to a sharp barking sound. No patterned sound production has been reported. In the few anecdotal observations available, it appears that the bursts of sound are produced as a defence ploy to startle predators (M. McIntyre, Wellington, 1993, personal communication).

Although most species of *Hemiandrus* have femoro-abdominal stridulatory structures and make rocking movements to engage the structures (Fig. 15.8F), recordings or descriptions of emitted sounds have yet to be made. In several species of *Hemiandrus*, males make drumming sounds against the substrate (D. Gwynne, Otago, 1994, personal communication). This has not been investigated in detail, but apparently is similar to the tibial drum-

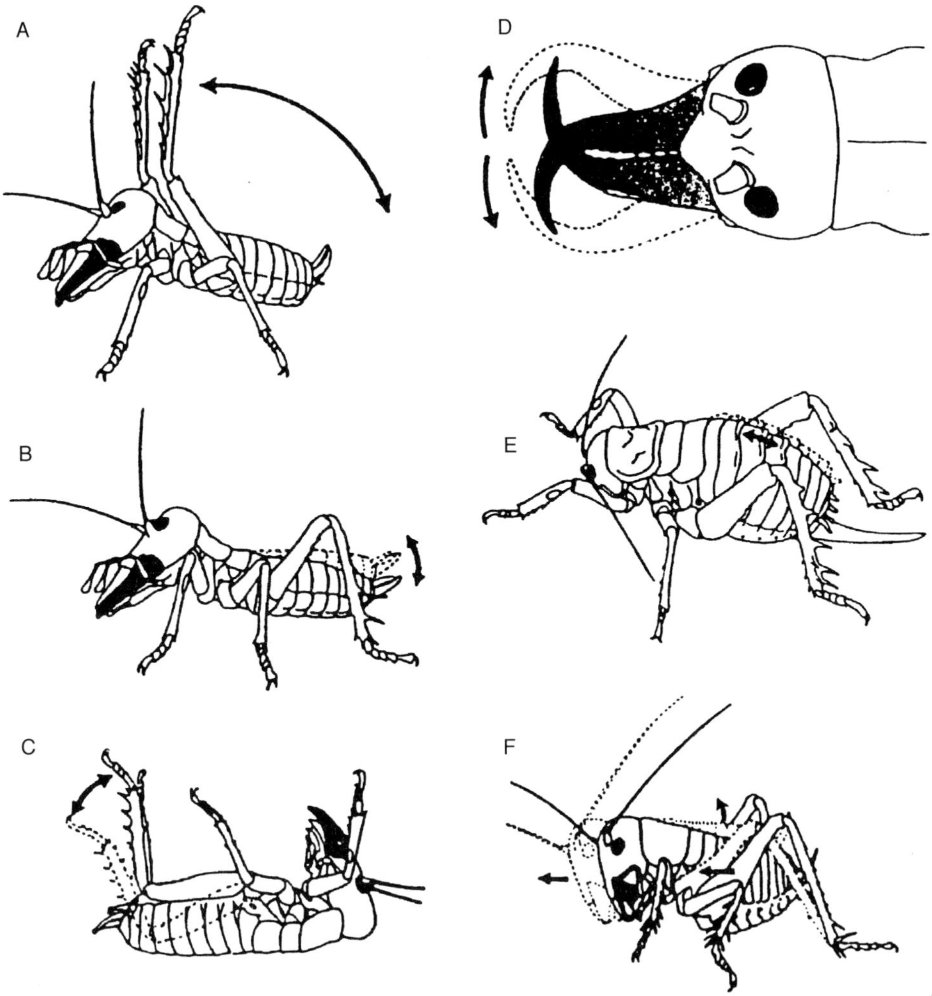

Fig. 15.8. Stridulatory behaviours found in New Zealand wetas. A. Defence stridulation typical of all species of *Hemideina* and *Deinacrida*. B. Rapid abdominal movements of the abdomen against the appressed femora is used for social signalling in *Hemideina* spp. C. Defence stridulation in *H. maori* after assuming the inverted alarm position using flip/splay behaviour. D. Sound production in tusked wetas (*Motuweta isolata* shown) by rapid opening of appressed tusks when alarmed. E. Tergo-tergal stridulation by the *D. rugosa* group of species is produced by telescoping the abdomen out repeatedly as the abdomen is raised and lowered. F. The *Hemiandrus* group of species produces defence stridulation by rocking forward and raising the abdomen against the appressed femora.

ming behaviour described for Australian gryllacridids (Field and Bailey, 1997). In the Gryllacrididae, both sexes rapidly tap the hind tibiae against the substrate in a species-specific pattern. Short trills are repeated erratically in a call-and-response duet, as males search through the dense underbrush for females in silk-lined nests. In contrast, *Hemiandrus* spp. occupy burrows in soil and both sexes aggressively defend their burrow entrances, particularly during the first few hours of darkness and just before dawn (Barrett, 1991). Males occupying burrows may send out vibratory signals in the substrate to announce territories.

A single intriguing observation sheds some light on the above suggestion. In a laboratory colony of *Zealandrosandrus gracilis*, drumming of one male has led to the appearance of other males from their galleries and subsequent drumming by all males. The drumming movements consisted of rapid vertical movements of a single hind tibia for 1–3 s (Cary, 1981). Although Cary interpreted these movements as acoustic stridulation of the femur against the abdomen, they almost certainly represented drumming behaviour, and in fact closely resemble the single hind-leg drumming observed in the Australian gryllacridid, *Ametrus* sp. (Field and Bailey, 1997). Substrate vibration communication in henicine wetas remains an area worthy of research. Much more detailed observations on vibratory communication in a related family of subterranean gryllacridoids, the Stenopelmatidae, have been made by Weissman (Chapter 19, this volume). There, drumming on the substrate is made by abdominal tapping.

Sound production in Deinacrida

Sound production by all of the stridulatory mechanisms described for this genus is essentially restricted to defence behaviour. The femoro-abdominal method is used in the upraised hind-leg defence posture, when the legs are swept downwards toward a predator and sound is emitted (Fig. 15.8A). In *D. connectens*, the same mechanism is used during walking when the insect is retreating from a source of disturbance (Field, 1980a). In both cases, the abdomen is apparently inflated to engage the sound-producing structures.

Acoustic social communication either is generally lacking or has not been discovered in the henicine ground wetas and in the giant *Deinacrida* spp. Only one report of patterned sound production in the latter has appeared for *D. mahoenui*, in which a subadult was heard to repeat three to five soft, short, scratching sounds, followed by an interval of silence over a period of 5 min (Sherley and Hayes, 1993). Anecdotal reports of similar calling by the rare *D. heteracantha* (D. Smith, Auckland, 1980, personal communication) on New Zealand's Little Barrier Island could be the basis of observed population clustering of this largest of giant wetas (G. Gibbs, Cairns, 1997, personal communication).

The tergo-tergal mechanism in the *D. rugosa* group of giant wetas is used to produce sounds during defence stridulation. Repeated telescoping movements of the abdomen result in bursts of hissing sounds as apposing pegs rub against each other (Fig. 15.8E). The sound has not been studied in detail.

In many of the giant wetas, an unusual tick sound has been observed, although the structural basis for its production is not proved. Ticks produced by *D. rugosa* are composed of four to eight extremely brief (250 μs) damped pulses. The weta produces about three ticks per second when disturbed or alarmed (Field, 1980b; Sherley and Hayes, 1993). Because the spectral frequency of the ticks (12 kHz peak) matches that of sounds produced by the femoro-abdominal mechanism used during defence stridulation, Field (1980b) proposed that minute movements of the hind leg against the abdomen cause one or a few peg releases, which accounts for tick production. It is possible that other undescribed stridulatory structures are involved, since the exact source of the sound has not been demonstrated.

Tremulation in Hemideina

During agonistic battles between male *H. femorata* and *H. crassidens* (probably other species as well), the intruder male often produces tremulation behaviour. This occurs while the intruder is attempting to evict the resident from a tree gallery by gripping a hind tibia with its mandibles and pulling. The movement involves rapid vertical oscillation of the abdomen in repeated bursts of about a second in duration. The abdomen does not contact the stridulatory ridges of the hind femur, and no stridulatory sounds are audible during tremulation (L.H. Field, Christchurch, 1996–1998, personal observations). All six tarsi remain on the substrate. Although no recordings have been carried out, it is probable that low-frequency vibrations are introduced into the tree-trunk during tremulation; it is also probable that the resident can detect such low-frequency energy, since its subgenual organs in each leg are extremely sensitive to substrate vibration induced into the tree surface during aggression stridulation (McVean and Field, 1996).

Sound production in Hemideina

Stridulatory behaviour and some associated sound patterns have been described in *H. crassidens, H. thoracica* and *H. ricta* in New Zealand (McVean, 1986; Field and Rind, 1992; Field, 1993b). Unpublished recordings have also been gathered for sound patterns of *H. femorata* (L.H. Field, Christchurch, 1993–1998, personal observations).

A complete repertoire has been established only for *H. crassidens* (Field and Rind, 1992), although the sound patterns and behaviours amongst all the above species appear remarkably similar. The patterns are named after the behavioural contexts in which they occur. All sounds are normally produced at night, except the defence and disturbance sounds, which are made any time wetas are threatened physically. An outstanding feature of the patterns is the extremely large variation in temporal parameters compared with similar measurements made for the highly stereotyped songs of crickets and bush crickets (e.g. Samways, 1977).

Although acoustic communication frequently appears in the social interactions of *Hemideina*, it is important to note that in no case has it been demonstrated that females are attracted to male sounds. Since such attraction plays a primary role in reproduction in other acoustic Orthoptera, it is crucial that future research demonstrates whether this role exists in tree wetas.

Aggression stridulation

This is produced primarily by males before, during and after agonistic interactions centred around the gallery entrance, and following interruption of mating by arrival of a second male. The pattern for aggression stridulation in *H. crassidens* (Fig. 15.6A) consists of a train (echeme) of four to seven syllables spaced about 82–121 ms apart at 20-22°C. In 3% of female aggressive behaviour towards male or female conspecifics (n = 207 interactions), the aggression stridulation pattern was produced by females (Sandlant, 1981).

Mating stridulation

This is produced by males after failed mating attempts with a female or after a male is rebuffed by a female. Contrary to the case in many other orthopterans, it is not a prerequisite for mating, nor is it produced during mating. The mating stridulation pattern is almost identical to the aggression pattern (mean of two to seven syllables), except that it is slower at similar temperatures (Fig. 15.6B). Syllables are 93–238 ms apart (Field and Rind, 1992). The allopatric sister species, *H. ricta,* has very similar timing for both sound patterns: aggression syllables are about 100 ms apart and mating syllables are 130 ms apart (Field, 1993b).

The movements used to produce aggression and mating stridulation are similar; the abdomen is moved up and down in a small arc (about 5–10°) against the appressed legs, while the insect is standing on the substrate (Fig. 15.8B). The abdomen is raised and stiffened prior to stridulation, and probably inflated to assist engagement of the ridges and pegs. A sound syllable is produced on each up-stroke of the abdomen. Occasionally, in aggression stridulation, the partially flexed hind legs are raised and vibrated through a small arc against the abdomen (Field and Rind, 1992).

Calling stridulation

This is made by single males (adults or juveniles) when in or near the gallery entrance (L.H. Field, personal observation; R. Ordish, Wellington, 1992, personal communication). Apparently it is not dependent upon the presence of other wetas of either sex, but males sometimes respond to each other in loosely timed alternation in the forest (L.H. Field, personal observation). Although it is often assumed that calling carries a territorial message, no documenting evidence has been published to this effect.

Calling stridulation probably involves the same abdominal vibratory movement in a vertical plane (not documented), but at a slower rate than that for aggression or mating sounds. It consists of a pattern of repeated singlet, doublet or triplet syllables occurring in an echeme (Fig. 15.6C), which varies considerably amongst males within a population. This variation is a subject of great interest, since it is atypical for Orthoptera and has not been recorded in any other species. Published measurements for *H. crassidens* indicate that syllables occur 177–258 ms apart at 11–14°C (Field and Rind, 1992).

Defence stridulation

This is produced at any time by either sex in response to intense or sudden visual, tactile or acoustic stimulation. It is normally used in the presence of predators and is only very rarely used between conspecifics. The movements are fully described in Field and Glasgow (Chapter 16, this volume). All species of *Hemideina* and *Deinacrida* raise the extended hind legs above the head and, when highly provoked, the legs are kicked down in an arc toward the source of stimulation (Fig. 15.8A). During each downward kick, a loud scratchy sound is produced (Fig. 15.6D). With less drastic stimulation, one or both legs are partially flexed, raised and kicked, with similar sound production.

Disturbance stridulation

This is made whenever tree wetas are mildly provoked by an outside disturbance while they are inside their galleries. This stridulation appears to involve similar movements to those for aggression, mating and calling to engage the stridulatory mechanism, but there is no clear pattern to the sound (Fig. 15.6E).

Although the five stridulatory behaviours are classified for *H. crassidens*, they are known to occur also in *H. femorata, H. thoracica* and *H. ricta* (McVean, 1986; Field, 1993b; L.H. Field, personal observation). The alpine *H. maori* has two kinds of defence stridulatory behaviour (Field and Glasgow, Chapter 16, this volume), but other stridulatory patterns are not known. *Hemideina broughi* has the same defence stridulation behaviour as the other species, but is not known to have social communication.

Variation in temporal parameters of Hemideina *patterns*

The large variation in timing of syllables within the patterns of *H. crassidens* is one of the major departures from the characteristically stereotyped sound patterns seen in singing Orthoptera. Normally, the pattern variation within any one species is much less than that observed amongst sympatric species (Alexander, 1962; Walker, 1962). A summary of the variation within individuals and within populations found in tree weta sound patterns is given in Table 15.2. The extraordinarily large values, expressed as ranges of coefficient of variation (CV), stand in marked contrast to values measured, for example, in the bush cricket *Platycleis*, where the CV of the syllable duration (SD) for proclamation songs of five species ranged from 2.94 to 6.24%. The expectation of a small CV value (hence a high degree of stereotypy) is based upon information theory, which states that a signal should have predictability for the message to be understood by the receiver. Hence, for a parameter to convey information usefully, it should have a narrow variation amongst individuals and each individual should have a narrow variation about its mean, as seen in the *Platycleis* example above. In Table 15.2, syllable period (SP) is the parameter with the smallest amount of variation; although this condition accords with data from cricket and tettigoniid (bush cricket) patterns, the large values in wetas should still have a deleterious effect on communication based upon this parameter alone (Field and Rind, 1992). Furthermore, aggression and mating patterns have heavily overlapping distributions of SP amongst populations and could only be discriminated, in principle, on the basis of the ratio of SP/SD, rather than on SP alone.

Field and Rind (1992) suggested that several factors underlie the lack of stereotypy seen in tree weta sound patterns. First, the species of *Hemideina* are mostly allopatrically distributed (see Morgan-Richards *et al.*, Chapter 7, this volume)

Table 15.2. Summary of within-individual variation (expressed as the range of coefficient of variation (CV = mean/standard deviation) in %), and within-population variation (range of mean values expressed as % of largest mean) measured for temporal parameters of *H. crassidens* sound patterns (after Field and Rind, 1992).

Parameter	Aggression	Mating	Calling (within echemes)	Defence
Syllables per echeme				
Individual	16.7–35.7	9.5–31.7	17.5–55.2	30.2–90.3
Population	37.9	64.2	56.5	63.3
Syllable duration				
Individual	25.5–76.3	21.7–49.0	18.0–30.4	39.2–77.8
Population	55.2	59.2	53.3	59.0
Syllable period				
Individual	4.5–28.0	7.5–20.6	8.3–20.5	25.0–82.4
Population	33.0	61.0	31.7	55.6

and hence may not have experienced strong selection pressure to evolve highly stereotyped sound patterns for species identification signals. This contrasts with many sympatrically distributed continental species of orthopterans, which have optimized species identification through stereotyped acoustic signals. Secondly, it is very likely that the information content of the weta sound patterns is strongly augmented by their behavioural context. This is especially pertinent to the aggression and mating patterns, where there is major overlap in the temporal parameters, with the result that they could not be discriminated on the basis of syllable period alone. Thirdly, in the calling pattern at least, there is evidence for highly individualistic patterning amongst males, which directly contrasts with the trend seen in crickets and bush crickets. This phenomenon opens the possibility of individual male recognition based upon individual calling patterns.

Evidence for unique individual male calling patterns

Recent studies have highlighted the importance of further investigation into intraspecific variation in orthopteran sound patterns, because such variation may be the mechanism whereby outcomes of mate choice and male competition are determined (Schatral and Bailey, 1991). Although species-specific patterns are clearly necessary, variation between individuals must work within the range of the species as a whole to allow female discrimination in systems based upon sexual selection. Attempts to isolate factors responsible for intraspecific call variation in crickets and tettigoniids have been reviewed by Zuk and Simmons (1997). While some cricket species show no relationship between body size and call parameters, *Gryllus bimaculatus* produces a higher chirp rate with larger body size, and females prefer higher chirp rates. Increased aggressiveness is correlated with increases in chirp length or chirp interval in the bush crickets *Platycleis albopunctata* and *Platycleis camellifolia*, while increasing age is correlated with a decrease in chirp variation and variability in *P. camellifolia*. Furthermore, variations in calling song parameters may be found in different parts of the geographical range of crickets (Walker, 1962) and bush crickets (Den Hollander and Barrientos, 1994).

All of the above pattern variation involves shifts in continuous temporal variables within a basic species-specific pattern. The situation found in tree weta calling is fundamentally and remarkably different. Because the syllable composition of any particular pattern differs greatly amongst individuals (Fig. 15.9), there is no specific pattern that can be ascribed to the species as a whole based upon any one individual (*H. crassidens* population sampled in Westland, New Zealand). For example, syllables may be repeated as singlets, doublets, triplets and even quadruplets within an echeme, as shown for calls from four males in Fig. 15.9A. Besides patterns of only singlets (Fig. 15.9Ai) or doublets (Fig. 15.9Aii), males may mix syllable groups, often commencing with triplets and doublets and drifting into doublets or singlets (e.g. Fig. 15.9Aiv).

The immediate question that arises is whether males call with patterns unique to each individual in a population. This possibility would allow, in principle, males to identify each other by acoustic signature alone, as often seen in birds and mammals. By logging calls over several weeks with voice-activated recorders attached next to galleries, a preliminary study has shown that call patterns of neighbouring males within a population differ markedly. Mean call patterns are shown for ten males (Fig. 15.9B), in which the mean number of syllables is plotted for sequential groups of syllables comprising the echemes of the entire data set for each male. The range of variation in Fig. 15.9B includes several males that progressively decreased the number of syllables per group (e.g. three to five syllables per group at onset, decreasing to a terminal group of two or three syllables), while others progressively increased or decreased the number of syllables per group and ended with a long train of doublets or singlets. Another male produced nearly all singlets.

The above analysis dealt with the mean calling pattern produced over the entire sampling period (males call erratically, or sometimes not at all on any given night). An additional analysis explored how consistent the pattern remained within single nights, compared with the variation over the entire sampling period, by accumulating mean nightly call patterns for each of 17 males in the population. Five examples of the range of differences are shown in Fig. 15.10. Some males showed highly consistent nightly patterns (e.g. Fig. 15.10ii, iv), while others varied but retained some essential feature of uniqueness in their pattern. Thus, in

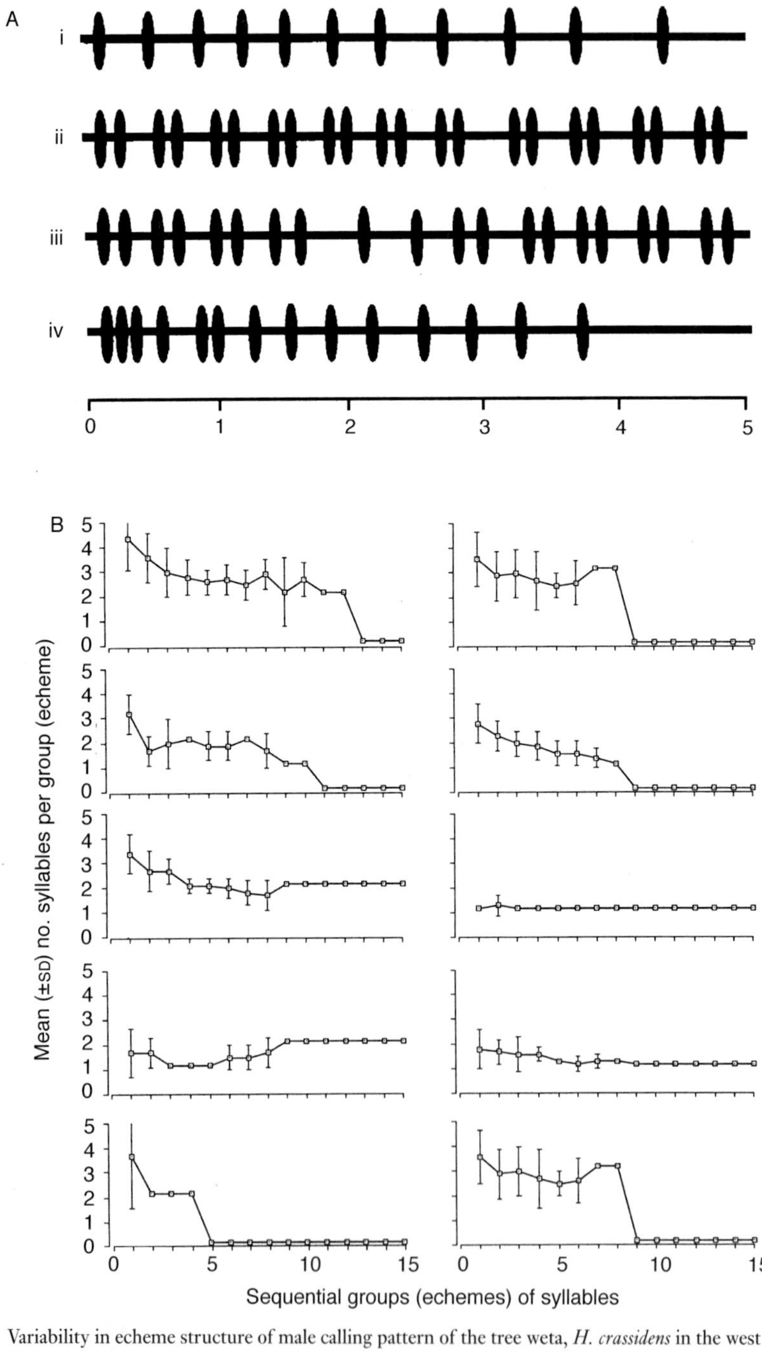

Fig. 15.9. Variability in echeme structure of male calling pattern of the tree weta, *H. crassidens* in the west coast bush of New Zealand. A. Four different males produced echemes of singlets (i), doublets (ii), doublets and singlets (iii) and triplets, doublets and singlets (iv), respectively. Note tendency for most echemes to decrease in rate towards the end of a call in some but not other males. B. Analysis of mean call pattern of ten different males over at least three nights of calling. Mean (± SD) number of syllables per echeme is plotted for repeated echemes over total recording period. Inter-individual differences occurred in the following variables; mean number of syllables per echeme, variance in number of syllables per echeme, increase or decrease in number of syllables per echeme and number of echemes per call.

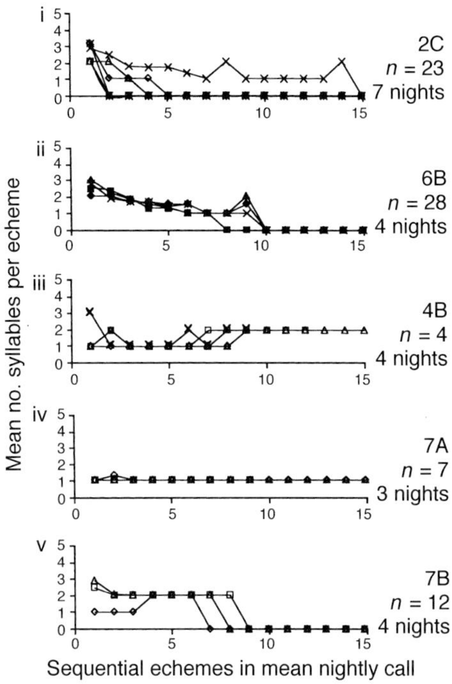

Fig. 15.10. Breakdown of consistency of call pattern in different males (*H. crassidens*) over three or more nights. Mean number of syllables per echeme is plotted against sequential echemes for each night of calling. Thus mean pattern is shown for each night. Each male retains features typical of its own pattern from night to night, even though some (e.g. i) have variability in call length.

Fig. 15.10i, the first syllable group always had the greatest number of syllables and, in Fig. 15.10iii, the call always ended in doublets preceded by several singlets, even though the first syllable group varied. Again, in Fig. 15.10v, the call always ended in doublets, even though its onset differed in one of the 12 echemes recorded. Thus there always appears to be some degree of predictability in the calls of different males within this population.

The greatest source of variation was seen in the length of the echemes (i.e. number of syllable groups within a call) during any one night. Such variability was seen, for example, in Fig. 15.10i. Thus, any estimate of information content relating to the identity of the sender must reside in the pattern of syllable groups, not in the total number of groups.

The existence of individual differences in insect calling signals has not been explored fully. Specifically, the possibility that unique patterns are maintained by individual males raises many questions about the sophistication of the tree weta signalling system. Are the pattern differences due to genetic differences within the population, or can males learn to vary their call pattern (perhaps to avoid overlap with immediate neighbours)? Are males aware of the acoustic signature of their neighbours and, if so, how does this affect interactions and territoriality? What message(s) can be inherent in call signals that show such high variability within a species? The tree weta mating system almost certainly involves sexual selection. Thus it is important to learn whether individual differences in male call patterns are used by females in the selection of mates.

Role of sound in male aggressive interactions

Male tree wetas show a striking propensity to engage in aggressive combat during their nightly activity cycle in the natural environment, as well as in laboratory conditions. Aggression stridulation forms part of the interaction and occupies a more or less prominent and predictable role, depending on the species. A detailed analysis of the role of sound in aggression behaviour is given by Field (Chapter 18, this volume). The following is a brief summary. Although most studies have been carried out with laboratory colonies, observations have been confirmed in natural habitats (Möller, 1985).

Most aggressive interactions occur at or around gallery entrances, presumably because males defend galleries as a resource in a defence-resource polygynous mating system (Field and Sandlant, 1983). Intruder males approach a resident perched in the gallery opening with spiny hind tibia blocking the entrance. The intruder instigates a fight by attempting to pull the resident out (see Field, Chapter 18, this volume, for description) and the ensuing interaction may range from early decamping by the intruder, if it cannot displace the resident, to escalation into full combat, involving biting, lunging and grappling by both wetas (Sandlant, 1981). The most stridulation during aggressive encounters occurs in *H. crassidens* (Field and Rind, 1992). In this species, the intruder shows a tremulation behaviour, and sometimes stridulation, while trying to dislodge the resident. If a battle follows, both males loosely alternate stridulation bouts while sparring with biting attacks, and tremulation is absent. The eventual winner produces more echemes during the battle than the loser, leading to the conclusion that sound production is proportional to aggressive motivation. Once the loser decamps, it immediately ceases to stridulate. The winner continues to stridulate for a time that is roughly proportional to the intensity of the battle. During this time, echemes are closely spaced initially but become progressively more separated over periods lasting up to 20 min. The most intense battles and greatest amount of stridulation occur during changes of winner/loser status of a dominance hierarchy (L.H. Field, observations made with laboratory colony). If, following a battle, the two combatants meet again, the previous winner produces a prolonged sequence of echemes, while the loser remains essentially silent (Field and Rind, 1992). In male aggressive encounters on Stephens Island, Möller (1985) also observed that most battles occurred at gallery entrances and that winner males stridulated more often than losers (21 encounters). The loser always became silent as soon as it departed, while the winner gained entrance to the gallery and continued to stridulate.

Other species of tree weta engage in similar battles for gallery space. However, *H. femorata* and *H. ricta* differ from the above scenario in that little sound is produced during a battle. Following a battle, sound production occurred in a waning series of echemes by winners, in the same fashion as in *H. crassidens*, and losers remained silent. If a winner encountered a loser within 30 min following a battle, it produced a series of echemes even if no further aggressive interaction occurred (Sandlant, 1981; Field, 1993b).

Semantics of Hemideina *stridulatory patterns*

Two problems arise when attempting to understand how receiver tree wetas could interpret meaning from conspecific acoustic signals. First, the extremely large variances (Table 15.2) indicate that the stridulatory patterns contain relatively little stereotypy and that information content (or predictability of message) based upon temporal parameters alone would be highly eroded. This stands in stark contrast to the more stereotyped and species-specific patterns found in crickets and bush crickets. Secondly, the large overlap in distributions of temporal parameters and the echeme similarity in aggression and mating patterns suggest that these two patterns would be difficult to discriminate by receiver wetas. Rather than relying on a traditional information theory framework to understand how message content might be interpreted by receiver wetas (i.e. information content and predictability should increase with greater stereotypy), Field and Rind (1992) found that a contextual interpretation was more useful in understanding the role of sound in weta behaviour. Thus, the set of conditions (referent) co-occurring with a sound signal gives the receiver some predictability of the intent of the sender. In aggressive encounters between male wetas, the referent for aggression stridulation could be: the sender is male, the receiver is male, the past history of wins and losses of fights for both males, and mandibular flaring and raised foreleg threat displays (see Field and Glasgow, Chapter 16, this volume). In mating encounters, the referent could be: the sender is male and capable of copulatory behaviour, the receiver is female, abdominal curving by the male and palpitation of the female and/or rejection of the male's mating attempts by the female.

The acoustic signal and its referent may not necessarily indicate precisely what the sender may do next, but at least they are likely to convey a message of probable behaviour. In agonistic interactions, the sender may actually be unclear about its next move, and may be biding time in the face of the two conflicting options of fighting or fleeing (Hind, 1981; Markl, 1985). Thus, in weta battles the message associated with the aggression sound pattern may depend upon the progress of the encounter. In the initial stage before biting and lunging attacks, it could be: 'I am a male and I am not going to leave my position' or 'I may attack, depending upon what you do.' Escalating threat displays of the opponent may alter the sender's message now to mean 'I am very likely to attack'. Very strong attacking behaviour of the opponent may force the message of the sender to mean 'I'm undecided whether to flee or to escalate, and I'm biding time until you commit yourself further'. Finally, a complete cessation of sound from the former sender would indicate 'Escape is imminent and attack probability has dropped to zero'. Following the battle, the winner's aggression stridulatory pattern would mean 'I now occupy this gallery and will attack any intruders' (Field and Rind, 1992).

The meaning of mating stridulation could be: 'I am a male and am likely to attempt (or re-attempt) copulation'. If another male arrives on the scene, the same sound could now contain the message 'I am a male and I will attack any intruding males'. In this case, the sender would experience two conflicting motivations (fighting and mating), and it could try to send two different sound patterns to the two receivers (male and female). But clearly a single pattern will suffice to carry two different messages, because each receiver has a different referent (Field and Rind, 1992). Many animals utilize this conservative approach of employing the same signal in a variety of referents to send simple, basic messages (Smith, 1979).

If a single acoustic pattern could carry different messages, the validity of differentiating an aggression and a mating pattern (based upon behavioural context) may be questionable. Furthermore, temporal parameters of these two patterns showed a great deal of overlap, raising the question of whether it is really just one highly variable pattern. An answer to this problem lies in the observation that the same approximately six-syllable pattern, which is produced in both behavioural situations, tends to have a longer syllable period in the mating context and a shorter period in the aggression and agonism context. Field and Rind (1992) postulated that a single basic echeme pattern is used, but that the syllable period varies in a graded fashion according to the excitation of the sender male. Their hypothesis assumes that: (i) the excitation level in a sender male in the presence of a female is less than that in the presence of a rival male; and (ii) the echeme syllable period becomes shorter as the excitation level increases. A slow echeme would usually indicate that a mating referent exists, while a faster echeme would develop in an aggression referent. The hypothesis also predicts that the longer mean syllable period in mating interactions would decrease, due to either active aggression by the female or the appearance of a rival male. In aggression interactions between males at gallery entrances, the syllable period should decrease rapidly if a battle escalates. The above model is easily testable, especially during observations of mating interactions interrupted by a rival male, but so far this has not been done.

Tree wetas are not the only insects in which the message of one acoustic signal changes depending upon the behavioural context in which it occurs. While in tree wetas the context dependency is found in aggressive and mating encounters, other behavioural referents appear to influence the meaning of physically undifferentiable signals in grasshoppers (Lewis and Gower, 1980). Thus, responding vs. rivalry behaviour in *Chorthippus brunneus* and copulation vs. courtship behaviour in *C. brunneus, Stenobothrus lineatus* and *Omocestus viridulus* utilized the same species-specific sound pattern in different behavioural contexts, with apparently different messages (Haskell, 1957; Young, 1971).

Comparison with other orthopteran sound patterns

In relatively recent families of the tettigonoid and acridoid orthoptera, sound patterns in social communication have reached a sophisticated level of complexity and diversity, including male attraction and courtship songs directed at females, songs used during and after mating, and agonistic and territorial songs, which can involve duetting, with complex messages about intermale spacing (Alexander, 1961; Elsner, 1974; Otte, 1977; Bailey, 1991). In contrast, tree wetas have sound patterns that are relatively simple, but which may give clues to the complexity in the evolution of acoustic signalling in the Orthoptera.

The outstanding difference in the tree weta acoustic system is the lack of a courtship sound pattern. The only sound made by males in the presence of females is the mating pattern but, because it is not a prerequisite for copulation and may even connote aggressive intent by the male sender, it cannot serve the courtship function often found in bush crickets, grasshoppers and crickets (Elsner, 1974; Bennet-Clark, 1989). Further support for this conclusion is the observation that the mating pattern occurs after attempts by males to mate with females, rather than before. In fact, the only difference between the mating and aggression patterns is a slightly slower time course in syllable repetition in the former.

Aggression stridulation occurs in similar circumstances to those which elicit rivalry songs in other orthopterans, but wetas do not engage in the highly organized phonoresponses found in tettigoniids and acridids. In the latter, accurate manipulation of timing, alternation and higher-order synchrony in rivalry duets is used to establish dominance and distance between males (Greenfield and Shaw, 1983; Sismondo, 1990), whereas *Hemideina* males only loosely alternate aggression sounds in correspondence with biting and fencing attacks during physical combat (Field and Sandlant, 1983). This situation more closely resembles the combat singing found in some crickets, which have a more ancient origin than wetas (Gwynne, 1995).

Evolution of sound patterns in tree wetas

The evolutionary development of song patterns in the Orthoptera has been the subject of much speculation, but published ideas have seldom been tested (W.J. Bailey, Perth, 1998, personal communication). In the crickets, and possibly in other winged Orthoptera, acoustic communication is thought to have commenced with the appearance of a courtship song associated with wing fanning by males for local broadcasting of pheromone(s) (Alexander, 1961, 1962). Once established, courtship sounds could have led to calling sounds that attracted females from a greater distance. The consequence of this could have been the development of male aggregations, in which territoriality and agonistic encounters led to specific sound signals associated with these two behaviours. Two mechanisms for the evolution of such a sequence have been proposed. In the first, diversification occurs when one signal, initially associated with one context, becomes associated with two contexts. This is eventually followed by structural alteration of the song pattern for the second context, yielding a second signal (Alexander, 1962). In the second, selection acts most strongly on acoustic signals that promote early intraspecific recognition (Alexander, 1967).

Could the above sequence of song diversification apply to wetas, or more generally to the evolution of acoustic communication in gryllacridoids? Did defence stridulation movements become associated with mating and eventually contribute to the evolution of courtship stridulation? And could this subsequently have led to calling and aggressive sound patterns, as in crickets?

The previous comparisons with other groups suggest that evolution of sound patterns associated with intraspecific behaviours in the gryllacridoid orthopteran line may have differed from that in crickets. It is noteworthy that few examples of acoustic defence signals exist amongst extant cricket species (Otte, 1977), where sound is produced by the wings. This contrasts with the almost universal production of defence sounds in the anostostomatids and gryllacridids, most of which utilize femoro-abdominal stridulating mechanisms. Defence stridulation occurs even in the absence of tympanal ears in these groups, suggesting that the defence pattern is plesiomorphic and was the precursor of other acoustic signals in the more primitive gryllacridoid line of ensiferans.

The lack of a courtship song suggests that tree wetas have not undergone the intense evolutionary selection for species-specific courtship songs found in continental orthopterans. In the latter, sympatric distributions and frequent interspecific encounters would necessitate an efficient means of mate recognition. In contrast, the tree weta species in New Zealand are primarily allopatric in distribution (see Gibbs, Chapter 2, this volume) and none of the species seem to have evolved a courtship song pattern (L.H. Field, personal observation). Therefore a crucial link in the sequence of song evolution proposed for crickets is missing in tree wetas. Perhaps this state represents the retention of an early, simple stage in the evolution of acoustic signalling in anostostomatid Orthoptera, wherein pheromones or other signals played a more important role than sound in the success of mating interactions between the sexes. The defence-resource polygyny, in which male wetas aggressively defend tree galleries and maintain female harems, may have precluded the need for acoustic attraction of females since females seek galleries for diurnal protection in any case. This situation is enhanced because wetas are apterous and tend to be restricted to trees as island habitats.

Although the deinacridine wetas display the raised-leg form of defence behaviour, many other gryllacridoid species, including those weta species lacking ears (Henicinae), utilize some form of vertical abdominal oscillation (e.g. Fig. 15.6F) to generate acoustic defence signals (Field, 1993a; Field and Bailey, 1997). Such a movement may have become associated with intraspecific behaviour in tree wetas and could have then given rise to the extant patterns used for intraspecific communication. Mating/aggression and calling sound patterns involve oscillatory movements of the abdomen against the hind femur, with the stridulatory structures engaging on the upward stroke. Tremulation also involves this movement, but the hind femora are not pressed against the abdomen. With the exception of calling, all of these patterns utilize the same neuronal circuitry for abdominal motor coordination and are therefore likely to have a common evolutionary origin, based upon ancestral oscillatory motor-control circuitry, which drove the defence movements of the abdomen.

The habit of occupying tree galleries would have exacerbated competition for refuges and for associated females. This, in turn, would have induced intense selection for successful agonistic behaviour patterns in males, including tremulation (which could produce substrate-borne signals) and, with little modification, aggression stridulation. The conclusions of Field and Rind (1992) about potentially aggressive messages in the mating pattern suggest that it did not evolve in the same way proposed for courtship stridulation in crickets, but instead may have been associated somehow with mild aggressive intent by males when rejected by females.

The calling pattern utilizes the same abdominal vertical oscillatory movements, but it is highly variable between males. Within this variability, calling echemes often contain short intersyllable intervals characteristic of doublets, which are a common component of the patterns from different male tree wetas. However, there is no obvious relationship between the timing of doublets and of mating/aggression patterns. It seems clear that calling is the most complex mode of signalling in tree wetas and is likely to have evolved as a major elaboration in the neuronal circuitry controlling the other patterns.

If the hypothesis of a single-variable mating/aggression pattern is accepted, it follows that the tree weta acoustic signalling system represents a simpler, intermediate stage in the evolution of orthopteran acoustic signalling, wherein defence led to a highly variable mating/aggression pattern, to a related tremulation pattern, also associated with agonistic interactions, and to another highly variable calling pattern, based upon rather different timing. The most interesting challenge for the future is to determine whether sound production by males in a mating referent affects female cooperativity and enhances mating success in males. Possibly species of tree wetas not yet examined show a more distinct difference in sound pattern between mating and agonistic fighting.

Sound and Substrate Vibration Propagation in Tree Weta Galleries

Because male tree wetas produce calling sounds from the gallery entrance, it is of interest to know whether the tree and/or gallery cavity has any effect on the quality and radiation of sound. Just as many calling insects utilize natural structures in their habitat (from burrows shaped like trumpets to leaves which act as baffles) to enhance sound output, it could be possible that the tree gallery aids male wetas in sound production. Another role for trees in weta sound production could be to distribute vibrations from one calling weta to others in the same tree. Insects such as plant hoppers detect vibrations induced on to the stalks of plants by the calls of conspecifics (Michelsen *et al.*, 1982). Although trees are more massive than any plants studied previously in regard to substrate vibration communication, a recent study has shown that male wetas are capable of inducing vibratory waves into tree-trunks. This raises questions about the use of substrate vibration as another method of communication by wetas.

The effect of the tree gallery on sound quality and radiation

Males calling from tree galleries are presumably broadcasting messages to other wetas in neighbouring trees. The calling male sits either in the gallery entrance or inside the gallery. Sound radiated from such an individual is likely to be affected by the tree acting as a dense acoustic baffle and also by the gallery acting as a potentially resonating acoustic cavity. To test whether the first effect occurs, Bailey and Field (2001) measured the intensity of a 'standard' tape-recorded weta calling sound generated by a small loudspeaker placed inside galleries. The sound intensity was set to a level comparable to that created by male wetas. The galleries were contained in sections of tree-trunks brought into a semi-anechoic room. Measurements were made by recording average sound intensity at three radial distances (10, 20, 40 cm) from a gallery entrance as the microphone was rotated to eight different directions around each trunk.

The results consistently showed that sound was radiated most intensely in the direction of the gallery entrance and least intensely from behind the trunk. The fall-off in intensity was 2.5 dB from 20 to 40 cm due to near-field effects of the log, and thereafter it was 6 dB loss with each doubling of distance. Therefore the tree acts as a mild directional attenuator of sound propagating from the gallery. The important question from a communication perspective is to ask from how far away the sound can be heard by other wetas. Bailey and

Field calculated this distance by using the threshold intensity of hearing for tree wetas (25 dB, *H. crassidens*: Field *et al.*, 1980) and by assuming a 6 dB loss decrement with distance. For a male calling from just outside the gallery at a typical intensity of 69 dB, the sound should just be heard by another male 25 m away. However, if the sender male is inside the gallery (which was found to attenuate the sound by 10–12 dB), a receiver male would need to be 3 m away to just detect the sound. Finally, if both wetas were in galleries, the effective threshold hearing distance would become 2 m. The conclusion was that calling by male wetas is probably limited to communication with other wetas in the same tree or with nearby neighbours, which could be potential gallery rivals. The call is likely to carry a territorial message indicating that the resident will fight to defend the gallery. More research is needed to elucidate the possibility of territoriality and the role of calling stridulation in natural populations of tree wetas.

The second effect of the tree on calling sounds could be to alter the sound through resonance effects imposed by the gallery. This was tested by comparing the frequency spectrum of a white noise signal broadcast from the speaker within the test galleries to a frequency spectrum of the same signal generated by the speaker in a free field outside the galleries. Any difference in the frequency content of the two recorded spectra should be due to effects from the gallery. In the free field, the speaker gave a flat response (± 5 dB) up to 15 kHz. Bailey and Field (2000) found that the galleries appeared to resonate at 2–3 kHz, as shown by a peak of enhanced sound at this frequency band. This is remarkably similar to the frequency peak of around 2 kHz in the spectrum of sound produced by a calling weta. Moreover, the spectral analysis showed that the galleries filtered out higher frequencies of sound and thus enhanced the signal-to-noise ratio of the emitted call. When a similar spectral analysis was made for the recording of a weta calling sound broadcast in the two conditions, a major peak at 2 kHz was found. Therefore it appears that tree wetas receive at least two acoustic benefits by calling from galleries in tree-trunks.

The peak sound frequencies produced by tree wetas are very similar to the peak frequency sensitivity of the tympanal organs (ears) (see Field, Chapter 22, this volume). Bailey and Field (2000) speculated that selection on the peak calling frequency is likely to have been driven by the properties of galleries created by wood-boring insects, notably cerambycid beetle larvae. Since the galleries occupied by tree wetas are chosen for a reasonably consistent internal diameter related to body size, it follows that the resonating properties of the galleries will be reasonably consistent from tree to tree. They further suggested that the hearing sensitivity in the tympanal organs may have evolved to match that of the calling sound, due to a process of sexual selection. The reasoning follows that of Endler (1992), who noted that, where the acoustic signal carries information on male quality, sexual selection will lead to the hearing system following frequencies close to the best male's call.

A further adaptation in tree weta calling stridulation comes from an unrelated study of the propagation of bird calls through forests. It was shown that the sound frequencies which propagate best through the lower levels of forests (i.e. tree-trunks and sparse foliage) are in the range of 1.5–2.5 kHz (Morton, 1975). Because tree wetas occupy galleries predominantly distributed below the foliage level (see Field and Sandlant, Chapter 13, this volume), they appear to be generating optimum frequencies for propagation through their habitat.

Communication by substrate vibration in trees

The remarkable sensitivity of some insects to vibrations carried in plant stems (e.g. Cokl, 1983) raised the question of whether tree wetas are able to induce vibrations in tree-trunks when calling. If so, could such vibrations be used for communication?

These questions were tested in a biophysical study by McVean and Field (1996) for a different species of weta, *H. femorata*, which lives primarily in manuka trees (*Leptospermum ericoides*). By first testing a tree for its vibratory response to a standard broad-band signal, they showed that the tree propagated a narrow-band peak of vibration, which increased in frequency as the trunk diameter decreased with height. This was shown to take the form of a bending wave travelling in the trunk (maximum diameter =13 cm). In addition, the tree severely attenuated other frequencies by acting as a selective filter. This indicates that the tree had a resonant frequency at which it oscillated when energized by any induced vibration, and that the resonant frequency for any level depended upon trunk diameter. The resonant frequency was around 500 Hz at the base and increased to 0.8–1.2 kHz over the rest of the trunk.

By next causing a weta to stridulate while on the tree, it was possible to demonstrate that the insect induces a vibratory signal into the tree-trunk, which had a narrow signature of 1.25–1.5 kHz, and, in strong stridulation, a minor peak at 7.5 kHz (Fig. 22.6, Field, Chapter 22, this volume). The vibratory spectrum was different from the spectrum of the stridulatory sound, which further underscored the existence of a vibratory resonance in the tree, which could be driven by a broad range of input frequencies. Thus, a stridulating *H. femorata* appears to excite bending waves at the natural frequency of the tree, but, in addition, it generates a forced vibration at a higher frequency, around 7.5 kHz. Both components would be modified as they travel through the tree and into the branches, but it is likely that the 7.5 kHz component would be damped more quickly, since it differs from the resonant frequency. Therefore, it should more accurately follow the temporal structure of the calling pattern. If another weta were able to detect this signature frequency as well as the temporal pattern of vibration, it would have the unequivocal identity of the sender, since only wetas produce such acoustic and vibratory signals at night in the forest habitat.

Can *H. femorata* detect the vibratory signals on the tree-trunk? An affirmative answer was provided by neurophysiological recordings from the subgenual organ, a vibration-detecting sense organ in the tibia of each leg. A restrained weta, prepared for recording, rested upon a suspended log in the laboratory. A second, stridulating weta, placed upon the same log, produced substrate vibrations well above threshold for detection by the subgenual organ. The weta subgenual organ responded to sinusoidal test vibrations ranging from 20 Hz to 3 kHz, with a threshold stimulus intensity of 0.015 ms^{-2} at 1.0 kHz (see Field, Chapter 22, this volume).

Calculations of the vibratory acceleration (approximately 0.034 ms^{-2}) generated by a calling weta indicated that the signal should propagate through the tree over a distance of 2–3 m above and below the weta's position (McVean and Field, 1996). Thus, all of the wetas in a manuka tree should be in vibratory communication with each other.

Armed with this knowledge, it is clear that there are distinct advantages to using vibrational as well as acoustic communication in tree weta social interactions. A receiver weta using only acoustic communication would have no means of determining whether a caller resided in the same tree or a neighbouring tree. Since a calling weta is likely to be a male occupying a gallery containing females, a receiving male interested in competing for the gallery would have to undertake lengthy journeys, including travel across the forest floor, where predator encounter is likely, in order to locate a caller in another tree. If the receive could detect substrate vibrations coupled to the acoustic signal, it could be sure that the caller resided on the same tree. If the calling weta was on another tree and a searching weta was moving to new trees, it could easily eliminate inappropriate trees by simply climbing on to the base of a trunk and testing for the presence of a vibrational signal accompanying the acoustic one.

References

Alexander, R.D. (1961) Aggressiveness, territoriality and sexual behaviour in field crickets (Orthoptera: Gryllidae). *Behaviour* 17, 130–223.

Alexander, R.D. (1962) Evolutionary change in cricket acoustical communication. *Evolution* 16, 443–467.

Alexander, R.D. (1967) Acoustical communication in arthropods. *Annual Review of Entomology* 12, 495–526.

Bailey, W.J. (1991) *Acoustic Behaviour of Insects.* Chapman and Hall, London, 225 pp.

Bailey, W.J. and Field, L.H. (2001) Are the calls of New Zealand wetas used in distance communication? An analysis of sound transmission in *Hemideina crassidens* (Stenopelmatidae: Orthoptera). *Journal of Natural History* (submitted).

Barrett, P. (1991) *Keeping Wetas in Captivity*. Wellington Zoological Gardens, Wellington, New Zealand, 60 pp.

Bennet-Clark, H.C. (1989) Songs and the physics of sound production. In: Huber, F., Moore, T.E. and Loher, W. (eds) *Cricket Behavior and Neurobiology.* Cornell University Press, Ithaca, New York, pp. 227–261.

Bennett, A. and Toms, R.B. (1995) Sexual dimorphism in the mouthparts of the king cricket *Libanasidus vittatus* (Kirby) (Orthoptera: Mimnermidae). *Annals of the Transvaal Museum* 36, 205–214.

Cary, P.R.L. (1981) The biology of the weta *Zealandrosandrus gracilis* (Orthoptera:

Stenopelmatidae) from the Cass region. MSc thesis, Department of Zoology, University of Canterbury, Christchurch, New Zealand.

Cokl. A. (1983) Functional properties of vibroreceptors in the legs of *Nezara viridula* (L.) (Heteroptera, Pentatomidae). *Journal of Comparative Physiology* 150, 261–269.

Den Hollander, J. and Barrientos, L. (1994) Acoustic and morphometric differences between allopatric populations of *Pterophylla beltrani* (Orthoptera: Tettigoniidae: Pseudophyllinae). *Journal of Orthoptera Research* 2, 29–34.

Dumortier, B. (1963) The physical characteristics of sound emissions in Arthropoda. In: Busnel, R.-G. (ed.) *Acoustic Behaviour of Animals*. Elsevier Publishing Company, Amsterdam, pp. 277–345.

Edmunds, M. (1974) *Defence in Animals*. Longman, London, 357 pp.

Elsner, N. (1974) Neuroethology of sound producing in gomphocerine grasshoppers (Orthoptera: Acrididae). I. Song patterns and stridulatory movements. *Journal of Comparative Physiology* 88, 67–102.

Endler, J.A. (1993) Signals, signal conditions, and the direction of evolution. *American Naturalist* 139, 125–153.

Field, L.H. (1978) The stridulatory apparatus of New Zealand wetas in the genus *Hemideina* (Insecta: Orthoptera: Stenopelmatidae). *Journal of the Royal Society of New Zealand* 8, 359–375.

Field, L.H. (1980a) Observations on the biology of *Deinacrida connectens* (Orthoptera: Stenopelmatidae), an alpine weta. *New Zealand Journal of Zoology* 7, 211–220.

Field, L.H. (1980b) The tick sound of a giant weta, *Deinacrida rugosa* (Orthoptera: Stenopelmatidae: Deinacrideinae). *New Zealand Entomologist* 7, 176–183.

Field, L.H. (1982) Stridulatory structures and acoustic spectra of seven species of New Zealand wetas (Orthoptera: Stenopelmatidae). *International Journal of Insect Morphology and Embryology* 11, 39–51.

Field, L.H. (1993a) Structure and evolution of stridulatory mechanisms in New Zealand wetas (Orthoptera: Stenopelmatidae). *International Journal of Insect Morphology and Embryology* 22, 163–183.

Field, L.H. (1993b) Observations on the stridulatory, agonistic, and mating behaviour of *Hemideina ricta* (Stenopelmatidae: Orthoptera), the rare Banks Peninsula weta. *New Zealand Entomologist* 16, 68–74.

Field, L.H. and Bailey, W.J. (1997) Sound production in primitive Orthoptera from Western Australia: sounds used in defence and social communication in *Ametrus* sp. and *Hadrogryllacris* sp. (Gryllacrididae: Orthoptera). *Journal of Natural History* 31, 1127–1141.

Field, L.H. and Rind, F.C. (1992) Stridulatory behaviour in a New Zealand weta, *Hemideina crassidens*. *Journal of Zoology, London* 228, 371–394.

Field, L.H. and Sandlant, G.R. (1983) Aggression and mating behaviour in the Stenopelmatidae (Orthoptera, Ensifera), with reference to New Zealand wetas. In: Gwynne, D.T. and Morris, G.K. (eds) *Orthopteran Mating Systems – Sexual Competition in a Diverse Group of Insects*. Westview Press, Boulder, Colorado, pp. 120–146.

Field, L.H., Hill, K.G. and Ball, E.E. (1980) Physiological and biophysical properties of the auditory system of the New Zealand weta *Hemideina crassidens* (Blanchard, 1851) (Ensifera: Stenopelmatidae). *Journal of Comparative Physiology* 141, 31–37.

Graber, V. (1874) Uber den Bau und die Entstehungeinige noch wenig bekannter Stridulationsorgane der Heuschreken und der Spinner. *Mitteilungen des Naturwissenschaften Vereins fur Steiermark* 1874, 1–15.

Greenfield, M.D. and Shaw, K.C. (1983) Adaptive significance of chorusing with special reference to the Orthoptera. In: Gwynne, D.T. and Morris, G.K. (eds) *Orthopteran Mating Systems: Sexual Competition in a Diverse Group of Insects*. Westview Press, Boulder, Colorado, pp. 1–27.

Gwynne, D.T. (1995) Phylogeny of the Ensifera (Orthoptera): a hypothesis supporting multiple origins of acoustical signalling, complex spermatophores and maternal care in crickets, katydids, and weta. *Journal of Orthopteran Research* 4, 203–218.

Haskell, P.T. (1957) Stridulation and associated behaviour in certain Orthoptera. 1. Analysis of the stridulation of, and behaviour between, males. *British Journal of Animal Behaviour* 5, 139–148.

Haskell, P.T. (1961) *Insect Sounds*. H.F. Withersby and G. Withersby, London, 189 pp.

Hind, R.A. (1981) Animal signals: ethological and games-theory approaches are not incompatible. *Animal Behavior* 29, 535–542.

Lewis, D.B. and Gower, D.M. (1980) *Biology of Communication*. Blackie and Sons, Glasgow.

McVean, A. (1986) The song of the New Zealand weta, *Hemideina thoracica* (Orthoptera: Stenopelmatidae). *Journal of Zoology, London (A)* 208, 171–190.

McVean, A. and Field, L.H. (1996) Communication by substratum vibration in the New Zealand tree weta, *Hemideina femorata* (Stenopelmatidae: Orthoptera). *Journal of Zoology, London* 239, 101–122.

Markl, H. (1985) Manipulation, modulation, information and cognition: some of the riddles of communication. *Forschritte Zoologische* 31, 163–194.

Michelsen, A., Fink, F., Gogala, M. and Traue, D. (1982) Plants as transmission channels for insect vibrational songs. *Behavioral and Ecological Sociobiology* 11, 269–281.

Möller, H. (1985) Tree wetas (*Hemideina crassicruris*, Orthoptera: Stenopelmatidae) of Stephens Island, Cook Strait. *New Zealand Journal of Zoology* 12, 55–70.

Morgan-Richards, M. (1997) Intraspecific karyotype variation is not concordant with allozyme variation in the Auckland tree weta of New Zealand, *Hemideina thoracica* (Orthoptera: Stenopelmatidae). *Biological Journal of the Linnean Society* 60, 423–442.

Morton, E.S. (1975) Ecological sources of selection on avian sounds. *American Naturalist* 109, 17–34.

Otte, D. (1977) Communication in orthoptera. In: Sebeok, T.A. (ed.) *How Animals Communicate*. Indiana University Press, Bloomington, pp. 334–361.

Ramsay, G.W. (1953) A supplementary sound-producing device in *Deinacrida rugosa* Buller (the Stephens Island weta). *New Zealand Entomologist* 1, 12–14.

Richards, A.M. (1972) A comparative study of the biology of the giant wetas *Deinacrida heteracantha* and *D. fallai* (Orthoptera: Henicidae) from New Zealand. *Journal of Zoology, London (A)* 169, 195–236.

Riek, E.F. (1970) Fossil history. In: CSIRO (eds) *Insects of Australia*. CSIRO, Melbourne University Press, Melbourne, pp. 168–186.

Samways, M.J. (1977) Song modification in the Orthoptera. IV. The *Platycleis intermedia/P. affinis* interaction quantified. *Physiological Entomology* 2, 301–315.

Sandlant, G.R. (1981) Aggressive behaviour of the Canterbury weta *Hemideina femorata* (Orthoptera: Stenopelmatidae): its adaptive significance in resource allocation. MSc thesis, The University of Canterbury, Christchurch, New Zealand.

Schatral, A. and Bailey, W.J. (1991) Song variability and the response to conspecific song and to song models of different frequency contents in males of the bush cricket *Requena verticalis* (Orthoptera: Tettigoniidae). *Behaviour* 116, 163–179.

Sherley, G.H. and Hayes, L. (1993) The conservation of a giant weta (*Deinacrida* n. sp. Orthoptera: Stenopelmatidae) at Mahoenui, King Country: habitat use, and other aspects of its ecology. *New Zealand Entomology* 16, 55–68.

Sismondo, E. (1990) Synchronous, alternating and phase-locked stridulation by a tropical katydid. *Science* 249, 55–58.

Smith, P.E. (1979) Behavioural and reflexive analysis of a defence response in *Hemideina maori* (Orthoptera: Stenopelmatidae). BSc (Hons) thesis, University of Canterbury, Christchurch, New Zealand.

Walker, T.J. (1962) Factors responsible for intraspecific variation in the calling songs of crickets. *Evolution* 16, 407–428.

Young, A.J. (1971) Studies on the acoustic behaviour of certain Orthoptera. *Animal Behavior* 19, 727–743.

Zuk, M. and Simmons, L.W. (1997) Reproductive strategies of the crickets. In: Choe, J.C. and Crespi, B.J. (eds) *Evolution of Mating Systems in Insects and Arachnids*. Cambridge University Press, Cambridge, pp. 89–109.

16
Defence Behaviour

Laurence H. Field and Stephen Glasgow
Department of Zoology, University of Canterbury, Private Bag 4800, Christchurch, New Zealand

Introduction

The hallmark of the New Zealand giant wetas and tree wetas is the spectacular display of vertically raised, spine-studded hind legs and loud rasping sounds given off when humans disturb or poke at the insects (Fig. 16.1). In popular books, this has been termed defence behaviour (from the human perspective). The important questions about this behaviour should be framed in ecological terms: what are its detailed components, is it produced in the presence of real weta predators, and how effective is it against these predators? In this chapter, we explore these questions and try to place the answers into a comparative view of the family as a whole.

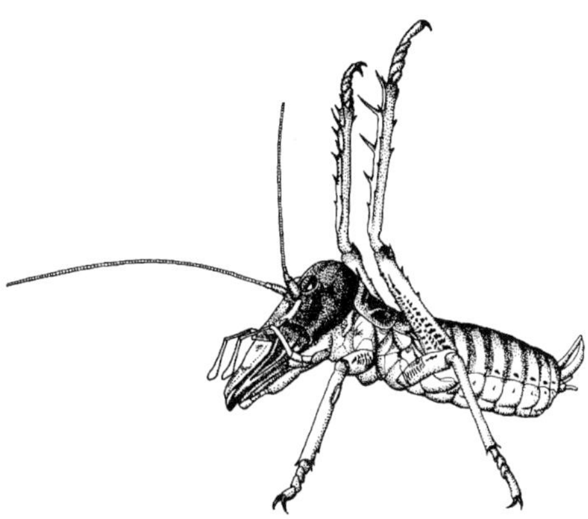

Fig. 16.1. Defence display of male tree weta *Hemideina crassidens* of New Zealand. When provoked, the hind legs are swept rapidly downwards to produce a stridulatory sound. (From Field, 1978, with permission.)

The general lack of wings in most anostostomatid genera has meant that modes of escape include only jumping or walking (running) away from predators. Inevitably having to face predators, anostostomatids appear to have evolved a variety of behavioural movements, postures and anatomical features adapted to prevent predation. The extent to which this had been successful only in an isolated island habitat with a limited suite of predators, excluding mammals, is shown by the disappearance of many of the New Zealand giant wetas from nearly all areas currently inhabited by introduced rats and feral cats. In those countries, including New Zealand, where anostostomatids survive in the presence of mammalian and avian predators, we see mechanisms designed to reduce chances of contact to a minimum, backed up by defence behaviours and protective morphology. However, experimental behavioural studies, as well as stomach and faecal analyses, show that such secondary adaptations (see below) may be of little value once the insects are faced by such predators. This, in turn, leads us to suggest later in this chapter that defence adaptations found in New Zealand anostostomatids evolved against reptilian, arthropod and perhaps some avian predators.

The defence system of any animal may comprise sensory components concerned with specific detection of predators, behavioural components concerned with avoiding detection, actively escaping from or battling against predators and morphological (sometimes physiological) components concerned with deterring predators. These defence mechanisms may be divided into two categories: primary and secondary defence (Edmunds, 1974). Primary defence operates regardless of whether or not there is a predator in the vicinity. Secondary defences operate as soon as a prey detects a predator (which may be prior to being detected by the predator) or after the predator has initiated a prey-capture attempt.

Primary Defence Mechanisms

Primary defences function to decrease the probability that an interaction will occur between a prey animal and a potential predator. Nearly all anostostomatids hide in the daytime and only emerge at night. This is a primary defence from diurnal predators, mostly birds, but is limited if the hiding-place is not robust against predators which can detect the smell of some anostostomatids. Occupation of galleries in trees by *Hemideina* wetas in New Zealand during the daytime is a robust primary defence ploy (termed anachoresis). Although the galleries serve other functions for tree wetas (see Field, Chapter 18, this volume), they are not obligatory: wetas also seek diurnal refuge in bolt-holes in fence posts, piles of firewood and hollow wooden traps placed by clever biologists. The tree wetas, and some ground wetas (*Hemiandrus*) which also inhabit tree holes, have survived well in the face of introduced predators in New Zealand. Other species, notably ground wetas (Henicinae) and the giant tusked wetas in New Zealand and the Jerusalem crickets of North America, live in burrows dug into the earth (Salmon, 1950; Meads, 1990a; see also Weissman, Chapter 3, this volume). For a review of burrowing adaptations, see Gorochov (Chapter 1, this volume). The giant wetas (*Deinacrida*) of New Zealand spend the daytime in refuges at ground level (depressions at the bases of shrubs, under rocks, in rotting logs, temporarily buried in loose soil) or hidden in overhanging dead fronds of tree ferns and under bark of trees (Richards, 1973; Meads, 1990b). They rely heavily upon camouflage coloration, the primary defence adaptation, termed crypsis. They are variously coloured in tan or brown tones, except for the 'Mahoenui' giant weta (*Deinacrida mahoenui*), which has two colour morphs: one blackish brown and the other mottled yellowish tan (Sherley and Hayes, 1993). Except for the alpine *Deinacrida connectens*, which lives under rocks above the timberline, most giant wetas have suffered heavily from introduced predators and are considered endangered species in New Zealand (see Sherley, Chapter 26, this volume). Immobility, as a primary defence behaviour, is apparently effective against reptiles. Moller (1978) noted that a female *Hemideina crassidens* next to a tuatara (*Sphenodon punctatus*) was not attacked until it moved and, in all experiments with geckos, described below, no attacks occurred unless the wetas were moving.

Secondary Defence Mechanisms in New Zealand Deinacridinae (Tree Wetas, Giant Wetas)

Secondary defence mechanisms are those behaviours and morphological adaptations brought into action once a predator is detected. The spectacular alarm response in deinacridine wetas provides three examples of secondary defence mechanisms. When provoked these wetas raise and kick the heavily spined hind legs, while producing a rasping sound and facing an antagonist with wide-opened mandibles (Hutton, 1897; Salmon, 1950; Field, 1978). Thus: (i) the leg raising/kicking and mandible opening constitute a visible behavioural defence; (ii) the sound production is an audible defence mechanism; and (iii) the hind tibial spination provides a morphological defence mechanism. Variations on this theme, as well as other defence behaviours occur throughout the Anostostomatidae. The following section will explore these secondary defence mechanisms in New Zealand Deinacridinae.

Defence behaviour of the tree weta, Hemideina crassidens

The following analysis of *H. crassidens* defence behaviour will serve as framework for comparison with other species (no other detailed studies exist). The experiments were carried out in the laboratory to maintain controlled conditions, but the same behavioural responses were also seen in interactions with predators (presented later). The behavioural responses consist of stereotyped components, which have been given descriptive, not functional, names. The names are used later for discussing interactions with predators and the evolution of defence behaviour.

Methods

Forty wetas (30 adults and ten juveniles) were collected from forests on the west coast of the South Island of New Zealand, and maintained on a reverse light cycle (16L : 8D) in moss substrate containers at 15°C. Food (apple, leaves from native trees) and water were provided every 2 days. A 25 W red lamp placed 30 cm from the experimental arenas provided illumination to enable observations during the artificial night period. An arena was a clear circular plastic container (11 cm × 23 cm), into which a weta was placed 2 h before each experiment and left undisturbed.

The experiment consisted of lightly prodding the weta with a hand-held 3 mm dia. wooden rod every 30 s until: (i) the weta escaped from the container; or (ii) the response showed no variation over ten consecutive stimuli. Previous experience showed that tactile stimulation, rather than acoustic or visual stimulation, was more effective in producing a full range of defence responses. For 15 males, 15 females (adults, tibia > 17 mm) and ten juveniles (tibia < 10 mm), one body region (antennae, pronotum or abdomen) was stimulated in each experiment and most animals were used only once.

The behavioural components comprising the defence response repertoire to the above tactile stimulus had been defined in previous tests using other wetas. These are presented below. The terms 'mildly excited' and 'highly excited' are subjective descriptions of the observed reaction levels of the wetas. 'Mildly excited' means that the insects gave one or two desultory responses, while 'highly excited' means that the response was rapid and, if phasic, repeated several times per stimulation.

Kick hind legs

From the normal resting position, one or both hind tibiae are raised and kicked posteriorad by upward femoral levation through about 25° and then rapid femoro-tibial extension from 80° to 110°–130° as the femur is depressed. This is a phasic movement, which thrusts the spines of the tibiae in the direction of a stimulus to the abdomen (Fig. 16.2A). Wetas often orientated their posterior end toward the stimulus.

Hind leg fend

The femur is raised to an angle of 70°–100° from the horizontal by anterior coxal rotation and the femoro-tibial joint is held at an angle of 80°–90°. This is a maintained posture with one or both legs, aiming the tibial spines toward an attack from above the abdomen (Fig. 16.2B).

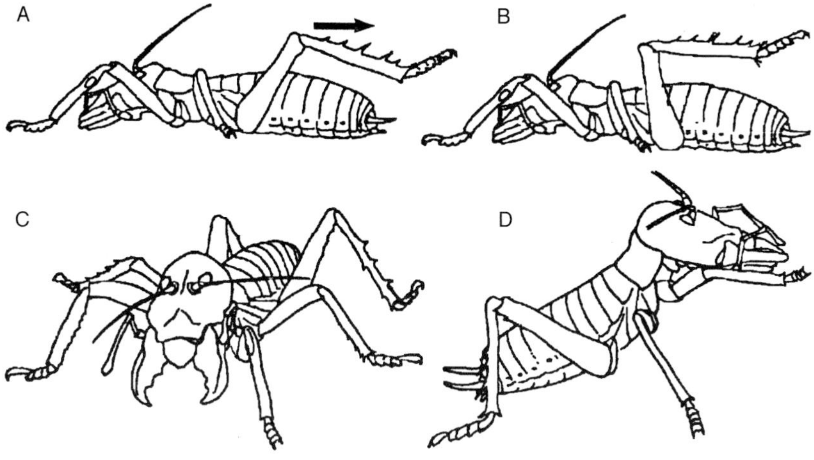

Fig. 16.2. Elements of defence behaviour of *H. crassidens*. A. 'Kick hind legs' behaviour is used to direct the sharp tibial spines at an attacker. B. 'Hind leg fend' is a display produced when mildly provoked. Greater provocation leads to 'raise hind legs'. C. 'Mandible gape' is a posture directed toward an attacker and may lead to biting. D. A more highly excited weta displays 'raise foreleg' behaviour, accompanied by 'raise head' and 'mandible gape' when stimulated on the head or antennae.

Raise hind legs

This is an exaggerated version of 'hind leg fend', in which one or both legs are extended fully and rotated to an extreme anterior or vertical position. The coxa may rotate the femur forward to an angle of 110°–170° to the horizontal, and the tibia is extended to a femoro-tibial angle of about 145° (Fig. 16.1). The response often occurs between stridulatory bouts and is characteristic of highly excited wetas. It is evoked by tactile stimulation to the antennae or pronotum. The tibial spines are aimed toward attack from above the thorax or, in the extreme anterior position, toward attack from the front of the weta.

Close hind leg

The hind legs are appressed medially to bring the tibiae parallel and close together, without raising the legs. This is evoked by mild stimulation to the abdomen. The tibial spines block attack of the abdomen from the rear.

Mandible gape

The mandibles are opened partially or fully in readiness for biting, which occurs if they are contacted. The clypeus is withdrawn, the maxillary palps are withdrawn to the level of the mandible tips, the labium is reflected downwards and the maxilla are held ventrad parallel to the mandibles (Fig. 16.2C). This may be a phasic movement or a maintained posture. The display is formidable in males, due to their enlarged mandibles: a 50 mm long male can have a maximum gape of 10 mm. 'Mandible gape' is evoked by stimulation of the anterior part of the body, including the antennae, and is characteristic of mildly to highly excited wetas.

Raise head

The head is raised and the front and middle legs are extended. The abdomen remains in contact with the substrate. This is a maintained posture evoked by stimulation of the antennae and pronotum, and is characteristic of highly excited wetas (Fig. 16.2D). 'Raise head' is usually accompanied by 'mandible gape', and it may serve to present 'mandible gape' as a more intimidating display. When escalated, 'raise head' involves arching the head and body back by extension of the middle legs. This always includes 'mandible gape' and 'raise foreleg' (see below) displays.

Raise foreleg

One or both forelegs are raised so that the femora are horizontal to the substrate and the tibia are extended maximally (Fig. 16.2D). This maintained posture is evoked especially by stimulation of the antennae. It often accompanies 'mandible gape' and 'raise head'.

Stridulation

Two modes occur during defence behaviour (Field, 1993). In the first (termed 'hind leg stridulation'), the hind legs are initially held in 'raise hind legs' position and then swept rapidly downward by posterior rotation through an arc of 70°–80°. The legs usually return to the raised position (see Fig. 15.7A , Field, Chapter 15, this volume). During 'hind leg stridulation', a patch of cuticular pegs on the medial femur surface is scraped against a file on the second abdominal tergite, producing an audible rasp (femoro-abdominal stridulation: Field, 1978, 1993; Field and Rind, 1992). The weta inflates its abdomen during each down-stroke of the legs, presumably to engage the file and pegs. This rapid behaviour occurs when a very highly excited weta is stimulated by touch to almost any part of its body or even by visual movement of nearby objects, such as a hand. A second mode of stridulation (termed 'abdominal stridulation') often occurs when wetas are disturbed in their galleries, by, for example, touching the abdomen. The abdomen is vibrated several times in a vertical plane against the hind femora, while the legs remain on the substrate. Each downstroke results in a weak rasping sound (Fig. 15.7B, Field, Chapter 15, this volume).

Move antennae

This response varies, depending on where the stimulus is applied to the weta. Anterior body stimulation evokes rapid antennal flicking against the substrate in front and to the side at a rate of about 2 Hz. Abdominal stimulation causes the antennae to be waved posteriorly over the abdomen. This is characteristic of a mildly excited weta. Also, 'move antennae' always occurs during 'escape' (see below).

Vibrate palps

The maxillary and labial palps are extended anteriorad to the level of the mandible tips and vibrated against the substrate at a rate of 4–6 Hz. This is often associated with 'move antennae' and always accompanies 'escape'.

Escape

An adult weta rapidly walks away from an applied stimulus, while antennae are tapped against the substrate (2–3 Hz) or waved laterally through an arc of 100°–110°. 'Vibrate palps' always occurs. 'Escape' is evoked by stimulation to all body areas. With abdominal stimulation, 'kick hind legs' or, less commonly, 'hind leg fend' occurs, while the weta continues to walk with the fore- and middle legs.

Jump

The legs are rapidly extended to propel the weta forward and upward to a distance of about 10 cm. This is more characteristic of early instars (Field, 1978), but occurs in adults as well. 'Jump' is evoked by abdominal stimulation once a weta has become highly excited.

Additional behaviours were evoked by handling wetas. This appears to cause greater excitement than the above tactile stimulation, but is likely to have artificial components included (e.g. warmth and pressure of fingers). Nevertheless, the behaviours are interesting. If a weta is grasped between thumb and forefinger, the sharp hind tibial spines are pushed upwards against these fingers; the 'raise hind legs' posture becomes an active prodding behaviour. If a grasped weta is allowed to bite (on protecting gloves), it often regurgitates its crop contents and/or defecates. Both the crop contents and faeces have a distinctive odour to humans and possibly weta predators; this is a common orthopteran secondary defence mechanism, described also for Australian and South African king crickets and the Chilean red cricket (respectively: Monteith and Field, Chapter 5, Toms, Chapter 4, and Angulo, Chapter 11, this volume).

Analysis of behavioural components

Analysis of the frequency of occurrence of the different behavioural components for each of the three stimulated body regions (Table 16.1) showed significant differences between body stimulation sites for wetas of the same sex, and between males and females of the same size (chi-squared association test, 5% significance, 8 d.f.).

The differences in patterns for each stimulation site are best seen by plotting the percentage occurrence of each component out of the total number of responses from adult males (Fig. 16.3). Antennal stimulation evoked 'mandible gape', 'raise head', 'raise forelegs' and the minor responses, as a suite of components (Fig. 16.3, top). Pronotum stimulation evoked a different suite: 'mandible gape', 'hind leg fend' and 'raise hind legs' (Fig. 16.3, middle), while abdominal stimulation evoked mostly 'kick hind legs', 'hind leg fend' and a high frequency of 'escape' (Fig. 16.3,

Table 16.1. Percentage occurrence of defence behaviour components associated with stimulation of different body regions in male, female, adult and juvenile (j) wetas.

Stimulation site	Sex	Percentage component occurrence out of total responses displayed									Total responses
		MG	RH	RF	KL	LF	RL	J	E	MR	
Antennae	Male	31.3	8.1	7.1	1.3	4.0	3.3	–	13.1	31.8	396
Antennae	Female	29.5	3.9	5.3	–	2.9	1.9	–	28.0	28.5	207
Pronotum	Male	27.6	1.1	0.4	1.1	7.5	10.3	–	27.6	24.0	279
Pronotum	Female	23.5	2.7	–	1.5	2.3	6.3	–	30.7	33.1	260
Pronotum	Male (j)	2.3	1.0	–	0.5	4.2	1.4	–	42.6	47.0	216
Pronotum	Female (j)	4.8	–	–	–	2.8	12.4	–	31.0	49.0	145
Abdomen	Male	3.7	–	–	27.5	2.5	–	1.2	38.6	26.5	321
Abdomen	Female	0.5	–	–	27.5	4.3	1.0	0.5	30.0	36.2	207

MG, mandible gape; RH, raise head; RF, raise foreleg; KL, kick hind legs; LF, hind leg fend; RL, raise hind legs; J, jump; E, escape; MR, minor responses (vibrate palps, move antennae, close hind legs).

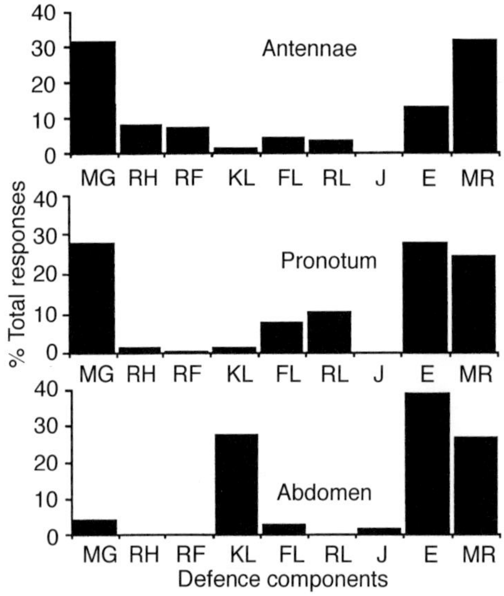

Fig. 16.3. Histograms showing how the suite of behavioural defence components alters according to site of tactile stimulation in male *H. crassidens*. MG, mandible gape; RH, raise head; RF, raise foreleg; KL, kick hind legs; LF, hind leg fend; RL, raise hind legs; J, jump; E, escape; MR, minor responses (vibrate palps, move antennae, close hind legs).

bottom). 'Escape' occurred about one-third of the time, within the termination criteria set for the tests, and 'jump' rarely occurred. Thus about two-thirds of the time the wetas stood their ground and continued to show defence behaviour. The minor responses (MR: 'vibrate palps', 'move antennae' and 'close hind legs') accompanied 'escape' with similar frequencies of occurrence, except in antennal stimulation, where 'escape' was reduced.

Sexual dimorphism partially accounted for differences between the sexes, and between adult and juvenile wetas. For example, adult females, with smaller mandibles, generally showed less 'mandible gape', 'raise head' and 'raise foreleg' behaviour for all three stimulated regions (Table 16.1). Also, adult females have heavier abdomens than males and may be less prone to escalate to the full defence behaviour repertoire (reduced percentage of 'raise hind legs'), in addition to showing greater minor responses and 'escape' behaviour. The juvenile tree wetas showed greatly reduced or no 'mandible gape', 'raise head' or 'raise foreleg' for pronotum stimulation. 'Kick hind leg', 'hind leg fend' and 'raise hind legs' were displayed as in adults, except that juvenile females showed a much greater percentage of 'raise hind legs' than males (Table 16.1). Juvenile wetas showed high percentages of minor response and 'escape' behaviour.

In summary, tree weta defence responses to repeated tactile stimulation varied, depending upon the sex and age of the individual and the body region stimulated. Compared with females, the enlarged head and elongated mandibles of males were associated with a greater frequency of threat display of the gaped mandibles, enhanced by raising the head and forelegs. The hind legs were used to fend, kick and protect the abdomen and pronotum when those body regions were stimulated. A mild defence response consisted mostly of antennal waving, palp vibration and closing the hind tibiae over the abdomen while in rest posture. As the display escalated with repeated stimulation, the hind legs were raised to protect the stimulated region of the body and to kick at the offending object. More escalation led to a fully extended hind leg posture. Hind–hind leg stridulation often accompanies the latter behaviour when humans 'attack' wetas, but this did not occur with the experimental paradigm used. Escalation was also marked by gaped mandibles, raised head and raised forelegs. Jumping, as a form of escape, occurred only rarely in the above experiments, but it is often produced when juvenile wetas are handled or extracted from tree galleries.

Comparison with other Hemideina *species*

A similar study to that described above was made with *Hemideina maori*, the only *Hemideina* species that does not live in tree galleries in New Zealand. Instead, it is an alpine species living under rocks in grassy fell fields (Salmon, 1950; Field, 1978; Miller, 1985). The same tactile stimulus was used on 30 adult specimens brought into the laboratory. Over 200 trials were performed in the quiet arena conditions described above, but stimuli were not consistently restricted to only one body region, and results were not numerically tabulated. The number of observations was sufficient to obtain a full appreciation of the defence repertoire of *H. maori* and to confirm an earlier description (Field, 1993).

The same components found for *H. crassidens* were observed in the responses of *H. maori*, but additional components were discovered as the defence behaviour escalated. For low-level excitation *H. maori* displayed 'hind leg fend', 'kick hind legs' and the suite of minor responses ('vibrate palps', 'move antennae' and 'close hind legs') which often accompanied 'escape'. Continued stimulation led to escalation, as indicated by 'mandible gape', 'raise foreleg', 'raise hind legs', and 'hind leg stridulation'. However, new components now appeared that are unique for *H. maori,* and possibly its sister species, *Hemideina ricta* (see below). These are termed 'recurve body', 'flip/splay', 'supine stridulate' and 'flex/bite'. Their descriptions follow.

Recurve body

In response to repeated stimulation of any body region except the head, the body is bent in an arc, thus presenting a concave surface in the direction of the stimulus (Fig. 16.4A). At the same time, the recurved body is tilted away from the stimulus and, while nearly lying on its side, the weta raises either or both the front and middle leg ipsilateral to the stimulus and displays 'mandible gape' toward the stimulating object.

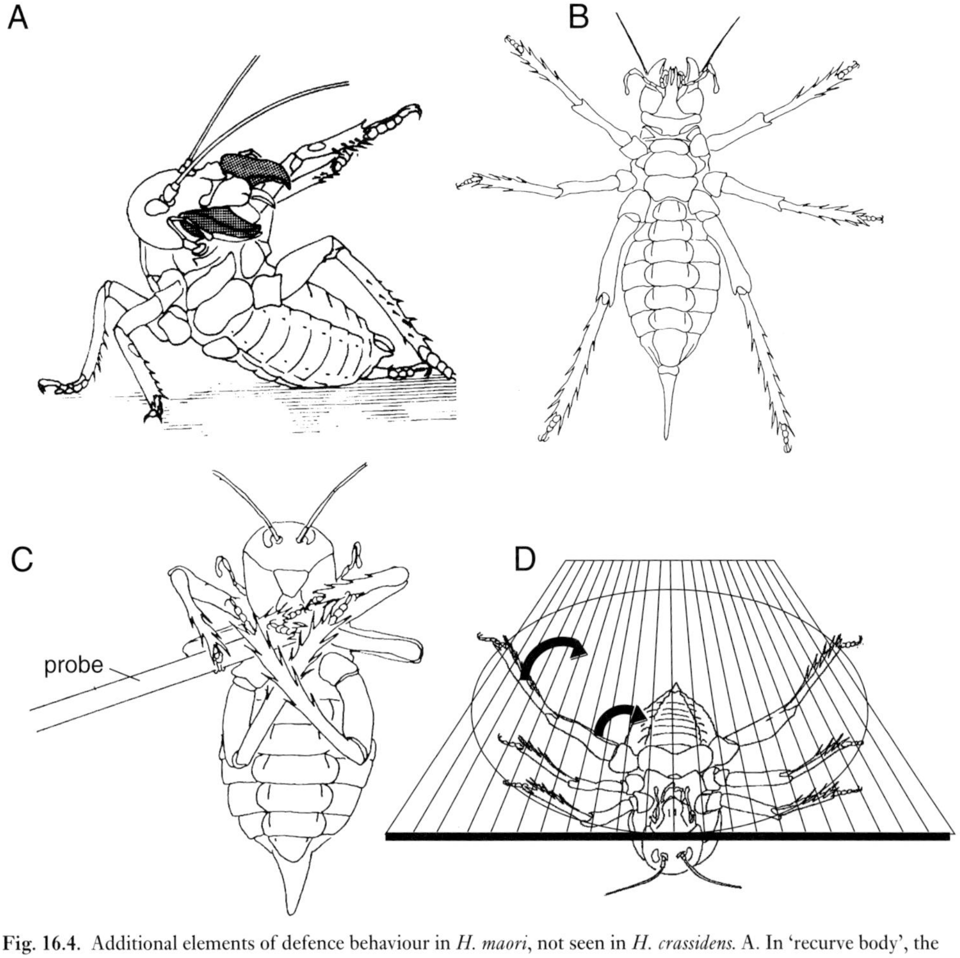

Fig. 16.4. Additional elements of defence behaviour in *H. maori*, not seen in *H. crassidens.* A. In 'recurve body', the weta bends the body and tilts away from the stimulus while raising one or more legs and displaying 'mandible gape'. B. The previous display leads to 'flip/splay', involving the weta flipping over on to its dorsal side with legs outspread and mandibles gaped in a highly alert and reactive state. C. Any tactile stimulus to the outspread legs or ventrum causes 'flex/bite' behaviour, shown after a pencil was lightly touched to a leg. D. Diagrammatic representation of the defence space of *H. maori* in the supine 'flip/splay' position. The circular intersection of the legs with the plane define a tactile stimulation space, including the ventral surface of the body, which will elicit the 'flex/bite' reflexive response. B, C, D after Smith (1979).

Flip/splay

From 'recurve body' position, any further provocation causes the weta to flip on to its back by collapsing the supporting legs and pushing itself over with the ipsilateral hind leg. The weta then assumes a splayed leg posture with mandibles fully opened (Fig. 16.4B). Although immobile in this supine position, the weta is highly alert and reacts very rapidly to any further stimulation with the behaviours described below.

Supine stridulate

While in the supine position, an additional stimulus (shadow, touch, sudden sound, sudden jet of air) causes one to five syllables of stridulatory sound. This is accomplished by rubbing the hind femora against the second abdominal tergites through a smaller arc than in 'hind leg stridulate', and simultaneously moving the abdomen opposite to each leg movement (Fig. 15.7C Field, Chapter 15, this volume).

Flex/bite

While in the supine position, tactile stimulation of the ventral surface of the weta causes a sudden synchronous flexion and adduction of the legs, bringing the tarsi to the tips of the mandibles. If the legs are contacted (the usual case), they sweep the object to the opened mandibles, which then bite it vigorously. To do this the body is folded ventrally and the head is directed posteriorly (Fig. 16.4C). The leg flexion is an automatic reflex, involving the giant fibre system in the ventral nerve cord (described in Field, Chapter 23, this volume). This reflex varies, depending upon what region is stimulated. If the head, thorax or (true) ventral surfaces of the legs are touched, all six legs will be flexed toward the mandibles. If the abdomen is touched, the hind legs will flex toward the mandibles, while the middle legs only partially flex. However, if the probe is contacted by any leg at this time, then all six legs rapidly sweep it to the waiting mandibles.

If stimulation was continued while these supine responses occurred, they eventually waned and the weta righted itself and 'escaped'.

In an earlier study (Field, 1978) of the defence behaviour of different *H. maori* instars, it was found that juveniles up through the third instar lacked femoral pegs in the stridulatory apparatus and hence could not produce sound. Correlated with this was the observation that they always showed 'jump' defence behaviour rather than the displays described above. As body mass increased in later instars, with proportionately less increase in the mass of the tibial extensor muscle, 'jump' was greatly reduced and the repertoire of additional defence behaviours occurred during tactile stimulation. The 'flip/splay' behaviour would not be effective against avian predators, which could stab from above with a long beak, but it would certainly seem to provide protection against reptilian predators, such as geckos and skinks, which inhabit the alpine habitat of New Zealand.

Anecdotal observations of defence behaviour of *H. ricta*, a lowland forest dweller with many morphological affinities to *H. maori*, indicated that it also shows 'flip/splay' and presumably the associated supine responses (Townsend, 1995). The other five *Hemideina* species in New Zealand are tree wetas, and their defence behaviour appears to resemble that of *H. crassidens*. They all show male secondary sexual megacephaly, except *Hemideina broughi*, which is thought to belong to a new genus (Ramsay and Bigelow, 1978).

Comparison with other deinacridine New Zealand wetas

Many anecdotal accounts describe the defence behaviour of *Deinacrida* species in New Zealand (Hutton, 1897; Richards, 1973; Meads, 1990a, b), but no analytical studies have been made. The displays are especially spectacular, because of the large size of the giant wetas, the robust, sharp spines on their hind tibiae and the loud crackling sounds produced during stridulation. The following summary is derived from personal observations of the *Deinacrida* species *D. rugosa, D. parva, D. heteracantha, D. connectens, D. mahoenui, D. fallai* and *D. elegans*, using the behavioural components identified for *H. crassidens*. None of the species show the secondary sexual characters found in *Hemideina*, which probably correlates with the lack of some defence components.

Visual, acoustic and especially tactile stimulation causes 'mandible gape', but not 'raise head' or 'raise foreleg', presumably owing to the lack of megacephaly. If handled while disturbed, most *Deinacrida* will bite. 'Hind leg fend', 'hind leg kick' and 'raise hind legs' are prominent components. 'Hind leg stridulate' is characteristic of all species, and in *D. rugosa, D. parva, D. heteracantha, D. mahoenui* and *D. fallai* an additional form of stridulation occurs (Ramsay, 1953; Richards, 1973; Field, 1993). This consists of telescoping movements of the abdomen, which cause stridulatory structures on the overlapping abdominal tergites to rub against each other (tergo-tergal stridulation), thereby producing a hissing sound (illustrated in Field, Chapter 15, this volume). The minor responses are present: 'vibrate palps', 'move antennae' and 'close hind legs'. In *D. rugosa* and *D. parva*, 'close hind legs' is dramatically assisted by the most proximal medial pair of tibial spines just distal to the femur–tibia joint. These are large and recurved medially; any object stimulating the abdomen from above is gripped by the tibia during 'close hind legs' and, if withdrawn, becomes gashed by the recurved spines (Fig. 8.28E, O'Brien and Field, Chapter 8, this volume). This could have important implications for attack by mammalian or reptilian predators (see 'Experimental analysis of defence against predators' below).

Secondary Defence Mechanisms in New Zealand Henicinae (Ground Wetas)

The henicine wetas are morphologically distinct from the deinacridine wetas. They have robust hind femora with large extensor tibiae muscles for jumping. The tibiae lack prominent spines and, except for the tusked wetas, there is never secondary sexual megacephaly. The tusked wetas present a remarkable departure from the above condition. Males have enlarged heads and a pair of stout elongate tusks protruding anteriorly from the mandibles (Fig. 10.1, Field and Deans, Chapter 10, this volume). In addition to these secondary sexual characters, stridulatory tubercles on the tusks contribute to defence behaviour (Field, 1993). Ground wetas and tusked wetas face outwards in their burrows, in contrast to the tree and giant wetas, which face inwards and present any intruder with the sharp spines of the hind tibiae.

When disturbed or prodded, most ground wetas usually 'jump' as the first means of escape. The minor responses 'vibrate palps' and 'move antennae' are low-level responses to any small disturbance. Some ground wetas, notably the alpine burrowers (*Hemiandrus focalis, Hemiandrus. maculifrons*), produce a stridulatory behaviour, 'rocking stridulation', which is very different from that described for the deinacridine wetas (Field, 1993). Here, the weta rocks forward, while simultaneously raising the abdomen against the hind legs, which are appressed against minute peg fields on the abdominal tergites. When provoked outside the burrow, this behaviour results in a hissing sound and is repeated about twice per second for several bouts (Fig. 15.7F, Field, Chapter 15, this volume). A very similar behaviour has been described for two species of Gryllacrididae in Australia (Field and Bailey, 1997).

The tusked wetas comprise three species: *Anisoura nicobarica* (= *Hemiandrus monstrosus*) (17 mm), the giant Mercury Island species *Motuweta isolata* (85 mm) and the recently discovered undescribed 'Raukumara tusked weta' (40 mm). *Anisoura nicobarica* lives in tree galleries, while the other two species are ground burrowers, and all three face outward while occupying the retreat (Bellingham, 1991; Gibbs, 1998b). When disturbed away from their burrow, the large tusked wetas rush for cover or 'jump' to escape (Meads, 1990b). This takes a remarkably different form in the Raukumara tusked weta. Owing to its preference for living beside streams, this species readily jumps or rolls from streamside rocks into the water when disturbed at night (Gibbs, 1998b). The wetas were observed by torchlight to cling to underwater substrates for at least 5 min and to move across cobbles on the stream bottom. One adult female was encountered by torchlight, quite wet and apparently emerging from the water's edge, while another juvenile readily sank to the bottom of a shallow pool and pressed itself against a rock crevice when it was dropped into the water. It remained for 3 min and then moved across the bottom to emerge with just the head and part of the thorax above water. After 15 min, it was still mostly submerged (Gibbs, 1998b).

If any of the tusked wetas are provoked further, they display 'mandible gape', 'raise foreleg' and 'raise head' as a suite of components, which closely resemble those of *Hemideina* species. When prodded at the burrow entrance, the males lunge forward with closed tusks and push at the threatening object (*M. isolata*: Meads, 1990a). In *A. nicobarica*, the body was reported to curve sideways during 'raise head', and the tusks were used as pincers during the forward lunges (Bellingham, 1991). A characteristic femoro-abdominal stridulatory mechanism is present in both sexes, and the forward lunging may be a form of 'rocking stridulation'. The question of femoro-abdominal stridulatory sound production requires further documentation in tusked wetas, since such sounds have not been described clearly. Bellingham (1991) described hissing sounds from *M. isolata* during the suite of threat displays described above. Male tusked wetas produce yet another form of defence stridulation, 'tusk stridulation'. This is a rapid opening movement of the mandibles, causing the crossed tusks to engage the stridulatory tubercles. The result in *M. isolata* is a sharp, loud, barking sound, produced when the male is disturbed at night (Meads, 1990b; Field, 1993; M. McIntyre and G. Gibbs, personal communication). In *A. nicobarica*, males raise their tusks and rasp them together to produce stridulatory sounds when disturbed (Bellingham, 1991). The Raukumara males lack stridulatory tubercles on the tusks and, presumably, they can only produce sound by a femoro-abdominal mechanism reported to be present (sounds not heard: Gibbs, 1998b). Females of all tusked wetas lack the tusks but show the typical ground weta threat displays 'mandible gape', 'raise foreleg' and 'raise head' when disturbed at the burrow entrance.

Secondary Defence Mechanisms in Other Anostostomatids

Behavioural analyses have not been conducted on king crickets and Jerusalem crickets of Australia, North and South America and Africa, but anecdotal accounts appear in entomology texts covering these regions. For example, the South African tusked king cricket threatens with gaped mandibles and emits foul-smelling faeces when threatened (see Toms, Chapter 4, this volume), while *Henicus monstrosus* is said to gnash its mandibles to make scratching sounds. In Australia, *Anostostoma australasiae* produces erratic jumping and stridulation behaviour when threatened and, like the New Zealand *H. ricta* and *H. maori*, flips over on its back with outspread legs and gaped mandibles if the threat persists (Monteith and Field, Chapter 5, this volume).

Experimental Analysis of Defence Against Predators

Although the human interpretation of the behaviours just described is that of a defence function, there are no published reports which document this with real predators. We have carried out experiments with New Zealand wetas faced by endemic and introduced predators with the aim of answering the following questions: (i) What responses are produced by wetas when attacked by a potential predator? (ii) How effective are the weta responses, i.e. how does the predator react to the weta? (iii) How does the predator overcome the weta defence behaviour? The tree weta *H. crassidens* was used in the experiments, because its 'defence' behavioural repertoire had already been established. Since it was first necessary to determine what species in New Zealand are predators of wetas, the following review was made.

Endemic predators of New Zealand wetas

At least one large megalomorph species of spider (similar to *Porothele*, but undescribed; A.D. Blest, Christchurch, 1991, personal communication) appears to prey upon the tree weta *Hemideina femorata*. This spider occupies large tree hollows and galleries made by cerambycid beetle larvae in kanuka trees (*Kunzia ericoides*), which are also used by the wetas. Wetas never cohabit galleries occupied by the spiders (Field and Sandlant, Chapter, 13, this volume) but are often found in adjacent galleries. In five such cases, the dried bodies of individual adult wetas, previously monitored over several months, were found in the silk of such spiders.

The analysis of gut and faecal contents has revealed that a number of endemic vertebrates feed upon wetas. Reptiles include the tuatara, skinks and geckos. The tuatara (*Sphenodon punctatus*) preys upon the tree weta *H. crassidens* on Stephens Island (Cook Strait) and upon the giant tusked weta on Mercury Island (Moller, 1978; M. McIntyre, Wellington, 1996, personal communication). The spotted skink (*Leiolopisma infrapunctatum*) has been reported to feed on juvenile wetas in the South Island of New Zealand (Sharell, 1966). The gecko *Hoplodactylus pacificus* lives on trees and on the ground of mountain forests throughout New Zealand, and preys upon ground wetas and other orthopterans (based upon gut analyses) (McIvor, 1972).

Diurnal as well as nocturnal native birds prey upon wetas. The South Island robin (*Petroica australis australis*) has been observed to attack and consume *H. femorata* and the ground weta *Hemiandrus* sp. in the South Island (R. Powlesland and G. Sherley, personal communication). Harriers, the only buteo hawk species (*Circus approximans*), feed upon *H. femorata* (Douglas, 1970) and the giant weta *D. fallai* of Poor Knights Islands (Richards, 1973). The long-tailed cuckoo (*Eudynamis taitensis*), the forest parrot (kaka, *Nestor meridionalis*) and the North Island saddleback (*Philesturnis carunculatus rufusater*) are reported to feed upon *Hemideina thoracica* in the North Island of New Zealand (Buller, 1871; Atkinson and Campbell, 1966, Gibbs and McIntyre, 1997). Nocturnal avian predators in New Zealand include the North and South Island subspecies of a large rail, the weka (*Gallirallus australis greyi* and *Gallirallus australis australis*), which are known to feed upon *H. crassidens* and *H. thoracica* (Carroll, 1963; L.H. Field, personal observation). The morepork owl (*Ninox novaeseelandiae*) has been reported to feed upon *H. thoracica* in the North Island (Lindsay and Ordish, 1964) and upon the endangered giant weta *D. heteracantha* on Little Barrier Island (Richards, 1973).

Introduced predators

New Zealand's only native mammals, two species of small bats, do not feed upon wetas. However, a number of introduced mammals appear to have been responsible for the demise or severe reduction of weta populations on the main islands (Gibbs, 1998a). The first introduced mammal was the kiore (*Rattus exulans*), brought by Maori voyagers over one thousand years ago (King, 1984). Within the last two centuries Europeans have brought the following potential predators of wetas: rats (*Rattus r. rattus*, *R. norvegicus*), mice (*Mus* spp.), stoat (*Mustela erminea*), ferret (*Mustela putorius*), cat (*Felis catus*) and hedgehog (*Erinaceus europaeus*). All of these mammals have been implicated anecdotally as weta predators (Ramsay and Bigelow, 1978; King, 1984; Moller, 1985). Studies that have verified this by faecal or gut analysis indicate, for example, that the tree weta *H. thoracica* forms an important food source in the diet of the ship rat in Orongorongo Valley, Wellington (Daniel, 1973), and of the feral cat (Richards, 1973). In the South Island, the tree weta *H. femorata* comprised 21% of the diet of stoats in Canterbury and 57% of the stoat diet in Fiordland (Marshall, 1963; Fitzgerald, 1964). A recent study has shown clearly that *Hemideina* population densities are significantly lower in offshore rodent-infested islands compared with rodent-free island populations (Ruffaut, 1995).

Predator–prey experiments

Methods

The following adult predators were selected: gecko (*H. pacificus*, $n = 8$), South Island robin (*P. a. australis*, $n = 10$), South Island weka (*G. a. australis*, $n = 1$), ferret (*M. putorius*, $n = 1$) and hedgehog (*E. europaeus*, $n = 2$).The experimental arenas differed for each predator. For the geckos (total body length 70–80 mm), glass terraria (30 cm × 15 cm × 15 cm) were used in the same red-light reversed day/night cycle under which the wetas were kept. The hedgehog trials were conducted in 50 cm × 33 cm × 30 cm steel-mesh cages with straw substrate. The weka and ferret trials were conducted in outdoor wire mesh cages (weka: 8 m × 5 m × 2 m; ferret: 2 m × 1 m × 1 m), which contained sparse vegetation. The robin trials were conducted in a forested area of the South Island (Kowhai Bush, Kaikoura), by making a 0.75 m clearing on the moss substrate. The last four predator experiments occurred in the dim light of the late afternoon. In all cases, except with the robins, the predator was left undisturbed and without food for 24 h before a trial. Except for the avian predators, a weta was introduced out of sight of the predator and the trial began when the predator sighted the weta. To begin a robin trial, a weta was placed on the ground arena 2 m from the experimenter after the predator had been attracted by shaking a branch or rustling leaves. All trials were terminated if the weta was killed or if the predator showed no action for 5 min (15 min for the geckos). The trials were timed with a stopwatch and ongoing observations were quietly dictated into a tape recorder. The observer was hidden behind a barrier, except for the weka and ferret trials, since those animals were relatively tame and fed readily in the observer's presence.

Descriptions of interactions

The first question addressed by these experiments was to determine which responses are produced by wetas when attacked by a potential predator. We first present descriptions of the encounters, followed by an analysis of the data.

In the gecko experiments, five adult wetas were tested first, but they were never attacked (presumably because their length of 30–40 mm was too large as a prey item). Thus all results are from 15 juveniles of 12–15 mm body length. These can be eaten by geckos, since a juvenile left in a gecko terrarium was consumed after 3 days. In the trials, the geckos would approach a weta with slow steps until about 3 cm away, at which point it would make a sudden lunge and attempt to grasp the weta. If grasped ($n = 8$), the weta responded by 'raise hind legs', 'fend hind leg' and 'kick hind legs', with attempts to push the tibial spines at the gecko's head (Fig. 16.5). The gecko immediately shook its head laterally and released the weta in seven cases ('escape' ensued); in the eighth, the weta was not released until it began biting the gecko's mouth. The gecko failed to secure a grasp in the other 15 attacks. None of the wetas were eaten.

Three juvenile wetas were tested with robins. All were carried away after being grasped from the side; one was first struck five times against a branch after being pecked in the head and regrasped in the robin's beak. No defence responses occurred, and they were presumably eaten. The seven adults only showed defence responses after first being approached from the side and pecked on the pronotum. This elicited 'raise hind legs' and 'hind leg stridulate' repeatedly. 'Mandible gape' and 'raise foreleg' occurred to a lesser extent. These behaviours typically caused the robin to withdraw 10–40 cm until the weta stopped stridulation movements of the hind legs. Then the robin approached, pecked and withdrew, or sometimes grasped the weta by the thorax or middle leg and carried it 8–90 cm before dropping it. If the weta landed overturned, the robin made one to three attacks to the underside and sufficiently injured the weta to grasp it by the pronotum and carry it away. Twice a robin grasped the weta and shook its head laterally before releasing the weta. Also twice a robin struck the weta held in its beak against the branch upon which it was perched. Only two wetas were able to 'escape' after the robin abandoned its attack in the face of vigorous defence by the wetas. The other eight were eventually eaten.

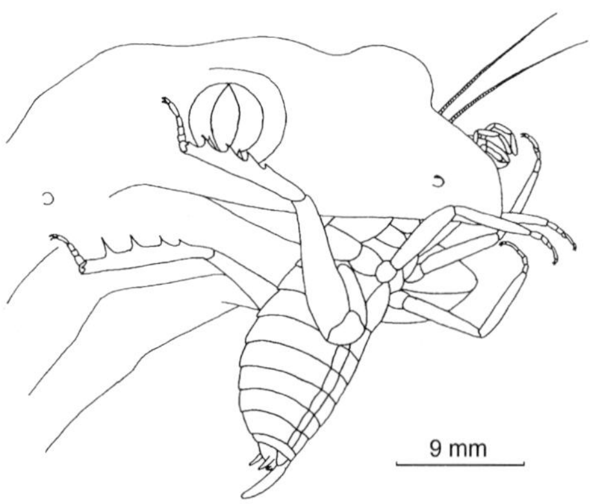

Fig. 16.5. Although tree wetas are susceptible to attack from the side (allowing a gecko to grab the insect), the hind leg spines are an effective second line of defence when pushed by the weta into the face or eyes of the predator. Geckos invariably released wetas when this happened.

The six wetas presented to the weka (the size of a small chicken) were all eaten. The weka would approach from the side, grasp the weta by the pronotum and carry it about 3 m away. This elicited 'mandible gape' and 'raise hind legs', with the tibial spines orientated toward the weka. The continued attack consisted of short, sharp pecks and lateral shaking of the weka's head before releasing the weta. Several times the insect's abdomen or legs were torn off by the shaking movements. The weka continued attacking the ventrum and leg bases with pecks, and then eventually swallowed the weta whole.

The hedgehogs approached the weta, lunged and grasped it in the mouth with rapid head shaking before releasing it. The wetas either 'jumped' or showed 'hind leg kick' and 'fend hind leg' during 'escape'. They also displayed 'mandible gape'. The hedgehog usually followed and tried to bite and repeat its previous attack, sometimes using its feet to hold the weta. In ten interactions involving 'hind leg kick' and 'fend hind leg', the tibial spines of the weta struck the hedgehog's snout and it withdrew before further attacks. Five of the wetas successfully defended themselves in this manner and one hedgehog discontinued attacks. The remaining wetas were killed by the other hedgehog using bites to the head, abdomen and thorax. This hedgehog learned to avoid the tibial spines as trials progressed.

The ferret killed and ate three of the four wetas presented. It typically approached and lunged at a weta to bite it and then (nine of 12 bites) violently shook its head laterally before releasing the weta. Although the wetas displayed 'mandible gape', this did not appear to deter further attacks. In one trial the ferret used its snout to overturn the weta. Despite the defence components of 'leg fend' and 'kick hind legs', the ferret still attacked the abdomen or thorax with biting and shaking, causing severe wounding and even dismemberment ($n = 2$). One weta landed about 10 cm away when the ferret released its bite, and was able to 'escape' beneath a log before the ferret made further attacks. Another was retrieved during 'escape' and soon killed by quick bites.

Analysis of weta responses

The results were first analysed in terms of the percentage of individual defence behaviour components shown by wetas when confronted by the different predators (Table 16.2). A surprising but convincing observation is that in no case were all the components displayed against a predator. In many cases, the attacks were too rapid or too violent to allow the weta time to defend itself or to advance through an escalation of display components. The ferret and the weka took the least time, as shown by the following list of mean kill times: ferret – 46 s ($n = 3$), weka – 77 s ($n = 6$), hedgehog – 94 s ($n = 10$), robin – 195 s ($n = 10$) for adult wetas, 25 s ($n = 3$) for juveniles, and geckos did not kill wetas, but took minutes during their trials. For the slower, endemic predators (gecko, robin), more defence components were presented (Table 16.2). This correlates with the facts that the individual attacks appeared less vicious and there were more interactions per trial. Presumably, then, the wetas escalated through more of the component repertoire without being seriously wounded initially. In all attacks by all predators, the minor responses 'vibrate palps' and 'move antennae' occurred and are not tabulated in Table 16.2.

The distribution of components displayed by wetas differs for each predator in Table 16.2. Based upon the data of Fig. 16.3, we might expect that the differences reflect differences in body region attacked by each predator. This is reasonably well borne out. For example, the hedgehogs often attacked the weta's abdomen by following and biting, and the wetas defended predominantly with 'kick hind legs', 'hind leg fend', 'mandible gape', 'jump' and 'escape' (as predicted by Fig. 16.3). The ferret's tendency to bite the abdomen also elicited 'kick hind legs', 'hind leg fend' and 'mandible gape', as predicted. On the other hand, the geckos and robins often attacked the pronotum, and the wetas defended mostly with 'raise hind legs', 'hind leg fend', 'mandible gape', 'escape' and sometimes 'raise front leg' and 'kick hind legs', also predicted by Fig. 16.3. Similarly, the weka's initial grasp of the pronotum elicited only two displays, 'raise hind legs' and 'mandible gape', plus 'escape', which are the most prominent components of pronotum attack in Fig. 16.3. 'Escape' was attempted during the attacks by all of the predators; its success was mainly limited by the extent to which the wetas were wounded by the attacker.

Differences from the predictions of Fig. 16.3 include the occurrence of 'bite' and 'hind leg stridulate'. The mandibles were brought into play only when the wetas had time to 'bite' while being grasped but not violently shaken, i.e. during attacks by the geckos. The other predators only

Table 16.2. Frequency of occurrence (%) of defence responses by wetas when attacked by predators.

	Percentage of wetas showing each component										Outcome (%)		
Predator	MG	B	RH	RF	KL	LF	RL	S	J	E	Eaten	Escaped	No. wetas
Gecko	–	6	–	13	20	27	20	–	–	100	0	100	15
Robin	30	–	10	–	–	–	60	50	–	30	80	20	10
Weka	33	–	–	–	–	–	50	–	–	33	100	0	6
Hedgehog	31	–	–	–	47	15	–	–	15	23	62	38	13
Ferret	100	–	–	–	25	25	–	–	–	50	75	25	4

MG, mandible gape; B, bite; RH, raise head; RF, raise foreleg; KL, kick hind legs; LF, hind leg fend; RL, raise hind legs; S, hind leg stridulate; J, jump; E, escape.

attacked the head with rapid pecks (robin) or quick bites and violent shaking after the weta was subdued (ferret), and thus avoided being bitten. As discussed below, the predators appeared to respect and avoid the mandibles of the wetas. The production of stridulation only occurred during the robin attacks. It is not clear why it was not produced during 'raise hind legs' against the other predators, although, in the case of the weka, the pecking attack movements were so fast and violent that the weta may have been unable to respond with stridulation. The question of whether stridulatory sounds deter predators is addressed in the next section.

Analysis of predator behaviour

Another aim of the experiments was to determine how the predators reacted to the weta defence behaviour. This was analysed by tabulating the direction from which each predator attacked the wetas relative to the weta body axis and the body region attacked (Table 16.3).

The most interesting result is that the avian predators and the ferret concentrated upon attacks from the side (60°–120°), mostly to the weta thorax. The implication is that the predators were avoiding the hind tibial spines and the mandibles of the wetas. This is substantiated by the low rate of attack from the front (0°–60°) for each predator, and the low rate of rear attack for all but the geckos and hedgehogs. Even with the hedgehogs, the rear attacks declined rapidly after the initial trials, in which the tibial spines were kicked at the hedgehogs' snouts. The only exception is the high rate of gecko attack at the legs of the wetas. As noted previously, most of these attacks were unsuccessful. The hind tibial spines may have deterred the geckos from obtaining a firm grasp of the leg, but apparently did not result in avoidance learning by the predator.

The few attacks to the head of male and female wetas occurred with about equal frequency, despite the larger mandibles of the male wetas (data only analysed for hedgehog and ferret, not shown). Presumably the rapidity of the attack bites by these mammals precluded biting by the wetas.

In several cases, the predators tried to overturn the wetas in order to attack the vulnerable underside of the abdomen and thorax ('ventrum' in Table 16.3). This area is apparently more difficult for *H. crassidens* to defend. The tibiae must be fully flexed and cover the abdomen in order for the spines to aim toward a predator and provide a defence. As noted for the alpine weta, *H. maori*, another very different defence ploy is to deliberately expose the ventrum, by flipping over, and to bite a predator if it touches the legs or body. Such a behaviour is quite inappropriate against a bird with a long beak, such as a weka (which tends not to live in the

Table 16.3. Frequency of occurrence (%) of predator attacks to different weta body regions, and from different directions relative to the body axis of the weta. Directional data were not obtained for the ferret, owing to the rapidity of the attacks.

Predator	Total number of attacks	Percentage attacks to each body region					Orientation of attacks	(%)
		Head	Thorax	Abdomen	Leg	Ventrum		
Gecko	23	9	–	–	–	–	0°–60°	9
		–	26	–	4	–	60°–120°	30
		–	–	–	61	–	120°–180°	61
Robin	91	1	13	–	3	–	0°–60°	18
		2	52	1	9	7	60°–120°	70
		–	8	4	–	–	120°–180°	12
Weka	154	1	–	–	1	–	0°–60°	2
		10	26	9	1	34	60°–120°	81
		1	–	5	1	10	120°–180°	17
Hedgehog	20	5	10	5	–	–	0°–60°	20
		5	5	15	5	–	60°–120°	30
		15	15	20	–	–	120°–180°	50
Ferret	12	17	42	33	8	–	–	100
Total	300	11%	41%	13%	11%	24%		

alpine habitat), but instead seems to be adapted for defence against the alpine geckos and skinks.

In summary, predators tended to approach wetas from the side during attacks. The thorax was the main target of initial attack and, when grasped, a universal behaviour of the predators was to shake their head laterally. With the ferret, this caused the legs or abdomen to fall off, but apparently dazed the wetas in all cases. The predators used repeated bites or pecks to subdue wetas, and sometimes tried to overturn wetas to attack the ventrum. With the smaller predators, the mandibles and hind tibiae of wetas seemed to act as deterrents, but less so with the larger, more powerful predators. The robin especially tended to avoid the mandibles and hind tibiae, and may have been deterred by stridulatory defence sounds of the wetas.

Defence space models for wetas

In principle, *H. crassidens* possesses fearsome secondary defence mechanisms against attack from predators. Any attack from the front can be met with a pair of sharp, powerful mandibles, which are especially formidable in males. This defence is amplified by 'mandible gape', 'head raise' and 'raise foreleg'. The defence responses using the spined hind legs can protect a large volume of space around the weta. This volume, the defence space of the weta, extends from the area above the head, above the body and behind the abdomen for *H. crassidens* (Fig. 16.6A, B). The size of such a space model depends upon the length of the hind legs. In an average adult with a tibial length of 22 mm and a femur length of 20 mm, the defence space extends about 10 mm anterior to the head, forms an arc about 30 mm above the body and extends about 20 mm posterior to the abdomen. In a juvenile weta with tibia and femur lengths of 9 mm, the potential defence space extends about 6 mm anterior to the head, about 12 mm above the body and about 8 mm posterior to the abdomen. In addition, the mandibles provide an anterior defence space forming a horizontal arc across the front of the head and extending to about 120° on each side (Fig. 16.6A, B). These dimensions are useful when considering the different modes of attack by predators. Theoretically, a predator must penetrate this defence space in order to injure the weta.

Despite possessing a reasonably large defence space in relation to its body size, *H. crassidens* and other tree wetas are potentially vulnerable to attack. They have no defence mechanism to protect against attacks directed from the side. Unless the weta orientates the longitudinal axis of its defence space in the direction of an attacking predator, it can be approached from the side without being able to use the tibial spines or the mandibles. If a predator takes advantage of this weakness in the weta's defence, it must still be subject to the possibility of being bitten or jabbed in the face by the hind tibial spines (Fig. 16.5).

The defence space of *H. maori* is different from that just described, once the 'flip/splay' behaviour occurs. Here the weta spreads an array of legs with tarsal claws ringing the perimeter. If a predator makes contact with the claws or with any part of the ventral surface of the body or legs, the weta's 'flex/bite' reflex behaviour ensures that the open mandibles will be brought to bear upon the

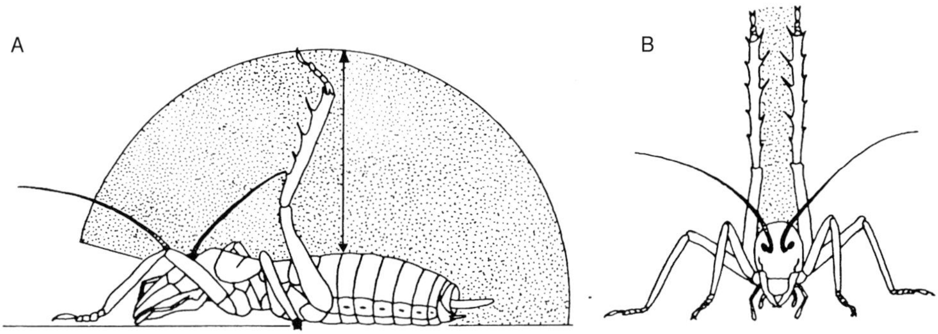

Fig.16. 6. Diagrammatic representation of the defence space of *H. crassidens* (as well as *H. femorata, H. thoracica, H. trewicki*). The stippled region is protected by the sharp tibial spines and is avoided by predators unless they happen to be birds with beaks longer than the height (arrow) of the defence space.

contacting object. Since the tarsal claws can grip a predator, it is as easy for the weta to lift and curl its body around the predator's face (e.g. that of a gecko) when biting, as it is for it to convey an object to its mandibles. Thus, its defence space takes the form of a broad, inverted truncated cone defined by the spread of the legs (Fig. 16.4D). There is a clear weakness in this defence space as well, since a bird with a beak longer than the depth of the defence space can attack directly from above without fear of the mandibles. The weta's bite against the beak is ineffective and the bird can kill the weta by pecking. Although predator–prey studies with *H. maori* have not been done, it appears that no long-beaked nocturnal birds live in the weta's habitat and that the major predators would be alpine geckos.

How, then, do predators overcome or avoid the defence spaces of wetas? This consideration is most cogent in relation to native predators of New Zealand, although interesting insights are gained from the behaviours of introduced ones. Geckos attack by stealth and a sudden lunge, before the defence display of a weta is produced. However, once grasped, the weta is effective in its use of tibial spines or mandibles and can often secure its release. Subsequent attacks must then face an established defence space by the weta and may be less successful. Thus, it appears that wetas produce a reasonably effective defence against geckos.

Our experiments indicated that, once wetas reach a certain size, they are not attacked by geckos. Although the early weta instars can be swallowed, once wetas grow to a tibial length of about 16 mm, they were either too large to swallow (*H. pacificus*) or appeared too large to represent prey items. This suggests that there exists a size refuge for *Hemideina* wetas against attack by geckos. The largest species of gecko in New Zealand grows to 20 cm body length (*Hoplodactylus duvauceli*) and the refuge size would be proportionately larger for wetas which it preys upon. The same arguments are likely to apply to skinks, which are known predators of wetas. However, the much larger tuatara simply grasps the weta and consumes it after several bites and is apparently oblivious of any secondary defence behaviour or size refuge of the insect (Moller, 1978).

Adult South Island robins have a beak length of 16 mm from culmen to tip (Oliver, 1930). When attacking adult *H. crassidens*, this beak length was too short to penetrate the defence space of the wetas without risk of injury from the tibial spines or mandibles. Instead, the robins made cautious, rapid attacks from the side. The birds were apparently deterred by the stridulatory behaviour, as indicated by the fact that they waited until stridulation stopped before recommencing attack. The success of robins was due to the rapidity and large number of attacks into the weak zone of the wetas' defence. This was not always successful, as shown by the 20% survival rate for wetas in the experiments (Table 16.2). Of 16 predatory interactions between *H. femorata* (the Canterbury tree weta) and robins, four resulted in the robin abandoning the attack (Field, 1978; R. Powlesland, personal communication). Further substantiation of the deterrent effect of the *Hemideina* defence behaviour comes from the time required by robins to subdue a weta: Powlesland estimated a mean time of 4 min 46 s, while the above experimental data gave a mean of 3 min 15 s. This contrasts markedly with Powlesland's estimated mean time of 10 s for robins to attack and consume the ground wetas *Hemiandrus* sp., which exhibit neither 'raise hind legs' nor 'hind leg stridulate'.

The concept of weta defence space also helps to explain the success of attacks by wekas. The South Island weka has a beak length of about 50 mm from culmen to tip (Oliver, 1930). Thus it was easily able to penetrate the defence space of wetas, and it appeared that the behaviours 'raise hind legs' and 'hind leg stridulate' had a minimal deterrent effect on the weka. The only indication of some deterrence was the observation that the weka initially tended to attack from the side at the pronotum. Otherwise, the weka took little heed of the weta defence responses, and even pecked at the pronotum between the upraised hind legs.

The introduced mammalian predators were clearly efficient predators. They were able to learn (hedgehogs) how to quickly kill wetas while avoiding or suppressing the wetas' defence mechanisms (hedgehog holding weta with its front paws to prevent 'raise hind legs' while biting the weta). Rapid and powerful biting was the main technique used. Weta secondary defence behaviour is ineffective against these predators, and they represent a formidable factor in altering the population ecology of New Zealand anostostomatids.

In conclusion, the defence responses of *Hemideina* tree wetas against native predators appear to be most effective in encounters with short-beaked birds and small reptiles. These would include known predators, such as the robin species, long-tailed and shining cuckoos, several gecko and skink species. Although morepork owls and harrier hawks have short beaks (< 32 mm), they have powerful, scaly talons that can hold and subdue wetas.

The effect of stridulatory sound in defence

The robins consistently avoided attacking the wetas during 'hind leg stridulate'. However, it is not clear whether the robins avoided the weta because of the hind-tibial down-stroke during 'hind leg stridulate', or because of the sound, or both. It is likely that robins learn to associate the sound with a dangerous prey (Haskell, 1961) and that the weta uses this behaviour as a form of acoustic aposomatism (Masters, 1979; Bailey, 1991). The effect of insect defence sounds on potential predators has seldom been studied, but several examples have shown that stridulation confers an advantage to the prey. The grasshopper mouse *Onychomys torridus* was less efficient in handling noisy cicadas compared with silenced ones and, in making more handling errors, allowed some of the noisy cicadas to escape (Smith and Langley, 1978). Similarly, stridulation by passalid beetles gave a similar advantage against predatory crows (Buchler *et al.*, 1981). Even invertebrate predators can be deterred by stridulation. Defence sounds produced by mutillid wasps (*Dasymutilla* sp.), water scavenger beetles (*Tropisternus* sp.) and round sand beetles (*Omophron labiatus*) significantly deterred attacks by lycosid spiders (Masters, 1979, 1980). Spiders are sensitive to vibratory stimuli and may pick up the stridulatory energy through the substrate or even the vibration of the prey while it is being handled.

In many cases of insect defence stridulation, including all examples from wetas, the frequency of the sound is broad-band and unpatterned (Haskell, 1961). Bailey (1991) pointed out that a broad-band sound is an efficient means of covering the hearing sensitivity ranges of a wide variety of predators, and is therefore not surprisingly found to be a widespread characteristic of defensive startle sounds.

Evolution of Defence Behaviour in Wetas

In view of the above conclusions, it is of interest to consider what predators may have influenced the evolution of defence behaviours in wetas. Anostostomatids are almost exclusively found in land masses that were once part of Gondwana: e.g. Australia, South America and Africa. The earliest fossilized ancestors related to wetas are three haglid species found in upper Triassic mudstones of Queensland, Australia (Riek, 1970). It is thought that ancestors of New Zealand anostostomatids probably migrated across the southern regions of Gondwana to New Zealand during the Triassic or upper Jurassic when land routes were still available. The ancestors were probably isolated from the Gondwana continent in the upper Jurassic, based upon a recent report of a fossil insect wing from a closely related family, the Prophalangopsidae, found in upper Jurassic siltstone of New Zealand (Grant-Mackie *et al.*, 1995).

The fossil record of terrestrial vertebrates in New Zealand is patchy. However a number of estimates have been suggested for the possible dates of colonization of New Zealand for some terrestrial vertebrates. These are included with the anostostomatid dates for comparison in the geological time sequence of events shown in Table 16.4. This sequence indicates that weta ancestors probably evolved with mainly reptilian predators in Gondwana, and later with tuatara ancestors in New Zealand. While the data for colonization of New Zealand are mainly based upon extrapolations rather than direct fossil evidence, Table 16.4 provides a tentative basis for developing working hypotheses about the evolution of defence behaviours in wetas.

Potential avian predators began colonizing New Zealand at least by the Cretaceous and continue to the present. Geckos probably arrived by rafting from Australia or Caledonia during the Miocene (Kuschel, 1975). Skinks are considered to have colonized more recently, during the Pleistocene (Table 16.4).

The genera *Hemideina* and *Deinacrida* are endemic to New Zealand, although Deinacridinae are also found in Australia. The 'raise hind legs' behaviour of the New Zealand wetas is also seen in the Australian genus *Australostoma* (Deinacridinae), although it is not accompanied by 'hind leg stridulate', as seen in the New Zealand deinacridine wetas (P. Johns, Christchurch, 1985, personal communication). Instead *Australostoma* rocks its body forward and backward to engage the femoro-abdominal stridulatory structures (Field, 1993), or else flips on to its back, as seen in *H.*

Table 16.4. Sequence of probable colonization events relating to weta ancestors and predatory vertebrate ancestors in New Zealand (data derived from Fleming, 1962; Kuschel, 1975; Stevens, 1980; Grant-Mackie *et al.*, 1995).

Million years ago (m.y.a.)	Events
225–190 (Triassic)	Ancestral relatives of wetas deposited in upper Triassic mudstone of Queensland, Australia (200 m.y.a.)
190–136 (Jurassic)	Ancestral relatives of wetas migrated to the New Zealand landmass after which it separated from Gondwana about upper Jurassic to Cretaceous Tuatara ancestors arrived in New Zealand
136–65 (Cretaceous)	Kiwi ancestors colonized New Zealand in upper Cretaceous
65–2.5 (Tertiary)	Saddleback ancestors probably colonized New Zealand during lower Tertiary Weka ancestors probably colonized during upper Tertiary Kaka ancestors probably colonized during upper Miocene Gecko ancestors probably colonized during Miocene Robin ancestors probably colonized during Pliocene
2.5–present (Quaternary)	Skink ancestors colonized New Zealand during Pleistocene Cuckoo and morepork (owl) ancestors colonized 20,000–1,000,000 years ago Harrier and kingfisher probably arrived comparatively recently

maori (see Fig. 5.7, Monteith and Field, Chapter 5, this volume). The rocking defence stridulation is seen in the Henicinae of both countries, in many other anostostomatids and also in the closely related, widespread family Gryllacrididae (Field and Bailey, 1997; L.H. Field and S. Glasgow, personal observation). The behaviour is associated with the plesiomorphic form of the femoro-abdominal stridulatory structures (Field, 1993), and probably predated the 'raise hind legs' behaviour. The sharing of 'raise hind legs' behaviour and the flipping defence behaviour between New Zealand and Australian deinacridines may indicate that these behaviours originated before the New Zealand land mass separated from Gondwana. At that time, squamate reptilian predators of ancestral anostostomatids (early birds and small dinosaurs) are likely to have predominated.

The present study indicated that 'raise hind legs' can be an effective mode of defence against reptilian predators if they are not greatly larger than the insect. Many orthoptera raise spine-studded legs as defence displays, but these are usually the forelegs (often with outspread wings), which may have aposomatic colour markings when displayed (Robinson, 1969; Sandow and Bailey, 1978; Bailey, 1991; Field and Bailey, 1997). In the Tettigoniidae, the Saginae are especially known for the prominent spines on the forelegs and accompanying displays when alarmed. These bush crickets are themselves fierce predators which also use the foreleg armament for grasping prey (Preston-Mafham, 1990; Rentz, 1993). Not so common in the Orthoptera is the development of large spines on the hind tibiae, as seen in the New Zealand deinacridine wetas and the Peruvian rainforest grasshopper *Agriacris trilineata* (Preston-Mafham, 1990). In these cases, the hind leg extensor tibiae muscle and hind femur are not massive, and the insects are not strong jumpers. Thus, unlike the majority of Orthoptera, which rely upon 'jump' as a means of escape, the deinacridine wetas have evolved an array of long and sometimes massive spines on the largest appendage of the body. These are presented to potential predators in a variety of effective ways. The deterrent function may involve: (i) increasing the apparent size of the insect prey; (ii) displaying to the predator potentially harmful spines; and (iii) hurting the predator by thrusting the spines when grasped. If this behaviour evolved under selection pressure from reptiles, the results from the present study suggest that (i) and (iii) were the most likely factors in operation.

The defence behaviours 'flip/splay' and 'flex /bite' of *H. maori* are unique additions to the above behaviour, and must have evolved in an environment free of long-beaked avian predators. In the alpine New Zealand habitat, a likely scenario is that *H. maori* ancestors evolved these behaviours under selective pressure from the invading ancestral geckos and skinks in the Miocene and Pleistocene (Table 16.4). The rise of the Southern Alps during this period may have provided a simultaneous alpine refuge for both predator and prey from larger predators in the lowlands. However, the occurrence of the same behaviour in other genera of two different orthopteran superfamilies (North American Jerusalem crickets (Stenopelmatidae) and an Australian king cricket (Anostostomatidae)) suggests a more complex background, probably including convergent evolution of a common defence pattern against reptilian predators. In New Zealand, the behaviour has been retained, in addition to 'raise hind legs', in spite of presumably intense selection pressure from long-beaked ancestral avian predators during periods when the repeated Pleistocene glaciations pushed the alpine fauna into lowland refugia (Fleming, 1962).

'Hind leg stridulate' is unique to New Zealand deinacridines, and it therefore probably evolved after the separation of New Zealand from Gondwana. It seems likely to have evolved against avian predators, which colonized New Zealand from the Cretaceous onward (Table 16.4). Of these, ancestors of the small, short-beaked robins from the Pliocene, and much later the cuckoos, would have provided the greatest selection pressure against the radiating *Hemideina* species, while the presumptive *Deinacrida* species would have received selection pressure from longer-beaked avians, such as saddleback ancestors. These conclusions are based upon the deterrent effect seen for this behaviour in the above experiments with South Island robins. As discussed by Field (Chapter 22, this volume), the file and peg form of the femoro-abdominal stridulatory mechanism also evolved during speciation of the Deinacridinae after New Zealand separated from Gondwana. In the case of *Hemideina*, this stridulatory structure became associated with the role of intraspecific acoustic communication, in addition to one of defence, and became more elaborated in the New Zealand species than anywhere else in the world (Field, 1993). The concomitant evolution of a mechanism of sound production for intraspecific communication (using different stridulatory movements) also led to the generation of louder sounds (as judged by comparison of extant species). A spin-off seems to be the enhancement of the effectiveness of 'hind leg stridulate' behaviour in a defence role.

References

Atkinson, I.A.E., and Campbell, D.J. (1966) Habitat factors affecting saddlebacks on Hen Island. *New Zealand Ecological Society Proceedings* 13, 35–40.

Bailey, W.J. (1991) *Acoustic Behaviour of Insects*. Chapman and Hall, London, 225 pp.

Bellingham, M. (1991) Field observations of two species of tusked weta. *The Weta (New Zealand Entomological Society)* 14, 30–32.

Buchler, E. R., Wright, T.B. and Brown, E.D. (1981) On the function of stridulation by the passalid beetle *Odontotaenius disjunctus* (Coleoptea: Passalidae). *Animal Behaviour* 29, 483–486.

Buller, W.L. (1871) Notes on the genus *Deinacrida* in New Zealand. *Transactions and Proceedings of the New Zealand Institute* 3, 34–37.

Carroll, A.L.K. (1963) Food habits of the North Island weka. *Notornis* 10, 289–300.

Daniel, M.J. (1973) Seasonal diet of the ship rat (*Rattus r. rattus*) in lowland forest in New Zealand. *New Zealand Ecological Society Proceedings* 20, 21–30.

Douglas, M.J.W. (1970) Foods of harriers in a high country habitat. *Notornis* 17, 92–95.

Edmunds, M. (1974) *Defence in Animals*. Longman, London, 357 pp.

Field, L.H. (1978) The stridulatory apparatus of New Zealand wetas in the genus *Hemideina* (Insecta: Orthoptera: Stenopelmatidae). *Journal of the Royal Society of New Zealand* 8, 359–375.

Field, L.H. (1993) Structure and evolution of stridulatory mechanisms in New Zealand wetas (Orthoptera: Stenopelmatidae). *International Journal of Insect Morphology and Embryology* 22, 163–183.

Field, L.H. and Bailey, W.J. (1997) Sound production in primitive Orthoptera from Western Australia: sounds used in defence and social communication in *Ametrus* sp. and *Hadrogryllacris* sp. (Gryllacrididae: Orthoptera). *Journal of Natural History* 31, 1127–1141.

Field, L.H. and Rind, F.C. (1992) Stridulatory behaviour in the New Zealand weta *Hemidiena crassidens*. *Journal of Zoology (London)* 228, 371–394.

Fitzgerald, B.M. (1964) Ecology of mustelids in New

Zealand. Unpublished MSc thesis, University of Canterbury, New Zealand.

Fleming, C.A. (1962) New Zealand biogeography: a paleontologist's approach. *Tuatara* 10, 53–108.

Gibbs, G.W. (1998a) Why are some weta (Orthoptera: Stenopelmatidae) vulnerable yet others are common? *Journal of Insect Conservation* 2, 161–166.

Gibbs, G. W. (1998b) Raukumara tusked weta; discovery, ecology and management implications. *Conservation Advisory Science Notes (New Zealand Department of Conservation)* 218, 1–14.

Gibbs, G.W. and McIntyre, M. (1997) *Abundance and Future Options for Wetapunga on Little Barrier Island.* Science for Conservation: 48, Department of Conservation, Wellington, 24 pp.

Grant-Mackie, J.A., Buckeridge, J.S. and Johns, P.M. (1995) Two new upper Jurassic arthropods from New Zealand. *Alcheringa* 19, 31–39.

Haskell, P.T. (1961) *Insect Sounds.* Witherby, London, 189 pp.

Hutton, F.W. (1897) The Stenopelmatidae of New Zealand. *Transactions and Proceedings of the New Zealand Institute* 29, 208–242.

King, C.M. (1984) *Immigrant Killers.* Oxford University Press, Melbourne, 224 pp.

Kuschel, G. (1975) *Biogeography and Ecology in New Zealand.* Junk, Amsterdam, 689 pp.

Lindsay, C.J. and Ordish, R.G. (1964) The food of the morepork. *Notornis* 11, 154–158.

McIvor, I.R. (1972) Ecology of a population of *Hoplodactylus pacificus,* the common New Zealand gecko (Reptilia: Gekkonidae). Unpublished MSc thesis, University of Canterbury, Christchurch, New Zealand.

Marshall, W.H. (1963) *The Ecology of Mustelids in New Zealand.* DSIR Information Series No. 38, DSIR Publishing, Wellington, New Zealand

Masters, W. (1979) Insect disturbance stridulation: its defensive role. *Behavioural Ecology and Sociobiology* 5, 187–200.

Masters, W.M. (1980) Insect disturbance stridulation: characterization of airborne and vibrational components of the sound. *Journal of Comparative Physiol*ogy 135, 259–268.

Meads, M.J. (1990a) *The Weta Book: a Guide to the Identification of Wetas.* DSIR Land Resources, Lower Hutt, New Zealand, 36 pp.

Meads, M.J. (1990b) *Forgotten Fauna.* DSIR Publishing, Wellington, New Zealand, 95 pp.

Miller, D. (1985) *Common Insects in New Zealand,* A.H. and A.W. Reed, Wellington, New Zealand, 139 pp.

Moller, H. (1978) Study of the wetas of Stephens Island. Unpublished report, File no. 4/15/9, Ecology Division, DSIR, Lower Hutt, New Zealand 38 pp.

Moller, H. (1985) Tree wetas (*Hemidiena crassicruris*) (Orthoptera: Stenopelmatidae) of Stephens Island, Cook Strait. *New Zealand Journal of Zoology* 12, 55–69.

Oliver, W.R.B. (1930) *New Zealand Birds.* Coulls, Somerville and Wilkie, Wellingon, 541 pp.

Preston-Mafham, K. (1990) *Grasshoppers and Mantids of the World.* Facts-on-File, Oxford, 192 pp.

Ramsay, G.W. (1953) A supplementary sound-producing device in *Deinacrida rugosa* Buller (the Stephens Island weta). *New Zealand Entomology* 1, 12–14.

Ramsay, G.W. and Bigelow, R.S. (1978) New Zealand wetas of the genus *Hemideina. The Weta* 1, 32–34.

Rentz, D.C.F. (1993) Do the spines on the legs of katydids have a role in predation? *Metaleptea* 14, 17.

Richards, A.M. (1973) A comparative study of the biology of the giant wetas *Deinacrida heteracantha* and *D. fallai* (Orthoptera: Henicidae) from New Zealand. *Journal of Zool*ogy 169, 195–236.

Riek, E.F. (1970) Fossil history. In: *Insects of Australia.* Melbourne University Press, Melbourne, pp. 168–186.

Robinson, D. (1990) Acoustic communication between sexes among bushcrickets. In: Bailey, W.J. and Rentz, D.F.C. (eds) *The Tettigoniidae: Biology, Systematics and Evolution.* Crawford House Press, Bathurst, pp. 110–129.

Ruffaut, C.G. (1995) A comparative study of the Wellington tree weta, *Hemideina crassidens* (Blanchard, 1851) in the presence and absence of rodents. Unpublished MSc thesis, Victoria University, New Zealand.

Salmon, J.T. (1950) A revision of the New Zealand wetas: Anastostominae (Orthoptera: Stenopelmatidae). *Dominion Museum Records Entomology* 1, 121–177.

Sandow, J.D. and Bailey, W.J. (1978) An experimental study of defensive stridulation in *Mygalopsis ferruginea* Redtenbacher (Orthoptera: Tettigoniidae). *Animal Behaviour* 26, 1004–1011.

Sharell, R. (1966) *The Tuatara, Lizards and Frogs of New Zealand.* Collins, London, 94 pp.

Sherley, G. and Hayes, L.M. (1993) The conservation of a giant weta (*Deinacrida* n. sp. Orthoptera: Stenopelmatidae) at Mahoenui, King County: habitat use, and other aspects of its ecology. *New Zealand Entomology* 16, 55–68.

Smith, P.E. (1979) Behavioural and reflexive analysis of a defence response in *Hemideira maori* (Orthoptera: Stenopelmatidae). BSc (Hons) thesis, University of Canterbury, Christchurch, New Zealand.

Smith, R.L. and Langley, W.M. (1978) Cicada stress sound: an assay of its effectiveness as a predator defence mechanism. *Southwest Naturalist* 23, 187–196.

Stevens, G.R. (1980) *New Zealand Adrift.* A.H. and A.W. Reed, Hong Kong, 442 pp.

Townsend, J.A. (1995) Distribution and ecology of the Banks Peninsula tree weta *Hemidiena ricta.* Unpublished MSc thesis, Massey University, Palmerston North, New Zealand.

17
Mating Behaviour

Laurence H. Field and Thomas H. Jarman
Department of Zoology, University of Canterbury, Private Bag 4800, Christchurch, New Zealand

Introduction

While much attention has been paid to elucidating the elaborate behavioural mechanisms for mating in common orthopteran groups such as the crickets (Gryllidae), equally fascinating, but different, mating behaviour is found in the gryllacridoid families of Ensifera (see Gorochov, Chapter 1, this volume, for phylogenetic details). For example, New Zealand tree weta males fight for possession of harems of females, and show associated sexual selection for exaggerated weaponry (such as enormous mandibles) developed as secondary sexual characters (described in Field and Deans, Chapter 10, this volume). In contrast to these, the closely related giant wetas have retained a primitive mating behaviour, and males aggregate around a female for mating, rather than showing aggressive competition. In addition, males of the giant tusked weta and at least several tree weta species show early maturation and can become imagines up to three instars earlier than normal. Such undersized males engage in sneak mating tactics. The ultimate form of male parental investment was discovered in Jerusalem crickets, where the male is eaten by the female in a bizarre act of postmating cannibalism.

Much of this information lies in unpublished studies. Some aspects have been reviewed by Brown and Gwynne (1997), while others are mentioned in related chapters in the present volume and will be placed in context here (Weissman, Chapter 3, and Stringer, Chapter 20, this volume). In this chapter, we review what is known about mating behaviour in New Zealand wetas and compare this with mating behaviour in other Ensifera. New material will be presented for quantitative behavioural analyses of mating in tree wetas and for evidence of chemical communication in tree wetas and giant wetas.

Overview of the Mating System of Tree Wetas

Most studies in New Zealand wetas have dealt with the tree wetas (*Hemideina*), in which males compete aggressively for females in a defence-resource polygyny mating system (Field and Sandlant, 1983). Males do not show any elaborate courtship towards females other than antennation, palpation and probing of the female with the genitalia. No acoustic courtship signals are given by males, although there is evidence that chemical signals from females elicit copulation by males. Mate choice is controlled by females, which allow a relatively low percentage of successful matings by males. Remarkably, there appears to be no species-specific recognition in this chemical communication system which triggers copulation within *Hemideina* (see Morgan-Richards *et al.*, Chapter 7, this volume). Copulation involves the evolutionarily derived end-to-end position, in contrast to the more primitive male below female position, which is employed by giant wetas (Brown and Gwynne, 1997). Males show aggressive behaviour towards females if rebuffed during mating attempts; this may be regarded as part of a guarding tactic. Guarding is much less obvious in tree wetas than in the giant wetas (*Deinacrida* spp.) or the ground wetas (*Hemiandrus* spp.).

Mating Systems

Resource defence polygyny in tree wetas

Hudson (1920) first suggested that the enlarged head and mandibles of *Hemideina* males should lead to sexual selection and aggressive competition between males for mating rights with females. This was confirmed and extended by Sandlant (1981) and reviewed by Field and Sandlant (1983) and Field (Chapter 18, this volume), who showed that tree weta males fight vigorously for and defend harems of females. Such harems are in galleries within trees or under boulders in the alpine habitat. Because females aggregate in a resource (tree galleries provide diurnal shelter from predators) for which males compete, and because dominant males mate with females within the defended resource, Sandlant (1981) showed that the mating system of tree wetas conforms to a resource defence polygyny (Emlen and Oring, 1977; Thornhill and Alcock, 1983).

In this mating system, males search nightly for galleries, which were originally excavated in trees by larvae of beetles or moths (see Field and Sandlant, Chapter 13, this volume) and which are secondarily occupied by females. Single adult males occupy galleries containing one or up to 12 females (Field and Sandlant, 1983; Gwynne and Jamieson, 1998; L.H. Field and T.H. Jarman, personal observation) and exclude other adult males. Both males and females make nightly feeding forays, but resident males spend most of their time occupying the gallery entrance. Unsuccessful males roam and engage in aggressive attempts to acquire galleries or to mate with females outside galleries (Sandlant, 1981). Resident males stridulate at the gallery entrance, using a femoro-abdominal mechanism (see Field, Chapter 15, this volume), mate with resident females at the entrance or within the gallery, if sufficient space exists, and aggressively repel intruder males. Daytime censuses of occupation of galleries in trees (or under-rock galleries for the alpine *Hemideina maori*) indicate that single adult males tolerate juvenile males and females but not other adult males.

The harem concept

Additional research is required to elucidate details of the formation, duration and identity of harems in the tree weta mating system. For example, the 'harem' concept implies exclusive mating rights to females that are defended by a dominant male (Krebs and Davies, 1978), and yet it is likely that some females in weta harems are mated outside the gallery by other males. This was recorded in a laboratory colony, where 70% of observed matings ($n = 57$) in *Hemideina crassidens* were at the gallery entrance, while the remainder occurred as females wandered about the enclosure (Spencer, 1995). Laboratory colonies, however, often have higher densities of wetas than naturally occurring populations. For the latter, it is not known how exclusively males dominate females within their respective harems.

Harems are not always common, nor are they necessarily large, in tree weta populations. The largest number of females reported with a single male is 13 (Asher, 1977), but the more typical harem size consists of one to three females. This may be seen in a comparison of four species of *Hemideina* (Fig. 17.1), each of which has a modal peak of one or two females per male. Some differences exist between species, which may be related to habitat and potential gallery size. For example, *H. maori* lives in an alpine environment devoid of trees, but often covered with grassy meadows littered with rocks. Retreats beneath large flat rocks, are likely to have more room for large harems than those in the more restricted tree galleries used by *Hemideina femorata*. On the other hand, *H. crassidens* sometimes occupies enlarged cavities in old trees and the enlarged space appears to be readily occupied by dominant males with large harems. These trends are reflected in the distributions shown in Fig. 17.1.

Details of how aggregations of females are formed, what factors allow them to be maintained and cost/benefit analyses of male vs. female investment in mating under harem conditions compared with paired male–female conditions would be instructive in understanding this mating system. A single study examining the relationship between harem size and male mandible length (as an index of male body size) has provided statistical evidence that larger males accumulate larger harems in *H. maori* (Gwynne and Jamieson, 1998).

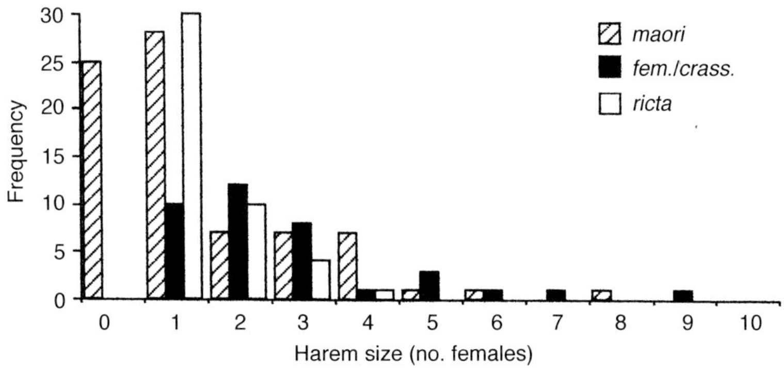

Fig. 17.1. Distributions of harem sizes of females cohabiting with single adult male in four species of New Zealand tree wetas (*H. maori, H. crassidens, H. femorata, H. ricta*) (additional unpublished data for *H. maori* were kindly contributed by D. Gwynne and I. Jamieson, with permission).

Early-maturing males and sneak mating strategy

A complexity is introduced into the system by the discovery that some male tree wetas (demonstrated for *H. crassidens*) mature at the eighth and ninth instar, while other males, and all females, mature at the tenth instar (Fig. 17.2A; Spencer, 1995). The earlier maturing males appear to adopt a sneak and wander strategy for mating with females when they are outside galleries, rather than fighting for gallery possession to access females. These smaller males inhabit small galleries with no females and, because they do not increase head width or mandible length after maturity, they are at a disadvantage in any contests for larger galleries containing the largest females (which, presumably, are most desirable for mating). This is due to the bias for tree weta males with greater mandible gapes winning battles against males with lesser gapes (data in Field, Chapter 18, this volume). Early-maturing males therefore attempt to mate with every female they encounter and their reproductive effort is spread widely, in contrast to that of tenth-instar males. Because females only spend 15% of the night outside of the gallery, tenth instar males would presumably have a greater overall monopoly of the females in the population. Spencer (1995) showed that eighth-instar males had a low mating success (22% of 23 attempts), but ninth-instar males had a greater success rate (61% of 14 attempts) than the largest (tenth-instar) males (44% of nine attempts). Ninth-instar males are able to gain possession of some galleries with resident females, and therefore have the ability to adopt either the gallery strategy or the wander strategy.

Ninth-instar males were more abundant than tenth-instar males in two *H. crassidens* populations studied by Spencer (1995) (Fig. 17.2B), so there is the possibility that intrasexual selection might be split into two directions in tree wetas. Heightened competition for galleries and for large females presumably leads to intrasexual selection for secondary sexual characters, such as the extremely long mandibles and broad head capsules in tenth-instar males. However, early maturation may split this selection pressure and lead to optimization of an alternate mating strategy, favouring smaller sneaking males that mate outside the galleries.

Mating systems in giant wetas, ground wetas and tusked wetas: scramble competition polygny

Although no mating behaviour studies have been carried out in other New Zealand genera, their lifestyles (reviewed by Gibbs, Chapter 2, this volume) suggest that they have a variety of mating strategies. For example, the arboreal giant wetas (*Deinacrida* spp.) of the North Island are nomadic and solitary, usually seeking temporary daytime shelter in loose bark, epiphytes or clusters of leaves in trees. Similarly, most of the other *Deinacrida* species are solitary and seek diurnal temporary refuges on the ground, amongst grass tussocks, in rock crevices or under rocks. Males of all these species do not appear to maintain mating territories and probably adopt a ‘scramble

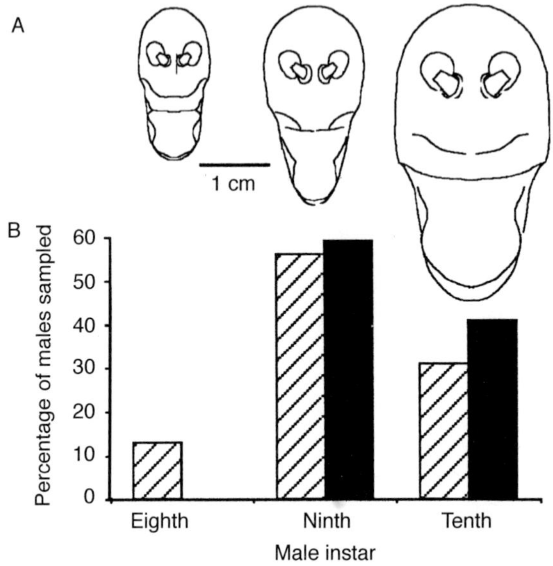

Fig. 17.2. Early maturation in eighth and ninth instars of tree wetas (*H. crassidens*) establishes sexually mature males in the population with different head sizes. A. The greatest secondary sexual character development occurs in the normal tenth-instar male (right side), while the two earlier instars have progressively smaller heads and mandibles. B. Histograms of the distribution of adult males in the last three instars in two populations (solid and hatched bars) in the North Island of New Zealand. (Modified from Spencer, 1995, with permission.)

competition polygyny' mating system (Thornhill and Alcock, 1983), involving non-aggressive searching for females. Radio-tracking of the ground-dwelling *Dienacrida rugosa* has established that apparently receptive females wander until one or more males are attracted to them. Trailing males pursue the female until she somehow makes a decision to mate with one male, and the other males decamp. The mating pair remain together for a period of days (M. McIntyre, Wellington, 1998, personal communication). On the other hand, *Deinacrida talpa*, ground wetas (*Hemiandrus*) and the tusked wetas *Motuweta isolata* and the undescribed 'Raukumara' tusked weta are burrow dwellers, and at least *M. isolata* males appear to defend these as resources. Anecdotal information suggests that *Hemiandrus* males also defend their burrows (van Wyngaarden, 1995). Female *Hemiandrus* oviposit in the burrows and remain with the developing eggs. Therefore it is crucial to know whether males stay with females in a monogamous system or whether females aggressively defend burrows containing eggs while males are polygynous. Ground weta mating systems provide a fertile area for future research in this subject.

Early maturing-male instars have also been reported for the giant tusked weta *M. isolata*, but nothing is known of the relative mating success or strategies used in this species (Johns, 1997). The phenomenon of early maturation apparently does not occur in the giant wetas or in ground wetas of New Zealand.

Mate Attraction

Wetas do not appear to exhibit the obvious mate attraction tactics found in other ensiferans. The scant evidence for possible acoustic and chemical attraction is reviewed below.

Acoustic signals

Male tree wetas produce stridulatory sounds from the gallery entrances at night (see Field, Chapter 15, this volume). Anecdotal reports suggest that much of the sound is made by early instar males rather than the tenth-instar adults (Ordish, 1992; L.H. Field and T.H. Jarman, personal observation). If confirmed, this could imply that smaller males call from their galleries to attract females, while the larger males do not call if they are defending a harem of females. Since no quantitative observations have been made of which instars call and whether they possess or attract females, it is not clear if the calling is directed at females or at rival males. The signalling could announce territories either to rival males (with the message that any intruders would be met aggressively) or to females (with the message the male holds a refuge from predators and/or is announcing his fitness).

Although anecdotal reports of calling by giant wetas (*Deinacrida heteracantha, Deinacrida mahoenui*) have been published, nothing is known about the behavioural context of this sound or if it is used as a means of attracting mates. Both sexes possess elaborate femoro-abdominal stridulatory mechanisms, but it is not even known which sex makes calling sounds at night. Ground wetas (*Hemidandrus* spp.), on the other hand, are known to produce drumming vibrations in the substrate (Brown and Gwynne, 1997; see also Field, Chapter 15, this volume) which could serve as signals to attract females to burrows occupied by males. Ground wetas possess femoro-abdominal stridulatory structures but lack tibial ears, so they presumably do not use acoustic transmission for intraspecific signalling.

Chemical signals

No experimental chemical analysis has shown the existence of volatile pheromones that facilitate pair formation in any New Zealand weta species as of this writing. However, most species of tree and giant wetas have odours that are easily detectable by humans, and common accounts interpret this as evidence for pheromones (Ramsay, 1955). The difficulties with this interpretation lie in the lack of demonstrated attraction of females to males (one exception described below indicates that females attract males), lack of chemoreceptor studies, lack of chemical analysis of volatile compounds from wetas and the possibility that the odours are non-communicatory metabolic byproducts.

However, the following behavioural observations of two species of giant weta (*Deinacrida*) show that males are attracted to and follow females. In favourable weather conditions, females of *D. rugosa* become active in the early evening and wander up to 50 m from their daytime shelters (most giant wetas do not use burrows). As they move, one or several males begin to follow (Fig. 17.3), and they eventually end up in a pocket of soil beneath grass tussocks, where copulation occurs. Male trailing behaviour has also been observed in *Deinacrida fallai* (M. McIntyre, Wellington, 1998, personal communication). It is probable that future studies may demonstrate that female giant wetas use volatile pheromones to attract males. Here, behavioural and chemical analyses are of crucial interest. A similar mechanism appears to exist for the giant tusked wetas. When a female Raukumara tusked weta was placed into a terrarium, a male emerged from its burrow and readily located the female in complete darkness and commenced mating (McIntyre, 1998).

In contrast to the use of volatile compounds for pair formation, clear evidence exists for a female contact pheromone that elicits copulation in tree wetas (discussed below). It is probable that such contact pheromones exist in the other weta genera as well.

Recent experiments with the South African king cricket *Libanasidus vittatus* have demonstrated that males and females can discriminate between sexes by the odours of the faeces. This

Fig. 17.3. Trailing behaviour in the giant weta, *Deinacrida rugosa*, in which one or more males follow a wandering, receptive female for extended periods of the evening before she eventually chooses to mate with one and remains with it for days.

facilitates avoidance of aggression between males, and mate-finding behaviour between males and females (Bateman and Toms, 1998).

Minimal Courtship and Female Mate Choice

Courtship behaviour in tree wetas

Because New Zealand wetas lack the elaborate courtship behaviour found in crickets and grasshoppers, it was initially thought that courtship does not occur in wetas (Richards, 1973). However, a number of studies showed that males engage in general antennation, palpation and genital probing of females prior to attempting to copulate. These more subtle actions are now recognized as elements of courtship behaviour (Field and Sandlant, 1983). Descriptions of this behaviour appear similar for those species of tree wetas which have been examined (*H. femorata, Hemideina ricta* and *H. crassidens*) (Sandlant, 1981; Jarman, 1982; Spencer, 1995; Townsend, 1995). The most complete description is for 128 male–female encounters of *H. femorata* (Sandlant, 1981; reviewed in Field and Sandlant, 1983). A summary of courtship behaviour consists of the following elements, as set out by Sandlant (1981) and Jarman (1982): 'approach' (male initiates the interaction), 'antennate–palpate' (male uses mandibular and maxillary palps and antennae), 'bite tibia' (male grips female's hind tibia and pulls her out of gallery), 'apply genitalia' (and possibly 'elevation' of head, 'gape' of mandibles and 'lunge', although these 'aggressive' behaviours may not be associated with courtship intention). As discussed below, the apparent aggressive response of males after being rebuffed by the females is a little understood facet, which requires more research.

In a typical encounter for *H. femorata,* a female rests in the gallery entrance and a male usually initiates an encounter by 'antennate–palpate' for a variable time, depending upon the female's reaction. If she remains passive, the male continues 'antennate–palpate' and often grips her tibia ('bite tibia') to pull her out of the gallery sufficiently far to copulate. If the female 'kick-pushes' the male (6% of encounters), she retreats forward into the gallery and (rarely) produces stridulatory sounds. If a female is outside the gallery, the male begins 'antennate-palpate' on the dorsal surface of her head, thorax, abdomen and legs. This lasts about 10 s in *H. femorata* and *H. crassidens* (Jarman, 1982). The male next flexes the abdomen ventrally into a C shape and probes the female's body with his genitalia. Probing allows contact against her head, proximal segments of the legs and ventral and lateral abdominal surfaces without any particular order in the sequence. At this stage, the female of all the above species usually prolongs the encounter by resisting (87% of encounters in *H. femorata*; n = 128; Sandlant, 1981). This includes stepping and manoeuvring her abdomen, kick-pushing and re-entering the gallery, remaining motionless with abdomen pressed against the substrate or walking away. Males are usually undeterred by this and continue to probe with the genitalia, even if dragged over the substrate when outside the gallery. In timed encounters of *H. femorata* (n = 38; Sandlant, 1981), female resistance duration was approximately similar for interactions with or without subsequent copulation (66 s and 85 s, respectively, but note the large SD, Table 17.1). If the female departs after kicking the male, he often produces a 'mating pattern' of stridulatory sound (see Field, Chapter 15, this volume) and occasionally shows behaviours associated with male–male combat ('elevation' of head, 'mandible gape' and 'lunging' with open mandibles).

In *H. femorata* only 23% of the male–female encounters ended in copulation (n = 128; Sandlant, 1981), while in *H. crassidens* 64% of observed mating attempts ended with copulation (n = 95; Spencer, 1995). Reasons for failure to achieve successful matings are given in Table 17.2. In a small percentage of the encounters, the male departed after 'antennate–palpate'. The majority of failures occurred when the male met female resistance and either sex subsequently departed without further courtship. Between one-third and one-sixth of the encounters failed due to departure of either sex after the male had continued courtship with 'apply genitalia' behaviour. In these cases, the male probed with his genitalia against the various parts of the female's body without ever arriving at the female's genitalia. It is not clear whether there are insufficient chemical cues to guide the males or whether this behaviour was an extended preparation of the female for copulation which was not met with appropriate cues from the female.

Table 17.1. Summary of existing data for durations (in min) of various phases of sexual behaviour (male courtship, female resistance and copulation) in New Zealand wetas. The second row of data for *Hemideina femorata* is for encounters where female resistance resulted in no successful copulation. Sample sizes for observed encounters: *H. crassidens* $n = 33$ (Jarman, 1982); *H. femorata* $n = 38$ (Sandlant, 1981); *Deinacrida connectens* $n = 1$ (Field, 1980); *D. mahoenui* $n = 13$ (Domett, 1996).

Species	Male courtship (min (mean ± SD))	Female resistance (min (mean ± SD))	Copulation (min (mean ± SD))
H. crassidens	7.5 (5.3)	–	1.5 (0.6)
H. femorata	1.3 (1.2) Median = 1.0	1.1 (0.9) Median = 1.0	2.0 (1.3) Median = 1.6
H. femorata	1.9 (1.5) Median = 1.4	1.4 (1.2) Median = 1.1	No copulation occurred
H. maori	–	–	Mean = 3.1 (range = 1.9–4.3)
D. connectens	–	–	35
D. mahoenui	–	–	Median = 324 (range = 1.2–810)

Table 17.2. Reasons for unsuccessful mating attempts by two species of tree wetas (*Hemideina*). Data given as percentage of encounters which terminated for the reasons given in the left column. *n*, Total number of male–female encounters observed. (After Sandlant, 1981; Spencer, 1995.)

Reason	*H. crassidens*	*H. femorata*
Male departs after 'antennate–palpate'	8%	18%
Female resistance and either sex departs	68%	27%
Either sex departs after male 'apply genitalia'	15%	30%
	$n =$ 53	128

Female mate choice

Female resistance is interpreted as a means of assessing the fitness of potential mates, because resistance was often followed by copulation. In principle, females could assess male fitness by testing the duration of male persistence in courtship as well as a male's ability to dislodge her from a gallery. Sandlant (1981) showed that male persistence times were similar whether copulation occurred or not (median test: $\chi^2 = 0$, $P > 0.95$, Table 17.1), suggesting that males may have an optimal persistence time or that females may have an optimal resistance time. However, in crickets, where courtship involves male stridulation, the variation in courtship duration appears to be due primarily to variation in female receptiveness, rather than to qualities of male fitness (Boake, 1984). Also, Simmons (1986) found that, in addition to female variation, at least some courtship duration variation was also due to differences in attractiveness of male crickets to females. Thus, the relationship between courtship duration and female mate choice in tree wetas is open to further investigation. Special attention should be paid to male and female past mating history and the fitness of each partner (body weight and size, winner/loser status in male combat). In summary, males make many mating attempts, but achieve a low rate of successful matings because of female resistance as a presumed means of making mate choices.

Aggression behaviour of rebuffed males towards females

In a study of acoustic communication between tree wetas (Field and Rind, 1992), it was shown that a 'mating' stridulatory sound pattern was produced by males that had attempted to mate with females but were actively rejected by females. The difference between the normal 'aggression' stridulatory pattern and the mating pattern was slight. The mating pattern was statistically somewhat slower, but its syllable rate could overlap with that of the aggression pattern (see Field, Chapter 15, this volume). The conclusion was that the mating pattern contained an element of aggression in the message sent by the male to both the female and any intruder males.

Interpretation of this apparently contradictory behaviour may be facilitated by comparison with a similar behaviour observed in the bark beetle *Dendroctonus*. Males make the same sound toward a courted female as when fighting a rival for access to a gallery system. Females apparently require this demonstration of male aggressiveness before mating can proceed (Ryker and Rudinsky, 1976). A likely hypothesis to explain the aggressive component in mating is that the female is assessing the male's potential ability to defend the gallery if he is allowed to mate with her. Actual courtship involves aggressive pushing, shoving and biting, which suggests that the female also physically assesses the male's strength and aggressive tendency before mating. Using the same general principle, it is possible to interpret male tree weta 'mating' stridulation as a demonstration or affirmation of aggressive ability to a female that initially rejected the male. This stridulation usually accompanies continued pursuit of a female by the rebuffed male.

An interesting discussion of female assessment of males based upon demonstrations of aggressiveness, dominance status and resource gathering abilities is presented by Thornhill and Alcock (1983). They suggest that, in principle, females should make assessments, while males should strive towards multimodal demonstrations of fitness, in order to provide the females with indicators about the males' genetic quality. The fact that female tree wetas have been shown to resist male mating attempts 87% of the time lends support to the idea that such testing of males is an inherent part of the tree weta mating system.

Other New Zealand weta species

In ground wetas (*Hemiandrus gracilis*), courtship appears to be brief. It commences outside the burrows with male approach and antennation of the female from the front. Often the female rapidly departs, but in successful matings, observed on four occasions, the pair remained motionless while the male continued antennation. This was followed by rapid male palpation and mounting of the female (Cary, 1981).

Courtship has been described as precopulatory behaviour for several species of giant wetas (*Deinacrida*). Two aspects contrast sharply with tree weta mating behaviour: (i) after the trailing behaviour described above, often several male giant wetas surround a female without engaging in intense aggressive exclusion (Richards, 1973); and (ii) mating behaviour usually occurs in the daytime (see Stringer, Chapter 20, this volume). Male behavioural elements include 'antennate–palpate' and prodding the female's body with forelegs for many minutes. Eventually, the female backs under the male and manoeuvres the abdomen, which seems to elicit copulation. If several males are present, the one that she has encouraged by her movements will be the one to mate, while the others either remain nearby or depart (*D. rugosa, D. fallai*; M. McIntyre, Wellington, 1998, personal communication).

In the tusked wetas *M. isolata* and the Raukumara tusked weta, males have been observed to mate as soon as they approach females. Preliminary 'antennate–palpate' behaviour is likely to occur but has not been reported (McIntyre, 1998; M. McIntyre, Wellington, 1998, personal communication). In *M. isolata,* the males only emerge in the dark phase of the moon (to take advantage of the inky darkness for protection from tuatara lizards, which are the main predators inhabiting the small island habitat). The males wander large distances (30–50 m) in search of females and, when found, the females are mated and intensely monopolized by the males. From radio-tracking studies, females are known to rotate amongst burrows each night over an area roughly 5 m across. If a wandering male encounters another male, they engage in intense battles near burrow entrances, involving head-to-head wrestling with the tusks locked together (see Fig. 10.14, Field and Deans, Chapter 10, this volume). This implies that males defend burrows that are possibly used as a mating resource in a manner similar to that of crickets.

Copulation

Brown and Gwynne (1997) concluded that copulation occurred in ancestral Ensifera with the female above the male, and that, while this position is still retained in many extant taxa, the derived ensiferan positions also include end-to-end, and end-to-end followed by female-above. Different genera of New Zealand wetas utilize different positions. The side-by-side or female-above position occurs in giant wetas (Deinacridinae) and ground and tusked wetas (Henicinae). The male stands beside or underneath the female, while both face the same direction. The male's abdomen reaches beneath that of the female to make genital contact while their bodies form an angle of 30° (*Deinacrida connectens*) or 45°–90° (*D. fallai, D. heteracantha*). In rare instances, the male mounts the female and curls the abdomen ventrally and forward to contact the underside of the female (Richards, 1973). This same mating position (Fig. 17.4A) has been documented for the large tusked weta, *M. isolata* (M. MacIntyre, Wellington, 1999, personal communication). *Motuweta isolata* even copulates in this position while in its underground burrows (C.J. Winks, Auckland, 1998, personal communication).

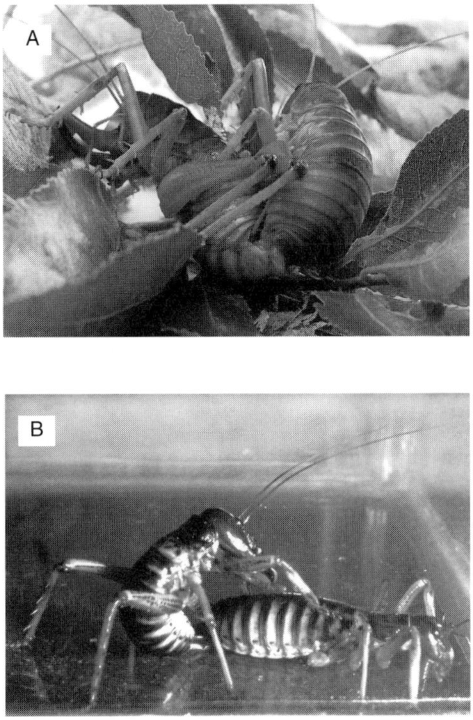

Fig. 17.4. Mating positions for New Zealand wetas. A. Giant wetas, ground and tusked wetas copulate with the male beneath the female while both face in the same direction, as shown for the tusked weta, *Motuweta isolata* (photo by Brett Robinson, Victoria University, Wellington, with permission). B. Tree wetas (*Hemideina* spp.) mate with the male curling the abdomen beneath the female while standing behind her.

Tree wetas (*Hemideina*) mate with the male behind the female. The male curls his abdomen ventrally and forward while grasping the female with his hind legs and (often) other legs (Fig. 17.4B). This is the usual position for mating near the gallery, as well as outside the gallery. Less frequently (18% of encounters, $n = 33$), the female-above position, described for giant wetas, has been observed on flat substrates for *H. crassidens* (Jarman, 1982). Male *H. maori* copulate when alongside the female by curving their abdomens towards the genitalia of the female (Gwynne and Jamieson, 1998).

Copulation in the ground weta *Zealandosandrus* (= *Hemiandrus*) *gracilis* occurs after the male approaches the female head-on and makes antennation contact with her. If the female is receptive, the male palpates for less than a minute and then turns to face parallel and in the same direction as the female. He then backs under her abdomen and arches the tip of his abdomen dorsally, while she flexes her abdomen ventrally to make genital contact (Cary, 1981).

Adult male *Hemideina* possess small, heavily sclerotized hooks on the anterior margin of the epiproct (dorsal lobe of the tenth abdominal tergite). These hooks are about 0.5 mm long, are 1–2 mm apart and are referred to as 'gin traps' (Hinton, 1946). The posterior margin of the ninth tergite projects a lip of thick cuticle midway between the gin traps to form a rudimentary grasping device. Gin traps are reduced in subadult males and lacking in females. The male uses the gin traps to open the female subgenital plate by sweeping the epiproct anteriorly over the female's subgenital plate. This is the direction in which the gin trap hooks point and in which the subgenital plate opens. The subgenital plate is opened when grasped between the hooks and the posterior lip of the male ninth tergite. When Jarman (1982) pulled copulating wetas apart prior to phallus intromission, the genitalia broke contact suddenly, confirming the nature of the gripping mechanism of the gin trap. The variety and roles of gin traps and similar copulatory grasping mechanisms in other Orthoptera are reviewed by Brown and Gwynne (1997), and in Stenopelmatidae by Weissman (Chapter 3, this volume).

The duration of copulation in wetas varies from about 1.5 min (*Hemideina*) to 2 h (*Z.* (= *Hemiandrus*) *gracilis*) to over 13 h (*Deinacrida*). Copulation times and frequencies for different species of wetas are summarized in Table 20.4 (see Stringer, Chapter 20, this volume). Long durations do not necessarily imply that this is the time required to transfer the spermatophore to the female. Jarman (1982) reported that a mean of 87 s (SD = 33 s, n = 6) was sufficient for transfer of single spermatophores in *H. crassidens.* Especially long times for mating are found in the various species of *Deinacrida.* Repeated copulation with the same individual occurs in both of the above New Zealand genera, and multiple spermatophores are transferred in the process (reviewed by Stringer, Chapter 20, this volume).

Spermatophores as Nuptial Gifts, and Multiple Matings: Parental Investment Theory

The spermatophores of New Zealand wetas are small in comparison with those of other Ensifera. In *H. crassidens* they are about 2 mm in diameter and weigh 2.6–3.6 mg (Jarman, 1982), while those of the giant weta, *D. heteracantha,* are of similar size (Richards, 1973). In *D. heteracantha,* the males are capable of producing eight spermatophores within 7.5 h during repeated copulation with the same female. The record is 15 spermatophores produced by a male *D. rugosa* in 12 h (Ramsay, 1955).

In tree and ground wetas (*Hemideina* and *Hemiandrus*, respectively), the female eats the spermatophore (Sandlant, 1981; Jarman, 1982; Brown and Gwynne, 1997), but this behaviour has not been seen in mated females of *Deinacrida.* In *H. crassidens* , the spermatophore is eaten on average 2.37 ± 0.98 (SD) min after genital contact is broken off (Jarman, 1982). The spermatophore of *H. crassidens* is relatively small, but it pushes out the subgenital plate sufficiently for it to be seen beneath. The female gets access to the spermatophore by bending around while holding on to her ovipositor with some of her legs and pulling open the subgenital plate with her mandibles (Jarman, 1982). Female *H. crassidens* also tend to leave the spermatophore in place longer if they are kept apart from males for some weeks or months before they are allowed to mate, compared with females that mate once or more a night (P.R.L. Cary, Christchurch, 1998, personal communication).

Gwynne (1997) and Brown and Gwynne (1997) have pointed out that, amongst the Orthoptera, ensiferans have specialized in producing large spermatophores, which are eaten after copulation. These are called nuptial gifts in so far as they constitute a male resource cost, and could serve as an ejaculate protection mechanism and/or an investment in the paternity of the offspring (Thornhill and Alcock, 1983; Gwynne, 1997). However, the function of the spermatophore meal has been open to much debate and hypothesis testing. The result has indicated that different orthopteran species use the structure for various solutions to problems raised by intersexual conflict. Gwynne (1997) concluded that the ancestral condition was a small spermatophore, which functioned as an ejaculate protection mechanism by distracting the female from mating with other males.

The small size of the spermatophore eaten by female *Hemideina* suggests that tree wetas have retained the primitive state. Indeed, the possibility that the tree weta spermatophore acts as a distracting mechanism from further mating with other males is diminished by Sandlant's (1981) observation that a female *H. femorata* consumed a spermatophore while being courted by another male, with which she subsequently mated within 1 min. A similar primitive condition is found in the Haglidae (*Cyphoderris*), two tettigoniids (*Decticus verrucivorus, Poecilimon veluchianus*) and a cricket (*Gryllodes supplicans*) (Gwynne, 1997). Because female giant wetas do not eat the spermatophore, a different tactic is used by males to protect their sperm investment by preventing the female from mating with other males (see 'Postcopulatory Behaviour', below).

Both tree wetas and giant wetas undergo multiple mating with the same female. This not only allows the male to fertilize more eggs in the female, but it possibly also acts to protect the male's investment by the action of chemical contents in the spermatheca. In a number of crickets and tettigoniids, such compounds act on the female nervous system to attenuate receptivity to male calling songs and to slow female movements (reviewed by Brown and Gwynne, 1997).

Chemical Releasers of Male Mating Behaviour

Abundant descriptions of the distinctive musky smell of tree wetas and giant wetas have been published, and commonly this smell has been described as a pheromone, without supporting evidence (e.g. Barrett, 1991; Gibbs, 1998). The odour is also characteristic of the faecal pellets, and it is not clear whether wetas detect and respond to this odour or to other(s) not detectable by humans. Nevertheless, there are two likely cases for pheromones. The first is a volatile substance produced by female giant wetas, which attracts males as the female wanders about (see 'Mate Attraction', above). There is no evidence that this specifically releases male mating behaviour. The second is a much less volatile (or non-volatile) substance present on female cuticle of all species of *Hemideina*, which appears to elicit male mating behaviour. The material's low volatility was shown in experiments using a container in which three live females were separated by a gauze partition (thus reducing the visibility of the females) from an adjacent arena into which single test males were introduced. By scoring the position of the ten males each minute for 30 min, with the females present or absent, Jarman (1982) found that there was no tendency for males to be attracted toward the female end of the arena ($\chi^2 = 1.42$, $P > 0.1$). Furthermore, he showed that there was no significant difference in the frequency of occurrence of male behaviour (duration of immobility, grooming, 'antennate-palpate') with females present or absent in the adjacent container ($\chi^2 = 3.44$, $P > 0.1$). When the partition was removed, males readily commenced mating behaviour upon contacting the females with their antennae (Jarman, 1982).

Jarman (1982) showed that this material is a chemical that can be washed off the female cuticle (*H. femorata*) with diethyl ether. Six freshly dead females were prepared: three were unwashed and three were triple-washed in fresh changes of ether and dried for 2 h. A washed or an unwashed female was presented to each of 12 males in random order, separated by 15 min intervals. The tests were carried out on two sequential days (48 trials in total). In 22 out of 24 trials with unwashed females, males showed copulatory behaviour ('apply genitalia') within 1 min, while, in 24 trials with the washed females, only four males showed 'apply genitalia' behaviour (three males responded only after the dead female was moved repeatedly next to the male). Clearly, the diethyl ether had removed the active compound(s) from the females. In further experiments with *H. crassidens* (L.H. Field and T.H. Jarman, unpublished data), we have found that filter-paper that lined shelter vials that housed females for 1 week accumulated a bioactive material detected by males. By testing male responses to filter-paper placed in the middle of a straight walking platform, we showed that 12 males stopped walking and spent an average of 4.2 min ('antennate–palpate' behaviour) on female-exposed filter-papers, while their behaviour towards the control (sterilized, clean) filter-papers was to either: (i) to continue walking over the paper; or (ii) to briefly stop and spend an average of 0.4 s in 'antennate–palpate' behaviour.

In experiments designed to determine which sensory appendages are used by males to detect females, Jarman (1982) excised antennae or maxillary and labial palps or cerci of males, either singly or in combination. The results (Table 17.3) indicated that no one pair of sensory appendages is responsible for male detection of the female pheromone. All males showed some degree of sensory deprivation, and all but one also showed post-operation copulatory behaviour when exposed to females. The latencies for the positive responses were longer than those for control males (which underwent CO_2 narcosis but no operation). The five out of 19 negative responses to females were from one doubly operated male lacking antennae and palps. This male reacted as if startled by the initial contact with females and responded aggressively ('mandible gape', 'elevate body'). It avoided the females thereafter. The conclusion from this series of experiments was that pheromone-sensitive chemoreceptor sensilla must reside on all of the sensory appendages tested, as well as elsewhere on the body. Detailed descriptions of the hair sensilla are found in Field (Chapter 22, this volume).

Tactile Component Contributes to Male Mating Behaviour

Once a male *Hemideina* has begun courtship behaviour towards a live female or a dead unwashed female, it is possible to induce 'apply genitalia' behaviour by applying gentle movements

Table 17.3. Effect of excision of sensory appendages from male tree wetas (*H. crassidens*) on their copulatory response when presented with females.

Appendages removed	No. animals	No. tests with females	No. copulatory responses
Antennae	3	12	12
Labial and maxillary palps	3	11	11
Antennae and palps	6	19	14
Cerci	1	6	6
Control (males anaesthesized only)	7	28	27

against his dorsal abdominal surface with a small camel-hair brush (Jarman, 1982). The resulting behaviour consisted of upward movements of the male genitalia against the brush. This purely tactile stimulus component was 100% effective in 20 trials performed on ten males (*H. crassidens*). When the brush was applied to the ventral side of aroused or non-aroused males, they briefly inspected the brush ('antennate–palpate') and departed. When the brush was applied to males that had not courted females, they showed defensive leg raising behaviour or 'kick–push' and departed. Thus the tactile stimulus facilitated release of copulatory behaviour only after the male had been aroused by presumably chemical contact with a female and only when applied dorsally.

Inappropriate Mate Choice: Interspecific and Homosexual Mating

Interspecific mating

Several workers have reported that cross-species mating readily occurs amongst species of *Hemideina* when kept in captivity (Richards, 1992; Field, 1993; see also Morgan-Richards *et al.*, Chapter 7, this volume). The phenomenon is likely to occur in the natural habitat as well, because it is not unusual to find members of two species sharing the same gallery in zones of parapatry (Trewick and Morgan-Richards, 1995). The genetic consequences and examples of hybrid offspring from such matings are discussed by Morgan-Richards *et al.* (Chapter 7, this volume). The implications for the present discussion relate to an apparent lack of species distinction in mating-releasing pheromones produced within the genus *Hemideina*. In selected cross-matings, Morgan-Richards *et al.* (Chapter 7, this volume) found that four of the possible five parapatric pairs of species mated, while the combination of *H. crassidens* and *H. femorata* did not. The significant feature of the latter pair is that *H. crassidens* is primarily a North Island species, while *H. femorata* is restricted to the South Island of New Zealand. These two species may have had a longer evolutionary history of separation than the other parapatric pairs, which presumably coevolved on one of the two islands.

The apparent lack of species-distinct chemical releasers for mating in tree wetas suggests that there is a potential for hybridization wherever species overlap. Some of the laboratory cross-matings (but not all) produced hybrids, and hybrids have also been found in the natural habitat. However, the hybrids were generally infertile and species integrity is apparently maintained in the natural habitat, because genetic studies do not show gene flow between species. Nevertheless, hybrid zones have been demonstrated in a number of orthopteran species (Chapco, 1997), so the question arises whether surface-borne compounds which act as mating signals are generalized amongst species in other genera besides *Hemideina*. Little is known of the diversity and function of contact pheromones within and between orthopteran species (Brown and Gwynne, 1997).

Apart from a single observation of a male *D. rugosa* unsuccessfully attempting to mate with a female *D. fallai* in captivity, nothing is known about interspecific mating in *Deinacrida* (Brown, 1995).

Homosexual mating behaviour

In tree wetas, male homosexual mating behaviour was often observed in *H. femorata* and *H. crassidens* (Sandlant, 1981; Jarman, 1982). The behaviour consisted of 'apply genitalia' to different regions of the recipient male's body. Usually the recipient male moved away or engaged in mild aggression behaviour, which discouraged the mounting male. Homosexual mating has also been observed in *D. mahoenui* and *D. heteracantha* (Richards, 1973; Richards, 1994).

Males in at least eight orders of insects besides the Orthoptera are known to engage in 'homosexual' mating behaviour (Thornhill and Alcock, 1983). Often this is due to an extremely low threshold for sexual excitation in males, combined with a response generalization to broad sensory (particularly visual) cues, rather than a response to specific details of female patterns. Such is the case in some butterflies and wasps (reviewed in Thornhill and Alcock, 1983). In other cases, males respond to general cues, such as larger size, since females may be larger than males, and hence large males are the subject of homosexual mating attempts (as in certain corixid bugs: Aiken, 1981). Finally, males may attempt to mate with objects that become contaminated with sexual pheromones from females. This may be a likely explanation for the homosexual mating observed in tree wetas, inasmuch as males and females move in and out of narrow galleries and the chances for transfer of female pheromone from the gallery walls on to males are certainly high.

While male homosexual mating attempts seem inappropriate at first glance, there may be evolutionary advantages for such a strategy. For example, the time spent by a male in waiting until a clear determination is made of the sex of a potential mate may be far longer than the time required for attempting to mate with any likely adult and rapidly receiving signals that clarify the sex of the receiver. Such a strategy is especially successful if females are more abundant than males. An example is provided by the digger bee, *Centris pallida*, where males emerge before females and therefore have a greater than 50% chance of encountering females when males dig up and try to mate with emerging individuals (Alcock *et al.*, 1977). The time spent in inappropriate mating attempts is kept minimal by curtailing the duration of contact, so males can afford to make mistakes. It would be enlightening to confirm this possibility in tree weta mating systems.

Postcopulatory Behaviour

Two features of postcopulatory behaviour in New Zealand wetas are of great interest. In tree wetas, males may become 'aggressive' towards females in what appears to be a tactic contradictory to principles of parental investment. In giant wetas, males remain with females for extremely long periods and appear to be protecting their sperm investment through their passive presence (since they do not fight with potential competitors).

Tree weta postcopulatory behaviour

Several different behaviours follow copulation in tree wetas, but the most consistent event observed was where the male stays with the female for a varying time. Jarman (1982) observed that such male guarding behaviour occupied a mean of 36.5 min (range 5.8–80 min, n = 11) in *H. crassidens*. During this period, the male stood over the female with his palpi and one or two pro- or mesothoracic legs in contact with her. This is consistent with the proposal by Gwynne (1997) that, when a large spermatophylax is absent, males protect their parental investment by guarding the female from insemination by other males. The unusual and little understood behaviour of apparent male 'aggression' towards the female occurs when the female attempts to depart during the guarding period. In *H. crassidens*, the males displayed aspects of behaviour normally seen in highly escalated agonistic battles with other males, as shown in Table 17.4. The behavioural units, characteristic of level IV male–male battles (see Field, Chapter 18, this volume), included 'stridulate', 'mandible gape', 'grapple' and 'lunge'.

Although the behaviour appeared to be the same as that seen in male–male agonistic encounters, care must be taken to allow other interpretations regarding male messages sent to females during postmating interactions. The result of such male behaviour usually involved departure of the female. The obvious conclusion is that the female was available for mating with other males and that the original male's paternal investment was not protected. Alternatively, male 'aggressive' behaviour towards females may involve a conflict of intention, possibly induced by crowded captive colony conditions. For example, Spencer (1995) found that the 'aggressive' behaviour was only displayed by large tenth (final)-instar males and that one such male attacked a female that had entered the gallery he had occupied. He pulled her out, 'stridulated' and 'lunged' at her, causing her to depart, and then occupied the gallery. The conflict

Table 17.4. Occurrence of 'aggressive' behaviour by male *H. crassidens* towards females before and following copulation. Such behavioural units normally occur in highly escalated battles between intruder and resident males for access to galleries containing females. (L.H. Field and T.H. Jarman, unpublished data from 30 male–female interactions.)

	Percentage occurrence (n = 30)	
Behaviour	Precopulatory	Postcopulatory
Grapple	0	13
Stridulate	3	43
Elevate	3	20
Gape mandibles	7	43
Lunge	3	23

between aggressively defending the gallery and guarding the female was apparently resolved to the detriment of his parental investment in this case.

Other postcopulatory behaviours in tree wetas were tabulated by Spencer (1995) for 42 mating interactions in *H. crassidens* (Table 17.5). The data emphasize a relatively low rate of male 'aggression' towards females (7%) and a much higher frequency of male guarding (45.5%, including remating attempts). A further illuminating feature is the high frequency of female or male departures immediately after mating. This implies that a conflict exists between a paternal investment strategy by males and a broadcast investment with as many females as possible. Detailed research into these aspects of tree weta mating systems is needed.

Giant weta and ground weta postcopulatory behaviour

The lack of female consumption of the spermatophore in the giant wetas is correlated with a male mate-guarding strategy to protect the ejaculate and to prevent sperm competition from other males. Males stay with the mated female for long times and undergo repeated copulations. Richards (1973) noted that *D. heteracantha* males normally do not part from the female during the 7–12 h pairing and that the male dominates the female to keep her in place. If the female attempts to wander once the pair is disturbed, the male immediately searches for her and remates. Radio-tracking studies showed that mated *D. rugosa* remain together for days, particularly in adverse weather conditions. A male remates with the female on subsequent nights and stays with her for up to 2 weeks. She presumably oviposits and the male breaks off during this period. The extent to which this ploy represents male guarding against mating by the female with other males is unknown, but such pairs are not found in association with other males during the postmating period (M. McIntyre, Wellington, 1999, personal communication). A similar prolonged mating period and possible mate-guarding behaviour was observed in captive Raukumara tusked wetas. One male remained with a female for 19 days, during which he copulated frequently and always remained with the female in the same cavity under a stone. The male then moved away from the female and died within 2

Table 17.5. Occurrence of different behaviours following successful copulation in *H. crassidens* (data after Spencer, 1995).

Behaviour		Percentage occurrence
Attempt to remate within 10 min		10
Aggression by male towards female		7
Female departs immediately		12
Male departs immediately		35.5
Male guards female		35.5
	Total matings	n = 42

days, and the female moved in with a second male. This male mated repeatedly with the female and remained with her for a further 37 days (McIntyre, 1998).

Very few data exist on the detailed interactions between male and female wetas in their natural habitat. Observations from radio-tracking is limited to large species of *Deinacrida* and *Motuweta*, discussed above, but it promises to be a useful means of uncovering many of the incompletely known aspects of mating behaviour in these remarkable insects. Many of the giant wetas are rare and endangered, so such knowledge is crucial to their effective conservation management in New Zealand.

References

Aiken, R.B. (1981) The relationship between body weight and homosexual mounting in *Palmacorixa nana* Walley (Heteroptera: Corixidae). *Florida Entomologist* 64, 267–271.

Alcock, J., Jones, C.E. and Buchanan, S.L. (1977) Male mating strategies in the bee *Centris pallida* Fox (Hymenoptera: Anthophoridae). *American Naturalist* 111, 145–155.

Asher, G.W. (1977) Ecological aspects of the common tree weta (*Hemideina thoracica*) in native vegetation. BSc thesis, Victoria University of Wellington, Wellington, New Zealand.

Barrett, P. (1991) *Keeping Wetas in Captivity.* Wellington Zoological Gardens, Wellington, New Zealand, 60 pp.

Bateman, P. and Toms, R.B. (1998) Olfactory intersexual discrimination in an African King cricket (Orthoptera: Mimnermidae). *Journal of Insect Behavior* 11, 159–163.

Boake, C.R.B. (1984) Male displays and female preferences in the courtship of a gregarious cricket. *Animal Behavior* 32, 690–697.

Brown, J. (1995) Behaviour of captive Poor Knights giant weta (*Deinacrida fallai*). BSc thesis, Victoria University of Wellington, Wellington, New Zealand.

Brown, W.D. and Gwynne, D.T. (1997) Evolution of mating in crickets, katydids and wetas (Ensifera). In: Gangwere, S.K., Muralirangan, M.C. and Muralirangan, M. (eds) *The Bionomics of Grasshoppers, Katydids and their Kin.* CAB International, Wallingford, pp. 281–314.

Cary, P.R.L. (1981) The biology of the weta *Zealandosandrus gracilis* (Orthoptera: Stenopelmatidae) from the Cass Region. Unpublished MSc thesis, University of Canterbury, New Zealand.

Chapco, W. (1997) Molecular evolutionary genetics in orthopteroid insects. In: Gangwere, S.K., Muralirangan, M.C. and Muralirangan, M. (eds) *The Bionomics of Grasshoppers, Katydids and their Kin.* CAB International, Wallingford, pp. 337–354.

Domett, E. (1996) Reproduction and behaviour of the Mahoenui Weta, *Deinacrida* n. sp. Unpublished MSc thesis, Massey University, New Zealand.

Emlen, S.T. and Oring, L.M. (1977) Ecology, sexual selection and the evolution of mating systems. *Science* 197, 215–223.

Field, L.H. (1980) Observations on the biology of *Deinacrida connectens* (Orthoptera: Stenopelmatidae), an alpine weta. *New Zealand Journal of Zoology* 7, 211–220.

Field, L.H. (1993) Observations on the stridulatory, agonistic, and mating behaviour of *Hemideina ricta* (Stenopelmatidae: Orthoptera), the rare Banks Peninsula weta. *New Zealand Entomologist* 16, 68–74.

Field, L.H. and Rind, F.C. (1992) Stridulatory behaviour in a New Zealand weta, *Hemideina crassidens. Journal of Zoology, London* 228, 371–394.

Field, L.H. and Sandlant, G.R. (1983) Aggression and mating behaviour in the Stenopelmatidae (Orthoptera, Ensifera), with reference to New Zealand wetas. In: Gwynne, D.T. and Morris, G.K. (eds) *Orthopteran Mating Systems – Sexual Competition in a Diverse Group of Insects.* Westview Press, Boulder, Colorado, pp. 120–146.

Gibbs, G.W. (1998) Why are some weta (Orthoptera: Stenopelmatidae) vulnerable yet others are common? *Journal of Insect Conservation* 2, 161–166.

Gwynne, D.T. (1997) The evolution of edible 'sperm sacs' and other forms of courtship feeding in crickets, katydids and their kin (Orthoptera: Ensifera). In: Choe, J.C. and Crespi, B.J. (eds) *The Evolution of Mating Systems in Insects and Arachnids.* Cambridge University Press, Cambridge, 1997, pp. 110–129.

Gwynne, D.T. and Jamieson, I. (1998) Sexual selection and sexual dimorphism in a harem-defending insect, the alpine weta (*Hemideina maori*, Orthoptera: Stenopelmatidae). *Ethology, Ecology and Evolution* 10, 393–402.

Hinton, H.E. (1946) The 'gin traps' of some beetle pupae; a protective device which appears to be unknown. *Transactions of the Entomological Society of London* 97, 473–496.

Hudson, G.V. (1920) On some examples of New Zealand insects illustrating the Darwinian principle of natural selection. *Transactions and Proceedings of the New Zealand Institute* 52, 431–438.

Jarman, T.H. (1982) Mating behaviour and its releasers in *Hemideina crassidens* (Orthoptera: Stenopelmatidae). Unpublished BSc Honours thesis, University of Canterbury, New Zealand.

Johns, P.M. (1997) The Gondwanaland weta: family Anostostomatidae (formerly in Stenopelmatidae, Henicidae or Mimnermidae): nomenclatural problems, world checklist, new genera and species. *Journal of Orthoptera Research* 6, 125–138.

Krebs, J.R. and Davies, N.B. (1978) *Behavioural Ecology: an Evolutionary Approach.* Blackwell Scientific Publications, Oxford, 494 pp.

McIntyre, M. (1998) *Raukumara Tusked Weta: Field and Captive Observations.* Conservation Advisory Science Notes No. 219, Department of Conservation, Wellington, 35 pp.

Ordish, R. (1992) Aggregation and communication of the Wellington weta *Hemideina crassidens* (Blanchard) (Orthoptera: Stenopelmatidae). *New Zealand Entomologist* 15, 1–8.

Ramsay, G.W. (1955) The exoskeleton and musculature of the head, and the life cycle of *Deinacrida rugosa* Buller, 1870. MSc thesis, Victoria University of Wellington, Wellington, New Zealand.

Richards, A.O. (1973) A comparative study of the biology of the giant wetas *Deinacrida heteracantha* and *D. fallai* (Orthoptera: Henicidae) from New Zealand. *Journal of Zoology* 169, 195–236.

Richards, G.E. (1994) Ecology and behaviour of the Mahoenui giant weta, *Deinacrida* nov. sp. MSc thesis, Massey University, Palmerston North, New Zealand.

Richards, M. (1992) Observations of matings between two species of tree weta. *The Weta. News Bulletin of the Entomological Society of New Zealand* 15, 15–16.

Ryker, L.C. and Rudinsky, J.A. (1976) Sound production in Scolytidae: aggressive and mating behavior of the mountain pine beetle. *Annals of the Entomological Society of America* 69, 677–680.

Sandlant, G.R. (1981) Aggressive behaviour of the Canterbury weta *Hemideina femorata* (Orthoptera: Stenopelmatidae): its adaptive significance in resource allocation. MSc thesis, University of Canterbury, New Zealand.

Simmons, L.W. (1986) Female choice in the field cricket, *Gryllus bimaculatus* (De Greer). *Animal Behavior* 34, 1463–1470.

Spencer, A.M. (1995) Sexual maturity in the male tree weta *Hemideina crassidens* (Orthoptera: Stenopelmatidae). MSc thesis, Victoria University of Wellington, Wellington, New Zealand.

Thornhill, R. and Alcock, J. (1983) *The Evolution of Insect Mating Systems.* Harvard University Press, Cambridge, Massachusetts, 547 pp.

Townsend, J.A. (1995) Distribution and ecology of the Banks Peninsula tree weta, *Hemideina ricta.* MSc thesis, Massey University, Palmerston North, New Zealand.

Trewick, S. and Morgan-Richards, M. (1995) On the distribution of tree weta in the North Island of New Zealand. *Journal of the Royal Society of New Zealand* 25, 1–9.

van Wyngaarden, F. (1995) The ecology of the Tekapo ground weta (*Hemiandrus* new sp.; Orthoptera: Anostostomatidae) and recommendations for the conservation of a threatened close relative. MSc thesis, University of Canterbury, Christchurch, New Zealand.

18
Aggression Behaviour in New Zealand Tree Wetas

Laurence H. Field
Department of Zoology, University of Canterbury, Private Bag 4800, Christchurch, New Zealand

Introduction

One of the most fascinating aspects of keeping tree wetas in captivity is to observe wandering males competing aggressively with other males occupying galleries. In this nightly scenario, an intruder male approaches a gallery entrance, which is blocked by the spiny hind tibiae of the resident male. After exploring the entrance with several ritualized behaviour patterns, the intruder deliberately opens his powerful mandibles and carefully secures a grip on one of the hind legs of the resident (Fig. 18.1). Then a persistent tug of war follows as the intruder pulls and strains in its attempts to extract the resident, who in turn kicks at the intruder and braces itself in the gallery. If the intruder is successful, the two males may then square off outside the gallery and engage in a ritual battle, which may escalate into grappling (Fig. 18.2), stridulation by either or both males and threat displays with wide-open mandibles. If neither yields, potentially injurious contact follows, as they lunge with open mandibles and attempt to outgrip the opponent's mandibles. Eventually, one male departs or is thrown off the tree-trunk; and the victor enters the gallery, where it stridulates repeatedly while occupying the entrance.

Many interesting questions are raised by this behaviour between males. What is its function and does it always occur in association with galleries? Is it driven by sexual selection and is it ritualized to reduce the risk of injury? How do the males decide who wins? In this chapter, these and other questions are addressed by reviewing the limited extant literature and by providing new data from studies of aggression behaviour in the tree wetas *Hemideina femorata* and *Hemideina crassidens*.

Aggression is defined as behaviour adapted to cause or to convey the threat of causing physical injury to another conspecific individual for the purpose of competing for a limiting resource (Hinde, 1974; Brown, 1975). In the case of tree

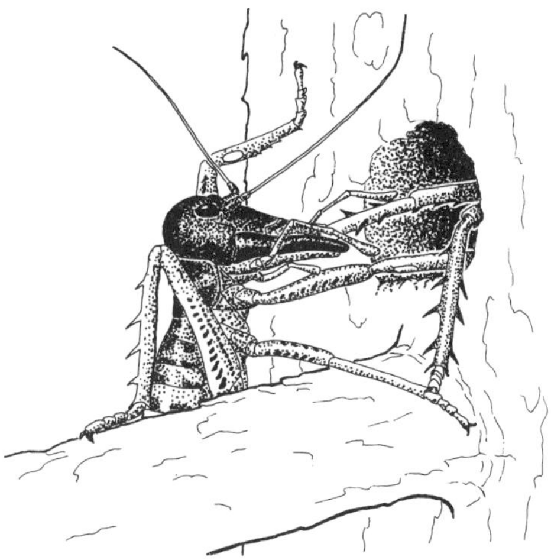

Fig. 18.1. Drawing from photograph of 'bite tibia' behaviour as intruder male tree weta (*H. femorata*) pulls on the hind leg of the resident in an attempt to evict it and gain access to the gallery (laboratory enclosure).

Fig. 18.2. Photograph of two male *H. femorata* engaged in 'grapple' behaviour. 'Grapple' appears to allow the contestants to assess physical factors, such as body weight and strength, which may enable decisions about winning to be made without risk of injury through escalation.

wetas, the gallery is a limited resource (made by cerambycid beetle larvae; see Field and Sandlant, Chapter 13, this volume), which provides diurnal refuge from predators. However, analyses of the frequency of gallery occupation by tree wetas, compared with total gallery availability in the forest habitat, showed that galleries are not a limiting resource; that is, available galleries are sometimes not occupied by wetas (Sandlant, 1981; Field and Sandlant, Chapter 13, this volume). Therefore, it becomes difficult to interpret aggressive behaviour of tree wetas in terms of competition for galleries solely as refuges from predators. In a previous review, Field and Sandlant (1983) showed that aggression occurring at night between adult male tree wetas represents agonistic competition for galleries that contain females. In this sense, the aggressive behaviour supports and drives sexual selection for enhanced secondary sexual characters, such as the enlarged head capsule and mandibles of males (reviewed in Field and Deans, Chapter 10, this volume). However, it was also shown that as dawn approaches, male and female tree wetas in crowded conditions display aggression towards each other as they compete for access to galleries. In this sense, aggression represents competition for galleries as diurnal refuges.

Aggression Behaviour Patterns

In addition to Sandlant's (1981) original laboratory study of a colony of *H. femorata*, we have carried out four more studies of aggression behaviour of *H. femorata* and *H. crassidens* in a similar laboratory enclosure (850 mm × 570 mm × 320 mm), where gallery availability could be altered and behavioural sequences could be recorded by video camera through the glass front with 15 W red light illumination. The wetas were maintained in a reversed light cycle (12D : 12L) at 15°C, and were provided with leaves of native shrubs, carrots and water. Personal observations (L.H. Field) of aggressive behaviour in the field at night and those of Möller (1985) have confirmed that similar behaviour occurs in the natural habitat.

Seventeen behavioural units have been identified as components included in, but not necessarily limited to, aggressive interactions. These are arranged below in order of increasing intensity of fighting and risk of injury:

- *Antennate–palpate*. Alternating movements of antennae (one to two per second) and mandibular and maxillary palps (three to five per second) to make contact with another weta.
- *Rest*. A resident remains motionless in the gallery entrance for up to several hours if undisturbed.
- *Emerge*. A resident emerges backwards from a gallery.
- *Depart*. The loser leaves the immediate scene of a fight.
- *Apply genitalia*. A male curls its abdomen forward and ventrad to make contact with another weta.
- *Step forward*. A resident weta moves further into the gallery.
- *Enter*. A weta enters a gallery.
- *Tremulate*. An intruder weta vibrates its abdomen dorsoventrally in short, rapid movements, which shake the whole body.
- *Stridulate*. A weta produces sound by short, rapid vertical movement of its abdomen against the hind femora.
- *Kick–push*. A resident weta braces its hind leg against an intruder and suddenly pushes it away in a single movement, or a resident gives several sharp kicks against an intruder.
- *Bite tibia*. The hind tibia of a resident is grasped by the mandibles of an intruder.

- *Pull*. An intruder weta pulls on the hind tibia of a resident in an attempt to extract the resident from a gallery.
- *Dislodge*. A weta is thrown off the tree-trunk as a result of grappling or kicking by the opponent.
- *Grapple*. Two males push against each other with a variable number of fore- and mid-legs, while gripping the vertical substrate with at least the hind legs.
- *Lunge*. A combatant raises itself slightly and suddenly thrusts its body forward 1–2 cm, directing its opened mandibles at the opponent.
- *Mandible gape*. A male opens its mandibles fully and displays them with slightly raised head towards the opponent. The gape is maintained for up to 15–20 s.
- *Bite head*. The mandibles of one male grip the head of the opponent above the subgenual suture or encompass the opponent's mandibles.

The above descriptions show remarkable similarity to behavioural acts of aggression in other Orthoptera. Kicking with the hind legs occurs in mild aggression interactions in crickets (e.g. *Acheta* sp. and *Teleogryllus oceanicus*) and grasshoppers (e.g. *Dissosteira carolina* and *Ligurotettix coquilleti*) (Alexander, 1961; Kerr, 1974; Otte and Joern, 1975; Burk, 1983). In higher-level aggressive interactions, the above insects all employ grappling with gaped mandibles, biting and lunging in a similar fashion to that seen in tree wetas. Crickets show much more intense aggressive chirping during a battle, while tree wetas show less stridulation during a battle and most stridulation by the winner after a battle.

Several of the behaviours in tree weta aggression could potentially cause physical injury to the combatants (e.g. 'kick–push', 'bite tibia', 'bite head'), while others could lead to accidental loss of grip and falling from the tree-trunk ('dislodge', 'lunge'). The resultant risk is that of exposure to nocturnal predators on the forest floor. Many of the other behaviours appear to be related to assessment of the opponent's size or ability to inflict injury. Suggested risk levels and potential for assessment of the opponent's aggressive intent are given in Table 18.1. Parker (1974) argued that natural selection should favour animals that can assess their opponents during a battle. If escalated aggressive behaviour brings with it greater risk of injury, assessment may allow the potential loser to withdraw unharmed. An example is 'mandible gape' behaviour, which appears to be ritualized to send a threatening signal to the opponent without actual injurious biting. If this is ignored, tree weta battles can potentially escalate to the point where

Table 18.1. Estimated levels of the risk of injury or of falling to the ground for behavioural units performed by a rival male combatant. The right-hand column indicates assessments that a male could make of the opponent's aggressive intent, and hence risk level, from each behavioural unit. The term 'crucial context of signal' means that, if the signal occurred at certain crucial points in the battle, a high risk level is signified. Although 'mandible gape' has the potential to cause serious injury if a rival bites the opponent, this rarely happens, and the behaviour instead appears to be a display that has become ritualized to prevent injury.

Risk of injury/falling	Behavioural unit	Possible assessment value
Low	Antennate–palpate	Rival's persistence
	Tremulate	Rival's persistence
	Stridulate	Crucial context of signal
Higher	Bite tibia	Rival's persistence
	Pull	Rival's persistence
	Grapple	Rival's mass/strength
Highest	Kick–push	Rival's mass/strength
	Dislodge	Rival's mass/strength
	Lunge	Mandible gape
	Bite head	Biting force
Ritualized display	Mandible gape	Crucial context of signal

high-risk behaviour, such as biting, is involved. The following section deals with a classification of fights into different levels according to the amount of escalation and risk of injury involved.

Levels of Aggression in Fights

Male tree wetas either guard the entrances of galleries containing females or they actively wander outside galleries, engaged in exploring, feeding, examining galleries and their occupants and fighting near gallery entrances. A study of *H. crassidens* showed that eighth- and ninth-instar males spend more time wandering than do large males in the tenth (final) instar (Spencer, 1995). Furthermore, those males living with females spent more time residing in the gallery than did males living alone. In contrast, females of both *H. femorata* and *H. crassidens* stayed mostly within galleries and only emerged to feed or defecate, after which they usually returned to their galleries (Sandlant, 1981; Field and Sandlant, 1983).

Male–male interactions away from galleries consisted of brief contact and departure and rarely included aggressive behaviour. The primary site of aggression was almost always the gallery entrance. Here, a typical interaction consisted of an intruder male attempting to extract a resident male that was facing inward and blocking the gallery entrance with its spiny hind tibiae. Sandlant (1981) arbitrarily classified these interactions into four levels of aggression, based upon successively greater escalation and rising risk of injury to the combatants (Table 18.2). Each level was characterized by several common behaviours, plus new ones that only occur upon escalation.

Behavioural sequences during fights at different levels

Flow diagrams were prepared from sequential analysis of unpublished data for 65 fights by *H. femorata*. In all encounters, the intruder initiated the action by 'antennate–palpate', if the resident was in the gallery entrance, or by 'enter', if the resident was inside the gallery. The flow diagram for level I (Fig. 18.3) shows that interactions consisted of behaviour with negligible risk of injury to the participants. Antennal contact by the intruder was usually followed by 'departure' of the intruder or by 'emerge' or 'stridulate' by the resident; occa-

Table 18.2. Percentage occurrence of behaviours associated with four arbitrary levels (I–IV) of increasing intensity of aggression between male tree wetas (*H. femorata*). The behaviours are arranged downwards in order of increasing risk of injury ($n = 347$ interactions). (L.H. Field, unpublished data pooled with those of Sandlant, 1981.)

	Aggression level			
Behaviour	I	II	III	IV
Antennate–palpate	100	100	100	100
Rest	86	88	86	87
Emerge	87	58	80	87
Depart	88	94	100	100
Apply genitalia	15	62	10	9
Step forward	4	20	9	50
Enter	6	12	14	50
Tremulate	5	7	22	10
Stridulate	5	10	14	13
Kick–push		45	88	100
Grapple		15	16	40
Pull		10	86	86
Dislodge			7	7
Bite tibia			79	49
Lunge			80	100
Mandible gape				13
Bite head				20

sionally the resident performed 'tremulate'. The resident's responses usually resulted in departure by the intruder, after which the resident 'stridulated'. Acoustic and vibratory signals were sent only by the resident. In level I interactions, the resident's intention to stay in the gallery is apparently high, while the aggressive intention of the intruder is apparently low.

In level II interactions the intruder's 'antennate–palpate' behaviour was usually answered by the resident's repeated 'kick–push', as its sharp tibial spines were jabbed at the intruder (Fig. 18.4). Residents often 'emerged' to 'kick–push' with added force, in which case the intruder was dislodged. Although not seen in this sample, residents sometimes 'stridulate' at this stage (Field and Sandlant, 1983). The residents usually retained possession of the gallery. However, the intruder showed more apparent interest than in level I, as evidenced by the repeated 'tremulate', 'enter', 'antennate–palpate' cycle shown in the upper part of Fig. 18.4.

Level III marks the development of aggressive behaviour by the intruder. The flow diagram (Fig. 18.5) is more complex, due to several cyclic sequences, in addition to a variety of outcomes, which arise at several critical decision points. In determined attempts to physically displace the resident, the intruder used its mandibles to 'bite tibia' and 'pull'. The resident usually resisted with 'kick–push' and 'step forward', which caused a loss of grip and repeat of the cycle through 'antennate–palpate' (the loop around 1 in Fig. 18.5 and sample fights in Fig. 18.6). A similar cycle at this stage included 'tremulate' (indicated by loop 2 in Fig. 18.5). The circle around 'kick–push' and resident 'stridulate' indicates that these behaviours occurred repeatedly in sequence. The first major decision point is marked by resident 'emerge' (being pulled out by the intruder). Three possible

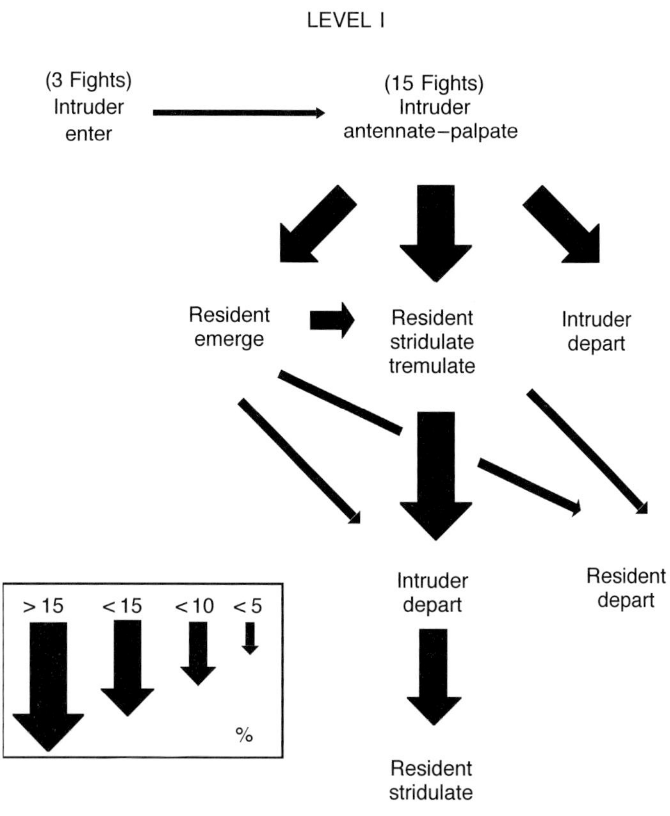

Fig. 18.3. Flow diagram for level I fights between male *H. femorata*. Behavioural units are defined in the text. The arrow thickness gives the percentage occurrence (inset) of each kind of behavioural transition out of all transitions between behaviour pairs recorded at this level; n = 18 fights.

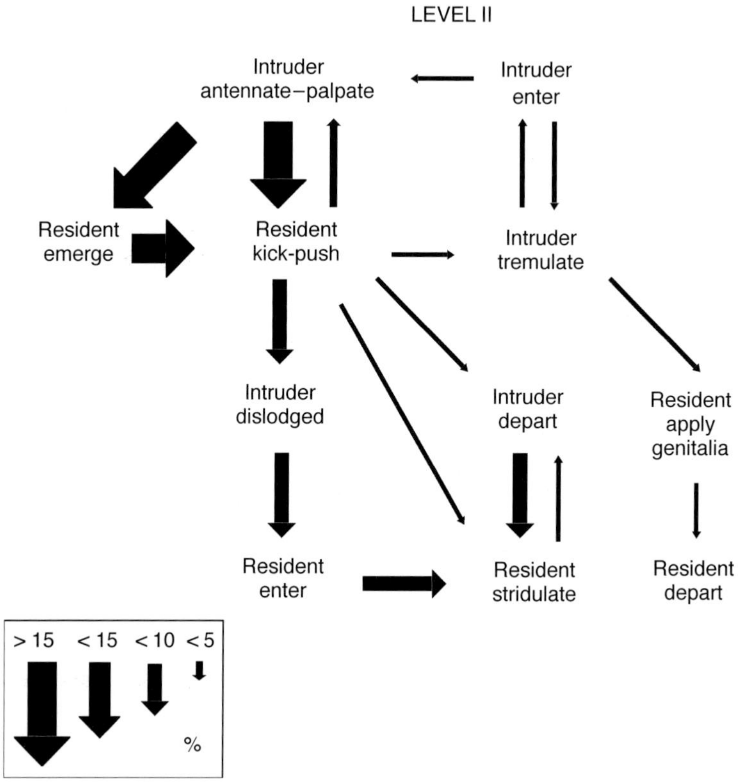

Fig. 18.4. Flow diagram for level II fights between male *H. femorata*. Behavioural units are defined in the text. The arrow thickness gives the percentage occurrence (inset) of each kind of behavioural transition out of all transitions between behaviour pairs recorded at this level; $n = 7$ fights.

outcomes resulted. The most frequent flow of events proceeded from resident 'kick–push' and/or resident 'stridulate', to intruder 'depart', resident 'enter' and 'stridulate' during 'rest in gallery'. This downward progression through Fig. 18.5 was often cycled through resident 'emerge', 'kick–push'/'stridulate', 'enter', intruder 'antennate–palpate' and 'bite tibia' more than once (the 'bite tibia' cycle). Nevertheless, in these cases the resident was the winner. An example of such a fight is shown in Fig. 18.6 (left side). A second, less frequent outcome involved intruder 'enter' and resident 'depart', in which case the intruder was the winner and 'entered' the gallery where it 'stridulated' (Fig. 18.6, right side). A third outcome involved intruder 'depart' following resident 'emerge'.

The second major decision point occurred at resident 'kick–push'. Either the fight was prolonged, directly or indirectly, through the 'bite tibia' cycle, as seen above, or it was brought to an end directly, by intruder 'depart' or 'dislodge' of intruder, or indirectly, by resident 'depart'. In either case, the winner 'entered' the gallery before or after 'stridulate'.

In level III fights, both participants appear to assess the rival's strength through behaviours such as 'pull', 'kick–push', 'step forward', 'bite tibia'. Each rival's aggressive intent is assessed through the degree of persistence in repeating the 'bite tibia' – 'pull' – 'step forward'/'kick–push' cycle and perhaps through the occurrence of 'stridulate' (discussed further in 'Role of Sound in Aggression' below).

Level IV interactions escalate from level III whenever the intruder 'lunges' at the occupant, and both opponents use their mandibles in the added behaviours 'mandible gape' and 'bite head'

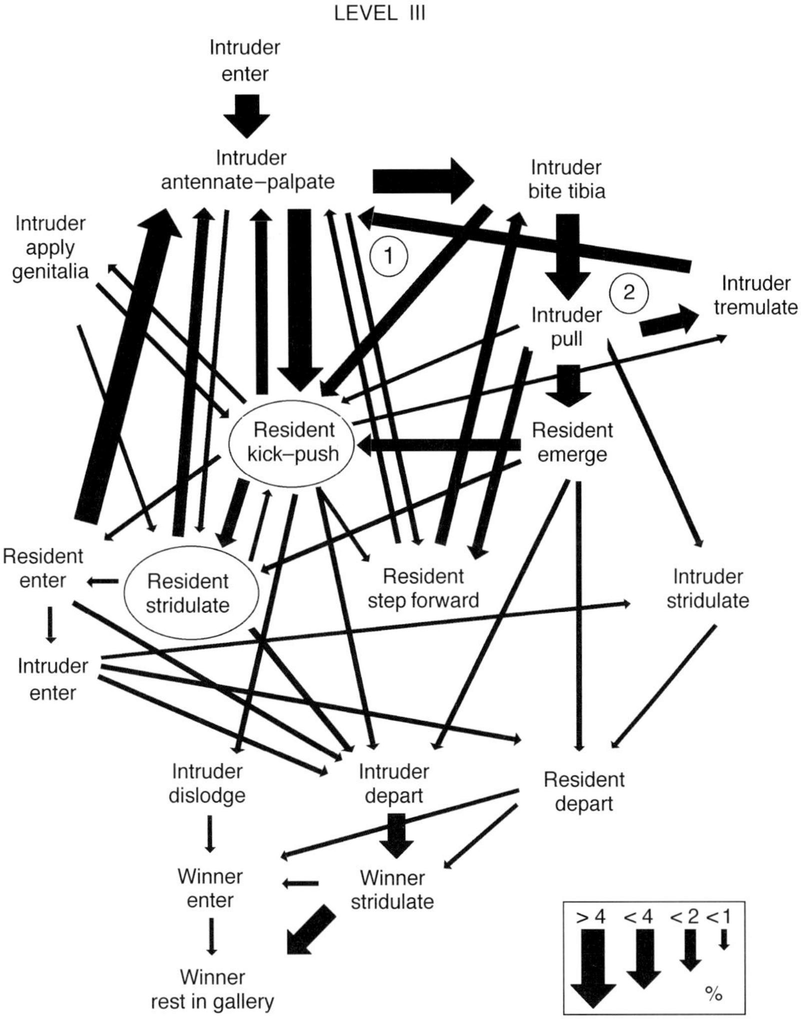

Fig. 18.5. Flow diagram for level III fights between male *H. femorata*. Behavioural units are defined in the text. The circled behaviours occurred repeatedly in sequence before a transition occurred. The arrow thickness gives the percentage occurrence (inset) of each kind of behavioural transition out of all transitions between behaviour pairs recorded at this level; $n = 27$ fights.

(Fig. 18.7). The same 'bite tibia' cyclic sequence occurs initially, as well as the intruder 'tremulate' cycle seen in levels II and III. The thick arrows at this level of the flow diagram confirm the persistence of the intruder in perpetuating the fight and the resistance of the resident to departure. Ultimately, the resident emerges and a major decision point is reached at 'grapple'. From here either combatant might depart, but the most common pathway leads into the complex 'mandible gape'/'lunge' interaction ('bite head' was not seen in the *H. femorata* sample of fights used for Fig. 18.7). This is also another decision point, since the outcome leads either to resident 'depart' (most frequent result), intruder 'depart', or both wetas may lose their grip on the tree and fall to the ground ('dislodge').

In level IV, there is potential for severe injury by the mandibles, since they are sharpened on the inner cutting edge and are powerful (see Fig. 10.8 in Field and Deans, Chapter 10, this volume). 'Mandible gape' is usually used as a threat display that can serve to reduce injury, but, when it escalates to 'bite head', the attacker appears to bite

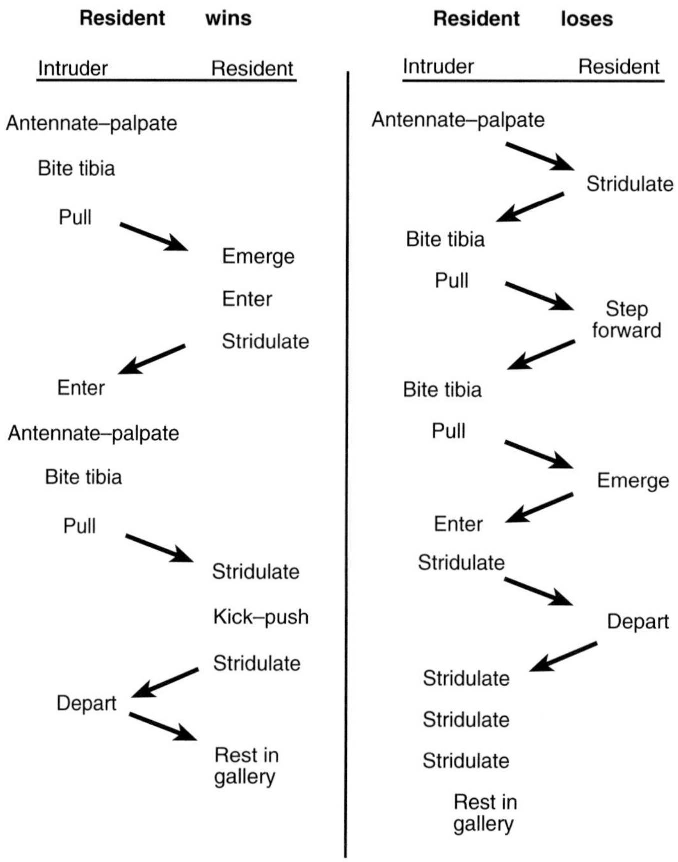

Fig. 18.6. Examples of level III fight sequences which occurred at gallery entrances. Each column indicates behaviour units performed by one or the other combatant. The arrows indicate the transition from one unit to the next. Left side: fight in which the resident wins. Note that the resident 'stridulated' and performed 'kick–push' repeatedly, apparently indicating strong intent not to vacate the gallery. Right side: fight in which the intruder wins. Note that the intruder stridulated and that the resident did not perform 'kick–push', suggesting greater aggressive persistence by the intruder.

with enough force to cause a clear withdrawal response by the defender. These battles occurred very rarely. In such escalation, the definitive signal to end the battle occurs when the attacker's mandibles enclose the defender's mandibles with 'bite head'. However, in three cases, one weta bit the head capsule of the other during violent jousting with gaped mandibles and caused puncturing of the head cuticle (Sandlant, 1981). The bitten weta always departed abruptly.

The homosexual attempts of the intruder to mate with the resident ('apply genitalia'), especially in level III interactions, may be an artefact of the higher than usual density of wetas in the laboratory colonies, wherein males could become marked with a female pheromone. Alternatively, it could point to a poor male–female signalling system for mating. The resident always responded to the intruder with aggressive behaviour ('kick–push', 'stridulate').

Departure rates of intruder vs. aggression level

A comparison of the intruder's frequency of departure versus the aggression levels of fights gives some idea of the intruder's persistence and, by implication, aggressive intent (Table 18.3). The data are taken from fights between *H. femorata* males. It is clear that intruders are less likely to

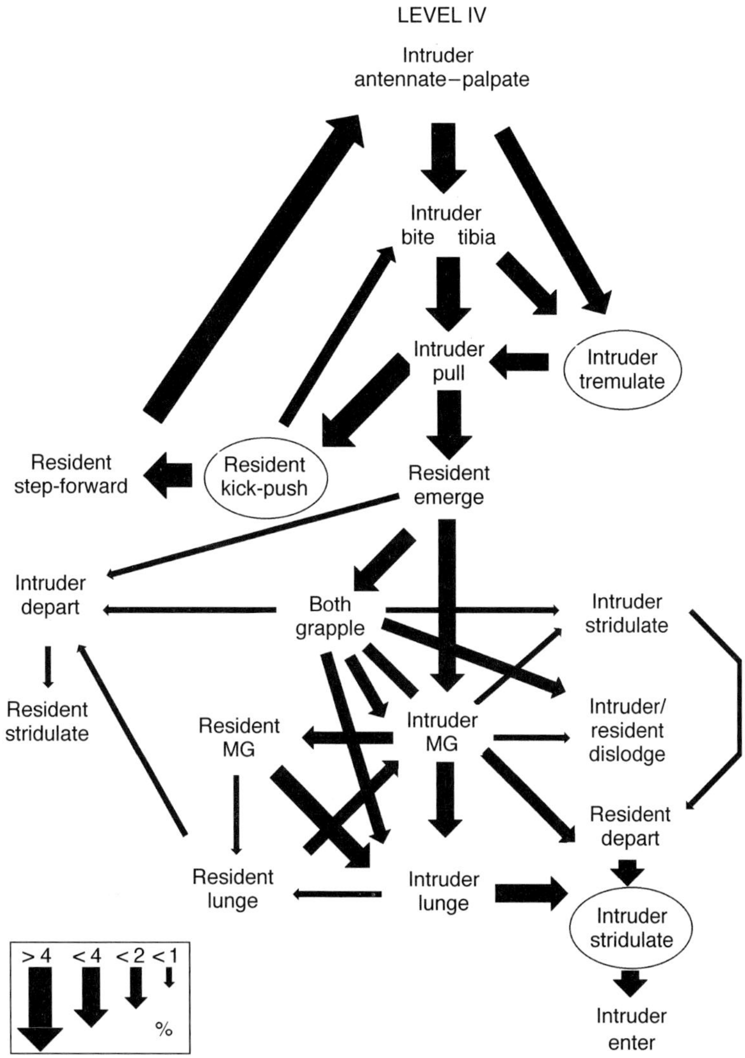

Fig. 18.7. Flow diagram for level IV fights between male *H. femorata*. Behavioural units are defined in the text. The circled behaviours occurred repeatedly in sequence before a transition occurred. The arrow thickness gives the percentage occurrence (inset) of each kind of behavioural transition out of all transitions between behaviour pairs recorded at this level; $n = 13$ fights. MG, mandible gape.

abandon a fight as the level of aggression increases. Correlated with this is the fact that larger males tend to win fights easily, while more evenly matched males tend to escalate fights. The same observation was made for fights between males in the South African tusked king cricket, *Libanasidus vittatus* (Bateman and Toms, 1998).

The unwillingness of an intruder to depart and its high aggressive intent appear to be driven by the intruder's search for females, rather than searching for available galleries. After winning a fight and evicting the resident, the intruder would explore the gallery briefly and if it did not contain a female, the intruder would depart. If a female was present in the gallery, the intruder usually attempted to pull the female out and copulate. This observation suggests that the basis of male aggression at the gallery site lies in competition

Table 18.3. Frequency of intruder departure for fights at levels I–IV, given as mean percentages for pooled data from studies of *H. femorata* (Sandlant, 1981; L.H. Field, unpublished data).

Aggression Level	Percentage of fights at each Level	Mean percentage of intruder departures	Number of interactions
I	32.10%	87%	119
II	22.40%	59%	111
III	35.30%	36%	120
IV	19.20%	45%	28

for females during the normal nightly activity period.

Role of Sound in Aggression

Many Orthoptera produce sounds during aggressive encounters, but the function and diversity of sound patterns differ (Loher and Dambach, 1989; Bailey, 1991). In the tree weta *H. crassidens*, winners produce an aggression sound pattern during a battle and after a battle has been won. The post-battle stridulation appears to announce the victorious status of that individual (Field and Rind, 1992). The following analysis extends to *H. femorata* as well, and broadens our understanding of the role of sound during a battle, as well as differences between species.

It is clear that the two species of tree weta differ markedly in the amount of aggression stridulation associated with fights. Only 10.9% of all fights by *H. femorata* had stridulation (pooled data, $n = 350$), while 55.8% of all fights by *H. crassidens* involved stridulation ($n = 39$). Within those fights which included stridulation, there was a trend to produce more stridulation as the aggression level of the fight increased. This is shown in Table 18.4 for the large data set from *H. femorata* fights.

Based upon the preceding discussion of the flow diagrams (Figs 18.3–18.5, 18.7) the source of stridulation should arise from both the resident (signalling increased resistance to removal from its gallery) and the intruder (signalling increased aggressive intent). However, analysis of stridulation by winner and loser males showed that it is primarily the winner that stridulates during fights (left half of Fig. 18.8). The trend is similar for both species of *Hemideina*, although there is some indication that *H. crassidens* males stridulate more during fights than do *H. femorata* males.

Two very clear patterns characterize sound production once a winner–loser decision has been made. First, the loser becomes silent once it has decided upon its status. This has now been observed for *H. femorata* and *Hemideina ricta*, in addition to previous reports for *H. crassidens* (Field and Rind, 1992). Secondly, a prolonged period of stridulation by the winner occurs once the loser has made its decision to decamp (Fig. 18.8, 'Winner After', right side). In this way, the winner apparently proclaims its victory status over the ensuing period of up to approximately 20 min, and may be sending the message that it is prepared to fight aggressively again if challenged. This surmise is borne out by the observation that subsequent meetings of the two combatants result in 'mandible gape', 'lunge' and 'stridulate' by the winner and 'depart' by the loser. A graphic demonstration of all three phenomena is shown in the running log of a fight in Fig. 18.9 (*H. crassidens*). The prospective loser (open circle

Table 18.4. Comparison of rates of stridulation by male *H. femorata* in fights of increasing risk of injury.

Aggression level	Percentage of fights with stridulation	*n*
I	5.0%	110
II	10.4%	107
III	15.6%	107
IV	21.3%	23

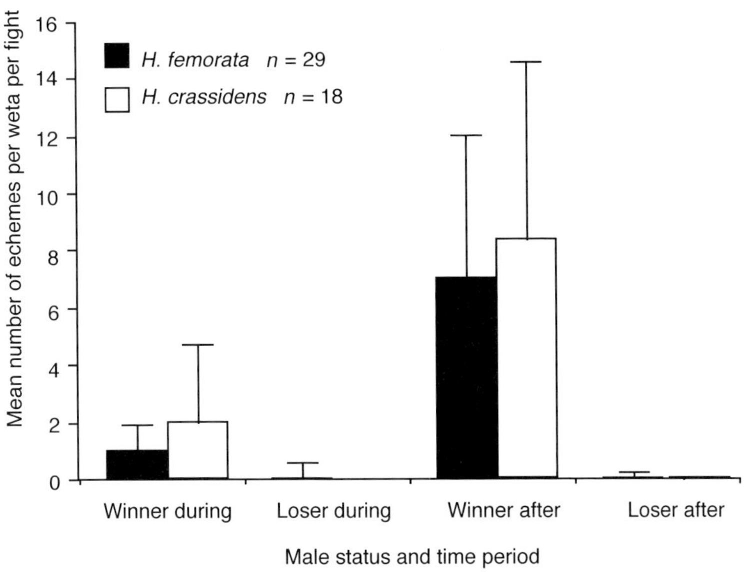

Fig. 18.8. Mean rate (± SD) of sound production by winner and loser males during and after fights for two species of tree weta. An echeme is a series of syllables produced as one stridulatory bout and repeated erratically. Stridulation is very rarely produced by losers after a fight.

symbols) initially made three threats, which consisted of stridulatory bouts prior to the actual fight at time = 160 s. The prospective winner then rapidly escalated the battle and stridulated vigorously until, at time = 220 s, the loser made a decision to flee and immediately became silent thereafter. The winner then began the prolonged phase of stridulation following the fight. At time = 300 s, upon approach by the loser, the winner displayed 'mandible gape', 'lunge' and 'stridulate' and the loser fled.

If the previous allusion to the message inherent in stridulation by the winner is correct, it should be possible to show that stridulation is correlated with some measure of aggressive intent of the winner. This was done for the data set from Fig. 18.8, using the aggression level of fights as the measure of aggressive intent of the winner, for the two

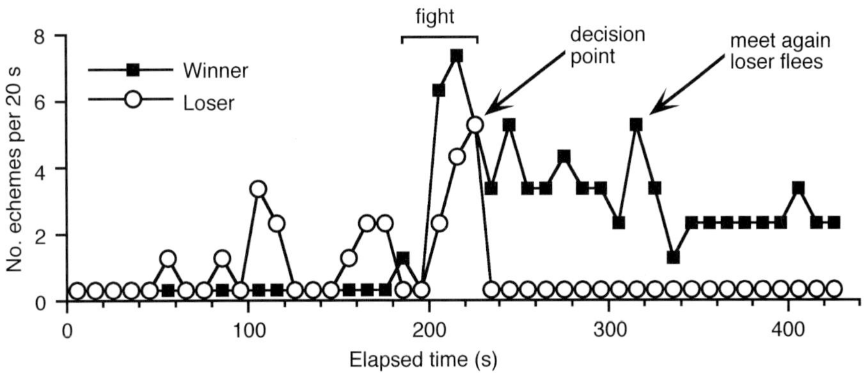

Fig. 18.9. Running log of a fight by two male *H. crassidens* plotted as rate of echeme production vs. elapsed time. Note abrupt cessation of all stridulation by the loser once it fled, and the prolonged stridulation by the winner following the fight. A subsequent meeting of the two wetas resulted in threat displays and increased rate of stridulation by the winner and flight of the loser. (After Field and Rind, 1992, with permission.)

Table 18.5. Mean number of stridulatory bouts (echemes) made by winners and losers of fights, which were classified into four levels of aggression (I–IV), according to the risk of injury. Two species of tree wetas, *H. femorata* and *H. crassidens,* were compared. Note that the potential losers were usually silent during and after fights in both species, while winners were more vociferous during fights, especially in *H. crassidens.* A clear increase in winner sound production occurs as the level of fight increases in *H. crassidens.*

	Hemideina femorata (*n* = 29 fights): Mean number of echemes				*Hemideina crassidens* (*n* = 18 fights): Mean number of echemes			
	During fight		After fight		During fight		After fight	
	Winner	Loser	Winner	Loser	Winner	Loser	Winner	Loser
Level I	0.22	0	3.67	0.44	1	0	2.75	0
Level II	0.0	0	4.0	0	0	0	5.5	0
Level III	1.1	0.38	6.81	0	1	0	7.8	0
Level IV	0.33	0	3.33	0	6	0	17.2	0.1

species of *Hemideina* (Table 18.5). The trend is for winners to produce an increasing amount of stridulation as the aggression level of the fight progresses from I to IV. This is particularly pronounced in *H. crassidens* for stridulation after fights. Therefore, it appears that a winner stridulates during and after fights to convey the level of aggressive intent it is likely to show towards an opponent.

Another way to uncover the underlying message(s) sent by stridulating males in fights is to analyse the frequency of behavioural units that precede and follow bouts of stridulation. By studying the occurrence of sound in its behavioural context, it should be possible to gain insight into the meaning of messages from the actions preceding and the responses to stridulation. This was done for 15 fights by *H. crassidens* (Fig. 18.10).

Considering first the sender's behaviours (Fig. 18.10A), the most prominent units were 'antennate–palpate', 'pull', 'lunge', 'rest in gallery' (and 'stridulate', which occurred in repeated fashion). The first three units represent aggressive behaviour by the intruder as it tried to evict a resident during a fight. They occurred typically before stridulation (filled bars, Fig. 18.10A). 'Rest in gallery' represents the behaviour preceding and following stridulation by the winner of a fight (open bars, Fig. 18.10A); this may be either the resident, which has successfully repelled an intruder, or an intruder, which has evicted a resident. In either case, since there was no longer any interaction at this point, the stridulation signal was likely to convey a message relating to the sender's status or motivation based upon the previous interactions in the fight. Since we know that this stage of stridulation increases in intensity and duration in proportion to the aggressiveness of the preceding fight, the message must contain an indication of the aggressive tendency of the sender.

The receiver's behaviours are equally revealing. If the receiver was an intruder, its most common units, before it received stridulation from the opponent, were 'antennate–palpate', 'bite tibia' and 'grapple' (filled bars, Fig. 18.10B). These are aggressive behavioural units, which evoked the stridulatory reply from the resident. If the receiver was a resident, its prevailing behaviours, which evoked stridulation by the intruder, were 'kick–push' and 'rest in gallery'. Finally, 'depart' by either combatant signified that it acquired loser status and therefore received the victory stridulation by the winner.

The receiver's behaviour following the sender's stridulation confirmed the aggressive nature of the sender's message (open bars, Fig. 18.10B). There are two aspects of this interpretation. First, if the opponent stridulated during a fight, the receiver responded by 'enter', 'bite tibia', 'mandible gape', and 'lunge'. These are aggressive responses, which led to escalation after receiving stridulation from the opponent, when the sender was either a resident stridulating from the gallery entrance or one of the combatants fighting outside the gallery. Second, the sender's stridulation caused the additional behaviours 'emerge', 'step back' and 'depart', which are all responses given by the receiver if it was a loser resident or the loser of a

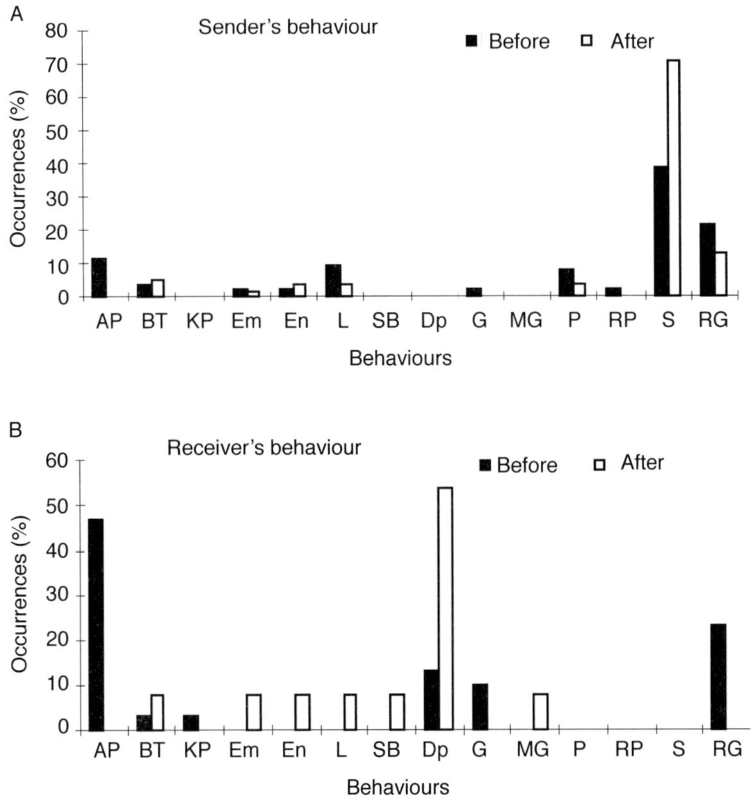

Fig. 18.10. Analysis of behavioural contexts for stridulation in *H. crassidens*. A. Frequency of occurrence of each behavioural unit that the sender executed immediately before (solid bars) or after (empty bars) its own stridulation behaviour. B. Frequency of occurrence of each behavioural unit executed by a receiver weta immediately before (solid bars) or after (hatched bars) it received a stridulation signal from its opponent. AP, antennate–palpate; BT, bite tibia; KP, kick–push; Em, emerge; En, enter; L, lunge; SB, step back; Dp, depart; G, grapple; MG, mandible gape; P, pull; RP, rest in place; S, stridulate; RG, rest in gallery.

fight outside the gallery. If the resident voluntarily emerged following the intruder's stridulation, the former weta invariably departed, since the usual tendency was to remain in the gallery until pulled out by an intruder. Thus the interpretation is that stridulation by the intruder either caused the receiver to depart (typical) or to escalate the fight (less common).

To summarize, stridulation by male wetas during fights appears to carry either a message of aggressive intent, if the sender is an intruder, or a message of resistance to displacement from the gallery, if the sender is a resident. Thus the message is context-dependent and carries different meanings depending upon the status of the sender. A similar conclusion was reached by Field and Rind (1992) in attempting to distinguish message differences between the aggression sound pattern and the rather similar pattern produced in mating stridulation by tree wetas (see Field, Chapter 15, this volume). The semantic analysis of Field and Rind confirmed that the male stridulatory pattern carried different messages depending upon the behavioural context (referent), including sex of the receiver. The use of a basic pattern to carry a variety of messages appears to reflect a highly conservative evolution of stridulatory signalling in tree wetas. This has also been described for some grasshoppers and for ants (Lewis and Gower, 1980; Markl, 1985), and presents a contrast to the elaborate diversity of signalling patterns in crickets (reviewed by Haskell, 1961).

Do Somatic Features Affect Battle Outcome?

According to the application of game theory to animal conflicts, when contests are asymmetric (characterized by factors such as one opponent being a resource owner, a larger opponent, an opponent with larger weapons), the asymmetry is used as a cue to settle the contest (Maynard-Smith, 1974). In male tree weta fights, body size, weight, mandible gape and force exerted during lunging are likely to be cues that enable a weta to evaluate the opponent's ability to inflict injury. This is supported by the observation that, when male *H. femorata* opponents obviously differed in size, fights were often settled quickly, whereas more evenly matched individuals engaged in prolonged fights (Field and Sandlant, 1983). Tactile behavioural units were used at all stages of fights, especially at levels III and IV, but it did not appear that great force was used. Presumably tactile behaviours, such as 'grapple', 'bite tibia', 'lunge' and 'bite head', provide accurate information concerning the size and strength of the opponents.

In recent experiments in our laboratory, these ideas were tested by calculating a relative body index for males of two species (*H. femorata, H. crassidens*), which were then scored for subsequent fighting success. The body index (BI) was a value between 0.0 and 1.0 calculated as:

$$BI = ((Wt + HW)/2)^2$$

where Wt is body weight (g) and HW is head width (mm). The results were similar for the two species (Fig. 18.11). Fighting success increased as a roughly logarithmic proportion to body index ($y = 90.7 + 229.8\ \log(x)$, with the larger males having greater success. Interestingly the body index of smaller males (BI between 0.4 and 0.55) gave less predictability for success than that for the larger males.

Since tree wetas are nocturnal insects, it is logical that the behavioural cues used in gauging the abilities of an opponent male are either acoustic or tactile in nature. However, there is the challenging question of how successfully combating males can perceive 'mandible gape', a visually mediated display behaviour. Aggressive displays evolve to pro-

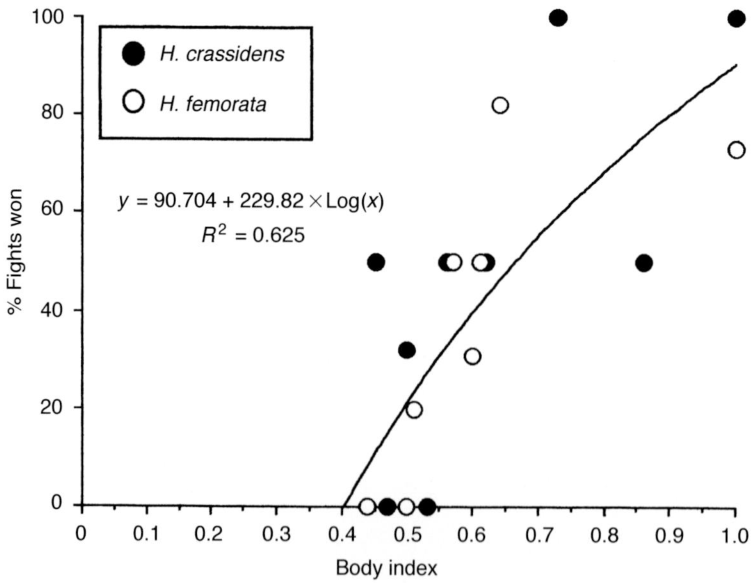

Fig. 18.11. Relationship between fighting success (percentage of fights won) and body index of male *H. crassidens* and *H. femorata*. The body index (BI) is given by body weight (Wt) and head width (HW) in the following formula: $BI = ((Wt + HW)/2)^2$. The logarithmic relationship is described by $y = 90.7 + 229.8\log(x)$. For *H. femorata*, $n = 8$ males (61 fights); for *H. crassidens*, $n = 9$ males (20 fights, unpublished data from R. Ewers and G. Cowley, with permission).

vide reliable information about the apparent injury potential of an opponent, thus making it advantageous for a less aggressive combatant to withdraw based upon its perception of the display. Clearly, sexual selection has driven the evolutionary development of mandible length, and combating males readily display them as a threat during level IV fights. The combatants must be able to see the mandibles at least some of the time, but it remains a question for future research to determine the real effectiveness of the display.

The Effect of Artificially Limiting Gallery Space

In observations of wetas (*H. crassidens*) in their natural habitat, Möller (1985) recorded an increased rate of stridulation (and, presumably, an increased rate of aggressive interactions) around 05.00, before dawn. While observing tree wetas in captivity, Sandlant (1981) also noticed that aggressive interactions of *H. femorata* seemed to increase as dawn approached, even in artificial lighting conditions, when 'dawn' consisted of a sudden switch from lights off to lights on. Later experiments showed that, under normal laboratory colony conditions, the frequency of fights for the 2.5 h period preceding dawn increased by 260% for females and by 350% for males, compared with the frequency of fights in previous hours. This gave rise to the possibility that aggressive behaviour in males was driven not only by competition for females, but also by competition for refuge space, regardless of the presence or absence of females.

Sandlant devised two experiments to test the above concept with *H. femorata*.. In the first, six males and no females were placed in an artificial habitat, with six galleries available initially. After 1 week, the gallery availability was progressively reduced from six to three to one over two cycles on successive nights, and fighting was monitored for 2.5 h before dawn. In the second, the same experiment was repeated with two males and four females, using the same enclosure and conditions.

The results indicated that male fighting frequency in both conditions (pooled data) increased dramatically as gallery availability decreased (Fig. 18.12A). Initially, even though one gallery was available for each weta, a baseline level of fighting occurred (column '6', Fig. 18.12A). When three galleries were closed off on subsequent nights, the

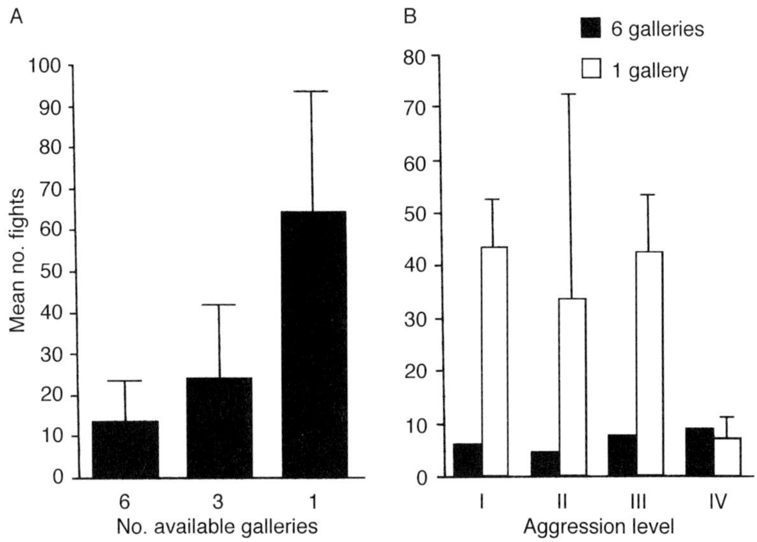

Fig. 18.12. The effects of limiting gallery availability on the frequency and nature of fights between male tree wetas (*H. femorata*). A. Mean (± SD) number of fights observed in 2.5 h period before dawn increases over threefold as available galleries are blocked off on successive nights (n = 454 fights, 10 males). B. Comparison of aggression levels (I–IV) observed in fights with six galleries available (filled bars) vs. one gallery available (open bars) (n = 10 males, 307 fights). Note lack of effect on level IV fights.

frequency of fighting increased by 176% in the predawn period (column '3', Fig. 18.12A). Finally when all six wetas had to compete for a single gallery on further subsequent nights, the rate of fighting increased by 476% compared with the original six-gallery condition.

Furthermore, there was a significant increase in the number of level I (χ^2 =22.4, $P < 0.001$), level II (χ^2 = 40.7, $P < 0.001$) and level III fights (χ^2 = 26.9, $P < 0.001$) when one gallery was available (Fig. 18.12B, open bars), compared with when six galleries were available (Fig. 18.12B, filled bars) (Sandlant, 1981; pooled data from both experiments). The large standard deviation in level II fights in Fig. 18.12B resulted from a difference between the presence and absence of females. When females were absent, the frequency of level II fights was low (males increasingly pulled each other out of galleries (level III) without the intruder being easily repulsed by 'kick–push' and 'stridulate', as in level II fights). On the other hand, when females were present, resident males (with females in galleries) increasingly showed vigorous 'kick–push' and 'stridulate', as they strongly resisted eviction, and intruders were more often deterred from escalating to level III fights. This resulted in a high frequency of level II fights.

The most interesting result of these experiments was the lack of effect of gallery restriction on the frequency of level IV fights (Fig 18.12B). Apparently, the males were not interested in escalating to level IV when dawn approached, even with only one gallery available and with females present. The interpretation is that it was more important to spend time trying to enter galleries than to defend them. Support for this was the observation that, under such unnaturally crowded conditions, occasionally one male tolerated another in its gallery without attempting to repulse it (Field and Sandlant, 1983). Such behaviour is almost never observed in the natural habitat.

Female Aggression Behaviour

By limiting gallery availability, Sandlant (1981) also showed that female aggressive behaviour can be induced. Normally, the females spend little time outside the galleries and they readily tolerate other females within galleries. But, as gallery number was reduced, female *H. femorata* began to compete aggressively for the refuge resource in the hour before dawn (laboratory population).

Female *H. femorata* show only three levels of aggression. Level I is characterized by 'antennate–palpate'; level II consists of 'antennate–palpate' and 'butt' by the intruder, and 'kick–push' and 'step forward' by the resident (the latter to accommodate the intruder); and level III involves the additional behaviours 'bite tibia', 'pull', 'emerge' and, rarely, 'grapple' and 'mandible gape' (Field and Sandlant, 1983).

Female *Hemideina maori* in captivity also showed infrequent aggressive behaviour, involving 'antennate–palpate', 'mandible gape' and 'bite' (usually incomplete), when kept in the same container. Mutual 'antennate–palpate' and often the other behaviours lasted for up to half a minute before one of the females turned away and departed. The second female (winner) occasionally followed and reinitiated the interaction (O'Brien, 1984). These interactions apparently resulted from the restricting conditions of captivity (e.g. small containers). In light of the crowded daytime conditions observed for females under rock slabs in their alpine habitat, it is probable that crowding is tolerated in the daytime, provided greater individual distances are maintained during the nightly activity period.

Functions of Aggression Behaviour in Tree Wetas

Intraspecific male aggression occurs at gallery entrances as intruder males fight for possession of galleries. It is argued here that sexual selection drives this behaviour as a primary tactic in male competition for females during the nightly activity cycle, but that, as dawn draws near, the function of aggression behaviour switches to competition for galleries as refuge space against diurnal predators. Exaggerated morphological features, such as male mandibular hypertrophy and cranial megacephaly (see Field and Deans, Chapter 10, this volume), are also the product of sexual competition, and they allow larger males to have greater success in winning battles, which escalate during competition for females. Supporting evidence for the postulation that mate competition underlies aggressive fighting by males for gallery access derives from the following facts: (i) resident males remain near the entrance of galleries, while females are always further inside; (ii) resident males behave aggressively toward intruder males but not towards intruder females; (iii) males make forays from galleries lacking females and try to gain access to other galleries, which might have females; (iv) once a winner male evicts a resident, it inspects the gallery and departs if females are absent; and (v) in the natural habitat, one male is found in a gallery with several females, but virtually never with other adult males. Thus the gallery serves as a resource for which males compete aggressively to gain access to females for mating purposes. This explanation applies particularly to intruder males.

A slightly different strategy is likely to underlie the tactics of resident males. In defending the gallery resource (with occupant females) against intruder males, they can be viewed as guarding their reproductive investment against sperm competition from other males. This conclusion follows because female receptivity is such that they are likely to mate with other males if given the opportunity (Field and Sandlant, 1983). A male able to guard one or more females from mating by other males would enjoy the advantage of greater reproductive output. Because the gallery configuration allows a male to easily monopolize females by guarding the entrance, the guarding behaviour appears to have led to the development of harems in the tree weta mating system (see Field and Jarman, Chapter 17, this volume).

References

Alexander, R.D. (1961) Aggressiveness, territoriality and sexual behaviour in field crickets (Orthoptera: Gryllidae). *Behaviour* 17, 130–223.

Bailey, W.J. (1991) *Acoustic Behaviour of Insects.* Chapman and Hall, London, 225 pp.

Bateman, P.W. and Toms, R.B. (1998) Mating, mate guarding and male–male relative strength assessment in an African king cricket (Orthoptera: Mimnermidae). *Transactions of the American Entomological Society* 124, 69–75.

Brown, J.L. (1975) *The Evolution of Behaviour.* Norton, New York, 761 pp.

Burk, T. (1983) Male aggression and female choice in a field cricket (*Teleogryllus oceanicus*): the importance of the courtship song. In: Gwynne, D.T. and Morris, G.K. (eds) *Orthopteran Mating Systems – Sexual Competition in a Diverse Group of Insects.* Westview Press, Boulder, Colorado, pp. 120–146.

Field, L.H. and Rind, F.C. (1992) Stridulatory behaviour in a New Zealand weta, *Hemideina crassidens. Journal of Zoology, London* 228, 371–394.

Field, L.H. and Sandlant, G.R. (1983) Aggression and mating behaviour in the Stenopelmatidae (Orthoptera, Ensifera), with reference to New Zealand wetas. In: Gwynne, D.T. and Morris, G.K. (eds) *Orthopteran Mating Systems – Sexual Competition in a Diverse Group of Insects.* Westview Press, Boulder, Colorado, pp. 120–146

Haskell, P.T. (1961) *Insect Sounds.* H.F. and G. Witherby Ltd, London, 189 pp.

Hinde, R.A. (1974) *Biological Bases of Human Social Behaviour.* McGraw Hill, New York, 462 pp.

Kerr, G.E. (1974) Visual and acoustical communicative behaviour in *Dissosteira carolina* (Orthoptera: Acrididae). *Canadian Entomologist* 106, 263–272.

Lewis, D.B. and Gower, D.M. (1980) *Biology of Communication.* Blackie and Sons, Glasgow, 239 pp.

Loher, W. and Dambach. M. (1989) Reproductive behaviour. In: Huber, F., Moore, T.E. and Loher, W. (eds) *Cricket Behavior and Neurobiology.* Cornell University Press, Ithaca, pp. 43–82.

Markl, H. (1985) Manipulation, modulation, information, cognition: some of the riddles of communication. *Fortschritt der Zoologie* 31, 163–194.

Maynard-Smith, J. (1974) The theory of games and the evolution of animal conflicts. *Journal of Theoretical Biology* 47, 209–221.

Möller, H. (1985) Tree wetas (*Hemideina crassicruris*, Orthoptera: Stenopelmatidae) of Stephens Island, Cook Strait. *New Zealand Journal of Zoology* 12, 55–70.

O'Brien, B. (1984) Mandibular movements and their control in the weta, *Hermideina maori* (Orthoptera: Fusifera: Stenopelmatidae). PhD thesis, University of Canterbury, Christchurch, New Zealand.

Otte, D. and Joern, A. (1975) Insect territoriality and its evolution: population studies of desert grasshoppers on creosote bushes. *Journal of Animal Ecology* 44, 29–54.

Parker, G.A. (1974) Assessment strategy and the evolution of fighting behaviour. *Journal of Theoretical Biology* 47, 223–243.

Sandlant, G.R. (1981) Aggressive behaviour of the Canterbury weta *Hemideina femorata* (Orthoptera: Stenopelmatidae): its adaptive significance in resource allocation. MSc thesis, University of Canterbury, Christchurch, New Zealand.

Spencer, A.M. (1995) Sexual maturity in the male tree weta *Hemideina crassidens* (Orthoptera: Stenopelmatidae). MSc thesis, Victoria University of Wellington, Wellington, New Zealand.

19
Communication and Reproductive Behaviour in North American Jerusalem Crickets (*Stenopelmatus*) (Orthoptera: Stenopelmatidae)

David B. Weissman
Department of Entomology, California Academy of Sciences, Golden Gate Park, San Francisco, CA 94118, USA

Introduction

No group of animals displays the variety of mechanically based communication systems seen in insects. Those airborne techniques, widely employed by various orthopteroid (crickets, grasshoppers and katydids) and cicada species, are well studied because of their pervasiveness, loudness and obvious importance in uniting the sexes through a genetically determined, species-specific calling song. Recent research has focused on the widespread but more subtle drumming (where the abdomen strikes the substrate) and tremulation (where the abdomen vibrates but does not contact the substrate) communication systems of various insect groups (e.g. Plecoptera, Neuroptera, Isoptera, Homoptera, Psocoptera, Coleoptera, Mecoptera, and Megaloptera (Bell, 1980; Morris, 1980; Henry, 1984, 1985a; Zeigler and Stewart, 1986; Maketon *et al.*, 1988; Stiedl and Kalmring, 1989; Stewart *et al.*, 1995; Henry *et al.*, 1999b). In almost all cases (see Stewart *et al.*, 1991, for exception), sympatric species have different substrate transmitted calling songs, which serve as effective premating barriers (Henry, 1984, 1985b, 1993; Stewart and Zeigler, 1984a; Henry *et al.*, 1999a). These studies have shown that many supposedly widespread species are actually assemblages of morphologically similar taxa (or cryptic species: see Walker, 1964), easily separated by their unique calling songs.

The Jerusalem crickets (JCs) (*Stenopelmatus* Burmeister, *Ammopelmatus* Tinkham and *Viscainopelmatus* Tinkham) can now be added to this list of insect groups (along with the taxa in Table 19.1) with subtle, and heretofore undiscovered, extensive mechanical communication systems. Indeed, this discovery has been the cornerstone to understanding their systematics. When I began my revisionary studies 20 years ago, there were 16 JC names in the literature for US and Canadian taxa, far short of the actual 60–80 species I now recognize (D.B. Weissman, unpublished). Sexually receptive adults of both sexes of all species produce, in many cases, species-specific 'calling' drums, with some individuals producing sounds audible 20 m away. In more than 85% of species, these calling drums are the same for conspecific males and females; sympatric species have different calling songs at 28 of 30 localities. Immatures of both sexes of a few species also produce their species-specific calling pattern ('nymphal' drum) sometimes a year prior to the final moult to adult. A calling drum received by adult males of many species is answered by a 'sex clarification' drum (SCD) announcing the receiver's sex. 'Courtship' drums are non-species-specific and made preceding mating.

Table 19.1. US and Mexican Orthoptera taxa which drum, either as 'calling' songs or interspersed during tegminal singing.

Taxa	Sex	Family	Body part drummed	Reference
Cnemotettix Caudell all species	Male	Anostostomatidae	One rear leg at a time	This report
Glaphyrosoma Brunner, ? all species, Mexico	Male	Anostostomatidae	One rear leg at a time	This report
Stenopelmatus Burmeister, all species	Male, female	Stenopelmatidae	Abdomen	Tinkham, 1968 Tinkham and Rentz, 1969 This report
Gammarotettix Scudder all species	Male	Rhaphidophoridae	Abdomen	Tinkham and Rentz, 1969 This report
Neduba (N.) castanea (Scudder)	Male	Tettigoniidae	Both rear legs together	This report
Neduba (*N.*) species, Monterey County, CA	Male	Tettigoniidae	Abdomen	This report
Daihinibaenetes arizonensis (Tinkham)	Male	Rhaphidophoridae	Both rear legs together	D.C. Lightfoot, New Mexico, 1998, personal communication

Calling-song drumming rates vary linearly with temperature. Information presented by Walker and Weissman (Appendix A of Chapter 19, this volume) allows calculation of the approximate drumming rate at any temperature, for any species, when a drumming rate at another temperature is known.

The orientation of JCs to each other during copulation is different from that of other orthopteroid groups. Male JCs pass a significant spermatophylax, which is almost never eaten by the female; its function is unknown, but it may serve as a mating plug or contain a chastity pheromone. After mating, the female occasionally consumes the male, who makes no effort to escape. Such consenting postcoital cannibalism is previously unreported and is discussed in terms of sexual selection theory.

The reader is referred elsewhere (Weissman, Chapter 3, this volume) for background information on the taxonomy, distribution, life cycle, ecology and related biology of this group, which extends from southern Canada east to the Great Plains of the USA and south to Panama. Undescribed species are referred to by number, with locality information given in Table 19.2.

Recording Techniques

Jerusalem crickets were recorded in the 225 cm^3 (8 oz.) margarine tubs they were raised in, but with most sand removed. Recordings were made on a Uher 4000 Report IC automatic tape-recorder and, suspended over the tub, initially a Uher M517 microphone and later a Sennheiser ME 40 and then ME 64 microphone with a cardioid pick-up pattern. Recordings were analysed on various machines, the latest a Tektronix 2214 digital storage oscilloscope. Hard copies were made on a Kay Elemetrics Model 5500 Sonagraph.

Laboratory temperatures were controlled as described in Weissman (Chapter 3, this volume).

Drumming rates were determined directly from the oscilloscope screen by counting the number of drums in several seconds and averaging. I usually started at the 15th drum because earlier ones were frequently more rapid than when the cadence 'settled down'. For those series with fewer than 20 drums (e.g. Fig. 19.1), counting was started near the middle, after the initial fast period and before the rate slowed toward the end.

Mating Techniques

All matings involved field-captured nymphs raised to adult in the laboratory, except where noted. Because of potential cannibalism, all JCs were always kept separate. Spermatophore weights were determined by weighing male and female JCs immediately before and after mating on an electronic Sartorius Handy Model H51 scale. Only

Table 19.2. Undescribed and described species of Jerusalem crickets referred to in text and their collection localities.

Species	Collection locality
1	USA, California, Riverside County, Yucca Valley
2	USA, California, Riverside County, San Jacinto Mountains
3	Mexico, Baja California Norte, La Rumorosa
4	USA, California, Tuolumne County, 24 km east of Oakdale
5	USA, California, El Dorado County, Lake Tahoe
6	USA, California, Riverside County, Winchester
7	Mexico, Baja California Sur, Sierra Agua Verde
8	Mexico, Baja California Norte, Descanso Dunes
9	USA, New Mexico, Hidalgo County, 5 km east of Cloverdale
10	USA, California, Los Angeles County, Santa Catalina Island
11	USA, California, Monterey County, Big Sur
12	USA, California, Alameda County, Oakland; *Stenopelmatus intermedius*
13	USA, Nevada, Carson City County, Carson City
14	USA, Arizona, Santa Cruz County, 16 km west of Nogales
15	USA, California, San Diego County, Tijuana Estuary
16	USA, California, Santa Barbara County, Santa Barbara
17	USA, California, Inyo County, Big Pine
18	USA, California, Santa Clara County, Los Gatos
19	USA, California, Santa Clara County, Mountain View
20	USA, California, Santa Clara County, Los Gatos; *Stenopelmatus pictus*
21	USA, California, Fresno County, Jacalitos Canyon; *Stenopelmatus nigrocapitatus*
22	USA, California, Riverside County, Calimesa
23	USA, New Mexico, Chaves County, Mescalero Sands; *Stenopelmatus mescaleroensis*
24	USA, Arizona, Cochise County, Portal
25	USA, California, Santa Barbara County, Lake Cachuma
26	USA, California, Kings County, Tar Canyon
27	USA, California, Riverside County, Glen Avon
28	Canada, British Columbia, Osoyoos

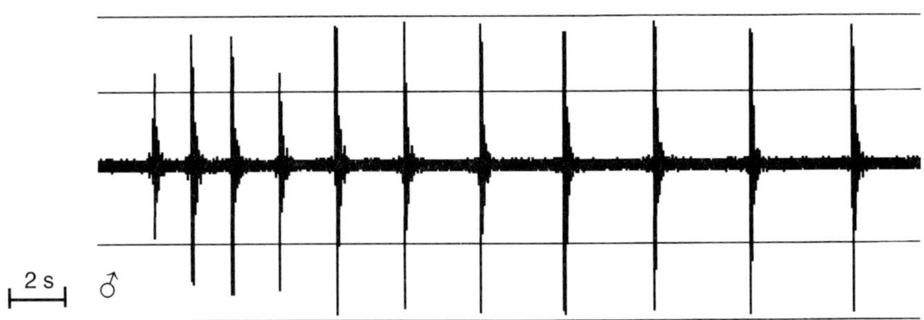

Fig. 19.1. Calculating drums per second for species with brief calling songs and few total drums. For this male of species no. 1, I arbitrarily did not count the first four 'rapid' drums and the last 'slow' one, and measured six drums in 5.5 s at 21°C.

male weight differences are reported. Females gained less weight during mating than males lost because parts of the transferred spermatophore stuck to the container.

Drumming and Communication

Jerusalem crickets have no auditory tympana to detect airborne sounds, nor do any species have wings with a sound producing apparatus (there are Mexican and Central American species that have fully formed wings of unknown function). Jerusalem crickets are generally nocturnal and largely subterranean, thus limiting the effectiveness of visual and olfactory communication. Information is exchanged via ground transmission of impulses produced by the abdomen (and sometimes the thorax–abdomen unit) striking the substrate, producing a sound wave. These ground impulses are detected by other JCs through their subgenual organs, which have been demonstrated in several orthopteroid families (Dambach, 1989; Ewing, 1989), including the related anostostomatids (Hutchings and Lewis, 1983; McVean and Field, 1996), but not specifically in stenopelmatids. These organs are located in the proximal tibia of all six legs (Devetak and Amon, 1997; review by Field and Matheson, 1998) and, in the orthopteroids, they are the most vibration-sensitive organs in any insect group (Kalmring and Kuhne, 1983), able to detect small amplitude displacements of less than 10 Å (1 Å = 10^{-10} m; Autrum and Schneider, 1948, cited in Brownell and Farley, 1979). Subgenual organs also respond to high-intensity airborne sounds (Hutchings and Lewis, 1983; Shaw, 1994). On several occasions, I have heard conspecifics in their containers in the laboratory, each on a separate bench isolated by 4 m of carpeted floor, duet (alternating) repeatedly with each other. The lids of these containers have either a few 'microholes' or no holes, and their drums were audible at 20 m.

Kinds of drumming

There are four kinds of drumming.

'Calling' drums

'Calling' drums are, in most cases, species-specific and spontaneously made by isolated adult males and adult females, the latter usually virgins. Spontaneous calling in female insects, while rare, is known elsewhere (Henry, 1984; Saxena and Kumar, 1984; Stewart and Zeigler, 1984a; Kraus, 1989; Hunt, 1993). Jerusalem cricket duets also occur between adults of the same species or different species with similar calling patterns, or in response to finger tapping if similar to the calling drum. This drumming is as valuable a taxonomic character as the species-specific calling songs of field crickets and katydids (see, for example, Weissman *et al.*, 1980; Rentz and Weissman, 1981; Ragge and Reynolds, 1998), because it appears to serve in uniting the sexes. Information on an individual's size may also be sent (De Luca and Morris, 1998).

Individual JCs respond only to drumming similar to their own calling song (as in other arthropods: Zeigler and Stewart, 1986; Hassage *et al.*, 1988; Stewart *et al.*, 1988; Barth, 1990; Henry *et al.*, 1999a). Different species drum with different regularity in the laboratory, but the drum is the same for the same individual on different days; for both sexes of the same species (see below for exceptions); and between populations of the same species. These drums vary in complexity among species and range from a series of single substrate strikes at rates of 0.5 (Fig. 19.2c) to 9 drums s^{-1} (Fig. 19.3a) to groupings of strikes (trills) with rates of over 40 drums s^{-1} (Fig. 19.4), or to combinations of both.

Length of calling song is the most common difference between males and females of the same species. Both sexes drum at the same rate, but a male's burst may last for seconds while the female drums for minutes. I suspect these differences are not artefacts of captivity but reflect true biological behaviours, as seen elsewhere (Field and Bailey, 1997; Stewart, 1997). Females may remain stationary and buried while drumming. Their long song gives males time to locate them. Males have short songs to trigger females; between drumming, they move about listening for drumming females.

In eight species, males and females have distinctly different calling drums (Table 19.3). I consider only those species where males drum at least 50% faster than females, to eliminate any possibility of individual or temperature variation. For all JC species where the sexes have a simple drum pattern, males always drum as fast as or faster than females. Differences in drumming patterns include species no. 2 (Fig. 19.2) and no. 5, where the male has single drums and the female has grouped drums. In species no. 3, the male (Fig. 19.3a) has a short series of rapid drums, while the female (Fig. 19.3b) combines a rapid and a slower section. In species nos. 6–9, the male drums at least twice as fast as the female. In species no. 10 (Fig. 19.5), the male's drum consists of three different cadences, while the female has one or two. This latter species is endemic on Santa Catalina Island near Los Angeles, California, and is the only JC taxon on the island. I do not know why it

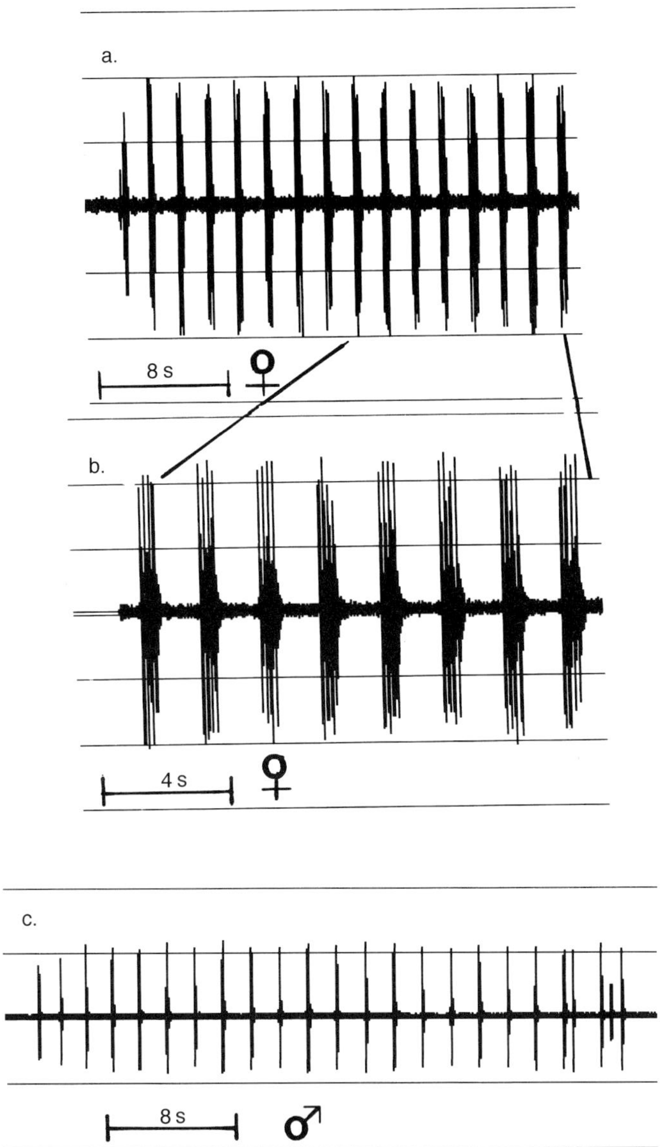

Fig. 19.2. Calling song of species no. 2: (a, b) Female song comprised of five trills (four drums per trill) in 8.5 s at 18°C. (c) Male song comprised of ten single drums in 16 s at 20°C.

has the most complex calling song of any JC species.

Calling-song drumming rate in JCs varies linearly with temperature (as in Walker, 1962, 1975; Prestwich and Walker, 1981; Henry, 1989; but see Henry, 1983), but the increase in rate with increasing temperature, is not constant for all species (Fig. 19.6a). Those species with fast rates increase their drumming speed more for a given temperature rise than does a slower-drumming species, as seen by the greater slope of the former's regression line.

Figure 19.6a hints at regression lines converging near a common point. Thus, not surprisingly, when I plot the slopes of these four regression lines versus the respective drumming rate at 20°C

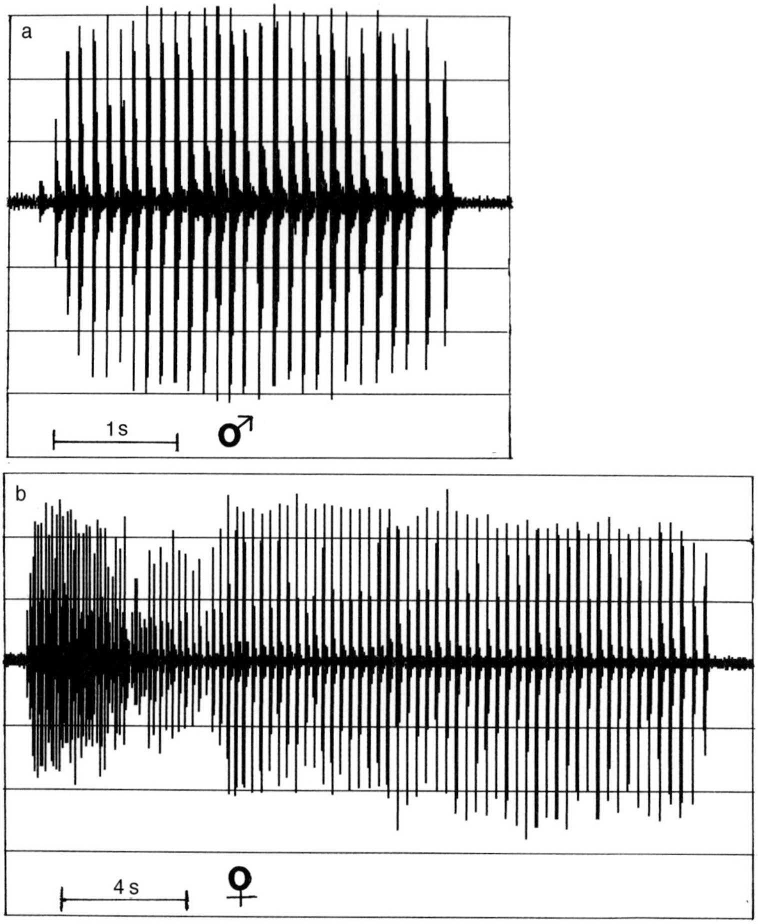

Fig. 19.3. Calling song of species no. 3: (a) Male calling at nine drums per second at 19°C. (b) Female calling with two cadences: fast at nine drums per second and slower at 3.5 drums per second at 19°C.

Table 19.3. List of Jerusalem cricket species where males and females drum differently. See Table 19.2 for species' localities. Drumming rates at 20°C.

Species no.	Male drum rate (drums s^{-1})	Female drum rate (drums s^{-1})	Text figure
2	0.5	Groups of 3 to 5	Fig. 19.2
3	7 to 9	Fast at 7 to 9, Slow at 3 to 4	Fig. 19.3
5	0.5	Groups of 3 to 5	–
6	20	10	–
7	10.5	5	–
8	6	3	–
9	2 parts: 6 to 8; 1.5 to 2.5	2 parts: 3; 0.5	–
10	3 parts: slow, paired, faster	2 parts: irregular, slow	Fig. 19.5

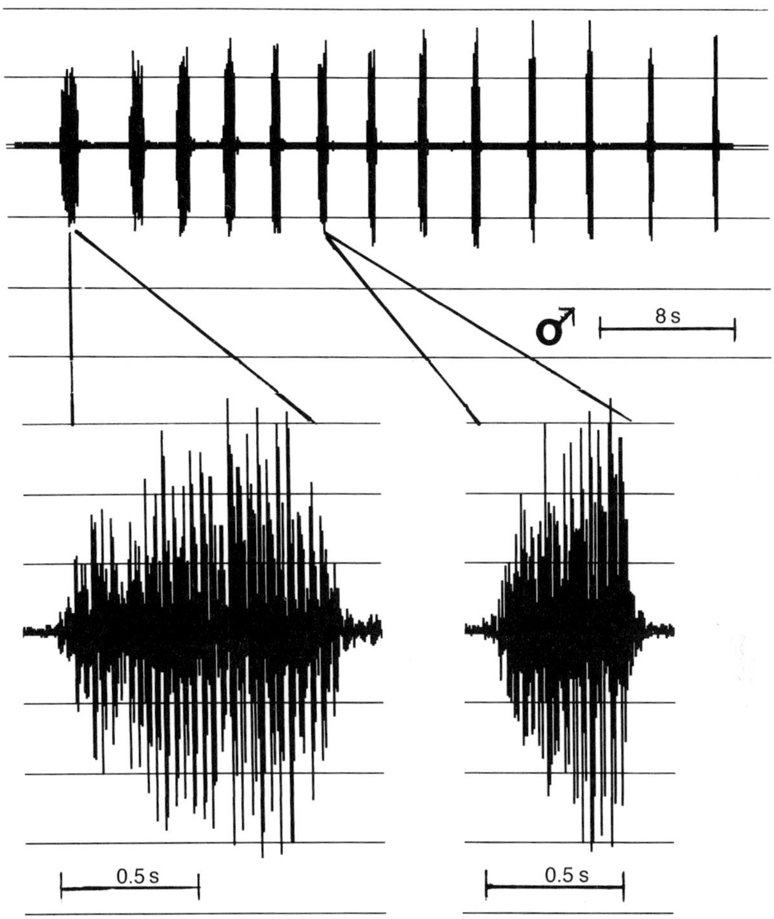

Fig. 19.4. Calling song of species no. 4 consists of a series of trills, the drumming rate within a trill exceeding 40 times s^{-1} at 20°C. This is the fastest-drumming Jerusalem cricket.

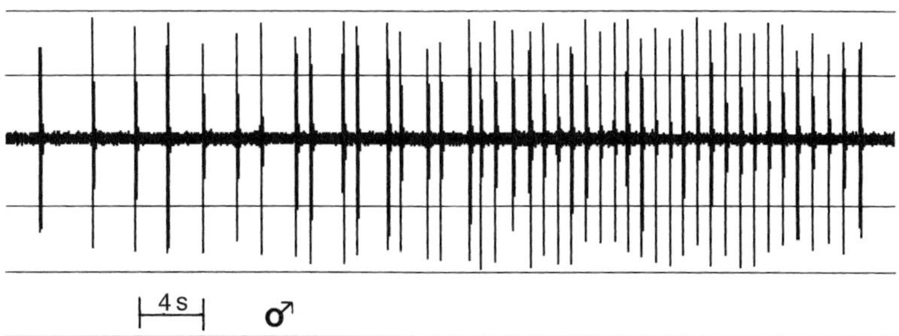

Fig. 19.5. Calling song of species no. 10, the most complex of any Jerusalem cricket taxon. Sonograph shows five slow drums (I ignore the first two drums) over 9 s, four pairs over 11 s and then 28 drums over 28 s, at 18°C.

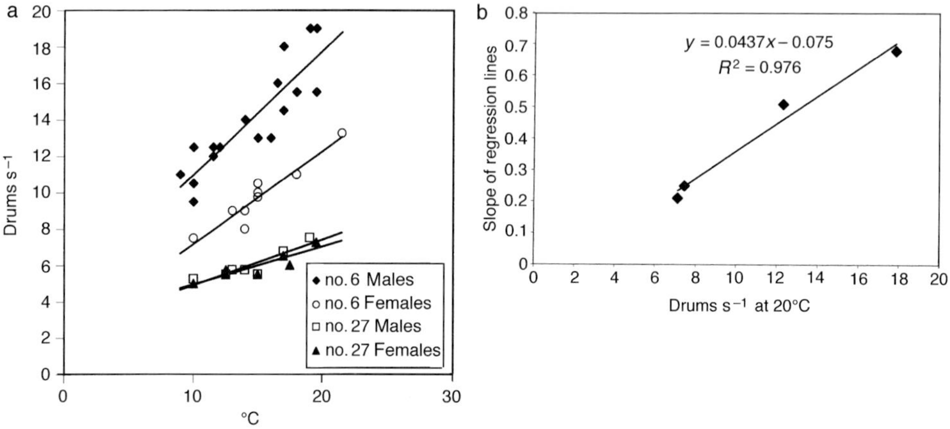

Fig. 19.6. (a) Effect of temperature on adult Jerusalem cricket drumming rate, data combined for several individuals of four species. The regression lines for the species are: ◆ drums per second (D/S) = $0.68t + 4.21$, $r^2 = 0.77$; ○ D/S = $0.51t + 2.07$, $r^2 = 0.90$; □ D/S = $0.25t + 2.43$, $r^2 = 0.82$; ▲ D/S = $0.21t + 2.90$, $r^2 = 0.82$. (b) Relationship between slope of the regression lines in (a) and drumming rate at 20°C.

(or at any other temperature as calculated from the regression equation), for each species, I discover a linear relationship between the two quantities (Fig. 19.6b). This association is so predictable ($r^2 = 0.98$) that, simply by measuring the drumming rate at 20°C, I can determine the slope of the drumming rate versus the temperature regression line, and thus the drumming rate at any other physiological temperature. As expected, this relationship is also true for an individual JC recorded at several temperatures (Fig. 19.7a, b). I show this approximation to apply to these three species, whose drumming rates, at 20°C, vary from 3.2 to 17.8 drums s^{-1} (while the known range for the genus is 0.5 to 40 drums s^{-1}), and over a temperature range of 8 to 28°C. Other JC species need to be investigated to examine the universality of this correlation but, should this relationship hold true across the genus, standardization of drumming rates between species to one uniform temperature

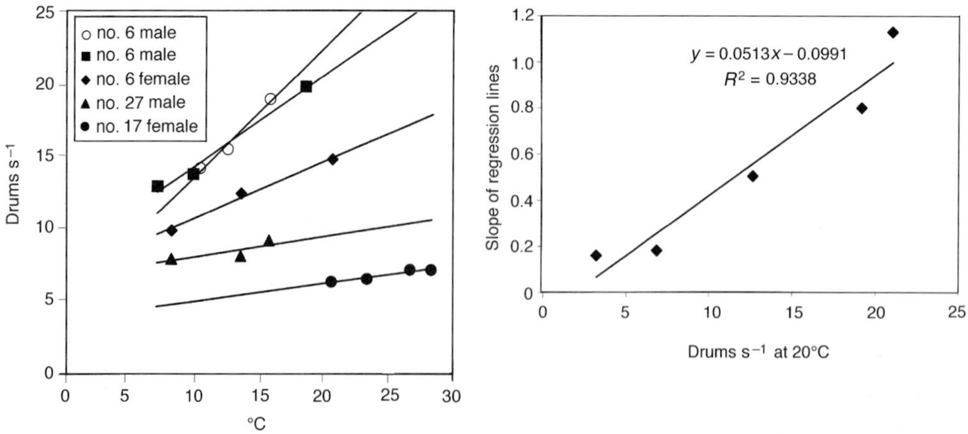

Fig. 19.7. (a) Effect of temperature on individual Jerusalem cricket drumming rate, data for five adults of three different species. The regression lines for the individuals are: ○ drums per second (D/S) = $1.12t–1.20$, $r^2 = 0.98$; ■ D/S = $0.79t + 3.47$, $r^2 = 0.99$; ◆ D/S = $0.50t + 2.72$, $r^2 = 0.99$; ▲ D/S = $0.18t + 3.28$, $r^2 = 0.87$; ● D/S = $0.16t + 0.02$, $r^2 = 0.94$. (b) Relationship between slope of the regression lines in (a) and drumming rate at 20°C.

will be greatly facilitated, since no taxonomic decisions need be made (see Appendix A). Also to be investigated is how temperature affects the calling song of species that trill (as in Figs 19.2a, b and 19.4), and species that combine single and grouped drums (as in Fig. 19.5).

Drumming speed appears to be unrelated to body size. There are large-bodied (5 cm) species that drum fast – at ten times per second – and small-bodied (2 cm) species that drum slowly – once per second. There are three species (nos 4, 11 and 12) that drum faster than 20 s^{-1} – all are less than 3 cm long. Most sand-dune-inhabiting taxa, several of which are less than 2 cm in length, drum slowly, perhaps in response to the sound transmission properties of sand when compared with harder substrates (see Michelsen, 1979; Stewart and Zeigler, 1984b; Kalmring *et al.*, 1990; Romer, 1998).

In thousands of recordings, I have only one presumptive hybrid: a female collected at Mt. Tehachapi State Park, Kern County, California, who drummed at four times per second, while her microsympatric parents drummed once and ten times per second. This hybrid would be at a disadvantage in nature, because she probably would not respond to the drum of either parent, nor would they to hers (as in Wells and Henry, 1994).

'Sex clarification' drums (SCD)

SCD are produced only by adult males (see below for exceptions) of some 35 US species (Mexican and Central American taxa are not well known). These are a series of fast (rates over 15 drums s^{-1} at 20°C – see species no. 13, Fig. 19.8), loud (the result of faster abdominal movements and the need to be heard by the female while she is drumming?), non-species-specific drums produced when these males detect their calling drum. SCD occur only in species where the male and female have the same calling drum rate and one sex cannot tell who they are answering. In the laboratory, if the first drummer is a male, he frequently ceases his calling drum upon detecting the second male's SCD. Males can also answer each other with SCD and may be displaying territoriality or aggression. If the initial drummer is a female, she usually continues or repeats the calling drum after sensing the SCD. The male may repeat his SCD as she continues her calling drum. Once she stops, he frequently makes a calling drum or may combine an SCD and a calling drum (Fig. 19.8). In the field, this answering strategy should enable males to conserve energy and avoid possible injury or predation by not searching for another male. SCD, besides telling females that there is a male nearby, may also give an indication of his size (e.g. De Luca and Morris, 1998) and that she should continue drumming while he searches for her (males usually start their SCD as soon as they sense the calling drum). SCD are unknown in any species that has a calling drum faster than ten drums per second, possibly because of the close similarity between the two drums and the likelihood of confusion. SCD are unknown in any species that trills, except for one male from Sutter Buttes, Sutter County, California. In this species, the calling song 'trills' made by males and females usually contain only one or two drums, rarely three, so that an SCD will be distinctive, because it contains more and faster drums. The variable percentage of males of any one taxa that make SCD is interesting: 6% of individuals of species no. 14 (one of 17 males recorded) to 100% in species no. 15 (four males recorded), with most species less than 50%. Species from all four karyotype groups (see Weissman, Chapter 3, this volume), and from both *Stenopelmatus* and *Ammopelmatus* utilize an SCD, perhaps indicating the antiquity of this behaviour.

Jerusalem crickets are rarely microsympatric, two species having been found together at only 30 localities involving 20 different species-pair combinations. At only one locality (Pt Conception sand-dunes, Santa Barbara County, California), do both species use an SCD. They have slightly different calling drums: at 20°C: 2.8 to 3.4 drums s^{-1} versus 3.8 to 4.6 drums s^{-1}. One species in this pair is an 'adventive,' normally found off sand-dunes (see Weissman, Chapter 3, under 'Ecology', this volume), in that area. Where two species with SCD occur together, I expect confused females to be unable to distinguish if their conspecific male is replying, unless males also make calling drums temporally close (as in Fig. 19.8) to their SCD. Incidentally, both JC species were common at Point Conception, but no hybrids were found.

No species in Table 19.3, where males and females drum significantly differently, makes an SCD, as expected, since each sex can identify the other sex by drumming rate. Although I have raised thousands of juveniles over the years, only three male nymphs (one each of species nos 16, 17 and 28) were heard to make an SCD, each on only

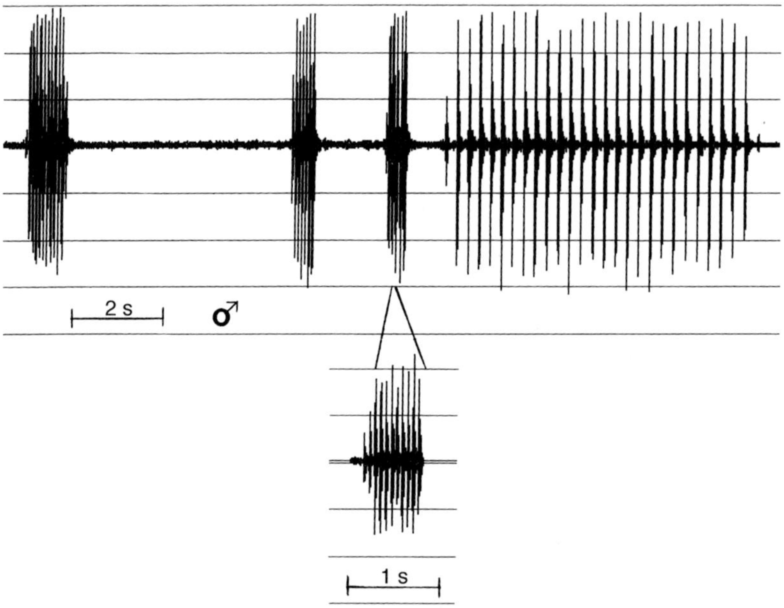

Fig. 19.8. Three series of 'sex clarification' drums made by a male of species no. 13, followed by a calling song drum, all at 20°C. Sex clarification drum rate around 20 s^{-1}, calling drum rate 4 s^{-1}.

one occasion. I can frequently elicit an SCD from a male by tapping my finger on the surface his container rests on, at a rate approximating the calling drum.

In summary, conspecific adults may be able to determine each other's sex at long distances: (i) by employing an SCD; (ii) where drumming rates differ between the sexes; (iii) where the sexes have calling songs of different lengths; (iv) where the drumming pattern differs between the sexes; (v) where one sex may be polite (that is, waits until the first drummer finishes before answering) and the other sex is not polite; and (vi) where one sex may be louder than the other. Females are typically heavier than males, but I have no data to show they are louder. Because males usually drum faster than females, they frequently sound louder. Loudness to the receiver, of course, may be a measure of proximity. Maybe the interval between the ending of the drum and initiation of the answering drum is integrated by the receiver to get a composite of size and distance.

Long-range sex discrimination behaviour, as found in JCs, is also known in male water-striders (Wilcox, 1979; Spence and Andersen, 1994). Because JC SCD are effective at long range, they are functionally distinct from the close-range rivalry sounds (Saxena and Kumar, 1984) made by male leafhoppers in the presence of other males and the aggressive or defensive tremulation (Rupprecht, 1981) made by male Plecoptera after being mounted by another male.

'Courtship' drums

'Courtship' drums are frequently made when two adults of either sex are within 6 cm of each other. This drumming consists of short series of barely audible, non-species-specific abdominal strikes or tremulations at a rate of two to four per second. Individuals also occasionally employ thoraco-abdominal stridulation during this period, despite the fact that they have no tympana (they can probably 'hear' each other (see McVean and Field, 1996)).

'Nymphal' drums

'Nymphal' drums were heard in at least 25 species in the laboratory, although performed by less than 25% of the individuals of each species. From any one nymph, this drumming was rarely heard for longer than a 2-week period (versus months for the adult calling drum). For any one species, this drumming was not always heard in both sexes, possibly because of the few captive individuals and the less than intense effort directed toward finding this behaviour. Nymphal drums have the same pattern as the adult species-specific calling drum, but they are produced less frequently, are usually of shorter duration and are rarely heard during more than one instar. These subadult JCs do duet with other individuals (both nymphs and adults), but less frequently than adults do. Nymphs that drum display no mating behaviour when enclosed with a sexually receptive adult of the opposite sex, and nymphs younger than the penultimate instar have no gonadal maturation.

While it is not clear why nymphs drum, two opposing hypotheses deserve discussion:

1. Nymphs drum to cause aggregation and so facilitate subsequent adult mating. Population densities of JCs are usually low and somewhat restricted (D.B. Weissman, unpublished), but their contiguous potential habitat (e.g. sand-dunes, open fields) may be extensive. Although many nymphs drum over 1 year before they become adult, drumming and answering may serve to keep them cohesive until mating begins.
2. Nymphs drum to cause segregation. It is my impression that, in the field, adult and subadult JCs are spaced – one per rock or one per sand dune plant. Segregation could decrease both intraspecific cannibalism (easily observed in the laboratory) and competition for food (as in Thanthianga and Mitchell, 1987; Moring *et al.*, 1991). In the laboratory, nymphs drum almost exclusively during the winter months. Food for these omnivorous insects may be scarcer during winter, when JC population densities are highest because of recent egg hatchings. Thus, the probability of a chance encounter with another (hungry) JC is increased. This danger of cannibalism of nymphs is increased because many adult female JCs are still alive during the autumn and early winter months. Adults have tougher integuments and larger mandibles and should prevail in most fights with nymphs. Under this hypothesis, nymphs sensing another drum will retreat. By answering each other, nymphal JCs can clarify position and ensure movement in opposite directions.

The above contrasting hypotheses are readily testable using a nearest-neighbour analysis and experiments that observe nymphal reaction (attraction versus repulsion) to other drums. Of particular interest will be the reaction of nymphs of sympatric species to each other's species-specific drum. In any case, I believe that nymphal drumming, which is rarely reported in insects (Wilcox, 1979; see also Sivinski, 1983, p. 149), has a function (although unknown), because it is potentially risky behaviour: it could attract an adult JC, which could eat the nymph, or attract a scorpion, also a JC predator (Polis, 1979), because the latter are extremely sensitive to substrate-transmitted vibrations (Brownell and Farley, 1979). Also owls and other predators could locate JCs from their drumming.

Mating Behaviour

The act

Jerusalem cricket orientation during mating appears to be unique among the orthopteroid insects (compare with Alexander and Otte, 1967). Laboratory-reared virgin adults introduced into an arena require several minutes to adjust. Initial aimless exploring changes to mutual antennal palpation when the sexes meet. There may be calling drums, courtship drums and tremulation or stridulation at that time, but none appeared necessary in the 68 matings involving 15 species that I observed (Table 19.4). Since 38 of these copulations involved species no. 18, all comments below refer to that taxon unless otherwise noted. The male initiates mating by rolling on to his side. To proceed further, the female must roll on to her side. If she does not roll voluntarily, or cannot be 'coaxed' by the male (although in some cases he physically forces her), mating will not occur. Once she is on her side, the male assumes a lateral position beside her but facing in the opposite direction (Fig. 19.9a). He bites one of her hind tibia (doing no damage; she may also engage in biting the male), positions his hind leg tarsi near her coxae and curls his abdomen between his hind legs and her hind legs toward her subgenital plate (Fig. 19.9b), a manoeuvre that often requires several minutes of adjusting. As the male's telescoping abdomen nears the female's subgenital plate, he positions his hooks (see Weissman, Chapter 3, this volume) underneath her antepenultimate or penultimate ventral plate. This provides the necessary anchor as he everts his phallic lobes (Figs. 19.9b and 19.10a) and makes contact with her subgenital plate area. Immediately upon contact, a bilobed ampulla containing sperm is passed to the female (Fig. 19.10b) and the everted phallic lobes then empty and deposit a sticky bilobed spermatophylax lateral to each ampullar lobe (Fig. 19.10c), the complete structure comprising the spermatophore (Fig. 19.10d). Transfer takes 10–20 s and does not leave the male's abdomen empty (*contra* Tinkham and Rentz, 1969), because the transferred mass is only 3–7.5% of the male's body weight (Table 19.5).

After mating, the female tries to disengage. Some 20% of males immediately run and another 15% continue to hold her rear tibia, sometimes (as in species no. 9) for as long as 1 h. In those cases where both disengaged simultaneously, the female usually antennated the motionless male and then walked away, gathering dirt on the dragging spermatophore. The spermatophylax can disappear (by dehydration) in a few hours or in a few days. Only one (species no. 18) of 68 females ate the spermatophylax, first starting 3 h after mating. The ampulla was never eaten after any mating, hardened in a few hours, and disappeared in 1–4 days, apparently flaking off. There appears to be no absorption of any part of the spermatophore, except possibly of ejaculatory substances in the genital tract (as in Loher, 1984).

The props

The male's abdominal hooks are essential for successful, expeditious mating. I removed both hooks from four virgin males of species no. 18, 2 months after they became adults, and isolated each with a virgin female. All pairs were in appropriate position within 20 min, when the males extruded their lobes and initiated repeated sweepings of their abdomens along the female's abdomen, trying to anchor their hooks. Male no. 1 passed the ampulla in 36 min but the spermatophylax was stuck to him; male no. 2 passed the ampulla between 60 and 90 min, but a spermatophylax was never seen; male no. 3, after several hours, managed to pass an ampulla on to the female's abdomen several segments cephalad from its intended position; male no. 4 was still in proper orientation on his side with his female 20 h later attempting to mate!

I subsequently removed only one hook using four other virgin males of species no. 18, 2 months after they became adults, and isolated each with a virgin female. Two males mated within 15 min and passed a normal spermatophore; one male mated normally in 20 min and the fourth male had mated within several hours. All these males experienced some difficulty in hooking properly, but all mated successfully.

Table 19.4. Species of Jerusalem crickets involved in 68 laboratory matings. See Table 19.2 for species' localities.

Species no.	Number of matings	Species no.	Number of matings
2	3	20	1
3	1	21	1
4	1	22	1
9	9	23	1
11	1	24	2
14	3	25	2
18	38	26	1
19	3		

Table 19.5. Spermatophore weights in four species of Jerusalem crickets as percentage of male body weight. See Table 19.2 for species' localities.

Species no.	Number of males	Premating weight (g) of males ± SD (range)		Spermatophore weight as % male body weight ± SD (range)	
9	5	1.09 ± 0.18	(0.77–1.22)	6.14 ±1.5	(3.6–7.3)
18	5	2.05 ± 0.37	(1.69–2.47)	5.90 ±1.6	(3.3–7.5)
24	1	1.97		6.4	
25	1	4.38		3.0	

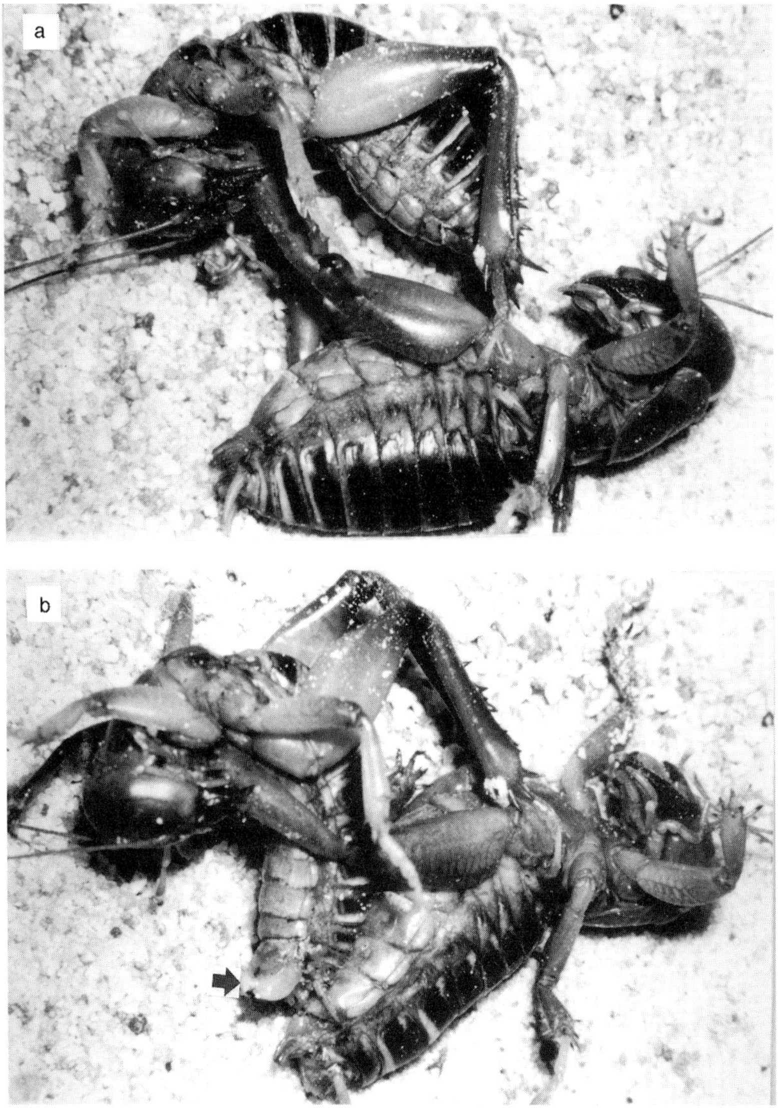

Fig. 19.9. Early mating stages in species no. 18. (a) Male (top) is biting female's left rear tibia while positioning his rear legs; then (b) curling his abdomen between her rear legs toward her subgenital plate. Note everted phallic lobes (arrow).

Interestingly, one male of species no. 6 had two hooks on one side and the normal single hook on the other side. Unfortunately, I made no mating trials with him but suspect he would have done fine.

The consequences

After six of 68 laboratory matings, the female ate the male (Table 19.6). In four of these cases, both males and females were virgins. The fifth mating involved adults who had mated once weeks previously, but not with each other. After this second mating, she attempted to eat him but I separated them. They mated with each other again one week later and she ate him. Two days later I paired this latter female with a male who had mated 4 weeks earlier. He ignored her but the status of her attached spermatophore (see below) was not noted. The sixth case involved a field-caught adult (and

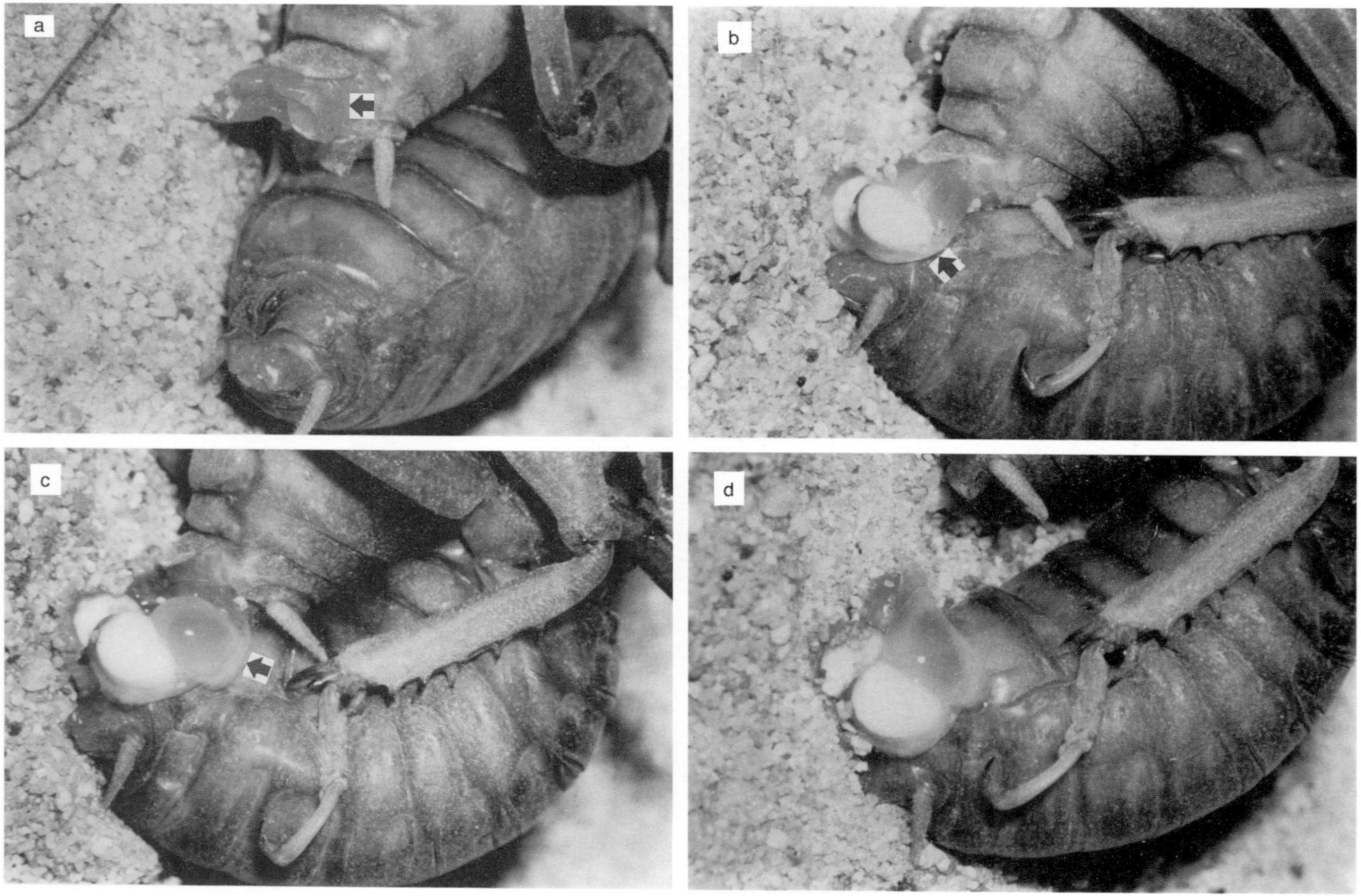

Fig. 19.10. Actual mating, here in species no. 2. (a) Male (top) everting his phallic lobes (arrow) and ready to anchor his hooks. (b) Male passing bilobed ampulla (arrow) and (c) bilobed spermatophylax (arrow). (d) Mating finished, complete spermatophore passed, and male withdrawing his abdomen (top).

therefore probably not a virgin) female. She was mated to a virgin male 8 weeks after her capture and ate him. In all cases of cannibalism, the male remained motionless after mating and separation, while the female spent 3–10 min probing for a vulnerable location (usually the neck or genital area (Fig. 19.11a)) to start eating. One female of species no. 18 walked away for 15 min before returning to the male. No male offered any resistance or made any effort to escape while being eaten alive (Fig. 19.11b).

Postcoital cannibalism in laboratory-mated JCs is documented elsewhere (Davis, 1927; Tinkham and Rentz, 1969; S.J. Arnold, in Halliday, 1980, p. 11), but its significance is questionable, because the mating, ovipositional and nutritional status of the individuals was unknown. Fortunately, I have discovered three field matings, all involving species no. 18, with all three females having fresh, intact, uneaten spermatophores and adult males nearby. In one case (on 18 December), the female was eating the male. This finding of field sexual cannibalism validates the veracity of laboratory observations.

Multiplicity

Males can mate repeatedly (Table 19.7), the shortest interval being some 24 h, although a second mating was not attempted sooner. Mating trials started with a virgin male paired with three successive virgin females. He was allowed to get into position with each female but not to consummate, thus establishing receptivity of all involved. This male was then allowed to mate with a fourth female. On successive nights, this original male was presented with one of the three original females until he mated a second time. He was then presented with the third female on the next night, and so on. Male no. 2 mated four times over 10 days. For the four trial males, a normal-sized spermatophore was passed only on the first mating. In subsequent matings, both ampulla and spermatophylax were sequentially reduced in size, the ampulla being 66–75% normal size on the second mating and 30–66% normal size on the third mating. The spermatophylax is reduced to 50–66% normal size on the second mating and 25% normal size on the third and fourth matings.

Table 19.6. Life history of cannibalized males. See Table 19.2 for species' localities. All males listed were eaten on date mated.

Species no.	Male captured	Male adult	Female captured	Female adult	Mated
2	28/12/84	4/9/85	14/3/85	11/9/85	19/11/85
3	18/12/82	13/8/83	18/12/82	13/8/83	5/10/83
18	12/12/82	12/9/83	12/12/82	15/9/83	9/12/83
18*	Spring/89	24/9/89	Spring/89	24/10/89	14/1/90
18†	Spring/89	24/9/89	19/10/89	–	21/12/89
26	18/12/87	29/7/89	17/10/87	29/7/89	10/10/89

* Male mated 1/11/89 with different female; mated 7/1 and 14/1/90 with this female. She attempted to eat him on 7/1/90 but separated; ate him on 14/1/90. Female mated 21/12/89 with another male and did not eat, before mating with this male twice. She was ignored by a third male 2 days after eating second male.
† Female captured as adult.

Table 19.7. Male multiple mating trials in species no. 18. All dates in 1990.

Male no.	When captured	When became adult	Mating trial results				
1	24/5	4/10	Mated:	23/11	24/11*	27/11	
			Did not mate:	25/11	26/11		
2	27/5	26/9	Mated:	23/11	25/11	29/11	2/12
			Did not mate:	24/11	26/11	27/11	
3	20/5	22/9	Mated:	27/11	29/11		
4	3/6	18/9	Mated:	30/11	2/12		

*Mated 24 h after first mating.

Fig. 19.11. Postmating. (a) Female of species no. 18 (note fresh spermatophore – arrow) biting male's posterior end. (b) Same female several hours later having devoured her mate.

Females can also mate repeatedly (Table 19.8), one in 2 days (she rejected a male 24 h after the first mating) and another in 4 days. Trials consisted of mated females being paired with virgin males on successive nights. Before males will show an interest in these non-virgin females, the latter's external spermatophore remnants must disappear, which takes 1–3 days. Males will then get into proper position with the female (indicating her receptivity), but they may proceed no further. Another 1–3 days must then elapse until a male will mate with this female. This time schedule indicates that the second male apparently recognizes some substance from the first male, since the mating is aborted without a physical barrier (e.g. the hardened ampulla) interfering. One could test this hypothesis by removing spermatophore traces from the mated female and observing a second

Table 19.8. Female multiple mating trials in species no. 18. All dates in 1990.

Female no.	When captured	When became adult	Mating trial results					
1	3/6	8/10	Mated:	25/11	27/11*	2/12		
			Did not mate:	26/11†	27/11†	29/11†	30/11	
2	20/3	4/10	Mated:	23/11	27/11			
			Did not mate:	24/11§	25/11‡	26/11†		
3	10/6	9/10	Mated:	24/11	29/11			
			Did not mate:	25/11‡	26/11§	27/11†	2/12†	
4	6/8	21/9	Mated:	23/11	30/11			
			Did not mate:	24/11§	25/11‡	26/11‡	27/11	29/11†

* Rejected first male, mated with second male.
† No spermatophore remnants visible (if no symbol, remnants not checked).
‡ Ampulla remnants only visible.
§ Ampulla and spermatophylax remnants both visible.

male's response. In some trials, it was the female that showed a refractory period and rejected a receptive male.

Discussion

In JCs, the spermatophore (see Heller *et al.*, 1998, for detailed chemical analysis of katydid spermatophore) is comprised of two parts: a sperm-containing ampulla and a gelatinous, protein-rich spermatophylax, the latter usually the same size as to two times larger than the former. Near JC relatives (Gwynne, 1995), the katydids (Tettigoniidae) and sagebrush crickets (Haglidae), both have spermatophores structured similar to those of JCs, but not the king crickets or wetas (Anostostomatidae), which both lack a spermatophylax (Bateman and Toms, 1998). In the katydids and sagebrush crickets, the spermatophylax serves various functions including providing nutrition (male investment) for the female when she eats it and serving as a decoy (ejaculate protection) while the ampulla empties (Bowen *et al.*, 1984; Gwynne, 1988). Female katydids and sagebrush crickets eat the entire spermatophore after mating; neither king crickets nor wetas eat the ampulla (Bateman and Toms, 1998).

Male JCs invest between 3% and 7.5% of their wet body weight in each spermatophore. In 71 matings (68 laboratory and three field, involving 15 species), only one female JC ate the attached spermatophylax (see also Tinkham and Rentz, 1969), and then starting 3 h after transfer. The spermatophylax is still visible on some females after 96 h, although it normally dehydrates and disappears in less than 6 h. I do not see how spermatophylax nutrients could be physically absorbed by the female through her reproductive tract (see Bowen *et al.*, 1984). Thus, the function of the spermatophylax is unknown in JCs, unless its getting hard and dirty somehow prolongs fidelity in the female. The spermatophore complex, through some pheromone, may inhibit another male from copulating with this female until the structure falls off or is dislodged, as during oviposition. In female multiple mating trials (see Table 19.8), males rejected receptive females that still had spermatophore remnants visible. Unfortunately, I have been unable to test if oviposition will reverse a male's rejection of a particular female, because I cannot get females to lay eggs. In any case, the male rejection period is short, because all spermatophore vestiges are usually gone within 2–6 days, and males will again copulate with these females.

After mating, the ampulla quickly becomes hard and covered with dirt and could thus function as a physical barrier (= mating plug of Thornhill and Alcock, 1983; Loher and Dambach, 1989; reviewed in Simmons and Siva-Jothy, 1998) to further matings, until it falls off. Thus, the plug could confer a few days of forced fidelity if males were interested, although they seem not to be. In JCs, this chastity device, or the chemicals associated with it, appears at least as important in preventing remating, as a chemically induced female refractory period (*contra* katydid examples (Gwynne, 1997)).

In these 71 matings, the female ate the male

seven times. These, I believe, are the first described cases of male complicity in postcoital cannibalism, because the male lets the female attack him after completed spermatophore transfer. These six laboratory males had the opportunity to escape but made no such effort. In one laboratory mating, the female left the male for a full 15 min, during which time he did not move, before returning to consume him. Once attacked by their female, no male made any effort to escape or repel her onslaught.

I distinguish this true postcopulatory cannibalism from the situation found in the Australian redback spider, where male sacrifice occurs during sperm transfer (Andrade, 1996, 1998). Consumed male redbacks gain two advantages over non-cannibalized males: if the female was a virgin, she remates less frequently; if the female was previously mated, the second male's paternity is increased. Male redbacks are short-lived in comparison with the female and part of the male's intromittant organ always breaks off inside the female – two events not seen in JCs. Elgar (1992, p. 139) discusses possible male complicity in sexual cannibalism in the spider *Argiope aemura*, where the male was always killed sometime during the second insertion. Because of size differences between the sexes, male contribution to female fecundity is probably minimal, and the 'why' of the whole process is a mystery (Elgar, 1992).

Male complicity in postcoital cannibalism could be selected for under certain conditions (see Buskirk *et al.*, 1984). Specifically, if adult males greatly outnumber adult females, it is unlikely that males would mate more than once. Natural selection would act on males to optimize this one mating opportunity. The male's sacrifice (himself) appears to be the ultimate parental investment in his offspring. He can simultaneously raise the female's fecundity and his own fitness. Nevertheless, I am not sure what the male JC gets in return for his body. Because the adult female JC emergence period is several months long, a male might have repeated opportunities to mate with a virgin. If the female's eggs are already produced, this male may fertilize them, but the nutrients from his body can only assist in producing the next clutch (as in Vahed, 1998), which he may not fertilize, because females, at least those who do not eat their mates, are not faithful. Mating may be necessary to trigger some reproductive process such as egg formation or oviposition (Loher, 1984), which somehow ensures the eaten male's paternity.

Do female JCs that eat their mates become so physically large that other males are not able to mate with them for an extended time period? It is, in fact, just the opposite situation that makes for difficult matings: males of species no. 18, that are 50–100% larger than their mates, have trouble obtaining the proper orientation and may be unable to couple after 12 h of trying. Are old males senescent and more likely to permit females to eat them? Table 19.6 would indicate otherwise (see also Elgar and Nash, 1988), as four eaten males were virgins and adult for only 7–14 weeks. Although all looked healthy and robust during their first and only mating, these males may be 'old' as gauged by already severely declining numbers of field adult males. Most adult males of species no. 18 easily live 6 months or more in the laboratory. Does the scarcity of adult males in the field in all species by early winter, then, reflect widespread sexual cannibalism? I am now trying to ascertain if males copulating soon after becoming adult are more likely to run away than those mating for the first time after several months as adults. Are females that eat their mates shunned by subsequent males because of their recognized infidelity?

Could these males contain parasites that 'drive' them to be complicit (as in Hoek *et al.*, 1997) and thus ensure that the parasites are passed on to a new host or the host's eggs? Is this process similar to horse hair-worms 'driving' JCs to water, where the worms emerge from their host (see Weissman, Chapter 3, this volume), killing it, and complete the next stage of their cycle in the water (Poinar, 1991)?

When eaten, the male's investment in his offspring is greater than the female's; he is expected to reverse roles and be the discriminating mate (Thornhill and Alcock, 1983). Unfortunately, there may be confounding factors in this apparently paradoxical behaviour as it relates to JCs: these laboratory-raised males are sexually deprived and they lack female JCs from which to select, so they appear to mate with the first female they find. Still, I was surprised to find that these males can mate repeatedly, as often as twice in 24 h and four times in 10 days, though they pass both smaller ampulla and smaller spermatophylax with these short intervals (as opposed to Sakaluk and Smith, 1988; Gwynne and Simmons, 1990). Two weeks after capture as adults (and I assume not virgins), five males of species no. 9 each passed almost no spermatophylax on mating. Subsequently, four laboratory-raised virgin males of species no. 9 passed a normal-sized spermatophylax on their first copulation, indicating that repeat matings in the field also result in shrinking spermatophylaces.

Before the question 'Why do male JCs let themselves be eaten?' can be answered, several areas will need to be investigated. For example, the female reproductive cycle needs clarification. How soon after becoming adults will females mate? When are eggs present? Are virgin adult females always sexually receptive? How many clutches are possible? When and where are eggs laid? Is there ever parental care? Are any nutrients absorbed from the spermatophore?

What is the situation of sperm precedence? Is it first-male precedence (Gwynne, 1984; Gwynne and Brown, 1994)? Last-male precedence (Parker, 1984)? Or a random mixing (Wedell, 1991)? Is there cryptic female choice (LaMunyon and Eisner, 1993; Eberhard, 1996)? Can JC males deplete their sperm with repeat matings? Can they sense that they are 'low on sperm' and thus have little to lose by being eaten?

How important is female nutritional status (Simmons and Bailey, 1990; Fagan and Hurd, 1991; Lawrence, 1992; Gwynne, 1993; Hurd *et al.*, 1994; Simmons, 1994; Andrade, 1998; Maxwell, 1998) as to whether or not she eats the male? Is male nutritional status (senility) important (as in Mappes *et al.*, 1996)? Can females tell if males are virgins?

Is the age of the female JC important to the male? In situations with first-sperm precedence, older females are more likely to be mated and male protandry may be a common (Simmons *et al.*, 1994) strategy as males try to secure young (i.e. virgin) females. Studies (Lynam *et al.*, 1992; references in Simmons *et al.*, 1994) in insects demonstrate that males can distinguish between virgin and non-virgin females.

Can males 'predict' paternity (Simmons, 1995)? Can they tell if a female has mated and eaten her first mate? Do males somehow stimulate females to oviposit (as in Loher, 1984; Spence and Andersen, 1994)?

Despite some difficulties in working with these insects (e.g. long time to maturity; necessity to raise separately, inability to obtain ovipositions), JCs present some unique research opportunities. It is apparently advantageous for certain male JCs to let their mates eat them. We just have to figure out why.

Acknowledgements

I thank V.F. Lee, D.C. Lightfoot and M.S. Caterino for assistance above and beyond the call of duty in the field; and L.F. Baptista for sonagraph usage. T.J. Cohn, R.V. Dowell, D.T. Gwynne, B. John, V.F. Lee and D.C. Lightfoot provided valuable comments on an earlier draft, and T.J. Walker gave valuable discussion on temperature responses.

References

Alexander, R.D. and Otte, D. (1967) The evolution of genitalia and mating behavior in crickets (Gryllidae) and other Orthoptera. *Miscellaneous Publications Museum Zoology, University Michigan* 133, 1–62.

Andrade, M.C.B. (1996) Sexual selection for male sacrifice in the Australian redback spider. *Science* 271, 70–72.

Andrade, M.C.B. (1998) Female hunger can explain variation in cannibalistic behavior despite male sacrifice in redback spiders. *Behavioral Ecology* 9, 33–42.

Autrum, H. and Schneider, W. (1948) Vergleichende Untersuchungen uber den Erschutterungssinn der Insekten. *Zeitschrift für Vergleichende Physiologie* 31, 77–88.

Barth, F.G. (1990) Spider courtship: male vibrations, female responsiveness and reproductive isolation. In: Gribakin, F.G., Wiese, K. and Popov, A.V. (eds) *Sensory Systems and Communication in Arthropods.* Birkhauser Verlag, Basle, Boston, pp. 161–166.

Bateman, P.W. and Toms, R.B. (1998) Mating, mate guarding and male–male relative strength assessment in an African king cricket (Orthoptera: Mimnermidae). *Transactions of the American Entomological Society* 124, 69–75.

Bell, P.D. (1980) Transmission of vibrations along plant stems: implications for insect communication. *Journal of the New York Entomological Society* 88, 210–216.

Bowen, B.J., Codd, C.G. and Gwynne, D.T. (1984) The katydid spermatophore (Orthoptera: Tettigoniidae): male nutritional investment and its fate in the mated female. *Australian Journal of Zoology* 32, 23–31.

Brownell, P. and Farley, R.D. (1979) Detection of vibrations in sand by tarsal sense organs of the nocturnal scorpion, *Paruroctonus mesaensis*. *Journal of Comparative Physiology* 131, 23–30.

Buskirk, R.E., Frohlich, C. and Ross, K.G. (1984) The natural selection of sexual cannibalism. *American Naturalist* 123, 612–625.

Dambach, M. (1989) Vibrational responses. In: Huber, F., Moore, T.F. and Loher, W. (eds) *Cricket Behavior and Neurobiology*. Cornell University Press, Ithaca, New York, pp. 178–197.

Davis, A.C. (1927) Studies of the anatomy and histology of *Stenopelmatus fuscus* Hald. *University of California Publications in Entomology* 4, 159–208.

De Luca, P.A. and Morris, G.K. (1998) Courtship communication in meadow katydids: female preference for large male vibrations. *Behaviour* 135, 777–793.

Devetak, D. and Amon, T. (1997) Substrate vibration sensitivity of the leg scolopidial organs in the green lacewing, *Chrysoperla carnea*. *Journal of Insect Physiology* 43, 433–437.

Eberhard, W.G. (1996) *Female Control: Sexual Selection by Cryptic Female Choice*. Princeton University Press, Princeton, New Jersey, 501 pp.

Elgar, M.A. (1992) Sexual cannibalism in spiders and other invertebrates. In: Elgar, M.A. and Crespi, B.J. (eds) *Cannibalism: Ecology and Evolution among Diverse Taxa*. Oxford University Press, New York, pp. 128–155.

Elgar, M.A. and Nash, D.R. (1988) Sexual cannibalism in the garden spider *Araneus diadematus*. *Animal Behaviour* 36, 1511–1517.

Ewing, A.W. (1989) *Arthropod Bioacoustics, Neurobiology and Behaviour*. Cornell University Press, Ithaca, New York, 260 pp.

Fagan, W.F. and Hurd, L.E. (1991) Late season food level, cannibalism, and oviposition in adult mantids (Orthoptera: Mantidae): sources of variability in a field experiment. *Proceedings of the Entomological Society, Washington* 93, 956–961.

Field, L.H. and Bailey, W.J. (1997) Sound production in primitive Orthoptera from Western Australia: sounds used in defence and social communication in *Ametrus* sp. and *Hadrogryllacris* sp. (Gryllacrididae: Orthoptera). *Journal of Natural History* 31, 1127–1141.

Field, L.H. and Matheson, T. (1998) Chordotonal organs of insects. *Advances in Insect Physiology* 27, 1–228.

Gwynne, D.T. (1984) Male mating effort, confidence of paternity, and insect sperm competition. In: Smith, R.L. (ed.) *Sperm Competition and the Evolution of Animal Mating Systems*. Academic Press, New York, pp. 117–149.

Gwynne, D.T. (1988) Courtship feeding and the fitness of female katydids (Orthoptera: Tettigoniidae). *Evolution* 42, 545–555.

Gwynne, D.T. (1993) Food quality controls sexual selection in mormon crickets by altering male mating investment. *Ecology* 74, 1406–1413.

Gwynne, D.T. (1995) Phylogeny of the Ensifera (Orthoptera): a hypothesis supporting multiple origins of acoustic signalling, complex spermatophores and maternal care in crickets, katydids, and weta. *Journal of Orthoptera Research* 4, 203–218.

Gwynne, D.T. (1997) The evolution of edible 'sperm sacs' and other forms of courtship feeding in crickets, katydids and their kin (Orthoptera: Ensifera). In: Choe, J.C. and Crespi, B.J. (eds) *The Evolution of Mating Systems in Insects and Arachnids*. Cambridge University Press, New York, pp. 110–129.

Gwynne, D.T. and Brown, W.D. (1994) Mate feeding, offspring investment, and sexual differences in katydids (Orthoptera: Tettigoniidae). *Behavioural Ecology* 5, 267–272.

Gwynne, D.T. and Simmons, L.W. (1990) Experimental reversal of courtship roles in an insect. *Nature* 346, 172–174.

Halliday, T. (1980) *Sexual Strategy*. University of Chicago Press, Chicago, 158 pp.

Hassage, R.L., Stewart, K.W. and Zeigler, D.D. (1988) Female response to computer-simulated male drumming call variations in *Pteronarcella badia* (Plecoptera: Pteronarcyidae). *Annals of the Entomological Society of America* 81, 528–531.

Heller, K.-G. Faltin, S., Fleischmann, P. and von Helversen, O. (1998) The chemical composition of the spermatophore in some species of phaneropterid bushcrickets (Orthoptera: Tettigonioidea). *Journal of Insect Physiology* 44, 1001–1008.

Henry, C.S. (1983) Temperature-induced changes in the calls of the green lacewing, *Chrysoperla plorabunda* (Neuroptera: Chrysopidae). *Psyche* 90, 343–360.

Henry, C.S. (1984) The sexual behavior of green lacewings. In: Canard, M., Semeria, Y. and New, T.R. (eds) *Biology of Chrysopidae*. Dr W. Junk, Boston, pp. 101–110.

Henry, C.S. (1985a) The proliferation of cryptic species in *Chrysoperla* green lacewings through song divergence. *Florida Entomologist* 68, 18–38.

Henry, C.S. (1985b) Sibling species, call differences, and speciation in green lacewings (Neuroptera: Chrysopidae: *Chrysoperla*). *Evolution* 39, 965–984.

Henry, C.S. (1989) The unique purring song of *Chrysoperla comanche* (Banks), a green lacewing of the *rufilabris* species group (Neuroptera: Chrysopidae). *Proceedings of the Entomological Society, Washington* 91, 133–142.

Henry, C.S. (1993) *Chrysoperla johnsoni* (Neuroptera: Chrysopidae): acoustic evidence for full species status. *Annals of the Entomological Society of America* 86, 14–25.

Henry, C.S., Brooks, S.J., Johnson, J.B. and Duelli, P (1999a) Revised concept of *Chrysoperia mediter-*

ranea (Holzel), a green lacewing associated with conifers: courtship songs across 2800 kilometers of Europe (Neuroptera: Chrysopidae). *Systematic Entomology* 24, 335–350.

Henry, C.S., Wells, M.L.M. and Simon, C.M. (1999b) Convergent evolution of courtship songs among cryptic species of the *carnea* group of green lacewings (Neuroptera: Chrysopidae: *Chrysoperla*). *Evolution* 53, 1165–1179.

Hoek, R.M., van Kesteren, R.E., Smit, A.B., de Jong-Brink, M. and Geraerts, W.P.M. (1997) Altered gene expression in the host brain caused by a trematode parasite: neuropeptide genes are preferentially affected during parasitosis. *Proceedings of the National Academy of Sciences, USA* 94, 14072–14076.

Hunt, R.E. (1993) Role of vibrational signals in mating behavior of *Spissistilus festinus* (Homoptera: Membracidae). *Annals of the Entomological Society of America* 86, 356–361.

Hurd, L.E., Eisenberg, R.M., Fagan, W.F., Tilmon, K.J., Snyder, W.E., Vandersall, K.S., Datz, S.G. and Welch, J.D. (1994) Cannibalism reverses male-biased sex ratio in adult mantids: female strategy against food limitation? *Oikos* 69, 193–198.

Hutchings, M. and Lewis, B. (1983) Insect sound and vibration receptors. In: Lewis, B. (ed.) *Bioacoustics: a Comparative Approach*. Academic Press, San Francisco, pp. 181–205.

Kalmring, K. and Kuhne, R. (1983) The processing of acoustic and vibrational information in insects. In: Lewis, B. (ed.) *Bioacoustics: a Comparative Approach*. Academic Press, San Francisco, pp. 261–282.

Kalmring, K., Keuper, A. and Kaiser, W. (1990) Aspects of acoustic and vibratory communication in seven European bushcrickets. In: Bailey, W.J. and Rentz, D.C.F. (eds) *The Tettigoniidae: Biology, Systematics and Evolution*. Crawford House Press, Bathurst, Australia, pp. 191–216.

Kraus, W.F. (1989) Surface wave communication during courtship in the giant water bug, *Abedus indentatus* (Heteroptera: Belostomatidae). *Journal of the Kansas Entomological Society* 62, 316–328.

LaMunyon, C.W. and Eisner, T. (1993) Postcopulatory sexual selection in an arctiid moth (*Utetheisa ornatrix*). *Proceedings of the National Academy of Sciences, USA* 90, 4689–4692.

Lawrence, S.E. (1992) Sexual cannibalism in the praying mantid, *Mantis religiosa*: a field study. *Animal Behaviour* 43, 569–583.

Loher, W (1984) Behavioral and physiological changes in cricket females after mating. *Advances in Invertebrate Reproduction* 3, 189–201.

Loher, W. and Dambach, M. (1989) Reproductive behavior. In: Huber, F., Moore, T.F. and Loher, W. (eds) *Cricket Behavior and Neurobiology*. Cornell University Press, Ithaca, New York, pp. 43–82.

Lynam, A.J., Morris, S. and Gwynne, D.T. (1992) Differential mating success of virgin female katydids *Requena verticalis* (Orthoptera: Tettigoniidae). *Journal of Insect Behavior* 5, 51–59.

Macdonald, J.A. (1981) Temperature compensation in the peripheral nervous system: Antarctic vs. temperate poikilotherms. *Journal of Comparative Physiology A, Sensory Neural and Behavioral Physiology* 142, 411–418.

McVean, A. and Field, L.H. (1996) Communication by substratum vibration in the New Zealand tree weta, *Hemideina femorata* (Stenopelmatidae: Orthoptera). *Journal of Zoology, London* 239, 101–122.

Maketon, M., Stewart, K.W., Kondratieff, B.C. and Kirchner, R.F. (1988) New descriptions of drumming and evolution of the behavior of North American Perlodidae (Plecoptera). *Journal of the Kansas Entomological Society* 61, 161–168.

Mappes, J., Alatalo, R.V., Kotiaho, J. and Parri, S (1996) Viability costs of condition-dependent sexual male display in a drumming wolf spider. *Proceedings of the Royal Society, London, B* 263, 785–789.

Maxwell, M.R. (1998) Lifetime mating opportunities and male mating behaviour in sexually cannibalistic praying mantids. *Animal Behaviour* 55, 1011–1028.

Michelsen, A. (1979) Insect ears as mechanical systems. *American Scientist* 67, 696–706.

Moring, J.B., Hassage, R.L. and Stewart, K.W. (1991) Behavioral interactions among male and female nymphs of the predatory stonefly *Hydroperla crosbyi* (Plecoptera: Perlodidae). *Annals of the Entomological Society of America* 84, 207–211.

Morris, G.K. (1980) Calling display and mating behaviour of *Copiphora rhinoceros* Pictet (Orthoptera: Tettigoniidae). *Animal Behaviour* 28, 42–51.

Parker, G.A. (1984) Sperm competition and the evolution of animal mating strategies. In: Smith, R.L. (ed.) *Sperm Competition and the Evolution of Animal Mating Systems*. Academic Press, New York, pp. 1–60.

Poinar, G.O. (1991) Hairworm (Nematomorpha: Gordioidea) parasites of New Zealand wetas (Orthoptera: Stenopelmatidae). *Canadian Journal of Zoology* 69, 1592–1599.

Polis, G.A. (1979) Prey and feeding phenology of the desert sand scorpion *Paruroctonus mesaensis* (Scorpionidae: Vaejovidae). *Journal of Zoology, London* 188, 333–346.

Prestwich, K.N. and Walker, T.J. (1981) Energetics of singing in crickets: effect of temperature in three trilling species (Orthoptera: Gryllidae). *Journal of Comparative Physiology* 143, 199–212.

Ragge, D.R. and Reynolds, W.J. (1998) *The Songs of the Grasshoppers and Crickets of Western Europe*. Harley Books, Essex, England, 591 pp.

Rentz, D.C.F. and Weissman, D.B. (1981) Faunal affinities, systematics, and bionomics of the Orthoptera of the California Channel Islands. *University of California Publications in Entomology* 94, 1–240.

Romer, H. (1998) The sensory ecology of acoustic communication in insects. In: Hoy, R.R., Popper, A.N. and Fay, R.R. (eds) *Comparative Hearing: Insects*. Springer-Verlag, New York, pp. 63–96.

Rupprecht, R. (1981) A new system of communication within Plecoptera and a signal with a new significance. *Biology of Inland Waters* 2, 30–35.

Sakaluk, S.K. and Smith, R.L. (1988) Inheritance of male parental investment in an insect. *American Naturalist* 132, 594–601.

SAS Institute (1989) SAS *User's Guide*, 4th edn. SAS Institute, Cary, North Carolina, 1686 pp.

Saxena, K.N. and Kumar, H. (1984) Acoustic communication in the sexual behaviour of the leafhopper, *Amrasca devastans*. *Physiological Entomology* 9, 77–86.

Shaw, S. (1994) Detection of airborne sound by a cockroach 'vibration detector': a possible missing link in insect auditory evolution. *Journal of Experimental Biology* 193, 13–47.

Simmons, L.W. (1994) Courtship role reversal in bush crickets: another role for parasites? *Behavioural Ecology* 5, 259–266.

Simmons, L.W. (1995) Relative parental expenditure, potential reproductive rates, and the control of sexual selection in katydids. *American Naturalist* 145, 797–808.

Simmons, L.W. and Bailey, W.J. (1990) Resource influenced sex roles of zaprochiline tettigoniids (Orthoptera: Tettigoniidae). *Evolution* 44, 1853–1868.

Simmons, L.W. and Siva-Jothy, M.T. (1998) Sperm competition in insects: mechanisms and the potential for selection. In: Birkhead, T.R. and Moller, A.P. (eds) *Sperm Competition and Sexual Selection*, Academic Press, San Diego, California, pp. 341–434.

Simmons, L.W., Llorens, T., Schinzig, M., Hosken, D. and Craig, M. (1994) Sperm competition selects for male mate choice and protandry in the bushcricket, *Requena verticalis* (Orthoptera: Tettigoniidae). *Animal Behaviour* 47, 117–122.

Sivinski, J. (1983) Predation and sperm competition in the evolution of coupling durations, particularly in the stick insect *Diapheromera veliei*. In: Gwynne, D.T. and Morris, G.K. (eds) *Orthopteran Mating Systems*. Westview Press, Boulder, Colorado, pp. 147–162.

Spence, J.R. and Andersen, N.M. (1994) Biology of water striders: interactions between systematics and ecology. *Annual Review of Entomology* 39, 101–128.

Stewart, K.W. (1997) Vibrational communication in insects. Epitome in the language of stoneflies? *American Entomologist* 43(2), 81–91.

Stewart, K.W. and Zeigler, D.D. (1984a) Drumming behavior of twelve North American stonefly (Plecoptera) species: first descriptions in Peltoperlidae, Taeniopterygidae and Chloroperlidae. *Aquatic Insects* 6, 49–61.

Stewart, K.W. and Zeigler, D.D. (1984b) The use of larval morphology and drumming in Plecoptera systematics, and further studies of drumming behavior. *Annals of Limnology* 20, 105–114.

Stewart, K.W., Szczytko, S.W. and Maketon, M. (1988) Drumming as a behavioral line of evidence for delineating species in the genera *Isoperla*, *Pteronarcys*, and *Taeniopteryx* (Plecoptera). *Annals of the Entomological Society of America* 81, 689–699.

Stewart, K.W., Bottorff, R.L., Knight, A.W. and Moring, J.B. (1991) Drumming of four North American euholognathan stonefly species, and a new complex signal pattern in *Nemoura spiniloba* Jewett (Plecoptera: Nemouridae). *Annals of the Entomological Society of America* 84, 201–206.

Stewart, K.W., Abbott, J.C. and Bottorff, R.L. (1995) The drumming signals of two stonefly species *Cosumnoperla hypocrena* (Perlodidae) and *Paraperla wilsoni* (Chloroperlidae); a newly discovered duet pattern in Plecoptera. *Entomological News* 106, 13–18.

Stiedl, O. and Kalmring, K. (1989) The importance of song and vibratory signals in the behaviour of the bushcricket *Ephippiger ephippiger* Fiebig (Orthoptera, Tettigoniidae): taxis by females. *Oecologia* 80, 142–144.

Thanthianga, C. and Mitchell, R. (1987) Vibrations mediate prudent resource exploitation by competing larvae of the bruuchid bean weevil *Callosobruchus maculatus*. *Entomologia Experimentalis et Applicata* 44, 15–21.

Thornhill, R. and Alcock, J. (1983) *Evolution of Insect Mating Systems*. Harvard University Press, Cambridge, Massachusetts, 547 pp.

Tinkham, E.R. (1968) Studies in Nearctic desert sand dune Orthoptera. Part XI. A new arenicolous species of *Stenopelmatus* from the Coachella Valley with key and biological notes. *Great Basin Naturalist* 28, 124–131.

Tinkham, E.R. and Rentz, D.C. (1969) Notes on the bionomics and distribution of the genus *Stenopelmatus* in central California with the description of a new species (Orthoptera: Gryllacrididae). *Pan-Pacific Entomologist* 45, 4–14.

Vahed, K. (1998) Sperm precedence and the potential of the nuptial gift to function as paternal investment in the Tettigoniid *Steropleurus stali* Bolivar (Orthoptera: Tettigoniidae: Epphippigerinae). *Journal of Orthoptera Research* 7, 223–226.

Walker, T.J. (1962) Factors responsible for intraspecific variation in the calling songs of crickets. *Evolution* 16, 407–428.

Walker, T.J. (1964) Cryptic species among sound-producing ensiferan Orthoptera (Gryllidae and Tettigoniidae). *Quarterly Review Biology* 39, 345–355.

Walker, T.J. (1975) Effects of temperature on rates in poikilotherm nervous systems: evidence from the calling songs of meadow katydids (Orthoptera: Tettigoniidae: *Orchelimum)* and reanalysis of published data. *Journal of Comparative Physiology* 101, 57–69.

Wedell, N. (1991) Sperm competition selects for nuptial feeding in a bushcricket. *Evolution* 45, 1975–1978.

Weissman, D.B., Rentz, D.C.F, Alexander, R.D. and Loher, W. (1980) Field crickets (*Gryllus* and *Acheta*) of California and Baja California, Mexico (Orthoptera: Gryllidae: Gryllinae). *Transactions of the American Entomological Society* 106, 327–356.

Wells, M.M. and Henry, C.S. (1994) Behavioral responses of hybrid lacewings (Neuroptera: Chrysopidae) to courtship songs. *Journal of Insect Behavior* 7, 649–662.

Wilcox, R.S. (1979) Sex discrimination in *Gerris remigis*: role of a surface wave signal. *Science* 206, 1325–1327.

Zeigler, D.D. and Stewart, K.W. (1986) Female response thresholds of two stonefly (Plecoptera) species to computer-simulated and modified male drumming calls. *Animal Behaviour* 34, 929–931.

Appendix A: Effect of Temperature on Drumming Rates of Jerusalem Crickets (*Stenopelmatus*: Stenopelmatidae: Orthoptera)

Thomas J. Walker and David B. Weissman

Internal temperatures of most insects closely follow the surrounding, external temperatures. As the internal temperature rises or falls, biochemical processes speed up or slow down. Therefore, it is not surprising that the rates at which crickets and katydids rub their wings as they produce their calling songs is a function of ambient temperature, with higher wing-stroke rates corresponding to higher ambient temperatures. What is surprising, when first observed, is that the wing-stroke rate in calling crickets and katydids is a linear function of temperature. After all, physiologists generally reduce rate vs. temperature effects to Q_{10} values, thus implying that rate is an exponential function of temperature. Furthermore, if wing-stroke rates are linearly regressed against temperature for the calling songs of a number of closely related species, the regression lines tend to converge at a wing-stroke rate of 0 (Walker, 1962, 1975). The effect of these regularities is that, if the wing-stroke rates of the calling songs of a group of related species of crickets or katydid species have been measured at a range of temperatures, a point of convergence at $y = 0$ can be estimated. Then, if the wing-stroke rate of another species of that group is known at a single temperature, its rates at all other temperatures can be estimated. For example, the x when $y = 0$ for the extrapolated regressions of wing-stroke rates of nine species of trilling *Oecanthus* (Gryllidae) ranges from 2.9 to 5.7 (Walker, 1962). The mean, 4.3, is the estimated convergence at $y = 0$.

In this study, the drummings of three species of Jerusalem crickets (JCs) were recorded at a range of temperatures, making it the first study of the effects of temperature on rates within stenopelmatids.

Methods

Weissman (Chapter 19, this volume) describes how temperatures were measured and how drummings were recorded and their rates determined.

For each species and sex, drum rate was regressed against temperature. When the regressions were not significantly different, the data were pooled and the pooled data regressed against temperature. For the four linear regression equations that resulted, a non-linear model was used to force them to converge at temperature x and rate y, with both x and y being unknowns. The non-linear model was fitted using PROC NLIN of the SAS System (SAS Institute, 1989).

Results

Regression of drum rate on temperature was significant for each sex and species, but the regressions for males and females were not significantly different for species no. 27 (Fig. 19.A.1). When the lines were forced to converge with PROC NLIN, x was -4.2 ± 8.2°C and y was 1.2 ± 4.4 (95% confidence intervals).

Discussion

The wide confidence intervals for the forced convergence of the four regression lines for JCs show that their tendency to converge when extrapolated is weak. If convergence needs to be assumed to predict lines for other species (or sexes) based on a single datum, two estimates compete: the empirically most likely convergence point is $x = -4.2$, $y = 1.2$, whereas the point $x = -4.2$, $y = 0$ is a better choice based on the results of more extensive studies on a variety of poikilothermic groups, including gryllids and tettigoniids (Walker, 1975; Macdonald, 1981). In either case, the formula for predicting the rate at any other temperature from a single x, y pair can be derived algebraically.

Let T_1 = temperature of single datum, D_1 = drums s^{-1} of single datum, T_c = temperature of point of convergence, D_c = drum rate of point of convergence, T_2 = temperature for which drum rate is to be estimated and D_2 = estimated drums s^{-1} at temperature T_2. Then:

$$D_2 = D_1 + (T_2 - T_1)\left(\frac{D_1 - D_c}{T_1 - T_c}\right)$$

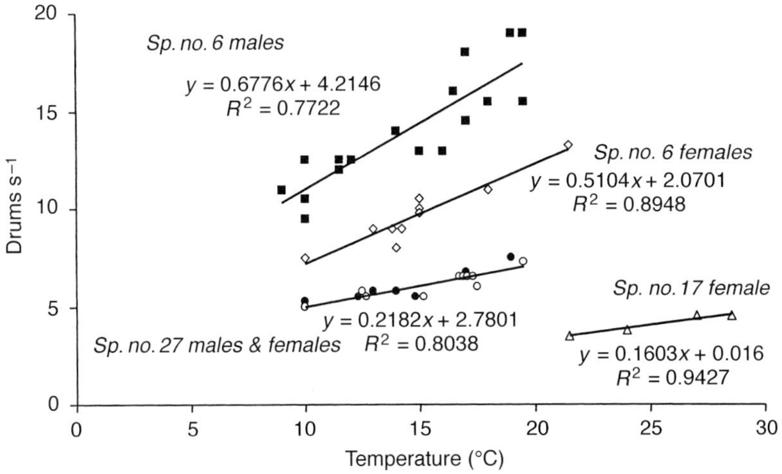

Fig. 19.A.1. Drum rate as a function of temperature in three species of Jerusalem crickets. Males are represented by closed symbols and females by open ones. In species no. 6, but not in species no. 27, regressions for males and females were significantly different. When points were identical, they were jittered horizontally for displaying on the graph (but not for calculations). Such points were $x = 14$, $y = 9$ ($n = 2$); 12.5, 5.5 ($n = 2$); 15, 5.5 ($n = 2$); and 17.0, 6.5 ($n = 4$).

Q_{10} values are defended theoretically on the basis of the effects of temperature on *in vitro* chemical reactions, whereas the observed linearity of rate vs. temperature regressions in many studies of poikilotherms long had no theoretical basis. However, in 1981, J.A. Macdonald addressed the question and described two parameters likely to limit the velocity of spike propagation in nerve fibres, both of which approximate linear functions of temperature. These were axoplasmic conductivity and maximum membrane conductivity during sodium activation.

Part V
Reproduction and Development

20
The Reproductive Biology and the Eggs of New Zealand Anostostomatidae

Ian A.N. Stringer
Ecology Group, Institute of Natural Resources, Massey University, PO Box 11222, Palmerston North, New Zealand

Introduction

Three main areas of interest relate to the reproductive biology of weta (Anostostomatidae). First, these insects appear to be typically *k*-strategists because many are large, all appear to have long developmental periods and some, at least, are long-lived as adults. In addition, many have a low reproductive rate in relation to other insects (Ramsay, 1978; Gibbs, 1994). Their size and longevity allow them to occupy the niches in New Zealand that mice and rats have elsewhere in the world. This has happened over a long period of isolation in the absence of mammals (Daugherty *et al.*, 1993). Secondly, adult male tusked weta and species of *Hemideina* have, respectively, mandibular horns and conspicuously enlarged heads with mandibles that are used for fighting other males (Field and Deans, Chapter 10, this volume). Male *Hemideina* fight to defend harems of females in cavities in wood or under rocks (Sandlant, 1981; Field and Sandlant, 1983; Gwynne and Jamieson, 1998; see also Field, Chapter 18, this volume). This is a life-history trait that is more typical of certain vertebrates (Gwynne and Jamieson, 1998). Fights between male tusked weta seem to be associated with defending a good refuge and it is possible that such refuges may be more attractive to females (Gibbs, 1998).

The third area of interest relates to the Anostostomatidae being a primitive orthopteran family. It follows that some aspects of their reproductive biology are quite likely to show ancient features, so further investigation could provide insight into the evolution of other Orthoptera. One area of particular interest is that of sexual selection and mating systems. Here, New Zealand giant weta (species of *Deinacrida*) appear to have no courtship behaviour. This is in marked contrast to most other orthopterans (Gwynne, 1995, 1997). Male giant weta will even attempt homosexual copulations (Richards, 1994; see also Field and Jarman, Chapter 17, this volume) and such behaviour could be either primitive or derived, but in any event it bears more detailed investigation.

Much of the information on weta reproductive biology is fragmented or incomplete. Some is incidental material scattered through unpublished scientific reports and student theses, which are often difficult to access outside New Zealand. In this chapter, I attempt to bring together all of the written information on the reproductive biology of weta which is not included elsewhere in this book (see Field and Jarman, Chapter 17; Stringer and Cary, Chapter 21, this volume). I have included information on weta eggs but not on the histology of the reproductive organs. The reader is referred, instead, to Maskell (1927), on the tree weta *Hemideina crassidens* (Blanchard), to Domett (1996), on the giant weta, *Deinacrida mahoenui* Gibbs, and to O'Brien and Field (Chapter 8, this volume).

Reproductive Traits

Sex ratio

The sex ratio of adult weta is generally not significantly different from a male to female (M : F) ratio of 1 : 1 (Table 20.1), although it can vary depending on a number of factors. In adult *Hemiandrus,* for example, it changes with the time of year. From mid-November to early January, it is between 1 : 0.56 and 1 : 1.56 M : F in the field, but the proportion of males decreases thereafter until only females remain by early March (Wahid, 1978). Hamilton (1991) also suggested that the time of year might affect the sex ratio of *Hemideina maori* (Pictet & Saussure) and Brown (1995) found that male *Deinacrida fallai* Salmon become adult earlier than females in captivity and this affects their sex ratio (Table 20.1).

The apparent sex ratio of weta in the field can also be affected by the method of collection or observation. Pitfall traps catch significantly fewer male *Hemiandrus maculifrons* (Walker) than females (1 : 3.57 M : F), even though there is no significant

Table 20.1. Sex ratio of adult weta.

Species	Sex ratio male : female	*n*	Sig. diff. from 1 : 1	Comments	Author
Deinacrida fallai	1 : 4	–	–	Captive reared	Richards, 1973
Deinacrida fallai	1 : 0.07	15	$P < 0.01$	Captive: March	Brown, 1995
	1 : 0.8	34	NS	Captive: June	
Deinacrida heteracantha	1 : 4	–	–	Captive reared	Richards, 1973
Deinacrida rugosa	1 : 0.81	29	NS	Mana Is.	Meads and Moller, 1977
Deinacrida mahoenui	1 : 0.81	646	NS	Mahoenui	Sherley and Hayes, 1993
Deinacrida mahoenui	1 : 0.99	392	NS	Mahoenui	Richards, 1994
Hemideina crassidens	1 : 1.56	45	NS	Near Upper Hutt	Asher, 1977
Hemideina crassidens	1 : 0.85	252	NS	Stephens Is./ by mark–recapture	Moller, 1978a
Hemideina crassidens	1 : 1.63	21	NS	Arapawa Is.	Moller, 1978b
	1 : 1.58	24	NS	Maud Is.	Meads and Moller, 1978
Hemideina crassidens	1 : 1.86	–	$P < 0.05$	In galleries	Field and Sandlant, 1983
Hemideina crassidens	1 : 0.85	235	NS	Stephens Is.	Moller, 1985
Hemideina crassidens	1 : 0.72	82	NS	Near Hutt valley	Spencer, 1995
Hemideina crassidens	~1 : 1	–	–	Pine logs, Nelson	Lemke, 1994
Hemideina ricta	1 : 0.77	211	NS	Banks Peninsula	Townsend, 1995
Hemideina femorata	1 : 0.96	122	NS	In galleries	Sandlant, 1981
Hemideina femorata	1 : 1.84	71	$P < 0.05$	Banks Peninsula	Townsend, 1995
Hemideina maori	1 : 0.89	124	NS	Rock and Pillar range	Hamilton, 1991
Hemiandrus maculifrons	1 : 0.93	–	NS	Cass	Cary, 1981
Hemiandrus subantarcticus	1 : 103	–	NS	Snares Is.	Butts, 1983
Hemiandrus sp.	1 : 1.63	–	–	Horotane Valley	Wahid, 1978
Hemiandrus sp.	1 : 2.23	–	–	Captivity	Wahid, 1978
Raukumara tusked weta	1 : 1.52	34	NS	Raukumara	McIntyre, 1998

difference between the numbers of males or females collected by hand (Table 20.1). The proportion of male and female weta found during night-time searches can be affected by a number of factors, including their 'apparency' to the observer (Rufaut, 1995). Males of *H. crassidens* and *Hemideina femorata* Hutton spend more time in and around galleries in comparison with females (Barrett and Ramsay, 1991; L.H. Field, Christchurch, 1998, personal communication), so, if tree-trunks are searched predominantly, this may favour males, but, if the search concentrates more on vegetation, this may favour finding females. The time of night may also play a part, because more adult male *H. crassidens* can be active just before dawn (Field and Sandlant, 1983; Rufaut, 1995).

Artificial galleries are being increasingly used for surveying tree weta, but harems are often formed in them and this biases the sex ratio towards females. Such galleries may, nevertheless, show seasonal changes in sex ratio. For example, Rufaut (1995) reported that fewer male than female *H. crassidens* occupied them in December (1 : 1.24 M : F), whereas fewer females occupied them in March (1 : 0.79 M : F). However, these sex ratios did not differ significantly from 1 : 1.

The presence of introduced mammalian predators may affect the apparent sex ratio, because more active female *H. crassidens* were observed per hour on Maud Island and Stephens Island, which are rodent-free, than on locations where rodents are present (Rufaut, 1995). This is consistent with the suggestion that ground predators should increase female mortality during oviposition, thus affecting the sex ratio, age and sex structure and also the behaviour of tree weta (Moller, 1985). However, in one test of the effects of Polynesian rat (*Rattus exulans* (Peale)) eradication on Nukuwaiata Island, no change was found in the sex ratio of *H. crassidens* between samples taken 7 months and 19 months after the eradication (Rufaut, 1995). Differences were, however, observed in the activity patterns of male and female weta which may relate to the presence of these rats. The proportion of male weta that were active at night increased in relation to the proportion of active females 1 year following the eradication (Rufaut, 1995).

Sexual dimorphism

Apart from the external genitalia, which are described by Stringer and Cary (Chapter 21, this volume), the most noticeable sexually dimorphic feature amongst all weta is the enlarged heads of adult male *Hemideina* and the mandibular horns of tusked weta. The head of male *H. crassidens* can be up to 45% of the body length, whereas the corresponding proportion for the head of adult females is about 25% (Asher, 1977; see also Field and Deans, Chapter 10, this volume). Head size – in particular, the gape of the mandibles – plays an important role in male–male agonistic encounters in *H. crassidens*, *H. femorata* and *H. maori* and this influences the outcome of disputes over access to galleries and harems, which, in turn, influences reproductive success in some species of *Hemideina* (Hudson, 1920; Field, 1980; Sandlant, 1981; Deans, 1982; Field and Sandlant, 1983; Gwynne and Jamieson, 1998). In general, males of *H. crassidens* and *H. femorata* occupy galleries just large enough to allow their heads to enter (Sandlant, 1981; Rufaut, 1995; Spencer, 1995) and this influences their success in keeping harems. If a male with a head larger than that of the resident male can enter, it usually displaces the resident male (see Field, Chapter 18, this volume).

The mating strategy of *H. crassidens* males is particularly interesting. Males of this species can become adult at any instar between seven and ten (see Table 21.1, Stringer and Cary, Chapter 21, this volume) and their mating strategy varies according to when they become adult (Spencer, 1995). Males that mature in the eighth instar have relatively small heads and tend to search outside at night and mate with any females they find wandering about. In contrast, tenth-instar males with the largest heads are not as successful at mating in the open but spend most of their time in galleries, where they usually mate. The most frequently found adult males are ninth instars. These can adopt either strategy but have better mating success in the open than either eighth- or tenth-instar adults (Spencer, 1995).

Male tusked weta also mature over a range of instars. The body size of male Raukumara tusked weta varies by an equivalent length of one or possibly two instars (e.g. metafemur length range 29–41 mm) (McIntyre, 1998). Adult males of the giant tusked weta *Motuweta isolata* Johns vary in size. The largest individuals possess long (20–30 mm), robust mandibular horns, which curve towards each other so that the tips cross over, while other adults have shorter, non-crossing tusks (comparative discussion in Field and Deans, Chapter 10, this volume).

Other sexually dimorphic features are not usually as obvious. Adult female *Deinacrida* are generally larger and weigh more than males. Some of their sclerites are larger than those of males, whereas other sclerites are approximately the same size in both sexes (Richards, 1973; Field, 1980; Sherley, 1992). Adult female *Hemideina* also usually weigh more and are larger than males, even though many of the sclerites and limbs of females are smaller than those of males. This follows because the abdomens of females can be much larger than those of males (Asher, 1977; Moller, 1985). Females of *Hemiandrus subantarcticus* (Salmon) are also larger than males (Butts, 1983). The cerci of male *Hemideina* and some *Deinacrida* are more elongated and curve inwards, whereas those of females are straight and relatively shorter (Barrett and Ramsay, 1991). This can be a useful feature for distinguishing adults from immature males (see Stringer and Cary, Chapter 21, this volume). The only author who described a function for such curved cerci was Ramsay (1955), who suggested that they are used to grip the female during copulation in *Deinacrida rugosa* Buller.

Only one polymorphic feature, the colour morphs of Mahoenui weta *D. mahoenui*, is reported to show sexual differences in frequency. Mahoenui weta have two colour morphs: one is dark brown and the other is yellow with dark brown areas and speckling (Richards, 1994). The ratio of yellow to brown morphs in males is 1 : 3.35 ($n = 320$) and this differs significantly from the ratio in females (1 : 1.56 ($n = 261$)) (Sherley and Hayes, 1993). Richards (1994) reported even higher significantly different ratios of yellow to brown morphs of 1 : 5.67 ($n = 195$) in males and 1 : 2.33 ($n = 190$) in females. Nothing else is known about these colour differences. Interestingly, *H. maori* has yellow, black and intermediate-colour polymorphs, but none of these differ significantly from a 1 : 1 ratio of M : F. However, black males have significantly longer mandibles than intermediate-colour morphs (Hamilton, 1991).

Longevity and seasonality of reproduction

Observations on seasonality of mating and ovipositing are given in Table 20.2. Little information is available about maturation period and reproductive lifespan. Total lifespans, in which maturation was followed until ecdysis to the adult, are only known for giant weta (Table 20.3). Richards (1973) reported that the first matings by

Table 20.2. Time of year when reproductive behaviour has been reported.

Species	Mating	Oviposition	Comments	Author
Deinacrida fallai	Apr–Jan	Jun–Nov	Captive	Richards, 1973
Deinacrida fallai		May	Poor Knights Is.	Watt, 1984
Deinacrida fallai	Mar–Jun (max. April)	Apr–May	Captive	Brown, 1995
Deinacrida heteracantha	Nov–Mar	Nov–Mar	Captive	Richards, 1973
Deinacrida rugosa	Nov–May	Oct–May	Captive	Ramsay, 1955
Deinacrida rugosa	–	Apr–May	Stephens Is.	Meads, 1976
Deinacrida mahoenui	–	Mar–May	Captive	Barrett and Ramsay, 1991
Deinacrida mahoenui	–	Jan–May	Captive	Richards, 1994
Hemideina crassidens	Sept–May	Apr–May	Captive	Barrett and Ramsay, 1991
Hemideina crassidens	–	Apr	Arapawa Is.	Moller, 1978b
Hemideina crassidens	–	Jan	Stephens Is.	Moller, 1985
Hemideina maori	–	Apr	Captive	Barrett and Ramsay, 1991
Hemideina ricta	Apr–May, Nov	Apr–Jul	Captive	Townsend, 1995
Hemiandrus maculifrons	–	All year	Cass	Cary, 1981
Hemiandrus pallitarsis	–	Dec–Feb	Captive	Barrett and Ramsay, 1991
Motuweta isolata	Mar–Aug	Mar–Aug	Middle Mercury Is.	McIntyre 1998
Motuweta isolata	–	Sep	Middle Mercury Is.	I. Southey in McIntyre 1992
Motuweta isolata	–	Oct	Captive	Meads, 1995
Raukumara tusked weta	Jun–Aug	Jun–Aug	Raukumara	McIntyre, 1998
Raukumara tusked weta	–	Mar–May	Captive	Gibbs, 1998

Table 20.3. Reported lifespans of adult weta (given as mean ± SE and/or range).

Species	Male	Female	Comment	Author
Deinacrida fallai	309 days	210–293 days	Captive (*n* = 1, 4)	Richards, 1973
Deinacrida fallai	~35–70 days	~35–49 days	Captive (*n* = 3, 3)	Brown, 1995
Deinacrida heteracantha	122–358 days	112–160 days	Captive (*n* = 4, 6)	Richards, 1973
Deinacrida rugosa	1.2–14 months	1.2–14 months	Captive	Ramsay, 1955
Deinacrida mahoenui	44–331 days (149 ± 27days)	92–356 days (189 ± 31 days)	Captive (*n* = 10, 10) (NS difference between sexes)	Richards, 1994
Deinacrida mahoenui	~3–4 months	~3–4 months	Mahoenui	Richards, 1994

four male *Deinacrida heteracantha* White occurred 6–35 days after the imaginal ecdysis and five females of this species first mated 4–35 days after moulting. These females mated after their weight had increased up to four times their weight at ecdysis. Four of the females laid their first eggs between 36 and 149 days after ecdysis and the fifth female never laid eggs. Richards (1973) reared only one male *D. fallai,* which first mated 100 days after ecdysis, and three females, which first mated after 89–125 days as adults. These females had doubled their weight after ecdysis when they first oviposited at 94–126 days. Brown (1995) reported that three female *D. fallai* mated 8 days after the adult ecdysis and the earliest a male mated was 13 days after ecdysis. Oviposition occurred approximately 4.5–13 weeks after mating. However, males became adult sooner than females and had to delay mating until females were available. Once oviposition commences, it continues throughout the remainder of the life of both *D. heteracantha* and *D. fallai* (Richards, 1973). The above account suggests that copulation acts as a releaser for oviposition, as it does in many insects (Sehnal, 1985). The only other reported evidence that suggests a releaser action is by Richards (1994), who noted that Mahoenui weta oviposited 2–3 weeks following the first observed copulation. No postreproductive periods have yet been reported for weta, although some period of senescence is likely to occur. This is certainly implied for female *Hemiandrus*, which are reported to guard their eggs and sometimes also the first-instar nymphs in their burrows for some time before the adults die (Salmon, 1950; Miller, 1971; Wahid, 1978; Cary, 1981; Barrett and Ramsay, 1991; van Wyngaarden, 1995).

Courtship and Copulation

Mating behaviour

The following account is restricted to an overview of mating behaviour in New Zealand stenopelmatids. The reader is referred to Field and Jarman (Chapter 17, this volume) for a detailed description of the mating behaviour of *H. crassidens* and information on the female pheromone that elicits male sexual behaviour.

Mating (and possible courtship) is observed at night in most weta (Table 20.2), but it has not been documented for *Hemiandrus*, which apparently mates within subterranean galleries. Mating can also take place in the tree galleries of *Hemideina*, although these weta will also mate during the night out in the open (Jarman, 1982; Field, 1993; Rufaut, 1995; Townsend, 1995). Such activity can account for up to 20% of the total activity of the weta during the first two-thirds of the night in *H. crassidens* and up to 10% of the activity during the first half of the night in *Hemideina ricta* Hutton (Rufaut, 1995; Townsend, 1995). There is also an indication that in *H. crassidens,* less mating activity occurs outside galleries when mammalian predators are present (Rufaut, 1995). Courtship and copulation in *H. subantarcticus* also take place outside at night (Butts, 1983) and these behaviours were reported to occur 4–6 h after dark in captive *H. maculifrons* (Walker) (Cary, 1981). Both of the latter species occupy tunnels during the day.

In contrast, *M. isolata* and Raukumara tusked weta pairs stay together for long periods during the day, while they mate repeatedly (Gibbs, 1998; M. McIntyre, Wellington, 1998, personal communication), and species of *Deinacrida* appear to mate primarily during the day. *Deinacrida rugosa*, *D. fallai* and *D. heteracantha* were all reported to start copulating in the morning and continue mating throughout the daytime (Ramsay, 1955; Richards, 1973). Mating behaviour was also observed during the day in both captive and wild *D. mahoenui*, but it could commence just before dawn or soon afterwards and continue throughout the day (Sherley, 1992; Richards, 1994; Domett, 1996) or even right through to the following day (Barrett and Ramsay, 1991). Such behaviour often involved multiple copulations. In the field, initial contact between Mahoenui weta was only observed to take place at night in summer. Those Mahoenui weta that did mate at night in the field did so most often during the first half of the night. During this period, mating behaviour accounted for up to 5% of the total activity budget (Richards, 1994; Domett, 1996). Mahoenui weta were reported to remain copulating for between 70 min and 13.5 h (median 5.42 h, $n = 13$) and, in captivity, they can spend from 10 to 17% of their total time either in contact with each other or copulating. These behaviours occur most frequently in summer, when the males spend more time pursuing females. In contrast, captive Mahoenui weta exhibit two maxima in their mating activity at night, one during the first 10–25% of the night and another from 70 to 80% of the night. During these maxima, up to 40% of the activity budget of adults can be spent in contact with the opposite sex (Richards, 1994; Domett, 1996).

Most observations of copulation were made on weta that were outside their galleries or daytime retreats, but *H. maori* has been found copulating under rocks (G. Monteith, personal communication, in Gwynne and Jamieson, 1998). Species of small tusked weta and weta in the genus *Deinacrida* appear to have no precopulatory behaviour, except that males follow or chase females (Richards, 1994; Brown and Gwynne, 1997; McIntyre, 1998). *Motuweta isolata*, however, have been reported to slowly walk around each other above ground and touch each other with their antennae and palpi prior to copulating in their burrow. There is no evidence that stridulation plays a part in mating behaviour. Female *D. rugosa* produce a musky odour, which has led to the suggestion that chemical communication may play a role in courtship (Ramsay, 1955; Brown and Gwynne, 1997), and McIntyre (1998) suggested that the odour of tusked weta might also allow males to detect females from a distance (see Field and Jarman, Chapter 17, this volume, for discussion of pheromones). Males of most weta species seem to initiate most matings, but their approach is by no means always successful, because the female often resists or moves away (Sandlant, 1981; Jarman, 1982; Brown, 1995; Spencer, 1995; Domett, 1996; McIntyre, 1998). Mating success is low, ranging from 25% to 64% in *Hemideina* species (see Table 17.2, Field and Jarman, Chapter 17, this volume). Homosexual male mating attempts, mating with immature females and mating with sibling species are reviewed and discussed by Field and Jarman (Chapter 17, this volume).

Spermatophore transfer

Most weta stay together after the initial mating and copulate repeatedly. The male usually transfers one spermatophore on each occasion (Table 20.4), although over a dozen spermatophores can be transferred in a single long copulation (Brown and Gwynne, 1997). Mate guarding has been reported in only one species of *Hemiandrus* by Brown and Gwynne (1997). Here the male always remains with the female after mating. Females of *Deinacrida* do not eat the spermatophores (Brown and Gwynne, 1997), as also observed for the sister family Stenopelmatidae (Weissman, Chapter 3, this volume). Old spermatophores from multiple copulations are left as clusters ventrally at the base of the ovipositor in *D. fallai* and *D. heteracantha* where they eventually dry and fall off (Richards, 1973; Brown, 1995). In contrast, male *D. rugosa* may remove a spermatophore that is already in place by working it out from beneath the subgenital plate of the female before copulating (Ramsay, 1955).

The spermatophore

The spermatophores of weta have a single sperm cavity (ampulla), in contrast to the double ampulla of most tettigoniids, although the ampulla in *D. rugosa* has a ridge which could be the remnant of a dividing wall (Ramsay, 1955; Gwynne, 1995). In addition, the spermatophore of *Deinacrida* lacks a spermatophylax, whilst the spermatophylax is very reduced in *Hemideina* (Brown and Gwynne, 1997; Gwynne and Jamieson, 1998). The spermatophylax in other Ensifera forms a postcopulatory meal for the female, which may have a variety of posi-

Table 20.4. Mating behaviour of weta. Individual times are given as numbers, other times are given as the range (mean) or (mean ± SD).

Species	No. genital contacts	Total time mating	Duration of each genital contact (min)	No. spermatophores exchanged	Author
Deinacrida connectens	1	35 min	–	–	Field, 1980
Deinacrida fallai	Many	6.5–13.5 h	~60	1–6 (3)	Richards, 1973
Deinacrida fallai	Many	–	28–51 (41 ± 10)	0–6	Brown, 1995
Deinacrida heteracantha	Many	6.5–13.5 h	~60	3–8 (5)	Richards, 1973
Deinacrida rugosa	Many	Up to 45 min	–	1	Ramsay, 1955
Deinacrida mahoenui	Many	2 h	~6–34	–	Domett, 1996
Deinacrida mahoenui	Many	1.2–>13.5 h	30–163	Up to 7	Richards, 1994
Hemideina crassidens	Many	–	2	–	Spencer, 1995
Hemideina crassidens	Many	–	(1.62 ± 0.55)	–	Jarman, 1982
Hemideina femorata	–	0.5–6 min (1.98 ± 1.38 min)	0.33–5.98	–	Sandlant, 1981
Hemideina ricta	Many	–	~2	Not visible	Townsend, 1995
Hemiandrus subantarcticus	–	–	35–47	Not observed	Butts, 1983
Raukumara tusked weta	Many	19–62 days	–	Not observed	McIntyre, 1998

tive effects on the fecundity and fitness of the offspring (see reviews by Brown and Gwynne, 1997; Gwynne, 1997). The benefit to offspring correlates with whether or not the female weta eats the spermatophore, as mentioned above.

Oviposition

Oviposition behaviour

There are few reported observations of when weta oviposit (Table 20.2). Mahoenui weta were most often seen ovipositing from early in the morning to mid-afternoon and only once was it observed at night (Richards, 1994). One specimen of *D. rugosa* was observed ovipositing in daylight. On this occasion, the weta had not mated for some time and it died soon after ovipositing (Ramsay, 1955). In all other weta, oviposition typically takes place at night. Moller (1985) also reported one female *H. crassidens* that stopped ovipositing at 04.16 in January, when it was quite light.

Prior to oviposition the female palpates the ground and female *H. subantarcticus* and *M. isolata* also sweep their antennae over it. When a suitable place is located, *H. subantarcticus* continues this behaviour for 2–4 min before raising and arching the abdomen and beginning to probe with the ovipositor. Female *D. mahoenui* frequently also push their heads partially into the soil and dig or scrape a small depression into which they oviposit (see Butts (1983) for *H. subantarcticus*; Richards (1994) for Mahoenui weta; C.J. Winks (Auckland, 1998, personal communication) for *M. isolata*).

Descriptions of weta ovipositing have been published for *D. heteracantha* and *D. fallai* (Richards, 1973), *D. rugosa* (Ramsay, 1955), *H. crassidens* (Blanchard) (Moller, 1985), *H. maculifrons* (Cary, 1981) and *H. subantarcticus* (Butts, 1983). All of these species raise their body up by stretching their legs, bend or arch their abdomen until their ovipositor is approximately vertical and then lower their body again to thrust the ovipositor into the ground. Usually, the ovipositor is first partially inserted or used to probe a few times in slightly different places. Once an apparently suitable position is found, the weta inserts the ovipositor to most or all of its length. It then stays in this position for 1–5 min in *Deinacrida* (Ramsay, 1955; Richards, 1973; Richards, 1994), 10–15 min in *Hemiandrus* (Cary, 1981; Butts, 1983) and from about 15 min to 1 h 14 min in *H. crassidens* (Moller, 1978a, 1985). Whilst the ovipositor remains in the ground, the body of *D. rugosa* remains motionless, except for its abdomen, which is twisted and convulsed (Ramsay, 1955). The abdomens of *D. heteracantha* and *D. fallai* are also periodically convulsed and they make rhythmic telescoping contractions (Richards, 1973). In *H.*

subantarcticus, the whole body is swayed slightly, causing the ovipositor valves to slide back and forth (Butts, 1983).

At the conclusion of egg laying, the ovipositor is withdrawn rapidly and then the female often moves a few centimetres away and begins probing with the ovipositor again prior to laying more eggs. One Mahoenui weta repeated this behaviour seven times within 6 h 40 min (Richards, 1994), whereas *D. heteracantha* may spend 50–60 min in an ovipositing session (McIntyre, 1997). *Deinacrida fallai, D. heteracantha* and *D. rugosa* were also reported to lay eggs within small areas of about 15 cm in diameter in captivity (Ramsay, 1955; Richards, 1973; Brown, 1995).

The oviposition behaviour of *M. isolata* is similar to that of other weta, according to C.J. Winks (Auckland, 1998, personal communication). Once the ovipositor is inserted into the ground, the female remains still, except for rhythmic pulsations of the abdomen. These occur every 2–3 s and become more rapid after several minutes. The whole body then appears to tense and may give several shudders, before the ovipositor is withdrawn. The female may then move several centimetres and repeat the behaviour.

Oviposition is probably accompanied by a large weight loss, and this is especially likely in weta such as *D. heteracantha*, which lay large numbers of eggs over a short time (Richards, 1973). So far, however, such a weight loss has only been measured after oviposition in Mahoenui weta (Richards, 1994).

Meteorological factors that may influence oviposition

Moller (1978b) found eight female *H. crassidens* ovipositing on Arapawa Island during the only night when it rained during his study. This night was also warm and humid (17–19°C, 86% RH). According to Richards (1994), Mahoenui weta were most likely to be found on the ground in summer at night and especially around the middle of the night.

Placement of eggs

The eggs of most weta are laid in soil where they are placed vertically or almost vertically, some distance below the surface (Table 20.5). Because some captive weta will oviposit in potting mix and damp sawdust, it is likely that some arboreal species may also lay eggs in leaf mould associated with epiphytes or hollows at the bases of branches (Maskell, 1927; Richards, 1994). Arboreal weta, however, are generally assumed to come down to the ground to lay their eggs. Both sexes of at least one arboreal weta, *D. mahoenui*, may spend up to 12% of their time on the ground, although this species has never been observed ovipositing in the field (Richards, 1994). Such behaviour increases the chance of them being preyed on by ground-dwelling birds and introduced mammals such as rats, mustelids, hedgehogs and cats (Glasgow, 1982; Rufaut, 1995; reviewed by Field and Glasgow, Chapter 16, this volume). Except for *H. subantarcticus*, which oviposits on the open ground on Snares Island, all henicine species of *Hemiandrus* with short ovipositors lay their eggs in underground galleries (Butts, 1983).

Three authors report searching for weta eggs in soil. Moller (1985) found up to 75 eggs of *H. crassidens* per 5.75 cm^2 core sample within 1.2 m of the bases of trees on Stephens Island and Richards (1994) examined four 0.0625 m^2 soil samples from beneath ten gorse bushes with Mahoenui weta but found no eggs. Gibbs (1998) found empty chorions of Raukumara tusked weta in silt on a log in a stream bed where he had previously observed a female ovipositing.

Wahid (1978) investigated soil preferences for oviposition sites in henicine weta. He found no significant association between the soil texture or ground cover and the places where underground galleries of an undescribed species of *Hemiandrus* were located, in which females laid their eggs. However, more galleries with eggs were found under total grass cover (14 batches m^{-2}) than in the open (1–2 batches m^{-2}). The highest density of galleries containing eggs also occurred where the soil had a high proportion of sand (46%) and low proportion of clay (24%), whereas the lowest density of such galleries was where there was 37% sand and 30% clay. In addition, more galleries were found in well drained areas than in depressions, where waterlogging could occur for long periods.

Postoviposition behaviour and egg guarding

Most female weta simply walk away after oviposition, but some ground weta remain with their eggs for some time. Eggs of several species, including *H. maculifrons*, are laid in a clump of up to 36 on

Table 20.5. Positioning of eggs in soil.

Species	Depth (cm)	Orientation	Grouping	No. laid (mean)	Comments	Author
Deinacrida fallai	2.6–3	–	1–5	15–39 (27)	Captive	Richards, 1973
Deinacrida heteracantha	2.6–3	–	1–5	15–39 (27)	Captive	Richards, 1973
Deinacrida rugosa	2–2.5	Vertical to ~horizontal	1–40 (usually 3–7)	–	Captive	Ramsay, 1955
Deinacrida mahoenui	~0.2–3 (usually 2–2.6)	Vertical	1–30	–	Captive	Richards, 1994
Deinacrida mahoenui	0.9–2.3	Vertical	–	–	Captive	Barrett and Ramsay, 1991
Hemideina crassidens	~1	–	1	21	Captive; in sawdust	Maskell, 1927
Hemideina crassidens	~0.2–2	Vertical	–	–	Captive	Barrett and Ramsay, 1991
Hemideina maori	~0.2–2	Vertical	–	–	Captive	Barrett and Ramsay, 1991
Hemideina thoracica	~0.2–2	Vertical	–	–	Captive	Barrett and Ramsay, 1991
Hemiandrus subantarcticus	2–3	–	–	–	Snares Is.	Butts, 1983

the sides of the female's underground gallery or in a side chamber of the gallery (Salmon, 1950; Miller, 1971; Wahid, 1978; Cary, 1981; Barrett and Ramsay, 1991; van Wyngaarden, 1995). Sometimes the eggs are partially embedded in the walls, as described for burrowing Australian anostostomatids (see Monteith and Field, Chapter 5, this volume). Females of *Hemiandrus* sp. from Tekapo and *H. maculifrons* remain in their galleries with their eggs for some time (Cary, 1981; Barrett and Ramsay, 1991; van Wyngaarden, 1995), while an unnamed species of *Hemiandrus* from the Horotane Valley (Christchurch) stays with its eggs over winter and dies about the time the eggs hatch (Wahid, 1978). Some *Hemiandrus* species may even remain in their burrows with the young nymphs (Salmon, 1950; Miller, 1971; Barrett and Ramsay, 1991). Some degree of maternal care for eggs is also known from representatives of four other ensiferan families, all of which use burrows (see review by Gwynne, 1995).

Fecundity

Fecundity of female weta in captivity has been recorded for *M. isolata* by C.J. Winks (Auckland, 1998, personal communication) and for *D. heteracantha* and *D. fallai* by Richards (1973). Two female *M. isolata* oviposited a total of 153 eggs, one female *D. heteracantha* mated and laid 209 eggs and two mated *D. fallai* females laid 300 and 216 eggs. An unmated female *D. fallai* laid eight normal and 42 small eggs but none hatched. Female *Hemiandrus* from the Horotane Valley laid only 18–84 eggs each (mean 50 ± 8 SE) (Wahid, 1978). Each female of this ground weta is likely to lay only a single batch of eggs in its gallery, because only one batch of eggs was ever found with a female and the females die about the time that the eggs hatch.

Some information is available on the number of fully formed eggs found in adult weta (Table 20.6), but this does not provide an accurate estimate of the number of eggs that weta lay. Some of the weta containing eggs listed in Table 20.6 had died after ovipositing, whereas others had been collected in the field and could have already oviposited. This could be why Moller (1985) found no apparent relationship between the number of fully formed eggs within female *H. crassidens* and the body weight, body length or length of the hind tibia. Cary (1981) suggested that arboreal giant weta, such as *D. heteracantha* and *D. fallai*, are able to store more fully formed eggs in the calyces of their ovaries than ground weta, so that the eggs are readily available to be laid quickly. This could allow these species to reduce their exposure to ground dwelling predators by rapid oviposition and reduction in time spent on the ground.

The Egg

Morphology

Eggs of weta are elongate and rounded. The size and appearance of those that have been described are summarized in Table 20.7. In *H. crassidens*, egg size may vary between females (Moller, 1978a, b). The egg can be slightly C-shaped, because one side can be slightly more curved, as in *D. rugosa* (Ramsay, 1955). The anterior pole of the eggs of *D. rugosa*, *D. heteracantha*, *D. fallai* and *D. mahoenui* is rounded and the posterior pole is slightly pointed (Ramsay, 1955; Richards, 1973; Richards, 1994). The posterior pole of *D. heteracantha* and *D. fallai* can also be darker in colour (Richards, 1973), whereas that of *D. rugosa* has a circular pale-coloured zone, with either a cap-like thickening or a flattened apex (Ramsay, 1955). The presence of micropyles has only been noted in *D. rugosa*. This species usually has six (range four to ten) arranged around the equator of the egg (Ramsay, 1955). Most weta eggs are white, brown or black (Table 20.7) but the colour of *D. rugosa* eggs, which are normally black, can change to grey when they dry out (Ramsay, 1955). The colour can also change during development, as described below.

The surface of the chorion has fine sculpturing (Fig. 20.1), which differs among species. Although not thoroughly explored, this led to the suggestion that the sculpturing could have some classificatory value (Ramsay, 1955; Cary, 1981; Butts, 1983). G. Gibbs (Wellington, 1998, personal communication) recently found that the type of pits with minute pores is a synapomorphy of *Deinacrida* and *Hemideina*, whereas the eggs of *Hemiandrus* and tusked weta form two other distinct groups.

Table 20.6. Numbers of fully formed eggs found in female weta.

Species	No. weta	Range	Mean ± SE	Comments	Author
Deinacrida fallai	1	373	–	Total eggs, captive, died after laying	Richards, 1973
Deinacrida heteracantha	2	135, 366	–	Total eggs, captive, died after laying	Richards, 1973
Deinacrida heteracantha	1	449	–	Total eggs, captive, unmated	Richards, 1973
Deinacrida rugosa	1	236	–	Died in captivity	Ramsay, 1955
Deinacrida mahoenui	1	185	–	Mahoenui	Meads, 1988
Hemideina crassidens	16	0–53	11.7±2.8	Eggs per calyx. Stephens Is.	Moller, 1985
Hemideina crassidens	16	2–39	22.7±1.4	Eggs per ovariole. Stephens Is.	Moller, 1985
Hemideina crassidens	7	29–96	53.6±10.0	Eggs per female. Stephens Is.	Moller, 1985
Hemideina crassidens	8	0–42	24.9±3.58	Arapawa Is., eggs per calyx, female ovipositing	Moller, 1978b
Hemideina crassidens	8	4–79	51.9±10.3	Eggs per ovipositing female, Arapawa Is.	Moller, 1978b
Hemiandrus maculifrons	33	0–14	7	Total eggs, Cass	Cary, 1981
Hemiandrus subantarcticus	13	5–57	52.6	Total eggs, Snares Is.	Butts, 1983
Motuweta isolata	1	234	–	Total eggs	McIntyre, 1992

Fig. 20.1. Scanning electron micrographs of weta eggs. (A) End of egg of *Deinacrida fallai* Salmon; (B) higher-power view of central portion of A; (C) high-power view of chorion of *D. fallai* about halfway along egg to show details of sculpture; (D) opposite end of egg of *D. fallai* shown in B; (E) entire egg of *Hemiandrus subantarcticus* (Salmon) (from Butts, 1983, by courtesy of the author); (F) higher-power view to show chorionic sculpture of *H. subantarcticus*; (G) entire egg of *Hemiandrus maculifrons* (Walker); (H) higher-power view of end of the egg of *H. maculifrons*; (I) light photomicrograph of the egg of *H. maculifrons* showing eye-spot and indications of the limbs and antennae of the embryo within. (Photographs E and F from Butts, 1983, by courtesy of the author; G, H and I from Cary, 1981, by courtesy of the author).

Table 20.7. Size and appearance of newly laid weta eggs.

Species	Length (mean ± SE) (mm)	Width (mean ± SE) (mm)	Mass (mean) (mg)	Sculpture: colour	Author
Deinacrida fallai	*c.* 0.7	*c.* 2.5	–	Uniform fine overall: mid-brown	Richards, 1973
Deinacrida heteracantha	*c.* 0.7	*c.* 2.5	–	Uniform fine overall: light brown	Richards, 1973
Deinacrida rugosa	6.25–7.15	2.1–2.4	5.3–18.6	Circular ridges: black, some brown to white	Ramsay, 1955
Deinacrida mahoenui	*c.* 0.7	*c.* 2.5	–	Brown	Richards, 1994
	–	–	–	Deep brown to black	Domett, 1996
Hemideina crassidens	5–6	–	–	With spines: white to grey	Maskell, 1927
Hemideina crassidens	4.8–6.0 (5.46 ± 0.02)	1.6–2.4 (2.14 ± 0.01)	–	–	Meads and Moller, 1978
Hemideina crassidens	(5.45 ± 0.02)	(2.02 ± 0.01)	–	Dark brown	Moller, 1978b
Hemideina crassidens	*c.* 5	*c.* 2	9–13	Hexagonal ridges	Jarman, 1982
Hemideina ricta	Mean 5.91	–	(16.8)	–	Townsend, 1995
Hemiandrus maculifrons	3.0–3.2 (3.1)	1.0–1.2 (1.0)	1.8–2.5	Longitudinal ridges: translucent	Cary, 1981
Hemiandrus subantarcticus	(4.03 ± 0.04)	(1.43 ± 0.02)	–	Rectangular ridge pattern: pale cream	Butts, 1983
Hemiandrus sp.	(2.94 ± 0.02)	(1.40 ± 0.01)	(3.3)	Smooth: transparent yellow	Wahid, 1978
Raukumara tusked weta	5.1–5.5	–	–	Translucent to light brown	McIntyre, 1998

Hemiandrus eggs have pores but also have quite different features from the eggs of the *Hemideina–Deinacrida* group, whereas the surfaces of tusked weta eggs lack pores and are decorated with knob-like structures.

Little is known about the chorion of weta eggs, although it appears to be similar to that of the Chilean *Cratomelus* (see Angulo, Chapter 11, this volume). It accounts for about 68% of the weight of Mahoenui weta eggs (Richards, 1994) and its internal structure is described only for *H. crassidens* (Maskell, 1927). In the latter, openings occur in depressions in the centre of the hexagonal surface areas. Canals lead inward from these openings and branch deeply within the exochorion (outer region of the chorion) (Maskell, 1927). Such a structure could act as a plastron to facilitate respiration when the egg is immersed in water, because of its similarity to the plastrons of some other insect eggs (Hill, 1985). However, this function has not yet been investigated. Beneath the exochorion of *H. crassidens* lies an endochorion, which consists of three dense layers, separated by trabecular canals (Maskell, 1927).

Incubation Period

The incubation period of weta eggs varies widely among different species (Table 20.8) and between eggs from the same female laid on the same date. Different batches of *D. mahoenui* eggs that were laid on the same day hatched over a range of 4–14 days, even though each batch was kept together (Richards, 1994).

Environmental effects on incubation

Water is essential for embryonic development. The eggs swell during incubation and their appearance may also change (Table 20.9). The eggs of *Hemiandrus* sp. require water only during the initial months of development (April to July) and thereafter they will develop even if the soil is

Table 20.8. Incubation times of weta eggs.

Species	Date laid	Date of eclosion	Incubation period (mean)	Comments	Author
Deinacrida fallai	Early Sep	–	147 d – ~3 mo	Captive	Richards, 1973
Deinacrida fallai	Sep–Dec	Feb–Mar	95–182 d (125 d)	Captive	Richards, 1973
Deinacrida heteracantha	Late Oct	–	141 d – ~3 mo	Captive	Richards, 1973
Deinacrida heteracantha	Late Oct	–	85–258 d (125 d)	Captive	Richards, 1973
Deinacrida heteracantha	Oct–Dec	Mar–Apr	4->5 mo	Captive	Richards, 1973
Deinacrida rugosa	Apr–May	Late Dec +	9–10 mo	Captive	Ramsay, 1955
Deinacrida rugosa	Early Mar	Late Dec +	Peak mid-Jan	Captive	Ramsay, 1955
Deinacrida rugosa	Apr	Early Nov +	11+ mo	Captive	Ramsay, 1955
Deinacrida mahoenui	Feb–Mar	Jan–Feb	~10 mo	In unheated room	Richards, 1994
Hemideina crassidens	–	Early Dec	~7 mo	Captive	Maskell, 1927
Hemideina crassidens	May	Nov	8 mo	Under house	Spencer, 1995
Hemideina crassidens	Apr–May	Oct–Dec	~6–7 mo	Captive	Barrett and Ramsay, 1991
Hemideina thoracica	Oct	–	–	Captive	Barrett and Ramsay, 1991
Hemiandrus maculifrons	–	–	12–18 mo	Lab ~20°C	Cary, 1981
Hemiandrus subantarcticus	Early Dec–early Jan	–	82–118 d (106.5 d)	In lab	Butts, 1983
Hemiandrus sp.	Mid Apr	Dec–?	8 + mo	Horotane Valley	Wahid, 1978
Hemiandrus sp.	Apr	–	56–83 d (70 d)	In lab	Wahid, 1978
Motuweta isolata	–	–	< 1.8 mo	In lab	Winks, 1995
Motuweta isolata	–	–	6–8 mo	In lab	C.J. Winks in McIntyre, 1998

relatively dry (Wahid, 1978). In contrast, the eggs of *D. rugosa* show their greatest increase in size during the last 3 months of development (Ramsay, 1955). No comparable data are available for other species.

Temperature may affect the development of weta eggs. Diapause has not yet been demonstrated in any weta egg, but there are suggestions that the eggs of some species may have a mild diapause, during which low temperature speeds their development. Thus, diapause may explain why eggs of *H. crassidens* failed to hatch when kept moist and inside (Spencer, 1995) and why some eggs of *D. heteracantha* took much longer than normal to develop in the laboratory (up to 258 days) (Richards, 1973). The eggs of *D. rugosa* are quiescent in winter, but they may have a very mild diapause, because eggs kept outside during winter hatch sooner when brought into the laboratory than those kept continuously in the laboratory (Ramsay, 1955). However, exposing eggs of *D. rugosa* to 8 days of cold, 4–8 months after they

Table 20.9. Change in appearance and average measurements of weta eggs during development.

Species	Incubation time	Length increase	Width increase	Weight change	Colour change	Author
Deinacrida fallai and *D. heteracantha*	127 days	137%	137%	–	No change	Richards, 1973
Deinacrida rugosa	9 months	140%	142%	228%	Paler colour	Ramsay, 1955
Deinacrida mahoenui	–	–	*c.* 114%	–	–	Richards, 1994
Hemiandrus maculifrons	2.1 months	(Swells: in field)	(Swells: in field)	(Swells: in field)	Milky white	Cary, 1981
Hemiandrus maculifrons	4–6 months	(Swells: in lab)	(Swells: in lab)	(Swells: in lab)	Milky white	Cary, 1981
Hemiandrus subantarcticus	–	110.9%	125.7%	–	–	Butts, 1983
Hemiandrus sp.	–	112.5%	130%	up to 167%	Pale cream	Wahid, 1978

were laid had no effect on development time (Ramsay, 1955). Finally, Cary (1981) reported that females of the henicine *H. maculifrons* lay their eggs in underground galleries over a month or so, but the eggs subsequently hatch within 4 days of each other. Such synchronization could be produced by a diapause, which is terminated by a period of cold and then followed by a period of quiescence until the return of favourable temperatures in spring (Tauber and Tauber, 1976).

Few studies have been done on the development of weta eggs at different temperatures. Eggs of *Hemiandrus* sp. develop slowly and absorb water at 15°C but show no sign of development at 5.5°C after 40 days. Normal development occurs when these eggs are kept at 20°C, but none hatch when kept at 25°C or higher (Wahid, 1978). Under field conditions, these eggs become quiescent during cold weather, because they show a negative linear relationship between time to eclosion and date when they are brought into the laboratory from the field (mostly over the winter) (Wahid, 1978). Mahoenui giant weta eggs take 7.5–8 months to hatch when kept at a constant temperature of 18 ± 1°C but they hatch after about 10 months if exposed to cold winter temperatures (Richards, 1994). However, one batch of eggs, laid in December 1992, hatched 24–26 months later. These had been kept inside a warm house and they hatched after being placed outside for several weeks during the second winter (Domett, 1996). This suggests that they could have been in a diapause, which was broken by exposure to cold. However, no controls were kept, because the eggs were thought to be dead (Domett, 1996). Interestingly, all 104 of these eggs hatched, whereas Mahoenui eggs generally have a much higher mortality (Table 20.10).

Table 20.10. Hatching success of weta eggs.

Species	Hatching rate (%)	No. eggs	Comments	Author
Deinacrida fallai	23 (3.2–30)	–	In captivity	Richards, 1973
Deinacrida heteracantha	36 (2.3–63)	–	In captivity	Richards, 1973
Dinnacrida mahoenui	48.6	186	Caged outside	Richards, 1994
Dinnacrida mahoenui	100	104	Inside plus cold period	Dommett, 1996
Hemideina crassidens	99.0	962	In captivity	Barrett and Ramsay, 1991
Hemiandrus sp.	85 (48–96)	–	Collected Sep	Wahid, 1978
Hemiandrus sp.	82 (62–98)	–	Collected Oct	Wahid, 1978
Motuweta isolata	33	153	In captivity	Winks, 1995

Development and embryology

The eggs of *H. maculifrons* and *H. subantarcticus* go through three stages of development, based on their external appearance (Cary, 1981; Butts, 1983). The chorion is hard and translucent in stage I, whereas in stage II it is still hard but the egg is swollen and a milky white substance appears within, which eventually fills the entire egg. Dark eye-spots appear, followed by other embryonic features, in stage III. The durations of stages I, II and III in the laboratory are 7–10 months, 4–6 months and 7–8 weeks, respectively, for *H. maculifrons* and the approximate means are 20 days, 20 days and 53 days, respectively, for *H. subantarcticus* (Cary, 1981; Butts, 1983). There is also some evidence that eggs of *H. maculifrons* may progress faster through stages I and II in the field than in the laboratory (Cary, 1981).

The embryology of weta has not been investigated in detail, but the mature embryo of *D. rugosa* has been described (Fig. 20.2) (Ramsay, 1955). A pair of well developed pleuropodia (segmental appendages of the second abdominal segment) is present, as in other orthopteran embryos, and the embryonic cuticle only loosely envelops the pronymph within. The almost completely developed pronymph possesses the mottled pigmentation of the first-instar insect. The embryonic cuticle is covered dorsally and laterally with numerous, small, scale-like structures. These are arranged in longitudinal rows, with their tips directed posteriorly. They possibly help the insect escape from the chorion during eclosion, as well as aiding it when working its way to the surface. In addition, there is a swollen cervical ampulla and a median hatching ridge is present on the frons. This ridge is dark brown, serrated and blade-like. It presumably helps burst the chorion during eclosion (Ramsay, 1955).

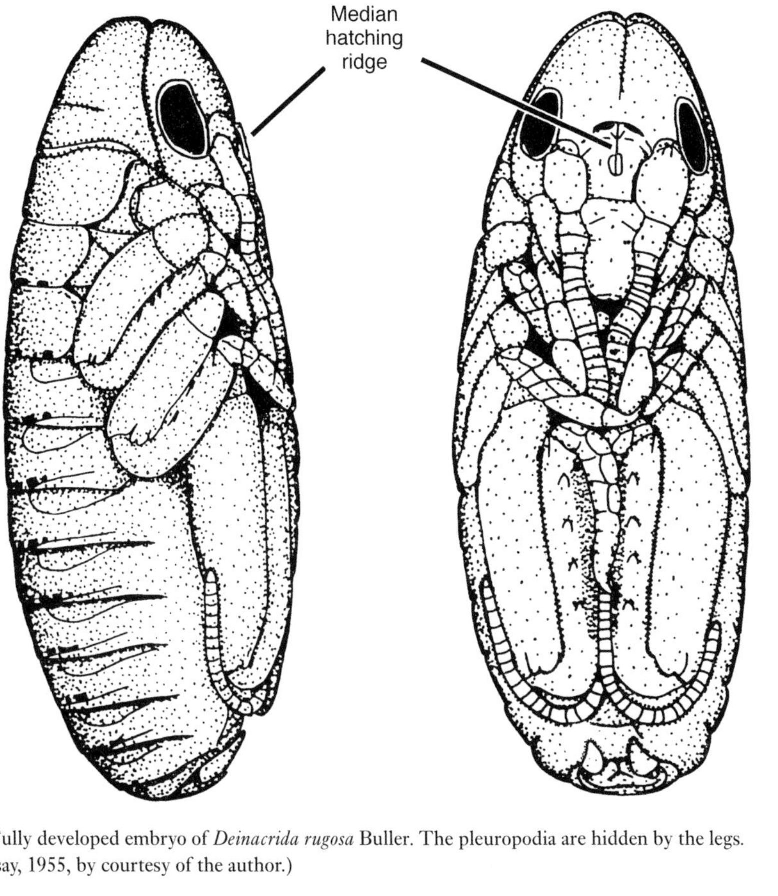

Fig. 20.2. Fully developed embryo of *Deinacrida rugosa* Buller. The pleuropodia are hidden by the legs. (From Ramsay, 1955, by courtesy of the author.)

Richards (1973) and Wahid (1978) opened eggs of *D. fallai, D. heteracantha* and *Hemiandrus* sp., both early and late in development. No embryos were reported from early eggs and only fully formed embryos were found late in development. A hatching ridge or spine and a cervical ampulla are present in advanced embryos of both *D. fallai* and *D. heteracantha* (Richards, 1973). The chorion of some weta eggs becomes more fragile and is easily split close to hatching (Ramsay, 1955; Richards, 1973; Wahid, 1978; Richards, 1994). In at least four weta species (*Hemiandrus* sp., *H. maculifrons, H. subantarcticus,* Raukumara tusked weta), black eye-spots become visible through the chorion as embryonic development nears completion (Fig. 20.1I). Abdominal segmentation and the outline of the antennae become visible just before eclosion (Wahid, 1978; Cary, 1981; Butts, 1983; McIntyre, 1998).

Mortality and parasitism of the egg

The hatching success for weta eggs is given in Table 20.10. The causes of mortality are generally not known. Egg parasites are reported only for *H. maculifrons* (Cary, 1981). One is an undescribed scelionid wasp (tribe Psilanteridini), which is found in up to 50% of egg clutches and parasitizes 15.8% of eggs taken into the laboratory. The second known egg parasite is a species of *Platypatasson* (Mymaridae). This occurs in fewer than 1% of eggs, but ten to 12 wasps emerge from each affected egg. The scelionid leaves one large rounded hole at the anterior of the egg, whilst *Platypatasson* leaves four to five smaller holes. No eggs with such damage were, however, found in the field (Cary, 1981).

There are few reports of direct egg mortality. Ramsay (1955) observed a small nymph of *D. rugosa* eating an egg and Winks (1995) reported that 27 eggs of *M. isolata* were eaten. Richards (1994) noted that fungi can attack Mahoenui weta eggs in captivity and Winks (1995) reported that four eggs became infected with fungi. Lack of water is another probable cause of mortality, because most authors presume that eggs die if they dry. This has not, however, been tested. Eggs of *D. heteracantha* and *D. fallai* can swell up again in water if they have previously lost some weight, but it is not known if such eggs are viable (Richards, 1973).

The environment seems to have a strong influence on hatching success in captivity. When eggs of *Hemiandrus* sp. were kept on moist plaster of Paris, 20–26% hatched, whereas 40–71% of those on moist soil hatched (Wahid, 1978). The number of *D. fallai* and *D. heteracantha* eggs that hatch in captivity can depend on where they are kept. Hatching success in *D. fallai* varied from 3.2% in an insectary to 30% in an outside shed, whereas in *D. heteracantha* it varied from 2.3% for eggs kept outside to 63% for eggs held in an insectary (Richards, 1973).

Eclosion

Normally, eclosion occurs underground in most weta, but many eggs of *D. fallai, D. heteracantha* and *D. mahoenui* appear at the surface of the soil prior to this (Richards, 1973; Richards, 1994; Domett, 1996). The eggs of Mahoenui weta can even remain on the surface from a few days to several weeks before they hatch (Richards, 1994). Empty chorions of *D. rugosa* also occasionally appear on the surface, but this is apparently caused by emerging nymphs, which pull their chorions to the surface after they have been unable to escape completely from them underground (Ramsay, 1955).

Eclosion from the egg usually occurs at night in *H. maculifrons* (Cary, 1981), *D. fallai* and *D. heteracantha* (Richards, 1973), *D. mahoenui* (Richards, 1994; Domett, 1996) and *Hemideina* (Barrett and Ramsay, 1991). In *D. rugosa*, eclosion usually occurs in the early evening, but it may occasionally take place in the late afternoon, late at night or early morning (Ramsay, 1955). Interestingly, the time of eclosion in *D. rugosa* is apparently unaffected by artificial illumination (Ramsay, 1955).

Eclosion takes a little over 20 min to complete in *D. rugosa* (Ramsay, 1955), from 13 to 17 min ($n = 6$) in *D. mahoenui* (Domett, 1996), and 25–35 min in *H. subantarcticus* (Butts, 1983). Eggs of the latter took longer to hatch if they were on the surface (Butts, 1983). After eclosion, first-instar *D. mahoenui* remain by their eggs for a further 9 min before they are sufficiently sclerotized to walk away (Domett, 1996).

Eclosion is similar in all weta. The abdominal segments undergo anterior-directed peristaltic waves. These direct pressure on to the hatching spine and cause the cervical ampulla to swell (Richards, 1973; Wahid, 1978; Cary, 1981; Barrett and Ramsay, 1991). The chorion splits over the anterior pole of the egg and this split continues posteriorly, along either the ventral or the dorsal surface of the egg for up to about half the length of the egg. The split may also extend a short distance over the opposite side of the egg (Ramsay, 1955; Richards, 1973; Wahid, 1978; Butts, 1983; Barrett and Ramsay, 1991; McIntyre, 1998). In *H. maculifrons*, the head emerges first, followed in sequence by the thorax, fore- and mid-legs, abdomen, hind legs and ends of the antennae (Cary, 1981). In *D. rugosa*, the swollen cervical ampulla causes the head to bend forwards under the body, but the head later adopts its normal position when the insect reaches the surface of the soil and the ampulla withdraws (Ramsay, 1955).

Weta usually hatch as pronymphs, which moult again soon after eclosion (Ramsay, 1955; Richards, 1973; Butts, 1983; Barrett and Ramsay, 1991). It is also possible that in some weta this moult may take place within the chorion, as it does in most exopterygotes (Sehnal, 1985). In *Hemideina* and *D. rugosa* the moult to first instar nymph takes place soon after the pronymph reaches the surface of the soil (Ramsay, 1955; Barrett and Ramsay, 1991). The pronymphal moult has been described in detail only for *H. subantarcticus*, by Butts (1983). This species takes 45–73 min to ecdyse and the process is the same as that described for subsequent moults by Stringer and Cary (Chapter 21, this volume). First-instar nymphs of *D. rugosa* and *H. subantarcticus* usually eat their pronymphal exuviae (Ramsay, 1955; Butts, 1983) and this behaviour also occurs frequently after subsequent moults (see Stringer and Cary, Chapter 21, this volume).

Acknowledgements

I am grateful for editorial help from Rachel Standish and Murray Potter (Institute of Natural Resources (INR), Massey University, New Zealand).

References

Asher, G.W. (1977) Ecological aspects of the common tree weta (*Hemideina thoracica*) in native vegetation. BSc thesis, Victoria University of Wellington, Wellington, New Zealand.

Barrett, P. and Ramsay, G.W. (1991) *Keeping Wetas in Captivity.* Wellington Zoological Gardens, Wellington, 60 pp.

Brown, J. (1995) Behaviour of captive Poor Knights giant weta (*Deinacrida fallai*). BSc thesis, Victoria University of Wellington, Wellington, New Zealand.

Brown, W.D. and Gwynne, D.T. (1997) Evolution of mating in crickets, katydids, and wetas (Ensifera). In: Gangwere, S.K., Muralirangen, M.C. and Muralirangen, M. (eds) *The Bionomics of Grasshoppers, Katydids and their Kin.* CAB International, Wallingford, pp. 279–312.

Butts, C.A. (1983) The biologies of two species of weta endemic to the Snares Island *Zealandrosandrus subantarcticus* Salmon (Orthoptera: Stenopelmatidae) and *Insulanoplectron spinosum* Richards (Orthoptera: Rhaphidophoridae). BSc (Hons) thesis, University of Canterbury, Christchurch, New Zealand.

Cary, P.R.L. (1981) The biology of the weta *Zealandosandrus gracilis* (Orthoptera: Stenopelmatidae) from the Cass Region. MSc thesis, University of Canterbury, Christchurch, New Zealand.

Daugherty, C.H., Gibbs, G.W. and Hitchmough, R.A. (1993) Mega-island or micro-continent? New Zealand and its fauna. *Trends in Ecology and Evolution* 8, 437–442.

Deans, N.A. (1982) The functional morphology of the mandibles in the genus *Hemideina.* BSc (Hons) thesis, University of Canterbury, Christchurch, New Zealand.

Domett, E. (1996) Reproduction and behaviour of the Mahoenui weta, *Deinacrida* n.sp. MSc thesis, Massey University, Palmerston North, New Zealand.

Field, L.H. (1980) Observations on the biology of *Deinacrida connectens* (Orthoptera: Stenopelmatidae), an alpine weta. *New Zealand Journal of Zoology* 7, 211–220.

Field, L.H. (1993) Observations on stridulatory, agonistic, and mating behaviour of *Hemideina ricta* (Stenopelmatidae: Orthoptera), the rare Banks Peninsula weta. *New Zealand Entomologist* 16, 68–74.

Field, L.H. and Sandlant, G.R. (1983) Aggression and mating behaviour in the Stenopelmatidae (Orthoptera; Ensifera), with reference to New Zealand wetas. In: Gwynne, D.T. and Morris, G.K. (eds) *Orthopteran Mating Systems: Sexual Competition in a Diverse Group of Insects.* Westview Press, Boulder, Colorado, pp. 120–146.

Gibbs, G. (1994) The demon grasshoppers. *New Zealand Geographic* 21, 90–117.

Gibbs, G. (1998) *Raukumara Tusked Weta: a Report to Department of Conservation on its Discovery, Ecology and Management Implications.* Department of Conservation, Gisborne, New Zealand.

Glasgow, S. (1982) Defence behaviour of *Hemideina crassidens* (Orthoptera: Stenopelmatidae) and its effectiveness against potential predators. BSc thesis, University of Canterbury, Christchurch, New Zealand.

Gwynne, D.T. (1995) Phylogeny of the Ensifera (Orthoptera): a hypothesis supporting multiple origins of acoustical signaling, complex spermatophores and maternal care in crickets, katydids, and weta. *Journal of Orthoptera Research* 4, 203–218

Gwynne, D.T. (1997) The evolution of edible 'sperm sacs' and other forms of courtship feeding in crickets, katydids and their kin (Orthoptera: Ensifera). In: Choe, J.C. and Crespi, B. (eds) *Social Competition and Cooperation in Insects and Arachnids*, Vol. I, *The Evolution of Mating Systems.* Cambridge University Press, Cambridge, pp. 110–129.

Gwynne, D.T. and Jamieson, I. (1998) Sexual selection and sexual dimorphism in a harem-defending insect, the alpine weta (*Hemideina maori*, Orthoptera: Stenopelmatidae). *Ethology, Ecology and Evolution* 10, 393–402.

Hamilton, S.A. (1991) The role of sex ratio, spatial distribution, and head size in the mating system of the Rock and Pillar weta, *Hemideina maori*. Diploma in Wildlife Management thesis, University of Otago, Dunedin, New Zealand.

Hill, P.J. (1985) Structure and physiology of the respiratory system. In: Kerkut, G.A. and Gilbert, L.I. (eds) *Comprehensive Insect Physiology Biochemistry and Pharmacology*, Vol. 3, *Integument, Respiration and Circulation.* Pergamon Press, Oxford, pp. 517–593.

Hudson, G.V. (1920) On some examples of New Zealand insects illustrating the Darwinian principle of sexual selection. *Transactions and Proceedings of the New Zealand Institute* 52, 431–438.

Jarman, T.H. (1982) Mating behaviour and its releasers in *Hemideina crassidens* (Orthoptera: Stenopelmatidae). BSc thesis, University of Canterbury, Christchurch, New Zealand.

Lemke, D. (1994) *Distribution of the Common Tree Weta (*Hemideina crassidens*) in Pine Plantations (*Pinus radiata*) at Hira Forest, Nelson.* Manaaki Whenua Landcare Research, Nelson, New Zealand.

McIntyre, M. (1992) *The Status and Habitat of the Middle Island (Mercury Group) Tusked Wetas, with Implications for Management.* Department of Conservation, Wellington, New Zealand.

McIntyre, M. (1997) Cover picture. *The Weta* 20.

McIntyre, M. (1998) *Raukumara Tusked Weta*, Part II. Department of Conservation, Gisborne, New Zealand.

Maskell, F.G. (1927) The anatomy of *Hemideina thoracica*. *Transactions and Proceedings of the New Zealand Institute* 57, 637–670.

Meads, M.J. (1976) *Visit to Stephens Island, Cook Strait, from 22 April to 3 May 1967: Preliminary Report.* DSIR Ecology Division, Lower Hutt, New Zealand.

Meads, M.J. (1988) *A Brief Review of the Curent Status of the Giant Wetas and a Strategy for their Conservation.* Report No. 11, DSIR Ecology Division, Lower Hutt, New Zealand.

Meads, M. (1995) *Captive Breeding the Mercury Tusked Weta to Establish New Populations.* World Wide Fund for Nature, Wellington, New Zealand.

Meads, M.J. and Moller, H. (1977) *Report of a Visit to Mana Island in September 1977.* DSIR Ecology Division, Lower Hutt, New Zealand.

Meads, M.J. and Moller, H. (1978) *Introduction of Giant Wetas (*Deinacrida rugosa*) to Maud Island and Observations of Tree Wetas, Paryphantids and Other Invertebrates.* DSIR Ecology Division, Lower Hutt, New Zealand.

Miller, D. (1971) *Common Insects in New Zealand.* A.H. and A.W. Reed, Wellington, 178 pp.

Moller, H. (1978a) Study of the wetas of Stephens Island. Unpublished report, DSIR Ecology Division, Lower Hutt, New Zealand.

Moller, H. (1978b) A weta and rodent study on Arapawa Island. Unpublished report, DSIR Ecology Division, Lower Hutt, New Zealand.

Moller, H. (1985) Tree wetas (*Hemideina crassicruris*) (Stenopelmatidae: Orthoptera) of Stephens Island, Cook Strait. *New Zealand Journal of Zoology* 12, 55–69.

Ramsay, G.W. (1955) The exoskeleton and musculature of the head and the life-cycle of *Deinacrida rugosa* Buller, 1870. MSc thesis, Victoria University of Wellington, Wellington, New Zealand.

Ramsay, G.W. (1978) Seasonality in New Zealand Orthoptera. *New Zealand Entomologist* 8, 357–358.

Richards, A.O. (1973) A comparative study of the biology of the giant wetas *Deinacrida heteracantha* and *D. fallai* (Orthoptera: Henicidae) from New Zealand. *Journal of Zoology* 169, 195–236.

Richards G.E. (1994) Ecology and behaviour of the Mahoenui giant weta, *Deinacrida* nov. sp. MSc thesis, Massey University, Palmerston North, New Zealand.

Rufaut, C.G. (1995) A comparative study of the Wellington tree weta, *Hemideina crassidens* (Blanchard, 1951) in the presence and absence of rodents. MSc thesis, Victoria University of Wellington, Wellington, New Zealand.

Salmon, J.T. (1950) A revision of the New Zealand wetas Anostostominae (Orthoptera: Stenopelmatidae). *Dominion Museum Records, Entomology, Wellington* 1, 121–177.

Sandlant, G.G. (1981) Aggressive behaviour of the Canterbury weta *Hamideina femorata* (Orthoptera: Stenopelmatidae): its adaptive significance in resource allocation. MSc thesis, University of Canterbury, Christchurch, New Zealand.

Sehnal, F. (1985) Growth and life cycles. In: Kerkut, G.A. and Gilbert, L.I. (eds) *Comprehensive Insect Physiology Biochemistry and Pharmacology*, Vol. 2, *Postembryonic Development*. Pergamon Press, Oxford, pp. 1–86.

Sherley, G.H. (1992) Gorse and goats: are pest species necessary for the conservation of a new giant weta (*Deinacrida* sp.) at Mahoenui, King Country? In: Heath, A.C.G. (ed.) *Proceedings of the 41st Annual Conference of the Entomological Society of New Zealand*. Entomology Society of New Zealand, Heretaunga, pp. 26–36.

Sherley, G.H. and Hayes, L.M. (1993) The conservation of a giant weta (*Deinacrida* n. sp. Orthoptera: Stenopelmatidae) at Mahoenui, King Country: habitat use and other aspects of its ecology. *New Zealand Entomologist* 16, 55–68.

Spencer, A.M. (1995) Sexual maturity in the male tree weta *Hemideina crassidens* (Orthoptera: Stenopelmatidae). MSc thesis, Victoria University of Wellington, Wellington, New Zealand.

Tauber, M.J. and Tauber, C.A. (1976) Insect seasonality: diapause maintenance, termination and postdiapause development. *Annual Review of Entomology* 21, 81–107.

Townsend, J.A. (1995) Distribution and ecology of the Banks Peninsula tree weta, *Hemideina ricta*. MSc thesis, Massey University, Palmerston North, New Zealand.

van Wyngaarden, F. (1995) The ecology of the Tekapo ground weta (*Hemiandrus* new sp.; Orthoptera: Anostostomatidae) and recommendations for the conservation of a threatened close relative. MSc thesis, University of Canterbury, Christchurch, New Zealand.

Wahid, M.B. (1978) The biology and economic impact of the weta, *Hemiandrus* sp. (Orthoptera: Stenopelmatidae) in an apricot orchard, Horotane Valley. MHortSc thesis, Lincoln College, Lincoln, New Zealand.

Watt, J.C. (1984) A note on the Poor Knights weta *Deinacrida fallai* (Orthoptera: Stenopelmatidae). *The Weta* 7, 64.

Winks, C.J. (1995) Captive breeding of Middle Island tusked weta. Unpublished interim investigation summary, Department of Conservation, Wellington, New Zealand.

21
Postembryonic Development and Related Changes

Ian A.N. Stringer and Paul R.L. Cary
Ecology Group, Institute of Natural Resources, Massey University, PO Box 11222, Palmerston North, New Zealand

Introduction

This chapter summarizes the wealth of mostly unpublished information on the postembryonic development of New Zealand weta. Much of this information is incidental in student projects and theses and in reports to government agencies. In addition, we include new data by one of us (P.R.L. Cary) on the gross morphology of the internal reproductive organs for *Hemideina crassidens* (Blanchard).

Many details about weta life histories are incomplete and several interesting aspects have not yet been explored. For example, we still do not have enough data to prepare a life-table for any weta species, so we cannot begin to calculate the strength of selection that acts on life-history traits and, in particular, on postembryonic development. One application of this could relate to how mammalian predators might affect the life histories of weta. Because New Zealand was isolated from other land masses and only acquired most of its mammalian predators with the arrival of humans, it can serve as a natural laboratory for such studies (Daugherty *et al.*, 1993). For example, we might expect, from life-history theory, that small eggs and small adults should be favoured when vertebrate predation is high (Stauffer and Whitman, 1997), but so far the only reports on how mammals might affect weta concern behavioural changes by these insects (Meads and Moller, 1978; Moller, 1978a, b; Rufaut, 1995).

The primary focus of this chapter is on the changes that occur during postembryonic development, but we also include a short section on dispersal behaviour of first-instar nymphs once they have hatched. Eclosion from the egg is treated in Stringer (Chapter 20, this volume), but the term eclosion also refers to the process of emerging from the exuviae during a moult which is covered here.

Posteclosion Behaviour

First-instar nymphs of most weta probably disperse quickly after emerging from their eggs, but this behaviour has only been described for *H. crassidens* (Ordish, 1992). Many weta will eat other weta when they are vulnerable (see below), and Richards (1994) suggested that losses from cannibalism are minimized when Mahoenui weta (*Deinacrida mahoenui* Gibbs) hatch, because eclosion is spread over several days to allow the emerging young time to disperse. In ground weta species (*Hemiandrus*), the first instars remain initially with the adult female in her burrow (Salmon, 1950; Miller, 1971; Barrett and Ramsay, 1991; Gibbs, 1994). Such behaviour occurs in other burrow-using ensiferans in three other families (see review by Gwynne, 1995). In the case of an undescribed species of *Hemiandrus* from the Horotane Valley, the first four nymphal instars have a highly aggregated distribution in the field, because they remain in the mother's burrow. These nymphs tend to form groups of three to seven in the laboratory when removed from the burrow. They subsequently tend to become less aggregated as second instars and eventually become randomly distributed on reaching the fifth instar (Wahid, 1978).

Moulting

Introduction

Insects grow by a series of moults, whereby the cuticle is replaced by new cuticle. This is the only time when sclerotized parts of the cuticle increase in size. The stages between moults are referred to as instars and the durations between moults are referred to as stadia. The process of moulting is termed ecdysis and the term eclosion here refers to the actual process of shedding the old cuticle.

The time of day when ecdysis occurs

Most weta ecdyse at night (Ramsay, 1955; Butts, 1983; Barrett and Ramsay, 1991; Brown, 1995; Meads and Notman, 1995; McIntyre, 1998) or sometimes in the early morning (Butts, 1983). There are, however, few detailed records of the times when it has been observed. In *Hemiandrus* sp. (Horotane Valley), ecdysis usually occurred between 12.00 and 17.00 h but it was observed as early as 10.00 h (Wahid, 1978). Richards (1973) reported that moulting usually occurred between 22.00 and 03.00 h, and occasionally between 04.00 and 06.00 h, in captive *Deinacrida heteracantha* White and *Deinacrida fallai* Salmon. Brown (1995) found that it occurred after 50% of the night had elapsed and finished before 75% of the night had passed (n = 17). Moulting was also observed once in *D. heteracantha* at 21.44 h on Little Barrier Island (Meads and Notman, 1995). Richards (1994) found that both captive and wild *D. mahoenui* moulted during the first 40% of the night-time period in summer, but during the winter (in captivity) it occurred during the first 75% and was most frequent between 50% and 70% of the night-time period.

Changes prior to moulting

Weta, like other insects, lose weight before moulting and subsequently gain weight after ecdysis. Richards (1994) observed that a large decrease in the weight of a *D. mahoenui* nymph usually indicated that it was going to moult within 1–2 weeks. Prior to moulting, this species is relatively inactive, often does not emerge at night and tends not to eat for 4–14 days (although one specimen drank water just before ecdysis) (Richards, 1994). Tree weta species (*Hemideina*) are also inactive 1–2 days before moulting and they become slightly paler in colour (Barrett and Ramsay, 1991). Other observations before ecdysis are reported by Butts (1983), who noted that *Hemiandrus subantarcticus* (Salmon) had a puffy appearance just before eclosion, particularly in the region of the pronotum, and that their cuticle appeared dull.

Ecdysis

When moulting, most tree weta and giant weta hang head downward with their hind legs, and sometimes also their middle legs, firmly hooked on to vegetation. The only reported comparable observations on ground weta are by Butts (1983), who noted that *H. subantarcticus* lie to one side on their dorsal surface when moulting. The Raukumara tusked weta and *Motuweta isolata* Johns moult while sealed in underground burrows (McIntyre, 1998; C.J. Winks, Auckland, 1998, personal communication). Moulting has not been described for these weta, but they are reported to stay in their underground burrows for up to 56 days and 4 months, respectively, when this occurs.

Ecdysis begins in all weta when the abdomen starts undergoing rhythmic contractions. Usually, the end of the abdomen and cerci visibly pull away a small amount from inside the old cuticle. Contractions and arching movements cause the cervical and thoracic regions to swell and split the cuticle along an ecdysial line, which extends from the top of the head to the first abdominal tergum. Bouts of contractions and arching or wriggling movements are followed by periods of rest. Usually, the thorax and anterior of the abdomen emerge first from the split (Fig. 21.1). These are followed by the head and the remainder of the abdomen, then first the forelegs and finally the hind legs are pulled out. As the legs become free, they usually reattach to the outside of the exuviae so that the weta does not fall. The antennae are usually last to be pulled out, with the aid of the mandibles (Fig. 21.1) (Ramsay, 1955; Richards, 1973; Butts, 1983; Moller, 1985; Sherley and

Fig. 21.1. Moulting in *Hemideina femorata* Hutton. (A) Thorax and abdomen emerging from the exuviae. (B) 42 min later the body and limbs emerged except for the end of the abdomen and tips of the antennae. (C) 53 min later the weta separated and dropped from the exuviae. (Photographs by courtesy of L.H. Field.)

Hayes, 1993; Richards, 1994; Brown, 1995; Domett, 1996). The exuviae is either colourless (Richards, 1973; Sherley and Hayes, 1993) or pale brown (Butts, 1983; Richards, 1994).

Duration of ecdysis

After emerging, the weta usually rests, hanging from its exuviae (for about 30–50 min in *D. mahoenui* (Richards, 1994; Domett, 1996) and 9–27 min in *H. subantarcticus* (Butts, 1983)). The entire process of ecdysis takes about 90 min in *D. heteracantha* (Richards, 1973). Brown (1995) found that ecdysis in *D. fallai* took 1.9 ± 0.3 h ($n = 9$) in males and 2.0 ± 0.9 h ($n = 8$) in females, although one female took 8.5 h. Wahid (1978) noted that *Hemiandrus* sp. took between 1 and 2 h to moult. Other information is given by Butts (1983), who found that five specimens of *H. subantarcticus* moulted in 74–97 min, and Moller (1985), who noted that three specimens of *H. crassidens* moulted in 1.8–5.8 h.

After moulting, the cuticle becomes fully sclerotized after 14 h in *Deinacrida rugosa* Buller (Ramsay, 1955) and within 24 h or so in *D. fallai, D. heteracantha* and *Hemideina* (Richards, 1973; Barrett and Ramsay, 1991). Butts (1983) reported that instars 7–12 of *H. subantarcticus* were still somewhat soft 48 h after ecdysis.

Colour changes during ecdysis

Newly moulted weta, like other insects, are much lighter in colour than when fully sclerotized (Fig. 21.1). Most are almost white, except for the eyes, which are black, and the mandibles, which are often brown. Colour can return within a few hours as sclerotization of the cuticle proceeds (Ramsay, 1955), but it can take as long as 7–33 h in *H. subantarcticus* (Butts, 1983). The colour can also return faster in small instars than in large ones (Butts, 1983). Newly moulted *D. fallai* and *D. heteracantha* are greenish (Richards, 1973), whereas *D. fallai* are pale rose-pink (Brown, 1995). Newly moulted *H. crassidens* are milky pink (Moller, 1985), and *Hemiandrus* sp. from Horotane Valley near Christchurch are pale blue (Wahid, 1978). These colours disappear as the cuticle darkens and becomes brown. The blue colour in *Hemiandrus* is even apparent after eclosion from the egg. It becomes more intense in later moults, when a purple tinge also appears, so that the colour changes almost to violet in newly moulted older nymphs (Wahid, 1978). Although Wahid (1978) suggested that the colour is probably Tyndall blue, which is due to light scattering by small particles, a more likely explanation is that it comes from the colour of blood pigments and possibly haemocyanin.

Behaviour following ecdysis

Once sufficiently sclerotized, newly moulted weta generally turn around and eat at least some of their exuviae. Younger instars frequently consume more than older individuals (Ramsay, 1955; Richards, 1973; Wahid, 1978; Little, 1980; Butts, 1983; Moller, 1985; Barrett and Ramsay, 1991; Sherley and Hayes, 1993; Richards, 1994; Spencer, 1995; Domett, 1996), while tusked weta consume all of their exuviae (McIntyre, 1998; C.J. Winks, Auckland, 1998, personal communication). The time required to consume exuviae varied from 1 h to 2.5 h in *D. mahoenui* (Richards, 1994). The only weta that has not been observed to eat exuviae is *D. heteracantha* (Richards, 1973).

Feeding resumes within 24 h of ecdysis for instars 1 to 5 in *Hemideina*, whereas larger individuals may not feed for 2–3 days, (Barrett and Ramsay, 1991). McIntyre (1998) noted that *Hemideina* and *Deinacrida* resume feeding within 2–3 days, and Richards (1994) reported that large instars of *D. mahoenui* started feeding again on the third night after a moult and that their first faecal pellets appeared after the fifth night.

Regeneration of appendages and deformities that occur during growth

Weta often lose appendages or parts of them, but they rarely autotomize a damaged or trapped appendage, as do some other Orthopterans (Ramsay, 1964b; Richards, 1973). However, they regenerate parts of their limbs and antennae, or even entire appendages, if the damage is followed by a sufficient number of moults (see review by Ramsay, 1964b) (Fig. 21.2). Regeneration usually commences a few moults after the loss and the regenerating appendage then grows at a faster rate than undamaged limbs over successive instars. However, the growth rate returns to normal before the limb reaches the size it would have been if it had not been damaged. Regenerated parts therefore never quite reach the size of their undamaged counterparts (Ramsay, 1955, 1964b; Richards,

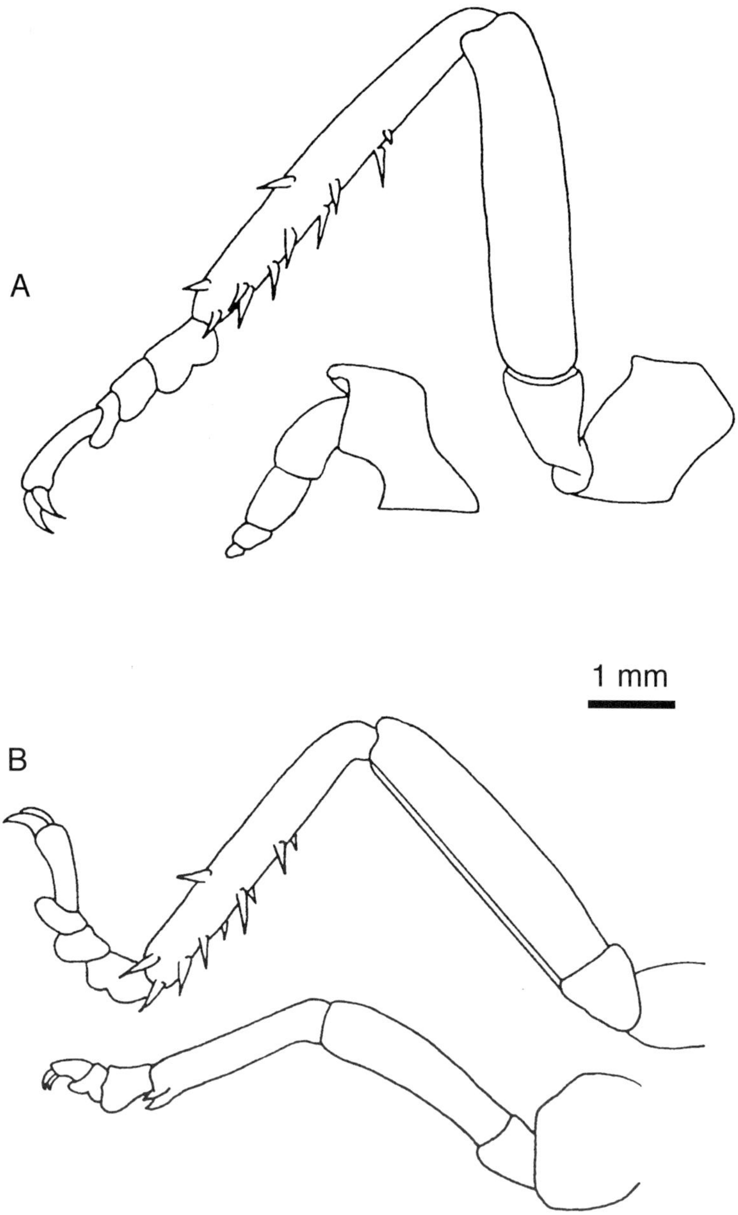

Fig. 21.2. Normal and partially regenerated limbs of *Hemiandrus maculifrons* (Walker). (A) Prothoracic legs from an adult female. (B) Prothoracic legs from an eighth-instar female. (From Cary, 1981, by courtesy of the author.)

1973). The antennae are exceptional in this respect, because they regenerate readily and may apparently reach their full length over only one or a few moults (Ramsay, 1955, 1964b; Barrett and Ramsay, 1991; Richards, 1994).

Weta with regenerating limbs have been reported occasionally from the field (Meads and Moller, 1978; Moller, 1978b; Sherley and Hayes, 1993; Richards, 1994). This suggests that either weta do not lose portions of appendages very often or that instances of regeneration are either overlooked or not reported. There are three published instances of the proportions of damaged weta found in the field. Richards (1994) reported that 5.7% of Mahoenui weta had damaged or missing appendages (n = 405). Of these, 45% had damaged or missing legs, 35% had damaged antennae, 26% had damaged thoracic or abdominal tergites and 23% of females had damaged or bent ovipositors. Slightly more females (57%) than males had damage. Moller (1978b) found damage in 23% of female *H. crassidens* (n = 13) and 12% of males (n = 8) on Arapawa Island. The damage comprised a missing front leg, a broken ovipositor, missing tarsi and broken spines. Amongst *H. crassidens* on Stephens Island, a significantly higher proportion of females (42%, n = 41) were damaged than males (17%, n = 36), and most damage was to limbs, especially tarsi (Moller, 1985). Other damage included a broken mandible, a broken ovipositor and a damaged head. Interestingly, Butts (1983) reported that individuals of *H. subantarcticus* were more likely to be cannibalized in captivity if they had lost limbs.

In captivity, frequent damage and hence regeneration result either from the actions of other individuals or from moulting, if the insect is unable to escape successfully from its exuviae (Richards, 1973; Richards, 1994). Also bent or distorted legs, antennae and ovipositors are relatively common occurrences in captivity. These may result if the containers holding the weta are cramped (Spencer, 1995) or when weta are prevented from hanging in the normal way during ecdysis. Other types of deformations often result when appendages regenerate (Ramsay, 1955, 1964b; Richards, 1973; Cary, 1981; Barrett and Ramsay, 1991; Richards, 1994; Spencer, 1995). Typical examples are bent antennae or annuli of uneven lengths, clubbed cerci and misshapen tibiae and tarsi (Fig. 21.2). Some spines, claws or tarsal segments may also be missing after a limb has regenerated. Reduplication is rare, but cases associated with regeneration were reported in *D. rugosa*, *Hemideina thoracica* White and *H. crassidens* by Ramsay (1955). All such reduplications consisted of an additional small portion of a tarsus that branched alongside a regenerated tarsus. Richards (1973) also reported one additional claw on the pretarsus of one specimen of *D. heteracantha* and Barrett and Ramsay (1991) found a third cercus on one *H. crassidens*.

Growth Through the Instars

Number of instars

The number of moults in weta varies between six and 11 in different species (Table 21.1). Within other Gryllacridoidea, six to 11 instars are known and six to nine instars are usual (Ramsay, 1964a). Males and females of most species undergo the same number of moults, but four species, which also have unusual developmental patterns, are known to have sexual differences in moult number (Table 21.1). Usually, within the Gryllacridoidea, when sexual differences in moult number arise, females have more instars than males. This follows because the larger females generally undergo more moults (Ramsay, 1964a). The first three exceptions occur in species of *Hemiandrus*, where the males undergo fewer moults than females. In the first, from Tekapo, males have more moults than females, even though the male is the smaller sex (van Wyngaarden, 1995). The second exception is in *Hemiandrus* sp. (Horotane Valley, Christchurch), where males can have seven to nine moults, while females have eight (Table 21.1). Wahid (1978), however, suggested that males of the latter species probably undergo seven moults in the field and that the supernumerary moults only occur in laboratory-reared insects. The third exception is *Hemiandrus maculifrons* (Walker). Here males undergo fewer moults to maturity than females (Table 21.1) and yet the imagines of both sexes are approximately equal in size. In this case, the average growth ratio between instars is greater for males than for females (Cary, 1981). The last unusual developmental pattern occurs in the tree weta, *H. crassidens*. Males of this species can mature after six to 11 moults, whereas females mature after nine moults (Barrett and Ramsay, 1991; Spencer, 1995; P.R.L. Cary, unpublished data). In contrast, males and females of other species of *Hemideina* that have been reared become adult at the tenth instar (Barrett and Ramsay, 1991). Interestingly, the size of the imago of male *H. crassidens* increases with the number of moults it has undergone, and this affects the shape of the head, because of allometric changes that occur towards the end of development, as described below (Spencer, 1995; also discussed in Field and Deans, Chapter 10, this volume). *Hemideina crassidens* is the first confirmed case of a male orthopteran maturing at different instars, although

Table 21.1. Number of moults reported during development in weta. The moult from pronymph is excluded.

Species	No. moults male	No. moults female	Comments	Author
Deinacrida fallai	9	9	Reared through	Richards, 1973
Deinacrida heteracantha	10	10	Reared through	Richards, 1973
Deinacrida rugosa	9	9	Reared overlapping instars	Ramsay, 1955, 1965
Deinacrida mahoenui	9	9	Reared through	Richards, 1994
Hemideina crassidens	–	9	Field measurements	Asher, 1977
Hemideina crassidens	7–9	9	Field measurements	Spencer, 1995
Hemideina crassidens	6–9	9	Captive rearing	Barrett and Ramsay, 1991
Hemideina crassidens	8–11	11	Reared overlapping instars	P.R.L. Cary (unpublished data)
Hemideina femorata	> 6	–	Field measurements	Little, 1980
Hemideina maori	> 7	> 5	Field measurements	Little, 1980
Hemiandrus maculifrons	7	9	Reared overlapping instars	Cary, 1981
Hemiandrus subantarcticus	11	11	Reared overlapping instars	Butts, 1983
Hemiandrus sp. (Tekapo)	9	8	Field and captive	van Wyngaarden, 1995
Hemiandrus sp. (Horotane)	7–9	8	Field and captive	Wahid, 1978

some female Orthoptera, Phasmatodea and Mantodea are known to become mature at different instars (Spencer, 1995).

Identification of instars

Laboratory observations

The best way to determine the numbers of weta instars is to rear them in the laboratory and collect exuviae. In most cases, missed instars (from consumed exuviae) were detected by plotting the logarithms of measured data against instar number. So far, measurements have been taken from individuals reared from egg to adult in five species: females of *D. rugosa* (Ramsay, 1955, 1965) and both sexes of *D. fallai* (Richards, 1973), *D. heteracantha* (Richards, 1973), *D. mahoenui* (Richards, 1994) and *Hemiandrus* sp. (Wahid, 1978). In all cases five or fewer individuals of each sex completed their entire development; thus limited data are available for older nymphs. Other species of weta have been reared successfully in captivity for more than one generation, but no details of their development have been published (Meads, 1988; Barrett and Ramsay, 1991). An alternative approach was to obtain information on all instars by rearing a number of different individuals of different ages until they overlapped the instars of other individuals. This method was employed for three student research projects, although time constraints prevented weta from being reared through their entire life cycle (Little, 1980; Cary, 1981; Butts, 1983), while, for *H. crassidens*, difficulties occurred in rearing these weta from egg to adult (P.R.L. Cary, unpublished data). Moults were signalled by size changes (Cary, 1981; Butts, 1983) and also by the disappearance of marks painted on the cuticle (P.R.L. Cary, unpublished data).

Methods for successfully rearing a variety of weta species in captivity have now been published (Barrett and Ramsay, 1991) and this opens the way to obtaining complete information on the growth rates, numbers of instars and identification of instars for several more weta.

Morphometric determination of instars

Morphological measurements that have been taken from weta, given in order of frequency of use, are pronotum length; ovipositor length; maximum head width and hind femur length; pronotum width; pro- and mesofemur lengths; pro-, meso- and metathoracic tibiae and tarsi lengths; length from anterior edge of pronotum to anus; mass; head length, including mandibles, and interocular distance; mandible length; and antennal length (Ramsay, 1955; Richards, 1973; Asher, 1977; Meads and Moller, 1978; Cary, 1981; Butts, 1983; Sherley, 1992; Sherley and Hayes, 1993; Richards, 1994; Meads and Notman, 1995; Townsend, 1995; van Wyngaarden, 1995). Less common measurements include the maximum head width at the point where the mandibles articulate, the total body length from the tip of the mandibles to the anus, the length from the anterior edge of the pronotum to the anus and the ovipositor length from the base of the subgenital plate (Rufaut, 1995; Spencer, 1995). Body length and weight, especially, can vary too much during a stadium to be useful for determining the instar number (e.g. Richards, 1973; Field, 1980; Richards, 1994), although weight was used for one weta species to help distinguish age (Meads and Moller, 1978). The antennae are particularly likely to break during ecdysis, so they are not useful for identifying instars either.

Morphometrics do not generally provide an accurate estimation of the number of instars, unless large sample sizes are available that span the entire size range of the species. When available, the instars may often be determined by plotting frequency distributions of a particular sclerite or structure. This was done successfully for the larger instars of *H. crassidens, Hemideina femorata* Hutton, *Hemideina maori* (Pictet and Saussure) and *D. mahoenui* (Asher, 1977; Little, 1980; Sherley, 1992; Sherley and Hayes, 1993). Relatively large, sclerotized structures were usually measured for ease and accuracy (Cary, 1981; Butts, 1983; van Wyngaarden, 1995). While efficient statistical methods are available that can improve the recognition of each instar grouping using the above approach (MacDonald and Pitcher, 1979), these have not yet been applied to weta. In general, plotting the length of one sclerite against another has not improved discrimination between the different instars (Meads and Moller, 1978; Field, 1980; Little, 1980; Townsend, 1995). While the above approach does not always allow reliable instar placement of individuals, it can help by providing size ranges for probable instars. In general, the size of a sclerite usually shows distinct differences between the first three or four instars of weta, while measurements of older nymphs show increasing overlap and hence generate increasing uncertainty (Asher, 1977; Meads and Moller, 1978; Sandlant, 1981; Butts, 1983; Richards, 1994).

A number of other morphometric approaches have been explored to separate different instars. Butts (1983) applied different cluster analysis methods to measurements from ten different structures plus the ovipositor length in females. She found that Euclidean distance was best, although it was still unsatisfactory, because it resolved 14 male and 16 female size groups when both sexes have 12 instars when reared. Spencer (1995) found that discriminant analysis gave relatively good separation into instars 5 to 10, using a variety of measurements taken from *H. crassidens* (Fig. 21.3). This method provided a better than 82% chance of correctly assigning an individual to its instar, and in most instances the accuracy was 97% or better. She found that the best features for discriminating instar number, listed in decreasing importance, were as follows: in males, head length, femur length and tibia length; and in females, ovipositor length followed by tibia length and femur length (Spencer, 1995).

Methods for checking the number of instars

Most authors checked if there were missing instars by plotting the logarithm of the means or modes of a part of the body against the instar number (Ramsay, 1955; Wahid, 1978; Little, 1980; Cary, 1981; Butts, 1983; Richards, 1994). If all instars are present, the measurements of most structures lie more or less along a straight line. A step in the plot indicates that an instar is missing or two

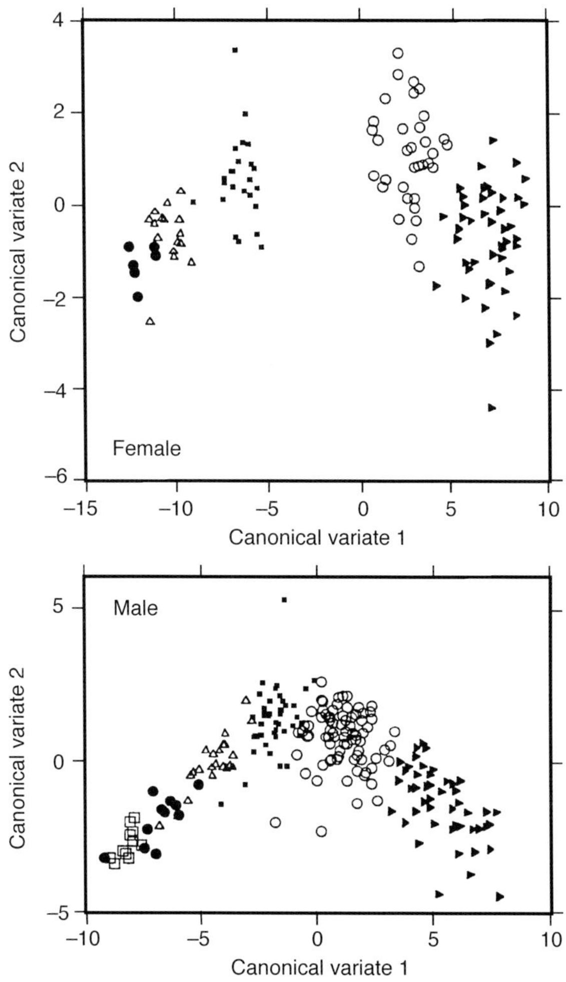

Fig. 21.3. Discriminant analysis of measurements taken from instars 5 to 10 of male and female *Hemideina crassidens* (Blanchard) (from Spencer, 1995, by courtesy of the author).

instars are amalgamated together. The measurements plotted may be means from captive animals or modes obtained from histograms of data. If the constant proportional increase in size between successive instars has a coefficient close to 1 : 1.4 ($\text{instar}_n : \text{instar}_{n+1}$), growth follows Dyar's rule. If this coefficient is close to 1 : 1.26, it follows Prizbram's rule, but it is quite common to have any intermediate ratio between these (Sehnal, 1985). The growth coefficients reported for weta range from almost 1 : 1 to 1 : 1.3, with most falling close to Prizbram's ratio (Table 21.2). Little (1980) and Butts (1983) also used the growth ratio rules of Brown and Davies (1972) and Crosby (1974) to test if all instars were present. According to these rules, if two successive growth ratios differ by more than 10–20%, it is likely that an instar was missed.

Some structures in some weta grow allometrically. Here a plot of the logarithm of a body measurement against instar number results in an upward curve. A straight line is achieved by plotting the logarithm of the measurement against the logarithm of the instar number. Well-known

Table 21.2. Growth ratios of parts of weta that show a constant proportional increase when the logarithm of the measurement is plotted against instar number.

Species	Body part	Sex	Growth ratio	Author
Deinacrida rugosa	Head length, pronotum length	Female	1.222	Ramsay, 1955
	Pronotum length	Male	1.222	Ramsay, 1955
	Head length,	Male	1.016–1.325	Ramsay, 1955
	max. weight	Both	~1.26	Ramsay, 1955
Deinacrida heteracantha	–	Female	~1.25	Richards, 1973
Deinacrida heteracantha	–	Male	~1.24	Richards, 1973
Deinacrida fallai	–	Female	~1.27	Richards, 1973
Deinacrida fallai	–	Male	~1.23	Richards, 1973
Hemiandrus subantarcticus	Most parts	Female	1.06–1.27	Butts, 1983
Hemiandrus subantarcticus	Most parts	Male	1.13–1.26	Butts, 1983
Hemiandrus maculifrons	Most parts	Female	1.15–1.30	Cary, 1981
Hemiandrus maculifrons	Most parts	Male	1.13–1.35	Cary, 1981

examples of allometric growth in weta are the head – in particular the mandibles (described by Field and Deans, Chapter 10, this volume) – and the cerci of male *Hemideina* over the last few instars (Ramsay, 1955; Asher, 1977; Field, 1980; Little, 1980; Hamilton, 1991; Spencer, 1995; Townsend, 1995). The ovipositor of most weta also grows allometrically in the few instars that precede the penultimate instar but there is usually a relatively small change in length between the penultimate instar and the imago (Ramsay, 1955, 1965; Hamilton, 1991; Richards, 1994; Spencer, 1995; Townsend, 1995). The ovipositor of tusked weta, in contrast, shows greatest elongation during the final moult (McIntyre, 1998).

Change in size between instars was investigated in most detail in *H. crassidens*, by Spencer (1995). She found that most structures showed some variability in their proportional growth rate. The most uniform were the lengths of the tibia and femur, which were similar in males and females. The pronotum length and head width also increase at an approximately constant proportional rate in the female between instars 6 and 10 (adult), but have different proportional rates in males after the seventh instar. The head width of males is also greater than that of females from the seventh instar onwards. Finally, *H. crassidens* males become mature at the seventh to twelfth instars, as mentioned above, but Spencer (1995) found no predictive relationship between measurements taken from earlier instars and the instar numbers of male adults.

The external genitalia as an aid to identifying instars

The state of development of the external genitalia seems to provide one of the best features for determining the instar, once males and females can be recognized. This usually occurs after the first few instars, as described under genital morphology below. Because genital structures vary in first appearance, shape and relative size in different instars, measurements of a genital component can sometimes also help to identify some instars. In the Mahoenui giant weta *D. mahoenui*, for example, the length of the ovipositor is useful for distinguishing between female instars 4 to 8 but it does not distinguish between the ninth instar and adult, because these measurements overlap (Richards, 1994). In general, all instars except the first few can be identified in this way, if the development of the external genitalia has been previously documented. There are, however, practical difficulties in the field, because the genitalia are often too difficult to observe when weta are very small (e.g. *H. crassidens* nymphs before the sixth instar (Spencer, 1995); *Hemiandrus* nymphs (Salmon, 1950)). In general, determination of instar is best done by using a combination of measurements from large sclerotized structures and the developmental state of the external genitalia (Cary, 1981; Butts, 1983; van Wyngaarden, 1995).

Features of the imago

Adult weta are apterous, so it is sometimes difficult to distinguish them from large nymphs until one is familiar with the particular species. Adult male tree weta (*H. crassidens*) were identified as those that attempted to copulate with females, and females were identified as adults by dissecting and searching for mature eggs within their reproductive tracts (Spencer, 1995). Morphologically, adult female *Hemideina* can be recognized by the upward curve of the ovipositor, whereas it is straight in last-instar nymphs. The adult ovipositor is also sharper, finer and more sclerotized (darker) than that of nymphs (Meads and Moller, 1978; Sandlant, 1981; Moller, 1985; Barrett and Ramsay, 1991; Hamilton, 1991; Richards, 1994; Spencer, 1995; Townsend, 1995). Brown (1995) used mating to confirm imagines of *D. fallai*. The best indicators of maturity in the Raukumara tusked weta are ovipositor length (adult range 14–17 mm; last-instar nymph range 6–8 mm) and the third valvulae (dorsal), which overlap the first valvulae (ventral) by 2–3 mm in adults but not in juveniles (McIntyre, 1998). A similar dorsal overlap was also reported in adult female *H. maori* (Gwynne and Jamieson, 1998).

Male weta are often hard to recognize. Adult Mahoenui weta have more pointed cerci than nymphs (Richards, 1994), whereas in *Hemideina* the cerci are curved inwards and are relatively longer than the straight cerci of nymphs (Barrett and Ramsay, 1991; Spencer, 1995; Townsend, 1995). The tips of the cerci in adult male *H. maori* are flexible, in contrast to those of nymphs (Gwynne and Jamieson, 1998). In addition, adult *Hemideina ricta* Hutton and *H. femorata* can be recognized by the presence of a pair of small black hooks on their paranotal plates (P.M. Johns, Christchurch, 1998, personal communication; see also Field and Jarman, Chapter 17, this volume). Adult male Raukumara tusked weta can be recognized by their mandibular horns, which usually curve slightly downward and are asymmetrical, so that they either come close to each other or cross at the tips, in contrast to the straight, short, mandibular horns of larger nymphs (Gibbs, 1998). In contrast, the mandibular horns of adult *M. isolata* curve upwards and have ridges that form a set of stridulatory pegs. Adult males can often be distinguished from large juveniles, because the mandibular horns of the latter are symmetrical, do not meet and are smooth-surfaced (McIntyre, 1998). Despite this, some adult male *M. isolata* are still difficult to recognize using the appearance of their tusks alone (McIntyre, 1998).

Other features of adult weta are as follows: the imagines may be shinier than the nymphs, their dark areas or markings may be more pronounced and darker and their general coloration may also be darker (Meads and Moller, 1978; Sandlant, 1981; Barrett and Ramsay, 1991; Richards, 1994; Spencer, 1995; compare with *Cratomelus* in Angulo, Chapter 11, this volume). There may also be other colour changes, as, for example, in *H. subantarcticus*, where adults are reddish brown in comparison with the dull brown of nymphs (Butts, 1983), and in Raukumara tusked weta, where distinctive bright orange–tan marks on the pronotum of nymphs fade in adults (Gibbs, 1998; McIntyre, 1998). The spines on the hind tibia of *D. rugosa* are generally less robust in nymphs (Meads and Moller, 1978).

In male *Hemideina*, the head is much larger in proportion to the rest of the body, as described in Field and Deans (Chapter 10, this volume). However, even this relative size difference may not always provide a good indication of maturity, as in the case of male *H. crassidens*, which can become mature between the seventh and twelfth instars (Spencer, 1995; P.R.L. Cary, unpublished data). In this species the head is noticeably elongated in the seventh instar so adult male head length is particularly variable (Sandlant, 1981; Barrett and Ramsay, 1991; Spencer, 1995; see also Fig. 17.2A, Field and Jarman, Chapter 17, this volume).

Size criteria have sometimes been used to identify adults. For example, Moller (1985) considered that male *H. crassidens* with a hind tibia length of 24 mm or greater were probably adult, whereas adult females had a hind tibia length of 20–26 mm and their ovipositor was longer than 19 mm. Sandlant (1981) also noted that most adult *H. femorata* are 30 mm or greater from the anterior edge of the pronotum to the anus. In addition, he reported that adult males had a head length to body length ratio of 0.4 or greater, and that adult females had an ovipositor to body length ratio of 0.35 or more. However, no single dimension can be used to identify the imago unambiguously, as discussed above, and even using ratios is likely to provide some uncertainty.

Morphological Changes

The external genitalia

First instar male and female weta cannot be distinguished (Ramsay, 1955; Cary, 1981; Butts, 1983). In *Hemiandrus* sp. from Tekapo, South Island, the sexes are hard to distinguish in the second instar (van Wyngaarden, 1995), and in the first three instars of *D. heteracantha* and *D. fallai* (Richards, 1973); Mahoenui weta (Richards, 1994), and *Hemiandrus* sp. from the Horotane Valley, Christchurch (Wahid, 1978). Once the external genitalia begin to differentiate, they show proportional changes between successive instars and, in the female, new structures may arise or change their position, while others may fuse together or disappear in subsequent moults.

The most detailed account of the development of the genitalia for any weta is by Ramsay (1965), who followed each sclerite from its first appearance in female *D. rugosa*. Good descriptions of postembryonic development are available for both male and female *H. maculifrons* (Cary, 1981), *H. subantarcticus* (Butts, 1983) and *Hemiandrus* sp. from Tekapo (van Wyngaarden, 1995). In addition, descriptions are available for the first four instars of male *D. rugosa* (Ramsay, 1955) and for some early instars of *Hemiandrus* sp. from the Horotane Valley, Christchurch (Wahid, 1978). The following account represents a synthesis from all of the above descriptions, together with some unpublished results (P.R.L. Cary) for *H. crassidens*.

The only genital structures present in the first instars, when sex is indeterminate, are a pair of styles, which are each attached to a posterior lobe of the ninth sternum (Figs 21.4 and 21.5). The lobes represent coxites, which usually appear to be fused with the sternum, although they are distinct entities in at least *D. rugosa* (Ramsay, 1965). Parameres are present as a pair of distinct sclerites, which are ventral and medial to the cerci. They are sternal sclerites of the eleventh segment, whereas the epiproct, the dorsal plate posterior to the cerci, represents the eleventh tergum (Ander, 1970). Sclerites of the other terminal segments are unmodified, except for the eighth sternum, which may be slightly narrower than the preceding sterna. The posterior margin of this latter sclerite may bear a pair of very slight posterior expansions in female *D. rugosa* (Ramsay, 1955, 1965), but these have not been reported in other weta.

The male

The male genitalia develop entirely from structures already present in the sexually indeterminate stages. Development occurs through change in relative size and shape between each successive instar (Figs 21.4 and 21.5). The styli become relatively smaller as the insect grows, but they do not disappear (such retention is neotenic rather than primitive (Matsuda, 1976)), as they do in some Orthoptera (Ander, 1970). As development continues, the fused sternum and coxites of the ninth segment enlarge and elongate posteriorly to become the subgenital plate. This plate generally covers the end of the abdomen, as seen from below, in later instars, and it may even extend past the end of the abdomen in the adult. The paraprocts enlarge relative to the other terminalia and undergo a change in shape. This is most pronounced in later instars. The cerci gradually elongate and finally become curved in adult *Hemideina* (Fig. 21.5). Lastly, paired hook-like modifications of the tenth tergite (a ‘gin trap’) appear at the adult moult in *Hemiandrus* (Cary, 1981; van Wyngaarden, 1995; Fig. 21.4).

The female

The changes that take place in the female genitalia during development are much more complex than those that occur in the male (Figs 21.6 and 21.7). In addition, some of the components of the female genitalia develop directly from structures that are present in the sexually indeterminate instars, whereas others appear *de novo* as development proceeds. Many components of the ovipositor in insects have been given alternative names by different authors, as listed by Ander (1970), whose terminology is used below.

The paired coxites of the ninth segment acquire distinct articulations with the ninth ster-

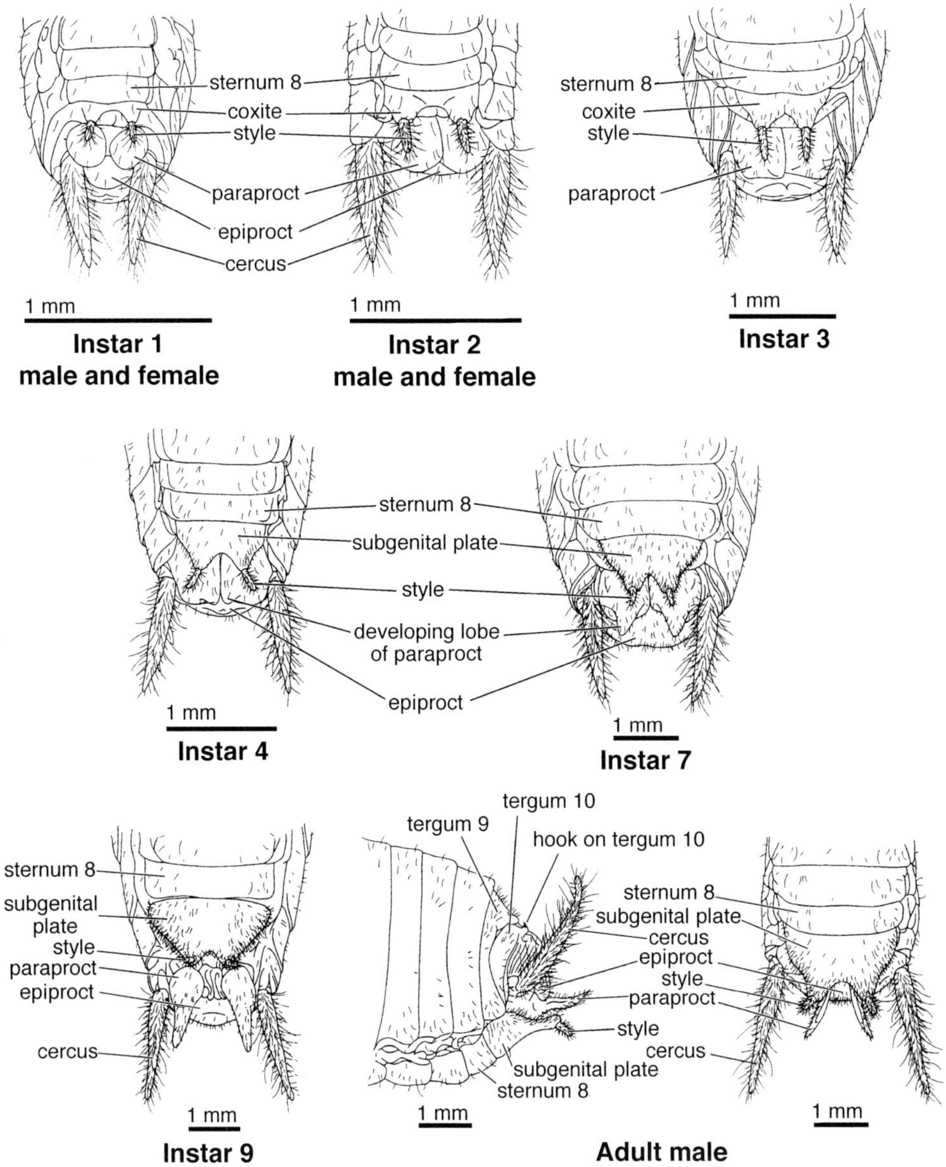

Fig. 21.4. Development of the external genitalia of male *Hemiandrus* sp. (from van Wyngaarden, 1995, by courtesy of the author).

num at the moult from the last sexually indeterminate stage. These coxites eventually develop into the third valvulae, but their styli gradually become relatively smaller and disappear about halfway through development. Ramsay (1965) considered that the styli become incorporated into the third valvulae in *D. rugosa*. New structures that appear at the moult from the last indeterminate stage or at the next moult are a pair of first-valvulae rudiments and a pair of lobes that are rudimentary second valvulae. The first valvulae appear from the posterior sternal region of the eighth segment, whereas the second valvulae develop initially as lobes medial to the bases of the third valvulae.

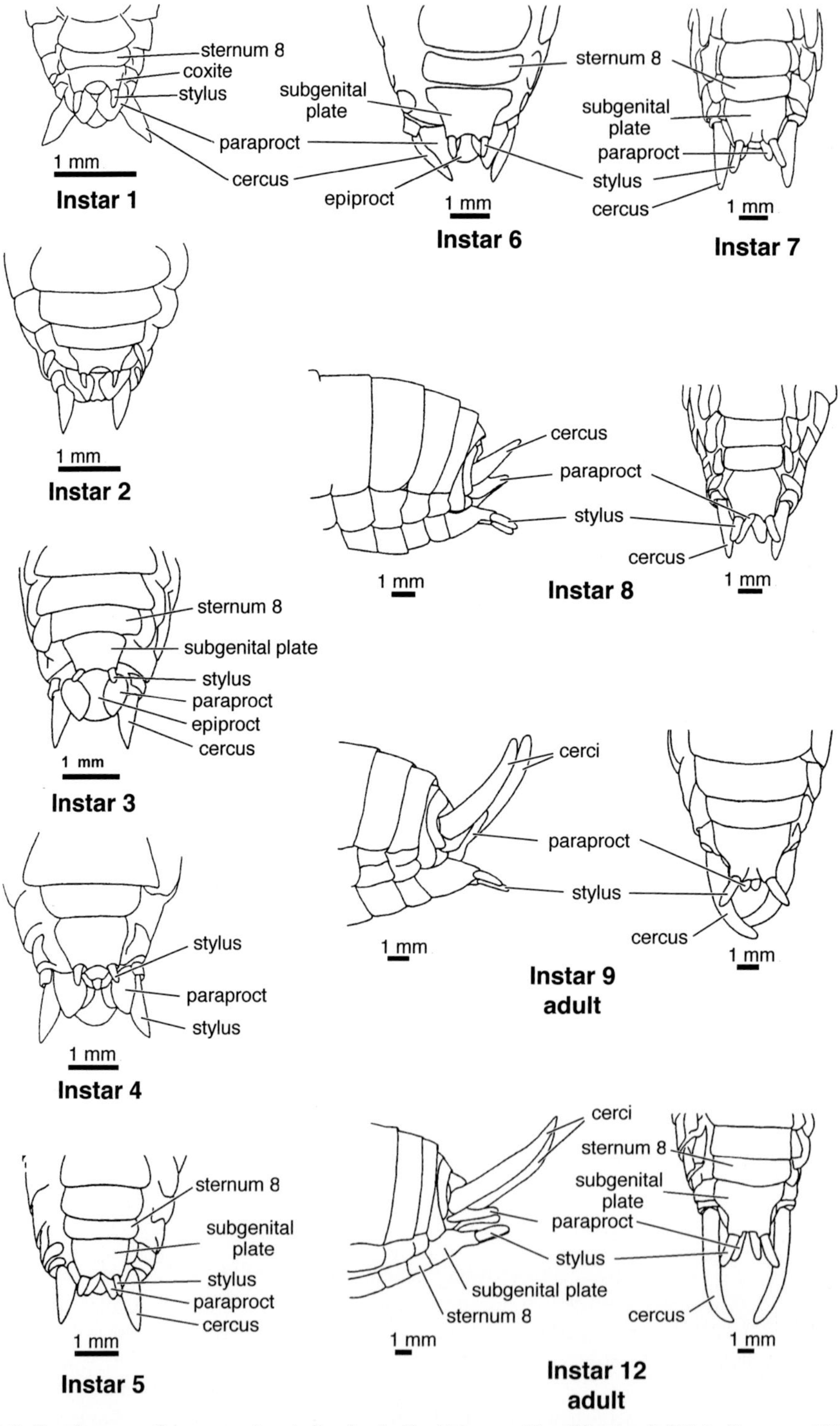

Fig. 21.5. Development of the external genitalia of male *Hemideina crassidens* (Blanchard) (P.R.L. Cary, unpublished data).

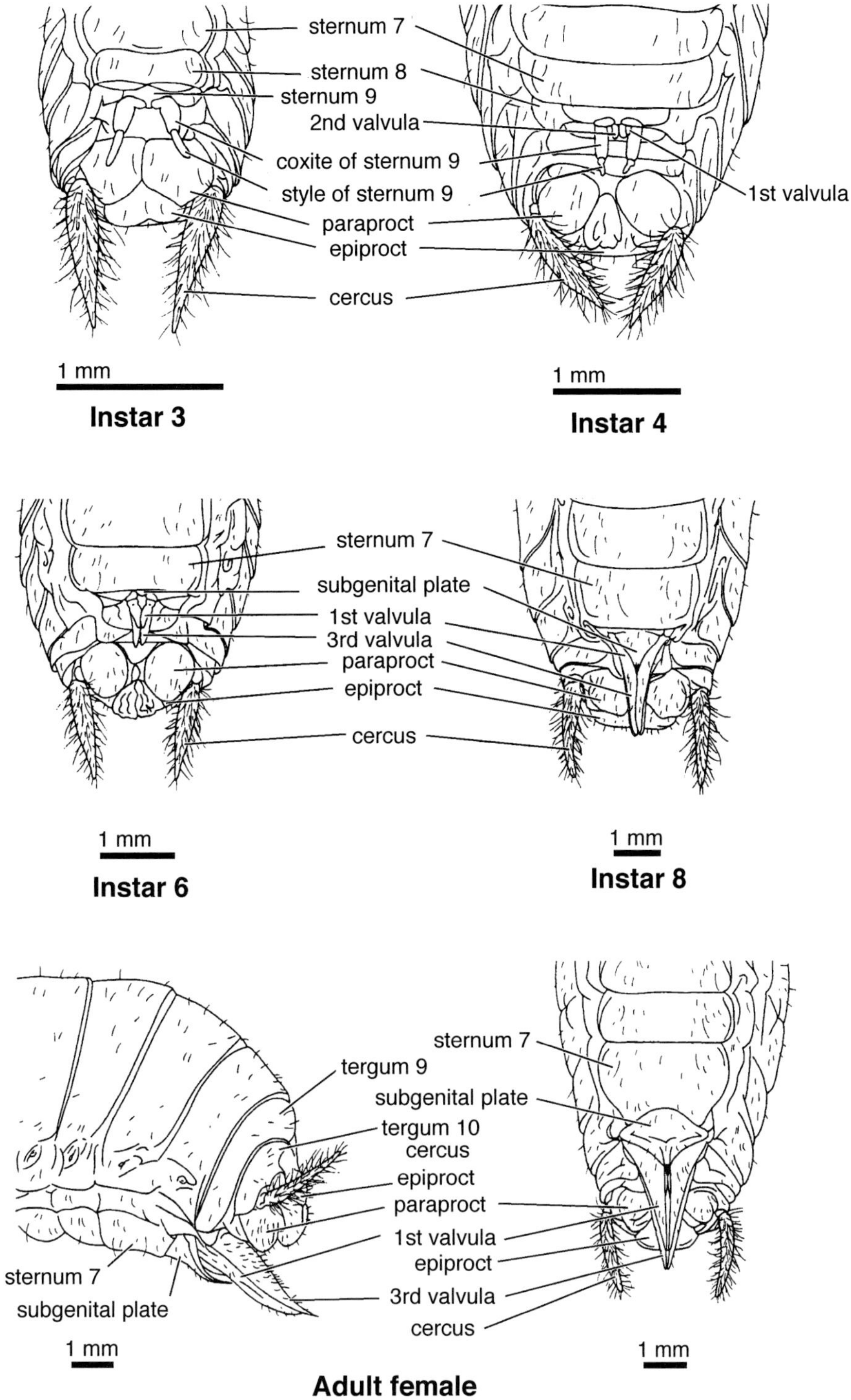

Fig. 21. 6. Development of the external genitalia of female *Hemiandrus* sp. The first two instars are shown in Fig. 21.4. (From van Wyngaarden, 1995, by courtesy of the author.)

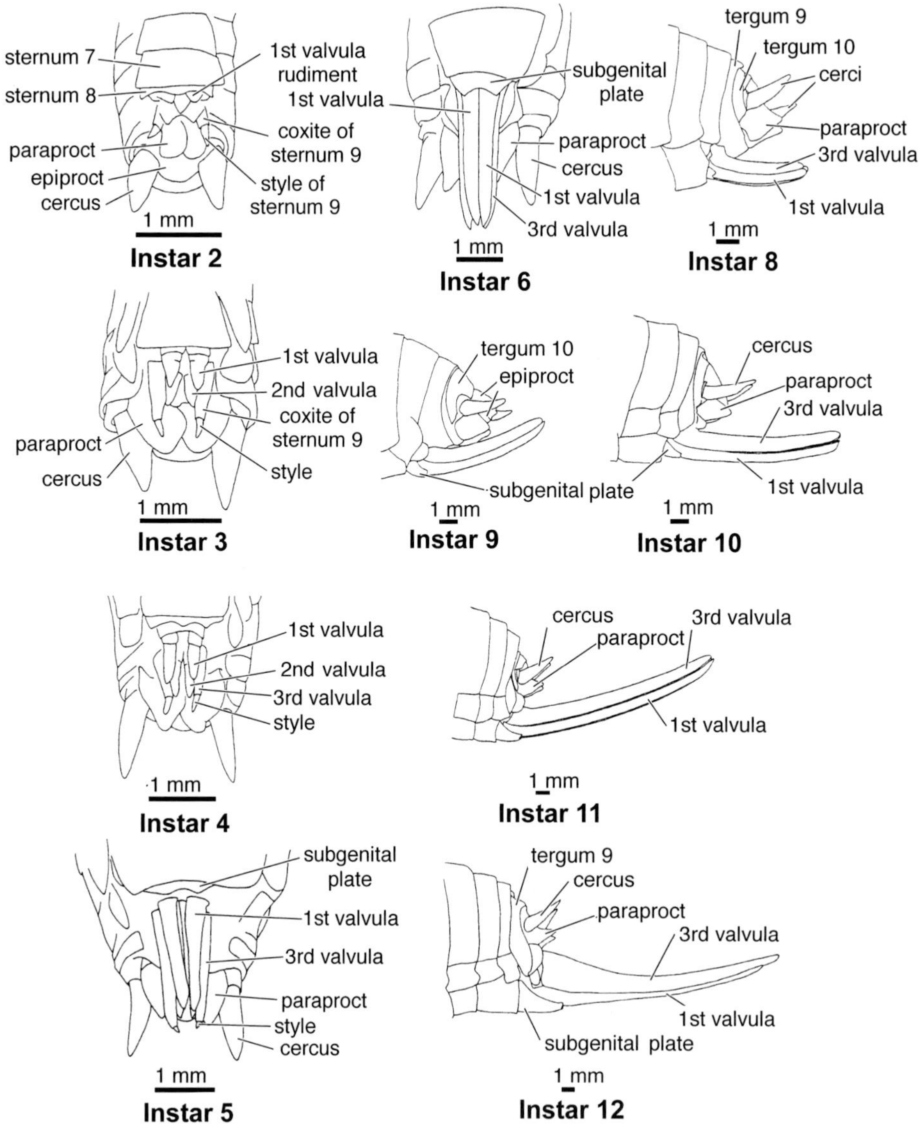

Fig. 21.7. Development of the external genitalia of female *Hemideina crassidens* (Blanchard). The first instar is shown in Fig. 21.5. (P.R.L. Cary, unpublished data.)

Both first and second valvulae usually appear to become two-jointed a few moults after they appear, but the joints disappear again later in development. The third valvulae clearly develop from the ninth coxites and possibly also their styli, so they are derived from parts of segmental appendages. The origins of the first and second valvulae, however, are not at all clear so there is debate about their homologies (see Scudder, 1971).

The rudimentary first valvulae are spaced well apart when they first appear (in instars 2–4, depending on the species), but they migrate medially during successive moults until they lie close together in the midline. The developing second valvulae elongate and become distinctly separated from the third valvulae. A rudimentary ovipositor forms when the second valvulae become enclosed ventrally by the first valvulae and laterally and dorsally by the third valvulae. Further growth to the adult ovipositor is largely by elongation of the valvulae.

The subgenital plate first appears as a lobe from the posterior margin of the seventh sternum (Figs 21.6 and 21.7). At first it may be weakly bilobed, as in *D. rugosa* and *H. crassidens* (Ramsay, 1965; P.R.L. Cary, unpublished data), or it may be a single lobe, as in *Hemiandrus* (Cary, 1981; van Wyngaarden, 1995). Eventually it expands to the entire width of the sternal region and elongates to cover the base of the ovipositor.

The seventh and sometimes also the sixth sterna gradually become elongated in proportion to the preceding sterna. The fates of the coxites and sterna of the eighth and ninth segments, however, have been followed only in *D. rugosa* (Ramsay, 1965). Ramsay describes the eighth coxites as appearing in the second instar as narrow basal joints of the first valvulae. These fuse together in the third instar, but become separate valvifers, which later fuse with their respective valvulae in the seventh instar. Eventually, they become reduced projections near the base of the first valvulae in the imago. The ninth coxites first enlarge and then become fused with two lateral sternal sclerites to form compound second valvifers in the eighth instar. The second valvifers are small sclerites that lie laterally by the bases of the second valvulae. They fuse with the eighth valvulae at the next moult and become reduced to small lateral projections at the bases of the second valvulae in the adult. The eighth and ninth sterna both become relatively smaller as development proceeds but the eighth is eventually lost entirely by the fifth instar. The ninth sternum first becomes divided into three sclerites at the fourth instar, but only two lateral sclerites remain at the fifth instar, when the medial sternal sclerite is lost. The lateral sclerites later fuse with the coxite bases in the eighth instar to form small compound second valvifers. A pair of new sclerites, the interbasivalvulae, appear between the bases of the first valvulae in the seventh instar. These fuse at the next moult to become a small rounded interbasivalvula in the adult. Another new sclerite, the intervalvula, first appears as a transverse thickening between the bases of the third valvulae in the eighth instar. This thickens into a stout rod in the adult (Ramsay, 1965).

Development of the internal reproductive organs

The morphological development of the internal reproductive organs has been followed from first instar to adult in *H. crassidens* (P.R.L. Cary, unpublished data). There is as yet no histological information on the development of the internal reproductive organs for any weta. In *H. crassidens*, the male and female reproductive organs are very similar in the first instar. Both consist of a pair of elongated gonad rudiments connected to a pair of fine tubes. These tubes pass posteriorly and unite ventrally in the region of the ninth and tenth abdominal sterna. The only difference between first-instar males and females is that these tubes swell slightly before uniting in males. The reproductive organs subsequently become increasingly larger with each moult and differences between male and female become evident.

The male

The swellings at the bases of the developing vasa deferentia change little at each moult, except for an increase in size. The greatest increase occurs when the insect becomes adult, and then the vasa deferentia become coiled and the accessory glands acquire their numerous diverticulae (Fig. 21.8). The testis begins to lose its elongated appearance and starts to round off in the third instar but its enlargement to the adult size is spread more evenly over the last instars than in Fig. 21.8.

The female

The first sign of a spermatheca becomes apparent in females as a small sac in the fifth instar (Fig. 21.9). This elongates and bends into a C shape by the seventh and eighth instars. In subsequent instars, it moves posteriorly and unites with the common oviduct. The first signs of developing oocytes appear in the fourth instar. These gradually enlarge with each moult, but fully formed eggs do not appear until the adult (Fig. 21.9). Developing oocytes were also visible in the ovaries of large nymphs of *H. maculifrons*, but none had chorions (Cary, 1981). Last-instar nymphs of *H. maculifrons* may also have slightly more ovarioles per ovary than adults (nymph 23–28, mean 26; adult 17–23, mean 22), but the difference was not significant (Cary, 1981). The situation may be

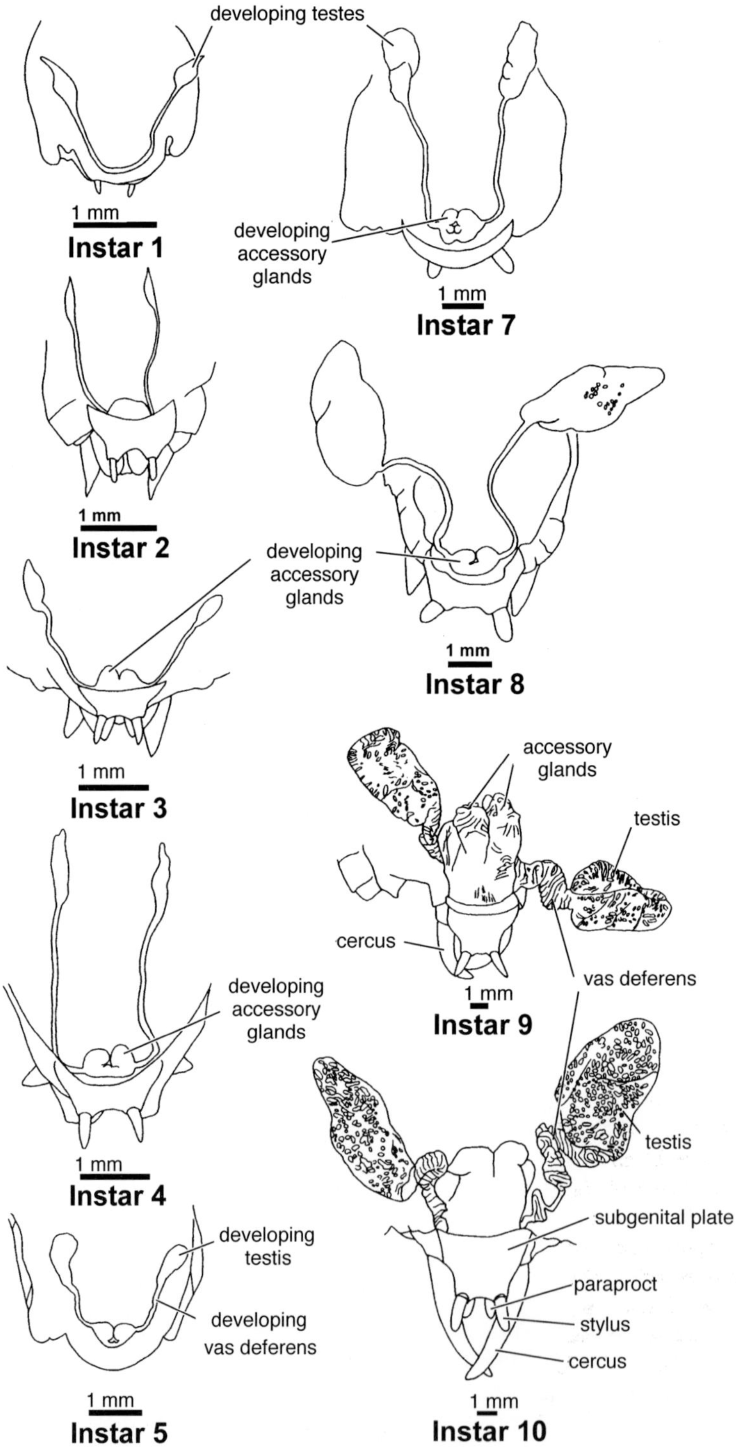

Fig. 21.8. Development of the internal reproductive organs of male *Hemideina crassidens* (Blanchard) (P.R.L. Cary, unpublished data).

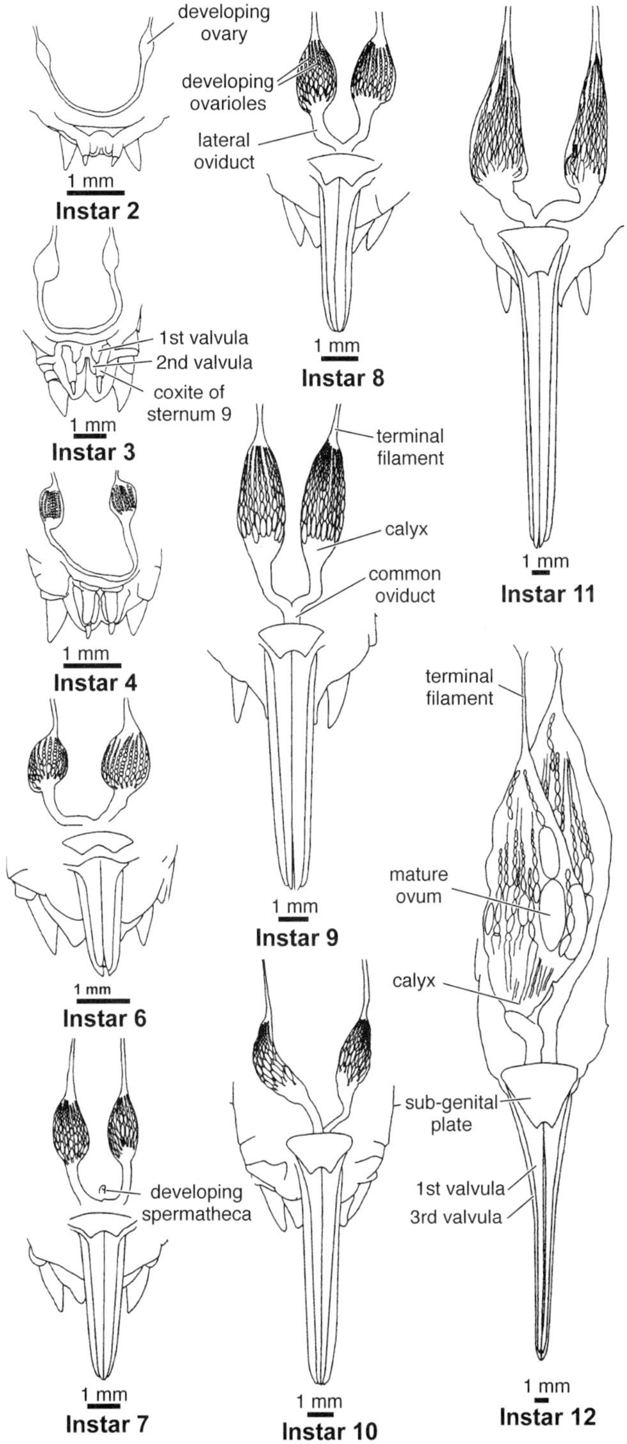

Fig. 21.9. Development of the internal reproductive organs of female *Hemideina crassidens* (Blanchard) (P.R.L. Cary, unpublished data).

quite different in *H. subantarcticus*, because no sign of visible oocytes was reported in the nymphs by Butts (1983).

Other Changes that Occur during Development

Diet

The diets of only two species are reported to change with age. Less animal matter is eaten during the first four instars of the omnivorous ground weta *H. subantarcticus* than in later instars, although, in general, females of this species eat a lower percentage of animal matter than males (Butts, 1983). *Hemiandrus maculifrons*, in contrast, is carnivorous, but the diet of females changes with age. Early-instar females and all stages of males primarily eat insect larvae, whereas later-instar females eat more adult insects (Cary, 1983).

Habitat

In *D. rugosa*, on Mana Island, juveniles of about 10–20 mm in length are found exclusively on vegetation 0.5 m or higher above the ground, whereas adults can be found nearer the ground (Ramsay 1955). In captivity, young nymphs also excavate burrows in soil, if no shelter or suitable cracks are available, but larger nymphs or adults will not.

Colour change

Some weta undergo colour changes during growth (Richards, 1973). The first instars of *D. fallai* and *D. heteracantha* are jade green or pale green, with some brown to black pigmentation on the edges of the body sclerites and on the ends of leg segments. The green colour gradually becomes less obvious in subsequent instars and disappears by the third instar (*D. mahoenui*; Richards, 1994) or the seventh instar in other species. Ramsay (1955) noted that the first-instar *D. rugosa* is pale brown, with dark brown mottling and a conspicuous notched pattern of alternate dark brown and white areas on the posterior edges of the abdominal terga. The mottling becomes coarser in subsequent instars and the notched pattern becomes less conspicuous, until it is finally lost by the fifth instar. Other subtle colour changes also occur in *D. rugosa* during the later instars, and the body surface also becomes rough at the third instar and then develops more pronounced sculpturing in subsequent instars (Ramsay, 1955). The Raukumara tusked weta has distinctive, bright, orange–tan marks on the pronotum of mid-sized to large juveniles, but these are not present in small juveniles and they fade in adults. Adult *M. isolata* are a lighter tan colour than the juveniles (Gibbs, 1998; McIntyre, 1998).

Body size

In addition to head size in *Hemideina*, the overall size of giant weta is a sexually dimorphic feature. Adult females appear larger than males, even though many of their sclerites are approximately the same size (see Stringer, Chapter 20, this volume). This size difference is mostly due to the females having larger abdomens and being heavier than adult males. This dimorphism appears at different stages of development in different giant weta species. It is first noticeable in the seventh instar in Mahoenui weta, although the difference is only significant in instars 9 and 10 (Richards, 1994); it becomes apparent in the sixth instar in *D. heteracantha* and *D. fallai* (Richards, 1973) and in the fourth instar in *D. rugosa* (Ramsay, 1955).

Tusks

The complete development of tusked weta has not been followed, but the first indications of the mandibular horns in male Raukumara tusked weta are 'knobs' which appear on the front of the mandible of slightly less than mid-sized juveniles (McIntyre, 1998). There is no trace of mandibular tusks in 30 mm long male nymphs of *M. isolata*, but tusks increase from 8 to 10 mm long in last-instar juveniles to 20–30 mm long in adults (Gibbs, 1998; Meads, 1995). The mandibular tusks of these juveniles are smooth-surfaced and symmetrical, and they do not meet each other (Meads, 1995).

Frass size

Finally, there is also a good correlation between frass length and weta age for *H. crassidens* (Rufaut, 1995), so the distribution of frass length can be used as an approximate estimate of the age distribution of these weta (Meads and Notman, 1995; Rufaut, 1995). An interesting example of such an analysis was provided by Rufaut (1995). She inferred, from a comparison of frass collected 7 months after the eradication of kiore (native rats) and again 12 months later, that the population of *H. crassidens* from Nukuwaiata Island in the Marlborough Sounds had increased and that there was an overall younger population structure of weta at the later date.

Differences between Captive and Wild Weta

The giant weta species *D. rugosa and D. mahoenui* were smaller when reared in captivity than those in the field and they matured faster than those in the wild (Ramsay, 1955; Richards, 1994). However, this difference in maturation rate was only slight in Mahoenui weta, because the overall growth rates of weta in the field were similar to those in captivity (Richards, 1994).

Growth and Factors that Affect Growth Rate

Seasonality

Most weta species show a more or less distinct seasonal developmental pattern, although there is usually much overlap between generations. Emergence of the ground weta *H. maculifrons* is particularly poorly synchronized with the seasons and all developmental stages occur throughout the year, despite the fact that it lives where the leaf litter can be frozen in winter. Cary (1981, 1983) suggested that the lack of a seasonal pattern is probably related to the carnivorous diet of this weta. It primarily feeds on leaf-litter invertebrates, which are always abundant, even under snow (Cary, 1983).

The developmental patterns of most other species appear to be better synchronized seasonally. Most are omnivorous to some extent, although their main food is vegetation. Asher (1977), for example, noted that there was a lack of small instars of *H. crassidens* in July. The adults of many other species are reported to occur from late spring or early summer to autumn and few of these adults appear to live into winter (Table 21.3), again suggesting that they are seasonal. In addition, some species are found in the field in two (*D. rugosa* (Meads, 1989a), *Hemiandrus* sp. (Wahid, 1978)) or three (*D. rugosa* (Ramsay, 1955; Meads and Moller, 1977), *H. maori* (Sutherland, 1964), *Hemiandrus* sp. (van Wyngaarden, 1995), *D. rugosa* (Meads, 1989a), *Deinacrida tibiospina* Salmon (Meads, 1989b), Raukumara tusked weta (McIntyre, 1998)) distinct age classes, at least at some time of the year. This also suggests that they have a total life cycle of about 2–3 years. Even seasonal species, however, show some overlap between adjacent instars, as shown in the example given in Fig. 21.10. Mahoenui weta also show pronounced seasonality, even though three cohorts, including almost all life-history stages, occur throughout most of the year. Some stages may be present in extremely low numbers during the colder months (Sherley and Hayes, 1993; Richards, 1994). In captivity, development of *D. heteracantha* and *D. fallai* is not rigidly tied to season (Richards, 1973), but three more or less distinct age classes of *D. heteracantha* were reported on Little Barrier Island in March (Meads and Notman, 1995).

Temperature

During winter, the growth of most weta species slows and stadia can extend over several months. This is clearly shown in captive *D. fallai* and *D. heteracantha*, where the nymphal stadia are much longer in winter than at other times of the year (Richards, 1973). Interestingly, Mahoenui weta developed fastest and gained most weight when kept at a constant 18°C, compared with weta kept in the variable conditions of a laboratory (Richards, 1994). Different individuals of this species were reared at either one constant temperature or at fluctuating temperature in a laboratory, so their developmental rates could not be modelled.

Temperature may directly cause some of the considerable variability in developmental rates shown by some species at different times of year. For example, individuals of *D. heteracantha* that hatch from eggs in spring may become adult 14 months later and pass through only one winter, whereas others, hatching in autumn, pass through two winters and become adult 21 months later (Richards, 1973). It is also possible for *D. mahoenui* to become adult in about 14 months if the eggs hatch early (late autumn), although the average developmental time for this species is about 27 months (Richards, 1994). *Hemideina crassidens* may take either 1 or 2 years to mature after hatching from the egg in captivity. This depends on when the eggs hatch and how far development has proceeded before the first winter (P.R.L. Cary,

Table 21.3. Time of year when nymphs and adults occur.

Species	Instar no. per adult	Time of year*	Comments	Author
Hemiandrus sp.	1	Nov–Dec	Tekapo	van Wyngaarden 1995
Hemiandrus sp.	4–5	~Jun–Aug	Tekapo	van Wyngaarden 1995
Hemiandrus sp.	5–7	~Sep–May	Tekapo	van Wyngaarden 1995
Hemiandrus sp.	7 + 9	~Jun–Aug	Tekapo	van Wyngaarden 1995
Hemiandrus sp.	Male imago	Nov–Mar	Tekapo	van Wyngaarden 1995
Hemiandrus sp.	Female imago	All year	Tekapo	van Wyngaarden 1995
Hemiandrus sp.	1	L. Nov–L. Jan	Christchurch	Wahid 1978
Hemiandrus sp.	6 + 7	~Jun–Aug	Christchurch	Wahid 1978
Hemiandrus sp.	Male imago	M. Nov–E. May (peak L . Jan)	Christchurch	Wahid 1978
Hemiandrus sp.	Adult female	M. Dec–L. Mar + (max M. Jan)	Live longer, guarding eggs	Wahid 1978
Raukumara tusked weta	Imago	Aug., Mar–Apr	Raukumara	McIntyre 1998
Motuweta isolata	3–4	Nov	Middle Mercury Is.	McIntyre 1992
Motuweta isolata	5–8	Mar, Oct, Nov	Middle Mercury Is.	McIntyre 1992
Motuweta isolata	Imago	Mar, Oct, Nov	Middle Mercury Is.	McIntyre 1992
Deinacrida fallai	1–4	~Mar–May	Captivity	Richards 1973
Deinacrida fallai	5–6	Jun–Oct	Captivity	Richards 1973
Deinacrida fallai	Imago	Jan–May +	Reach adult, captivity	Richards 1973
Deinacrida fallai	Imago	Aug	Poor Knights Is.	Richards 1973
Deinacrida heteracantha	1–2	Mar–May	Captivity	Richards 1973
Deinacrida heteracantha	3	May–Sep	Captivity	Richards 1973
Deinacrida heteracantha	4–9	Sep–May	Captivity	Richards 1973
Deinacrida heteracantha	10	May–Oct	Captivity	Richards 1973
Deinacrida heteracantha	Imago	Sep–Oct+	Reach adult, captivity	Richards 1973
Deinacrida heteracantha	Imago	E. Sep, Mar	Little Barrier Is.	Richards 1973
Deinacrida mahoenui	4–5	~Jun–Aug	Mahoenui; highest nos.	Sherley and Hayes, 1993
Deinacrida mahoenui	6–7	~Dec–Feb	Mahoenui; highest nos.	Sherley and Hayes, 1993
Deinacrida mahoenui	8–9	~Jun–Nov	Mahoenui; highest nos.	Sherley and Hayes, 1993
Deinacrida mahoenui	Imago	~Mar–Feb	Mahoenui; highest nos.	Sherley and Hayes, 1993
Deinacrida mahoenui	1	~Dec–Feb	Mahoenui; highest nos.	Richards 1994
Deinacrida mahoenui	4, 8	~Mar–May	Mahoenui; highest nos.	Richards 1994
Deinacrida mahoenui	6, 8–9	~Jun–Aug	Mahoenui; highest nos.	Richards 1994
Deinacrida mahoenui	6–7, 9	~Sep–Nov	Mahoenui; highest nos.	Richards 1994
Deinacrida mahoenui	Imago	~Dec–Feb	Mahoenui; highest nos.	Richards 1994
Deinacrida rugosa	Juvenile	Sep–Dec	Kaikoura	Meads 1989a
Deinacrida rugosa	Imago	Nov–Dec, Apr	Kaikoura	Meads 1987, 1989a
Deinacrida tibiospina	Juvenile	Feb–Mar	North-west Nelson	Meads 1989b
	Imago	Feb–Mar	North-west Nelson	Meads 1989b

* L, late, M, mid, E, early; ~ indicates that time of year was given as a season. These have been replaced with approximate months as follows: summer = Dec–Feb, autumn = Mar–May, winter = Jun–Aug, spring = Sep–Nov.

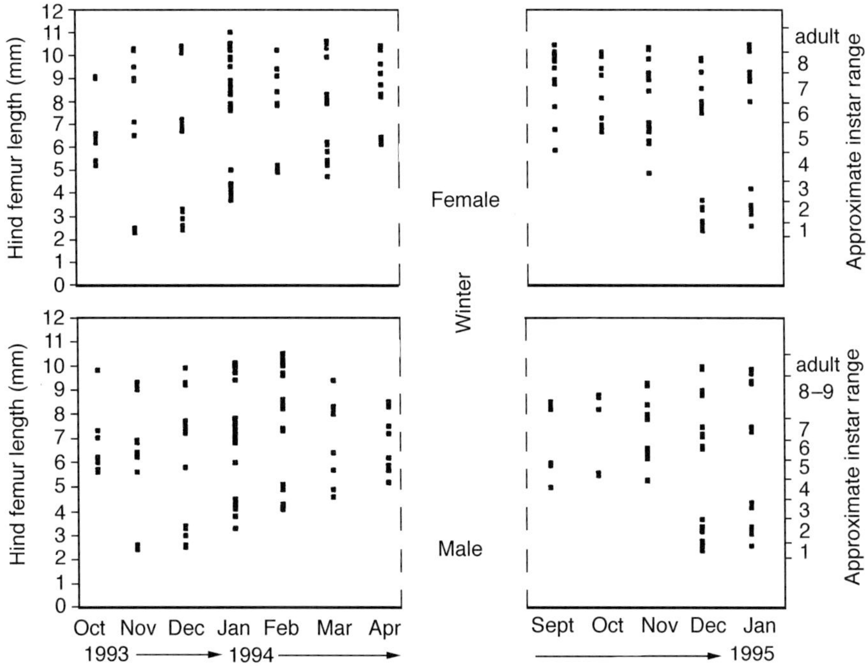

Fig. 21.10. Size distribution of a species of *Hemiandrus* in the field at Tekapo over a year (from van Wyngaarden, 1995, by courtesy of the author).

unpublished data). The duration of stadia, both in the field and among reared weta, can be variable (Table 21.4). This can lead to variation in the time taken to reach a certain stage (Table 21.5). For example, Ramsay (1955) found that the length of each stadium of *D. rugosa* was so variable that individuals reared in captivity were spread between the fifth and ninth instars 11 months after eclosion from the egg.

Nutrition

Food also has an effect on the growth rates of at least some weta species. Poor-quality food slows down growth in *Hemideina* (Barrett and Ramsay, 1991), and *D. mahoenui* gain weight faster when fed gorse and pasture plants than when fed native vegetation (Richards, 1994). The weight gain in a species of *Hemiandrus* was reported to depend on food by Wahid (1978). The addition of sorbic acid, streptomycin or ethanol to an artificial diet had a slightly negative effect on their weight gain, although there was no significant effect on the size of the individuals.

Sex

Males of *H. maculifrons*, *D. mahoenui* and *Hemiandrus* sp. develop faster than females of the same species (Cary, 1981; Wahid, 1978; Richards, 1994). In addition, males of *H. maculifrons* have a greater average growth ratio between successive instars than females, as mentioned above, but the males pass through fewer instars before they mature (see Table 21.1). The result is that both sexes mature at about the same size (Cary, 1981).

Lifespan

The size distribution of cohorts of nymphal *D. heteracantha* found at different times of year, plus the length of time this species takes to mature in captivity (Richards, 1973; Meads and Notman, 1995), suggests that *D. heteracantha* lives for at least 2 years after eclosion from the egg in the field. Most of the other weta species probably have total life cycles of 2–3 years, but the durations are only known for a few (Table 21.6).

Table 21.4. Duration of some stadia (durations of male and female imaginal stadia are given in Stringer, Table 20.3, Chapter 20, this volume).

Species	Instar	Duration of each stadium (days)*	Comments	Author
Hemiandrus sp.	All nymphs	~14–91	Large instars take longer	Wahid, 1978
Hemiandrus subantarcticus	5	59	Captive female	Butts, 1983
Hemiandrus subantarcticus	6	61	Captive	Butts, 1983
Raukumara tusked weta	Imago	~395–456	Captive	McIntyre, 1998
Hemideina sp.	1–5	~21–35	Captive	Barrett and Ramsay, 1991
Hemideina sp.	6–imago	~61–92	Captive	Barrett and Ramsay, 1991
Hemideina crassidens	1–5	~31	Captive	Spencer, 1995
Hemideina crassidens	6–imago	~92	Captive	Spencer, 1995
Hemideina ricta	(n – 2)–(n – 1)	101–120	Captive	Townsend, 1995
Hemideina ricta	(n – 1)–imago	112–202	Captive	
Hemideina ricta	(n – 1)–imago	~274	1 specimen, Banks Peninsula	Townsend, 1995
Deinacrida mahoenui	1–5	~30	Captive	Barrett in Sherley and Hayes, 1993
Deinacrida mahoenui	6–9	~61	Captive	Barrett in Sherley and Hayes, 1993
Deinacrida mahoenui	1–3	~30	Captive	Richards, 1994
Deinacrida mahoenui	4–5, 7	~30–122	Captive	Richards, 1994
Deinacrida mahoenui	6	~61–91	Captive	Richards, 1994
Deinacrida mahoenui	8	~30–61	Captive	Richards, 1994
Deinacrida mahoenui	9	~62–122	Captive	Richards, 1994
Deinacrida mahoenui	Imago	~91–122	Mahoenui	Richards, 1994
Deinacrida rugosa	1	17–24 (mean ~20)	Captive	Ramsay, 1955
Deinacrida heteracantha	1, 2, 4–6	27–58	Captive	Richards, 1973
Deinacrida heteracantha	3	94–97	Captive, during winter	Richards, 1973
Deinacrida heteracantha	7	17–91	Captive, during winter	Richards, 1973
Deinacrida heteracantha	8–9	38–76	Captive, during winter	Richards, 1973
Deinacrida heteracantha	10	95–166	Captive, in autumn and winter	Richards, 1973
Deinacrida fallai	1–3	22–29	Captive	Richards, 1973
Deinacrida fallai	4	45–50	Captive	Richards, 1973
Deinacrida fallai	5–9	38–79	Captive	Richards, 1973
Deinacrida fallai	Male imago	~35–70	Captive (n = 3)	Brown, 1995
Deinacrida fallai	Female imago	~35–49	Captive (n = 3)	Brown, 1995

*~indicates that the original time was given in weeks or months.

Table 21.5. Time taken to reach an instar after eclosion from egg.

Species	Instar reached	Sex	Time (months)	Comments	Author
Hemiandrus maculifrons	Imago	Male	4–10	Captive	Cary, 1981
Hemiandrus maculifrons	Imago	Female	5.7–13.8	Captive	Cary, 1981
Hemiandrus sp.	Imago	Both	~24	Tekapo	van Wyngaarden, 1995
Hemiandrus sp.	Imago	Both	~12	Christchurch	Wahid, 1978
Hemideina sp.	Imago	–	13–18	Captive	Barrett and Ramsay, 1991
Hemideina crassidens	Imago	–	15+	Captive	Spencer, 1995
Deinacrida rugosa	5–9	–	11	Captive	Ramsay, 1955
Deinacrida heteracantha	Imago	–	14	Captive, from spring egg hatch	Richards, 1973
Deinacrida heteracantha	Imago	–	21	Captive, from autumn egg hatch	Richards, 1973

Causes of Mortality

Age

In general, early instars experience high mortality in captivity and the mortality rate reduces with increasing age. Ramsay (1955) reported that 16.2% of *D. rugosa* reached the seventh instar, whereas Richards (1973) found that 89.9% of *D. fallai* and 82.7% of *D. heteracantha* died during the first five instars and only 1% of *D. fallai* and 2.4% of *D. heteracantha* reached maturity. Richards (1994) also reported that 80% of *D. mahoenui* died by the time they reached instar 6 and 40% of these subsequently died between instars 7 to 9.

One cause of mortality is a failure to successfully separate from the exuviae at ecdysis (Ramsay, 1955; Richards, 1973; Wahid, 1978; Cary, 1981; Butts, 1983; Richards, 1994; Brown, 1995). This also occurs occasionally in some species of *Deinacrida* when the first instar moults from the pronymph during eclosion (described in Stringer, Chapter 20, this volume), but apparently it can

Table 21.6. Duration of total life cycle.

Species	Duration	Comments	Author
Deinacrida fallai	2 years and 7 months	Average in captivity	Richards, 1973
Deinacrida heteracantha	2 years and 2 months	Average in captivity	Richards, 1973
Deinacrida rugosa	~3 years	Suggested: in field	Ramsay, 1955
Deinacrida mahoenui	2 years and 3 months	Average in captivity	Richards, 1994
Deinacrida mahoenui	25 months–37+ months	Range in captivity	Richards, 1994
Deinacrida mahoenui	~3 years	At Mahoenui	Richards, 1994
Hemiandrus sp.	~1 year and 11 months	1 male, Christchurch	Wahid, 1978
Hemiandrus sp.	~2 years and 8 months	1 female, Christchurch	Wahid, 1978
Raukumara tusked weta	~27–31 months	Suggested; in field	McIntyre, 1998
Motuweta isolata	2+ years	Middle Mercury Is.	McIntyre, 1992
Motuweta isolata	3+ years	In laboratory	C.J. Winks, in McIntyre, 1998

happen at any subsequent moult in most weta. It appears to be more frequent in later instars of one species of *Hemiandrus*, because Wahid (1978) only observed it at the sixth and subsequent instars. Mortality rates due to moulting included: 12.7% of *H. maculifrons* in captivity (failure to free their tibiae or tarsi (Cary, 1981); three out of 75 *D. mahoenui* moults in captivity (Richards, 1994) and one of ten moults in the field (Richards, 1994). Wahid (1978) attributed such moulting failure in *Hemiandrus* sp. fed artificial diets to be due to insufficient cholesterol.

Diet may also affect mortality rate during development, but the only published information on this is for *D. mahoenui*, where one out of nine weta reared on gorse and pasture plants died, whereas nine out of 11 died when fed native plants (Richards, 1994).

Cannibalism

Most weta are cannibalistic, but this has only been reported from weta kept together in captivity. Pieces of weta are frequently found in the field, but these could be the remains of partially eaten exuviae or weta that were eaten by a different predator. Cannibalism usually occurs during ecdysis, while the moulting insect is still unsclerotized and cannot move away or defend itself (Maskell, 1927; Ramsay, 1955; Richards, 1973; Little, 1980; Butts, 1983). The result is that limbs or antennae are damaged or lost, and death can result if parts of the head or body itself are attacked. Weta are especially likely to be cannibalistic if they are kept under crowded conditions, but cannibalism can also occur if the insects are not fed for several days, if they are trapped or have missing limbs, if they are kept with much larger individuals or if they are in apparent poor health (Maskell, 1927; Ramsay, 1955; Richards, 1973; Cary, 1981; Butts, 1983; McIntyre, 1998). Adult *D. fallai* and *D. heteracantha* have also been reported to be cannibalistic during mating or if one has to choose between two individuals of the opposite sex (Richards, 1973).

Disease and other factors

Other causes of mortality listed by Wahid (1978) for a species of *Hemiandrus* include high susceptibility of early instars to low humidity and to the presence of added mould inhibitor in their diet. There are few published cases of fungal diseases, except for infections in *D. rugosa* and Mahoenui weta. Here the fungi infected captive nymphs, as well as adults in the field (Ramsay, 1955; Meads, 1988; Richards, 1994). Only one of the fungi was identified, as *Cordyceps kirkii* G. Cunningham, and this was reported attacking *D. rugosa* (Ramsay, 1955). As far as parasites are concerned, gordian worms (Nematomorpha) are known to parasitize *Hemideina* and *Deinacrida* (Buller, 1871; Ramsay, 1955; Miller, 1971; Gander, 1976), one nematode is reported from *H. thoracica* White (Dale, 1967) and a species of *Gregarina* was reported from *H. maculifrons* (Cary, 1981). Finally, weta can also be infested with mites (Miller, 1971), but these probably do not kill the insect. Further information on the parasites of weta is given by Wharton *et al.* (Chapter 14, this volume).

Acknowledgements

We are grateful for editorial help from Rachel Standish and Murray Potter (Institute of Natural Resources (INR), Massey University, New Zealand).

References

Ander, K. (1970) Orthoptera Saltatoria. In: Tuxen, S.L. (ed.) *Taxonomist's Glossary of Genitalia in Insects.* Munksgaard, Copenhagen, pp. 61–71.

Asher, G.W. (1977) Ecological aspects of the common tree weta (*Hemideina thoracica*) in native vegetation. BSc thesis, Victoria University of Wellington, Wellington, New Zealand.

Barrett, P. and Ramsay, G.W. (1991) *Keeping Wetas in Captivity.* Wellington Zoological Gardens, Wellington, 60 pp.

Brown, J. (1995) Behaviour of captive Poor Knights giant weta (*Deinacrida fallai*). BSc thesis, Victoria University of Wellington, Wellington, New Zealand.

Brown, V. and Davies, R.G. (1972) Allometric growth in two species of *Ectobius* (Dictyoptera: Blattidae). *Journal of Zoology (London)* 166, 97–132.

Buller, W.L. (1971) Notes on the genus *Deinacrida* in New Zealand. *Transactions and Proceedings of the New Zealand Institute* 3, 34–37.

Butts, C.A. (1983) The biologies of two species of weta endemic to the Snares Island *Zealandrosandrus subantarcticus* Salmon (Orthoptera: Stenopelmatidae) and *Insulanoplectron spinosum* Richards (Orthoptera:

Rhaphidophoridae). BSc (Hons) thesis, University of Canterbury, Christchurch, New Zealand.

Cary, P.R.L. (1981) The biology of the weta *Zealandosandrus gracilis* (Orthoptera: Stenopelmatidae) from the Cass Region. MSc thesis, University of Canterbury, Christchurch, New Zealand.

Cary, P.R.L. (1983) Diet of the ground weta *Zealandosandrus gracilis* (Orthoptera: Stenopelmatidae). *New Zealand Journal of Zoology* 10, 295–298.

Crosby, T.K. (1974) Studies on Simuliidae (Diptera), with particular reference to *Austrosimulium tillyardianum*. PhD thesis, University of Canterbury, Christchurch, New Zealand.

Dale, P.S. (1967) *Wetanema hula* n. gen. et sp., a nematode from the weta *Hemideina thoracica*. *New Zealand Journal of Science* 10, 402–406.

Daugherty, C.H., Gibbs, G.W. and Hitchmough, R.A. (1993) Mega-island or micro-continent? New Zealand and its fauna. *Trends in Ecology and Evolution* 8, 437–444.

Domett, E. (1996) Reproduction and behaviour of the Mahoenui Weta, *Deinacrida* n.sp. MSc thesis, Massey University, Palmerston North, New Zealand.

Field, L.H. (1980) Observations on the biology of *Deinacrida connectens* (Orthoptera: Stenopelmatidae), an alpine weta. *New Zealand Journal of Zoology* 7, 211–220.

Gander, P.H. (1976) A model for the circadian pacemaker of *Hemideina thoracica* derived from the effects of temperature on its activity rhythm. PhD thesis, University of Auckland, Auckland, New Zealand.

Gibbs, G. (1994) The demon grasshoppers. *New Zealand Geographic* 21, 90–117.

Gibbs, G. (1998) *Raukumara Tusked Weta: a Report to Department of Conservation on its Discovery, Ecology and Management Implications*. Department of Conservation, Gisborne, New Zealand.

Gwynne, D.T. (1995) Phylogeny of the Ensifera (Orthoptera): a hypothesis supporting multiple origins of acoustical signalling, complex spermatophores and maternal care in crickets, katydids, and weta. *Journal of Orthoptera Research* 4, 203–218.

Gwynne, D.T. and Jamieson, I. (1998) Sexual selection and sexual dimorphism in a harem-defending insect, the alpine weta (*Hemideina maori*, Orthoptera: Stenopelmatidae). *Ethology, Ecology and Evolution* 10, 393–402.

Hamilton, S.A. (1991) The role of sex ratio, spatial distribution, and head size in the mating system of the Rock and Pillar weta, *Hemideina maori*. Diploma in Wildlife Management thesis, University of Otago, Dunedin, New Zealand.

Little, G.A. (1980) Food consumption and utilisation in two species of weta (*Hemideina femorata* and *H. maori*: Stenopelmatidae). BSc thesis, University of Canterbury, Christchurch, New Zealand.

MacDonald, P.D.M. and Pitcher, T.J. (1979) Age-groups from size-frequency data: a versatile and efficient method of analysing distribution mixtures. *Journal of the Fisheries Research Board of Canada* 36, 987–1001.

McIntyre, M. (1992) *The Status and Habitat of the Middle Island (Mercury Group) Tusked Wetas, with Implications for Management*. Department of Conservation, Hamilton, New Zealand.

McIntyre, M. (1998) Raukumara tusked weta part II. Unpublished report, Department of Conservation, Gisborne, New Zealand.

Maskell, F.G. (1927) The anatomy of *Hemideina thoracica*. *Transactions and Proceedings of the New Zealand Institute* 57, 637–670.

Matsuda, R. (1976) *Morphology and Evolution of the Insect Abdomen*. Pergamon Press, Oxford, 534 pp.

Meads, M.J. (1987) *The Giant Weta* (Deinacrida parva) *at Puhi Puhi, Kaikoura: Present Status and Strategy for Saving the Species*. Report No. 8, DSIR Ecology Division, Lower Hutt, New Zealand.

Meads, M.J. (1988) *A Brief Review of the Current Status of the Giant Wetas and a Strategy for their Conservation*. Report No. 11, DSIR Ecology Division, Lower Hutt, New Zealand.

Meads, M.J. (1989a) *An Evaluation of the Conservation Status of the Giant Weta (*Deinacrida parva*) at Kaikoura*. Report No. 20, DSIR Ecology Division, Lower Hutt, New Zealand.

Meads, M.J. (1989b) *The Conservation Status of the Giant Weta* Deinacrida tibiospina *in Northwest Nelson: Report on a Field Visit, and Notes on Other Invertebrates*. Report No. 21, DSIR Ecology Division, Lower Hutt, New Zealand.

Meads, M.J. (1995) *WWF-NZ Final Report on the Mercury Tusked Weta*. World Wildlife Fund, Wellington, New Zealand.

Meads, M.J. and Moller, H. (1977) *Report of a Visit to Mana Island in September 1977*. DSIR Ecology Division, Lower Hutt, New Zealand.

Meads, M.J. and Moller, H. (1978) *Introduction of Giant Wetas (*Deinacrida rugosa*) to Maud Island and Observations of Tree Wetas, Paryphantids and Other Invertebrates*. DSIR Ecology Division, Lower Hutt, New Zealand.

Meads, M. and Notman, P. (1995) *Surveys of Giant Weta. Little Barrier Island, Pig Island (Foveaux Strait), and Mt Faraday and Price's Basin (Southern Alps)*. Science for Conservation Series No. 16, Department of Conservation, Wellington, New Zealand.

Miller, D. (1971) *Common Insects in New Zealand*. A.H. and A.W. Reed, Wellington, 170 pp.

Moller, H. (1978a) Study of the wetas of Stephens

Island. Unpublished report. DSIR Ecology Division, Lower Hutt, New Zealand.

Moller, H. (1978b) A weta and rodent study on Arapawa Island. Unpublished report, DSIR Ecology Division, Lower Hutt, New Zealand.

Moller, H. (1985) Tree wetas (*Hemideina crassicruris*) (Orthoptera: Stenopelmatidae) of Stephens Island, Cook Strait. *New Zealand Journal of Zoology* 12, 55–69.

Ordish, R.G. (1992) Aggregation and communication of the Wellington weta *Hemideina crassidens* (Blanchard) (Orthoptera: Stenopelmatidae). *New Zealand Entomologist* 15, 1–8

Ramsay, G.W. (1955) The exoskeleton and musculature of the head, and the life-cycle of *Deinacrida rugosa* Buller, 1870. MSc thesis, Victoria University of Wellington, Wellington, New Zealand.

Ramsay, G.W. (1964a) Moult number in Orthoptera (Insecta). *New Zealand Journal of Science* 7, 644–666.

Ramsay, G.W. (1964b) Regeneration of appendages in some New Zealand wetas (Insecta: Orthoptera). *Transactions of the Royal Society of New Zealand (Zoology)* 4, 139–144.

Ramsay, G.W. (1965) Development of the ovipositor of *Deinacrida rugosa* Buller (Orthoptera: Gryllacridoidea: Henicidae), and a brief review of the ontogeny and morphology of the ovipositor with particular reference to the Orthoptera. *Proceedings of the Royal Entomological Society of London* 40, 41–50.

Richards, A.O. (1973) A comparative study of the biology of the giant wetas *Deinacrida heteracantha* and *D. fallai* (Orthoptera: Henicidae) from New Zealand. *Journal of Zoology* 169, 195–236.

Richards G.E. (1994) Ecology and behaviour of the Mahoenui giant weta, *Deinacrida* nov. sp. MSc thesis, Massey University, Palmerston North, New Zealand.

Rufaut, C.G. (1995) A comparative study of the Wellington tree weta, *Hemideina crassidens* (Blanchard, 1951) in the presence and absence of rodents. MSc thesis, Victoria University of Wellington, Wellington, New Zealand.

Salmon, J.T. (1950) A revision of the New Zealand wetas Anostostominae (Orthoptera: Stenopelmatidae). *Dominion Museum Records, Entomology, Wellington* 1, 121–177.

Sandlant, G.G. (1981) Aggressive behaviour of the Canterbury weta *Hemideina femorata* (Orthoptera: Stenopelmatidae): its adaptive significance in resource allocation. MSc thesis, University of Canterbury, Christchurch, New Zealand.

Scudder, G.G.E. (1971) Comparative morphology of insect genitalia. *Annual Review of Entomology* 16, 379–406.

Sehnal, F. (1985) Growth and life cycles. In: Kerkut, G.A. and Gilbert, L.I. (eds) *Comprehensive Insect Physiology Biochemistry and Pharmacology*, Vol. 2. *Postembryonic Development.* Pergamon Press, Oxford, pp. 1–86.

Sherley, G.H. (1992) Gorse and goats: are pest species necessary for the conservation of a new giant weta (*Deinacrida* sp.) at Mahoenui, King Country? In: Heath, A.C.G. (ed.) *Proceedings of the 41st Annual Conference of the Entomological Society of New Zealand.* Entomology Society of New Zealand, Heretaunga, pp. 26–36.

Sherley, G.H. and Hayes, L.M. (1993) The conservation of a giant weta (*Deinacrida* n. sp. Orthoptera: Stenopelmatidae) at Mahoenui, King Country: habitat use, and other aspects of its ecology. *New Zealand Entomologist* 16, 55–68.

Spencer, A.M. (1995) Sexual maturity in the male tree weta *Hemideina crassidens* (Orthoptera: Stenopelmatidae). MSc thesis, Victoria University of Wellington, Wellington, New Zealand.

Stauffer, T.W. and Whitman, D.W. (1997) Grasshopper oviposition. In: Gangwere, S.K., Muralirangan, M.C. and Muralirangan, M. (eds) *The Bionomics of Grasshoppers, Katydids and their Kin.* CAB International, Wallingford, UK, pp. 231–280.

Sutherland, O.R.W. (1964) Alpine wetas in New Zealand. *New Zealand Entomologist* 3, 16–17.

Townsend, J.A. (1995) Distribution and ecology of the Banks Peninsula tree weta, *Hemideina ricta.* MSc thesis, Massey University, Palmerston North, New Zealand.

van Wyngaarden, F. (1995) The ecology of the Tekapo ground weta (*Hemiandrus* new sp.; Orthoptera: Anostostomatidae) and recommendations for the conservation of a threatened close relative. MSc thesis, University of Canterbury, Christchurch, New Zealand.

Wahid, M.B. (1978) The biology and economic impact of the weta, *Hemiandrus* sp. (Orthoptera: Stenopelmatidae) in an apricot orchard, Horotane Valley. MHortSc thesis, Lincoln College, Lincoln, New Zealand.

Part VI
Physiology

22
Sensory Physiology

Laurence H. Field
Department of Zoology, University of Canterbury, Private Bag 4800, Christchurch, New Zealand

Introduction

Most biological studies of the little-known stenopelmatoid groups have concentrated on taxonomy rather than on the biology of the living insects. However, in New Zealand, a concerted effort has been broadly directed towards the latter pursuit, due to the ready availability of wetas, and also to issues arising from the unique island country's isolation. Interest in the physiology of the weta nervous system derives from the insects' relatively primitive and evolutionarily isolated status compared with more modern insects, such as the locust, in which neuronal mechanisms have been well studied. Inasmuch as the wetas of New Zealand are considered to have been isolated since the separation of Gondwanaland 75 to 100 million years ago, it is possible that the whole group has retained some primitive features of nervous system structure and function. In addition, the radiation of species in the isolated island ecosystems of New Zealand presents interesting mechanisms and adaptations for the particular lifestyles of these insects. Virtually nothing has been published about the sensory physiology of other anastostomatids elsewhere, except the suggestion, from communication studies in Australian gryllacridids, that they are extremely sensitive to substrate vibrations (Field and Bailey, 1997).

The search for interesting and unusual sensory phenomena in wetas has paid off handsomely. As explained in the following sections, wetas (primarily tree wetas) show a number of hitherto unknown physiological features, some of which are related to primitive origins and others of which are presumably adaptations to their particular way of life. Certainly, many of the sense organs and features of the nervous system anatomy are typical of the Orthoptera, but surprises became evident in many areas that have been investigated. For example, in the area of auditory physiology, an inflated tracheal system was discovered in each foreleg of tree wetas, which has proved to be a remarkable adaptation for improved hearing ability. Hence tree wetas are endowed with a feature unique in the entire insect world. The ears themselves (tibial tympana), also on the forelegs, are the largest ever reported in any orthopteran family. Furthermore, even though male tree wetas can hear the sounds produced by other males calling from the entrances of tree galleries, it was found that males are also extremely sensitive to the surface vibrations, which are imparted into the tree-trunk by the stridulatory energy of the calling male and propagated throughout the tree (McVean and Field, 1996). Another interesting discovery regarding nervous system organization is that the major stretch receptor in the femur–tibia joint in wetas has reflex input to all other muscles of the leg, rather than controlling only reflexes and posture in the femur muscles of the leg (as known for other insects) (Field and Rind, 1981). This contradicts the usual pattern of neural cirsuits, where the receptor of one joint controls reflexes only to muscles which operate that joint.

This chapter reviews the knowledge of those sense organ systems which have been investigated in tree wetas, including auditory receptors, air current detectors, joint movement receptors, muscle tension receptors, vibration detectors and the visual system. The anatomy of the weta nervous system is described in O'Brien and Field (Chapter 8, this volume).

Overview

Sensory structures in tree wetas, as in most insects, are situated either externally on the cuticle or internally, where they are usually associated with joints or muscles. External receptors (except for the compound eyes) usually consist of sensory hairs and circular and oval plates or cupolas innervated with one or more sensory neurons. A single such structure, together with its associated neuron(s) and accessory cells, is known as a sensillum. Many hair sensilla (= plural) serve as mechanoreceptors or as chemoreceptors. Internal receptors comprise either clusters of bipolar (spindle-shaped) sensory neurons with ciliated dendrites, known as chordotonal organs, or single multipolar (branched dendrite) sensory neurons, often innervating thin 'receptor' muscles dedicated to the sense organ's function. Most internal receptors are mechanoreceptors and, as chordotonal organs, are found in each joint of the insect exoskeleton. Some chordotonal organs are highly specialized mechanoreceptors, which detect the minute movements of a weta eardrum (e.g. the tympanal organ) or of the fluid vibrations within a leg (e.g. subgenual organ) (for review, see Field and Matheson, 1998). In all chordotonal organs, the basic sensillum is termed a scolopidium: a group of four cells, including the neuron, which comprise one micromechanical receptor.

In wetas, detailed studies have been carried out on auditory and vibratory detectors, joint chordotonal organs and muscle receptor organs. Less well-known are the compound eyes, cercal sensory systems and nearly all external hair sensilla.

Auditory and Vibration Detection Systems

In ensiferan orthopterans, each prothoracic leg bears an auditory sense organ (tympanal organ), which is intimately associated with a complex of other sense organs primarily concerned with detecting substrate vibrations rather than sound (reviewed in Field and Matheson, 1998). The associated organs (subgenual organ, intermediate organ and accessory organ) have no external manifestation, while the tympanal organ comprises two eardrums (tympana) and an internal linear array of auditory sensory neurons (the crista acoustica, which resembles an unrolled cochlea). In the other legs, the tympanal ear is lacking, but the internal complement of sense organs remains, including the subgenual organ, with its associated sense organs, and the crista acoustica. The number of sensory neurons in each meso- and metathoracic crista acoustica is reduced compared with that in the prothoracic crista acoustica, and the former are not associated with any internal specialization (Fornusek, 1993). In other ensiferan groups, the meso- and metathoracic crista acoustica are termed 'tracheal organs' (Field and Matheson, 1998). The neuroanatomy and organization of this complex group of sense organs will be described together, while the relevant physiological studies will be treated separately. The following treatment deals only with tree wetas (*Hemideina crassidens* and *Hemideina femorata*), since the other anostostomatids have not been studied.

Tympanal organ

External structure of the tympanal organ (ear)

The two tympana of a weta ear are thick oval membranes on the flattened anterior and posterior sides of each prothoracic tibia (Fig. 22.1A). Contrary to the condition in crickets (Gryllidae), where there is one large tympanum and one small one on the foreleg, both tympana in the weta ear are equally large. The tympana of wetas are larger than any other orthopteran tympana found on the forelegs. For comparison, the giant cricket, *Macrogryllus ephippium*, is of similar size to a tree weta (40 mm body length) but has tympana only about half the size of those found in *H. crassidens* (Ball and Field, 1981). The largest of all tympana found in the suborder Ensifera (Stenopelmatoidea, Tettigonoidea, Grylloidea) is that of the giant weta *Deinacrida heteracantha*, which has tympanal dimensions of 4.0 mm × 2.1 mm (Fig. 22.1A) and attains a body weight of over 70 g.

Each tympanum is divided into two regions: a dorsal, thicker depression of stiff cuticle, ellipsoidal in shape, and a surrounding thinner, rather

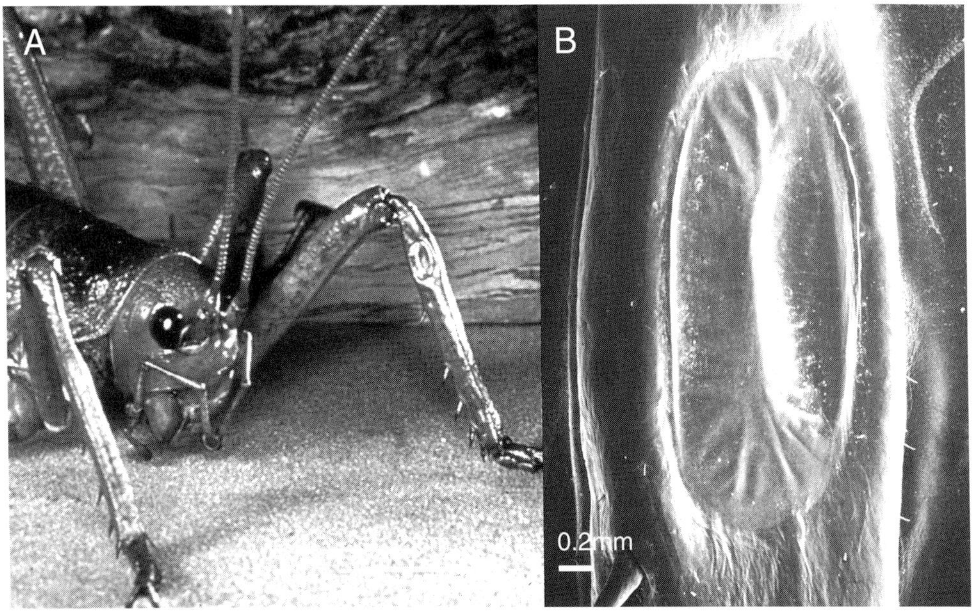

Fig. 22.1. Large size and a loose rippled surface are unusual characters found in the tympanal membranes of New Zealand wetas. A. Tibial ear of the giant weta *Deinacrida heteracantha*, measuring over 4 mm in length. This is the largest tibial ear of the insect world. B. Photograph of the tympanal membrane of the ear of the tree weta *Hemideina crassidens*, showing the rippled appearance of the region surrounding the depressed oval of hard cuticle. Presumably, the more dense oval area acts as an acoustic resistance, which swings like a door on the loose membrane.

transparent membrane (Fig. 22.1B). The weta tympanum stands apart from that of all other Orthoptera in lacking a taut, drum-like appearance. Instead, the membrane portion is rippled, with small folds, and appears to be loosely suspended (Ball and Field, 1981). Although the depressed cuticular region also occurs in the tettigoniid (katydid) tympanum, no other insect shows the folded surface of the membrane region in the weta. In addition, the weta tympanum is much thicker than those of other Orthoptera (60–110 μm), as seen in Fig. 22.2C. Both conditions would seem to place a severe restriction on efficient coupling of sound energy to the internal sensory receptors; however, the ear is surprisingly sensitive, as shown below. It is not known how such a rippled membrane can transduce sound energy in the weta ear.

Tracheal system associated with the tympanal organ

In orthopterans that have tibial tympana, the posterior of the two leg tracheae becomes enlarged to form vesiculae acousticae and tympanal vesicles associated with the ear. However, in tree wetas, only the tympanal vesicles are formed. In addition, a hitherto unknown modification of the posterior trachea occurs, in the form of two inflated chambers, one proximal and one distal to the tympanal vesicles associated with the tympana. Thus, the trachea in the foreleg is divided into three regions (Fig. 22.2A). Immediately upon entering the tibia, the posterior trachea becomes greatly inflated and occupies nearly the whole of the cross-section of the tibial lumen. The resulting bulbous proximal posterior inflated chamber (PPIC) is invaginated dorsally and extends to the proximal margin of the tibia, before becoming laterally compressed into the posterior tympanal vesicle (PTV, Fig. 22.2A). Distal to the PTV, the trachea again expands into a large, elongated distal posterior inflated chamber (DPIC), which is also invaginated dorsally. This region of the tibia is enlarged, giving the foreleg a swollen appearance. The muscles and nerves are restricted to a narrow ventral channel along the tibia. The anterior trachea is not enlarged except at the level of the tympanum, where it forms the

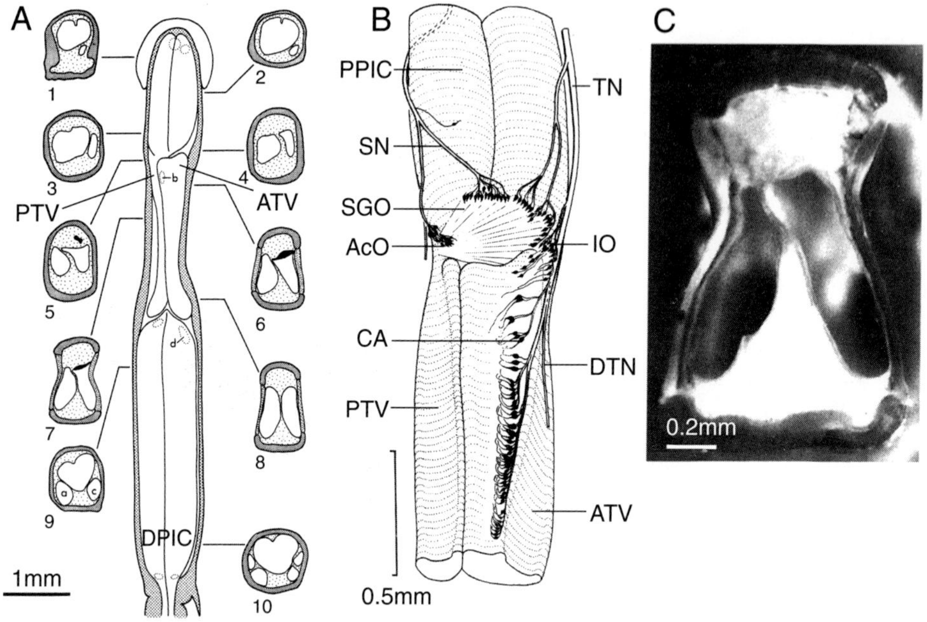

Fig. 22.2. Anatomy of the tracheal system in the tree weta prothoracic leg associated with the tympanal organ. A. Dorsal view of the right prothoracic leg showing the posterior trachea inflated into three chambers, the proximal posterior inflated chamber, the posterior tympanal vesicle and the distal posterior inflated chamber. The anterior trachea forms the anterior tympanal vesicle only, and otherwise remains restricted in size. Insets show transverse sections at each level. Dotted circles indicate foramina allowing air communication between chambers and tracheae. The black structure in insets 5, 6 and 7 represents the crista acoustica. B. Dorsal view of the innervation and sensory neurons of the tympanal organ complex in the foreleg. C. Transverse section through the tympanal area of the prothoracic tibia, showing the extraordinarily thick cuticle of the tympanal membrane and the tympanal vesicles closely applied to the tympana. AcO, accessory organ; ATV, anterior tympanal vesicle; CA, crista acoustica; DTN, distal tibial nerve; IO, intermediate organ; PPIC, proximal posterior inflated chamber; PTV, posterior tympanal vesicle; SGO, subgenual organ; SN, subgenual nerve; TN, tympanal nerve. (From Ball and Field, 1981, with permission.)

laterally compressed anterior tympanal vesicle (ATV, Fig. 22.2A) and, as with the PTV, is closely applied to the tympanum (Fig. 22.2C). However, unlike the PTV, the ATV is broadened and flattened dorsally to form a platform, upon which lies the crista acoustica (Fig. 22.2B). Both tracheae and inflated chambers are interconnected via small foramina (b, d in Fig. 22.2A).

The unusual inflation of the tibial tracheal system in *Hemideina* serves to enhance the physiological sensitivity of the auditory system, apparently by providing a resonant airspace, which matches the characteristics of the cuticular structure of the leg (Field *et al.*, 1980). Although not illustrated, Ander (1939) mentioned a possibly analogous feature in the legs of *Lezina*, in a related subfamily, Lezinae.

Tympanal innervation and sensory neurons

The main leg nerve (nerve 5B) gives rise to two sensory nerves upon entering the tibia from the femur. The tympanal nerve rises dorsally on the anterior side of the PPIC to innervate part of the subgenual organ, and the entire intermediate organ and crista acoustica (Fig. 22.2B). The subgenual nerve rises on the posterior side of the PPIC to innervate the accessory organ and remainder of the subgenual organ. The projection of the tympanal nerve into the prothoracic ganglion of the central nervous system (CNS) is almost indistinguishable from that seen in crickets (Gryllidae) and katydids (Tettigoniidae), and in all cases the tympanal nerve contains numerous axons from cuticular hair receptors as well as those from the tympanal organ (Ball and Field, 1981).

Crista acoustica of the tympanal organ complex (prothoracic leg)

The morphology and organization of the crista acoustica are much more similar to the tettigoniid plan than to that of the gryllids. The crista in the prothoracic leg of *H. crassidens* and *H. femorata* contains about 50 sensory neurons, which decrease in soma size distally in an array along the dorsal platform of the ATV (Ball and Field, 1981; H. Nishino, Christchurch, 1999, personal communication) (Fig. 22.2B). The dendrites of the scolopidial neurons each contain an internal cilium and are reflected proximally before terminating in attachment cells (Fig. 22.3D) (Ball, 1981). The neuron cell bodies and attachment cells are embedded in a gelatinous matrix, triangular in cross-section, which is attached proximally to the dorsal wall of the tibia. The gelatinous, and hence solid, nature of the crista acoustica has never been described in other insects, although it is likely to be found generally. It appears that the gelatinous matrix is secreted by the attachment cells, inasmuch as it lacks a bounding membrane and is built in layers around the cells. The attachment cells are highly unusual and of two types. The first is of a stellate morphology and is found in the proximal crista neurons (Fig. 22.3A). The second is a lamellate type, never described in other insects, which resembles the blade of a shovel arranged at an angle to the surface of the tympanal vesicle platform (Fig. 22.3B, C). As explained below, the stellate attachment cells are also found in the intermediate organ and the subgenual organ, while the lamellate attachment cells are unique to the distal half of the crista acoustica. There is no physiological correlate for the difference in attachment cell type, and this presents an interesting problem in functional cell morphology (Ball and Field, 1981).

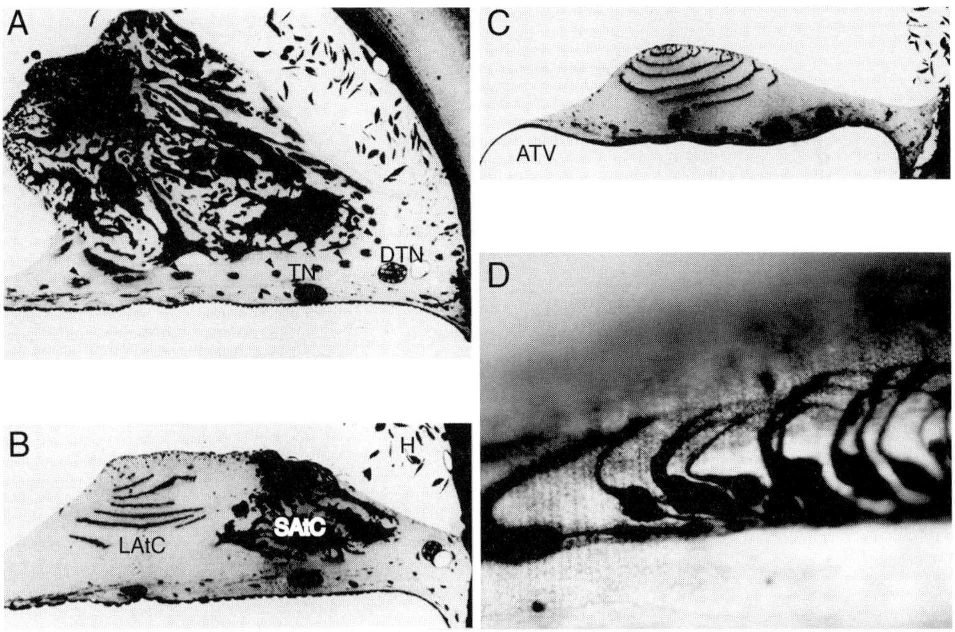

Fig. 22.3. Histology of the prothoracic crista acoustica and intermediate organ in the tympanal organ. A–C depict proximal to distal sections showing a decrease in size as the gelatinous mass of the intermediate organ and crista acoustica tapers distally on the platform of the anterior tympanal vesicle (ATV). A. Section through the intermediate organ showing darkly staining stellate attachment cells and scolopidia (arrowheads) of sensory neurons along the bottom. B. Further distally in the crista, some stellate attachment cells are replaced by lamellar attachment cells. C. As the crista narrows toward its distal end, all attachment cells are lamellar and parallel, but rising up at an angle from the base of the crista where the neurons are lying. D. Methylene blue stain of scolopidial neurons near distal end of crista, showing dendrites reflected dorsally and proximally from cell bodies along the ventral side. DTN, distal tibial nerve; H, haematocytes; LAtC, lamellar attachment cells; SAtC, stellate attachment cells; TN, tympanal nerve. (A–C from Ball and Field, 1981, with permission.)

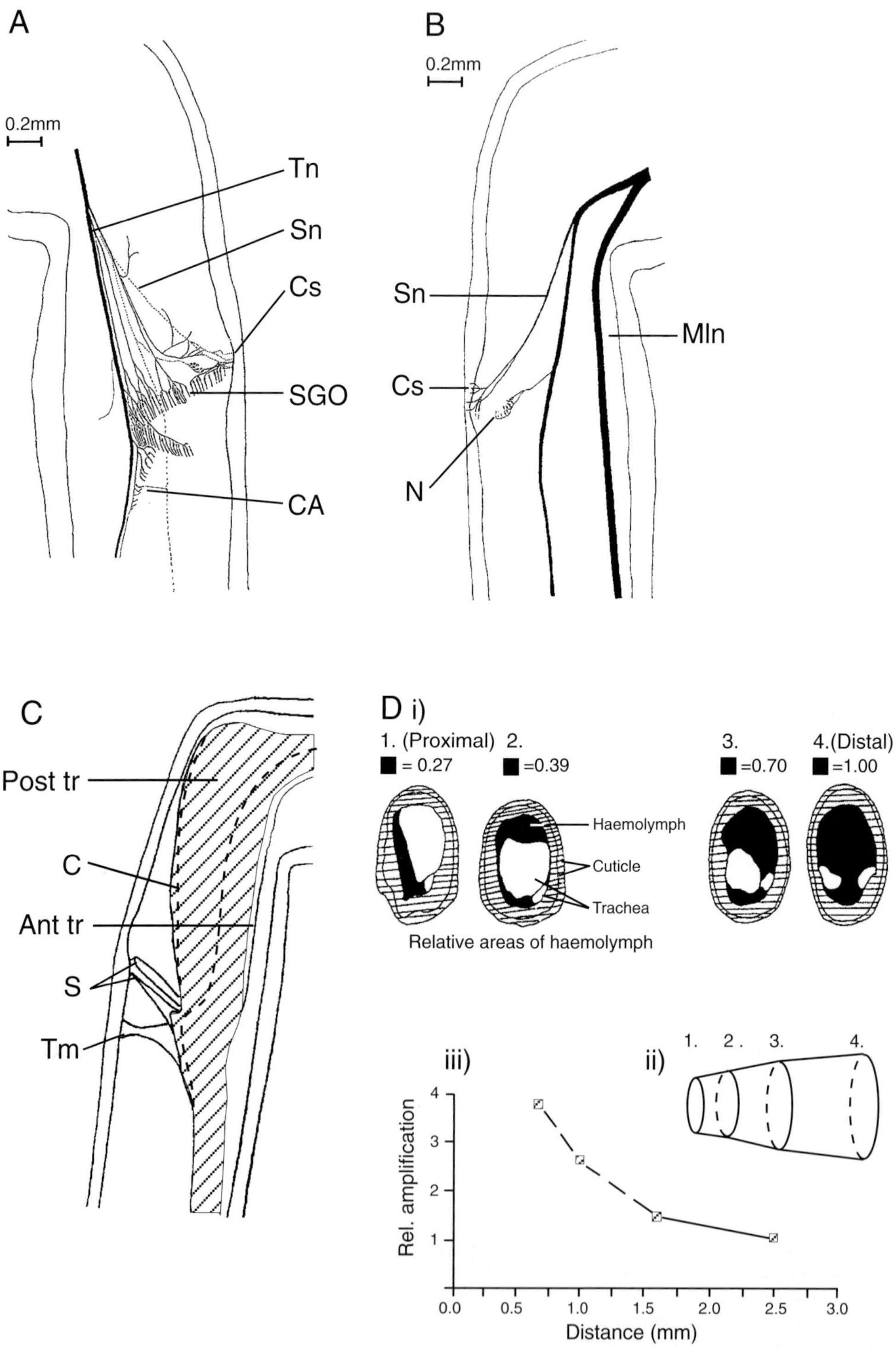
A
0.2mm
Tn
Sn
Cs
SGO
CA
B
0.2mm
Sn
Cs
N
Mln
C
Post tr
C
Ant tr
S
Tm
D i)
1. (Proximal)
= 0.27
2.
=0.39
3.
=0.70
4.(Distal)
=1.00
Haemolymph
Cuticle
Trachea
Relative areas of haemolymph
ii)
1. 2. 3. 4.
iii)
Rel. amplification
Distance (mm)
0.0 0.5 1.0 1.5 2.0 2.5 3.0
1 2 3 4

Crista acoustica of meso- and metathoracic legs

The crista acoustica also occurs in the meso- and metathoracic legs, as a graded series of scolopidial neurons resting along the dorsal surface of the anterior trachea (Fig. 22.4A), but it is not associated with any structural modification for sound detection (Fornusek, 1993). In gryllids and tettigoniids, this sense organ has been termed the 'tracheal organ' (reviewed in Field and Matheson, 1998), but it is considered homologous to the crista acoustica in the prothoracic leg and the latter term is used here. The number of scolopidial neurons decreases from the prothoracic leg to the metathoracic leg; in *H. femorata*, a mean count of 50 neurons ($n = 2$) was found in the pro-leg while the meta-leg had a mean of 22 neurons ($n = 6$) (Fornusek, 1993). As in the prothoracic crista, the dendrites are reflected dorsoproximally within a gelatinous matrix.

Intermediate organ

This collection of approximately 19 scolopidial sensilla in the tympanal organ complex of the prothoracic leg is essentially a proximal extension of the crista acoustica, and both organs arise from the same nerve. It has been termed the 'distal organ' in tettigoniids (Schwabe, 1906). The number of neurons is similar in the two species studied: *H. crassidens*: 19; *H. femorata*: 18. In the prothoracic leg, the sensilla form an arc against the anterior wall of the leg distal to the subgenual organ. The dendrites terminate in stellate attachment cells, embedded in the thick proximal section of the crista acoustica (Fig. 22.3A). In the meso- and metathoracic legs, the intermediate organ sensilla are more numerous (in *H. femorata*, a mean of 26 occurred ($n = 11$)). They insert on to a small, but thick, transverse septum distal to the large septum (described below) of the subgenual organ (Fig. 22.4C). The small septum appears to attach to the posterior trachea as well as to the dorsal wall of the tibia (Fornusek, 1993).

Subgenual organ

In most insects, the subgenual organ (SGO) is considered to be a detector of substrate vibrations (Field and Matheson, 1998). In wetas, it takes the usual orthopteran form of a transverse fan-shaped septum, which occludes the dorsal lumen in the proximal end of each tibia. Its position is marked externally by the presence of a small indentation containing eight campaniform sensilla (minute, button-shaped cuticular stress detectors, which do not have scolopidial sensory neurons). In the prothoracic leg, the SGO occurs just proximal to the tympanal organ (Fig. 22.2B), while in the meso- and metathoracic legs it is located at the constriction point on the posterior trachea (which is somewhat inflated, in a manner similar to the prothoracic PPIC of the prothoracic leg) (Fig. 22.4C). In all legs, from 50 to 70 scolopidial neurons are arrayed in a semicircle around the circumference of the septum, with dendrites projecting radially inwards, towards the septum

Fig. 22.4. Anatomy of subgenual organ (SGO) and associated sense organs in the metathoracic leg of the tree weta, *H. femorata*. A and B. Cobalt chloride backfilling demonstrates innervation and organization of sensory neurons. A. Anterior side of the hind tibia (proximal end) showing the neurons of the SGO in a semicircular array lining the inner wall of the tibia (SGO septum not shown). The intermediate organ forms an almost linear array of neurons projecting dendrites dorsoposteriorad on to a more distal septum (not shown) from the SGO. The crista acoustica is the most distal organ, comprising a linear array of neurons graded in size. Dotted lines show the position of nerves and sensilla on the posterior side of the tibia. B. Posterior side of the hind tibia showing the accessory organ and the group of campaniform sensilla located in a thin, slightly raised area of cuticle. C. Anterior view of hind tibia indicating tracheal anatomy and the position of the septa on to which the SGO and the intermediate organs insert dendrites. D. Analysis of fluid compartment in proximal region of metathoracic tibia containing the SGO complex. i. Transverse sections of tibia, indicating relative area occupied by haemolymph (black), tracheae (white) and cuticle (hatched). Normalized relative areas of haemolymph given beside each drawing. ii. Equivalent truncated cone model of fluid volume in the first 3 mm of hind tibia, estimated from cross-sectional data. iii. Graph of fluid displacement amplification for a given displacement pulse presented at the large (distal) end of model. Progressive proximal constriction results in increased fluid displacement as a pulse travels proximally. Ant tr, anterior trachea; CA, crista acoustica; Cs, campaniform sensilla in the tibial cuticle; Mln, main leg nerve; N, *neben* organ; Post tr, posterior trachea; SGO, subgenual organ; S, septa for SGO (upper) and for intermediate organ (lower); Sn, subgenual nerve; Tm, tectorial membrane attaching the crista acoustica to the cuticle; Tn, main tibial nerve (N5).

centre. The attachment cells into which the dendrites insert are elongated, highly branched and of complex structure, embedded within a matrix (Ball and Field, 1981).

On the posterior side of the SGO, a group of 13 (*H. femorata*) to 15 (*H. crassidens*) scolopidial neurons insert into the base of the SGO septum (Figs 22.2B and 22.4B). These have been termed the accessory organ (Ball and Field, 1981) or '*neben* organ' (Schnorbus, 1971).

The septum of the SGO is a thick sheet attached around its circumference to the wall of the tibia (Fig. 22.4C). Fine filaments connect the septum surface to the anterior trachea, as well as to the wall of the tibia. These may provide a mechanism allowing sensitivity to airborne sound energy within the trachea, as well as to haemolymph vibratory displacement within the tibia. The actual pathway of energy detection in the SGO is not well understood in any insect, but it is known that the organ can detect acoustic energy as well as substrate vibration energy in cockroaches (Shaw, 1994; Field and Matheson, 1998).

The proximal location of the SGO complex of organs appears to be an optimal position for intercepting fluid vibrational energy, while allowing the leg tracheation to function without interfering with leg muscles and joint movements (Rössler, 1992). Furthermore, Fornusek (1993) noted an interesting mechanism in tree wetas, whereby longitudinal fluid displacement within the tibia is likely to be amplified at the position of the SGO, due to the progressively increasing diameter of tracheation towards the proximal end of the tibia (Fig. 22.4Di). This has the effect of progressively occluding the haemolymph space proximally within the tibia. In a simplified model based upon measurements of tibial cross-sectional areas of trachea and haemolymph space (Fig. 22.4Dii, iii), Fornusek showed that the constriction of haemolymph space at the level of the SGO will cause a fourfold amplification of a given longitudinal displacement introduced into the distal end of the tibia. The mechanism could serve to enhance the sensitivity of the SGO.

Physiology of sound and vibration detectors

A detailed study has been carried out on the sensory responses of the tympanal organ of *H. crassidens* (Field *et al.*, 1980), and two studies have dealt with responses of the SGO of *H. femorata* (Fornusek, 1993; McVean and Field, 1996). A major outcome from the hearing experiments is that the weta ear is tuned for relatively low sound frequencies (matching the peak frequencies produced by conspecific stridulation) and that the remarkable inflated chambers of the forelegs are adaptations to enhance only the low frequency part of the hearing spectrum. The vibration sensitivity studies showed that tree wetas can readily detect the vibrations induced by stridulating males as they call from tree galleries, and that this is a second mode of communication amongst wetas within any one tree.

Tympanal organ sensitivity

Recordings from the tympanal nerve of *H. crassidens* in an anechoic chamber showed that the tympanal organ has its greatest sensitivity to sound frequencies around 2.5 kHz (1 Hz = 1 cycle s^{-1}), at threshold intensities of 20–35 dB sound pressure level (relative to the threshold of human hearing, 2×10^{-5} N m^{-2}). The threshold curve showed a reduction of sensitivity toward frequencies below the peak sensitivity frequency of about 15 dB per octave, while the decrease above peak sensitivity was about 27 dB per octave (Fig. 22.5A). The peak frequency of sensitivity in tree wetas is low compared with many other orthopterans; however, it is consistent with the unusual thickness and the large area of the tympana plus associated tracheal air spaces. The tuning of the ear is due either to the mechanical structure of the tympanum and tympanal vesicles or to mechanisms intrinsic to the scolopidial neurons of the crista acoustica, since the peak frequency of the tympanal organ is not shifted by experimentally altering the tracheal air spaces in the tibia. Such experiments demonstrated that resonance of the tracheal air spaces, such as the tympanal vesicles, does not contribute to tuning of the ear (Field *et al.*, 1980).

Unlike those of crickets, the two tympanal membranes in the tree weta ear contribute equally to sound reception. This was shown by sequentially occluding first one and then the other tympanum and observing only a small reduction in sensitivity; however, occlusion of both tympana at once caused a dramatic loss of sensitivity, by 20–25 dB (Field *et al.*, 1980).

The unique inflated tracheal chambers in the tree weta foreleg were shown to enhance low-

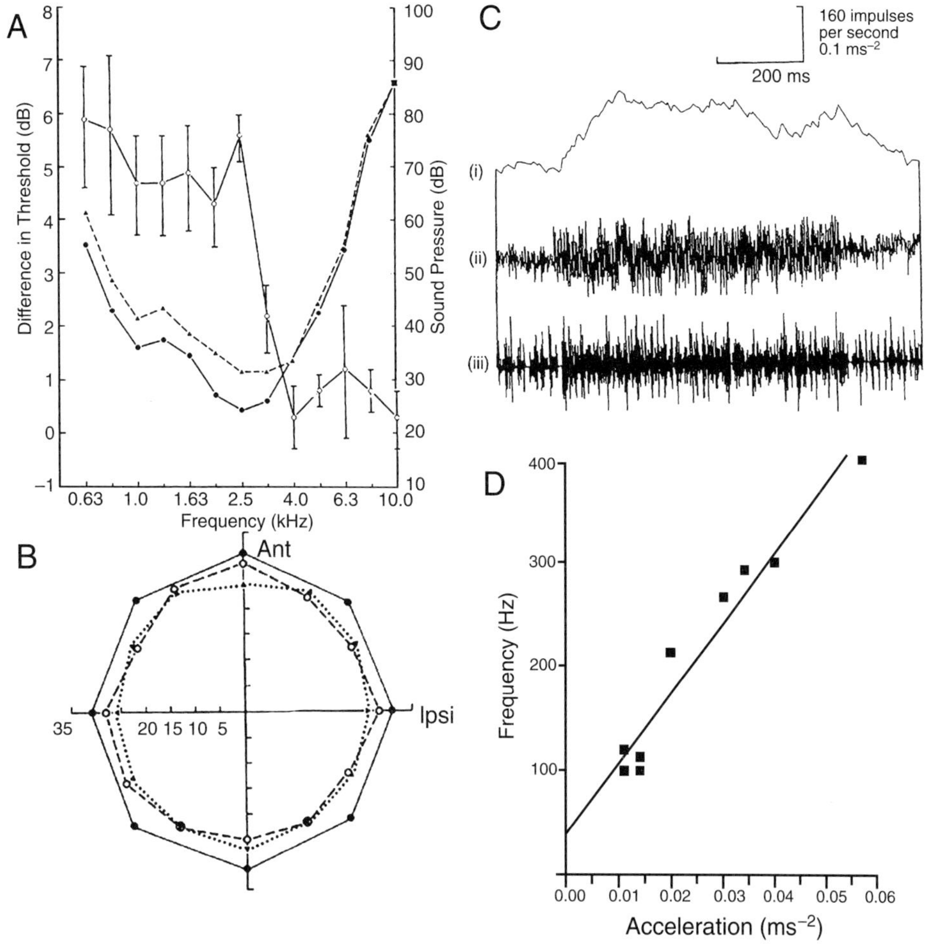

Fig. 22.5. Physiological characteristics of tympanal and subgenual organs. A. Auditory threshold curve (right-hand ordinate) for the tympanal organ, indicating that the greatest sensitivity (lowest threshold) occurs at 2.5 kHz (solid line and circles). When the distal posterior inflated chamber of the tibia was filled with saline, the sensitivity to sounds below 2.5 kHz decreased by approximately 6 dB (broken line; difference given by open circles, left-hand ordinate). B. Lack of directional sensitivity demonstrated by threshold curves for sound presented through 360° for three preparations. C. Extracellular responses from the SGO nerve in the metathoracic leg of weta, which was resting upon a log, to 1000 Hz vibration applied to the surface of the log. i. Integrated frequency of nerve impulses rises during application of vibratory stimulus. ii. Vibratory signal recorded from log by accelerometer. iii. Nerve impulses in response to stimulus. D. Linear relationship between vibration intensity (= acceleration amplitude) and frequency of nerve impulses over acceleration range of 0.012–0.06 m s^{-2}. (A and B from Field *et al.*, 1980, with permission; C and D from McVean and Field, 1996, with permission.)

frequency hearing sensitivity of the ear by increasing the mechanical compliance of the tympana. By selectively injecting the DPIC with saline, it was demonstrated that a hearing loss of 6 dB resulted for frequencies at or below the peak of 2.5 kHz (dashed line in Fig. 22.5A), while little or no effect occurred above 2.5 kHz. Thus the ear's sensitivity is adapted to match the peak sound frequencies produced by the broad-band (non-resonant) stridulation in *H. crassidens* (sound production described in Field, Chapter 15, this volume). Furthermore, it has been noted that non-resonant singers can optimize their broadcast distance if they use low frequencies (< 4 kHz) broadcast from a point above ground level (Michelsen, 1978). This is due to the tendency of ground herbage to absorb sounds above 4 kHz, as well as to the attenuation effect within a 'shadow zone', caused when sound is broadcast from ground level (Embleton *et al.*, 1976; Michelsen and Larson, 1983). An additional physical feature of forests that enhances tree weta communication is a differential absorption of sound frequencies at various vertical levels. Morton (1975) showed that forests transmit sound maximally at about 1.5–2.5 kHz if it is broadcast from an elevation of 0.3 to 1.5 m above ground level. It is known that tree weta galleries occur primarily within 1.5–2.5 m above ground (see Field and Sandlant, Chapter 13, this volume). Thus, male wetas call from gallery entrances that are located within the zone of maximum sound transmission and are not found calling high in the canopy zone.

Wetas resemble tettigoniids in sharing equal sized tympana, but differ from gryllids by lacking directional sensitivity in each tympanal organ. With non-directional hearing, sound from any direction elicits an equal threshold discharge from the tympanal nerve (Fig. 22.5B). This reflects a fundamental difference in the biophysics of the ear in crickets compared with that in wetas. In gryllids, the ear acts as a differential pressure-gradient receiver, in which sound waves act on the tympana from within the trachea, as well as from the outside (a two-input system). But, in wetas, the ear is a one-input pressure receiver, in which sound only acts on the external surface of the tympanum. This was demonstrated by occluding the contralateral tympana and showing no change in sensitivity, even though an internal tracheal pathway connects both ears (Field *et al.*, 1980).

Subgenual organ sensitivity

Extracellular recordings from the SGO nerve of *H. femorata* allowed measurement of the threshold to vibrational stimuli, of responses to intensity increases and of frequency sensitivity (McVean and Field, 1996). In experiments designed to assess the ability of the weta to detect vibrations carried along a tree-trunk, the insects were immobilized while the recorded leg rested on a 1 m, suspended section of a tree-trunk, to which were applied artificial and natural vibratory stimuli. As found for SGO responses in other insects, the sense organ is maximally sensitive to acceleration in the vibratory signal, rather than to displacement. The *H. femorata* preparations gave threshold responses to sinusoidal accelerations as low as 0.012 m s^{-2} (Fig. 22.5C). This is slightly higher than the threshold (0.01 m s^{-2}) measured for the cydnid bug *Nezara viridula* by Cokl (1983). The response increased linearly with a linear increase in acceleration (Fig. 22.5D). The SGO preparations also responded to vibrations induced in the tree-trunk by stridulation from another weta resting upon the trunk. In this case, both sound and vibration energy reached the preparation, but the SGO response was related only to the latency and frequency characteristics of the vibrational waves, and not to the sound. McVean and Field (1996) determined that the effective acceleration induced in the substrate by a male weta stridulating at the gallery entrance is approximately 0.34 m s^{-2}, which means that wetas should be able to communicate to others within a distance of 2–3 m above and below the gallery. Of several frequencies tested, the peak sensitivity occurred at vibrations of 20 Hz and 1 kHz (McVean and Field, 1996). In general, peak SGO responses in other insects occur at low frequencies, ranging from 1 to 3 kHz (Autrum and Schneider, 1948).

The discovery that wetas are sensitive to substrate vibrations opened a new avenue of research into weta communication systems. When male *H. femorata* stridulate at the gallery entrance, they generate a vibration spectrum with a peak frequency quite different from that of the acoustic spectrum (Fig. 22.6A). Therefore, it was possible that tree wetas communicate by vibration signals sent through a tree, as well as sound signals sent through the air. In order to test this hypothesis, it is necessary to determine whether trees inhabited by wetas are able to propagate vibrations or whether trees filter out the energy induced by stridulating wetas. McVean and Field (1996) investigated the response characteristics of manuka trees, known to be inhabited by *H. femorata*, to a standard broad-band impulsive stimulus (a steel ball-bearing accelerated by gravity down a channel at a constant angle over a constant distance). Manuka trees (*Leptospermum ericoides*) have a trunk around 13 cm in diameter and they grow to about 6 m in height. Accelerometers glued to a tree at various positions allowed measurement of the tree's transverse and longitudinal propagation of bending waves, as well as the frequency-filtering property of the tree. They found that the tree readily propagated vibrations well beyond 5 m from the point of impact, and that the tree acted as a filter to the broad-band energy

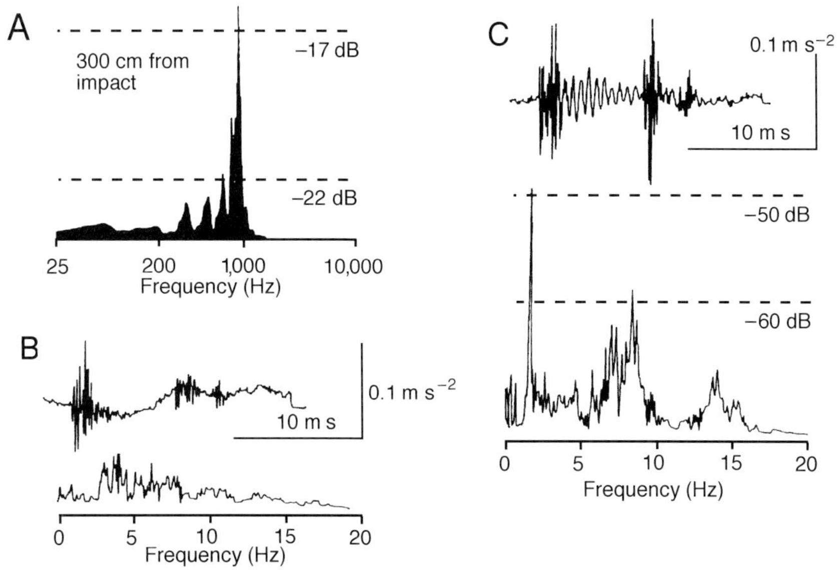

Fig. 22.6. Frequency (spectral) characteristics of vibratory energy produced by stridulating male weta, compared with resonant characteristics of tree inhabited by wetas. A. Acceleration wave-form induced into tree-trunk by stridulating weta, and recorded by accelerometer glued to tree. B. Power spectrum of frequencies in bending wave vibrations carried in tree-trunk during stridulation. Major peak is centred around 1000 Hz. C. Resonant frequency characteristics of tree measured using a standard test impulse to induce a vibration into the tree. The tree filters out most frequencies, except a peak at around 1000 Hz. (From McVean and Field, 1996, with permission.)

and oscillated with a resonant peak of around 1000 Hz (Fig. 22.6A). When male wetas were induced to stridulate on the tree containing accelerometers, their vibrational spectrum also contained a major peak at 1000 Hz, as well as a broad lesser peak around 7.5 kHz (Fig. 22.6C). This signal differed from the simultaneous acoustic signal and its spectrum (Fig. 22.6B). Thus it appears that wetas could induce vibrations at not only the natural frequency of the tree, but also forced vibrations at a signature frequency of around 7.5 kHz. The latter should allow receiver wetas to discern whether a vibration is generated by another weta, and in this way wetas within a single tree could remain in acoustic and vibratory communication. A weta on the forest floor could determine whether a given tree contained stridulating (hence potentially rival) males by detecting matching acoustic and vibratory signals.

Hair sensilla

As in all insects, wetas are covered with an astounding variety of sensory hair sensilla. These are undoubtedly used as mechanoreceptors (touch, or tactile, receptors and air-current detectors), and chemoreceptors (olfactory detectors of airborne volatiles, which give the sense of smell, or gustatory contact receptors, which give the sense of taste). No physiology has been carried out on weta hair sensilla, except for the cercal air-current detectors. However, sensilla on the antennae, mandibles, maxillae, labia, labrum and associated palps of tree wetas (*Hemideina*) have been examined with light and scanning electron microscopy (Jarman, 1982; O'Brien, 1984). All hair sensilla are articulated structures, presumably innervated by non-scolopidial bipolar sensory neurons.

Antennal hair sensilla

Hemideina crassidens possesses filiform antennae, which average 212 ± 15 (mean ± SD, $n = 6$) segments. With the exception of the hair-plate organs at the base of the antennae, few sensilla were present on the first 15–18 segments. The density increased thereafter, initially on the ventral surface and more distally over the full surface of each segment. The maximum density occurred at about segments 26–28 (Jarman, 1982).

Two types of chemoreceptor sensilla are found on the antennae, both of which are most likely to function as detectors of odours from a distance, i.e. the antennae serve as the sense of smell for wetas (Jarman, 1982). Thin-walled hairs were identified as chemoreceptors, based upon their rounded tips, transparent and colourless walls and the ability to uptake crystal violet throughout the lumen of the sensilla. The latter feature suggests that one or more pores occur at least near the tip and possibly along the hair shaft, and is characteristic for chemoreceptor sensilla (Slifer, 1960). The second type, coeloconic sensilla, consisted of short, peg-like hairs within cuticular pits. The roof of each pit was thin, with a small circular opening, as seen in other orthopterans and shown to be involved in olfaction (Slifer *et al.*, 1959).

Antennal tactile hair sensilla consisted of two types, neither of which showed uptake of crystal violet, which suggests that these sensilla are mechanoreceptors. First, a pair of hair-plate organs occurred at the base of the scape (first segment) and the base of the pedicel (second segment). Each consisted of a field of 40–50 short, stout hairs in slightly submerged sockets. The second type was the most abundant sensillum observed on all antennal segments. Each hair was narrow-tipped and slightly curved and had a thick wall, with no terminal pore (Jarman, 1982). Tactile sensilla allow the weta to use the antennae to detect the presence of physical objects. The great length of the antennae in all weta genera, plus the behaviour of moving the antennae in broad sweeping circles encompassing the front and side regions of the animal, serves to build a tactile impression of the surrounding physical environment. This may be seen especially well when a tree weta is walking along a thin branch or twig: the antennae keep constant contact with circular brushing sweeps over the branch as the weta walks forward.

Labial and maxillary palps

The single chemoreceptor type on the palps consisted of a thick-walled, truncated sensilla with an axially ridged sculpture on the shaft and a flattened tip. These sensilla were abundant at the bulbous tips of the palps, but did not occur elsewhere. Crystal violet was taken up at the hair tip only, confirming the preliminary identity as a chemoreceptor (Fig. 22.7A). Scanning electron microscopy confirmed the presence of at least one pore at the tip.

Tactile mechanoreceptor hairs were abundantly mixed with the chemoreceptor hair sensilla (Fig. 22.7C). They consisted of curved hairs which narrowed towards the tip and had spiral sculpture (Fig. 22.7B). In addition, hair-plate organs occurred at each joint of each palp.

J. Wynyard (Christchurch, 1999, personal communication) has shown that the palps are usually in constant alternating tapping motion against the substrate as a tree weta walks or investigates its surroundings. The frequency of tapping is around 5–7 cycles s^{-1}. The function of the chemoreceptors is to allow contact detection of the chemical nature of the substrate being probed. This serves to discriminate food items, as well as the sex of conspecific wetas. The tactile receptors have not been studied, but they presumably allow coordination of movements against the substrate objects being probed.

Maxilla, mandibles, labium and labrum

The mandibles contain apparent chemoreceptive truncated hairs in shallow open pits on the distal surfaces, particularly near the carina (Jarman, 1982; O'Brien, 1984). These also occurred on the distal surfaces of the galea of the maxilla, the labrum and the labium. Scanning electron microscopy showed an indentation at the hair tip, appearing as a pore (Fig. 22.7D). No thin-walled hair sensilla occurred on these appendages (Jarman, 1982).

Tactile hair sensilla (thick walls, narrow tips, no pores) occurred widely on the labium, labrum and the medial surface of the maxilla. Much longer (*c.* 200 μm) tactile hairs are found as a brush border along the mandibular cusps, and especially on the inner margin of the mandible immediately proximal to the cusps (Fig. 22.7F, G). A dense group of long hair sensilla occurs on the ventral surface of the mandible near the posterior articulation, and in lesser numbers along the ventrolateral regions (O'Brien, 1984).

The function of mandibular and palp tactile sensilla is likely to be in detecting contact with food and with other mouth-parts. The latter function would aid in coordinating activities such as chewing and manipulating food, drinking and cleaning mouth-parts.

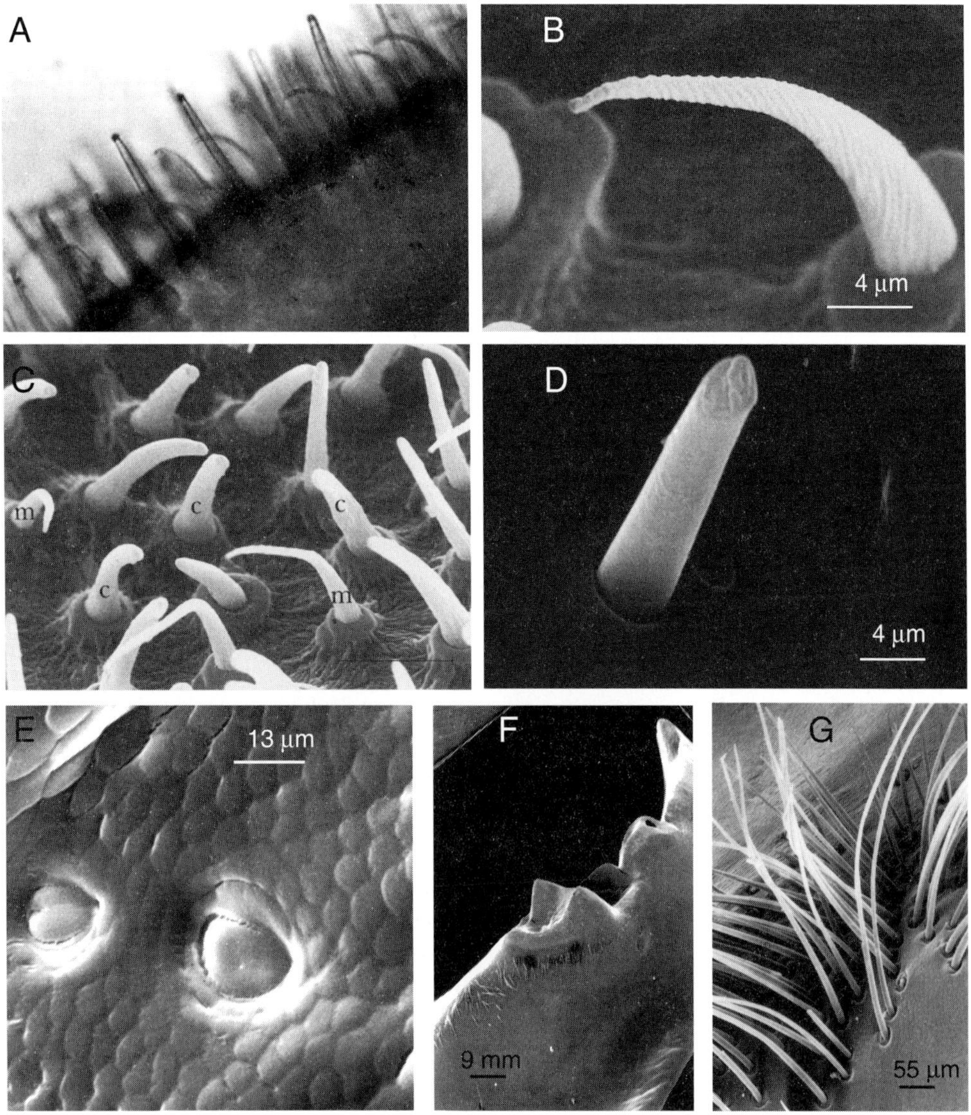

Fig. 22.7. Morphology of cuticular hair sensilla and campaniform sensilla. A. Light micrograph of labial palp hair sensilla fixed in Bouin's solution and stained in dilute crystal violet. Truncated chemoreceptor sensilla take up crystal violet at the tips (arrows), suggesting presence of one or more pores, while thinner, curved tactile mechanoreceptors do not. B. Enlarged view of spiral sculpture on thin tapered mechanoreceptor sensillum of labial palp. C. Scanning electron micrograph (SEM) of bulbous tip of labial palp showing thick-walled truncated chemoreceptors (c) with terminal pores, and thin curved mechanoreceptor sensilla (m). D. Short truncated chemoreceptor sensilla (SEM) on distal surface of mandible showing terminal pore. E. Campaniform sensilla (SEM) in the ventral group on the ventral surface of mandible. Central indentation on dome indicates point of underlying neuron insertion. Long axis of ovate dome lies parallel to slight folds in field of sensilla, and parallel to base of mandible. F. Low-magnification SEM view of cusp region of mandible, showing brush of long mechanoreceptor hair sensilla. G. Closer view of long mechanoreceptor hair sensilla at base of cusp. Similar sensilla found on ventral side of mandible. (A–D from Jarman, 1982, with permission; E–F from O'Brien, 1984, with permission.)

Air-current Detection by Cercal Sensilla, and Giant Axon Responses

The cerci of many insects bear long filiform hair sensilla, which are sensitive to air currents, as well as other, shorter, hair sensilla, which function as contact receptors and gravity receptors (Edwards and Palka, 1974). The physiology and morphology of such sensilla have been well studied in crickets and cockroaches, but comparative information for the Stenopelmatoidea is lacking. The following summarizes an unpublished study, by J.S. Edwards and L.H. Field, of cercal sensilla in the tree weta *H. femorata*.

The abdominal cerci of *H. femorata* are heavily sclerotized, but lightly pigmented, unsegmented, cornute appendages inserted ventrolaterally on the eleventh abdominal segment. They are sexually dimorphic, with the male cerci larger and more strongly recurved (Fig. 22.8A, B). Two types of sensilla occur on tree weta cerci, while other types are absent. Thus, clavate hairs, which occur on the basal part of cockroach and cricket cerci and which serve to orientate with respect to gravity, are absent in the tree weta. The first sensilla present includes filiform, socketed hairs, each of which is inserted on a flexible base, so that it is sensitive to small air movements (e.g. Fig. 22.9B). The patterned orientation of filiform hairs in other orthopteroid insects, such as crickets (Edwards and Palka (1974) and cockroaches (Ritzmann 1984), where a particular filiform hair can oscillate in one plane only, is not evident in those of the weta. Input from the pattern of orientated filiform hairs serves to localize a source of air movement and thus of a potential predator in cockroaches and crickets. The absence of orientated filiform hairs on the cerci of the tree weta suggests that the cerci do not function in predator location. The second sensilla type consists of stout appressed bristles, which are distributed over the entire cercus (Fig. 22.9C, D). Campaniform sensilla, which are prominently associated with the sockets of filiform hairs in crickets and cockroaches, also appear to be lacking, as in locusts.

Central projections of cercal sensilla

Afferent axons from the cercal sensory neurons terminate predominantly in the terminal abdominal ganglion, but a small projection of primary axons continues anteriorly in the ventral nerve cord in two bundles, a larger median and a smaller bundle, occupying a more lateral and central position (Fig. 22.10A). The anterior terminations of these fibres are not known; a few extend beyond the next anterior ganglion.

Within the terminal ganglion, cercal sensory axons terminate in a compact ovoid glomerulus situated predominantly in the ventral region of the neuropile, medial to the site of entry of the major tracheae. Two other ipsilateral ventral sites of termination lie adjacent to the major glomerulus. One of these lies close to the midline and slightly posterior to the major glomerulus, while a second lies mesial and anterior to the major glomerulus. A contralateral projection of six to eight axons, closely resembling those of *Acheta domesticus*, terminate near the midline of the ganglion in the ventral neuropile of the contralateral major glomerulus. The glomeruli appear to occupy regions of neuropile that receive processes from at least five of the eight giant neurons found in the terminal abdominal ganglion.

Physiological responses

Recordings from the cercal nerve with extracellular silver electrodes showed that the filiform hairs were extremely sensitive to air movement. Gently blowing from a distance of 30–40 cm elicited a discharge of many small sensory axons. Stronger stimuli elicited greater discharge. The extreme sensitivity of the hair sensilla was shown by reliably eliciting a response to a hand clap at 1 m distance from the preparation. Using a calibrated air-stream pulse as a stimulus, it was found that at least two classes of sensory axons responded to air movement. At air-flow rates less than 3.3 ml s^{-1},

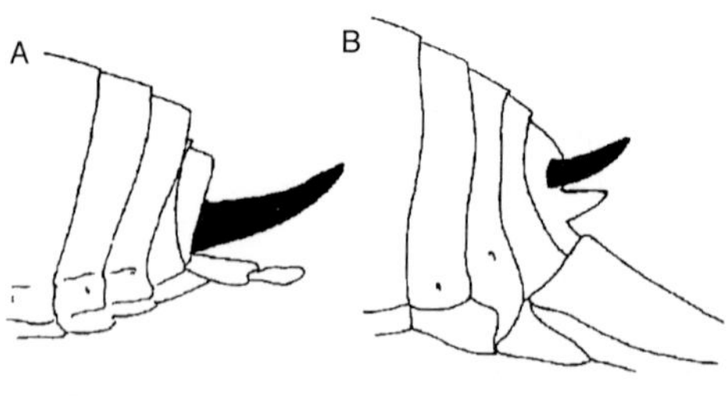

Fig. 22.8. Lateral views of terminal abdominal segments of adult male (A) and female (B) tree wetas (*H. femorata*), showing position and relative size of abdominal cerci (black). Sensilla not shown. (Courtesy of J.S. Edwards, with permission.)

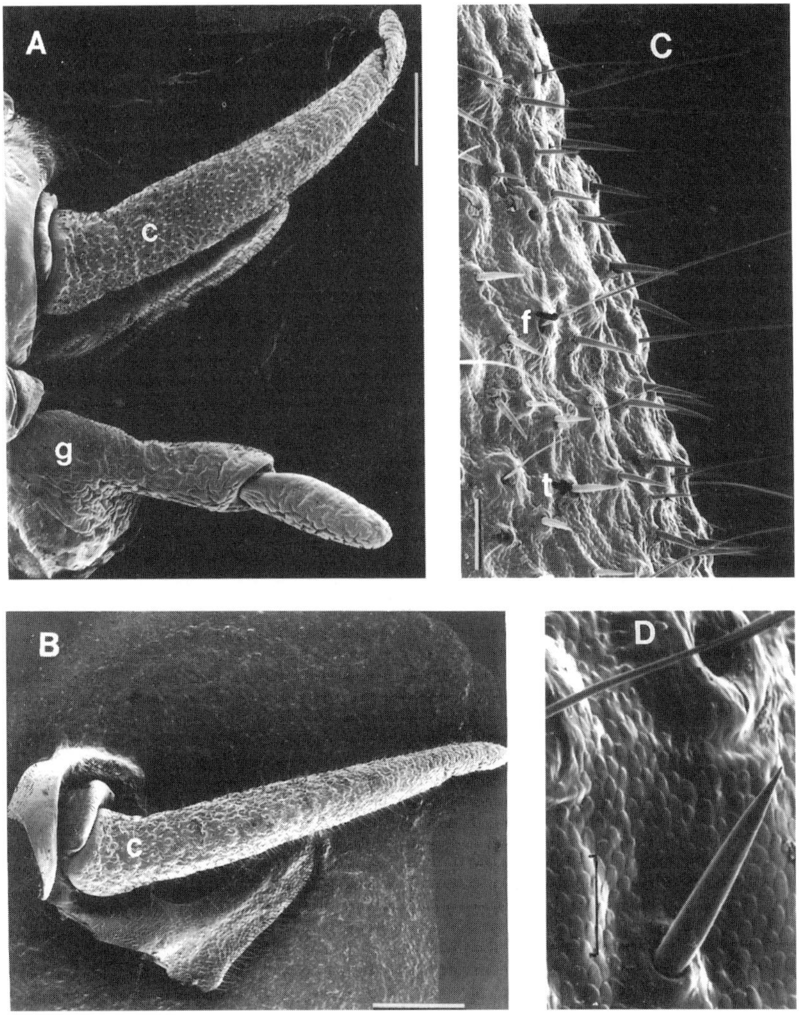

Fig. 22.9. Scanning electron micrographs of abdominal cerci of *H. femorata*. A. Male cercus (c) and gonapophysis (g). Scale bar, 1 mm. B. Abdominal cercus of female at same magnification as for male. Note that angle of view is not precisely transverse and thus actual length of cerci is not apparent. Scale bar, 1 mm. C. Lateral view of male cercus showing filiform hairs (f) interspersed with trichoid sensilla (t). Scale bar, 100 μm. D. Detail of filiform hair socket (above) and appressed trichoid sensillum (below). Scale bar, 50 μm. (Courtesy of J.S. Edwards, with permission.)

only small sensory spikes of many amplitudes were observed, often discharging for 30–60 ms after the stimulus termination. At rates above this value, spikes of twice the amplitude responded phasically for the duration to the air stimulus. Both kinds of response are shown in Fig. 22.10B. An intense burst of giant axon spikes followed the cercal discharge, with a latency of 8–10 ms, and showed no after-discharge. The responses readily habituated with repeated stimulation at 10 s intervals.

Campaniform Cuticular Stress Detectors

Campaniform sensilla appear as oval or round dome-shaped structures, embedded in the cuticle and surrounded by a small collar. They are innervated by a single bipolar neuron, which detects stress exerted on to the cuticle. Such sensilla have been observed on the legs (see Fig. 22.4B), as well as on the mandibles. O'Brien (1984) has described

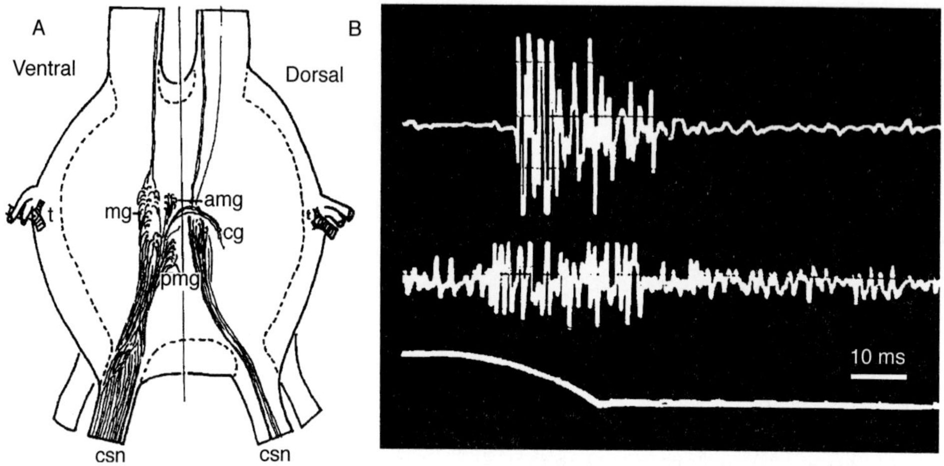

Fig. 22.10. Central projection and physiological response of cercal afferents. A. $CoCl_2$ axonal fill of cercal nerve (csn) projection to the neuropile of the sixth abdominal ganglion. Composite drawing showing ventral cercal projections on the left and more dorsal projections on the right. The largest area of projection forms an ovoid major glomerulus (mg) near the midline and medial to the site of entry of the major tracheal supply (t). Two smaller glomeruli, an anterior midline (amg) and a posterior (pmg) lie close to the midline. A small bundle of about six axons contributes to the contralateral major glomerulus (cg). B. Recordings from cercal nerve and ipsilateral connective of ventral nerve cord, showing cercal afferents (middle trace) firing in response to a puff of air delivered from a syringe (downward movement, lower trace), and driving a burst of spikes from several giant axons (upper trace).

in detail the external appearance and innervation of the mandibular sensilla in *Hemideina maori*. The most prominent group (12–20 sensilla) occurs in a differentiated region of thinner cuticle on the ventral surface, midway between the insertion of the largest tentoro-mandibular muscle (TM-1) and the base of the mandible. The thinned region is developed into folds, containing the campaniform sensilla, which lie with their long axis parallel to the fold axis. The ovate dome of each sensillum has a slight indentation at the point of insertion of the neuron dendrite from beneath, and is surrounded by a cuticular collar (Fig. 22.7E). This specialized region of campaniform sensilla lies almost in line with the mandibular hinge axis and the insertion of the apodeme from the major mandibular adductor muscle (M-21) (see internal anatomy in O'Brien and Field, Chapter 8, this volume).

A few sensilla lie around the outside of the thinned ventral region, rather closer to the TM-1 muscle insertion. All mandibular campaniform sensilla are innervated by the same nerve, a branch of cranial nerve I, which also contains axons innervating the main adductor muscle (M-21) (O'Brien, 1984).

Two other groups of campaniform sensilla occur next to the two mandibular articulations. They are not in regions of specialized cuticle. The frontal group near the anterior articulation is innervated by a branch of cranial nerve II, and gives responses during mandibular adduction (O'Brien, 1984). A small group also occurs near the base of the posterior articulation.

In serving as cuticular stress detectors, it is clear that the campaniform sensilla are strategically placed on the mandibles. During strong bites, the extremes of stress and strain are likely to occur at four places: the cusps, the two articulations and the apodeme insertion point of the M-21 adductor muscle. Campaniform sensilla occur at three of these sites, while the rich innervation of cusp receptors monitors the fourth. It is especially noteworthy that the greatest concentration of campaniform sensilla occurs at the M-21 apodeme insertion. This muscle develops a powerful force in adult male tree wetas (described in Field, Chapter 23, this volume), and it is likely that the campaniform sensilla serve as a safety mechanism to inhibit the adductor muscle when the strain reaches an extreme limit during strong biting. In cockroaches, the campaniform sensilla are known to inhibit motor neurons causing high levels of strain (Zill *et al.*, 1981), and the same mechanism may well operate in wetas (O'Brien, 1984).

Mandibular Cusp Receptors

The distal regions of the mandible (investigated only in *H. maori*) are innervated by cranial nerve III, which also supplies the cusp region. This region is supplied with an extremely rich innervation of tiny, unidentified sensory neurons, apparently associated with the inner cuticle of the cusps and not evident externally. Thin sections of the nerve branch to the cusps showed approximately 2700 tiny axons, ranging in diameter from 0.3 to 2.5 μm. Recordings from this nerve with thin silver hook electrodes gave 'on' and 'off' responses to gentle pressure applied to the cusps, as would occur during biting. Sustained pressure gave low-level continuous firing of tiny impulses and greater pressure resulted in recruitment of larger impulses (O'Brien, 1984). The histology and ultrastructure of these receptors are completely unknown in wetas, although they may represent the mandibular chordotonal organs described in other insects. Their function, as judged from their sensory response, is undoubtedly associated with monitoring and controlling any adjustments in position or force involved in the final stages of biting. The receptors are also likely to assess the consistency of materials being bitten or chewed; in principle, they could even act as a safety mechanism to prevent the build-up of excessive force on the mandibles (O'Brien, 1984)

Proprioceptor Systems: Chordotonal Organs and Muscle Receptor Organs

Proprioceptors are internal sense organs that usually detect joint position (chordotonal organs) or muscle tension (muscle receptor organs, strand receptors). These serve as stretch receptors, in which mechanical distortion is converted into electrical impulses in sensory neurons, which often have specialized dendrites (e.g. terminal cilia) and are associated with accessory cells comprising the sensillum. Chordotonal organs occur in nearly every joint in the insect body, while muscle receptor organs are found in a few specialized locations in the legs, abdomen and head.

Joint chordotonal organs

Joint chordotonal organs consist of a cluster of sensory neurons associated with an elastic strand, or ligament, which spans a joint between segments of appendages or body segments. In tree wetas, the only well-known representative is the femoral chordotonal organ (FeCO).

Femoral chordotonal organ

The FeCO is the largest and most complex joint chordotonal organ in the legs of wetas. It consists of a large cluster of scolopidia, formed into a bulbous structure, attached proximally to the femur wall, close to the femur–trochanter articulation (Fig. 22.11A). The organ tapers distally into an elastic ligament, spanning almost the entire femur as it attaches on to a long, thin, cuticular apodeme, which inserts on to the dorsal side of the tibia at the femoro-tibial joint. This insertion point ensures that flexion of the tibia stretches the FeCO. At the point of ligament insertion on to the apodeme, a curious, coiled, spring-like structure is found, which is homologous to that described for the cricket (Nowel *et al.*, 1995) and which apparently serves a viscoelastic function. This coiled structure is not found in the caeliferan orthopterans, and it further serves to establish the affinity of tree wetas with the ensiferan insects. The bulbous FeCO is separated into two scoloparia (groups of scolopidia), which are fused proximally, while distally the connective tissue ligaments remain separated until joining the apodeme (Fig. 22.11B). Of the two scoloparia, the dorsal and ventral, only the dorsal is attached at the proximal wall of the femur. The attachment forms a basal band of connective tissue anchoring the bulbous scoloparium to the hypodermis. The axons from the various groups of neuron cell bodies within the scoloparia join to form the FeCO nerve, which is a branch of nerve 5B1. Unlike the FeCO in the caeliferan legs in grasshoppers and locusts, the FeCO in wetas is similar in all three pairs of legs. It is located between the flexor and extensor muscles on the anterior side of the leg, running parallel to the main leg nerve 5B (Rind, 1976).

The sensory neurons are arranged into identifiable groups, very similar to those of the cricket FeCO (Nishino and Sakai, 1997), except for increased numbers of neurons in the weta. In the dorsal scoloparium, approximately 50 small

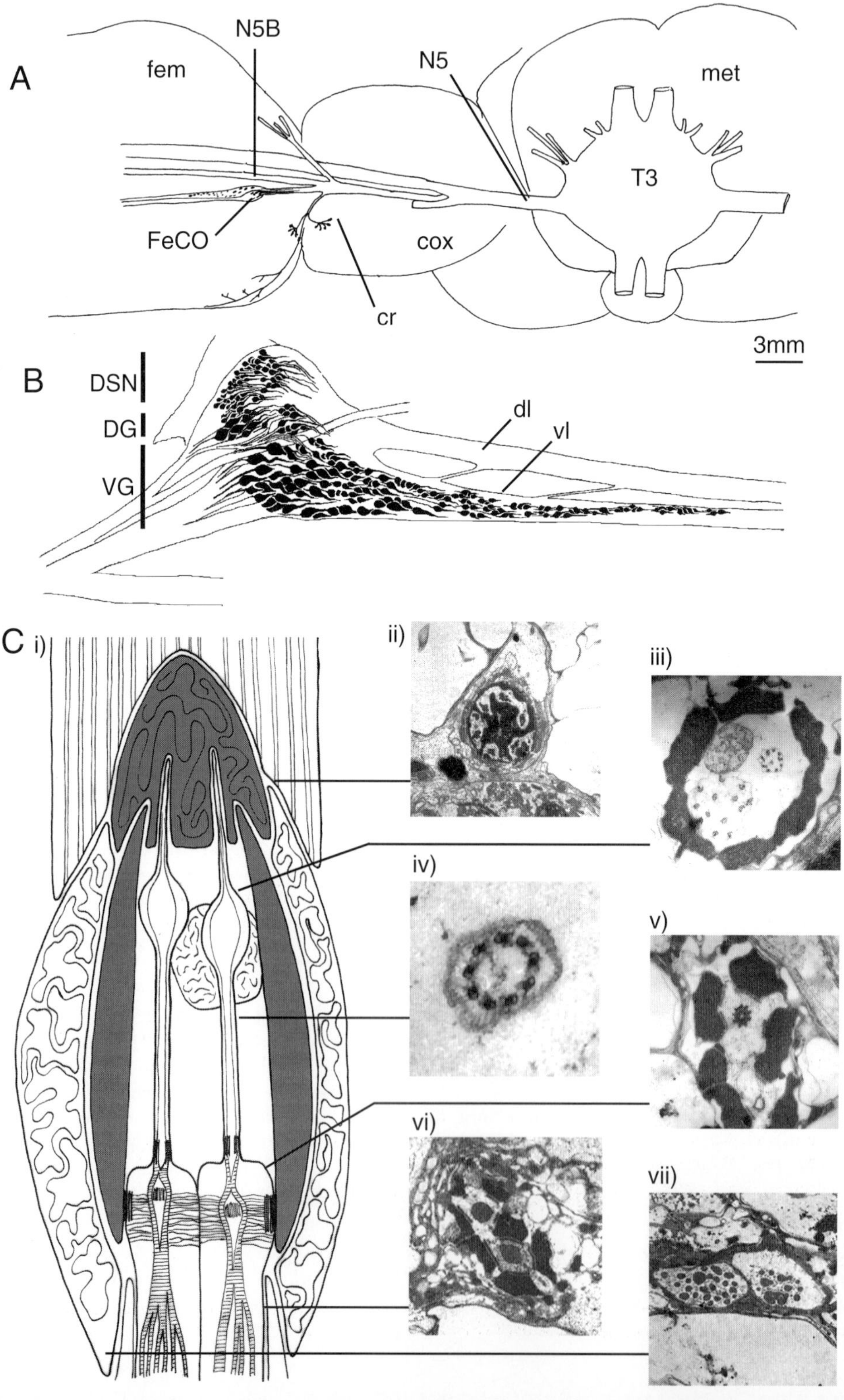
A
fem
N5B
N5
met
T3
FeCO
cox
cr
3mm
B
DSN
DG
VG
dl
vl
C i)
ii)
iii)
iv)
v)
vi)
vii)

neurons, designated as group DSN, form a tightly packed group in a conical array pointing distally into the dorsal ligament (Fig. 22.11B). These are all of similar size, as seen also in the apparently homologous 'proximal' scoloparium of locusts and tettigoniids (Field and Pflüger, 1989). In the ventral scoloparium, two groups are recognizable. Along the ventral extent of the scoloparium is an array of large neurons, proximally grading into smaller ones reaching the distal tip of the organ, where it merges into the ligament. This ventral group (VG) includes approximately 130 neurons, which is much greater than that described in crickets (43 neurons). A third, newly recognized group in the ventral scoloparium is the dorsal group (DG), which forms a cluster of large to intermediate-sized neurons dorsal to the VG, but still inserting into the ventral ligament.

The division of the FeCO into anatomical groups appears to have a functional basis, reflected by their different regions of projection into the CNS of related insects (Nishino *et al.*, 1999). The VG projects mainly to the lateral association centre (LAC) in crickets, which, in the locust, is a reflex centre for leg motor neurons and interneurons (Burrows *et al.*, 1988). Thus the VG neurons appear to be primarily responsible for controlling leg reflexes. The DSN projects into the median ventral association centre (mVAC), which is known to receive projections from the tympanal organ in crickets and tettigoniids (Kalmring *et al.*, 1978; Eibl and Huber, 1979), as well as from vibration-sensitive axons of the FeCO proximal scoloparium in the locust (Field and Pflüger, 1989). The DG also projects to the mVAC in an area known to receive input from the SGO in the cricket (Eibl and Huber, 1979). Thus, accumulating evidence suggests that the FeCO may be a vibration detector (acting possibly indirectly), in addition to its typical function as a joint movement and position detector for control of leg reflexes.

The ultrastructure of the FeCO has been studied (Rind, 1976) and the scolopidium unit has been reconstructed, based upon transmission electron micrographs (Fig. 22.11Ci). Each scolopidium contains two bipolar sensory neurons, which give rise to a cilium at the tip of each dendrite (Fig. 22.11Ci, vii). The axons of the neurons are wrapped in Schwann cell layers, while the dendrites are surrounded by layered glial cells. Each dendrite contains a collagen-like set of rootlets extending down from the cilium (Fig. 22.11Cvii). These rootlets fuse distally into a single large ciliary root, just below a basal body structure, resembling a centriole (Fig. 22.11Cvi). At this level, the dendrites are surrounded by a highly vacuolated scolopale cell, which secretes a barrel-shaped set of seven or eight scolopale rods,

Fig. 22.11. Morphology, neuroanatomy and fine structure of the femoral chordotonal organ (FeCO) of the hind leg (*H. femorata*). A. Position of the FeCO in the hind femur, and innervation from the main leg nerve, N5B. B. Camera lucida drawing of meta-FeCO neuron groups backfilled by cobalt/nickel chloride diffusion through the chordotonal organ nerve (drawn with kind permission from preparation by Dr H. Nishino). Three neuron groups are recognized and labelled. C. Ultrastructure of a scolopidium of the FeCO. (i) Longitudinal reconstruction based upon electron micrographs, showing dendrites giving rise to cilia, which are surrounded by scolopale rods, and which insert into the cap secreted by one viscoelastic attachment cell of the ligament. (ii-vii) Transmission electron micrographs of transverse sections through scolopidium. (ii) Section through base of cap, composed of amorphous electron-dense material, surrounded by attachment cell containing microtubules. Scolopale rods insert around the periphery of cap, while cilia insert into middle. (iii) At level of ciliary dilatation, the nine axoneme microtubule doublets are partially surrounded by membrane crescents and show attachments to cilia plasma membrane. Dynein-like arms are absent. Additional tubular membrane profiles occupy the central dilatation area. (iv) Within receptor lymph space formed by partially fused scolopale rods, both cilia show $9 \times 2 + 0$ microtubule doublets in axoneme surrounded by cilium plasma membrane. (v) This section through the proximal basal body of upper dendrite shows encompassing finger-like strands of ciliary root, desmosomes connecting dendrites to unfused scolopale rods, surrounded by vacuolated scolopale cell. (vi) Proximal to the basal bodies, the ciliary root becomes fused, scolopale rods are connected to dendrites by desmosomes and scolopale unit is surrounded by scolopale cell. (vii) More proximally, ciliary root has divided into many rootlets, and dendrites are surrounded by glial (sheath) cell wrappings. cox, coxa; cr, cuticular hair receptors; DG, dorsal group neurons of ventral scoloparium; dl, dorsal scoloparium ligament; DSN, dorsal scoloparium neurons; FeCO, femoral chordotonal organ; fem, femur; met, metathoracic segment; N5, fifth nerve arising from the thoracic ganglion; N5B, sensory nerve receiving FeCO axons; T3, third thoracic ganglion; VG, ventral group neurons of dorsal scoloparium; vl, ventral scoloparium ligament. (A and electron micrographs of C from Rind, 1976, with permission.)

which are intimately attached to the pair of dendrites by desmosomes (Fig. 22.11Cvi). Moving towards the dendrite tip, the root becomes hollow and then splits into nine strands, which expand outward to enclose the proximal basal body, containing a centriole (Fig. 22.11Cv). The desmosomes firmly link the dendrites to the scolopale rods. Extending inwards from the desmosomes is filamentous material connecting to the ciliary root around the basal body. Septate desmosomes connect infoldings of the scolopale cell (Rind, 1976).

Just distal to this region, each dendrite tapers rapidly and gives rise to a cilium, which extends from the dendrite tip via a distal basal body. The ciliary root material coalesces into the basal body. The cilium contains a $9 \times 2 + 0$ configuration of microtubules in the axoneme (Fig. 22.11Civ). The A microtubule is hollow, while the B microtubule is electron-dense. Dynein-like arms extend from the B subfibre, giving rise to the suggestion that the cilium is motile. The entire cilium is surrounded by the plasma membrane of the dendrite, and remains in an extracellular space (receptor lymph space) formed by the barrel-like configuration of the scolopale rods. The rods have partially fused together. At a more distal point, the cilia become dilated to at least three times their diameter, and the axoneme changes dramatically. The microtubule doublets (A and B subfibres) remain adjacent to the plasma membrane but are semi-surrounded by C-shaped structures, which may be membrane material (Fig. 22.11Ciii). Additional material links the microtubules to the cilium plasma membrane. In addition, four to five tubular membrane structures appear inside the dilatation and are connected by fibrous material to the other neighbouring structures. The B subfibres have lost their dynein arms. The scolopale rods are almost completely fused to form a receptor lymph space containing a membrane-bound region of more densely staining material (possibly extensions of the scolopale cell). At the apex of the scolopidium, the cilia and scolopale rods insert into the cap, which is made of infolded amorphous material (Fig. 22.11Cii). The cap is surrounded by the attachment cell, containing an abundance of microtubules. The ligaments are therefore composed of attachment cells, each of which connects to two sensory neurons through the scolopidial apparatus (Rind, 1976).

Nothing is known about the mechanism of transduction of stretch, directed through the ligaments, into electrical impulses in the neuron, although small unitary voltage events have been recorded from dendrites of tympanal scolopidial neurons of the locust, which have given rise to one hypothesis about the generation of impulses in such neurons (Hill, 1983).

Other chordotonal organ proprioceptors

Essentially every joint in the insect leg contains at least one chordotonal organ, which (presumably) reports the movement and position of the joint to the CNS. However, in wetas, only the FeCO has been carefully studied. In view of interesting differences in neuron numbers in the tree weta, it would be worthwhile investigating the morphology and anatomy of the other leg joint chordotonal organs. For example, the tibio-tarsal chordotonal organ appears to have an unusual morphology, with a large ligament inserting on the retractor unguis apodeme (L.H. Field, personal observation). This is not found in other orthopterans or in the cockroach, in which the sense organ has been most thoroughly studied.

Physiology of the femoral chordotonal organ

Two unpublished studies have analysed the physiological sensitivity of the FeCO in response to flexion and extension movements, as well as to set positions of the femur–tibia joint (Field and Rind, 1976; Rind, 1976). The aims of such studies, described below, were to determine whether the FeCO could code for: (i) the angular position of the femur–tibia joint; (ii) different directions of movement; (iii) different rates (velocities) of movement; (iv) the arc over which a movement occurs; and (v) acceleration. Recordings of joint position sensitivity were taken 60 s following each new position setting of the tibia, to allow for adaptation to occur (adaptation was 90% complete at this time). To minimize hysteresis in the responses, new joint-angle settings were approached from the extended (relaxed chordotonal organ) to the flexed (stretched organ) extreme femur–tibial angle (FTA). Movement sensitivity was measured by sinusoidal and ramp displacement of the tibia from set positions. Angular velocity was determined by the slope of the ramp and expressed in degrees s^{-1} (Rind, 1976).

As in most joint chordotonal organs studied in

orthopterans, the tree weta FeCO contains sensory units that respond with sustained firing to static joint position, known as tonic units, and others that fire only to joint movement, known as phasic units. The tonic units were generally smaller than the phasic units, suggesting that the distal neurons of the VG may be responsible for such responses. The large phasic units may belong to the proximal VG or DG groups.

Tonic responses

The overall sensitivity of the FeCO to joint position was greatest at both extremes of the FTA and least at the normal resting FTA of 60° (Field and Rind, 1976; Rind, 1976). This corresponds to similar results found in the locust leg (Usherwood *et al.*, 1968). In the weta, joint position sensitivity was encoded in a graded manner, related to neuron size. The smallest spikes (4.1–4.7 mV) showed very high sensitivity to extreme tibial flexion (FTA 10°–20°) and a decreasing sensitivity over the remaining FTA range (Fig. 22.12A). The next size class showed a steadily decreasing sensitivity over the whole FTA range as 140° was approached (Fig. 22.12A). In opposite manner, the intermediate- and large-sized spikes showed increasing sensitivity to increasing FTA. Specifically, the intermediate-sized ones (Fig. 22.12B) responded over the whole or most of the range, while the largest ones only fired at FTAs greater than 40° (11.3–11.7 mV) or 60° (10.4–10.6 mV). The latter showed either a peak sensitivity at 90° (11.3–11.7 mV) or two peaks at 90° and 140° (Fig. 22.12C).

The distribution of responses of individual classes of position-sensitive neurons with peaks of sensitivity at different FTAs is known as range fractionation. This reflects the different position tuning of each individual sensory neuron, and is a common mechanism for position discrimination. In range fractionation, a unique combination of firing frequencies will be sent to the CNS for any given joint angle (Field and Matheson, 1998).

Phasic responses

The FeCO is very sensitive to tiny movements and to vibrations. In general, the FeCO responds more strongly to extension movements (increasing FTA) than to flexion movements; nevertheless the chordotonal organ codes for both directions of movement (Rind, 1976). The movement-sensitive neurons give larger spikes than the position-sensitive ones, which suggests that the movement-sensitive neurons comprise the largest ones in the VG and DG of the ventral scoloparium.

Directionality is shown by all phasic neurons – that is, they are unidirectional in response and conform with the same characteristic seen on other insect chordotonal organs (Field and Matheson, 1998). Extension-sensitive neurons included the largest size classes, while flexion-sensitive neurons included the smaller and intermediate phasic units. Thus, over the same arc of movement at the same velocity, intermediate units fire strongly on flexion and the largest units fire most strongly on extension (Rind, 1976).

Velocity sensitivity is again related to neuron size. Furthermore, two distinct types of responses were observed: (i) the unit increased firing with increased velocity; and (ii) the unit decreased firing as velocity increased. The smallest phasic units showed the second type of response, with greatest sensitivity to low velocities in the range 0.05–0.15° s^{-1} and a decreased response at higher velocities (Fig. 22.12D). This contrasts with the responses of intermediate and large units, which all increased firing frequency as velocity increased and, in addition, showed range fractionation (Fig. 22.12E). The latter is seen in two classes of units, 11.3–12.3 mV and 12.3–15.2 mV, which steadily increased the response to increasing velocity, while another two classes, 8.8 mV and 12 mV, had no sensitivity to low velocities and only began to fire at velocities exceeding 0.3° s^{-1} (Fig. 22.12E). Finally, the most extreme high-threshold velocity detector and largest unit only fired at velocities exceeding 0.55° s^{-1} (15.2–16.3 mV) (Fig. 22.12E) (Rind, 1976).

At least one acceleration-sensitive unit fired only to rapid extension and never to rapid flexion. This was the largest unit observed (17 mV), and it is presumably one of the VG group.

Muscle receptor organs

Muscle receptor organs (MROs) contain one (rarely, many) sensory neurons, each with a branched dendrite, which inserts into a thin muscle and which detects the tension built up by that muscle. The muscle is usually too weak to develop significant tension at a joint, but is used instead to

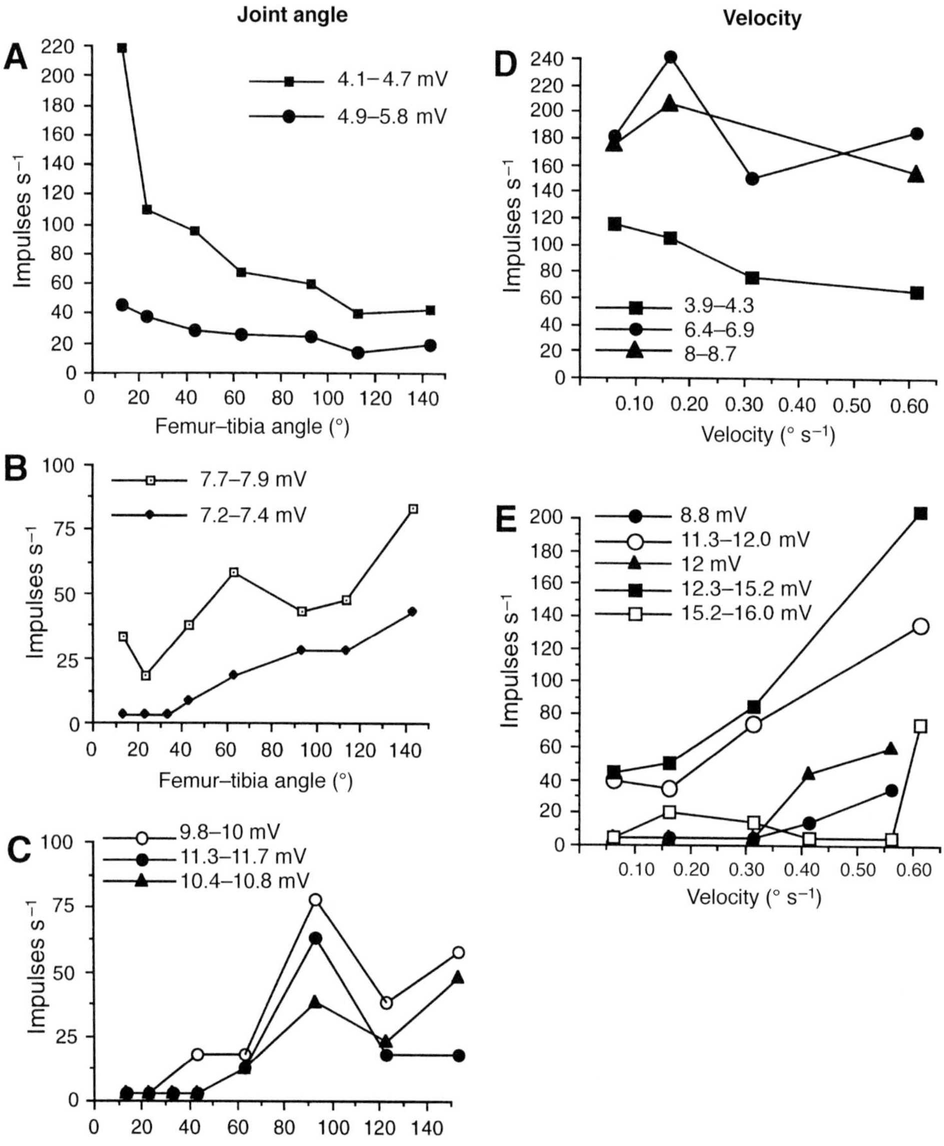

Fig. 22.12. Sensory responses of the metathoracic femoral chordotonal organ (FeCO) to position (A–C) and movement velocity (D, E) of the femur–tibial joint. A–C. Examples of position-sensitive unit response profiles plotted against femur–tibia joint angle (FTA). Units segregated into classes according to spike amplitude in mV. A. The smallest units were most sensitive to flexed positions and showed responses across the entire range of joint angles. B. Two intermediate size classes responded with increasing sensitivity to extended joint angles, and one did not fire at FTA less than 40°. C. The largest units showed peak sensitivities at an intermediate FTA, 90°, and a secondary peak at the extreme of extension. D, E. Responses of velocity-sensitive units as tibia was moved in constant velocity ramps over 20° arcs. D. The smallest units were maximally sensitive to low velocities, and decreased firing at higher velocities. They responded across the entire range of velocities tested (maximum, 0.6° s^{-1}). E. Intermediate and large units all gave increased response firing as velocity increased. Some showed sensitivity across the entire range of velocities tested, while other high-threshold units only responded to high velocities. (After Rind, 1976, with permission.)

regulate tension experienced by the sensory neuron, in parallel with tension development of adjacent main muscles (reviewed in Matheson and Field, 1995). While known in the head and abdomen in other insects, in tree wetas the only ones investigated are two MROs associated with muscles in the head, where they monitor and assist with control of the mandibles through reflexes to mandibular muscles. Both are contained within the mandibles, one placed ventrally and the other placed dorsally. The ventral muscle receptor organ (VMRO) is strikingly different from others described in insects, due to the presence of a very large number of sensory and motor neurons (O'Brien, 1984).

Ventral muscle receptor organ (VMRO) morphology

Of the two muscle receptor organs found associated with the mandibles, the VMRO is the most complex and physiologically the most important. It is a spindle-shaped stretch receptor, consisting of a thin muscle slip innervated by many sensory neurons, with branched dendrites intermingling amongst the individual muscle fibres along the length of the receptor organ. In addition, its muscle receives motor innervation, indicating that the sense organ is under central modulatory control. The VMRO lies in parallel with the non-sensory muscle TM-2a, both of which form part of the tentoro-mandibular complex (Fig. 22.13A). They span the joint between the mandible and head capsule, and therefore this stretch receptor is in a position to monitor mandible movement. The pair of muscles lies adjacent to the main tentorial muscle TM-1, which joins the strut-like tentorium of the head capsule to the ventral mandibular floor. Because the VMRO spans the mandibular joint at an oblique angle, it undergoes a 40% length change from fully closed to fully opened mandible (O'Brien, 1984).

Histological study has shown that the VMRO comprises five main components. The outside is surrounded by a layer of sheath cells, covered by a thick extracellular matrix of collagen-like material. Within the organ, there are sensory neuron cell bodies, a tract of sensory axons, a tract of motor axons and the muscle fibre tract. The thick sheath encapsulating the organ is probably made of elastic material, which is consistent with its role as a stretch receptor rather than as ordinary skeletal muscle. The sensory nerve tract occupies about one-quarter of the cross-sectional area of the VMRO. The cell bodies along this tract give off dendrites, which end in tiny branches embedded in connective tissue among the muscle fibres. These are naked branched endings, lacking accessory cells, and are characteristic of the multipolar type found in other MROs. Up to 170 sensory axons have been counted in the sensory nerve to the VMRO, making this the most richly innervated MRO ever reported. Also, one large multipolar sensory neuron occurs at the junction of the sensory nerve and the muscle. Its dendrites invade both the muscle and the sensory nerve, but its function is unknown. The motor nerve supply to the VMRO is a thin branch from cranial nerve I, which supplies muscles TM-1 and TM-2a. The VMRO branch contains seven axons, at least four large ones of which branch profusely to make synapses within the muscle tract. The muscle tract occupies about one-third of the cross-sectional area of the VMRO, and it consists of a loosely anastomosing network of disorganized tiny muscle fibres. This is unlike the normal skeletal-muscle histology. The VMRO muscles probably serve to stiffen the sense organ to allow changes in sensitivity of the sensory neurons to imposed stretch or position (O'Brien, 1984).

Normally, insect MROs have far fewer sensory neurons associated with the muscle and, while mandibular MROs have the highest count, no other insect has the huge numbers found in the tree weta *H. maori*. In the beetles *Dermestes* and *Oryzaephilus*, about ten and eight sensory neurons, respectively, have been found (Honomichl, 1978). Elsewhere in the insect body, both skeletal and specialized receptor muscles occur as MROs, but they invariably have only one or two multipolar sensory neurons. These are found in the thorax, abdomen and legs (Finlayson, 1976; Matheson and Field, 1995).

Physiology of the VMRO

The VMRO is primarily sensitive to opening movements and open positions of the mandible. The greater the angle of opening, the greater is the firing of the sensory neurons. However, the response is not simple: at the closed position, it consists of tonic (continuous) firing of small units, plus low frequency bursts of larger units at about

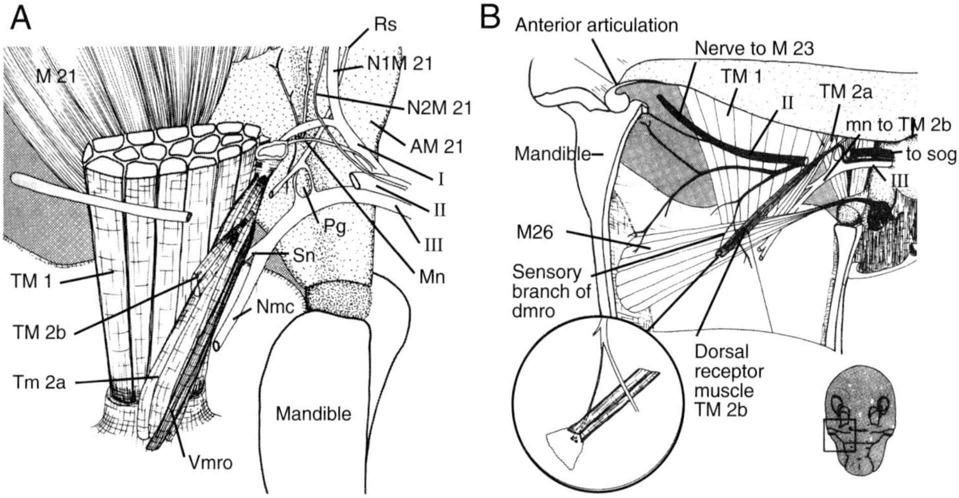

Fig. 22.13. Anatomy of mandibular muscle receptor organs in the weta *Hemideina maori*. A. Ventral muscle receptor organ (VMRO) lying in close association with the tentoro-mandibular muscle TM-2a. The VMRO inserts on the apodeme of the TM-2 muscle group, which, together with the main tentorium muscle TM-1, attach to the tentorium (removed). All three muscles have origins on the ventral floor of the mandible. B. Anatomy of dorsal muscle receptor (DMRO). The DMRO originates on the dorsal wall of the mandible, where its single receptor neuron (from cranial nerve II) innervates connective tissue in series with the receptor muscle. It inserts on to the tentorium and receives motor innervation from cranial nerve III (left inset). Right inset shows mandibular regions dissected for illustration. AM 21, apodeme to M 21; M 21, main mandibular adductor (closer) muscle; M 23, mandibular abductor (opener) muscle; M 26, hyperpharyngeal retractor muscle; mn, motor nerve; Nmc, nerve to cusps and distal mandible region; NM 21, motor nerves to M 21; Pg, sensory cell bodies supplying the apodeme receptor strand; Rs, apodeme receptor strand; Sn, sensory nerve from VMRO; sog, suboesophageal ganglion; TM 1, main tentoro-mandibular muscle; TM 2a, one of two branches of tentoro-mandibular muscle TM 2; II, cranial nerve II; III, cranial nerve III.

four bursts per scond. With increasing opening of the mandible, the tonic firing increased, more units were recruited in the bursts and the burst rate increased. Superimposed on this was an occasional long burst of still larger units. This mode of bursting also increased as mandible opening increased. With full opening, the bursts of the large units fused into continual firing (O'Brien, 1984).

The nature of the bursts was uneven and could be stopped altogether by section of the motor nerve to the VMRO. Myogram recordings from both the VMRO muscle and the other tentorial muscles showed that the bursting of sensory discharges closely followed the bursting discharges of the muscles as they contracted cyclically. Therefore, O'Brien (1984) concluded that the VMRO sensory units are monitoring the muscle tension developed by the VMRO muscle, rather than monitoring passive stretch-induced tension or muscle length. That is, the VMRO is a tension receptor that responds to simultaneously developed contraction of the tentoro-mandibular skeletal muscles via contraction of its own receptor muscle. The receptor muscle seems to be driven in synchrony with the skeletal muscles, with which it is connected in parallel, and the rate of sensory firing upon opening is modulated by the contractile tension developed by the VMRO muscle. Sectioning of the motor nerves to either the VMRO or the tentoro-mandibular muscles eliminates the bursting of the sensory units, as well as most of the tonically firing units (O'Brien, 1984).

The response of the VMRO is made more complicated by the discovery that its firing is inhibited by sensory discharges from the VMRO of the contralateral mandible. Thus, opening of the contralateral mandible inhibits firing of the small tonic units and the larger bursting units of the ipsilateral mandible. But closing of the contralateral mandible excites firing of the large bursting units of the ipsilateral mandible (O'Brien, 1984).

Investigation of the function of the VMRO involved two kinds of experiments: ablation by release of the attachment point of the VMRO to the mandible, and loading one mandible with weights and pulling it open via a fixed-head and pulley system. Two consistent results were seen by O'Brien (1984). First, the ablation always caused a disruption in the weta's correct positioning of the operated mandible. In ablated preparations, the opening angle of the mandible was always reduced and regular mismatch of the mandibles occurred upon closing contact. Secondly, unilateral loading caused wider opening (due to the outward pull of the load) of the mandible during the opening phase of chewing, but the mandibles always met correctly at contact during the closing phase. Thus, although it was possible to disrupt mandibular position during chewing, clear load compensation was still accomplished in the intact animal. VMRO ablation caused this compensation to disappear.

O'Brien concluded that the VMRO serves both as a positional control sensor and as an error detector in a load-compensating servo control circuit, which regulates tension development in the mandibular and tentoro-mandibular skeletal muscles during chewing. Position control was not met during the whole chewing cycle, as shown by disruption of position during loading, but it was clearly in action during the later part of the closing phase, just before the mandibles made contact. Thus, the mandibles always engaged correctly if the left and right VMROs remained intact. The load-compensating system involves the crossed inhibitory feedback from both VMROs, in allowing accurate detection and overcoming of a tension mismatch during mandibular chewing. Such systems have been described for mandibular control systems in the locust (Seath, 1977). The sensory physiology of the weta VMRO differs from that of abdominal MRO responses reported in the hemipteran *Rhodnius* (Anwyl, 1972) and the moth *Antheraea* (Weevers, 1966), in that the weta VMRO does not respond consistently to muscle length change.

Dorsal muscle receptor organ (DMRO) anatomy

The DMRO is much simpler than the VMRO. It consists of a thin slip of a few muscle fibres, only one motor neuron and only one multipolar sensory neuron (Fig. 22.13B). The muscle is TM-2b, a tentoro-mandibular muscle that arises as a branch from TM-2a and extends from the tentorium to the dorsal wall of the mandible, running almost parallel with the cuticle of the mandible. The muscle is so thin that it could not develop sufficient tension to act on the mandible. However, unlike the tiny, disorganized VMRO fibres, the DMRO fibres had the ultrastructural appearance of normal fibres of the surrounding skeletal muscles. The muscle is innervated by the single motor neuron that branches from cranial nerve III to adjacent tentoro-mandibular muscles (O'Brien, 1984).

The sensory neuron of DMRO branches from the distal extent of cranial nerve II, and the cell body lies adjacent to the distal end of the muscle. It is not incorporated into the receptor organ, as seen in the VMRO. The dendrite of the DMRO branches profusely into the collagenous matrix of the distal insertion of the receptor muscle, where it attaches to the mandibular hypodermis (Fig. 22.13B, inset). Thus, the neuron is clearly in series mechanically with tension development by the muscle (O'Brien, 1984).

Physiology of the DMRO

In a few experiments only with this tiny receptor, O'Brien showed that it responds to opening of the mandible, as seen for the VMRO. The response is tonic, with increasing firing as the mandible opens further, and there was no bursting observed. In addition, phasic responses could be elicited by directly stretching the muscle by short displacements. Pressure on the cuticle near the insertion of the receptor muscle caused inhibition of the firing of the sensory neuron (O'Brien, 1984). The normal role of the DMRO in mandibular control is not known, but it is likely to serve a similar function to that of the VMRO, probably augmenting the feedback from the latter.

Apodeme strand receptor

A third stretch receptor occurs at the mandibular joint, and is termed the apodeme strand receptor. This is formed from a thin strand of cuticle, which connects the apodeme of the main adductor, M-21, to the main tentoro-mandibular muscle, TM-2, close to the tentorium (Rs in Fig. 22.13A). The apodeme strand receptor is only elastic at the site of tentorial attachment, and it is stretched by movement of the adductor apodeme. The receptor is innervated by a small cluster of about 24 sensory cell bodies saddling nerve III and projecting through two nerve branchlets on to the receptor strand. Unlike the MROs discussed above, the apodeme strand receptor is fully stretched in the closed position of the mandibles, and it relaxes as the mandibles open. Nothing is known about its physiological responses to the opening and closing of the mandibles.

Visual Physiology

Very little is known about the eyes of wetas or king crickets. External examination of the head capsule reveals two types of photoreceptors, characteristic of virtually all orthopterans: compound eyes and ocelli. The compound eyes are the main visual sense organs, composed of hundreds of individual photoreceptor units, known as ommatidia. In New Zealand wetas, the eyes are large, pear-shaped (*Hemideina*) or slightly oval (*Deinacrida*) and black, with no pseudopupil. In *Hemideina* compound eyes, there is a mediodorsal region that lacks pigment and is pale cream-coloured (Fig. 22.14A). In crickets, as well as a number of other insect orders, the ommatidia in this region are sensitive to polarized light (Nilsson *et al.*, 1987), and the same function is likely to be found in wetas. This region of the eye is also prominent in Australian and African anostostomatids, where, in some species, it is bright white. The two lateral ocelli appear as minute, circular, clear areas on either side of the epicranial suture, while the median ocellus occurs on a raised projection below the lateral ocelli and between the antennae (Maskell, 1927).

Physiology of compound eyes

Electroretinograms (ERGs) have been recorded from the compound eyes of the tree weta *H. femorata* and the giant weta *Deinacrida mahoenui*, using a chlorided silver electrode in saline contact with the corneal surface (L. Field and I. Stringer, unpublished data). This technique, though gross for determining details of the compound eye's physiology, allows assessment of spectral sensitivity and intensity response functions, as well as allowing information to be gained on the temporal resolution of the eyes' response to light onset. Eyes were dark-adapted for 5 min before recordings were made.

ERG recordings from *Hemideina* and *Deinacrida* showed similarities but also marked differences. In both genera, a rapid initial deflection marked the response to a 200 lux white light onset ('1' in Fig. 22.14B, C). This deflection was always negative (downward in the records), and presumably represents the discharge of optic ganglion neurons, as described in classical studies (Autrum, 1950). A slow potential ('2' in Fig. 22.14B, C) followed the 'on' deflection and lasted for the duration of the stimulus. This plateau potential was of opposite polarity in the two genera: positive in *H. femorata* and negative in *D. mahoenui.* Slow 'on' potentials are usually representative of the discharge of the primary light receptor neurons in the compound eye, although their shape, polarity and time course differ with stimulus intensity, wavelength and the depth of recording in the eye or underlying optic ganglia (Mazokhin-Porshnyakov, 1969). In addition, differences appeared in the 'off' responses of both genera. While the 'off' responses moved in the positive direction, that of *H. femorata* showed an initial positive deflection away from the direction of the baseline, while that of *D. mahoenui* simply drifted back toward the baseline. These 'off' responses also represent primary visual receptor responses, but both differ in wave-form shape and reversal of polarity from the usual ERG recorded from diurnal orthopterans (e.g. locust) and from that of the cricket *Gryllus campestris.*

The relationship between magnitude of the ERG and stimulus light intensity differs between the two genera (Fig. 22.14E). Usually, in insect eyes, the ERG amplitude increases proportionally to the logarithm of the stimulus light intensity, over some narrow range of intensity (Mazokhin-Porshnyakov, 1969). In Fig. 22.14E, this is seen over the major portions of each curve, but it is not seen at low light intensities (inset, Fig. 22.14E). The outstanding difference between the two genera is the slopes of their intensity–response curves. *Deinacrida mahoenui* is more sensitive to light than *H. femorata*, inasmuch as it requires less light to reach an equal amplitude of response, by a factor of approximately five. Part of the reason for the larger response of *D. mahoenui* at any given intensity may be due to the larger eye of this species. However, the greater slope of its intensity–response function, even at very low light intensities (inset, Fig. 22.14E), suggests that it has greater sensitivity over the whole range tested. Both species are known to be nocturnal.

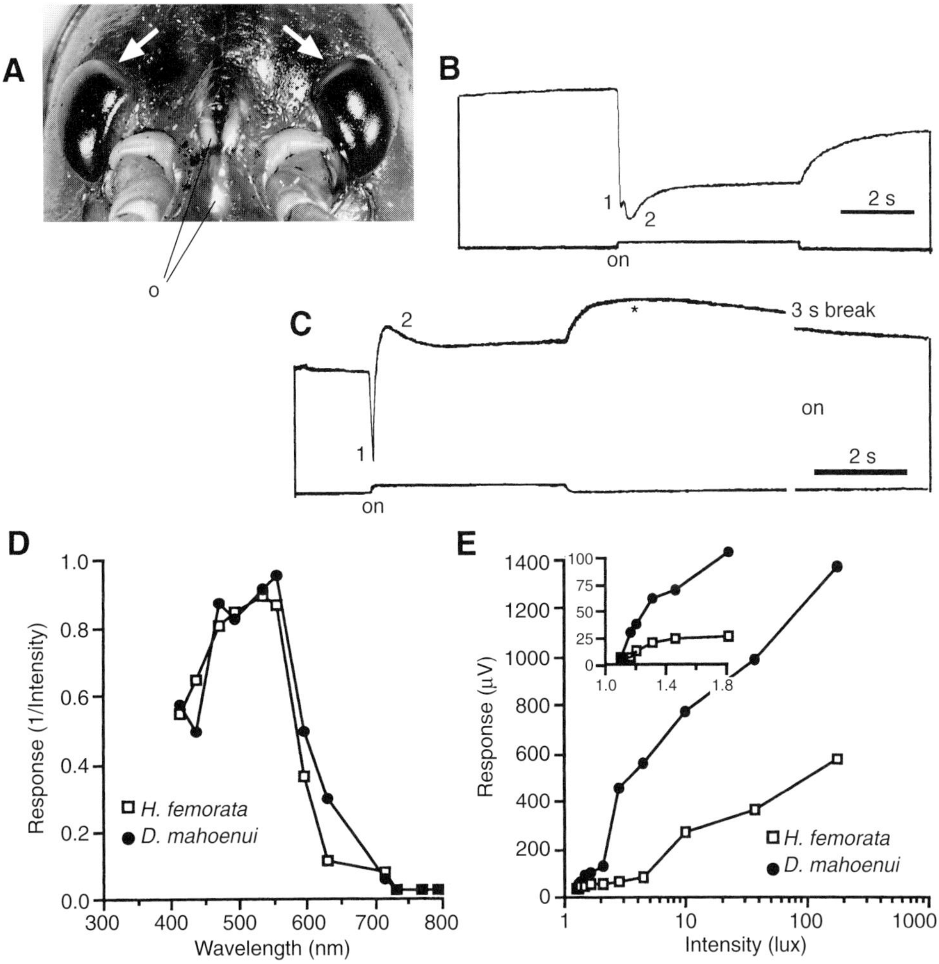

Fig. 22.14. Visual physiology of compound eyes of tree and giant wetas. A. Photograph of compound eyes of tree weta *Hemideina femorata*, showing white dorsal region (arrow), as well as ocelli (o) between antennae. B. Electroretinogram (ERG) from eye of the giant weta *Deinacrida mahoenui*, showing 'on' response consisting of initial rapid negative deflection (1) and following slow negative wave (2), which adapts to a negative plateau. The recording returns to baseline on 'light off'. C. ERG from eye of *H. femorata*, showing initial rapid negative deflection (1), and slow wave which is positive (2), and adapts on a positive plateau. The 'off' response is a second positive deflection (*), in contrast to lack of a deflection in the *Deinacrida* ERG. D. Examples of spectral response curves of both species studied above, plotting equal energy intensities required to achieve a criterion ERG response against wavelength of light. Both genera have peak sensitivity in the green and lack of sensitivity in the red band. E. Examples of intensity–response curves, using white light, for both species studied above. Linear portions indicate a logarithmic response function at higher intensities (note semi-log plot), although at lower intensities the response is non-linear. The response of *D. mahoenui* is considerably greater, even at low intensities (inset), presumably indicating a greater sensitivity to light.

Colour sensitivity of the tree weta and giant weta is shown in Fig. 22.14D, which presents equal energy curves required to achieve a supraliminal response. Both genera have a peak sensitivity to wavelengths around 520–530 nm, in the green region of the visible spectrum. Sensitivity drops to about half in the red and violet, and is nil at wavelengths greater than 620 nm in the deep red. This pattern reveals nothing out of the ordinary, compared with colour sensitivity in other opthopterans. It confirms previous observations made in other orthopterans that the insects are virtually blind at the red end of the spectrum and have decreased sensitivity in the blue (Mazokhin-Porshnyakov, 1969). This method of estimating colour sensitivity tells nothing of the spectral sensitivity of various ommatidial receptors, or of the number of colour-sensitive visual pigments present in weta eyes.

Practically, the above results indicate that behavioural experiments conducted with wetas at night should incorporate red-filtered light as close to the red (> 600 nm) end of the spectrum as possible, rather than using 'red' cellophane tissue, which transmits much light in the orange and yellow as well as the red band.

Summary

For the comparative physiologist, the Anostostomatidae, and indeed the other stenopelmatoid families, provide a rich source of physiological mechanisms, which have only been glimpsed through the initial investigations described above. The unusual findings accumulated to date suggest that many variations on the main orthopteran themes are to be uncovered, including some that herald modern adaptations to local evolutionary pressures, while others may well be primitively retained features, which can give insight into the early phylogenetic history of these interesting families, as well as the Stenopelmatoidea as a whole.

References

Ander, K. (1939) Vergleichend-anatomische und phylogenetische Studien über die Ensifera. *Opuscula Entomologica* Supplement 2, 1–306.

Anwyl, R, (1972) The structure and properties of an abdominal stretch receptor in *Rhodnius prolixus*. *Journal of Insect Physiology* 18, 2143–2154.

Autrum, H. (1950) Die Belichtungspotentiale und das Sehen der Insecten (Untersuchungen an *Calliphora* und *Dixippus*). *Zeitschrift für Vergleichende Physiologie* 32, 176.

Autrum, H. and Schneider, W. (1948) Vergleichende Untersuchengen über den Erschütterungssinn bei locustiden. *Zeitschrift für Vergleichende Physiologie* 28, 580–637.

Ball, E.E. (1981) Structure of the auditory system of the weta *Hemideina crassidens* (Blanchard 1851) (Orthoptera, Ensifera, Gryllacridoidea, Stenopelmatidae). 2. Ultrastructure of the auditory sensilla. *Cell and Tissue Research* 217, 345–359.

Ball, E.E. and Field, L.H. (1981) Structure of the auditory system of the weta *Hemideina crassidens* (Blanchard 1851) (Orthoptera, Ensifera, Gryllacridoidea, Stenopelmatidae). 1. Morphology and histology. *Cell and Tissue Research* 217, 321–343.

Burrows, M., Laurent, G.J. and Field, L.H. (1988) Proprioceptive inputs to non-spiking local interneurones contribute to local reflexes in a locust hindleg. *Journal of Neuroscience* 8, 3085–3093.

Cokl, A. (1983) Functional properties of vibroreceptors in the legs of *Nezara viridula* (L.) (Heteroptera, Pentatomidae). *Journal of Comparative Physiology* 150, 261–269.

Edwards, J.S. and Palka, J. (1974) Insect neural evolution – a fugue or an opera? *Seminars in the Neurosciences* 3, 391–398.

Eibl, E. and Huber, F. (1979) Central projections of tibial sensory fibers within the three thoracic ganglia of crickets (*Gryllus campestris* L., *Gryllus bimaculatus* DeGreer). *Zoomorphologie* 92, 1–17.

Embleton, T.G.W., Piercy, J.E. and Olson, N. (1976) Outdoor sound propagation over ground of infinite impedance. *Journal of the Acoustical Society of America* 59, 267–277.

Field, L.H. and Bailey, W.J. (1997) Sound production in primitive Orthoptera from Western Australia: sounds used in defence and social communication in *Ametrus* sp. and *Hadrogryllacris* sp. (Gryllacrididae: Orthoptera). *Journal of Natural History* 31, 1127–1141.

Field, L.H. and Matheson, T. (1998) Chordotonal organs of insects. *Advances in Insect Physiology* 27, 1–228.

Field, L.H. and Pflüger, H.-J. (1989) The femoral chordotonal organ: a bifunctional orthopteran (*Locusta migratoria*) sense organ. *Comparative Biochemistry and Physiology* 93A, 729–743.

Field, L.H. and Rind, F.C. (1976) The function of the femoral chordotonal organ in the weta. *New Zealand Medical Journal* 86, 451.

Field, L.H. and Rind, F.C. (1981) A single insect chordotonal organ mediates inter- and intra-segmental leg reflexes. *Comparative Biochemistry and Physiology* 68A, 99–102.

Field, L.H., Hill, K.G. and Ball, E. (1980) Physiological and biophysical properties of the auditory system of the New Zealand weta *Hemideina crassidens* (Blanchard 1851) (Ensifera: Stenopelmatidae). *Journal of Comparative Physiology* 141, 31–37.

Finlayson, L.H. (1976) Abdominal and thoracic receptors in insects, centipedes and scorpions. In: Mill, P.J. (ed.) *Structure and Function of Proprioceptors in the Invertebrates*. Chapman and Hall, London, pp. 153–211.

Fornusek, C. (1993) An investigation into the subgenual organ of *Hemideina femorata*. BSc Honours thesis, University of Canterbury, Christchurch, New Zealand.

Hill, K.G. (1983) The physiology of locust auditory receptors. II. Membrane potentials associated with the responses of the receptor cell. *Journal of Comparative Physiology* 152, 483–493.

Honomichl, K. (1978) Feinstruktur eines zweier Propriozeptoren im Kopf von *Oryzaephilus surinamensis* (L.) (Insecta, Coleoptera). *Zoomorphologie* 90, 213–226.

Jarman, T.H. (1982) Mating behaviour and its releasers in *Hemideina crassidens* (Orthoptera: Stenopelmatidae). Unpublished BSc (Hons) thesis, University of Canterbury, New Zealand.

Kalmring, K., Lewis, B. and Eichendorf, A. (1978) The physiological characteristics of the primary sensory neurone of the complex tibial organ of *Decticus verrucivorus* L. (Orthoptera, Tettigoniidae). *Journal of Comparative Physiology* 127, 109–121.

McVean, A. and Field, L.H. (1996) Communication by substrate vibration in the New Zealand tree weta, *Hemideina femorata* (Stenopelmatidae: Orthoptera). *Journal of Zoology, London* 239, 101–122.

Maskell, F.G. (1927) The anatomy of *Hemideina thoracica*. *Transactions of the Royal Society of New Zealand* 57, 637–669.

Matheson, T. and Field, L.H. (1995) An elaborate tension receptor system highlights sensory complexity in the hind leg of the locust. *Journal of Experimental Biology* 198, 1673–1689.

Mazhokhin-Porshnyakov, G.A. (1969) *Insect Vision*. Plenum Press, New York, 306 pp.

Michelsen, A. (1978) Sound reception in different environments. In: Ali, M.A. (ed.) *Sensory Ecology*. Plenum Press, New York, pp. 345–373.

Michelsen, A. and Larsen, O.N. (1983) Strategies for acoustic communication in complex environments. In: Huber, F. and Markl, H. (eds) *Neuroethology and Behavioral Physiology*. Springer Verlag, Berlin. 412 pp.

Morton, E.S. (1977) Ecological sources of selection on avian sounds. *American Naturalist* 109, 17–34.

Nilsson, D., Labhart, T. and Meyer, E. (1987) Photoreceptor design and optical properties affecting polarization sensitivity in ants and crickets. *Journal of Comparative Physiology A* 161, 645–658.

Nishino, H. and Sakai, M. (1997) Three neural groups in the femoral chordotonal organ of the cricket *Gryllus bimaculatus*: central projections and soma arrangement and displacement during joint flexion. *Journal of Experimental Biology* 200, 2583–2595.

Nishino, H., Sakai, M. and Field, L.H. (1999) Two antagonistic functions of neural groups of the femoral chordotonal organ underlie thanatosis in the cricket *Gryllus bimaculatus* DeGeer. *Journal of Comparative Physiology A* 185, 143–155.

Nowel, M.S., Shelton, P.M.J. and Stephen, R.O. (1995) Functional organisation of the metathoracic femoral chordotonal organ in the cricket *Acheta domesticus*. *Journal of Experimental Biology* 198, 1977–1988.

O'Brien, B. (1984) Mandibular movements and their control in the weta *Hemideina maori* (Orthoptera: Ensifera: Stenopelmatidae). PhD thesis, University of Canterbury, Christchurch, New Zealand.

Rind, F.C. (1976) The metathoracic chordotonal organ and its involvement in the posture and locomotion in the weta (*Hemideina thoracica*). BSc Honours thesis, University of Canterbury, Christchurch, New Zealand.

Ritzmann, R.E. (1984) The cockroach escape response. In: Eaton, R.C. (ed.) *Neural Mechanisms of Startle Behaviour*. Plenum Press, New York, pp. 93–131.

Rössler, W. (1992) Functional morphology and development of tibial organs in legs I, II, and III in the bush cricket *Ephippiger ephippiger* (Insecta: Ensifera). *Zoomorphology* 112, 181–188.

Schnorbus, H. (1971) Die Subgenualen Sinnesorgane von *Periplaneta americana:* Histologie und Vibrationsschwellen. *Zeitschrift für Vergleichende Physiologie* 71, 14–48.

Schwabe, J. (1906) Beiträge zur Morphologie und Histologie der tympanalen Sinnesapparate der Opthopteren. *Zoologica (Stuttgart)* 50, 1–154.

Seath, I. (1977) The effects of increasing mandibular load on electrical activity in the mandibular closer muscles during feeding in the desert locust, *Schistocerca gregaria*. *Physiological Entomology* 2, 237–240.

Shaw, S. (1994) Detection of airborne sound by a cockroach 'vibration detector': a possible missing link in insect auditory evolution. *Journal of Experimental Biology* 193, 13–47.

Slifer, E.H. (1960) A rapid and sensitive method for identifying permeable areas in the body wall of insects. *Entomological News* 71, 179–182.

Slifer, E.H., Prestage, J.J. and Beams, H.W. (1959) The chemoreceptors and other sense organs on the antennal flagellum of the grasshopper (Orthoptera, Acrididae). *Journal of Morphology* 105, 145–191.

Usherwood, P.N.R., Runion, H.I. and Campbell, J.I. (1968) Structure and physiology of a chordotonal organ in the locust leg. *Journal of Experimental Biology* 48, 305–323.

Weevers, R. deG. (1966) The physiology of a lepidopteran muscle receptor. I. The sensory response to stretch. *Journal of Experimental Biology* 44, 177–194.

Zill, S.N., Moran, D.T. and Varela, F.G. (1981) The exoskeleton and insect proprioception. II. Reflex effects of tibial campaniform sensilla in the American cockroach *Periplaneta americana*. *Journal of Experimental Biology* 94, 43–55.

23
Neuromuscular Physiology and Motor Control

Laurence H. Field
Department of Zoology, University of Canterbury, Private Bag 4800, Christchurch, New Zealand

Introduction

In introducing the chapter on sensory physiology, it was noted that the unique island geography and evolution of New Zealand have been the justification for studies of many biological aspects of wetas. The expectation has been that isolation has either allowed retention of primitively derived characters or has promoted development of interesting features specific to the isolated fauna. This conjecture is supported by the finding of numerous primitive anatomical characters in wetas, no less so in the neuromuscular system than in other organ systems. For example, the coxal leg musculature is more highly divided and dispersed than that of modern orthopterans (L.H. Field, personal observation), the heart has 11 chambers, which project well into the prothorax (Maskell, 1927), and the mandibular musculature has retained primitive features of the tentorial apparatus (O'Brien, 1984). In addition to anatomical features, physiological features could be expected to show the same trends related to the island situation. Thus, although the leg muscles are driven by the typical orthopteran array of motor neurons, the discovery that these muscles (especially the tibial extensor) can be locked into a catch contraction (i.e. frozen) state, whereby no further neuronal excitation is required to maintain prolonged muscle tension during the unusual defence display, is unprecedented in any other insect species. Furthermore, contrary to the results found in more modern insects, such as locusts, tree wetas do not appear to have developed as great a capability to retain learned limb positions in conditioned-learning experiments. This may represent either a more primitive state in the weta or a specific difference in central versus peripheral control of the learned limb position between the two kinds of insects (Hoyle and Field, 1983b).

Tree wetas have attracted physiological study for another reason: they serve as models for studying walking motor control. They walk evenly and relatively slowly, supported equally by all six legs, in contrast to the locust or to the rapid walking of cockroaches (two other insect models). Furthermore, the hind legs are not modified for jumping in the tree wetas and giant wetas, even though these legs have the typically large and elongated orthopteran morphology. Thus, much of this chapter will deal with neuromuscular studies of walking, including the contraction properties and motor innervation of leg muscles, and the remarkable role of the neuromodulator, octopamine, on establishing the catch contraction condition in leg muscles. This will be explored in the context of thanatosis-like behaviour during defence displays of tree wetas and during their diurnal rest period in galleries. Also, the reflex physiology of leg motor neurons and the role of stretch receptors in the coordination of walking will be reviewed and compared with similar studies in model insects, such as the locust.

The unusual secondary sexual characters of male tree wetas, greatly enlarged mandibles and head capsule (megacephaly), have also attracted the attention of physiologists. These enlarged structures present an ideal opportunity to study the engineering and mechanical problems facing the insect brain's control of the two asymmetrical mandibles, which must be brought to bear correctly to allow the cusps to intermesh as biting force is rapidly increased. This has been little studied in other insects, presumably due to the usual small size of mandibles. This chapter will review the motor and sensory feedback control involved in mandibular movements.

Motor Control of Leg Muscles

Motor neurons to tibial muscles and muscle innervation patterns

Approximately 70 motor neurons innervate the 17 muscles of the insect leg (see musculature, in O'Brien and Field, Chapter 8, this volume). The coxal and trochanteral muscles are important for support of the body during walking, while the tibial flexor and extensor muscles in the femur play a major role in either pushing (joint extension) or pulling (joint flexion) the body during stepping. The small muscles that raise (levate) and lower (depress) the tarsus assist in providing pressure on to the substrate, as well as lifting the tarsus to prevent the claws catching on the forward swing stroke of stepping. The weta does not differ substantially in this general scheme, although there are differences within muscles, attributed to a more primitive status (discussed in O'Brien and Field, Chapter 8, this volume).

As in other arthropods, insect muscle fibres are innervated by different excitatory motor neurons, the axons of which cause either a barely discernible or a large twitch to a single motor impulse. These are known as slow and fast motor neurons, respectively. Intermediate motor neurons, with intermediate twitch effects, are described for some muscles. Other motor neurons, known as inhibitors, can cause relaxation of ongoing muscle contraction induced by excitors.

The motor neurons controlling the major leg muscles in *Hemideina femorata* have been stained to reveal positions and numbers in the metathoracic ganglion, by cutting nerves to each muscle, perfusing with 6% cobalt chloride ($CoCl_2$) and silver intensifying with the Timms method (Currie, 1988). Other muscles have been studied physiologically to determine their innervation patterns. The general pattern of motor neuron cell-body distribution and their axon pathways in the thoracic ganglia are remarkably similar to those of the more modern locust, thus indicating an early orthopteran ganglionic ground-plan. However, interesting variation occurs, which may be related to the leg function in the different orthopteran groups. The innervation patterns of the tree weta and the locust are shown in Table 23.1.

The trochanteral muscles levate and depress the trochanter. Each muscle contains at least two excitors, as seen in recordings during walking. More may be present, based upon the large numbers seen in the locust, but have not been documented. Inhibitory and dorsal unpaired midline (DUM) motor neurons have not been studied in these or in any of the other muscles linking the coxa to the body.

The extensor tibiae muscle is innervated by one fast excitatory (FETi) and one slow excitatory (SETi) motor neuron and one common inhibitor

Table 23.1. Comparison of motor neuron innervation patterns between the tree weta and the locust. Data for tree weta from Field and Rind (1981), Hoyle and Field (1983b) and Currie (1988). Data for locust from Burrows (1996).

	Tree weta: numbers of motor neurons			Locust: numbers of motor neurons		
Muscle name	Excitors	Inhibitors	DUM	Excitors	Inhibitors	DUM
Levator trochanteris	2	?	?	7	1	2
Depressor trochanteris	2	?	?	6	1	?
Extensor tibialis	2	1	4	2	1	1
Flexor tibialis	4–7(?)	1	2	~9	2	1
Retractor unguis	3	1(?)	1	2	2	1(?)
Levator tarsi	2	?	?	1	2	?
Depressor tarsi	> 2	?	?	~3	2	?

(CI), which has branches common to many of the leg muscles (Fig. 23.1A). The positions of both neurons are similar to that in the locust, except that SETi in the tree weta is near the midline, while that in the locust is lateral to the base of the anterior connective (Fig. 22.1A, D). While the FETi axon emerges from nerve 5 and that of SETi emerges from nerve 3 in the metathoracic ganglion (Fig. 22.1E), the reverse occurs in the meso- and prothoracic ganglia, just as in the locust nervous system (Wilson and Hoyle, 1978). This is further confirmation of an early common ancestor in which the switch occurred. In addition, Currie (1988) found four dorsal unpaired median neurosecretory motor neurons (DUMETi), which send axons to the extensor tibiae through thoracic nerves 3 and 5 (discussed under 'DUM Motor Neurons', below). This is in marked contrast to a single DUMETi reported for the locust (Hoyle, 1978), but is similar to the number reported for the tettigoniid (bush cricket) *Decticus albifrons* (Theophilidis, 1983).

The flexor tibiae muscle has the largest number of motor neurons, although the exact number for the weta is not well established. In the locust, nine excitors and three inhibitors have been identified (Phillips, 1981; Hale and Burrows, 1985). Cobalt back-fills of the flexor motor nerve (Fig. 23.1B) showed at least ten motor neuron cell bodies clustered on each side in the anterior region of the metathoracic ganglion, plus the bilateral pair of common inhibitors in the medial midline region (Currie, 1988). However, cobalt peripheral fills into the nerve supplying the flexor muscle only revealed four to five axons (Fig. 23.1F). Currie demonstrated six motor neurons from physiological record stimulation (Fig. 23.1G). An additional two dorsal unpaired median flexor tibiae (DUMFlTi) neurosecretory motor neurons were described by Currie (1988) for *H. femorata*. In the locust, nine excitor motor neurons innervate the flexor muscle.

The retractor unguis muscle is also found in the femur, although its function is to depress the claw (unguitractor plate) by means of a very long apodeme, which extends from the femur, through the tibia and tarsus to the claw. Its nerve contains three motor axons, as revealed by cobalt staining (Fig. 23.1C), plus a faint DUM neuron (not shown), which was confirmed by physiological means (Currie, 1988). It is not known to contain a branch from the CI, although it is probably based upon the pattern seen in other orthopterans.

The tarsal muscles each receive two excitors (the depressor probably receives more), based upon physiological recordings (Field and Rind, 1981). The remaining innervation awaits investigation.

Muscle contraction physiology

The relative force development of muscle activation by slow or fast excitors differs markedly. For a given rate of motor neuron impulse firing, slow units cause the muscle to develop force gradually, with a smooth slope and tiny twitch amplitude, whereas fast units cause large twitches and therefore allow the insect to develop maximum force rapidly in the muscle, with a few closely spaced motor impulses. Wetas, like other insects, utilize this difference to control the legs with either smooth, graded movements or rapid movements, as in jumping. In the tree weta *H. femorata*, the contraction properties of the tibial extensor and flexor and the retractor unguis muscles have been studied.

Stimulated tension development

The extensor tibiae muscle in the hind leg differs from that of the locust in being smaller and similar in mass to the flexor tibiae muscle. This is undoubtedly related to the difference in function of the hind leg in the two insects. The locust has a massive and powerful extensor muscle, used for jumping (also found in the henicine wetas, crickets and grasshoppers), while the metathoracic flexor muscle is not enlarged and contacts with only 5% of the force developed by the extensor. In the deinacridine tree and giant wetas, the extensor and flexor tibiae muscles are almost equal in mass and power and the hind leg is primarily used for climbing or walking. These insects are weak jumpers as adults.

However, the weta and locust have a remarkable similarity in the pattern of distribution of excitor and inhibitory axons within the extensor muscle, as demonstrated by intracellular recording from individual muscle fibres (Hoyle and Field, 1983a). Thus, the proximal fibres are innervated by both excitors and the inhibitor, with some fibres receiving SETi and CI only. The middle region has fibres primarily innervated by FETi. Most distal fibres receive all three axons again, but the extreme distal fibres receive SETi and CI only.

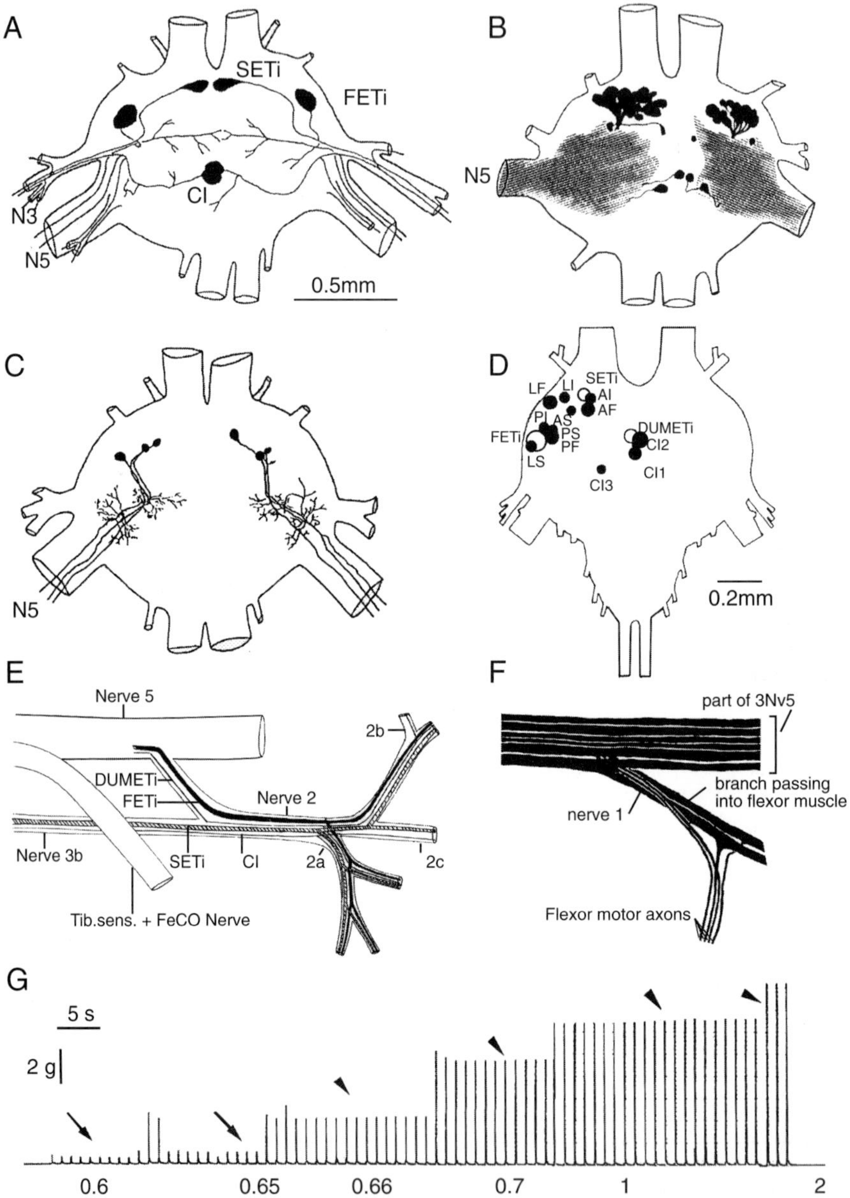

Fig. 23.1. Motor neuron morphology and innervation of tibial leg muscles in the tree weta, *H. femorata*. A. Somata and branching patterns of the slow (SETi) and fast (FETi) extensor tibiae motor neurons in the metathoracic ganglion, as revealed by axonal perfusion by hexamine cobalt chloride. CI, common inhibitors. B. Flexor tibiae motor neurons revealed by perfusion of flexor tibiae nerve. Shaded areas indicate regions of dense arborizations. C. Motor neurons of the retractor unguis muscle. D. Diagram of tibial extensor and flexor motor neuron somata positions in the locust metathoracic ganglion, for comparison with those of the weta. E. Origins of extensor tibiae motor neurons as they branch into the extensor muscle from nerves 3 and 5 (right). FETi and at least one DUM neuron travel in N5, while SETi and CI are in N3. F. Cobalt chloride filled axons of four flexor tibiae motor neurons branching from N5 into the flexor motor nerve. G. Progressive increase in stimulus voltage to the flexor tibiae nerve recruits six motor neurons. (A, B, C, G from Currie, 1988; E from de Ruiter, 1984; F from Smith, 1979. All figures with permission from the authors.)

Also, the excitatory electrical muscle response (known as the excitatory junction potential (EJP) to SETi stimulation has two forms. In the proximal and most distal fibres the EJP is small and summated and grows with repeated stimulation (facilitation). But in the extreme distal fibres the EJP is large and does not show facilitation (Fig. 23.2Ai-iii). The tension development for SETi stimulation is gradual and reaches a plateau with time (Fig. 23.2B). Individual twitches are very small compared with the very large single amplitude of a single twitch contraction from FETi stimulation (Fig 23.2B). The CI relaxes the contraction tension built up by SETi if stimulated simultaneously, thus allowing the weta to produce finely graded control of tibial extension (Fig. 23.2C).

Contraction profiles of the flexor tibiae muscle show mechanical summation of large twitches, which increase in size as more excitors are recruited by increasing stimulus voltage during nerve stimulation experiments (Fig. 23.2D). The

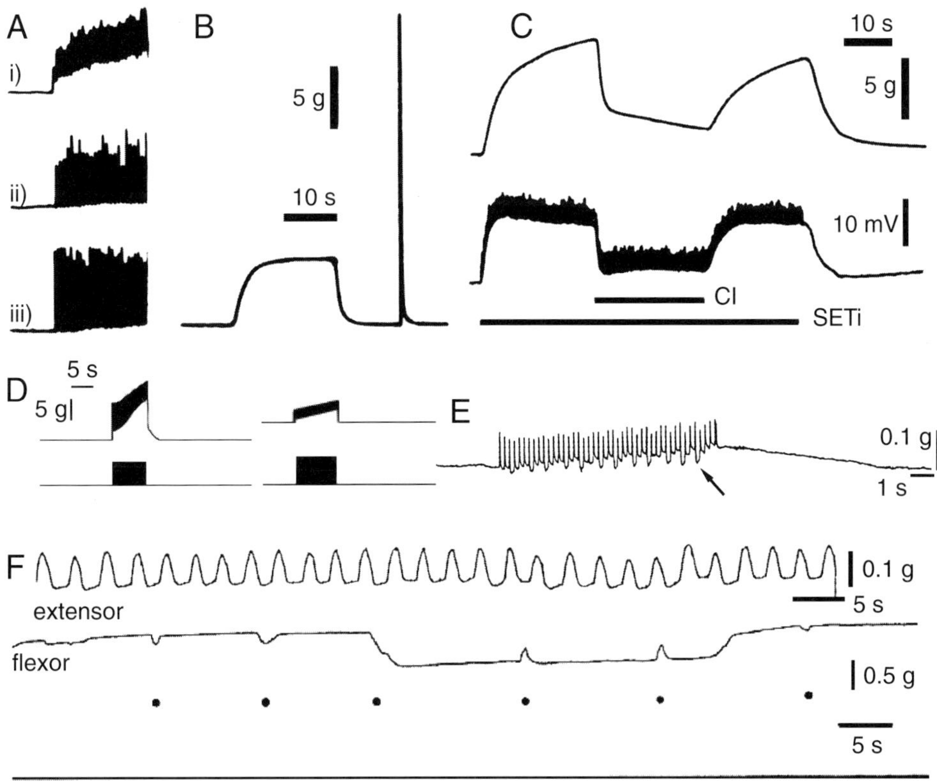

Fig. 23.2. Metathoracic tibial muscle responses to motor neuron stimulation. A. SETi stimulation generates different excitatory junction potentials (EJP) in different parts of the extensor tibiae. i. Proximal fibres show facilitation and summation. ii. Distal fibres show slight facilitation but little summation. iii. Extremely distal fibres show neither facilitation nor summation. B. Stimulation of SETi (20 Hz) produces smooth, gradual contraction with very tiny twitches, compared with very large single-twitch contraction (right) resulting from one stimulus pulse to FETi. C. Stimulation of the common inhibitor (bar, CI) produces relaxation of tension development resulting from stimulation of SETi (bar, SETi). Upper, tension in extensor tibiae muscle; middle, electrical muscle EJP recording. D. Responses of the flexor tibiae to stimulation (10 Hz) of flexor motor nerve at two stimulus voltages. E. Twitch contractions of retractor unguis muscle in response to stimulation of its nerve. Occasional recruitment of the common inhibitor (arrow) caused loss of tension. F. Tension recording of intrinsic rhythm (IR) in metathoracic extensor tibiae (upper) and (dots) in flexor tibiae (lower) muscles. Flexor also shows spontaneous changes in basic tonus (BT) levels, which reflect changes in IR. (A–C after Hoyle and Field, 1983b; D–F from Currie, 1988. All figures with permission.)

twitches of the retractor unguis summate in a similar way, but they are considerably smaller (compare tension scales in Fig. 23.2D and E). At high frequencies, the muscle undergoes tetanic contraction and shows little relaxation between each twitch (Currie, 1988).

Tonic tension phenomena: basic tonus and intrinsic rhythm

The maintained resting tension of muscle in the absence of neural excitation is known as basic tonus (BT). In insects, it is apparently utilized for maintenance of posture in antigravity muscles, and it is known to be under neurohumoral control (Hoyle, 1983). Recordings of resting tension from the extensor and flexor tibiae and retractor unguis muscles in *H. femorata* often showed spontaneous fluctuations in BT, together with the curious phenomenon known as intrinsic rhythm (IR). The latter is a slow, periodic increase and decrease in muscle tension, varying from 20 to 60 mg at rates of 3–30 min^{-1} in the case of the extensor tibiae (Currie, 1988). The most prominent IR occurs in the extensor tibiae muscle (Fig. 23.2F), but it also occurs with slower frequency (2–7 min^{-1}) in the other muscles investigated. The peak tension can reach 360 mg in the retractor unguis, which is about 11% of the maximum tetanic tension developed by that muscle. In the flexor tibiae, IR reaches 250 mg (about 1.9% of maximum tetanic tension), while in the extensor tibiae peak IR tension is about 1.7% of tetanic tension developed by SETi stimulation. In the locust, IR is reported from only a few proximal muscle fibres of the extensor tibiae, whereas it occurred in all three muscles tested in the tree weta.

The function of BT may well be for small adjustments of posture or for setting static limb positions, as demonstrated for the locust in associative conditioning experiments (Hoyle, 1983). The function of IR is obscure, but it may serve as an automatic exercise mechanism, in so far as it develops sufficient tension to extend the tibia to 120° in the locust hind leg (Hoyle, 1983). The mechanism of IR control is clearly influenced by that of BT, for an increase in BT causes a decrease in the IR peak of tension to the point at which large BT levels swamp any evidence of IR. When absent, IR can be initiated by SETi stimulation and, when present, a single SETi impulse can immediately terminate the IR contraction. Both BT and IR are affected by octopamine in the weta, as reviewed below.

Reflex control by the femoral chordotonal organ

Resistance reflex control of leg muscles

Almost all leg muscles are under proprioceptive feedback control, whereby they develop tension to resist passive movement of a joint. This resistance reflex circuit is activated by stretch receptors in the joints (proprioceptors) and it serves to allow the insect to stabilize posture against the effects of gravity or other passive perturbation forces in the environment.

From previous resistance-reflex research, mostly on locust leg muscles, it was thought that at least one chordotonal organ (defined in Field, Chapter 22, this volume) in each joint mediates resistance reflexes only to the muscles of that joint. However, it is now known that a major leg proprioceptor, the femoral chordotonal organ (FeCO), has resistance reflex control of other leg muscles, in addition to those of its own joint (the femur–tibia joint). This phenomenon was first discovered in the tree weta (*H. femorata*) and the principle probably applies to most other insects as well (Field and Rind, 1981). It was demonstrated by recording reflex activation of various muscles through implanted myogram electrodes during passive movement of the tibia, before and after ablation of the FeCO. In the classical resistance reflex, within the femur–tibia joint, passive extension of the tibia causes firing of the flexor tibiae muscle, while passive tibial flexion causes firing of the extensor tibiae (Fig. 23.3A). Thus, each muscle opposes a movement that would displace the joint. In addition, however, tibial extension also causes firing of the levator trochanteris muscle of the hind leg. This stiffens the base of the leg against opening of the coxa–trochanter joint in the event that the body is passively pushed forward (which would activate the FeCO by tibial extension) and is therefore a resistance reflex (Fig. 23.3B, lower myogram trace). Non-homonymous resistance reflexes have not been shown in other insects, but the same experiment could undoubtedly demonstrate them.

Locomotory assistance reflexes

The FeCO also drives other reflexes in the leg muscles that are not appropriate as resistance

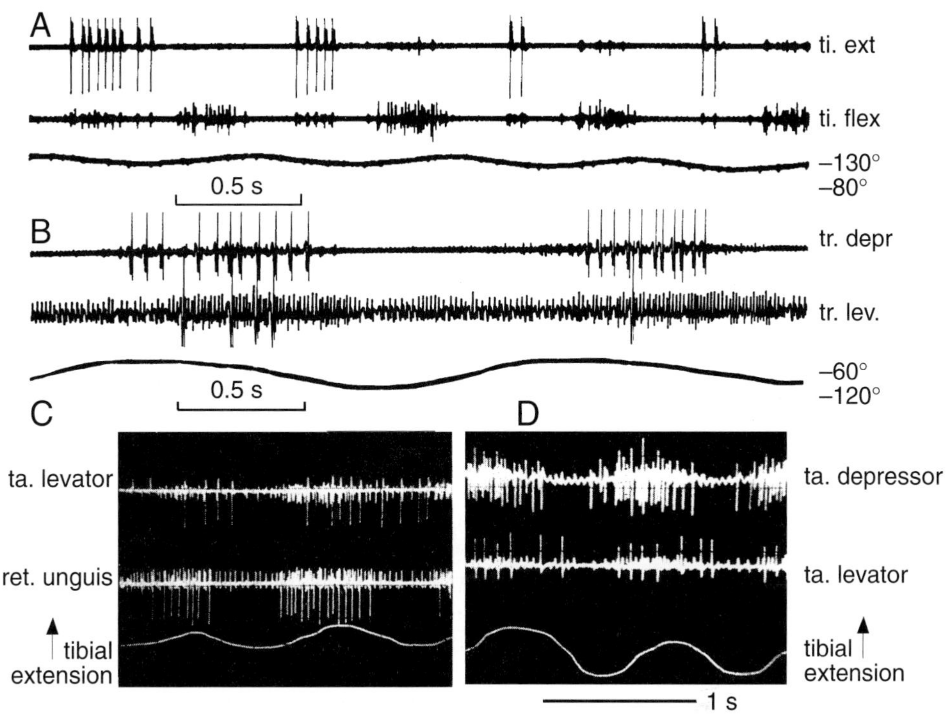

Fig. 23.3. Reflex responses in metathoracic leg muscles of different joints to passive imposed movement of only the femur–tibia joint. A. Resistance reflexes in the extensor (upper trace) and flexor (middle trace) muscles during tibial movement between 80° and 130° (lower trace). Some cross-talk occurs. B. Similar movement of the tibia (lower trace) causes firing in depressor and levator trochanter muscles, giving rise to non-homonymous reflexes. C and D. Tarsal levator and depressor, and retractor unguis, muscles are also activated by tibial movement to produce locomotory reflexes. (From Field and Rind, 1981, with permission.)

reflexes, in that they could augment locomotory movements of the hind leg during the extension (retraction, stance) phase of stepping, rather than resisting joint movements. These reflexes were demonstrated experimentally by passive tibial extension, and they are thought to serve as assistance reflexes in the active control of walking (Field and Rind, 1981). For example, tibial extension causes reflex firing of the trochanteral depressor (Fig. 23.3B, upper myogram trace). Such positive activation could enhance the ongoing depressor excitation that is normally generated by the walking motor programme (from the central nervous system (CNS)) as the hind leg extends to thrust the body forward. In the same fashion, the tarsal depressor is reflexly fired during tibial extension (Fig. 23.3D). The reflex could augment the downward force exerted by the tarsus on to the substrate during the stance phase of walking, thus ensuring a good grip. These potential assistance reflexes have only been demonstrated in fixed preparations, in which the tibia alone is moved, rather than during walking. Such assistance reflexes have not been demonstrated in other insects, but they are likely to exist as general mechanisms for the motor control of walking.

Two other reflexes act on remote joints, but they serve a different locomotory function. As the tibia approaches and leaves the extreme extended position, the tarsal levator is activated by the FeCO. The same phase is observed for reflex firing of the retractor unguis muscle (Fig. 23.3C). Both reflexes serve to augment the lifting of the hindleg claw and of the tarsus, as the leg nears the end of the stance phase and prepares to lift off from the substrate for the swing phase (protraction). They therefore assist to prevent the claws from catching on the substrate during protraction.

Motor Control of Walking

Because the tree weta walks relatively slowly and regularly, it serves as an ideal model for studying the motor control of walking. It is not difficult to attach fine wire leads to pairs of leg muscles and to record ongoing firing patterns while tethered wetas are walking. Interest has focused on the role of the FeCO in walking control, because it is easy either to tenotomize or to ablate this large chordotonal organ and because earlier studies implicated it as an important source of proprioceptive feedback for the walking motor programme (Usherwood *et al.*, 1968).

Normal walking – behavioural analysis

As for many hexapods, tree wetas walk with an alternating tripod gait, which is somewhat loosely coupled during slow walking and becomes more tightly coupled as walking speed increases (Fig. 23.4A). Each tripod consists of two legs on one side of the body and one on the other side, which move together in the swing and stance phases of stepping. The centre of gravity is always contained within the tripod of legs to prevent the insect from falling while in motion. The left-hand tripod indicated by the oval in Fig. 23.4A consists of the right meso-leg (R2), the left pro-leg (L1) and the left

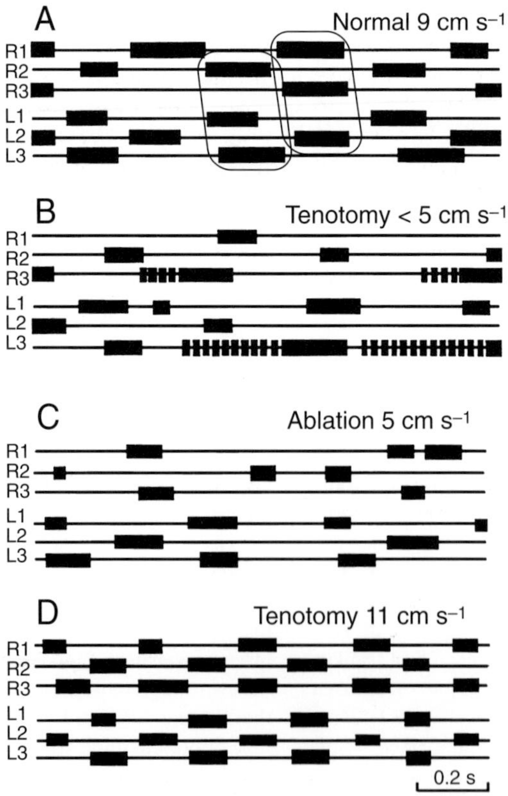

Fig. 23.4. Walking-gait patterns of tree wetas before and after interference with the femoral chordotonal organ (FeCO). Heavy bars indicate protraction (swing phase). Right and left front, middle and hind legs indicated by R1–R3 and L1–L3, respectively. A. Normal stepping at 9 cm s^{-1}. Oval lines indicate alternating leg tripods. B. Severing FeCO apodeme (tenotomy) causes uncoordinated gait with dragging of hind legs (dashed bars) at slow stepping speeds. C. With slightly higher stepping speed, partial coordination returns, even though FeCO was ablated. D. At high stepping following FeCO tenotomy, full coordination returns, indicating independence of motor programme from FeCO feedback at this speed. (Modified from de Ruiter, 1984, with permission.)

meta-leg (L3). The heavy bars represent protraction (swing forward) times, while the thin lines represent retraction (rearward movement during stance) times for each leg. Note that the coupling of the two tripods is not strict, even at the intermediate velocity of 9 cm s^{-1}.

Quantitative analysis of stepping sequences for wetas walking at different speeds showed that the retraction time (stance phase) decreases in a non-linear fashion as walking speed increases, while protraction time (swing phase) shows much less variation and remains essentially linear (Fig. 23.5A), as seen in walking patterns of other insects as well as vertebrates. In the locust, retraction time varies linearly with walking speed over a similar range of velocities measured in the weta (Burns, 1973). When plotted as a ratio of protraction to retraction times against walking speed, a linear relationship (correlation coefficient $r = 0.70$) is obtained, as also described, for example, in cockroaches (Delcomyn, 1971). A further characterization of walking may be made by measuring variation in step length as velocity increases. This relationship is linear in the weta (e.g. $r = 0.78$, slope 0.1, $n = 9$) over a wide range of velocities (Fig. 23.5C). Tree wetas walk at velocities that vary from about 2.5 to 15 cm s^{-1}, with stepping frequencies of about 1–6 Hz (Rind, 1976; de Ruiter, 1984).

The role of the femoral chordotonal organ in walking control

The largest proprioceptor in the insect leg is the FeCO, and interest in its role in the motor control of walking was inspired by the surgical interference experiments of Bässler (1967) and Usherwood *et al.* (1968). Both reported dramatic changes in normal walking motor coordination, but subsequent research, mainly on the stick insect, has shown the underlying neuronal control system to be complex and, as yet, unravelled. Experiments in the tree weta were easily accomplished, due to the ease of either cutting the FeCO apodeme (release) or the FeCO nerve (ablation). Because the hind legs provide the main propulsion force during walking, due to their larger muscle mass and greater length compared with the other legs, experiments on the role of the FeCO have been carried out on these legs. The effects of FeCO release, in which the FeCO apodeme was disconnected but the organ was left intact, were

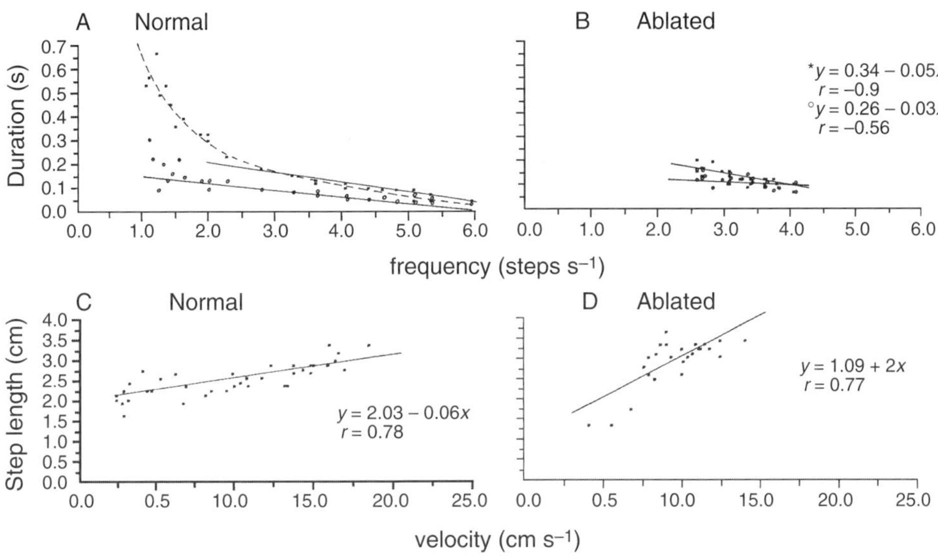

Fig. 23.5. Quantitative analysis of protraction/retraction times during walking in normal and FeCO-ablated tree wetas. A. In normal weta, retraction time (dashed line) decreases non-linearly as stepping speed increases, while protraction time remains almost linear (solid line). B. Following ablation, the insects often avoid slow walking speeds and at higher speeds protraction and retraction are little different from normal. C. Step length varies linearly with walking speed in normal wetas. D. Following ablation, the step length undergoes greater variation (but still varies linearly) as walking speed changes. (From de Ruiter, 1984, with permission.)

generally less severe than those of ablation (de Ruiter, 1984).

Qualitative changes

In general, a dramatic series of behavioural disruptions are seen during slow walking following the operations, but these tend to disappear as the weta enters fast walking. The most marked effects include changes in leg angles and poor coordination in stepping. The following qualitative changes in horizontal walking behaviour occur.

1. The alternating tripod gait becomes uncoupled, leading to erratic stepping and even repeated steps of one or more legs within a cycle at slow walking speeds (Fig. 23.4B, C). These effects disappear at high walking speeds (Fig. 23.4D). The same effect was reported for the locust (Usherwood *et al.*, 1968).
2. The femur–body angle increases, causing the hind legs to be more splayed out from the body (Fig. 23.6C). This is especially apparent during slow and halting walking, but is also seen with video analysis at fast walking speeds. This was also reported, though not quantified, for ablation of the FeCO in the locust.
3. Hind-leg hyperflexion (femur–tibia angle (FTA) = ~10°) occurs when a hind leg is lifted, often prior to walking or turning. At rest, the hind leg remains more flexed than normal and drifts into hyperflexion if lifted free of the substrate by the experimenter. During walking, the arc of FTA movement remains hyperflexed (Fig. 23.6B). Surprisingly, the opposite effect is described for ablated locusts, where the arc of FTA movement increases from 50°–80° to 80°–120° during walking.
4. As slow walking commences in the weta, both hind legs often become simultaneously hyperextended (FTA = 150°–170°) and are dragged, while the front and middle legs continue to step normally (Fig. 23.6A). Attempts by the weta to flex the hind legs often result in either partial flexion of one or simultaneous raising and lowering of both, without entering the walking cycle.
5. In the normal walking cycle, the tarsus and claw are raised just before protraction and depressed during the stance phase to assist propulsion. Following FeCO disruption in the weta, levation of the claw and tarsus does not occur and the claw drags on the substrate.
6. The hind tarsus of the weta often slips on the substrate, suggesting that the leg is not sufficiently

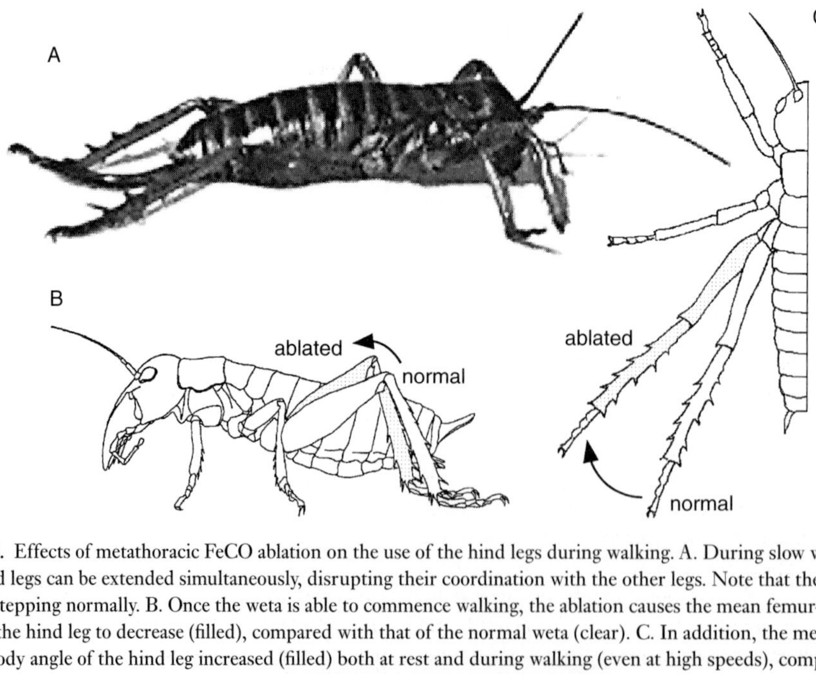

Fig. 23.6. Effects of metathoracic FeCO ablation on the use of the hind legs during walking. A. During slow walking, both hind legs can be extended simultaneously, disrupting their coordination with the other legs. Note that the middle legs are stepping normally. B. Once the weta is able to commence walking, the ablation causes the mean femur–tibia angle of the hind leg to decrease (filled), compared with that of the normal weta (clear). C. In addition, the mean femur–body angle of the hind leg increased (filled) both at rest and during walking (even at high speeds), compared with a smaller angle in normal wetas (clear). (A, after Rind, 1976, with permission.)

depressed by the trochanteral depressor and tarsal depressor during the stance phase. This shows up more clearly when the insect attempts to climb.

The above effects largely disappear as the walking speed increases over the first six to ten steps and the hind legs become more entrained into reciprocal stepping. Above velocities of 6 cm s^{-1} stepping appears to be almost normal in ablated wetas (Fig. 23.4D), although the FTA range is still hyperflexed. These results highlight the concept that the CNS requires proprioceptive feedback from the leg (especially the FeCO) during slow walking, but at fast walking speeds the more tightly phased motor pattern becomes generated almost entirely within the CNS, with little influence from peripheral sense organs (Pearson and Iles, 1970).

A remarkable adjustment has been seen in some wetas over a period of several weeks following FeCO ablation. They tend to walk only at high speeds and therefore do not enter the awkward initial stages of disrupted hind-leg coordination characteristic of slow walking. At times when they do commence with slow walking, the ablation deficits are similar to those of freshly operated animals. Film analysis showed that in fast walking the tarsi are often highly levated during the entire walking cycle in such wetas, or the leg is held abnormally high during the swing phase, in either case avoiding the problem of dragging the claws. The above observations suggest that the wetas are able to compensate for the proprioceptive deficit by altering their walking motor pattern. However, others do not show such remarkable alterations and continue to suffer from disruptions during slow walking attempts (de Ruiter, 1984).

Quantitative analysis

Because release or ablation usually causes the wetas to walk at intermediate and fast speeds, data comparable to slow normal walking are difficult to obtain, and the stepping-speed range is limited. Within the recorded ranges, no significant differences have been found for protraction (swing), retraction (stance) and protraction/retraction ratio of the stepping pattern (Fig. 23.5A, B). However, the slope of the relationship of step length vs. velocity is significantly steeper in ablated and tenotomized wetas, compared with that for normal wetas ($P < 0.001\%$, F test comparing regression slopes), as seen in Fig. 23.5C, D. In fact, at slow speeds, the step length is shorter than normal and, at high speeds, it is longer than normal. An explanation is provided by the weta's adjustment of swing phase duration, which controls step length. At slow speeds, protraction occupies a smaller fraction of the step cycle, while, at high speeds, protraction occupies a greater fraction of the cycle (de Ruiter, 1984). This correlates with the increased tendency to keep the leg more flexed when walking at higher speeds, as described in the following section.

Two leg angles are altered by either ablation or release of the FeCO during walking. The first, which correlates with alteration of step length, is the FTA, while the second is the femur–body angle. A clear decrease in the means of both maximum and minimum FTA during stepping occurs after both operations (Fig. 23.6B). For example, in nine ablated wetas, the mean arc of tibial movement narrowed from the normal FTA of 61°–126° to 49°–107°, causing the midpoint of the arc to decrease significantly in FTA from 93° to 78° ($P < 5\%$, t test). This reflects the tendency for the insect to walk with the hind legs at more flexed angles, with the appearance of having stiffer femur–tibia joints. The opposite effect occurred in experiments with locusts, and the basis for this radical difference has never been investigated (Usherwood *et al.*, 1968). In tree wetas, the mean maximum femur–body angle increased by 24° in the same experiments, while the minimum femur–body angle increased by 6.9° (Fig. 23.6C) (de Ruiter, 1984). These changes confirm the earlier observation of highly splayed legs in operated wetas and show that the effect is still present even at high walking speeds (Cadenhead, 1982; de Ruiter, 1984). The above effects occur in a single leg in the case of a unilateral ablation, as well as both legs with bilateral ablations.

In addition to disruptions in walking behaviour of wetas, the ablation also caused the loss of stridulatory behaviour and the loss of the raised hind leg defence display. In locusts, apparently the tendency to jump is lost also.

Myogram activity during normal walking

Myogram recordings of free-walking wetas before and after FeCO operations provided insight into mechanisms underlying the role of the FeCO in motor control of the leg (Cadenhead, 1982;

de Ruiter, 1984). The most immediate effect of release or ablation is the loss of resistance reflexes in the extensor and flexor tibiae muscles in the standing weta. During walking, SETi normally fires in an alternating pattern with the flexor motor neurons. Sometimes, the onset of the extensor motor neurons overlaps with the firing of the flexors, but the onset of flexor firing never overlaps with the extensor burst (insets, Fig. 23.7A, B). FETi is recruited as speed or substrate slope increases. Both SETi and FETi fire more strongly, and with greater duration, during walking on a vertical slope and at higher speeds, with more tightly controlled alternation (Fig. 23.7Ci). The flexor discharge represents firing of a number of motor units. It does not change markedly in duration or in spike frequency as slope or walking speed increases (Fig. 23.7A, C). This reflects the fact that the extensor provides power to the leg proportional to slope and walking speed, while the flexor only provides a rather constant protraction of the tibia under any walking condition. The tarsal levator and retractor unguis fire near the end of the extensor muscle burst, causing the claw and tarsus to be raised from the substrate before the swing phase of hind-leg movement. The

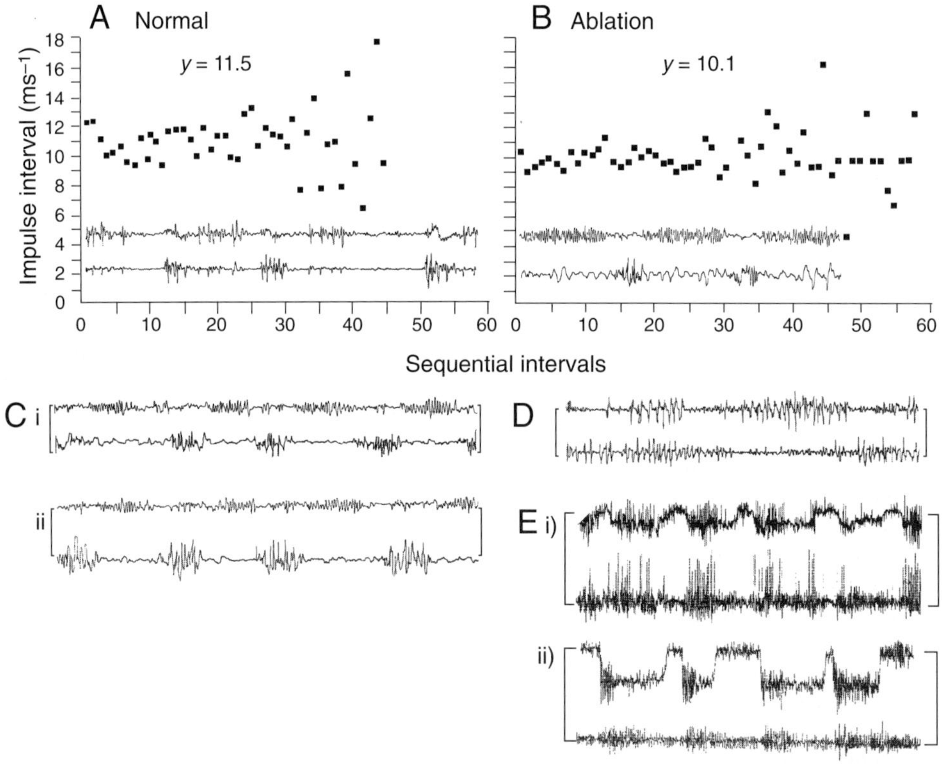

Fig. 23.7. Muscle-firing properties during walking in normal and FeCO-ablated wetas. A and B. Mean sequential impulse interval plots for slow extensor motor neuron (SETi) bursts before (A) and after (B) FeCO ablation during slow walking; $n = 30$ for each plot. Note prolonged burst duration and shorter intervals after ablation. Insets show myograms of the extensor muscle (upper trace, mostly SETi firing) and the flexor muscle (lower trace) during slow walking before (A) and after (B) the ablation. C. i. Extensor (upper trace) and flexor (lower trace) muscle bursts during fast walking in a normal weta. Slight cross-talk occurs in the extensor recording. ii. Following FeCO ablation, the SETi bursts are prolonged, but coordination is normal and the flexor bursts are unaffected. D. Semi-spastic co-extension of hind legs after ablation is reflected in simultaneous firing of SETi in right (upper) and left (lower) hind legs during attempt to commence slow walking. E. Effect of FeCO ablation on the trochanteral depressor muscle (lower trace). i. In the normal animal, a slow excitor fires tonically and at lease two fast units fire phasically during the stance phase of walking. ii. Following the ablation, at least one of the fast excitors fails to fire, while little effect is shown in the tonic excitor.

trochanteral depressor fires in synchrony with the extensors to provide downward force of the entire leg on to the substrate, while the trochanteral levator fires during the swing phase to raise the leg as it swings forward (Fig. 23.7Ei).

Effects of FeCO ablation on muscle activity

The most striking effect of FeCO release, as well as ablation, was an increase in burst duration and firing frequency of SETi during walking (Fig. 23.7A, B). This was especially pronounced for slow walking. Thus, it is clear that the abnormal hyperextension characteristic of ablated wetas is due to prolonged extensor firing, rather than to extreme tarsal depression causing the claws to catch on the substrate or to passive dragging of the hind leg by forward movement of the body. Plots of extensor burst duration against cycle time confirmed that a significant increase (*F* test: $P < 0.05$) in extensor burst duration occurred at slower walking frequencies (not shown) (de Ruiter, 1984). Another result of release and ablation is the loss of FETi recruitment in walking conditions where it normally occurs. Thus, a 30° substrate slope causes recruitment of the FETi in the normal animal, but, following ablation, no FETi firing occurs on a 60° or even vertical slope.

The ablation also causes disruption of coordination of extensor tibiae muscle bursting between the two hind legs. This bilateral disruption shows up in two ways. First, there is a phase advance in the burst onset of one side within the burst cycle of the other side. Normally, the legs alternate, with a mean phase of 0.52 (perfect alternation would predict that phase = 0.5), while ablated wetas showed a mean phase of 0.38 (de Ruiter, 1984). The second change is an increased variance in the phase of individual steps, causing larger variation in the coordination of the stance phase of both legs during walking. Furthermore, negative phase values occurred following the ablations, indicating that one leg can abnormally step twice and reverse its phase with the other.

The flexor muscles show little change in burst structure after ablation or release. Although plots of burst duration against cycle time normally show a slight increase with greater cycle time, no significant changes in the slope of the relationship occur following ablation. The only significant change in flexor firing is an increase in variance in burst duration after ablation or release.

Myogram records show that the trochanteral depressor burst becomes weaker and at least one large excitor is not recruited following ablation (Fig. 23.7Ei, ii). This may account for the lower body level during walking and frequent dragging of the abdomen, as shown when walking over smoked paper.

At higher walking speeds, the above effects are greatly diminished and muscle burst patterns in ablated wetas appear similar to those in normal animals (Fig. 23.7C). However, protraction/retraction ratios show that protraction (swing phase) takes up comparatively more time than retraction (stance phase). The muscle bursts of SETi during the stance phase are significantly shorter than normal at a given walking speed, which is a reversed condition to that observed in slow walking.

Model for role of FeCO in walking

The effects of FeCO ablation on details of muscle burst coordination in one or adjacent legs are complex. As also seen in the stick insect, they may disappear as the insect increases walking speed or may be masked by inputs from unablated sense organs (Bässler, 1983). In fact, Burrows (1996) concluded that the number of parallel receptors acting during walking make it almost impossible to ascribe any one leg action to just one chordotonal organ. Nevertheless, the experiments reported above show clearly that the FeCO is relied upon heavily for muscle burst coordination in most segments within a leg, as well as between contralateral legs, for slow walking in tree wetas. Furthermore, ablation experiments show that, upon careful analysis, the FeCO also contributes to fast-walking motor control, even though experiments indicate that much less proprioceptive feedback is required by the walking pattern generator at higher speeds.

SLOW WALKING. During slow walking, FeCO afference must normally feed back negatively on to SETi and FETi to attenuate the extensor tibiae muscle burst toward the end of the stance phase, since extensor firing is prolonged following ablation (Fig. 23.7B). Presumably, position-sensitive FeCO neurons fire as they detect the end of the extension phase of normal walking, and inhibit FETi and SETi. This is in agreement with results in the stick insect, where the FeCO acts as a position detector to control end-points of tibial movement during walking (Weiland and Koch, 1987). The flexor motor neurons of the weta do not normally fire until the extensors stop firing, since myograms show no overlap in the transition of firing in these antagonists at the end of the stance phase (Fig. 23.7A, C). The control of duration of flexor firing in the swing phase appears not to be influenced by the FeCO, since little change in firing is seen following ablation. However, the tibia often shows hyperflexion, both in postural and slow walking conditions, and the FeCO may normally act as a position detector to terminate the swing phase, as well as regulating postural control of the flexed leg. Similar results are reported for the stick insect, where experiments on perturbation of leg movements (which indirectly affected FeCO feedback) had no effect on the duration of the swing phase, but strongly affected the start both of the swing phase and of the stance phase (Cruse and Schwarz, 1988).

The FeCO also provides excitatory feedback on to the trochanteral depressor (Fig. 23.7Ei) and on to the tarsal depressor during the stance phase, to assist in firmly gripping the substrate. These are normal locomotory reflexes, described previously (Field and Rind, 1981).

FAST WALKING. In fast-walking ablated wetas, the extensor burst is shorter than normal and the stance phase terminates earlier, allowing the flexor burst to become relatively longer than the extensor burst. The effect is to move the whole arc of tibial movement into a more flexed position. This is interpreted as a loss of excitatory feedback from the FeCO on to the extensors (also suggested by the loss of FETi firing after ablation), as well as the loss of the end-point position detection function of the FeCO. The pattern generator normally receives FeCO excitatory input to keep the stance phase end point at an extended position. With less extensor drive, the stance phase terminates earlier and the constant flexor burst moves the entire swing phase to a more flexed position.

CONTRALATERAL COUPLING. The loss of reliable contralateral coupling of hind-leg coordination during slow walking highlights the dependence of the pattern generator of FeCO feedback. Details of the circuitry underlying this aspect of the walking-pattern generator are sketchy at best. Bässler (1983) concluded that the generator is a loosely coupled set of ganglionic oscillator networks, which are highly dependent upon proprioceptive feedback to control contralateral and ipsilateral leg coordination, as well as single-leg cycle coordination.

Learning and catch contraction of leg muscles

In the course of studying neuronal mechanisms underlying learning and memory in insects, Forman and Hoyle (1978) discovered that locusts could learn to position a leg joint at a specific angle. This phenomenon of conditioned learning and retention of the learned act represents the most sophisticated neuronal motor-control feature found in modern orthopterans. Since wetas are considered to have retained many ancestral traits and thus are considered to be more primitive, it was of interest to see if they could display the same learning abilities.

Negative-reinforcement experiments, in which a weta was required to turn off an aversive stimulus (loud sound and vibration) by placing its femur–tibia joint within a narrow window of FTA, were conducted on tree wetas (*H. femorata*) by Hoyle and Field (1983a). A photocell monitoring tibial movement was connected to a switch, which turned off the stimulus when the weta moved the tibia into the window. This task is readily learned, as the restrained weta rapidly flexes and extends its tibia through the FTA window upon onset of the aversive stimulus (Fig. 23.8A). Normally the leg is held at an FTA of about 76° while at rest. Learning required that the leg be set into windows > 120° and < 40° to ensure that active muscular contraction held the tibia in the set position. Wetas can learn to hold the tibia within window widths as narrow as 5°, positioned anywhere from full flexion (0° FTA) to almost full extension (160°) (Table 23.2).

Learning occurred within a mean of 1 min 18 s (n = 16 wetas of both sexes), and the leg was held in a stable position within the window (thus keeping the sound switched off) for a period of 10 min to 2h 40 min (examples in Table 23.2). This contrasts with locusts and grasshoppers, both of which allow the leg to drift in and out of the window, but always return it immediately to the correct setting.

While the set position of the weta tibia is very stable, once the tibia drifts out of the window wetas show an inability to rapidly relearn the window position. They must flex and extend the tibia

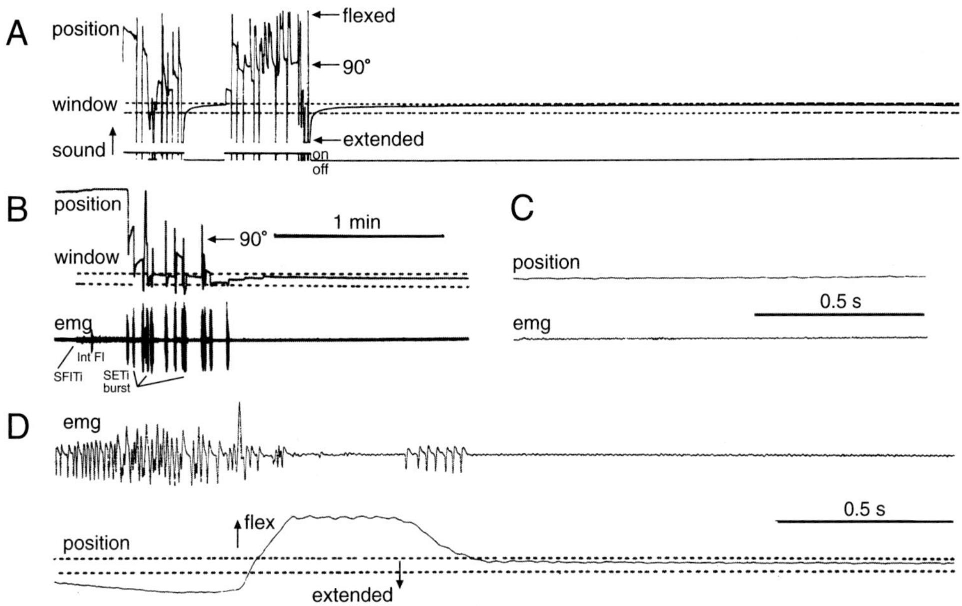

Fig. 23.8. Conditioned learning of hind-leg femur–tibia position in tree wetas. A. Record of tibial position (upper trace) during conditioning with aversive sound (lower trace). Sound started at the arrow. Dashed line indicates width of window in which tibia must rest to turn off sound (by photoelectronic switch). Leg struggling indicated by full flexion–extension movement at first, but soon leg rests within window, turning off sound. Note drift toward flexion within window. B. Demonstration of absence of electrical activity in extensor muscle myogram (lower trace) during maintenance of leg position (upper trace) within window following conditioning. emg, Electromyogram recording; IntFL, intermediate flexor tibiae motor neuron; SFLTi, slow flexor tibial motor neuron. C. A high-speed recording of the extensor myogram (lower) 10 min following B shows complete absence of electrical muscle activity during conditioned leg extension. D. Evidence that weta is using previously learned window boundaries to set leg into window. The mixed flexor/extensor myogram (upper trace) shows that the SETi fires a single burst, which drives the tibia into the window (dashed lines) but which terminates before the tibia enters the window.

Table 23.2. Examples from data (n = 16, Hoyle and Field, 1983a) showing variation in learning time (time to set tibia into window), time to first error (elapsed time before the tibia drifts out of the window), relearning time (time required to reset tibia into window). The arc of experimental window size and the angle of the proximal edge of the arc are also shown. Full extension = 170°, full flexion = 0°, normal rest FTA = 76°.

Size	Proximal FTA	Learning time		Time to first error		Relearning time	
		min	s	h	min	min	s
20°	120°	1	19	0	10	1	40
15°	120°	1	49	0	44	1	36
10°	160°	1	18	0	48	1	4
10°	140°	0	47	2	0	1	32
10°	40°	1	34	0	21	–	–
10°	30°	0	42	0	14	1	23
5°	0°	1	14	1	50	0	46

through the window to relocate its position before being able to reset the tibial position. The mean relearning time of 1 min 22 s (Table 23.2) was essentially as long as that of initial training. Two interesting conclusions arose from this observation. First, the weta apparently cannot retain a memory of the learned position, in direct contrast to locusts, which seem to retain a CNS representation of the required FTA. Secondly, if wetas do not retain a CNS memory of the required FTA, they must use some unknown peripheral muscular mechanism to hold the leg in the correct position, rather than a dynamic repositioning mechanism driven by the brain.

The clue to the mystery came from myogram recordings of the tibial extensor and flexor muscles during conditioning. These showed that the initial rapid tibial movements at the onset of the aversive stimulus were driven by slow and intermediate flexor motor neurons for flexion and by SETi for extension (Fig. 23.8B, D). However, once the weta locked the tibial FTA into the window, electrical activity of the myogram ceased and the muscles became silent. This remarkable observation could only mean that the muscles were being set into catch contraction, as seen in bivalve molluscan catch muscles, because the flexor and extensor motor neurons that move the tibia could be identified clearly by high-gain recordings during movements and ruled out by their absence after the tibia became set within the window (Fig. 23.8B, C).

There is further evidence for catch contraction based upon anatomical grounds. The femur–tibia joint contains special lubricated cuticular pads and flanges, which ensure a passive return of the joint to a mean rest position of 76° in the absence of muscular tension. The maintenance of the learned positions requires that the respective muscles produce a tension of 7 g for the 160° window and 5 g for the 0° window; therefore the joint is not held in the set positions by passive frictional joint resistance to movement (Hoyle and Field, 1983a).

Another intriguing aspect of weta leg-position learning is found in the mechanism by which SETi fires to cause the tibia to enter and remain in the window. Rather than firing impulses to extend the tibia until it enters the window and then terminating the muscle burst, SETi fires a burst, following which the tibia drifts into the window (Fig. 23.8D). The muscle firing ceases before the window is entered, suggesting that ongoing proprioception during movement into the window is not used to terminate the muscle burst when the correct FTA is obtained. Instead, the weta must be using information gained previously from frequency and duration of muscle firing to calculate the precise tension required to position the tibia correctly to switch off the aversive stimulus. Moreover, because the muscles normally relax once bursting ceases and yet the tibia remains within the window, resisting the passive return force to normal rest position, it follows that some additional motor output (not recorded as electrical muscle activity) must accompany or precede the electrical SETi output in order to produce the catch contraction. This output is very likely to be the release of the neuromodulator compound, octopamine, by DUM neurons in the CNS, as explained in the next section (Hoyle and Field, 1983b).

The conclusions from the above research posited that both wetas and locusts can undergo conditioned learning, but that wetas have much less ability to remember the learned task. Locusts can retain a memory of the learned behaviour within the CNS for at least an hour, while wetas do not retain a memory of window position after the tibia enters it. Therefore, the weta must be setting the position by a peripheral locking-mechanism catch contraction (which holds the tibia in the window for over 2 h), but is not able to reset the tibia into the window once it drifts outside; it must relearn the task. This appears to indicate that the more modern caeliferan Orthoptera (locusts and grasshoppers) have developed a sophisticated memory ability in the CNS, while the more primitive wetas have not.

The phenomenon of catch contraction, originally described in bivalve adductor and byssal retractor muscles, differs somewhat from that in the tree weta. No other insects exhibit this unique mode of operation, although other arthropods display a partial retention of tension following sporadic muscle firing or contraction followed by fast relaxation (crayfish dactyl opener muscle, cockroach coxal muscle, respectively (Hoyle, 1983)). Wetas appear to utilize catch contraction in maintenance of the defence posture and in thanatosis, where all limbs are frozen in position. The latter is easily elicited experimentally, but is likely to be a natural mode of resting vertically in tree galleries during the daytime without prolonged electrical neuromuscular activation. Little is known about this phenomenon and the area is in need of further research.

Dorsal Unpaired Midline (DUM) Motor Neurons and the Role of Octopamine

The observation of muscle tension being maintained in tree wetas without accompanying electrical activity led to the search for a chemical explanation of catch contraction. Earlier work in locusts had shown that the monoamine octopamine, a noradrenaline-like molecule that occurs in certain CNS interneurons and motor neurons, has modulatory effects on muscular contraction, IR and BT (Evans, 1985). Although this compound had diverse roles in different animal species, in insects it can act as a neurohormone, a neurotransmitter and a neuromodulator (reviewed in Burrows, 1996). Its role as a peripheral modulator of neuromuscular transmission and muscular contraction provided the rationale for an investigation of a possible role in catch contraction in tree wetas.

Octopamine typically occurs in motor neurons uniquely characterized by being unpaired and located in the ganglion midline. Hence they are named dorsal unpaired midline (DUM) cells. They differ from all other motor neurons, which are arranged in bilateral pairs within insect ganglia. DUM motor neurons still retain a bilateral function via a single axon, which bifurcates after emerging from the median cell body and projects symmetrically through one or more nerve roots on the left and right side of the ganglion. Other dorsal (and a few ventral) octopaminergic interneurons occur as paired and unpaired cells, which send branches locally or to other ganglia (Stevenson *et al.*, 1994). Octopamine has been conclusively demonstrated in the DUM cell to the locust extensor tibiae muscle (DUMETi), as well as in certain abdominal DUM cells supplying the oviducts (Orchard and Lange, 1985; Stevenson *et al.*, 1992).

DUM neuron position and morphology

In *H. femorata* and *Hemideina crassidens*, DUM neurons occur as median clusters along the midline of all three thoracic ganglia (Fig. 23.9A), as

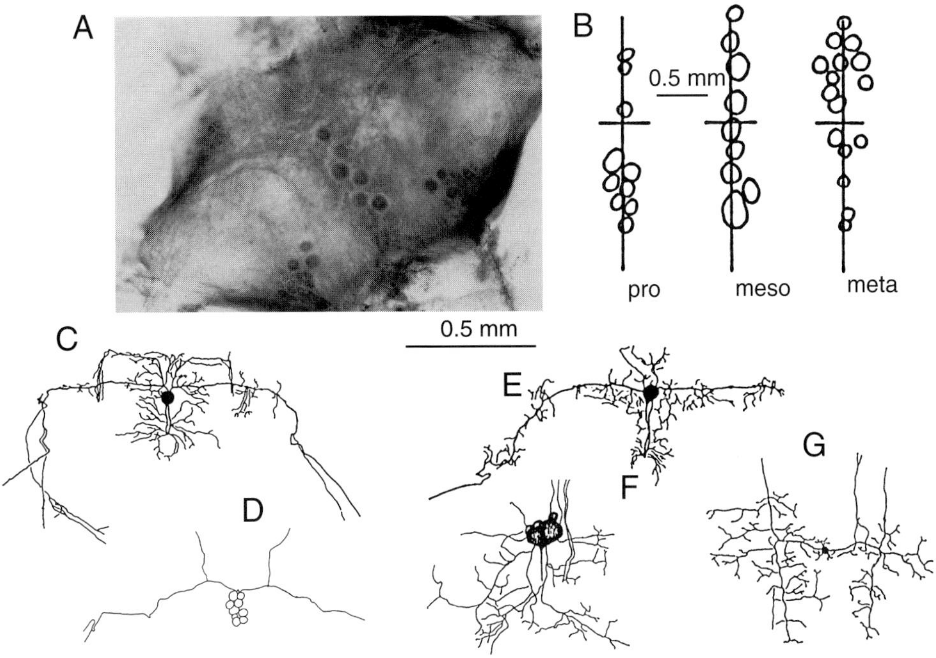

Fig. 23.9. DUM neuron position and morphology in tree wetas. A. Toluidine blue staining of cell bodies in metathoracic ganglion stains dorsal medial cluster of DUM neurons. B. Patterns of distribution of DUM neurons in the pro- (left), meso- (middle) and metathoracic (right) ganglia. Note clusters in pro- and metathoracic ganglia with doublet or triplets of cells as outliers, while the DUM neurons of the mesothoracic ganglion are arranged linearly. C. Morphology of motor neuron DUM3,4,5 with multiple axons exiting nerve roots 3, 4 and 5. D. DUM1,3 is a motor neuron giving off axons into roots 1 and 3, and has not been reported in other insects. E. Motor neuron DUM5 resembles the DUMETi, which innervates the extensor tibiae muscle in locusts and cockroaches. F. A spiking DUM interneuron, which sends an axon anteriorly through the meso–meta connective. G. Another DUM interneuron, with an H-shaped morphology and restricted to the metathoracic ganglion.

visualized with the non-specific stain toluidine blue. Specific staining of only the DUM cells is demonstrated with the monoamine-specific dye neutral red, which suggests that they contain octopamine (Currie, 1988). In tree wetas, as in locusts, the greatest number of DUM cells occurs in the metathoracic ganglion, while about half the number of cells occurs in the pro- and mesothoracic ganglia (Fig. 23.9B; Table 23.3). There are no differences in DUM cell number between species of *Hemideina*; however, cell-body position and grouping differ among ganglia within each species. For example, most DUM cells in the prothoracic ganglion are found in a posterior cluster, with a distinct doublet of cells always separated anteriorly. In the mesothoracic ganglion, the cells occur in a linear array along the midline, while in the metathoracic ganglion the majority of cells are clustered in an anterior group, with a doublet or triplet of cells separated posteriorly (Fig. 23.9B).

In the locust, as well as tree wetas, DUM neurons also occur in two size classes, of which the larger are motor neurons, which send axons out of ganglionic nerve roots (Evans and O'Shea, 1978). The smaller cell bodies are typically interneurons, either with processes locally restricted to a ganglion or with axons extending to neighbouring ganglia.

Three morphological types of DUM motor neurons were demonstrated in metathoracic ganglia of wetas by intracellular filling of living cells ($n = 30$) with the fluorescent dye Lucifer yellow, and a fourth type was found by $CoCl_2$ perfusion of the extensor tibiae nerve (Currie, 1988). In the first type, the primary neurite travels posteriorly first before turning anteriorly to make a T-junction. Dense processes are given off near the cell body. Each axon then gives off a fine branch, which makes a rectangular pattern anterior to each main axon. The latter divides near the ganglion periphery and exits through N3, N4 and N5 (Fig. 23.9C). This type is called DUM3,4,5 and has been described in the metathoracic ganglion of the locust (Watson, 1984) and of the cockroach (Arikawa *et al.*, 1984). It innervates the flexor tibiae, the pleuroaxillary and the tergocoxal remotor muscles in the locust.

The second type of DUM motor neuron occurs near the anterior of the ganglion. The primary neurite forms a T-junction close to the cell soma, with few central processes, and each lateral axon bifurcates to give off a branch through N1 and N3 (Fig. 23.9D). Dye coupling, in which dye leaks from one cell to its neighbours, was common in this type and prevented unique identification of cell bodies for each axon. This type, designated DUM1,3, has not been reported in other insects.

The third type gives off a posterior neurite, which doubles back anteriorly before forming a T-junction to give off left and right axons. Diffuse central processes are given off from the neurite and axons, which leave the ganglion through N5 (Fig. 23.9E). This type closely resembles DUMETi, identified as the single octopaminergic motor neuron innervating the extensor tibiae muscle in the locust (Evans and O'Shea, 1978), and designated DUM5 by Watson (1984). It has also been reported in the cockroach (Arikawa *et al.*, 1984).

The fourth type was only found by $CoCl_2$ nerve back-filling, and is apparently confined to the extensor tibiae muscle. It gives off a single left and right axon from the central T-junction, each of which divides and passes out N3 and N5, and thus is designated DUM3,5. This type has not been seen elsewhere.

Two morphological types of interneurons found in the weta metathoracic ganglion are also reported for the same ganglion of the locust (Hoyle, 1978). The first is a spiking DUM interneuron, which broadly projects processes in the mid- and posterior regions of the metathoracic ganglion, and sends an axon through one of the connectives to at least the mesothoracic ganglion

Table 23.3. Mean numbers of DUM neurons occurring in thoracic ganglia of *H. crassidens* and *H. femorata,* based upon neutral red staining (from Currie, 1988).

Ganglion	Mean number DUM cells	SD	*n*
Prothoracic	7.4	2.9	27
Mesothoracic	6.7	2.4	32
Metathoracic	13.6	4.7	26

(but not traced further) (Fig. 23.9F). The second is a DUM interneuron with an H-shaped set of processes, which are restricted to the metathoracic ganglion (Fig. 23.9G). This type has also been reported in the cockroach (Arikawa *et al.*, 1984).

DUM innervation to muscles

The innervation pattern to muscles of the hind femur has been determined by a combination of cobalt back-filling of specific nerves to each muscle, electron microscopy of each muscle's nerve and electrical stimulation of the contralateral nerve to each muscle, while recording from the homologous nerve to the ipsilateral muscle (Currie, 1988). The latter method takes advantage of the bilateral axon morphology of the DUM motor neurons to allow specific determination of innervation without suffering from contamination by stimulus spread contralaterally. The methods above showed that the extensor tibiae muscle is innervated by four DUM neurons, the flexor tibiae receives two DUM neurons and the retractor unguis receives a single DUM neuron (Table 23.1, Fig. 23.10D). The four neurons to the extensor are progressively recruited as stimulation voltage is increased in Fig. 23.10A, with the first firing at a threshold of 4 V. The two flexor DUM neurons and the single retractor unguis DUM neuron were recruited at threshold, and no further DUM neurons fired at higher voltages (Fig. 23.10B, C). By recording from small branches to different parts of the extensor and flexor tibiae muscle, it was found that at least two extensor DUM neurons occur in the ventral extensor nerve, while all four travel in the main extensor nerve. The two flexor DUM

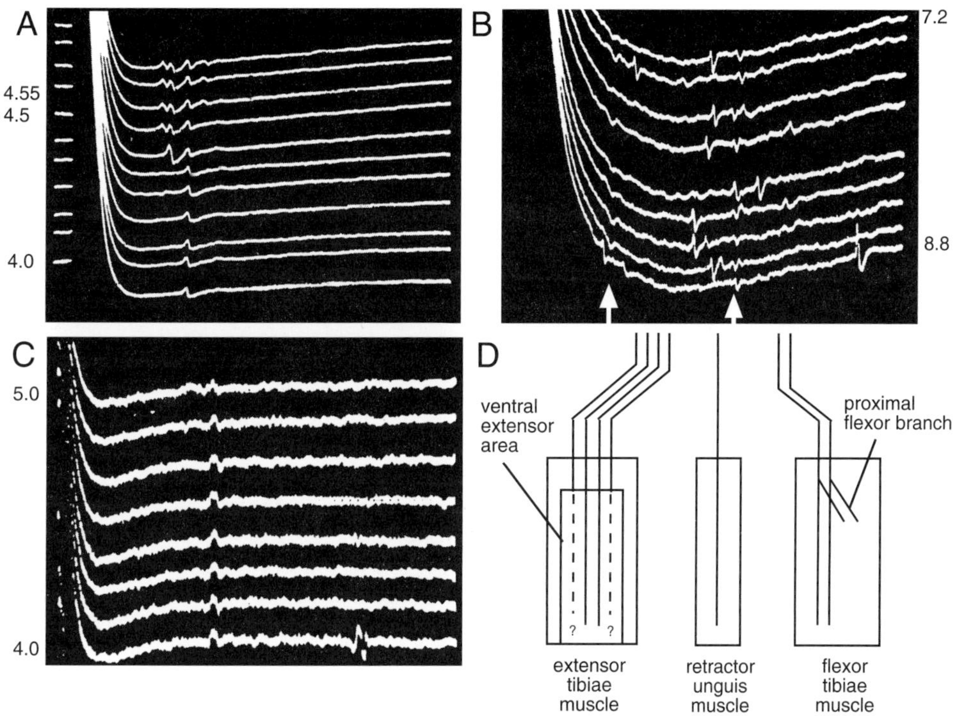

Fig. 23.10. Innervation patterns of hind leg muscles by DUM neurons in the tree weta, revealed by contralateral stimulation of motor nerves. A. Progressive recruitment (from 4 V) of four DUM motor neurons to the extensor tibiae muscle, recorded in the ipsilateral extensor nerve while stimulating the same nerve contralaterally. B. Recruitment of two DUM motor neurons (short and long latency, arrowheads) to the flexor tibiae muscle at a stimulus threshold of 7.2 V. These innervated both the proximal flexor branch and the main flexor nerve. C. A single DUM motor neuron is recorded in the motor nerve to the retractor unguis muscle, at a stimulus threshold of 4.0 V. D. Schematic diagram of the DUM innervation pattern to the leg muscles.

neurons travel in the tiny proximal branch, as well as in the main flexor nerve (Fig. 23.10D).

The DUM innervation to the flexor and retractor unguis muscles is similar to that described for the locust. However, the occurrence of four DUM neurons to the weta extensor tibiae, as well as multiple DUM innervation in the bush cricket extensor tibiae (*Decticus albifrons*, Tettigoniidae) (Theophilidis, 1983), suggests that there is a basic difference between ensiferan and caeliferan DUM innervation to this important jumping muscle.

Effects of Octopamine and DUM Stimulation on Leg Muscles

In establishing whether DUM neurons and octopamine play a role in the catch contractions observed during defence behaviour, Hoyle and Field (1983b) studied the contraction physiology of the weta metathoracic extensor tibiae muscle as modulated by octopamine solutions (10^{-10}–10^{-6} mol l^{-1}) and by stimulation of DUM neurons. Normal firing of SETi causes a rapid build-up of tension in the extensor, followed by immediate relaxation upon cessation of the SETi burst. The extensor muscle EJPs showed facilitation during progressive firing of SETi, much as seen in the well-known locust extensor tibiae muscle (Fig 23.11A). However, in some preparations, natural catch contraction occurred when SETi was stimulated, even in the absence of externally applied octopamine or DUM neuron stimulation. This exhibited all the properties of catch contraction elicited by the modulatory factors tested, e.g. the contraction plateau remained flat for 3–8 min

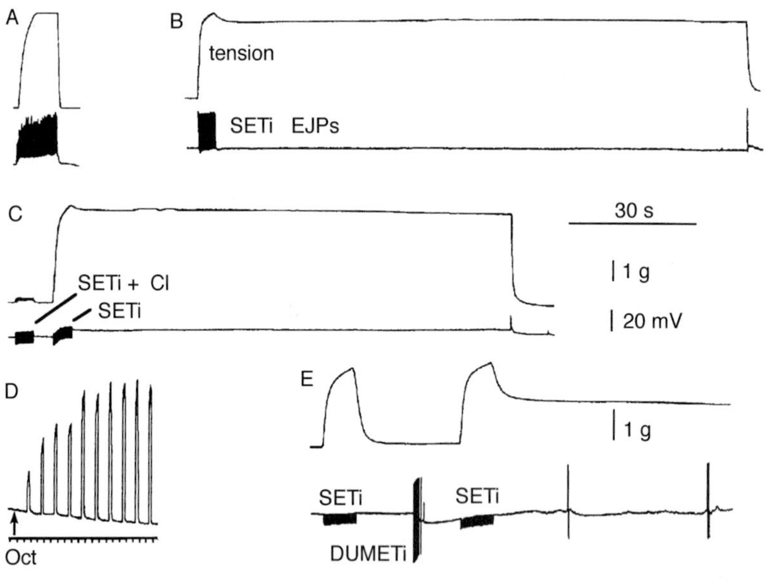

Fig. 23.11. Catch contraction and normal isometric contraction properties of the hind-leg extensor tibiae muscle. A. Normal isometric tension development (upper trace) during constant-frequency stimulation of the slow extensor tibiae motor neuron (SETi). Intracellular muscle response shown in lower trace. B. Natural catch contraction (upper trace) elicited by a short SETi burst at 10 Hz. Non-facilitating excitatory junction potentials (EJPs) in lower trace are characteristic of octopamine-promoted catch contraction, and differ from the normal facilitation shown in A. C. Catch contraction produced by SETi stimulation (8 Hz) following application of 10^{-8} mol l^{-1} octopamine to the extensor muscle for 20 min. SETi plus common inhibitor stimulation (8 Hz) did not induce the catch state. Catch contraction was abolished by single SETi impulse (right side, lower trace). D. Octopamine (10^{-8} mol l^{-1}, applied at arrow) initially causes potentiation of isometric tension before catch state is induced. SETi stimulated at 5 Hz for 10 s at 1 min intervals. Note relaxation of basic tonus (BT). Time marks 30 s. E. Appearance of isometric catch contraction (upper trace) during SETi stimulation following a brief burst of an identified, presumed octopaminergic, DUMETi-like neuron (intracellular recording, lower trace). The neuron was excited by passing depolarizing current into its soma; thick bars on record are SETi stimulation artefacts.

before relaxing over a period of up to 20 min and the muscle was electrically silent during contraction (Fig. 23.11B).

Octopamine modulated the contraction properties of the extensor muscle in two ways. First, when infused into the muscle as a saline solution, octopamine progressively potentiated the extensor's contraction tension, during SETi stimulation, within about 30–60 s following infusion (Fig. 23.11D). The growth of contraction tension was accompanied by a progressive increase in the amplitude of the extensor EJPs during SETi stimulation. This effect is also seen in the locust extensor muscle (Evans and O'Shea, 1978). Secondly, octopamine induced catch contraction upon SETi stimulation, starting at a concentration of 10^{-8} mol l^{-1} (Fig. 23.11C). This effect was delayed by an average of 12 min, but at 10^{-6} mol l^{-1} the delay was reduced to 4 min. The magnitude of catch contraction was proportional to the firing frequency of SETi, and it was especially sensitive to interpolated stimuli within a burst. The exact opposite effect is seen in the locust. Octopamine causes an increase in relaxation rate of twitch tension induced by SETi stimulation, and nothing resembling the catch contraction in the tree weta has been reported. Also, in the weta, octopamine appears to enhance the efficacy of the CI action on the extensor muscle, although this has not been well studied.

Rapid and complete termination of catch contraction is brought about by a single firing of SETi, FETi or CI (Fig. 23.11B, C). This remarkable phenomenon occurred any time after catch for the FETi, but required at least 10 s of maintained catch for termination to work with SETi stimulation. It is not known how the same stimulus that induces catch contraction can also terminate it.

Other muscles in the weta leg are affected by octopamine infusion, whereas this has not been demonstrated in the locust. The flexor tibiae shows potentiation of single twitches induced by the fast excitor as well as the slow excitor. Catch contraction is induced by 10^{-8} mol l^{-1} octopamine, although the very slow relaxation sets in sooner, after a catch duration of 1–2 min (Currie, 1988). The effect on both fast and slow units contrasts with the lack of octopamine effects on FETi-induced twitches in the locust extensor muscle. In addition, the weta retractor unguis shows potentiation of twitch contraction in 10^{-8} mol l^{-1} octopamine for both fast and slow motor neuron excitation of the muscle. Although catch contraction does not occur in this muscle, the relaxation phase of a tetanic contraction is greatly slowed (three times longer than normal) in 10^{-8} mol l^{-1} octopamine.

While octopamine has not conclusively been demonstrated in the DUM neurons in the weta, stimulation of single extensor DUMETi cells by an intracellular electrode in the cell body leads to the induction of catch contraction (Fig. 23.11E). DUMETi neurons could be identified by antidromic stimulation of their axons, which branch in the extensor muscle. While even a brief burst of DUMETi elicits a weak catch contraction, progressively larger catch tensions could be brought about by prolonged stimulation of DUMETi.

Recordings from DUMETi cells showed that the neurons fire more or less continually, but at different frequencies in different preparations. Such natural variation is undoubtedly due to differences in the diverse arrays of sensory inputs that drive the neurons (Hoyle and Dagan, 1978), and it could account for the large differences observed by Hoyle and Field (1983a) in leg-training experiments. Attempts to observe the natural firing of DUM neurons in semi-dissected crickets confirmed that many sensory pathways drive a variety of DUM motor neurons and intersegmental interneurons, and that these cells may naturally fire during walking to modulate the contractile responses of leg muscles (Gras *et al.*, 1990).

Much remains to be done to complete the story of catch contraction and DUM neuron control of behaviour in tree wetas. Certainly defence behaviour appears to be mediated by this neuromuscular phenomenon. Hoyle and Field (1983a) and H. Nishino (Christchurch, 1999, personal communication) have noted that tree wetas display thanatosis if held in a restricted position for a short period of time and then released. It is possible that this behaviour is also mediated by octopamine and DUM neurons. The natural resting position of tree wetas in galleries during the daytime is usually vertical. The catch mechanism could be a physiological method of locking the insect into this position without requiring the expenditure of neuronal activity to the muscles. Because there are many differences in the physiological responses to octopamine between locusts and wetas, it is important to investigate the extent to which octopamine plays a role in the behaviour of ensiferan Orthoptera compared with the caeliferan Orthoptera. Is the lack of catch contraction in the locust simply a reflection of the behavioural repertoire of this species, or is it a characteristic of the entire suborder of grasshoppers and locusts?

Motor Control of Defence Behaviour

In two species of New Zealand tree wetas, *Hemideina maori* and *Hemideina ricta*, and in the large Australian king cricket *Anostostoma australasiae*, defence behaviour includes flipping over on the insect's back with legs and mandibles outspread in a motionless but highly alert state. Any tactile stimulus to the ventral surface of the insect triggers a rapid flexion of the legs and biting of the offending object with the mandibles (described in Field and Glasgow, Chapter 16, this volume). The specific leg movement involves adduction of the coxo-trochanter joint and flexion of the femur–tibia joint which, when combined in all legs, brings any attacking object to the waiting open mandibles.

The rapid response of the legs to tactile stimulation of the ventral thorax and abdomen is known to be driven by local thoracic and abdominal reflex pathways, which do not require the brain or suboesophageal ganglion (Smith, 1979). The response to stimulation of the ventral thorax differs from that to stimulation of the ventral abdomen. Any stimulus to the thorax elicits flexion of all six legs, although the leg nearest to the stimulus locus usually gives the strongest response. In contrast, abdominal tactile stimulation results in flexion only by the hind legs, with weaker flexion by the middle legs (Fig. 23.12A). These observations prompted an investigation into the coordination pathways mediating the leg reflexes in the ventral nerve cord (Smith, 1979).

The neuroanatomy of tree wetas is described by O'Brien and Field (Chapter 8, this volume). Briefly, the thoracic and abdominal ganglia are linked by paired connectives (Fig. 23.12A). Each thoracic ganglion (T) gives off five pairs of nerves to the respective thoracic segment, of which the fifth (n5) is the main leg nerve. The metathoracic ganglion (T3/1) is fused with the first abdominal ganglion and therefore gives off an additional pair of nerves, which innervate the first abdominal segment. Another fusion occurs between the second and third abdominal ganglia (A2/3). Normally, each abdominal ganglion gives off one major pair of sternal nerves (stn) to its respective segment, however the fused A2/3 supplies one pair of nerves to A2 and a second pair of nerves to A3. Abdominal ganglion 8 (A8/11) represents a fusion of four terminal ganglia, and the resulting nerve supply is made more complex by sexual differences.

Connective ablations were made to determine CNS pathways for the reflex defence responses of the legs. The effects of single and double ablations were assessed by observation of reflex deficits, as well as by electrical recordings from the nerves (n5) to each leg. In the abdomen, for example, cutting the left connective between T3/1 and A2/3 resulted in loss of response to low-threshold (very weak) stimuli given to the abdominal sternites on the left side (Fig. 23.12B). This was reflected in the lack of motor axon firing in n5 to the left leg following ipsilateral abdominal tactile stimulation, while the right (contralateral) pathway remained intact and gave a normal motor response (Fig. 23.12E). However a high-threshold pathway was revealed to the ipsilateral side. Thus, strong stimulation to the left abdominal sternites caused a response in the left legs (Fig. 23.12C), which must have been mediated by a high-threshold pathway, which crosses the midline in A2/3 and crosses back to the ipsilateral side above the ablation (e.g. in T3/1).

Pathways in the thorax are more complicated. An example is indicated by a double ablation of the left connective between T1 and T2 and the right connective between T2 and T3 (Fig. 23.12D). Tactile stimulation of the second thoracic sternite caused a loss of response in the right front leg and the left hind leg. This means that within the mesothorax, ipsilateral stimuli activated a pathway to the nearest ipsilateral (middle) leg, as well as to the contralateral leg. In addition, pathways to the other thoracic segments must have remained ipsilateral and did not cross the midline.

As a result of specific combinations of ablations to the connectives and also to sternal abdominal sensory nerves, the following summary scheme has been worked out for the reflex pathway.

1. *Local branching.* The sensory pathway entering the CNS from tactile receptors on ventral cuticle immediately branches within any thoracic or abdominal ganglion innervating that area of

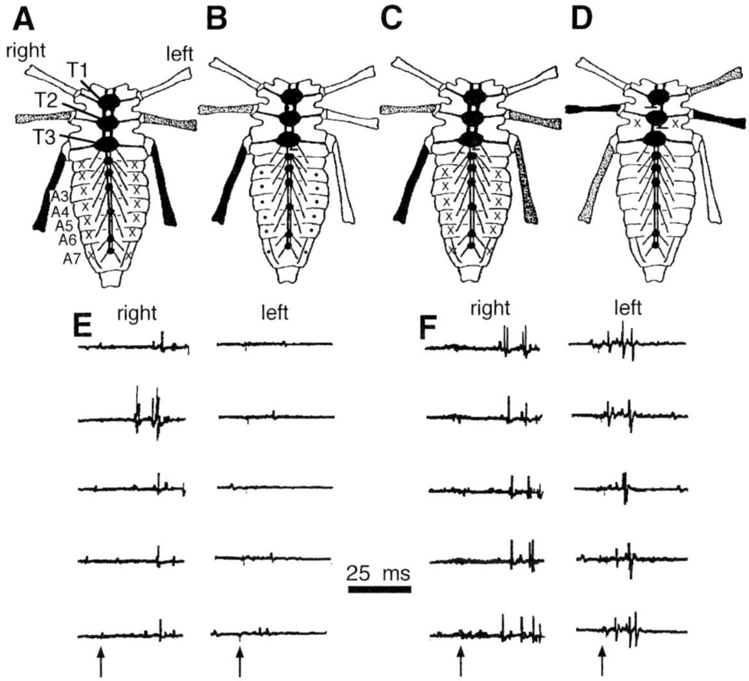

Fig. 23.12. Effects of cutting thoracic or abdominal connectives on the hind-leg defence reflex response. A. Normal intact weta. Stimulation to any abdominal sternite causes the strongest response in the ipsilateral hind leg (solid) and a weaker response in the middle leg (stippled). B. A left T3/1–A2/3 connective ablation removes all response from the left legs, when weak tactile stimuli are applied to the abdomen (dots). C. Strong tactile stimuli to the abdomen (x) reveal a high-threshold pathway, which crosses the midline to give left-leg responses. D. A double thoracic ablation (right T1–T2, left T2–T3) reveals local ganglion crossing only when strong stimuli (x) are applied to the mesothoracic sternite. E and F. Recordings from left and right metathoracic nerve 5 indicate motor neuron drive excited by reflex pathways to respective hind legs. Arrows indicate time of tactile stimulus application. E. Responses for ablation shown in B. Activity is lost on left side. F. With the same ablation (T3/1–A2/3), stronger stimulation (x) produces reflex response on both sides again, indicating that the pathway crosses from tactile stimulus site on left side of abdomen to right connective through abdominal ganglia, and then crosses the midline back to left thoracic legs via T2 and T3/1.

cuticle. One path crosses the ganglion midline, while the other remains ipsilateral. Then both pathways travel anteriorad in the case of the abdominal ganglia, and both anteriorad and posteriorad in thoracic ganglia. This means that, for a unilateral input to the thoracic ganglia, sufficient circuitry is available to activate a reflex in both legs of a segment within the single ganglion for that segment.

2. *Low-threshold pathways remain ipsilateral.* For inputs ascending from abdominal ganglia, low-threshold pathways do not cross the midline once they reach the thoracic ganglia. For inputs into any one thoracic ganglion, no crossing occurs after the pathway proceeds to neighbouring ganglia. Therefore tactile stimuli activate the reflex only in legs ipsilateral to weak (low-threshold) stimuli (Fig. 23.13A, C).

3. *High-threshold pathways cross repeatedly.* For high-threshold abdominal input ascending from A4 to A8, the pathway crosses the midline in A2/3 and again in T3/1 (Fig. 23.13B). Thus, in all ganglia posterior to A2/3, the abdominal pathway remains ipsilateral after branching within the ganglion of origin. The ascending abdominal pathway does not cross the midline in T2 or in T1. For high-threshold input to any thoracic ganglion, the pathway only branches within the ganglion of origin and does not cross the midline in any other thoracic ganglion (Fig. 23.13D).

In somewhat similar studies on the stridulatory

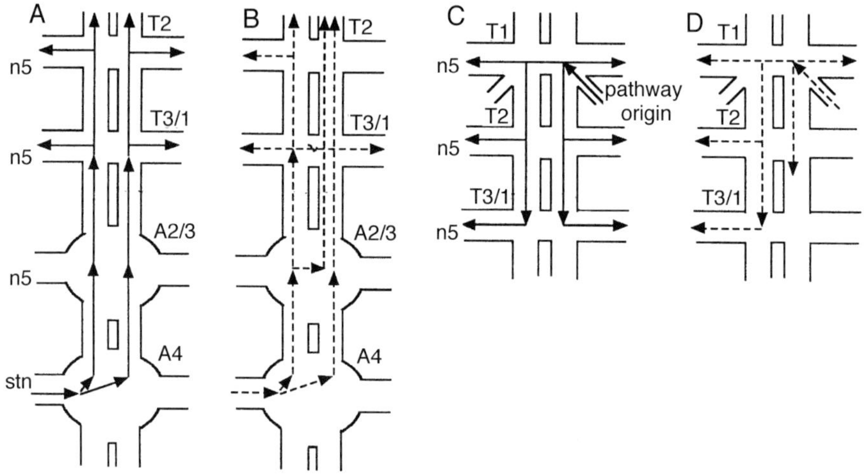

Fig. 23.13. Reflex pathways deduced for control of hind-leg defence flexion in the alpine tree weta, *H. maori*. Solid lines indicate low-threshold pathways; dashed lines indicate high-threshold pathways. A. Low-threshold pathways for sensory input originating in any abdominal ganglion below A2/3. Note only local midline crossing. B. High-threshold pathways for sensory input originating below A2/3. Note multiple abdominal and thoracic midline crossing. C. Low-threshold pathways for any thoracic sensory input; only local midline crossing occurs. D. High-threshold thoracic pathways resemble those for low-threshold stimuli. T1, T2, T3/1, pro-, meso- and metathoracic ganglia; A2/3, A4, abdominal ganglia 2/3 and 4; n5, fifth thoracic nerve root; stn, sternal nerve.

behaviour of the grasshopper *Gomphocerus rufus*, no midline crossing of commands occurred in T1 and T2 in descending pathways from the brain (Loher and Huber, 1966). Only coupling in T3 occurred, which was, furthermore, necessary for coordination of the hind legs during stridulation. This contrasts with the local ganglion branching observed in *H. maori*, where left–right coupling occurs in T1, T2 and T3. In the cockroach abdominal nerve cord, the pathways for escape behaviour mediated by the giant interneurons in the terminal ganglion are reminiscent of those described above for the weta. Local crossing of the midline occurs in the terminal abdominal ganglion and, from there onwards, strictly ipsilateral pathways are used to activate the legs by the ascending signals (Roeder, 1948).

Further central control of the defence reflex pathways occurs in the form of descending inhibition by the brain and suboesophageal ganglion. Ablation of the circumoesophageal connectives between the suboesophageal ganglion and T1 demonstrated that the strength of response increased significantly by a maximum of 163% (Smith, 1979). Also, the half-time to peak response decreased to a minimum of 90% after circumoesophageal connective ablation and 36% after T1–T2 ablation. The results not only show that descending inhibition from the brain or suboesophageal ganglion (or both) normally exerts an influence on the thoracic ganglia, but also that the first and second thoracic ganglia appear to exert an inhibitory influence posteriorad to the third thoracic ganglion. The inhibitory effect is greater in males than in females, probably owing to the stronger and larger hind legs, as well as the greater burst duration when the muscles are fired during the defence reflex.

In other insects, inhibition as well as excitation has been demonstrated between the brain/suboesophageal ganglion complex and the thoracic ganglia. The overall sum of influences depends upon specific behaviours and species. For example, classic experiments involving the removal of the brain (and descending inhibition) of the male praying mantis showed that mating behaviour is released (Roeder *et al.*, 1960). On the other hand, successive removal of the brain and suboesophageal ganglion in the locust *Schistocerca gregaria* causes first hyperactivity and then depression of cleaning behaviour by the front legs (Rowell, 1964). Furthermore, the meso- and metathoracic ganglia in the locust exert an inhibitory influence upon the prothoracic ganglion, which controls this behaviour.

Motor Control of Mandibular Movements

The mandibles are used for food-processing functions, such as chewing and biting off chunks of food for mastication, as well as for drinking (in association with the other mouth-parts). However, in New Zealand wetas, they are deployed in a variety of other behaviours. Thus, defence and agonistic behaviour of tree wetas (*Hemideina* spp.) involves a threat display with widely gaped mandibles (see Field and Glasgow, Chapter 16, and Field, Chapter 18, this volume), followed by defensive or offensive biting if attacked at close quarters. In males of tree wetas, as well as in Australian *Anostostoma* spp. and in many African king crickets, the mandibles have become enlarged through sexual selection and are used in agonistic behaviour during competition for females. In the New Zealand tusked wetas *Motuweta isolata* and *Hemiandrus monstrosus*, each mandible bears a forward projecting tusk containing a longitudinal row of stridulatory pegs or tubercles. These are used to produce sound when the closed tusks are rasped against each other during rapid opening of the mandibles (see Field and Deans, Chapter 10, and Field, Chapter 15, this volume). Another mandibular function, documented in *M. isolata* at least, involves interlocking and jousting with the mandibular tusks in males fighting for territorial rights. Only the alpine weta *H. maori* has been analysed neurophysiologically, in an elegant and detailed study of mandibular motor control (O'Brien, 1984). Mechanical analyses have been discussed by Field and Deans (Chapter 10, this volume).

Behavioural analysis of motor control

Feeding and drinking

Feeding behaviour in *H. maori* includes eating beetle larvae and chewing plant matter. In captivity, contact with *Tenebrio* larvae elicits wide opening of the mandibles, rapid grasping by the maxillae and a closing bite with the mandibles (the shearing action of which could sever the larva in a single bite). Several immediate and rapid bites (three to four cycles at a mean frequency of 2.3 Hz) are followed by a slower mastication (1–2 s^{-1}) of reduced amplitude until the food is ingested.

Normal chewing behaviour of plant leaves or fruit (e.g. apple chunks offered in captivity) is slower. Chewing begins with an initial sequence of four to five ineffective mandible closures, which progressively increase in amplitude until the chewing cycle is developed. The detailed movement of each mandible differs as a consequence of their asymmetric morphology. While holding the food piece with the maxillae, the longer overlapping left mandible reaches the open position in phase with the right, and then closes more or less smoothly until it reaches the normal rest position. At this stage, the right mandible is still closing (with a lag of 0.2–0.3 s in a 1.5 s cycle), but may have made contact with the food chunk as it approaches the midline. The left holds its position, while the right continues to close with a shearing action of the distal cusps over the food, to bite off a small piece. This plateau feature of the left mandible, which may occupy up to 45% of the chewing cycle, is clearly seen in Fig. 23.14A. Upon development of full force in the right closure, the left is slightly deflected. Usually, the right then opens before the left and both mandibles reach the fully opened position simultaneously. Smaller-amplitude masticating movements eventually follow, as the food is broken down against the proximal cusps, until the food is finally ingested. If a large food piece is obtained, the chewing movements may be interrupted by a prolonged closure to bite off a chunk of the food (arrow, Fig. 23.14A). For various food types, the amplitude and frequency of movement are adjusted to the task (bite off vs. mastication), indicating that peripheral proprioceptors are involved in motor control, but the frequency and coordination of movement appear to be centrally mediated. This is shown by loading one mandible with an outward-pulling (abduction sense) weight in a head-restrained weta, and observing normal frequency and mandibular synchrony during chewing (O'Brien, 1984). Drinking is accomplished by immersing the tips of the mandible, maxillae labrum and labium into water and cyclically opening and closing the mandibles with a low amplitude movement at about one-third masticating frequency.

Defence gaping and biting

Upon provocation or threat, the mandibles are widely gaped. Though apparently symmetrical, the two mandibles do not necessarily reach the same extreme opening angle (Fig. 23.14B). Maximum mean angles of 38° and 36° from the midline were measured in 15 wetas, with the greatest gape reaching 45° (90° between both mandibles) or 13 mm in a large weta. The extreme gape is only held for a short time (< 0.5 s) and is then followed by relaxation to a half-gape threat posture for up to a minute (Fig 23.14B, arrow). Alternatively, several rapid bites may follow if the animal is highly provoked (Fig. 23.14B, dots). Such defence bites are the most rapid action executed by the mandibles. Full closure from maximum gape may occur in 200 ms, which equals 150° s^{-1} (O'Brien, 1984). Any object grasped by the mandibles during a bite is forcefully held for 2–3 s, followed by reduced tension without releasing the object. Several cycles may occur, and the object may be grasped for 15–20 s before release. If the object moves or struggles, the biting cycle recurs with full force.

Motor neurons to mandibular and associated muscles

The anatomy of muscles and nerves in the head capsule is described by O'Brien and Field (Chapter 8, this volume). All of the muscles and sense organs associated with the mandibles are

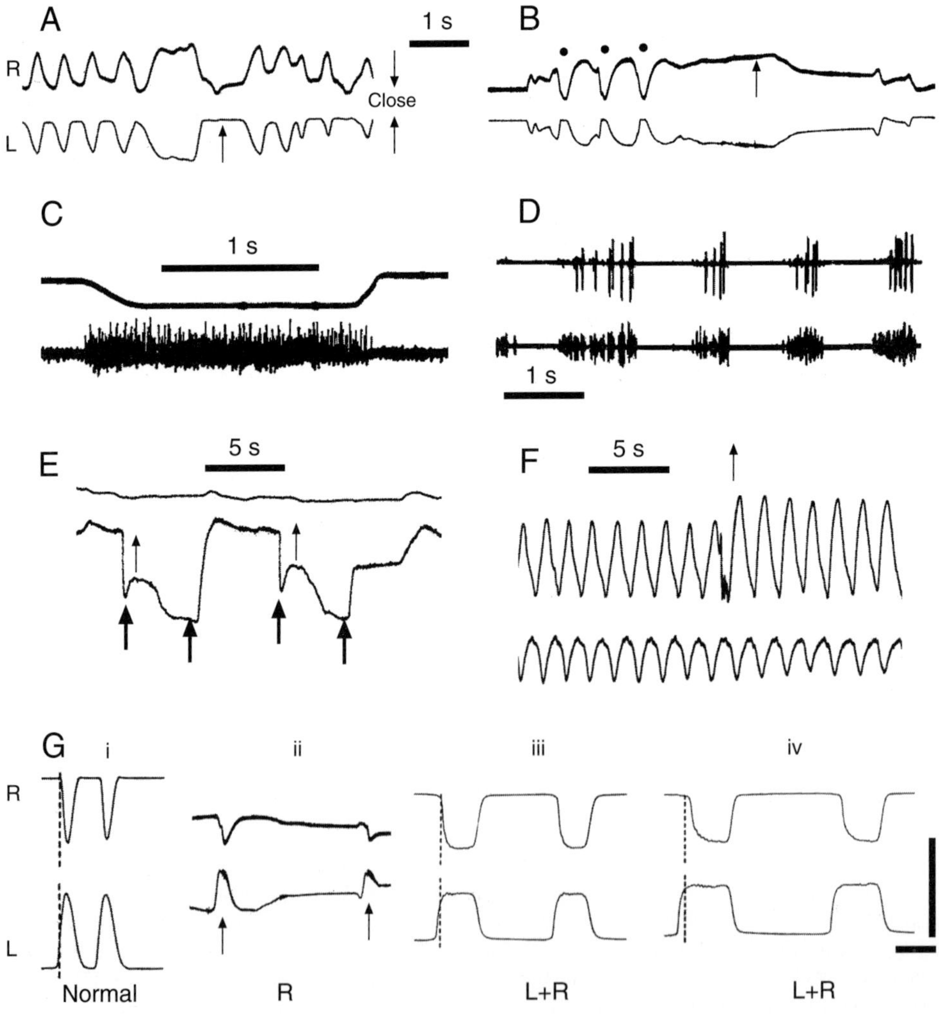

innervated by nerves arising from the suboesophageal ganglion, and not the brain.

The mandible is opened by the mandibular abductor muscle, M23. It receives at least six motor neurons (demonstrated by $CoCl_2$ back-filling; seven axon profiles observed in motor nerve sections). Only one or two of these units appear to fire as slow motor neurons in myogram recordings during mandible opening, and recruitment was not evident (O'Brien, 1984).

The mandible is closed by the massive mandibular adductor, M21, as well as the much smaller tentoro-mandibular muscle, TM1. Muscle M21 receives 18 motor neurons, which have been demonstrated both by sectioning the nerve and by axonal back-filling with $CoCl_2$ of cell bodies in the dorsal region of the suboesophageal ganglion. From myograms, these appear to include slow and fast motor neurons, which: (i) differentially innervate M21 in different regions of the muscle (Fig. 23.14D); and (ii) are recruited progressively, depending upon the strength of the movement. Fast units show marked facilitation and fire rapidly during strong bites.

Muscle M26 is the hypopharyngeal retractor. It is used to retract the hypopharynx during feeding and threat displays. Two excitatory motor neurons innervate M26; in myograms, one is continuously active during low levels of tension, while the other is recruited for higher tension levels.

A series of small muscles, collectively termed the tentoro-mandibular muscles, attaches from the tentorium to the mandible. They serve either to close the mandible or to monitor mandibular position and tension, as sensory receptor muscles (described in Field, Chapter 22, this volume). Muscle TM1 has not been studied histologically, but myogram recordings show that it is innervated by at least two excitatory motor neurons. They fire with the weak facilitation and high-frequency characteristic of slow motor neurons (Fig. 23.14C). This muscle provides weak force and is utilized in weak mandibular closing movements and bites. It is spontaneously active at a low rate, and is only inhibited during opening movements. Muscles TM2a and TM2b and the ventral muscle receptor organ (VMRO) are slender spindle-shaped muscles running parallel and located adjacent to TM1. The innervation of TM2a is unknown, while that of TM2b consists of a single excitor axon. The latter is part of the dorsal muscle receptor organ (DMRO). The motor patterns of firing in TM2a and b for normal mandibular behaviour are not known. They appear too weak to exert effective tension on the mandible.

Roles of the VMRO in mandible motor control

The sensory physiology of the VMRO, as well as the smaller DMRO and the apodeme strand receptor in the mandibles, is discussed by Field

Fig. 23.14. Motor control of mandibular movements in *H. maori*. A. Normal mandibular movements, with closing direction toward the midline. Left mandible (lower) arrives at closure first and holds a plateau, while right (upper) continues shearing action against left, clearly seen in biting off piece of apple (arrow). Note tiny deflection of left as right fully closes. B. Defence biting (dots) and gaping followed by half-relaxation (arrow) in male weta. C. Myogram recording of right muscle TM1 (lower trace) during imposed movement of right mandible (upper trace). At least two motor units fire with weak facilitation. Passive opening (downward trace deflection) of the mandible elicited resistance reflex firing of the muscle (absent in main mandibular adductor muscle). D. Myogram recordings from main adductor M21. Two recording sites (upper and lower traces) showed different motor units, indicating differential innervation of muscle. Fast units (upper) show strong facilitation compared with slow and intermediate units (lower). E. Effect of repeatedly applying load (13.1 g.cm) to the left mandible (movement, lower trace) while mandibles are at rest. Application and unloading of force shown by heavy arrows in opening direction (downward) and closing direction (upward), respectively. Upward arrows indicates initial closure response, which appears to represent the TM1 resistance reflex. The load then overpowers the TM1 action and opens the mandible. F. Effects of loading a mandible during mastication. Application of a 5.9 g.cm load to the right mandible (arrow) produces a grossly uneven opening position (upper trace) of the mandible compared with the left mandible position (lower trace), and yet a load compensation mechanism allows perfect synchrony and occlusion as the mandibles meet. G. Effects of ablation of VMRO sense organ on mandibular torque development during biting. Right mandible force: upper trace, left mandible force: lower trace. i. Normal torque (left mandible rests at 17° open from rest; right mandible rests at 14° open). ii. After right VMRO ablation, mismatched occlusion occurs (arrows) due to too rapid closure of left mandible. iii. After right and left (bilateral) VMRO ablations, the bite duration is prolonged and phase of closure is occasionally disrupted (mandible rest positions as in i). iv. With same bilateral ablation, but altered rest position (left, 17° open; right, 20° open), phase of closure is grossly disrupted, shown by dashed lines. Vertical scale: 1000 g.cm.

(Chapter 22, this volume). Through a combination of nerve and muscle recording, loading experiments and ablations of left and right VMRO receptors, O'Brien (1984) assessed the role of the receptors in control of the mandibles during active behaviours. Basically, two challenges are faced. First, the independently moving mandibles must be bilaterally coordinated during the closing phase to allow them to meet at the midline and occlude properly. The asymmetric and complex cusp surfaces require accuracy in bilateral control before the cusps are engaged. Secondly, once engaged against the cusps or a food particle, the mandibles must develop force proportional to the resistance encountered, with the limiting condition that force generation must be limited in the case of extreme resistance, which could pose danger to the integrity of the mandibles. Two major roles have been discovered: load compensation that provides control of mandibular position during active chewing and defensive biting, and load compensation to control adductor force against food. In addition, a resistance reflex operates to maintain passive (resting) mandibular posture.

Resistance reflexes

Passive opening of the mandibles (by an externally applied force) causes the tentorial muscle TM1 to fire both motor neurons as a resistance reflex (Fig. 23.14C). On the other hand, little response is seen in recordings from M21, unless the mandible is actively closing. Usually arthropods have evolved resistance reflexes in limbs to counteract perturbances induced by gravity and environmental forces against a steady postural stance. Thus, it is not surprising that little response is seen in the main adductor muscle of the mandible, since it is unlikely to be passively opened by environmental influences. The reflex seen in TM1, however, is characteristic of slow muscle resistance reflexes in motor neurons that aid in maintenance of posture. Loading experiments of single mandibles suggested that the reflex takes the form of a weak load-compensation reflex, which acts to maintain the rest 'posture' of the two mandibles in the midline, as discussed below (O'Brien, 1984).

Load compensation – weak postural control

By unilaterally loading the mandible with a weight attached via silk thread and led over a pulley, the perturbation to the resting posture of the closed mandibles in head-immobilized wetas could be studied. The first obvious result in a passive weta is that the mandible is pulled away from its midline rest position when the weight is imposed. Immediately after, the weta attempts to move the mandible back to the rest position, but is pulled by the weight toward the open position until unweighted by the experimenter (Fig. 23.14E). The initial closure response apparently results from the action of TM1 attempting to close the mandible, but is soon overpowered by the weight. Since TM1 and the VMRO are in parallel and act in concert, any change in length of TM1 is in direct proportion to that of VMRO. The tonic activity recorded in TM1 suggests that it normally operates to keep the mandible closed; it may be linked with a shared drive to the VMRO muscle, which is normally kept under tension by its own slow motor neuron drive. Any passive opening of the mandible would be detected by the VMRO, which would act as an error detector and could, in turn, enhance the ongoing motor drive to TM1. This is termed a weak load-compensation reflex, because TM1 is only capable of developing about 4 g tension, and it is not thought to participate in the forces used during biting and chewing (O'Brien, 1984).

Load compensation – chewing and biting

The dynamic voluntary behaviours of chewing and defensive biting involve complex peripheral feedback from the two VMROs, in addition to the main motor programme control from the CNS. Load compensation during mandibular movement is demonstrated by adding a horizontal load to one of the two mandibles during chewing (Fig. 23.14F). The open position of the loaded mandible is greatly increased for each chewing cycle (due to the outward pulling weight) and yet the two mandibles are always coordinated during the closure phase, without mismatching bites. This means that a load-compensating mechanism regulates mandibular position during the closing phase to allow the mandibles to reach the correct final midline position before they make contact. It is interesting to note that the opening phase of the cycle is not regulated by any obvious peripheral feedback, and the coordinated chewing may be achieved even with grossly differing opening positions (Fig. 23.14F).

An explanation of the compensation is provided by known VMRO interconnections. The two VMROs are linked by a reciprocal inhibitory circuit, whereby the opening of one mandible (stretch of its VMRO) inhibits firing in the VMRO of the other mandible. In this basic interaction, both VMROs fire evenly at a low level, unless the mandibles are asymmetrically opened. If the latter occurs, the VMRO output of the more widely opened mandible feeds back more strongly on to its own closer motor drive, while retarding the output of the contralateral VMRO on to its closer motor drive. The result is greater compensating drive to the more widely opened mandible during the closing phase, which would promote symmetry of mandible position during the closing phase before any mechanical contact is made.

Detailed support of this concept is provided by VMRO ablation experiments. Two major findings recurred in the effects of VMRO ablation on mandibular behaviours: (i) disruption of the perception of mandibular position; and (ii) disruption of symmetrical motor output to each mandible (O'Brien, 1984). The following effects occurred. VMRO ablation (unilateral and then bilateral) caused mismatched occlusion in biting, due to too rapid closure of the left mandible (Fig. 23.14Gii), prolonged duration of bites in both mandibles (Fig. 23.14Giii) and disrupted phase of closure, which could be enhanced by altering the relative positions of both mandibles (Fig. 23.14Giv). Additionally, the strength of bite could become uneven on both sides, the weta might sometimes bite with only one mandible and the angle of opening (amplitude of movement) could decrease on the operated side. The conclusion from these observations is that VMRO ablation disrupts the weta's estimation of the relative positions of both mandibles and that the effects are both unilateral and bilateral for a single VMRO ablation. Therefore, the receptors are acting as position error detectors.

Complexity is introduced by the observation that the effects of the right VMRO differ from those of the left VMRO. For example, ablation of the right VMRO causes an increased bite duration in both mandibles during bilateral chewing, and torque development and relaxation are out of phase between the two sides. On the other hand, ablation of the left VMRO caused only increased duration of the left mandible bite and had no effect on that of the right mandible. The right VMRO appears to exert a dominant effect over the left. This correlates with the different patterns of movement in the two mandibles in freely chewing wetas. The left reaches the midline sooner, while the right moves more in the final shearing phase of biting. This asymmetry of movement may have led to the evolution of different proprioceptive feedback effects in the two VMROs.

In addition to allowing bilateral coordination under the above experimental conditions, the same load-compensation mechanism should allow the weta to adjust closer muscle tension when the mandibles close asymmetrically upon a large food item during chewing. Obstruction of closure would be monitored by a VMRO error signal, whereby the movement of the mandibles does not correlate with programmed closure, and tension is built up in the VMRO muscle when its mandible is obstructed by resistance. It is postulated by O'Brien (1984) that the error signal would feed on to the ongoing drive of M21 and increase its output, in a similar fashion to that described in the locust (Seath, 1977). Similar autogenic reflexes have been described in many arthropod muscle receptors under efferent control (reviewed in Matheson and Field, 1995).

A final unusual conclusion arising from this investigation is that the rapid mandibular bite, evoked by stroking with a paint brush, appears to be a ballistically initiated manoeuvre, rather than being proprioceptively controlled during the closure phase, as seen in chewing. This follows from the observation that the rate of torque production during a defensive bite is uninfluenced by elimination of a variety of sense organs associated with the mandibles. Neither ablation of the ventral group of campaniform sensilla, nor presence or absence of cusp contact, nor ablation of VMROs had any effect on the rate or strength of muscle tension developed in rapid defensive biting.

However, VMRO ablation had one major effect on rapid biting: bite duration increased. Normally, rapid bites by intact wetas were of short duration against rigid force transducers. This apparently resulted from the inability of the restrained mandibles to close, and appeared to result from inhibition of the adductor muscle as it rapidly developed high tension against the immobile transducers. The inhibition must have been due to VMRO detection of a position error, rather than the high muscle tension, since removal of the VMRO had no effect on the strength of the bite.

The conclusion is that the bite was prolonged because the feedback that normally signals a position error and inhibits the adductor M21 was lacking.

Conclusions

The seemingly simple mandibular motor control system is shown to consist of an elaborate evolutionary response to solving the engineering challenge of controlling the position and force development of two opposing and independently articulated structures. The weta mandibles are asymmetrical both in their movements and in their morphology. By their apposable action, the movement of one directly affects the other. Biting requires the precise meeting of the mandibles and the synchronized production of strong forces.

The proprioceptors involved may not necessarily be involved in all of the movements made by mandibles. But, by imposing perturbations through loading and ablation experiments and by monitoring the position and tension of the mandibles, it was possible to see the expression of proprioceptor effects. The large array of proprioceptors in the weta mandible includes the VMRO, which is a complex stretch receptor under efferent control. The bilateral pair of these sense organs exerts differential influences on the mandibular control system, which may reflect the asymmetry and unique requirements of a biting mechanism.

References

Arikawa, K., Washio, H. and Tanaka, Y. (1984) Dorsal unpaired median neurones of the cockroach metathoracic ganglion. *Journal of Neurobiology* 15, 531–536.

Bässler, U. (1967) Zur Regelung der Stellen des Femur-Tibia-Gelenkes bei der Stabheuschrecke *Carausius morosus* in der Ruhe und der Lauf. *Kybernetik* 4, 18–26.

Bässler, U. (1983) *Neural Basis of Elementary Behaviour in Stick Insects*. Springer Verlag, Berlin, 169 pp.

Burns, M.D. (1973) The control of walking in Orthoptera. I. Leg movements in normal walking. *Journal of Experimental Biology* 58, 45–58.

Burrows, M. (1996) *The Neurobiology of an Insect Brain.* Oxford University Press, Oxford, 682 pp.

Cadenhead, L.C. (1982) The role of the metathoracic femoral chordotonal organ during walking in the weta. BSc Honours thesis, University of Canterbury, Christchurch, New Zealand.

Cruse, H. and Schwarz, W. (1988) Mechanisms of coupling between the ipsilateral legs of a walking insect (*Carausius morosus*). *Journal of Experimental Biology* 138, 455–469.

Currie, M.J. (1988) DUM neurones and the action of octopamine in the common New Zealand weta *Hemideina femorata*. MSc thesis, University of Canterbury, Christchurch, New Zealand.

Delcomyn, F. (1971) The locomotion of the cockroach *Periplaneta americana*. *Journal of Experimental Biology* 54, 453–469.

de Ruiter, A.F. (1984) The hind leg motor control system in wetas: anatomy, proprioceptive feedback, motor output. BSc Honours thesis, University of Canterbury, Christchurch, New Zealand.

Evans, P.D. (1985) Octopamine. In: Kerkut, G.A. and Gilbert, L.I. (eds) *Comprehensive Insect Physiology, Biochemistry and Pharmacology.* Vol. 11: *Pharmacology.* Pergamon Press, Oxford, pp. 499–530.

Evans, P.D. and O'Shea, M. (1978) Identification of an octopaminergic neurone and the modulation of a myogenic rhythm on the locust. *Journal of Experimental Biology* 73, 235–260.

Field, L.H. and Rind, F.C. (1981) A single insect chordotonal organ mediates inter- and intra-segmental leg reflexes. *Comparative Biochemistry and Physiology* 68, 99–102.

Forman, R. and Hoyle, G. (1978) Leg position learning by insects in behaviourally appropriate situations. In: *8th Annual Meeting of Society of Neuroscience.* Society for Neuroscience, Washington, DC, Abstract 591.

Gras, H., Hörner, M., Runge, L. and Schürmann, F.-W. (1990) Prothoracic DUM neurons of the cricket *Gryllus bimaculatus* – responses of natural stimuli and activity in walking behavior. *Journal of Comparative Physiology* 166, 901–914.

Hale, J.P. and Burrows, M. (1985) Innervation patterns of inhibitory motor neurones in the thorax of the locust. *Journal of Experimental Biology* 117, 401–413.

Hoyle, G. (1978) Distribution of nerve and muscle fibre types in locust jumping muscle. *Journal of Experimental Biology* 73, 205–234.

Hoyle, G. (1983) *Muscles and their Neural Control.* John Wiley and Sons, New York, 689 pp.

Hoyle, G. and Dagan, D. (1978) Physiological characteristics and reflex activation of DUM (octopaminergic) neurones of locust metathoracic ganglion. *Journal of Neurobiology* 9, 59–79.

Hoyle, G. and Field, L.H. (1983a) Defense posture and leg-position learning in a primitive insect utilize catch-like tension. *Journal of Neurobiology* 14, 285–298.

Hoyle, G. and Field, L.H. (1983b) Elicitation and abrupt termination of behaviorally significant catchlike tension in a primitive insect. *Journal of Neurobiology* 14, 299–312.

Loher, W. and Huber, F. (1966) Nervous and endocrine control of sexual behaviour in a grasshopper (*Gomphocerus rufus*, L., Acrididae). In: *Nervous and Hormonal Mechanisms of Integration. Symposium of the Society of Experimental Biology XX.* Cambridge University Press, Cambridge, pp. 381–400.

Maskell, F.G. (1927) The anatomy of *Hemideina thoracica. Transactions of the Royal Society of New Zealand* 57, 637–670.

Matheson, T. and Field, L.H. (1995) An elaborate tension receptor system highlights sensory complexity in the hind leg of the locust. *Journal of Experimental Biology* 198, 1673–1689.

O'Brien, B. (1984) Mandibular movements and their control in the weta, *Hemideina maori* (Orthoptera: Ensifera: Stenopelmatidae). PhD thesis, University of Canterbury, Christchurch, New Zealand.

Orchard, I. and Lange, A. B. (1985) Evidence for an octopaminergic modulation of an insect visceral muscle. *Journal of Neurobiology* 16, 171–181.

Pearson, K.G. and Iles, J.F. (1970) Nervous mechanisms underlying intersegmental coordination of leg movements during walking in the cockroach. *Journal of Experimental Biology* 58, 725–744.

Phillips, C.E. (1981) Organization of motor neurones to a multiply innervated insect muscle. *Journal of Neurobiology* 12, 269–280.

Rind, F.C. (1976) The metathoracic femoral chordotonal organ and its involvement in the posture and locomotion in the weta (*Hemideina thoracica*). BSc Honours thesis, University of Canterbury, Christchurch, New Zealand.

Roeder, K.D. (1948) Organization of the ascending giant fiber system in the cockroach (*Periplaneta americana*). *Journal of Experimental Zoology* 108, 243–261.

Roeder, K.D., Tozian, L. and Weiant, E.A. (1960) Endogenous nerve activity and behavior in the mantis and cockroach. *Journal of Insect Physiology* 4, 45–62.

Rowell, C.H.F. (1964) Central control of an insect segmental reflex. *Journal of Experimental Biology* 41, 559–572.

Seath, I. (1977) The effects of increasing mandibular load on electrical activity on the mandibular closer muscles during feeding in the desert locust, *Schistocerca gregaria. Physiological Entomology* 2, 147–156.

Smith, P.E. (1979) Behavioural and reflexive analysis of a defence response in *Hemideina maori* (Orthoptera: Stenopelmatidae). BSc Honours thesis, University of Canterbury, Christchurch, New Zealand.

Stevenson, P.A., Pflüger, H.-J., Eckert, M. and Rapus, J. (1992) Octopamine immunoreactive cell populations in locust thoracic–abdominal nervous system. *Journal of Comparative Neurology* 315, 382–397.

Stevenson, P.A., Pflüger, H.-J., Eckert, M and Rapus, J. (1994) Octopamine-like immunoreactive neurones in locust genital ganglia. *Cell Tissue Research* 275, 299–308.

Theophilidis, G. (1983) A comparative study of the anatomy and innervation of the metathoracic extensor tibia muscle in three orthopteran species. *Comparative Biochemistry and Physiology* 75a, 285–292.

Usherwood, P.N.R., Runion, H.I. and Campbell, J.I. (1968) Structure and function of the chordotonal organ in the locust leg. *Journal of Experimental Biology* 48, 305–323.

Watson, A.H.D. (1984) The dorsal unpaired median neurons of the locust metathoracic ganglion: neuronal structure and diversity, and synapse distribution. *Journal of Neurobiology* 13, 303–327.

Weiland, G. and Koch, U.T. (1987) Sensory feedback during active movements of stick insects. *Journal of Experimental Biology* 133, 137–156.

Wilson, J.A. and Hoyle, G. (1978) Serially homologous neurons as concomitants of functional specialization. *Nature* 274, 377–379.

24
Circadian Rhythms in Tree Wetas, *Hemideina thoracica*

Robert D. Lewis and Anna York
School of Biological Sciences, University of Auckland, Private Bag 92019, Auckland, New Zealand

Field Behaviour: the Question of Timing

All living organisms are subjected to a diverse set of geophysical oscillations, due to the ceaseless movement of the moon around the earth, the earth around the sun and the rotation of the earth on its own axis. Hence rhythmicity is a basic property of living systems (Saunders, 1977). This chapter will focus on the behavioural and physiological rhythms of the tree weta *Hemideina thoracica* (White), on their description and on the biological clocks that control them. The results, ideas and conclusions presented here are based on research projects carried out in the Department of Zoology (later the School of Biological Sciences) in the University of Auckland over almost three decades.

The approach we have taken to understanding weta rhythmicity is highly mechanistic, treating biological clocks as living oscillators and explaining their behaviour and regulation through analogies with the control systems of engineers. However, mechanistic models of behaviour are of little value unless they have a foundation in a knowledge of the natural behaviour of animals. We begin, therefore, with a brief consideration of the field behaviour of the tree weta.

Tree wetas, *H. thoracica*, in common with the majority of other terrestrial animals, are subjected to a range of rhythmic changes in their physical and biotic environments, and they have subsequently adapted their behaviour and physiology in order to survive. Their usual behaviour is to remain hidden during the day in holes in trees, under bark or in dense hedges, emerging at sunset to begin foraging on plant and animal matter. They return to the seclusion of their burrows before dawn. This typical nocturnal pattern of behaviour may be relaxed on wet, dull days (Smith, 1931). In winter, wetas may remain in their retreats, either in harems of a single male and several females or individually, for several days without activity (Morris, 1992). These temporal adaptations of behaviour enable them to avoid predation and harsh physical conditions and to take advantage of opportunities to catch prey or maintain their social structures (such as territoriality).

Bullivant (1990) and York (1998) have recorded the times of emergence of wetas from holes in logs in purpose-built cages, using direct observation (Fig. 24.1) or data loggers, which recorded weta movement, light intensity and temperature at regular intervals. In general, their results confirm the overall description of wetas' nocturnal behaviour, but it has become clear that the timing of activity may not be as precise as we first thought. In her recent study, York (1998) observed the emergence of wetas from holes in logs for over a week. She noted that the time of emergence advanced each day until it coincided with the evening twilight. Activity on the following day was delayed significantly, as if the timing had been reset by the exposure to the evening light. Although such an observation is consistent with findings in the flying

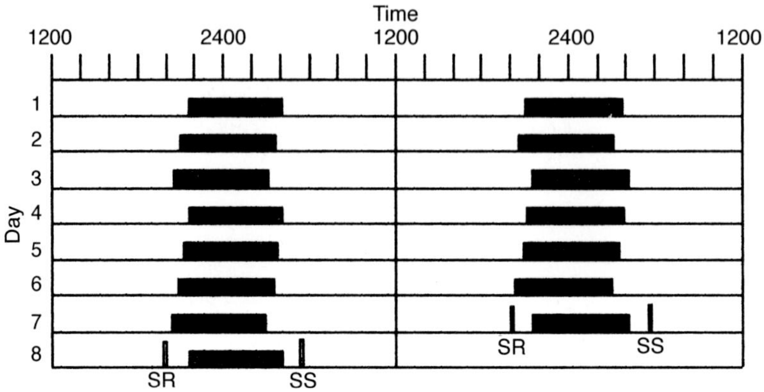

Fig. 24.1. The times of onsets and ends of weta activity in an outdoor arena in relation to light intensity/sunset. Note the progressively earlier onsets of activity until emergence coincides with evening twilight, at which point it is delayed by 2 h. SR, sunrise; SS, sunset.

squirrel, *Glaucomys volans* (DeCoursey, 1986, 1989), it is not what would be predicted based on laboratory results, where resetting is achieved on a daily basis. There is no substantial evidence that this species exhibits lunar phobia, although Bullivant (1990) noted the reduction of activity at the times of full moon in his Masters study of behaviour of wetas in the field.

These descriptions of the timing of activity prompt the question of the control of the initiation and cessation of the nocturnal bouts of weta activity. We could assume that the timing of these daily patterns of activity is triggered either by direct responses to the changes in light intensity at sunrise or sunset or the concurrent changes in temperature. Alternatively, there may be some element of endogenous control, as is the case for numerous other animals (Saunders, 1977). We favour the idea that the timing of locomotor activity is in part under the control of an endogenous circadian clock. It is pertinent to note that their daytime habitat shields them from immediate changes in the physical environment. We believe that a mechanism predicting evening or morning twilight would be of selective advantage over mechanisms that involve regular monitoring of the prevailing conditions.

Tests of the Endogenous Clock Model

The internal clock hypothesis has been tested in numerous animals by recording some aspect of behaviour or physiology in timeless conditions (for example, locomotor activity may be recorded in total darkness at constant temperature) over extended periods of time. Almost all of the animals that have been tested continue to show rhythmic activity (so called free-running rhythms), with timing similar to that exhibited in natural conditions. This has led to the development of the concept of internal clocks; endogenous timing systems, which, in the absence of external time cues, continue to run with frequencies approximating the natural geophysical equivalents. The clocks' functions in nature are to synchronize all of the rhythmic processes of the body and restrict activity to the most appropriate times of day (Enright, 1970).

Since endogenous biological clocks run in constant conditions, with a periodicity that differs slightly from the rhythms of the real world, they are termed circadian (about a day) clocks (Saunders, 1977). Circadian rhythms are therefore only exhibited in constant conditions, where they are free to drift away from the temporal constraints of the natural environment; in the field, the rhythms are strictly daily. In natural conditions, circadian clocks are synchronized to the 24 h time-scale of the solar day through regular adjustments, which also take into account seasonal changes in the times of sunset and sunrise (Aschoff, 1960).

To test the internal-clock hypothesis in wetas, the locomotor activity of many individuals has been recorded in constant laboratory conditions, where we believe they are effectively isolated from the rhythms of the environment. During this time, they are generally held in total darkness at 20°C ± 1°C and subjected to the minimum of disturbance, being cleaned and fed on carrot at irregular times at approximately 14-day intervals.

The locomotor activity is detected either with simple tilting actographs, in which movement of the subject within the pivoted tray activates a magnetic switch (Fig. 24.2a), or, alternatively, subjects are held in circular trays with bisecting infrared beams, which are invisible to the wetas (Fig. 24.2b). Activity events are recorded on slow-running event recorders, or are accumulated at 10 min intervals by computer-based data loggers. Results are pasted up from the event recorder traces as actograms, with consecutive days of activity below each other, or are plotted by computer. The height of each bar of the computer-produced actograms represents the amount of activity per 10 min. Actograms are generally 'double-plotted' to aid visualization of the patterns of the free-running rhythms (Fig. 24.3). In this presentation, the right panel of data is a repeat of the left side, raised up 1 day. Rhythms are analysed for periodicity and active phase length, either manually or with specialized methods of time-series analysis, such as

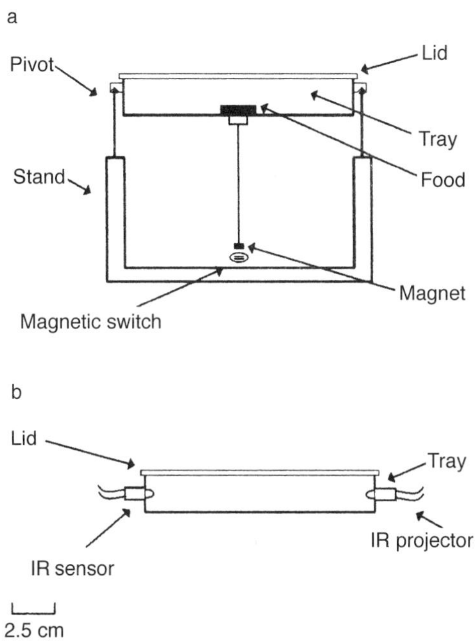

Fig. 24.2. Diagrams of (a) the tilting actograph apparatus and (b) the infrared (IR) light beam actograph used for recording weta locomotor activity in the laboratory.

the modified version of Enright's periodogram, and form estimates (Enright, 1965; Williams and Naylor, 1978).

The majority of freshly collected subjects display very clear circadian rhythms of locomotor activity, lasting for several days to many weeks in constant conditions. The longest record of consistent rhythmicity lasted over 6 months. The periods of the rhythms are generally less than 24 h during the first days in constant conditions, and then invariably lengthen, often increasing to 25 h or longer and displaying various kinds of lability, such as a change in period or breakdown of the rhythm (Lewis *et al.*, 1991). The longest steady-state period is about 28 h in total darkness. The active phases are typically about 6 h long, with abrupt onsets and ends of activity. These records are evidence for impressive timing mechanisms in these insects, and clearly support the endogenous clock hypothesis.

In many examples, the wetas are not uniformly active throughout the active phases, but exhibit short-period rhythms of activity and rest (Fig. 24.4). These high-frequency rhythms, often called ultradian rhythms in the literature (Lloyd and Rossi, 1992), could either arise simply from alternating bouts of activity and rest or could be the overt expression of high-frequency oscillators within the overall timing system. Alternatively, ultradian rhythms may be an additional layer of rhythmicity on top of the existing circadian system.

The Entrained Clock Hypothesis: Phase Control with Light Cycles

The free-running periods of the circadian clocks of almost all animals that have been tested are not exactly 24 h, but lie in the range of 23 h to 28 h. If endogenous circadian clocks were the sole controllers of the timing of activity in the field, it is obvious that the timing of natural behaviour would soon slip out of phase with the daily cycles. Hence there must be some adjustment of the endogenous rhythm to synchronize it with the 24 h cycles. The natural deviation of the free-running period from 24 h may seem, at first consideration, to be an error that has to be corrected for in field conditions. However, there may be an adaptive value to the deviant nature of endogenous clocks, relating to the optimal timing of behaviour in the field.

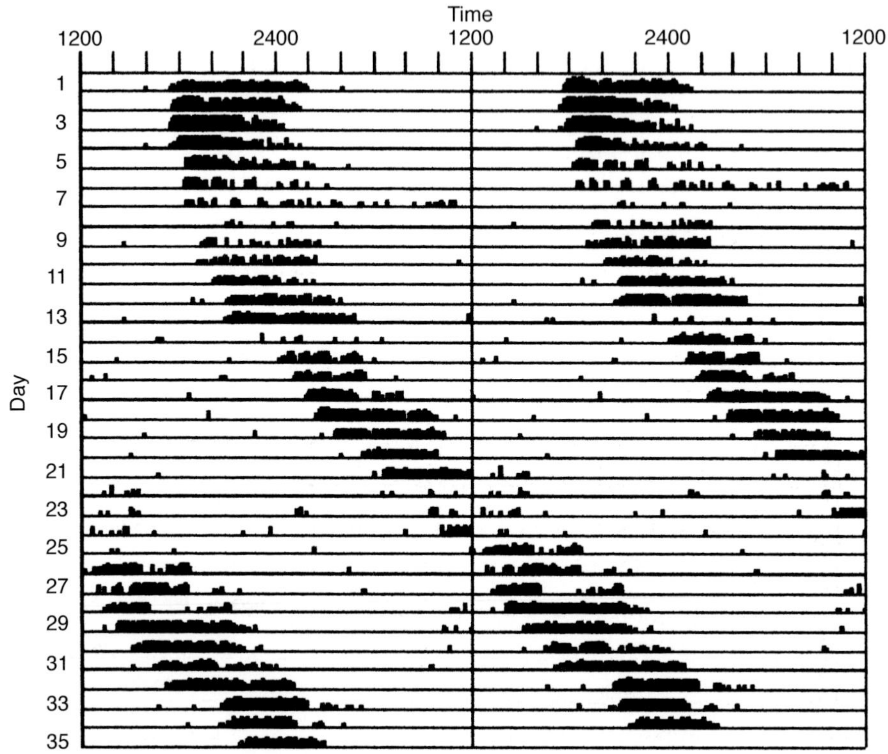

Fig. 24.3. A double-plotted actogram of the free-running locomotor rhythm of one weta in constant conditions. Note that the activity begins about half an hour earlier each day for the first few days (circadian period 23.5 h), and subsequently the period increases to 25 h for the remainder of the record. Results such as this support the endogenous clock model for wetas.

We propose that wetas' endogenous clocks are entrained in the field principally by natural light cycles. However, observations in the field support the notion that wetas are frequently cut off from the natural light cycles in the daytime as they lie hidden in holes in trees or under bark. Hence, the only light they perceive are the pulses of twilight as they move in and out of their cover. These twilight pulses are the main phase-controlling cues, adjusting the clock at regular intervals so that the active phase begins at the most appropriate time.

Field observations confirm that wetas begin to emerge from their daytime retreats soon after sunset, when the light intensity is about 200 lux. They remain with their posterior ends protruding from the holes before finally emerging to begin the active phase when the light intensity has fallen to 5 lux. They return to the holes before dawn (York, 1998). We propose that these daily patterns of activity exhibited in the field are the output of the entrained circadian system. This question of the mechanism of entrainment of the weta clock by light cycles has been extensively studied through observation of behaviour in the field and recording their locomotor rhythms under artificial light cycles and pulses in laboratory conditions (Lewis, 1976; Christensen, 1978).

In field conditions, weta clocks must be adjusted so that the free-running period adapts to the 24 h period of the solar day. But what is the natural period of circadian clocks of wetas in the field? The period of circadian clocks is only revealed in timeless conditions, since in the field it is constantly adjusted to ensure synchronization with the solar day. In constant conditions, the periods of free-running rhythms spontaneously change after days or weeks in isolation. None of these changes has any adaptive value. They are simply the product of the timing system after many days of experimental isolation, telling us

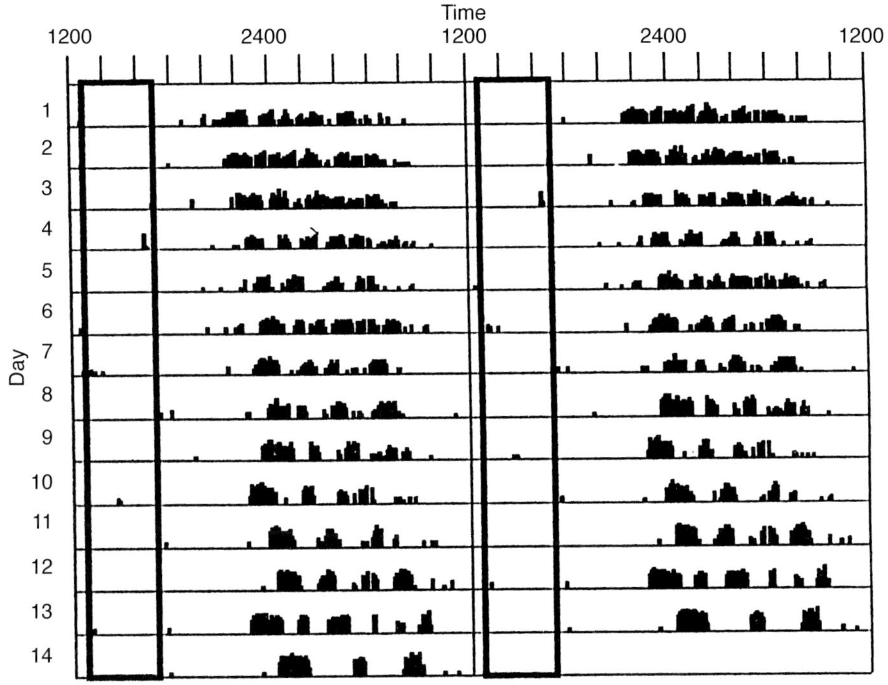

Fig. 24.4. A section of a weta rhythm entrained by light cycles (light : dark (LD) 1 : 23) in which the active phases subdivide into regular blocks of activity. This provides evidence for ultradian rhythms in *H. thoracica*.

something of the nature of the timing system, but having no ecological significance. We believe that it is the period of the rhythm in the first 2–3 days of timelessness that is relevant to the natural entrainment process. In the weta, this critical value is almost always close to 23 h (Lewis *et al.*, 1991).

The entrainment hypothesis has been tested in the laboratory by subjecting wetas to 24 h artificial light cycles of various pulse duration and intensity. In these tests, the subjects remain in the recording apparatus and are exposed to the full duration of the artificial light pulse. This is in contrast to the field situation, where we suppose that they only perceive light during evening twilight and avoid the full intensity of daylight. In a typical test of the entrainment hypothesis, wetas are subjected to light cycles of 8 h light (*c.* 200 lux) and 16 h dark (LD 8 : 16), ranging down to just 3 h of light in 24 h (LD 3 : 21) (Fig. 24.5).

Our results confirm that the locomotor rhythms of wetas may be entrained to laboratory light cycles when the light signal lasts from 12 h down to only 3 h per day. During entrainment, the light cycle has two functions. The obvious role is to adjust the period of the endogenous clock from the typical value of 23 h to the 24 h period of the solar day. This must be achieved by delaying the clock by 1 h, the difference between the length of the solar day and the natural period of the clock. The other role of the entraining cycles is to ensure that the activity begins after the onset of darkness, so that the phase relationship between the end of evening twilight and the onset of activity is appropriate (Pittendrigh, 1965).

It follows from this that a single light pulse administered during an otherwise free-running rhythm should advance or delay the rhythm, depending on the time of administration of the pulse in relation to the active phase. By plotting the phase changes produced by single pulses against the time of administration of the pulse, we can develop phase response curves (PRCs) for light pulses and test the entrainment model (Pittendrigh, 1965).

To develop PRCs for the effect of light pulses on wetas, individuals were subjected to single light pulses of either 12 h, 8 h, 4 h or 2 h duration, with

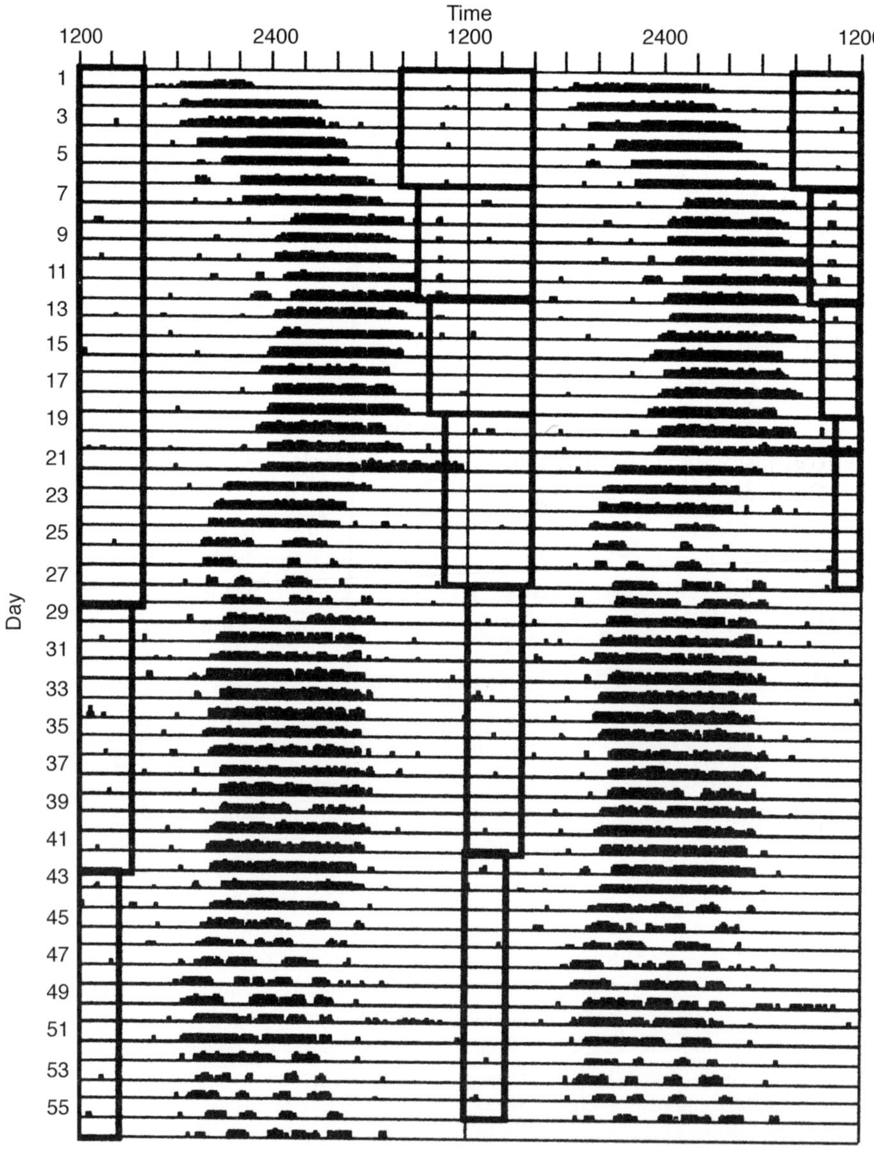

Fig. 24.5. A series of entrained weta rhythms in which the lengths of the light phases are systematically reduced from 8 h (LD 8 : 16) to 3 h (LD 3 : 21). The rhythms entrain to each light regime, with a steady-state phase relationship dictated by the pulse length.

each pulse falling at a different time of the circadian cycle, or circadian time (Fig. 24.6). The pulses either advanced the rhythm, delayed it or left it unperturbed. The PRC (Fig. 24.7) plots the phase changes produced, either advances or delays, by single light pulses falling at different times of the circadian cycle. The zero point of the circadian time-scale (CT 0) is defined as the onset of the activity phase preceding the pulse, whereas the time of the pulse is measured at its midpoint.

The resulting curves show that there are clearly times when single pulses advance the rhythm, and also times when they delay it. The amplitude of the curve depends on the duration of the pulse; 12 h pulses give larger phase changes than 2 h pulses. Examination of the curves confirms that the times when light pulses delay the rhythms are at about the times of the onset of the active phase, as we would expect from our field observations. The results of the tests of the entrainment model confirm its basic principles.

So we conclude that, in the field under natural entrainment, the *c.* 23 h clocks of the wetas are entrained to 24 h by the delaying action of some feature of the natural light cycle (such as evening twilight). This forces the clock into a 24 h period and ensures that activity begins at the correct time in relation to sunset. These 24 h cycles of light can be duplicated in the laboratory with similar results to the field situation.

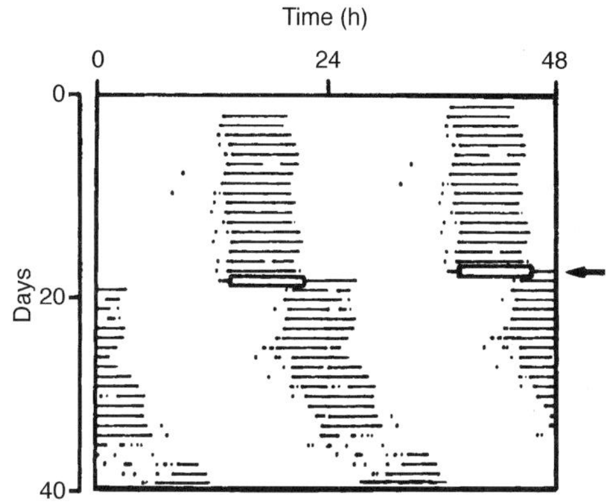

Fig. 24.6. The effect of a single 8 h light pulse on the phasing of a locomotor rhythm under otherwise constant conditions. The pulse in this example delays the rhythm by 5.5 h.

The Entrained Clock Hypothesis: Phase Control with Temperature Cycles

Whilst it is undoubtedly true that light cycles provide the most consistent entraining signals in the field, temperature cycles may well also provide synchronizing information, but with less day-to-day precision. At the times of the year when wetas are confined to the dark holes in the trees for several days or more, temperature cycles may well be the only source of entraining information. To test whether temperature cycles can entrain the locomotor rhythms, the activity of subjects was recorded in constant darkness with imposed temperature cycles (Gander, 1979).

Gander found that the circadian clock could be entrained by temperature cycles if the difference between the warm and cold phases exceeded about 7°C. The active phases started at the beginning of the cold phase, as we would expect from the timing of field behaviour. This may be relevant in natural conditions and may supplement entrainment by light cycles.

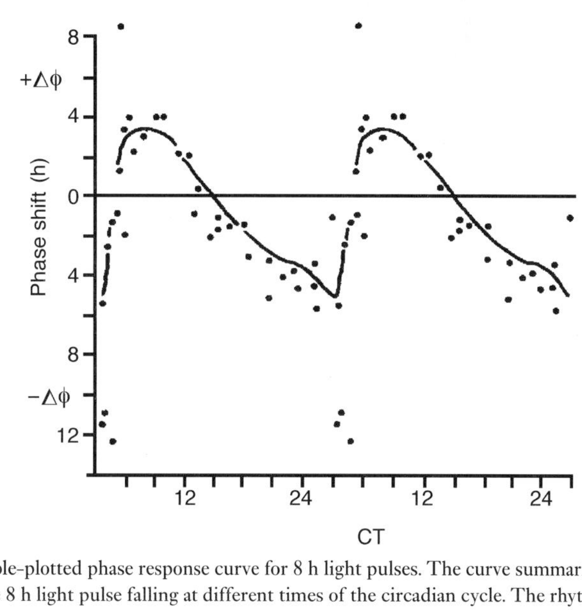

Fig. 24.7. The double-plotted phase response curve for 8 h light pulses. The curve summarizes the phase changes produced by a single 8 h light pulse falling at different times of the circadian cycle. The rhythms are either advanced or delayed, or remain unaffected.

The Entrained Clock Hypothesis: Phase Control with Cycles of Electromagnetic Fields (EMFs)

Light and temperature cycles are undoubtedly very obvious candidates for the natural entraining agents of the weta circadian clock, and it is not surprising that, in the laboratory, cycles of both light and temperature entrain rhythms with the appropriate period and phase relationships, as we have seen. Other entraining agents of circadian clocks include feeding times in honey-bees (Frisch and Aschoff, 1987) and house sparrows (Hau and Gwinner, 1992) and possibly electric fields for humans (see Palmer, 1976). Clearly, all these cues must be translated into some kind of information that directly perturbs the underlying clock to give the phase changes essential for entrainment.

During the course of our studies, one unexpected finding was that weta locomotor circadian rhythms may be entrained by electromagnetic fields (EMFs). Weta rhythms were recorded using the tilting actograph apparatus, coupled with a slow-running event recorder. They were subjected each day to 6 h pulse of EMFs created by a 230–6000 V transformer placed about 30 cm above the wetas. The locomotor activity was clearly entrained by the imposed cycles, with activity taking up the 24 h period of the EMF cycle, and beginning soon after the end of the 6 h pulse (Fig. 24.8). The locomotor rhythms remained entrained for many days, but occasionally broke away and re-entrained at a later time. There was very little doubt that our small sample of wetas was responding to the imposed field (King, 1988).

It is difficult to judge whether these unusual results have any relevance to the natural behaviour of wetas or whether the responses are non-adaptive effects of the EMFs on the mechanism of the circadian clock in this insect. The intensity of the earth's magnetic field changes on a daily basis, and could possibly entrain the clocks of animals that are sensitive to it. We know, for example, that transport of protein through the nuclear membrane is an essential process in the clock mechanism (Edery *et al.*, 1994). Maybe the electric fields upset the fine balance of ions or electrical charges that may be involved.

These results provide some of the first objective evidence that living organisms are affected by high-voltage fields, and this is undoubtedly relevant to the understanding of the effects of electric fields on other species.

Temperature Compensation

To be of value as biological timers, endogenous clocks should not be sensitive to long-term

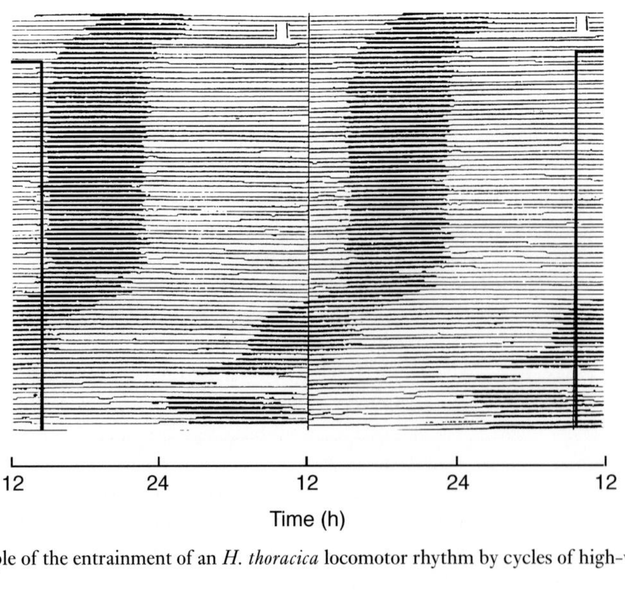

Fig. 24.8. An example of the entrainment of an *H. thoracica* locomotor rhythm by cycles of high-voltage electrical fields.

changes in temperature. They should not, for example, run faster in summer than in winter, as they would go beyond the limits of entrainment by the natural synchronizers. Normal biochemical processes, however, typically double their rate of reaction for a 10°C rise in temperature, but we should expect that the periods of all biological clocks are compensated for these effects of temperature, at least within the normally encountered range of biological temperatures (Balzer and Hardeland, 1988).

To test for temperature compensation of the weta circadian clock, the free-running rhythms of individuals were recorded at various constant temperatures between 15°C and 30°C over several days at each temperature. The locomotor rhythms were analysed for period and active phase length. In the temperature range between 18°C and 30°C, the period of the rhythms remained remarkably constant, often not changing more than an hour or so between the highest and lowest temperatures. However, the length of the active phases and the overall amount of activity depended on temperature, since both measures of activity are reduced at the lower temperature. Below about 15°C, the active phases became very reduced and faded away (Gander, 1979). These data support the generally accepted concept that biological clocks are temperature-compensated, at least over a limited range of temperatures, and that in natural conditions throughout the year the periods of the clocks lie within the range of values that may be entrained to the cycle of the natural environment.

Circadian clocks, however, are not independent of or insensitive to temperature, since 24 h temperature cycles can act as synchronizing agents for circadian rhythms (including the weta, as we have seen) (Gander, 1979). One of the major challenges in the development of models and explanations of the mechanisms of circadian clocks is the development of systems that account for all of these divergent responses of clocks to temperature.

Clock Location: the Role of the Optic Lobes

The reality of biological clocks would be more convincing if they could be identified in particular locations within the body. We need to be cautious, however, for, whilst it is often tempting to focus on the biological clock as a single structure, it is probably more realistic to consider the clock to be a complex system of interrelated parts scattered throughout the body. The search for the location of a single clock location may be inappropriate. Clock systems have, however, been modelled as hierarchies of master clocks and subclocks (Brady, 1982), in which we envisage the master clock both as the coordinator of an array of subsystems throughout the body and also the connection with the natural light cycles.

Proof that suspected tissues are the sites of the master clock in any species is difficult to obtain, since abolition of a rhythm following surgical removal of a site only shows that the tissue is vital for the expression of the rhythm (Colwell and Page, 1990). Even the recording of rhythmic activity from isolated tissues *in vitro* only demonstrates that the tissue is capable of self-sustained rhythmicity and could merely be a candidate for a subsystem.

The best test for a master clock is the transplantation of the suspect tissue, together with the characteristics of a donor's clock, to a clockless recipient. This has only been achieved in a handful of cases. Pineal glands in sparrows, whole brains in silk moths, optic lobes in cockroaches and crickets (reviewed by Saunders, 1977) and suprachiasmatic nuclei in neonatal hamsters (Ralph *et al.*, 1990) cover most of the published examples. Our work on wetas has only progressed in a limited fashion to answer the question of master clock location (Waddell *et al.*, 1990).

Surgical removal of both the left and right optic lobes destroys the locomotor rhythm (Fig. 24.9) (sham operations cause a minimum of disruption). Removal of either the left or right optic lobe inevitably leads to an increase in the period of the locomotor rhythm (as has been demonstrated in cockroaches). Cutting both optic tracts also abolishes the locomotor rhythm, at least for a few days, after which the neural connections re-form. So the optic lobes of the weta are vital for the expression of locomotor rhythms (Waddell *et al.*, 1990).

The rhythm of unilaterally lobectomized subjects continues even when the laminar region and much of the medulla is removed by cutting the optic lobe in half. Our interpretation of the anatomy of the optic lobe of the weta (Fig. 24.10) shows that the lobular region lies in the midbrain and not in the optic lobe itself. This observation, together with the results of the various surgical

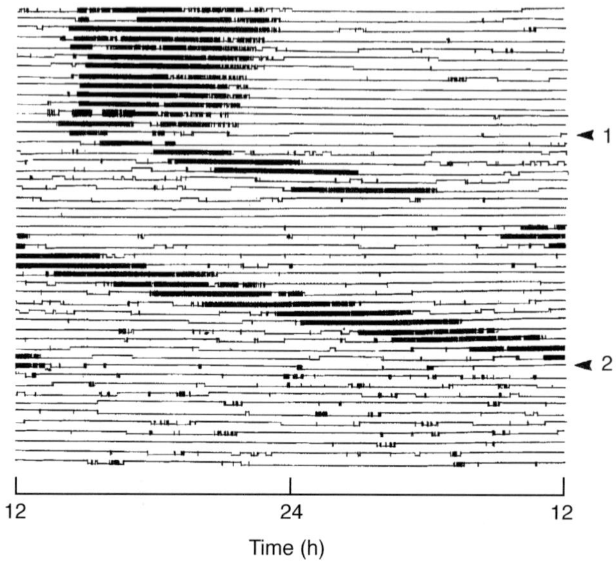

Fig. 24.9. A locomotor rhythm illustrating the effect of removal of first the right (arrow 1) and then the left (arrow 2) optic lobe. Note the increase in period following the first lobectomy, and total arrhythmicity after the removal of the remaining lobe.

procedures, suggests that the simplest interpretation is that the circadian clock of the weta lies in the lobular region of the brain. This, however, does not rule out the existence of other rhythmic centres in the midbrain or other regions of the body.

These results support the generalization that the optic lobes are essential for the expression of the circadian locomotor rhythm of exopterygote insects, but they do not prove this view without qualification. Transplant experiments are still required, as the structure of the optic lobes is different from that in other Orthopterans, where the lobular region is at the proximal end of the optic tract.

From its physical location, one would suspect that the pathway of light entrainment from imposed light cycles (natural or experimental) would be through the compound eye (or perhaps the ocelli) and to the lobular region via the retina and optic nerve. However, covering the ocelli and the compound eyes with black opaque material had only minor effects on entrainment; the phase relationship between the end of the light phase and onset of activity only increased slightly. Even surgical destruction of the compound eyes and ocelli failed to prevent entrainment by light cycles in the laboratory, and so we must presume that entrainment comes about either by direct illumination of the relevant neural tissue or through extraoptic photoreceptors (Waddell *et al.*, 1990).

Control Systems Models of the Circadian Clock: Single Oscillator

The support for the endogenous clock model of timing in the weta leads to the question of the nature of the endogenous clock in this insect. What is its chemical composition, cellular structure and mechanism of control? At first sight, the analyses of the free-running circadian locomotor rhythms reveal very little about the fundamental nature of endogenous clocks, even if we assume that the overt rhythm is a reasonable representation of the behaviour of the endogenous clock. It could be argued that the chain of events from the oscillator to the observable rhythm is too long to make any meaningful statements on the nature of the clock (Bünning, 1967).

However, all is not lost, since the oscillator theory for circadian clocks (Pittendrigh, 1974) proposes that the underlying mechanism of circadian clocks is based on oscillations in activity-regulating proteins (now identified in *Drosophila* as PER and TIM proteins (see Lewis *et al.*, 1997)), which

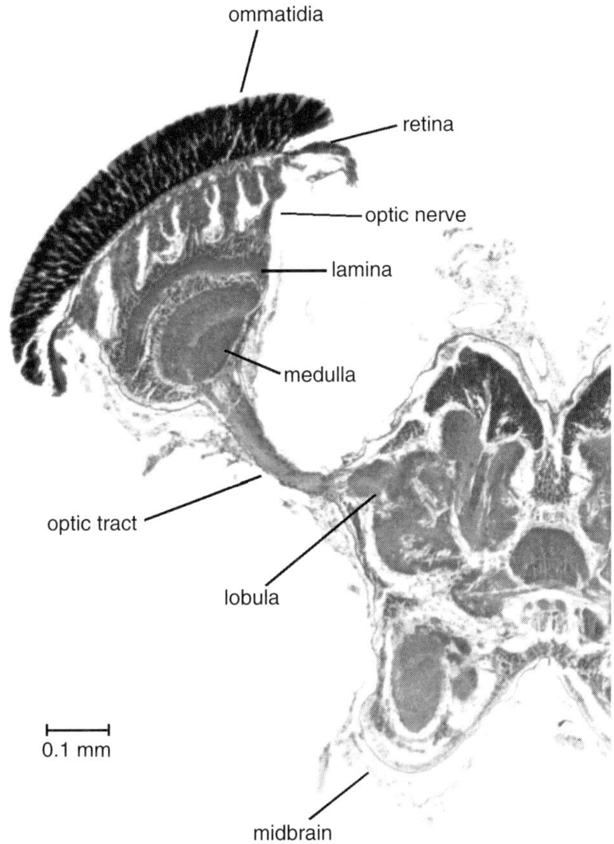

Fig. 24.10. Section through the head of *H. thoracica*, showing the compound eye (retina and ommatidia), optic lobe (lamina, medulla and lobula) and midbrain of *H. thoracica*.

have a period of about 24 h. We have adopted this model as the basis of the explanation of the mechanism of the circadian clock of the weta, and have elaborated on it by incorporating the concepts of negative feedback and time delay to explain the generation of the oscillations that lead to circadian locomotor rhythms.

Central to the mechanism of the clock is the synthesis of activity-promoting proteins. Locomotor activity is triggered as the concentration of the protein rises above a threshold, and ceases as the protein degrades to low levels (Fig. 24.11) (Gander and Lewis, 1979). The essential components of the model and the ways in which they interact can be summarized in a control systems diagram (Fig. 24.12). This depicts the flow of information between the functional components of the control system. Protein synthesis is switched on when the time-delayed value (time delay) of the oscillating protein falls below an innate reference value (Cref), so creating a positive error (Error). The synthesis increment for each time unit is determined by the synthesis function (SYN) and is added to the existing concentration. Protein is destroyed through constant loss or degradation (Loss). Time delay is vital for the generation of oscillations in the calculated values of the protein

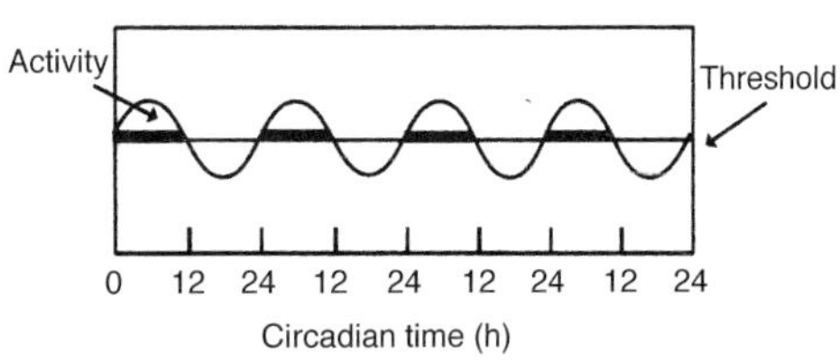

Fig. 24.11. The threshold model for circadian clock of *Hemideina thoracica*. Activity occurs if the concentration of the protein is above threshold.

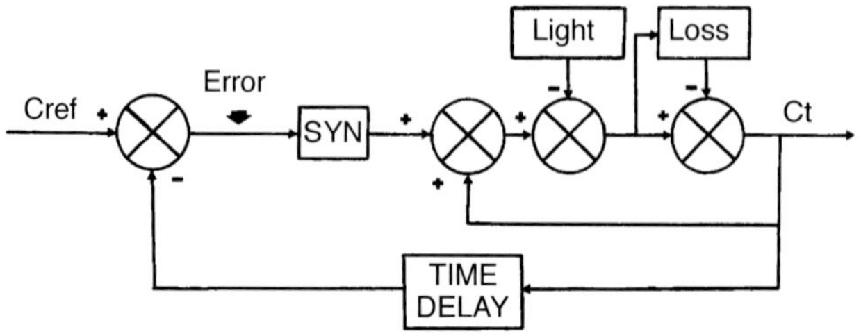

Fig. 24.12. Control systems diagram of the single-oscillator feedback model for the circadian clock.

(Ct), and in this control systems model the feedback information is delayed by 7 to 8 h. With the appropriate set of parameter values, the model simulates self-sustained oscillations, which account for the basic light and temperature properties of the weta clock.

From studies of the whole animal, it is clear that the biochemical oscillator must be sensitive to light. Our model takes light effects into account by proposing that light (Light) destroys the oscillating protein. This is shown in the control systems diagram, where the presence of light subtracts from the concentration of oscillating protein at the time the light is on. This convention allows us to simulate phase changes (Fig. 24.13) and the entrainment of the modelled rhythm with light cycles (Lewis, 1994).

Temperature, according to the model, only affects the rate of protein synthesis. As temperature rises, synthesis rate rises, with a temperature coefficient of 2.0. Loss and degradation processes are temperature-independent, with a temperature coefficient of 1.0. With these temperature conventions in place, the model accounts for the known properties of the weta circadian clock, covering temperature entrainment, temperature compensation and the stopping of the clock at low temperatures (Gander, 1979).

In its simplest form, we have envisaged the clock as a single oscillator, contained within a single cell or possibly an organelle. This model accounts for the basic properties of the weta clock, such as free-running rhythms, temperature compensation and light and temperature entrainment, and predicts other responses to light and temperature. For example, in constant bright light the oscillation damps, but in constant light of lower intensity, the period of the rhythm increases. These predictions on light effects are supported by experimental data (Lewis, 1994).

However, the single-oscillator model does not display the examples of lability, such as sponta-

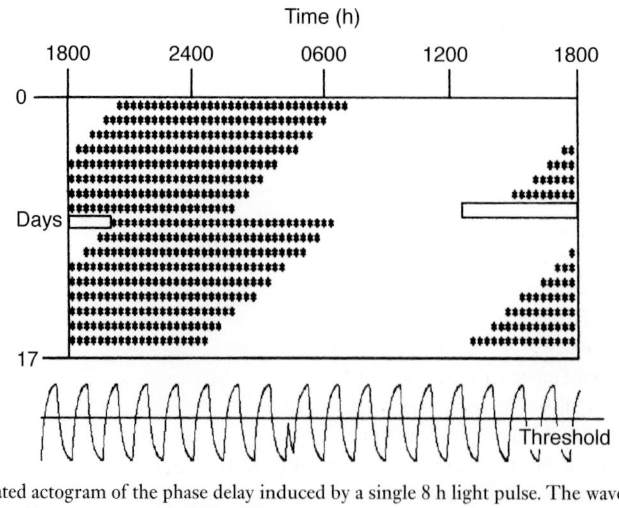

Fig. 24.13. Simulated actogram of the phase delay induced by a single 8 h light pulse. The wave-form below illustrates the effect of light on the oscillating protein.

neous change in free-running period and total breakdown of the free-running rhythm, observed in our results. For this, we have to elaborate the model into a population of similar oscillators (Christensen and Lewis, 1982) (Fig. 24.14).

The Population Models for the Circadian Clock of the Weta

Free-running rhythms of weta locomotor activity have been recorded in some cases for many months from individual subjects. For at least a part of the records, the majority exhibited very distinct locomotor rhythms, in which the timing of the onsets of the active phases are predictable to within a few minutes from day to day. Most rhythms, however, go through times of spontaneous change, arrhythmicity and recovery (Lewis *et al.*, 1991).

In typical actograms, the period over the first 10 days or so is generally less than 24 h, and it spontaneously increases to greater than 24 h as time passes. Thereafter, the records exhibit a variety of types of behaviour:

1. Regular rhythms with a constant, long free-running period.
2. Arrhythmicity and recovery.
3. Spontaneous changes in period.

In its simplest form, the population model states that the clock comprises a group of oscillators all having similar basic properties, but with a range of intrinsic free-running periods, stretching from *c.* 18 h to 30 h. The members of the population are weakly coupled, so that, during a normal free run, the majority of the oscillators are mutually coupled, but the fastest and slowest individuals are so deviant that they are not encompassed by the main group. They oscillate more or less independently in the background, with their own periodicity, and move in and out of phase with the main group (Christensen and Lewis, 1983).

This behaviour, together with other examples of free-run lability, such as spontaneous change in period, can be modelled by coupling together a number of feedback oscillators with periods ranging from about 18 h to 30 h. Coupling is achieved by partial sharing of the feedback information: a small portion comes from the mean value of all the oscillators and the remainder from each individual. In all other respects, the individual oscillators have the same properties as in the single-oscillator model.

Results of spontaneous breakdown and recovery are displayed in Fig. 24.15. We believe that the

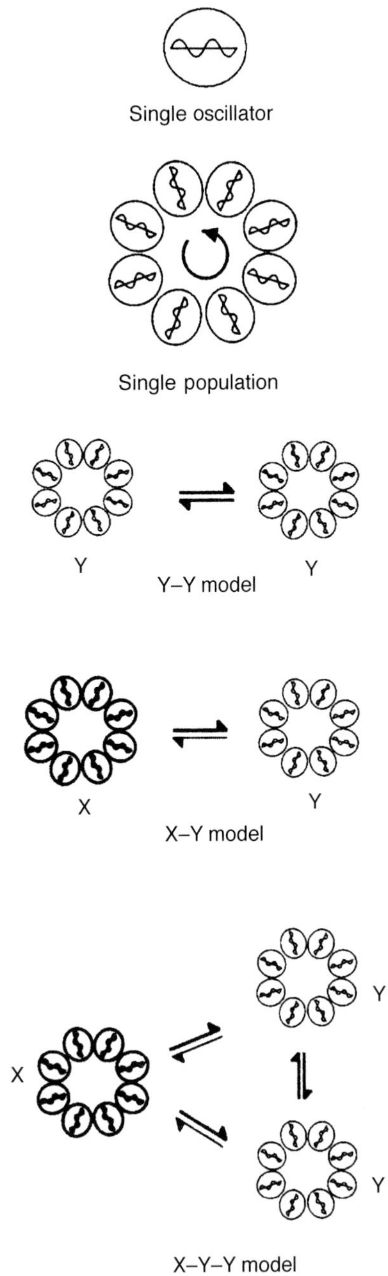

Fig. 24.14. Diagram showing the development of clock models from a single oscillator to the more complex X–Y–Y model.

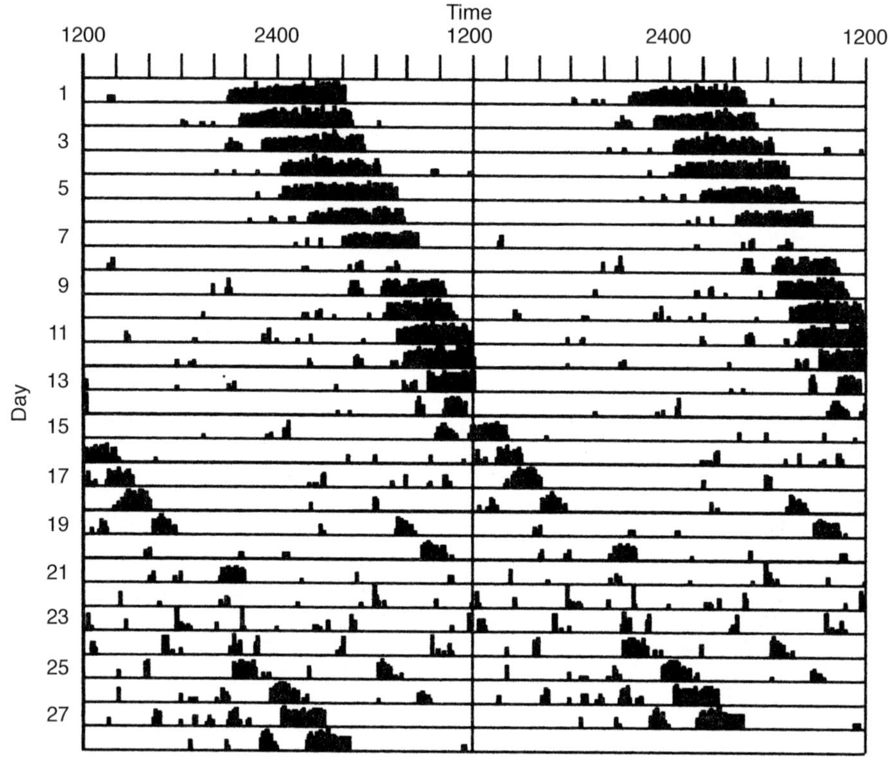

Fig. 24.15. Example of breakdown and spontaneous recovery of a locomotor rhythm.

elaboration of the model from a single oscillator to a population of loosely linked oscillators helps towards the understanding of an additional level of complexity of the weta clock.

If this model is appropriate, the population of oscillators should have physical validity and symbolize the master clock in the whole animal, residing (in wetas) in the lobular region of the optic lobes. It is clear that, as the brain is bilaterally symmetrical, there must be two populations of coupled oscillators and that the model summarizes the behaviour of both working together.

The Dual Pacemaker Model

The population model described above leaves two major categories of free-run lability unaccounted for. In examples of many free runs, the locomotor rhythm initially has a period of less than 24 h (as we have noted before) and then increases after about 10 days to greater than 24 h, followed by more or less systematic decreases and increases in period, which may be termed scalloping (Fig. 24.16). This type of period lability has also been described in mosquitoes (Clopton, 1984). In day-skipping, the active phase systematically reduces over several days to the point where it vanishes to nothing and then gradually reappears following several days of missed activity (Fig. 24.17).

These two interesting categories of rhythmicity cannot be explained by the simple population model, but require a further level of elaboration. The ease of recording locomotor rhythms, which we see as the obvious overt expression of the underlying clock, may distract us from looking further for evidence of rhythmic output from the weta. However, the cuticle of some newly moulted insects also shows daily patterns of deposition, even when cultured in constant conditions, and can be entrained by daily light cycles (Neville, 1983). Could the cuticle rhythm be part of a second major clock system, which runs independently of the locomotor rhythm in wetas?

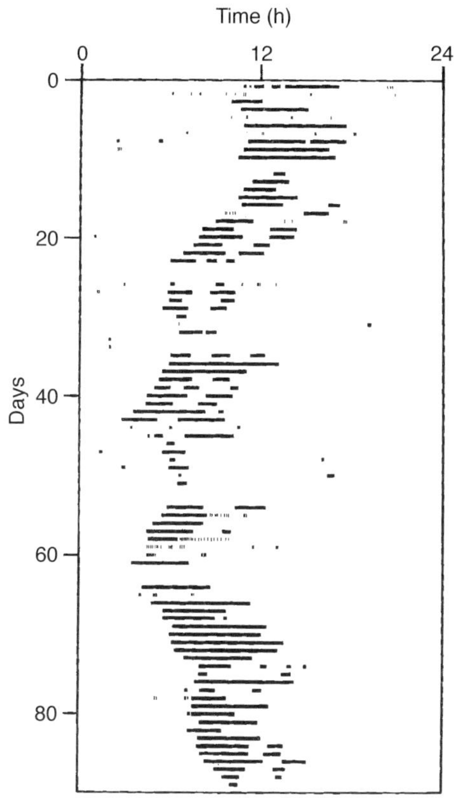

Fig. 24.16. An example of scalloping of a locomotor rhythm, providing evidence of an X–Y system.

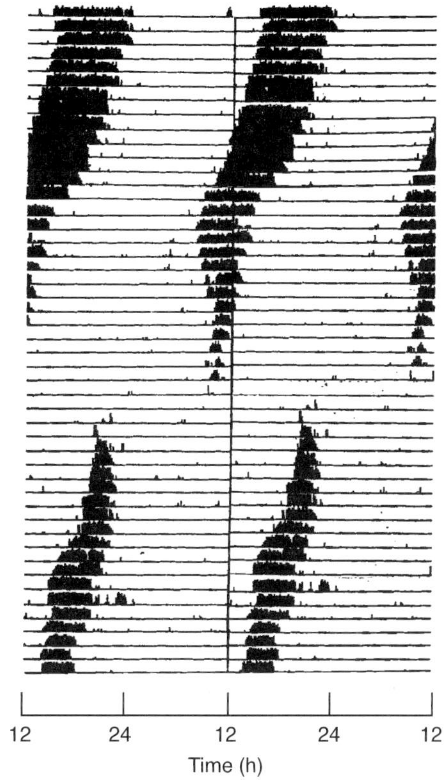

Fig. 24.17. Example of day-skipping in a locomotor rhythm, providing evidence of an X–Y system.

To answer this question, the number of cuticle bands in a sample of tarsal cuticle collected from a newly moulted weta was counted. The subject then free-ran for about 20 days, during which time the locomotor rhythms were recorded. To test the independence of the cuticle rhythm, the number of circadian days of locomotor activity was compared with the number of cuticle layers deposited over the same time. It was generally found that the number of cuticle layers was greater than the number of active phases, demonstrating that the cuticle clock was running faster (mean period 19 h) than the locomotor rhythm clock (mean period 25 h). This, we believe, supports a model of circadian control that incorporates two major clock centres, one having a long period and controlling the locomotor rhythm, and the other controlling, amongst other rhythms, the shorter-period cuticle deposition rhythm. In field conditions, these are both entrained by exogenous cycles to a 24 h period, but, when allowed to free-run, tend to interact only weakly. They pull apart to give the spontaneous patterns of change we observed in the experiments. It is interesting that in just two examples, where the cuticle rhythm had a period longer than the locomotor rhythm, the free-running period settled to a shorter value compared with the initial days in constant conditions. To quantify these ideas, we enhanced the population model to a dual pacemaker model.

The dual pacemaker model states that the circadian system of the weta comprises two major pacemakers, one (the Y system) regulates locomotor activity and the other (the X system) times the cuticle rhythm. These two major systems are loosely coupled and so interact to some extent. In natural conditions, these are entrained to the environmental cycles by the natural cycles of light and temperature, but in constant conditions in the laboratory they drift apart and, through their interactions, create the patterns of lability we observe in the real data.

So the single-population model has been expanded to encompass two populations of feedback oscillators, which are linked together to a small degree. The X system does not directly control locomotor activity but exerts its influence through its coupling with the Y pacemaker. When run with suitable parameter values, the dual pacemaker model mimics the variety of free-run lability observed in long records of weta locomotor activity in laboratory experiments. Small changes in coupling between the X and Y pacemakers lead to the range of variability demonstrated by the wetas.

We believe that, by treating the circadian clock of the weta as a living oscillator and by coupling oscillators together as populations, firstly single and finally dual populations, we can account for much of the behaviour of our wetas in constant conditions and shed light on the structure of the endogenous circadian system.

Conclusions and Discussion

In the field, wetas exhibit clear patterns of nocturnal behaviour, emerging from cover at twilight and returning well before dawn. We can speculate that the ecological significance of the timing of this behaviour is related to the avoidance of predation by birds and lizards, to foraging for food at optimal times and to the maintenance of territory. We see the role of the circadian clock as predicting the time of the onset of activity and preparing wetas, both physiologically and behaviourally, for forthcoming changes in the environment. The predictive power of the clock is particularly relevant and important in a species which, during its rest phase, is cut off from the outside world in the seclusion of holes in trees or under bark.

In tests of the internal circadian clock hypothesis, wetas exhibit free-running rhythms with remarkable precision, and clearly support the endogenous clock hypothesis for timing. The free-running period of the rhythms is initially less than 24 h, but spontaneously changes over time, and also exhibits a variety of distinctive patterns of free-run lability. The clock is entrained by light cycles, and we suspect that it is the light perceived at evening twilight that has the phase-changing role, as they are masked from the full intensity of daylight and return to the cover before dawn. Surgical removal of compound and simple eyes leads to the conclusion that light perception is extraoptic, and it is tempting to speculate that the time spent partially protruding from the retreat is to assess light intensity and fine-tune the exact time of emergence to suit the prevailing conditions.

Whilst we recognize that circadian clocks are in reality complex systems of interrelated parts, the lobular region of optic lobes seems a good candidate for the location of the circadian master clock for locomotor activity (the Y system of the dual pacemaker model) and matches the findings for crickets and cockroaches.

A single-oscillator feedback model incorporating time delay exhibits many of the basic properties of the weta circadian rhythm, but does not account for the complexities of behaviour evident in many of the free-running rhythms. However, expansion of the model first into a single population of loosely coupled oscillators and finally into two interacting pacemakers improves the realism of the model.

It is relevant that the molecular approach to the elucidation of the nature of the circadian mechanism of *Drosophila melanogaster* has also led to the development of time-delayed feedback models, in which the build-up of the protein feeds back on to the transcription of the *period* gene. We believe that the key to the understanding of the nature of the weta circadian clock also lies in the identification of the clock genes of this insect, the location of their expression and understanding the feedback pathways involved in the generations of circadian rhythms of clock proteins. Hand in hand with this is the understanding of the exact part played by light in the phase control of the rhythm.

Hemideina thoracica has proved to be an ideal subject for the study of insect circadian rhythms. It survives well in laboratory conditions, exhibits remarkably clear locomotor rhythms, which demonstrate a rich variety of rhythmic behaviour, which has proved ideal for analysis and modelling. These results and the related models not only help in the understanding of the control of the patterns of activity of this species in the field, but also hopefully may be generalized to the understanding of the nature of circadian clocks in other insects and, more widely, in other animal groups as well.

Acknowledgements

The authors would like to thank the University of Auckland Reseach Grants Committee for financial support over the years.

References

Aschoff, J. (1960) Exogenous and endogenous components in circadian rhythms. *Cold Spring Harbor Symposia on Quantitative Biology* 25, 11–28.

Balzer, I. and Hardeland, R. (1988) Influence of temperature on biological rhythms. *International Journal of Biometeorology* 32, 231–241.

Brady, J. (1982) Circadian rhythms in animal physiology. In: Brady, J. (ed.) *Biological Timekeeping*. Society for Experimental Biology Seminar Series 14, Cambridge University Press, Cambridge, pp. 121–142.

Bullivant, A.G. (1990) Long-term rhythms in the activity of the weta (*Hemideina thoracica*). MSc thesis, University of Auckland, Auckand, New Zealand.

Bünning, E. (1967) *The Physiological Clock*, 2nd rev. edn. Springer-Verlag, New York, 167 pp.

Christensen, N.D. (1978) The circadian clock of *Hemideina thoracica* as a population of weakly coupled oscillators. MSc thesis, University of Auckland, Auckland, New Zealand.

Christensen, N.D. and Lewis, R.D. (1982) The circadian locomotor rhythm of *Hemideina thoracica* (Orthoptera; Stenopelmatidae): the circadian clock as a population of interacting oscillators. *Physiological Entomology* 7, 1–13.

Christensen, N.D. and Lewis, R.D. (1983) The circadian locomotor rhythm of *Hemideina thoracica* (Orthoptera; Stenopelmatidae): a population of weakly coupled feedback oscillators as a model of the underlying pacemaker. *Biological Cybernetics* 47, 165–172.

Clopton, J.R. (1984) Mosquito circadian and circa-bidian flight rhythms: a two-oscillator model. *Journal of Comparative Physiology A* 155, 1–12.

Colwell, C.S. and Page, T.L. (1990) A circadian rhythm in neural activity can be recorded from the central nervous system of the cockroach. *Journal of Comparative Physiology A* 166, 643–649.

DeCoursey, P.J. (1986) Light-sampling behaviour in photoentrainment of a rodent circadian rhythm. *Journal of Comparative Physiology A* 159, 161–169.

DeCoursey, P.J. (1989) Photoentrainment of circadian rhythms: an ecologist's viewpoint. In: Hiroshige, T. and Honma, K. (eds) *Circadian Clocks and Ecology*. Hokkaido University Press, Sapporo, pp. 187–206.

Edery, I., Zwiebel, L.J., Dembinska, M.E. and Rosbash, M. (1994) Temporal phosphorylation of the *Drosophila* period protein. *Proceedings of the National Academy of Sciences of the United States of America* 91, 2260–2264.

Enright, J.T. (1965) The search for rhythmicity in biological time series. *Journal of Theoretical Biology* 8, 426–468.

Enright, J.T. (1970) Ecological aspects of endogenous rhythmicity. *Annual Review of Ecology and Systematics* 1, 221–238.

Frisch, B. and Aschoff, J. (1987) Circadian rhythms in honeybees: entrainment by feeding cycles. *Physiological Entomology* 12, 41–49.

Gander, P.H. (1979) The circadian locomotor activity rhythm of *Hemideina thoracica* (Orthoptera): the effects of temperature perturbations. *International Journal of Chronobiology* 6, 243–262.

Gander, P.H. and Lewis, R.D. (1979) The circadian locomotor rhythm of *Hemideina thoracica* (Orthoptera): feedback model for the underlying clock oscillation. *International Journal of Chronobiology* 6, 263–280.

Hau, M. and Gwinner, E. (1992) Circadian entrainment by feeding cycles in house sparrows, *Passer domesticus*. *Journal of Comparative Physiology A* 170, 403–409.

King, V.M. (1988) Advances in the model for the circadian organisation of the weta (*Hemideina thoracica*). MSc thesis, University of Auckland, Auckland, New Zealand.

Lewis, R.D. (1976) The circadian rhythm of the weta *Hemideina thoracica* (Orthoptera): free-running rhythms, circadian rule and light entrainment. *International Journal of Chronobiology* 3, 241–254.

Lewis, R.D. (1994) Modelling the circadian system of the weta, *Hemideina thoracica* (Orthoptera: Stenopelmatidae). *Journal of the Royal Society of New Zealand* 24, 395–421.

Lewis, R.D., Bullivant, A.G. and King, V.M. (1991) A dual pacemaker model for the circadian system of the insect *Hemideina thoracica*. *Journal of Interdisciplinary Cycle Research* 22, 293–309.

Lewis, R.D., Warman, G.R. and Saunders, D.S. (1997) Simulations of free-running rhythms, light entrainment and the light-pulse phase response curves for the locomotor activity rhythm in *period* mutant of *Drosophila melanogaster*. *Journal of Theoretical Biology* 185, 503–510.

Lloyd, D. and Rossi, E.L. (1992) *Ultradian Rhythms in Life Processes: an Inquiry into Fundamental Principles of Chronobiology and Psychobiology*. Springer-Verlag, London, 419 pp.

Morris, J.F. (1992) Tests of the dual pacemaker model of circadian organisation in the weta (*Hemideina thoracica*). MSc thesis, University of Auckland, Auckland, New Zealand.

Neville, A.C. (1983) Daily cuticular growth layers and the teneral stage in adult insects: a review. *Journal of Insect Physiology* 29, 211–219.

Palmer, J.D. (1976) *An Introduction to Biological Rhythms.* Academic Press, New York, 375 pp.

Pittendrigh, C.S. (1965) On the mechanism of entrainment of a circadian rhythm by light cycles. In: Aschoff, J. (ed.) *Circadian Clocks*. North-Holland, Amsterdam, pp. 277–297.

Pittendrigh, C.S. (1974) Circadian oscillations in cells and the circadian organization of multicellular systems. In: Schmitt, F.O. and Worden, F.G. (eds) *The Neurosciences Third Study Program*. MIT Press, Cambridge, Massachusetts, pp. 437–458.

Ralph, M.R., Foster, R.G., Davis, F.C. and Menaker, M. (1990) Transplanted suprachiasmatic nucleus determines circadian period. *Science* 247, 975–978.

Saunders, D.S. (1977) *An Introduction to Biological Rhythms*. Blackie, Glasgow and London, 170 pp.

Smith, W.W. (1931) Ants inhabiting Mount Egmont. *New Zealand Journal of Science and Technology* 13, 45–47.

Waddell, B., Lewis, R.D. and Engelmann, W. (1990) Localization of the circadian pacemakers of *Hemideina thoracica* (Orthoptera; Stenopelma-tidae). *Journal of Biological Rhythms* 5, 131–139.

Williams, J.A. and Naylor, E. (1978) A procedure for the assessment of significance of rhythmicity in time-series data. *International Journal of Chronobiology* 5, 435–444.

York, A. (1998) Mechanisms of entrainment of the circadian clock in the weta (*Hemideina thoracica*). MSc thesis, University of Auckland, Auckland, New Zealand.

25
Haemolymph Physiology

Doug Neufeld and John Leader
Department of Physiology, University of Otago, Dunedin 913, New Zealand

Considering that wetas make up such a conspicuous element of the New Zealand fauna, not only numerically but also in size, it is surprising that so little attention has been paid to their physiology. Fleming (1977) pointed out many years ago that these omnivorous insects seem to have evolved, in New Zealand, so as to inhabit niches which elsewhere in the world would have been occupied by rodents. Within New Zealand, weta is the term loosely given to species of two families of the order Orthoptera, the Rhaphidophoridae (cave wetas) and the Anostostomatidae (true wetas). This chapter is concerned with the Anostostomatidae, and principally with the physiology of one of the more interesting species, *Hemideina maori*, the New Zealand alpine weta. This insect lives in montane regions and subalpine valleys (altitude 1000 to 2000 m) in both the North and South Islands.

Haemolymph Composition

The composition of the haemolymph of wetas has been reported by Leader and Bedford (1978). Table 25.1 shows the ionic composition of the haemolymph of two herbivorous true weta species, *Hemideina femorata* and *H. maori*, and representatives of other Orthoptera. The inorganic composition of the haemolymph, rich in sodium and chloride, with small amounts of calcium and magnesium, is broadly similar to that of other orthopteran insects. Calcium levels are significantly lower than those reported for other related insects, but these data were obtained from sera after acid precipitation and thus represent the unbound fraction of the haemolymph calcium pool. Interestingly, the osmotic pressure of the haemolymph is high compared with that of most insects, reaching 479.9 mosmol l^{-1} in *H. maori*, even in insects collected in summer. In winter insects, the haemolymph osmotic pressure rises even higher than this.

As with most other insects, there is a substantial anion 'deficit' when the haemolymph is analysed as a simple inorganic solution. This must be satisfied in part by the characteristically high amino acid content of insect haemolymph. Table 25.2 gives data from Duchateau *et al.* (1953) for the amino acid content of the haemolymph of *Locusta migratoria* nymphs, for comparison with published data for *Hemideina* spp. (Leader and Bedford, 1978). Amino acids contribute about 50 mmol l^{-1} to the haemolymph, and in all three species glycine makes a significant contribution, between 20 and 30%. However, while glutamic acid (with glutamine) accounts for about 22% of the amino acid pool in *Locusta*, it is low in concentration in *Hemideina* haemolymph. In contrast, proline, present at 5.5 mmol l^{-1} in *Locusta*, is found in much higher concentrations in *Hemideina*, making up 33% of the amino acid pool in *H. maori*.

Malpighian Tubules

Like other Orthoptera, the primary excretory system consists of a large number (about 100) of short, blindly ending Malpighian tubules (described in O'Brien and Field, Chapter 8, this

Table 25.1. Composition of the haemolymph of representative Orthoptera.

Species	Na	K	Ca	Mg	Cl	References
Locusta migratoria	60	12	17.2	24.8	97.6	Duchateau *et al.* (1953) Hoyle (1953)
Schistocerca gregaria	81.3	5.8	17.8	34.6		Duchateau *et al.* (1953)
Chortophaga viridifasciata	108.9	3.4	2.8	21		Barsa (1954)
Gryllotalpa gryllotalpa	233.7	7.3	28.0	10.4		Duchateau *et al.* (1953)
Hemideina femorata	82.0	9.0	4.5	6.9	62.8	Leader and Bedford (1978)
Hemideina maori (summer)	95	5.5	4.0	9.5	65.6	Leader and Bedford (1978)

volume). The functional properties of the tubules in *H. maori* were investigated by Neufeld and Leader (1998a, b, c). The tubules were remarkable for their viability when isolated. At 4°C, tubules isolated in a glucose-containing saline were capable of continued function for up to 72 h, as judged by their capacity to concentrate phenol red in their lumina and to form a secretion. Like most herbivorous insects (Phillips, 1981), the secretory rate was low compared with that of bloodsucking insects after a blood meal. The maximal secretory rate found was 112 nl h^{-1}. At this rate, it is calculated that the water content of tubular cells turns over approximately every 17 min, compared with 15 s in *Rhodnius* tubules following a blood meal (Maddrell, 1991).

As in other insects, the tubular secretion is close to isosmoticity with the bathing fluid, and consists mainly of a solution of potassium chloride, with lesser amounts of sodium. Tubules incubated in normal saline (Na^+ 140, K^+ 10, Ca^{2+} 2, Mg^{2+} 1, Cl^- 137, HCO_3^- 18, $H_2PO_4^-$ 1 mmol l^{-1}, raised to an osmolality of 500 mosmol l^{-1} by addition of trehalose) secreted a fluid containing concentrations (calculated from determinations made with ion-selective electrodes) of potassium, sodium and chloride of 149, 131 and 197 mmol l^{-1}, respectively, at a rate of about 15 nl h^{-1}. The contribution of other anions to electroneutrality was not determined.

The rate and composition of the secretion was markedly affected, as in other insects, by the osmolality of the bathing fluid and by pharmacological agents. In a dilute bathing saline (osmolality about 300 mosmol l^{-1}), the rate of secretion rose to 75 nl h^{-1}, with no significant change in the concentration of potassium and chloride but a significant fall in sodium concentration. This results from a three- to fourfold increase in potassium and chloride secretion, but a doubling of sodium secretion. Two common secretagogues also increased the secretory rate. Both dibutyryl cAMP (10^{-4} mol l^{-1}) and 5-hydroxytryptamine (5-HT) (10^{-5} mol l^{-1}) caused a doubling of the secretory rate when added to the bathing solution under isosmotic conditions. In isosmotic saline, the increased rate of secretion in the presence of 5-HT or cAMP was achieved by an increase in both sodium and potassium secretion. However, in a dilute bathing medium, addition of 5-HT resulted in a decreased secretory rate compared with the untreated tubules, while dibutyryl cAMP caused a further increase in secretory rate, up to 115 nl h^{-1}. This increase appeared to be primarily due to an effect upon potassium secretion.

Analysis of the intracellular ionic activities and membrane potentials allowed determination of the transmembrane electrochemical gradients for ion movement. Under isosmotic conditions, the basolateral membrane potential was 58 mV (cell

Table 25.2. Amino acid content of the haemolymph of *Locusta migratoria, H. femorata* and *H. maori.*

	Locusta migratoria		*Hemideina maori*		*Hemideina femorata*	
	mmol l^{-1}	% total	mmol l^{-1}	% total	mmol l^{-1}	% total
Alanine	3.8	6.7	1.8	5.2	5.2	12.3
Arginine	1.4	2.4	0.5	1.4	0.4	0.9
Aspartic acid	1	1.7	–	–	–	–
Glutamic acid	11.2	19.6	0.8	2.2	0.6	1.5
Glycine	12.9	22.6	11.6	32.9	9.7	23
Histidine	1.9	3.3	1.6	4.7	1.0	2.3
Isoleucine	1.6	2.8	1.1	3.1	0.91	2.2
Leucine	1.6	2.8	1.2	3.5	0.81	1.9
Lysine	3.2	5.6	3.6	10.2	1.0	2.3
Methionine	0.4	0.7	0.2	0.6	0.4	1.0
Phenylalanine	0.4	0.7	0.2	0.5	0.5	1.1
Proline	5.5	9.6	7.4	22	14.3	34
Serine	4.7	8.2	1.9	5.4	2.0	4.7
Threonine	1.7	2.9	–	–	1.6	3.8
Tyrosine	1.5	2.6	0.2	0.6	0.7	1.7
Valine	4.7	8.2	3.3	9.3	3.3	1.3
Total	57	–	36	–	42	

interior negative) and the potential between bath and secreted fluid was 29 mV (secretion positive). Thus the potential across the apical membrane was 87 mV (cell interior negative). Such values are typical of those found in other insects (Fathpur *et al.*, 1983; Morgan and Mordue, 1983; Aneshansley *et al.*, 1988; Baldrick *et al.*, 1988; O'Donnell *et al.*, 1996). In the dilute bathing medium, these values did not change significantly. As in other tubules, the basal membrane potential was predominantly a potassium diffusion potential. A tenfold change in potassium activity in the bathing solution gave a change in apical membrane potential of 46 mV. Consistent with this finding, reduction of sodium in the bathing fluid from 116 to 16 mmol l^{-1} by replacement with choline caused a barely measurable change in basolateral membrane potential. Likewise, changes in chloride activity of the bathing fluid had no effect on membrane potentials.

Measurement of intracellular ionic activities demonstrated that the potassium ion was close to electrochemical equilibrium across the basolateral membrane, a condition common to all insect tubules investigated (Morgan and Mordue, 1983; O'Donnell and Maddrell, 1984; Baldrick *et al.*, 1988; Leyssens *et al.*, 1992). There was also a strong electrochemical gradient favouring inward sodium movement and a smaller one driving outward chloride movement. This distribution is commonly found in cellular systems, and is generally regarded as being maintained by an energy-dependent outward transport of sodium ions coupled to an inward movement of potassium ions, the so-called pump–leak hypothesis (Tosteson and Hoffman, 1960). Pannabecker *et al.* (1992) found that the basolateral membrane potential of the Malpighian tubules of *Aedes* is markedly affected by metabolic inhibitors, suggesting that it is maintained by metabolic processes. The basolateral membrane potential of tubules of *H. maori* was, however, unchanged after 15 min exposure to either dinitrophenol (10^{-4} mol l^{-1}) or cyanide (10^{-5} mol l^{-1}). Similarly, ouabain, a potent universal inhibitor of the membrane-bound ATPase, had no effect on either membrane potential or secretory rate. While this might suggest that the basolateral membrane potential is not maintained by active processes, it should be borne in mind that tubular cells are large and secretion rates relatively slow. A membrane potential established by metabolic processes will dissipate at a rate determined by electroneutral diffusion. If membrane permeability to sodium and chloride is low, the rate of decay of intracellular potassium activity and hence the fall in membrane potential will also be low.

In other insects, additional transport mechanisms have been implicated in ion transport across the basolateral membrane of Malpighian tubular cells. An inwardly directed $Na^+/K^+/2Cl^-$ transporter has been postulated on the basis of the effect of bumetanide, a specific inhibitor of this transporter, on tubular secretory rates (O'Donnell and Maddrell, 1984; Nicolson, 1993). In *H. maori*, bumetanide had no effect on the rate of secretion. However, the rate of secretion was markedly reduced by Ba^{2+} ions (1 mmol l^{-1}), a known blocker of potassium channels, and also by amiloride, which at the concentration used (10^{-5} mol l^{-1}) is known to block sodium channels (Kinsella and Aronson, 1981). Curiously, however, amiloride did not appear to affect the rate of sodium excretion by the tubules.

At the apical membrane, the secretion of a fluid rich in potassium must involve the movement of this ion against a large electrochemical gradient, a characteristic of most insect tubules (Phillips, 1981; Williams and Beyenbach, 1984). In general, it appears that the concentration of potassium secreted by herbivorous insects (e.g. *Locusta*: Morgan and Mordue, 1983) tends to be higher than that of stimulated tubules of bloodsucking insects (e.g. *Aedes aegypti*: Williams and Beyenbach, 1983; Hegarty *et al*., 1991). While the prime function of potassium secretion may be to generate a primary fluid flux, whose final composition is determined by selective reabsorption in the hind-gut and rectum (Phillips, 1977), its use in this role does serve the obvious function of secreting the excess potassium that inevitably arises from a vegetative diet. This is consistent with the fact that exposure to low osmolality or secretagogues appears to change the amount of potassium, rather than sodium, secreted. This is unlike *Locusta*, where stimulation of cation secretion appeared to be unselective (Morgan and Mordue, 1983). A summary diagram of the ion pathways and electrical potential gradients is given in Fig. 25.1.

Low Temperature

Most wetas live in temperate climates, where they are not exposed to temperatures far below zero. The montane weta, *H. maori*, however, is confined to relatively high altitudes, where temperature fluctuates over a large range. On the Rock and Pillar Range near Dunedin, for example, summer air temperatures may rise to 30°C during the day, while in winter they may fall at least to −8°C at night, warming in the day to temperatures often well above zero. However, the regime is variable and temperatures can fall below freezing at any time of the year (Bliss and Mark, 1974). This climate is more akin to those inhabited by insects that adopt a strategy of supercooling. Surprisingly, *H. maori* does not become supercooled, but can tolerate haemolymph freezing, an adaptation usually associated with insects that survive extremely low temperatures (Ring and Tesar, 1980; Churchill and Storey, 1989; Leather *et al*., 1993). Ramlow *et al*. (1992) showed that *H. maori* could tolerate freezing to at least −10°C for 30 min or more, provided that cooling was slow, of the order of 1°C min^{-1}. *Hemideina maori* is thus the largest insect known to survive freezing. This capacity is limited, however, since the insect never survives cooling to −15°C. The freezing point of an aqueous solution is lowered by 1.86°C by the addition of each mol l^{-1} of non-electrolyte. Thus when freezing begins, pure water separates out as ice and the remaining solution becomes progressively more concentrated as temperature falls. It follows that at −10°C the concentration of the unfrozen fraction of the haemolymph is about 5.3 molar. Wilson and Ramlow (1995) demonstrated an important feature of freezing tolerance in this insect, namely the existence of nucleating proteins, which accelerate the freezing of the haemolymph at temperatures close to freezing. The significance of this lies in the fact that dilute aqueous solutions, when cooled slowly, will often reach temperatures of 5° or more below zero before freezing commences. When ice crystals form and grow at this temperature, they do so very rapidly, creating not only the possibility of disruption of cell membranes by the sharp front of the growing crystal, but also mechanical distortion caused by the expansion of water in the solid phase. In addition, as the temperature is lowered, the probability of nucleation of ice crystals within cells increases steeply. With a few exceptions, intracellular ice formation in animals appears to be lethal. It is not surprising, therefore, that Sinclair and Wharton (1997) found that specimens of *H. maori* only rarely exhibited intracellular freezing, and confirmed that those cells in which ice had formed did not appear viable after thawing. Survival at −10°C therefore appears to involve the ability of the cells to tolerate the concentrated medium formed by the precipitation of water. Neufeld and Leader (1998b, c) investigated this further, using the Malpighian tubular cells as a

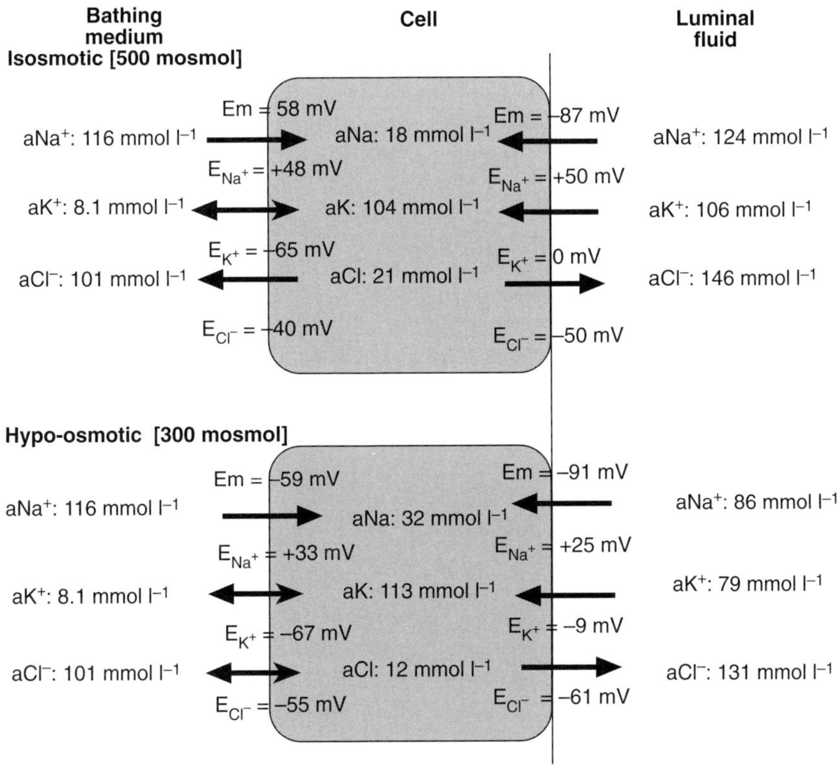

Fig. 25.1. Electrochemical characteristics of the Malpighian tubules of the weta *Hemideina maori*, bathed in either hypo-osmotic or iso-osmotic saline. Mean values are shown for the activities of the ions Na^+, K^+ and Cl^-, both extracellular and intracellular. Em is the transmembrane electrical potential, given as the potential of the cell interior relative to the basolateral or apical bathing fluids. Arrows indicate the direction of electrochemical driving forces for each ion, and E_{Na^+}, E_{K^+} and E_{Cl^-} are the equilibrium potentials for sodium, potassium and chloride, respectively, across each membrane. (Data from Neufeld and Leader, 1998a.)

model system. These have the advantage not only that they can be easily studied and cellular parameters accurately measured, but also that recovery of function can easily be determined by assessing the ability of the cells to concentrate phenol red in the tubular lumen and also by their ability to generate a secretion. Tubules were cooled at a rate close to 1°C min^{-1} usually supercooling to about −4°C before freezing was initiated. Isolated tubules frozen to −5°C in a saline containing only inorganic substances (hypo-osmotic saline) disintegrated on thawing. However, tubules bathed in a saline containing 200 or 400 mmol l^{-1} trehalose for either 1 or 18 h showed good recovery on thawing, as judged by their ability to maintain secretion, and possessed basolateral membrane potentials close to prefreezing values. There was partial recovery of function in tubules frozen to −15°C for 1 h in the presence of 400 mmol l^{-1} trehalose. To test whether temperature alone was the causative factor in tissue damage of those tubules which did not survive freezing, the effect of supercooling to −5°C was examined. Tubules exposed to these conditions, immersed in either hypo-osmotic saline or saline containing sucrose – solutions which result in considerable tissue damage if ice formation takes place – showed no change from control values for either the basolateral membrane potential or the secretion rate. In the converse experiment, the ability of tubules to survive chilling when suspended in a hyperosmotic solution was tested. At equilibrium at −5°C, the osmolality of the liquid phase of an aqueous solution is approximately 2690 mosmol kg^{-1}. A similar pattern of survival to that after freezing was observed. Tubules in the highest concentrations of trehalose survived best, while immersion in salines with no organic additives or containing sucrose resulted in tubular disintegration on rewarming. These experimental data correlate nicely with the environmental conditions the animals encounter in the wild. In summer, when temperatures can plunge rapidly to below freezing, trehalose concentrations in the haemolymph are low, generally about 40–50 mmol l^{-1}. This is sufficient to protect the tubules and presumably the other tissues from the mechanical and osmotic stresses resulting from mild freezing. In winter, the haemolymph concentration of trehalose rises to around 400 mmol l^{-1}, and this provides greater protection against lower temperatures. Trehalose has been shown to have a stabilizing effect on both membranes (Crowe *et al.*, 1990; Bakaltchava *et al.*, 1994) and proteins (Carpenter *et al.*, 1986), and its occurrence in *H. maori* in high concentrations reflects this protective role. Proline also occurs in the haemolymph of *H. maori* (Neufeld and Leader, 1998b), rising to high concentrations in winter (28 and 52 mmol l^{-1} in two samples). This amino acid is also thought to play a role in stabilizing membranes and proteins (Carpenter *et al.*, 1986; Crowe *et al.*, 1990), consistent with its occurrence in high concentrations in other freeze-tolerant insects (Ring and Tesar, 1980; Storey *et al.*, 1981).

An important feature, on which Neufeld and Leader (1998b) did not comment, is that high concentrations of neutral solutes in the body fluids of animals tolerant of low temperatures is generally associated with their ability to lower freezing-point and, in so doing, to reduce the temperature at which harmful levels of ionic strength are reached and to slow the rate of ice crystal growth, thus reducing the possibility that the sharp crystal fronts will penetrate and rupture membranes. In this case, however, nucleators ensure that, in life, freezing is initiated at relatively high temperatures and, in addition, the neutral solute sucrose was found to afford no protection against freezing damage. It seems likely, therefore, that at least part of the protective effect offered by trehalose (and proline) must be intracellular, as an osmolyte protecting against the damaging effects of intracellular dehydration. However, trehalose is normally poorly permeant, and it would be interesting to know whether this sugar is found in high concentration within the cells of the insect and, if so, how it reaches there. Conversely, if it is not, it would be equally interesting to know how the high extracellular osmolality caused by the presence of high trehalose concentrations is balanced intracellularly.

The haemolymph of *H. maori*, like that of many other freezing-tolerant insects, shows a large fluctuation in osmotic pressure between summer and winter. Since the membranes of animal cells are predominantly permeable to water and mechanically weak, the tissues of the animal must, if cell solute content remains unaltered, swell or shrink in concert with changes in extracellular osmolality. The alteration of cellular solute content in the face of an osmotic challenge so as to maintain cell volume is a widely observed phenomenon (Chamberlin and Strange, 1989). For invertebrates, the challenge to maintain cellular volume is complicated by the fact that temperature coefficients for the active and passive components of the cellular pump–leak system, which maintains a constant intracellular solute content, are different. Control or regulation of alterations in either active transport or passive membrane permeabilities to maintain a balance between ion and solute influx and efflux are thus one basis for cellular response to altered temperatures (Kamm *et al.*, 1979; Hochachka, 1986, 1988). It is known that the Malpighian tubules of *Rhodnius* can respond to changes in osmolality of the bathing fluid, an increase leading initially to osmotic shrinkage, followed by a restoration of cellular volume through active accumulation of intracellular solute. Conversely, reduction of osmolality of the bathing fluid leads to cell swelling, followed by recovery through loss of solute (O'Donnell and Mandelzys, 1988; Arenstein *et al.*, 1995). Like all other transport epithelia, this ability to resist challenges to cellular volume is superimposed on maintenance of a steady transepithelial flux of water and ions.

Since, at various times, the tissues of *H. maori* are subjected to a temperature regime from +25°C to at least −6°C and, in addition, the cells are exposed to an osmolality of about 400 mosmol l^{-1} in summer to 700 mosmol l^{-1} at supra-zero temperatures in winter and to several osmoles per litre when freezing occurs, Neufeld and Leader (1998c) examined the ability of Malpighian tubules of the insect to regulate cellular volume in response to temperature and osmotic challenges. Tubules transferred, at room temperature, from a saline containing 400 mmol l^{-1} of trehalose or glucose (700 mosmol l^{-1}) to a hypo-osmotic saline (400 mosmol l^{-1}) had an unchanged cellular volume (as estimated from cross-sectional area), although this change should result in a volume increase of 133% had the cells behaved osmometrically. Exposure, at 20°C, to an hyperosmotic, glucose-containing, saline of 2.1 osmol l^{-1}, equivalent to the osmotic concentration to which cells frozen at −4°C are exposed, resulted in a rapid decrease in cross-section of 33%, followed by a partial restoration of volume within 30 min. That this was achieved by accumulation of solute is confirmed by the overswelling of the cells when returned to normal saline. However, in the presence of hyperosmotic saline containing trehalose, the tubular cells behaved as osmometers, rapidly shrinking by about one-third and recovering to normal volume when returned to isosmotic saline. However, at 0°C, tubules showed no capacity to recover volume in either glucose- or trehalose-containing hyperosmotic solutions. Nor was there any apparent volume recovery by cells exposed at subzero temperatures to hyperosmotic solutions. Two conclusions may be drawn from these experiments. First, it is obvious that, if the function of cell volume regulation is to maintain integrity of spatial relationships between structures, then, at subzero temperatures, when metabolic functions are largely inoperative, other factors assume greater significance, such as the effect of increasing intracellular ionic strength on the structure of cellular constituents and the reduction of cellular volume by osmotic shrinkage bringing about their mechanical disruption. Secondly, it is clear that many questions remain unanswered. The significance of the absence of volume regulation in trehalose-containing solutions remains to be explained. Perhaps, however, the most curious feature of freezing tolerance in *H. maori* is that it occurs at all. Most freezing-tolerant insects live in environments where extremes of cold are common. In milder climates, where temperatures only a few (5–15°C) degrees below freezing are encountered, the accumulation of antifreeze agents and adoption of behaviours to avoid freezing are the most common adaptations. In this temperature range, cells in contact with ice experience the maximal risk of damage, since ice crystals may still grow, ionic diffusion and consequent redistribution may occur at significant rates and the chemical consequences of exposure to high ionic strength may become manifest. It may be that *H. maori* retains vestiges of an adaptation to the much lower temperatures which New Zealand experienced during the last Ice Age.

The family Anostostomatidae is a uniquely interesting group of insects, which merits serious further study, if only to clarify their ecological role, particularly in New Zealand, where they form such a dominant component of the endemic fauna.

References

Aneshansley, D.J., Marler, C.E. and Beyenbach, K.W. (1988) Transepithelial voltage measurements in isolated Malpighian tubules of *Aedes aegypti*. *Journal of Insect Physiology* 35, 41–52.

Arenstein, I.R., Caruso-Neves, C., Onuchic, L.F. and Lopes, A.G. (1995) Mechanisms of cell volume regulation in the proximal segment of the Malpighian tubule of *Rhodnius neglectus*. *Journal of Membrane Biology* 146, 47–57.

Bakaltchava, I., Williams, W.P., Schmitt, J.M. and Hincha, D.K. (1994) The solute permeability of thylakoid membranes is reduced by low concentrations of trehalose as a co-solute. *Biochimica Biophysica Acta* 1189, 38–44.

Baldrick, P., Hyde, D. and Anstee, J.H. (1988) Microelectrode studies on Malpighian tubule cells of *Locusta migratoria*: effects of external ions and inhibitors. *Journal of Insect Physiology* 34, 963–975.

Barsa, M.C. (1954) The behaviour of isolated hearts of the grasshopper, *Chortophaga viridifasciata*, and the moth *Samia walkeri*, in solutions with different concentrations of sodium, potassium, calcium, and magnesium. *Journal of General Physiology* 38, 79–92.

Bliss, L.C. and Mark, A.F. (1974) High-alpine environments and primary production on the Rock and Pillar Range, Central Otago, New Zealand. *New Zealand Journal of Botany* 12, 445–483.

Carpenter, J.F., Hand, S.C., Crowe, L.M. and Crowe, J.H. (1986) Cryoprotection of phosphofructokinase with organic solutes. *Archives de Biochimie et Biophysique* 250, 505–512.

Chamberlin, M.E. and Strange, K. (1989) Anisosmotic cell volume regulation: a comparative view. *American Journal of Physiology* 257, C159-C173.

Churchill, T.A. and Storey, K.B. (1989) Metabolic consequences of rapid cycles of temperature change for freeze-avoiding vs freeze-tolerant insects. *Journal of Insect Physiology* 35, 579–585.

Crowe, J.H., Carpenter, T.A., Crowe, L.M. and Anchordoguy, T.J. (1990) Are freezing and dehydration similar stress vectors? A comparison of modes of interaction of stabilizing solutes with biomolecules. *Cryobiology* 27, 219–231.

Duchateau, G., Florkin, M. and Leclerq, J. (1953) Concentrations des bases fixes et types de composition de la base totale de l'hémolymphe des insectes.

Archives Internationales de Physiologie et Biochimie 61, 518–537.

Fathpur, H., Anstee, J.H. and Hyde, D. (1983) Effect of Na^+, K^+, Ouabain, amiloride and ethacrynic acid on the transepithelial potential across Malpighian tubules of *Locusta*. *Journal of Insect Physiology* 29, 773–778.

Fleming, C.R. (1977) The history of life in New Zealand forests. *New Zealand Journal of Forestry* 22, 249–262.

Hegarty, J.L., Zhang, B., Pannabecker, T.L., Petzel, D.H, Baustian, M.D. and Beyenbach, K.W. (1991) Dibutyryl cAMP activates bumetanide-sensitive electrolyte transport in Malpighian tubules. *American Journal of Physiology* 261, C521–C529.

Hochachka, P.W. (1986) Defense strategies against hypoxia and hypothermia. *Science* 231, 234–241.

Hochachka, P.W. (1988) Channels and pumps – determinants of metabolic cold adaptation strategies. *Comparative Biochemistry and Physiology* 90B, 515–519.

Hoyle, G. (1953) Potassium in insect nerve and muscle. *Journal of Experimental Biology* 30, 121–135.

Kamm, K.E., Zatzman, M.L., Jones, A.W. and South, F.E. (1979) Maintenance of ion concentration gradients in the cold in aorta from rat and ground squirrel. *American Journal of Physiology* 237, C17–C22.

Kinsella, J.L. and Aronson, P.S. (1981) Amiloride inhibition of the Na^+-H^+ exchanger in renal microvillus membrane vesicles. *American Journal of Physiology* 241, F374–F379.

Leader, J.P. and Bedford, J.J. (1978) Haemolymph composition of two species of New Zealand weta, *Hemideina* (Orthoptera, Anostostomatidae). *Comparative Biochemistry and Physiology* 61A, 173–176.

Leather, S.R., Walters, K.F.A. and Bale, S.J. (1993) *The Ecology of Insect Overwintering*. Cambridge University Press, Cambridge, 255 pp.

Leyssens, A., Steels, P., Lohrmann, E., Weltens, R. and van Kerkhove, E. (1992) Intrinsic regulation of K^+ transport in Malpighian tubules (*Formica*): electrophysiological evidence. *Journal of Insect Physiology* 38, 431–446.

Maddrell, S.H.P. (1991) The fastest fluid-secreting cell known: the upper Malpighian tubule cell of *Rhodnius*. *BioEssays* 13, 357–362.

Morgan, P.J. and Mordue, W. (1983) Electrochemical gradients across *Locusta* Malpighian tubules. *Journal of Comparative Physiology B* 151, 175–183.

Neufeld, D.S. and Leader, J.P. (1998a) Electrochemical characteristics of ion secretion in Malpighian tubules of the New Zealand alpine weta (*Hemideina maori*). *Journal of Insect Physiology* 44, 39–48

Neufeld, D.S. and Leader, J.P. (1998b) Freezing survival by isolated Malpighian tubules of the New Zealand alpine weta *Hemideina maori*. *Journal of Experimental Biology* 201, 227–236.

Neufeld, D.S. and Leader, J.P (1998c) Cold inhibition of cell volume regulation during the freezing of insect Malpighian tubules. *Journal of Experimental Biology* 201, 2195–2204.

Nicolson, S.W. (1993) The ionic basis of fluid secretion in insect Malpighian tubules: advances in the last ten years. *Journal of Insect Physiology* 39, 451–458.

O'Donnell, M.J. and Maddrell, S.H.P. (1984) Secretion by the Malpighian tubules of *Rhodnius prolixus* Stål.: electrical events. *Journal of Experimental Biology* 110, 275–290.

O'Donnell, M.J. and Mandelzys, A. (1988) Cell volume maintenance and volume regulatory decrease in Malpighian tubule cells of an insect, *Rhodnius prolixus. Comparative Biochemistry and Physiology* 90B, 843–849.

O'Donnell, M.J., Dow, J.A.T., Huesmann, G.R., Tublitz, T.J. and Maddrell, S.H.P. (1996) Separate control of anion and cation transport in Malpighian tubules of *Drosophila melanogaster*. *Journal of Experimental Biology* 199, 1163–1175.

Pannabecker, T.L., Hayes, T.K. and Beyenbach, K.W. (1992) Unique electrophysiological effects of dinitrophenol in Malpighian tubules. *American Journal of Physiology* 263, R609–R614.

Phillips, J.E. (1977) Problems of water transport in insects. In: Jungreis, A.M., Hodges, T.K., Kleinzeller, A. and Schulz, S. (eds) *Water Relations in Membrane Transport in Plants and Animals.* Academic Press, New York, pp. 333–363.

Phillips, J.E. (1981) Comparative physiology of insect renal function. *American Journal of Physiology* 253, R241–R257.

Ramlow, H., Bedford, J.J. and Leader, J.P. (1992) Freezing tolerance of the New Zealand alpine weta, *Hemideina maori. Journal of Thermal Biology* 17, 51–54.

Ring, R.A. and Tesar, D. (1980) Cold-hardiness of the arctic beetle *Pytho americanus* Kirby Coleoptera, Pythidae (Salpingidae). *Journal of Insect Physiology* 26, 763–774.

Sinclair, B.J. and Wharton, D.A. (1997) Avoidance of intracellular freezing by the freezing-tolerant New Zealand alpine weta, *Hemideina maori* (Orthoptera: Anostostomatidae) *Journal of Insect Physiology* 43, 621–625.

Storey, K.B., Baust, J.G. and Storey, J.M. (1981) Intermediary metabolism during low temperature acclimation in the overwintering gall fly larva, *Eurosta solidaginis. Journal of Comparative Physiology* 144, 183–190.

Tosteson, D.C and Hoffman, J.F. (1960) Regulation of cell volume by active cation transport in high and low potassium sheep red cells. *Journal of General Physiology* 44, 169–194.

Williams, J.C. and Beyenbach, K.W. (1983) Differential effects of secretagogues on Na and K secretion in the Malpighian tubules of *Aedes aegypti*. *Journal of Comparative Physiology B* 149, 511–517.

Williams, J.C. and Beyenbach, K.W. (1984) Differential effects of secretagogues on the electrophysiology of the Malpighian tubules of the yellow fever mosquito. *Journal of Comparative Physiology B* 154, 301–309.

Wilson, P.W. and Ramlow, H. (1995) Haemolymph ice-nucleating proteins from the New Zealand alpine weta *Hemideina maori* (Orthoptera, Anostostomatidae). *Comparative Biochemistry and Physiology* 112B, 535–542.

Part VII
Conservation of Endangered Species

26
Conservation of Threatened Species of Weta (Orthoptera: Anastostomatidae) in New Zealand

Greg Sherley

Science and Research Division, Science Technology and Information Service, Department of Conservation, PO Box 10420, Wellington, New Zealand

Introduction

In New Zealand, the conservation of invertebrates has gained popularity and received an increased funding since 1987, when the New Zealand Department of Conservation was established. Most of this attention has been focused on the conservation of large-bodied, spectacular weta (Orthoptera: Anostostomatidae). These have acted as icons for invertebrate conservation generally (Sherley, 1989) and helped overcome the inherent reluctance of professional bodies and the general public to recognize the worth of conserving invertebrates and their habitats. A similar public and government awareness of the need to conserve flagship species is discussed for South African anostostomatids by Toms (Chapter 4, this volume). Large-bodied invertebrate species, such as weta, may be particularly threatened by predation from introduced rodents (Moors *et al.*, 1992) and possums (*Trichosurus vulpecula*), which live throughout New Zealand, including many offshore islands (Cowan, 1990).

In this chapter, I shall describe measures taken to protect threatened weta species in New Zealand. Measures include species recovery planning, translocation and captive breeding. Finally, I shall describe some of the priority research and management requirements for the conservation of weta.

Recovery Plans

Recovery plans encompass the research and management that are required to prevent extinction. They also set out priorities and timetables for specific tasks. In New Zealand, recovery plans are commissioned and sponsored by the Department of Conservation – usually for 3 to 5 years before being rewritten. Species recovery work is coordinated by a recovery group drawn from scientists, field managers and non-government organizations. New Zealand's indigenous people (Maori) are often represented.

The New Zealand Weta Recovery Plan

Part One of the draft plan provides background information relevant to weta conservation, including legislative protection, taxonomy, distribution and causes of decline, ecology, and species recovery. The latter describes what work has already been done to conserve individual species. Part One also deals with weta management, including: (i) describing management tasks for the restoration of threatened weta populations; (ii) a recovery strategy, which includes a long-term goal and specific objectives for all species mentioned in the plan; (iii) a ranked list of species requiring conservation work; and (iv) a ranked list of generic (applying to more than one species) research and management tasks. Finally, the recovery group is described, including its membership and how it should work. The latter includes a 'critical path', which is a schematic procedure for planning research and management of threatened weta.

Part Two of the draft plan details for each species its taxonomy, distribution and abundance, ecology, reasons for decline, current threats, captive breeding attempts, management tasks, contacts and a reference list for that species. Sixteen species (seven undescribed) are included in this section, including ten giant weta (*Deinacrida* spp.), three tusked weta (*Motuweta isolata, Hemideina monstrosa*, and an undescribed species), one tree weta (*Hemideina ricta*) and one cave weta (*Gymnoplectron giganteum*, Rhaphadiphoridae). Appendices describe survey and monitoring techniques, collection guidelines and captive breeding techniques.

Discussion of weta recovery planning

To achieve the best efficiency for conserving species, a recovery plan should be written prior to any conservation work. The draft weta recovery plan has been written well after the start of significant biosystematic and ecological research and protection work. While in some ways this was not ideal, recovery plans do require a priori knowledge of the biosystematics and conservation status of the group – things which are still an impediment to recovery planning for weta. Hence some research on biosystematics and distribution is necessary before writing a recovery plan.

Usually, species recovery plans for vertebrates deal with single species, but, because there were so many species of weta (at least 16) which needed conservation work, all of them were included in the one plan. One benefit of a multispecies plan is that some individual species that would otherwise not have warranted their own plan will now receive attention. This gain in efficiency is partly because many of the problems (e.g. biosystematics) are common to all the species. Similarly, the recovery group can operate more efficiently, as the problems facing each species can be tackled with knowledge gained from working with others. The advocacy role of the recovery plan and group in promoting the conservation of less spectacular species can benefit from the inclusion of the spectacular or icon species.

The biggest problem which has confounded ranking tasks for weta recovery is the uncertainty of the biosystematic status of many species. Consequently, there is a list of species appended to the recovery plan, which includes many deserving urgent work, but which are not included yet, simply because their biosystematics is completely unknown. As this knowledge is gained, they will be included in the recovery plan.

In general, it is important to be flexible in writing multispecies recovery plans, in order to avoid setting over-ambitious objectives and because of the reluctance of the administering department to authorize a plan with potentially huge commitments. Most of these problems can be overcome with a clear mandate for the recovery group, such as the ability to reorder species priorities as new knowledge comes to hand, vetting funding bids to reflect these priorities and setting achievable aims annually.

Translocation

Introduction

Translocation may be defined as 'the transfer of individuals from an endangered site to a protected or neutral one' (Anon., 1986). The reason most translocations are attempted is to establish more populations of the threatened species and therefore reduce the risk of extinction. Published accounts of weta translocations include Meads and Notman (1992) (Mana Island giant weta, *Deinacrida rugosa*, Anostostomatidae) and Sherley (1994) (Mahoenui giant weta, *Deinacrida mahoenui*). The following summary of weta translocation is based on experience gained from research and management of the latter species (Sherley and Hayes, 1993; Sherley, 1989, 1994) between 1989 and 1993 near Te Kuiti in the North Island of New Zealand. Since 1994, little management work has been conducted and much of the work needed to protect this species is incomplete.

Mahoenui giant weta are medium-sized, compared with others in their genus: males and females weigh up to 10.5 g and 19.0 g, respectively (Sherley and Hayes, 1993). Most individuals are coloured dark chocolate-brown with black markings, but a second colour morph occurs, which is dark yellow with black to brown markings. Their habitat today is mainly introduced gorse (*Ulex europeus*), although the endemic habitat was once probably mixed hardwood and podocarp forest. This weta species is primarily herbivorous, it may live to 2 years and populations have overlapping generations (Sherley and Hayes, 1993). Two populations have been located, although one is so small that only a few weta have ever been discovered. The research and management described here are based on the larger population, found near Mahoenui in the North Island of New Zealand (Sherley, 1994).

Methods and results

Site selection

Three mainland sites and an island were selected to receive introduced weta after a study of habitat requirements (Sherley and Hayes, 1993). Information is only available for the mainland translocations, so only these will be discussed. Sites were selected which most closely resembled the habitat where the weta naturally occurred. However, because gorse is a seral species throughout most of New Zealand (being shaded out by secondary native forest species), weta were introduced to gorse that was adjacent to native forest, with the hope that the weta would eventually establish in the forest.

Ideally, weta should have been translocated to new sites within their natural range. However, historical records of their range are non-existent. The next criterion considered was to select sites within the ecological region (*sensu* McEwen, 1987) within which the two natural populations occurred. Localities within an ecological region share similar geology, land-form and vegetation communities. Selecting sites within the ecological region would thus avoid exposing translocated weta to a new suite of selection pressures. Other considerations included land tenure (preferably a Crown-owned reserve), accessibility for monitoring, predator control and local people's attitude towards species translocations. In practice, all these criteria could not be met and weta ended up being translocated outside of their known range and ecological region.

Collection and transport

Weta were collected for translocation from all parts of the reserve established to protect the larger population (Sherley and Hayes, 1993) and from the adjacent habitat, which was being converted to exotic forestry. Groups of four to seven individuals of mixed-sex nymphs (instar 4 and over) were transported amongst fresh gorse foliage in darkened, well-aerated containers. The vegetation was used because it gave individuals more opportunity to separate themselves, thereby reducing stress, and because their natural rest position is a near vertical position in dense vegetation. Also, the death of the weta was prevented, in case ecdysis occurred during translocation, by providing vegetation to hang from. Adults found as pairs were transported separately.

Release

Weta were either released within 24 h of being collected or held in an enclosure at their new location.

IMMEDIATE RELEASE. Up to four translocations to three mainland sites (including the enclosure site) and one to an offshore island were made. Multiple translocations were made to two mainland sites (the Cowan and Maungaokewa reserves) to offset the expected weta mortality and dispersal (Sherley, 1994). Groups of 12–18 nymphs (similar age and sex ratio of 1 : 1) were released into the same gorse bush. Each bush and its neighbours were assessed, based on what is known about the habitat requirements of the species (Sherley and Hayes, 1993). Important criteria included high density of foliage, large bush size, close proximity to other high-quality bushes in an area approximately 10 m by 25 m and close proximity to native forest. Before release, all nymphs were measured, including pronotum maximum length and width, head capsule width and length of rear tarsus. All weta were released within 24 h of capture and most within 12 h. Adults found together as pairs were released together in their own bush.

Weta bred by a professional insectary were used in one translocation of 120 mixed-age and sex individuals in 1994. The occasion was used to involve media to promote weta conservation. Weta were transported singly in containers, except for pairs of adults, and released immediately in the Maungaokewa reserve in a similar manner to that described above. Thus the transferees were not acclimatized to the release site beforehand.

DELAYED RELEASE. The second method of translocation involved releasing wild-caught weta into an enclosure covering perceived high-quality gorse habitat at the new site. The enclosure protected the weta from predation. It was approximately 6 m wide by 10 m long by 3 m high and covered about 16 coppiced gorse bushes. These were kept clipped, much like a hedge, to retain the type of dense foliage preferred by weta and to prevent the gorse from pressing against the roof. The sides and roof of the enclosure are made from a synthetic polymer close-mesh weave, designed to allow through 60–70% of incident light. The sides were dug into the ground to prevent rodents (presumed predators of weta) from gaining access. At the time of writing, there has been one release into the enclosure of 12 adult pairs of wild-caught weta. At least one pair has been seen copulating and it is presumed that laying occurred. However, no juveniles have been found since (three monitoring surveys over 12 months).

Weta are expected to breed in the enclosure and their progeny will be allowed to disperse to the surrounding gorse when they reach about instar 4. It is hoped that the progeny from repeated releases into the enclosure will establish in the surrounding habitat, including, eventually, the adjacent forest. Predator control (mainly rodents) will be started, so that, by the time a population is establishing, risks from predators should be minimal. Rodent predator control will be achieved, using anticoagulant poisons in bait stations, immediately before nymphs are allowed to disperse from the enclosure.

A variation of the enclosure method will involve using captive-bred adults. The advantages of using an enclosure and captive-bred weta will be discussed below.

Monitoring released weta

While weta undoubtedly suffered mortality after release and did disperse, the extent could not be measured, despite repeated post-transfer monitoring (Sherley, 1994). The timing of monitoring and subsequent translocations at the Cowan and Maungaokewa reserves took into account the expected rate of development of transferees from the preceding translocation. The standard measurements taken from weta before they were released were used to estimate their instar, using data from Sherley and Hayes (1993). Thus, follow-up monitoring was timed so that the youngest weta from the previous translocation should have been dead and any weta found would be progeny from translocated animals.

Monitoring of weta released to the wild was carried out 10 to 26 months after release, to determine whether transferees had subsequently bred (Sherley, 1994). No monitoring of survival was done earlier because of: (i) limited time; (ii) no available systematic method; and (iii) the fact that casual observations the day after release showed that the transferees had quickly dispersed from the bushes they were released into and the chances of finding any again were slight. Monitoring to determine breeding was carried out by searching likely habitat in a radial pattern outward from the point of release to a distance of about 300 m. Any weta found were measured and their instar assessed and compared with the data on transferees. On the basis of Richards's (1973) observations on the time taken to reach a given instar, the age of the weta was assessed, thereby indicating whether it had hatched at the translocation site or was one of the group translocated.

Transferees from wild capture and captive breeding did breed (Sherley, 1994), but there are no self-sustaining populations established at either of the three mainland sites. No systematic monitoring has yet been attempted at the offshore island site.

Discussion

Many problems were encountered with translocating Mahoenui weta to the three mainland sites, and most of them have not been solved. Capturing wild weta and translocating them were often done under severe time constraints, because habitat was being destroyed either for forestry conversion or for creating fire-breaks. Under these circumstances, too many weta were on hand to afford them proper treatment and processing (e.g. measuring, careful containment). A system for processing translocated weta needs to be created well in advance, with sufficient personnel guaranteed.

Before translocation should be considered, one must be certain that that species does not already exist at the new site. For this, ideally, the biosystematics of the species needs to be clearly understood, so that surveys to determine its existence or not at the new site may be carried out with confidence. Pretranslocation surveys are also confounded by the difficulty of being able to detect animals at low densities – a problem that also confronts those monitoring transferees.

Finding transferees and their progeny presented a serious difficulty in monitoring Mahoenui giant weta after release. Observations of a few individual weta indicated that they did not remain in the chosen 'favourable' habitat. Clearly, research into the dispersal of translocated weta is needed. This may be facilitated in future with the use of portable harmonic radar transceivers, which can greatly assist in locating invertebrates (Lovei *et al.*, 1997). Using this technique involves gluing a transponder (diode in a closed circuit with a copper-loop aerial) on to the study animal and using a radar transceiver to locate it after release.

Measuring weta just prior to release was possibly unnecessary in most cases. Future weta releases need to simply estimate instar to three categories (6 and less, 7–8 or 9 and 10–adult) – something which we later discovered could be quite easily done by eye with practice. Measuring was unnecessary, because the timing of monitoring could have been planned to ensure that weta found must have been progeny of transferees and that all transferees would have died. The measuring also meant the weta were being handled three times (capture, measuring and release), which caused a lot of unnecessary stress. Although little mortality was evident between capture and release, there were signs that the weta were being subjected to considerable stress. For example, some adult females started ovipositing in their containers and most regurgitated a smelly fluid while being handled, which is, presumably, a defence against predation. While no data were systematically gathered, it appeared that younger instars were less stressed by the procedure than adults. Although no data were recorded, the individuals that died during handling and transfer were mainly adults.

Some of the problems with translocating wild-caught weta did not occur with captive-bred animals. These weta were simply transported in the pottles they were reared in, so the only handling necessary was releasing them into a gorse bush. Most animals were the same age and any measuring was carried out well in advance of the day of translocation. Hence, captive-bred animals appeared relatively unstressed, as evidenced by no deaths or aberrant behaviour during transfer. Captive rearing was conducted in controlled-climate rooms. The use of these in the future could speed up the rate of insect development and enable the timing of release to be manipulated to practical advantage, such as making it coincide with predator control and spring–summer, when survival is likely to be higher. However, before captive-bred animals are released, they should be allowed to acclimatize, either on site or in a controlled-climate facility that mimics the conditions of the release site.

Captive breeding of endangered invertebrates needs to be recognized as a formal area of study in its own right, which must be perfected before it can be integrated into translocation programmes. While captive breeding methods for some weta have been recorded (Barrett, 1991), other species, such as the Middle Island tusked weta (*M. isolata*), have proved difficult to breed in captivity (C. Winks, Auckland, 1997, personal communication). While no figures on costing have been kept, translocations involving the Mahoenui weta would have been far more efficient using captive-bred animals and the enclosure method, rather than live-capturing wild animals and releasing these directly to the new site. This is because huge numbers of person-hours were needed to search for animals in the wild, either for capture or for monitoring. In contrast, captive-breeding can involve the production of hundreds of weta employing one person part-time, especially when techniques are established.

The use of enclosures provides a central location from which to start monitoring searches for released progeny. Best use of an enclosure would involve introducing captive-bred adults or nymphs into it and monitoring their survival and breeding. This should minimize stress to transferees and the problems of acclimatization and dispersal, since the progeny to be released outside the enclosure would have been reared on site. Once a founder population is established, its genetic diversity will need to be monitored to ensure it resembles, as far as practicable, the parent population.

The success of a translocation plan is a question of coordinating the timing of the following activities: wild capture of animals to start a founder captive population; timing the production of captive offspring for translocation to the wild at a time of the year when survival is most likely and when similar instars are occurring naturally; building an enclosure and managing the plants in it at a suitable site well before transferees are due; controlling predators at the translocation site and any other habitat enhancement necessary; monitoring the success of each stage of the programme; and planning repeated translocations. While many of the above were done during the Mahoenui weta translocation programme, they were not planned in advance and the numerous parties were not coordinated. This project clearly demonstrated that translocation programmes are long-term ventures and need to be overseen by a single body, which should see the programme through a decision process such as that described in Meads (1994). Fortunately, a weta recovery plan has been prepared (Sherley, 1998) and a recovery group is to be formed that will coordinate future weta management in New Zealand.

Captive Breeding

Introduction

Captive breeding of threatened species of insects has great potential for helping to establish new populations or augmenting existing populations that are declining. This is because many species have high rates of reproduction, which may be manipulated in controlled-environment insectaries. The advantages of captive breeding programmes for weta have been described above and methods of breeding have been reported in Barrett (1991). Captive breeding of Middle Island tusked weta (*M. isolata*) has been carried out, on contract from the Department of Conservation, by Landcare Research New Zealand Ltd, but results have not yet been published. Other species of weta, mainly of the genus *Deinacrida*, have been kept in captivity by Dr George Gibbs (Victoria University, Wellington) and Mr Mike Meads (private consultant, Wellington). However, these attempts have not been reported either.

Methods and discussion

Captive breeding has been conducted: (i) with captive populations kept in enclosures and essentially exposed to outside weather conditions; (ii) using indoor populations in controlled environments, where conditions have been kept constant; or (iii) indoor populations with no controls over temperature, relative humidity or day/light regimes. No systematic comparison has been made of the relative advantages of each method. At the time of writing, Mahoenui giant weta are the only New Zealand species of insect that have been bred for translocation using the controlled-environment method (Sherley, 1994). The advantages and potential of this method for translocation have been described above. Further advantages of each method relate to research opportunities.

Field research on Mahoenui weta was limited to habitat use and an inferential study on the chronology of their life cycle (Sherley and Hayes, 1993). Essential details, such as the time taken between instars and information on laying and feeding ecology, are difficult to gather in field studies of New Zealand weta, but can be obtained from captive studies. Hence, field and captive studies and management (such as translocation) should be integrated together.

To date, field studies on the Mahoenui giant weta and Middle Island tusked weta have included monitoring environmental conditions, which have been copied in the attempts to breed these species in controlled environments. However, the studies (unpublished) have not been long enough in duration or sufficiently intensive to determine the specific factors (e.g. relative humidity, temperature, lunar cycle, photoperiod, laying substrate) that are influencing breeding and the emergence of nymphs. Similarly, the diet of weta, which are known to be omnivorous (Sherley and Hayes, 1993), needs adequate field study to ensure maximum growth rate in captivity. Understanding other aspects of their behavioural ecology is also necessary – especially breeding. Breeding attempts to date have been done in the complete absence of information, such as whether the species is monogamous or serially polygamous (males copulating with more than one female and vice versa).

Barrett (1991) has described methods he used to keep tree weta in captivity, including weta endemic to the following locations: Wellington (*Hemideina crassidens*), Stephens Island (*H. crassidens*), north of New Zealand (*Hemideina thoracica*) and South Island alpine zone (*Hemideina maori*). Other species covered included Mahoenui giant weta, ground weta (*Hemiandrus similis*) and cave weta (*Gymnoplectron longipes*). His accounts included methods of simply keeping weta (e.g. housing, diet, numbers, substrate, temperature, relative humidity), methods for rearing from eggs onwards and accounts of aspects of their behaviour (e.g. mating, aggression, oviposition, stridulation).

In future, integrated field and captive studies, which are, in turn, combined with a recovery plan, will be necessary to secure threatened weta against extinction. However, before captive breeding should be used, formal research is required to develop a method of breeding weta, as distinct from money required for the subsequent production of weta for translocation.

Research and management for weta conservation

Priority research and management topics for New Zealand weta include the biosystematics of giant (*Deinacrida*), ground (*Hemiandrus*) and tusked weta. Information on the distribution and population status of threatened taxa is also urgently required. To facilitate this and the monitoring of populations that have been manipulated (e.g. translocations, predator management), research into detecting individuals at low densities, such as using pheromone techniques for trapping, is required. Research is also required on captive rearing methods. Topics include housing weta (microclimate conditions, density); pairing (taking into account whether the species is polygamous or monogamous); laying substrate characteristics; microclimatic conditions for incubation; adult behaviour towards each other, eggs and nymphs; and extending longevity through manipulating microclimatic conditions, such as temperature.

More research is needed on translocation methods: for example, containing weta in enclosures in the new habitat until they become established, reducing threats at the release site (e.g. predators), identifying preferences in microclimate and other physical features of habitat preference. Introductions using weta from captive-bred stock need to be compared with those using wild-captured individuals. Finally, research is needed on how best to market (to the general public and government agencies) weta conservation and invertebrate conservation generally. This is essential in order to secure long-term commitment to protecting weta and invertebrate diversity.

Acknowledgements

I thank the following people for help in collecting weta for translocation and subsequent monitoring: Ray Scrimgeour, Phil Thomson, Gary Coles, Tony Green, Tony Williams, Andrew Campbell, Ross Barnes, Ian Flux and Linda Hayes. Thanks to Susan Timmins, Lisa Sinclair, Ralph Powlesland and Lynette Clelland for improving the text.

References

Anon. (1986) Insect re-establishment – a code of conservation practice. *Antenna* 10, 13–18.

Barrett, P. (1991) *Keeping Wetas in Captivity.* Wellington Zoological Gardens, Wellington, New Zealand, 60 pp.

Cowan, P.E. (1990) Brushtail possum. In: King, C.M. (ed.) *The Handbook of New Zealand Mammals.* Oxford University Press, Oxford, pp. 68–98.

Dodd, C.K. and Siegel, R.A. (1991) Relocation, repatriation, and translocation of amphibians and reptiles: are they conservation strategies that work? *Herpetologica* 47, 336–350.

Lovei, G., Stringer, I.A.N, Devine, C. and Cartellieri, M. (1997) Harmonic radar – a method using inexpensive tages to study invertebrate movement on land. *New Zealand Journal of Ecology* 21(2), 187–193.

McEwen, M. (1987) *Ecological Regions and Districts of New Zealand*, 3rd edn. New Zealand Biological Resources Centre Publication 5, New Zealand Department of Conservation, Wellington.

Meads, M.J. (1994) Translocation of New Zealand's endangered insects as a tool for conservation. In: Serena, M. (ed.) *Reintroduction Biology of Australian and New Zealand Fauna.* Surrey Beatty and Sons, Chipping Norton, pp. 53–56.

Meads, M.J. and Notman, P. (1992) *Resurvey for Giant Wetas* (Deinacrida rugosa) *released on Maud Island, Marlborough Sounds.* New Zealand DSIR Land Resources Technical Record 90, Lower Hutt, New Zealand.

Moors, P.J., Atkinson, I.A.E. and Sherley, G.H. (1992) Reducing the rat threat to island birds. *Bird Conservation International* 2, 93–114.

Richards, A.M. (1973) A comparative study of the biology of the giant wetas *Deinacrida heteracantha* and *D. fallai* (Orthoptera: Henicidae) from New Zealand. *Journal of Zoology London* 169, 195–236.

Sherley, G.H. (1989) Conservation research on New Zeland terrestrial arthropods: what is most important? *The Weta* 12, 40–46.

Sherley, G.H. (1994) Translocations of the Mahoenui giant weta *Deinacrida* n. sp. and *Placostylus* land snails in New Zealand: what have we learnt? In: Serena, M. (ed.) *Reintroduction Biology of Australian and New Zealand Fauna.* Surrey Beatty and Sons, Chipping Norton, pp. 57–63.

Sherley, G.H. and Hayes, L. (1993) The conservation of a giant weta (*Deinacrida* n. sp. Orthoptera: Stenopelmatidae) at Mahoenui, King Country: habitat use, and other aspects of its ecology. *New Zealand Entomology* 16, 55–68.

Index